TRIGONOMETRIC FUNCTIONS OF ACUTE ANGLES OF A RIGHT TRIANGLE

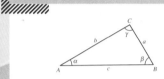

$$\sin \theta = \frac{\text{opp}}{\text{hyp}} \qquad \csc \theta = \frac{\text{hyp}}{\text{opp}}$$

$$\cos \theta = \frac{\text{adj}}{\text{hyp}} \qquad \sec \theta = \frac{\text{hyp}}{\text{adj}}$$

$$\tan \theta = \frac{\text{opp}}{\text{adj}} \qquad \cot \theta = \frac{\text{adj}}{\text{opp}}$$

LAW OF SINES AND LAW OF COSINES

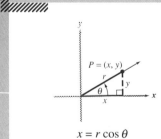

LAW OF SINES

$$\frac{a}{\sin \alpha} = \frac{b}{\sin \beta} = \frac{c}{\sin \gamma}$$

LAW OF COSINES

$$a^2 = b^2 + c^2 - 2bc \cos \alpha$$
$$b^2 = a^2 + c^2 - 2ac \cos \beta$$
$$c^2 = a^2 + b^2 - 2ab \cos \gamma$$

POLAR COORDINATES

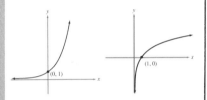

$$x = r \cos \theta$$
$$y = r \sin \theta$$
$$x^2 + y^2 = r^2$$

SLOPE OF A LINE

The slope of the line containing the points $P_1 = (x_1, y_1)$ and $P_2 = (x_2, y_2)$ is
$m = \dfrac{y_2 - y_1}{x_2 - x_1}$, if $x_1 \neq x_2$. If $x_1 = x_2$, the slope is undefined.

EQUATIONS OF LINES

$y = mx + b$	Slope-intercept form
$y - y_1 = m(x - x_1)$	Point-slope form
$ax + by + c = 0$	General form

EXPONENTIAL AND LOGARITHMIC FUNCTIONS

$f(x) = b^x, b > 1 \quad f^{-1}(x) = \log_b x, b > 1$ | $g(x) = b^x, 0 < b < 1 \quad g^{-1}(x) = \log_b x, 0 < b < 1$

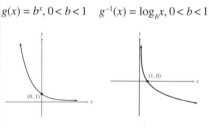

DEFINITIONS AND PROPERTIES OF LOGARITHMS

Assume a and b are positive numbers not equal to 1. Then

$$\log_a x = y \text{ if and only if } a^y = x.$$
$$\log_a xy = \log_a x + \log_a y$$
$$\log_a \frac{x}{y} = \log_a x - \log_a y$$
$$\log_a x^r = r \log_a x$$
$$\log_a a^x = x$$
$$a^{\log_a x} = x$$
$$\log_a x = \frac{\log_b x}{\log_b a}$$

PRECALCULUS

To Brett
With Best Wishes
Phillip
5/19/93

PRECALCULUS

Phillip W. Bean
Mercer University

Jack C. Sharp
Floyd College

Thomas J. Sharp
West Georgia College

PWS-KENT Publishing Company
BOSTON

PWS–KENT
Publishing Company

20 Park Plaza
Boston, Massachusetts 02116

PWS-KENT Publishing Company is a division of Wadsworth, Inc.

Library of Congress Cataloging-in-Publication Data

Bean, Phillip W.
    Precalculus / Phillip W. Bean, Jack C. Sharp, Thomas J. Sharp.
      p.   cm.
    Includes index.
    ISBN 0-534-93160-X
      1. Functions.   2. Algebra—Graphic methods.   I. Sharp, Jack C.
    II. Sharp, Thomas J.   III. Title
    QA331.B33   1993
    515—dc20   92-27803
                                                                    CIP

Sponsoring Editor: Timothy Anderson
Developmental Editor: Barbara Lovenvirth
Production Coordinator: Pamela Rockwell
Text and Cover Design: Julia Gecha
Manufacturing Coordinator: Marcia Locke
Typesetter: G & S Typesetters, Inc.
Printer/Binder: Arcata/Hawkins
Cover Printer: Henry N. Sawyer Company
Cover Photo: G. Kullenberg/Superstock, Inc.

Printed in the United States of America
93   94   95   96   97—10   9   8   7   6   5   4   3   2   1

This book is printed on recycled, acid-free paper.

CONTENTS

PREFACE

Our principal goal in writing this text is to provide a thorough, well-organized, and comprehensible presentation of the topics essential in preparing a student for the study of calculus. To attain this goal, we believe that two fundamental requirements must be met. First, the text must be readable. By employing an informal and conversational writing style, we hope to encourage students to interact with the material and to be actively involved in the study of mathematics. To further this interaction, we often use examples to introduce new concepts and pose questions that will be answered throughout each chapter. With this technique, we hope to direct students' thinking and heighten their interest, as many of these examples point up practical, as well as mathematical, aspects of the topic being covered.

Second, a text must be well-organized. We present topics in a natural and logical order, introducing ideas and concepts needed before the topic is discussed. In addition, topics introduced in earlier chapters are often revisited, extended, and expounded upon in later chapters. Throughout the text, we remind the student that the mathematics being studied is a unified, cohesive body of knowledge and not a series of independent or unrelated topics.

The cornerstone of this text is the study of functions and graphs. We believe a student must have a thorough understanding of elementary functions and their graphs to be successful in the study of calculus. A wide variety of problems—many related to everyday life situations—are presented. Solutions to these problems involve applications of functions and graphs.

To be successful in the study of calculus, a student must have a firm grasp of a variety of topics—algebra, trigonometry, and analytic geometry. This text is designed for students who have completed two years of high school algebra or an intermediate college algebra course, as well as a course in plane geometry.

CONTENTS

Chapter 1 begins with an introduction to set theory, emphasizing the central role of this topic in the study of mathematics. It defines the set of real numbers $\mathbb{R}$ and the set of complex numbers $\mathscr{C}$, and compares and contrasts properties of both sets. The chapter includes a discussion of the Cartesian plane and a brief review of equations and inequalities, with an emphasis on the manipulative skills needed in the study of calculus.

Chapter 2 includes the definitions of relation and function and a thorough study of linear and quadratic functions. Techniques for graphing functions are developed, using functions that students frequently encounter in calculus. Section 2.5 presents the algebra of functions, and the chapter concludes with a study of one-to-one functions and their inverses.

Following a brief historical overview of polynomial equations, Chapter 3 discusses both long division and synthetic division of polynomials. The Fundamental Theorem of Algebra is presented and used to obtain the full factorization of a polynomial of degree n over $\mathscr{C}$. The chapter concludes with a thorough discussion of polynomial and rational functions.

Chapter 4 contains a development of logarithmic and exponential functions. The number e is introduced by considering interest on an investment at a rate compounded continuously. The chapter includes a wealth of applications taken from the natural sciences, social sciences, and business.

Using the unit circle and wrapping function approach, trigonometric functions are developed in Chapter 5. Section 5.3 contains a careful treatment of graphing techniques for the six trigonometric functions. The chapter concludes by considering the inverse trigonometric functions.

Chapter 6 presents the trigonometric functions of angles and shows that the trigonometric functions defined for real numbers and those defined for angles are essentially the same. The chapter provides a thorough treatment of trigonometric identities and concludes with a study of trigonometric equations.

In Chapter 7, a wide variety of problems involving both right triangles and oblique triangles are solved using the tools of trigonometry. Following a careful development of the representation of complex numbers in trigonometric form and DeMoivre's Theorem, the chapter ends with a discussion of plane vectors and polar coordinates.

The conic sections are introduced in Chapter 8. The Cartesian equations of these curves are obtained from their definitions as sets of points in the plane. Throughout, many interesting applications of the conics appear. In Section 8.5, the rotation of axes and the general equation are studied, and the chapter concludes with a study of parametric equations.

Chapter 9 contains a thorough discussion of systems of equations and develops methods, such as Gaussian Elimination, for solving these systems. Section 9.3 contains a complete study of the algebra of matrices. Determinants are introduced in Section 9.5, and Cramer's Rule is used to solve systems of linear equations. The chapter includes a section (9.6) on solving systems of nonlinear equations. In Section 9.7, there is a study of linear inequalities showing how the method of linear programming is used in applications.

Chapter 10 includes a variety of topics intended to provide students with a solid background for the study of calculus and other, higher-level mathematics courses. The ideas discussed in this chapter include arithmetic and geometric sequences and series, mathematical induction, the Binomial Theorem, and partial fractions.

KEY FEATURES OF THE TEXT

Over 500 graphs and figures to enable students to visualize the concepts being presented with greater clarity and understanding.

Historical notes are provided to give students insight into the humanistic nature of the development of mathematics. One of the more interesting of these notes, found in Section 8.2, describes the Greek awareness of the reflecting properties of the parabola citing *On Burning Mirrors*, a book written by Diocles in the second century B.C. (For further information on the sources for the historical notes, see the list at the end of the Preface.)

A biographical sketch of a mathematician whose work has impacted on the material being studied is included in each chapter. For example, Chapter 2 features Leonhard Euler, who first introduced the functional notation $f(x)$, which is a vital part of this chapter.

Highlighted descriptions of formulas and properties are interspersed throughout the text.

Carefully selected examples introduce new topics and are accompanied by clear and precise explanations. These examples follow a logical sequence and serve as an excellent tool to enhance student understanding of the concept being introduced.

Cautions warn the student of commonly made errors and appear throughout the text wherever appropriate.

Exercises found at the end of each section are directly related to the subject and reinforce the material discussed in that section. Each exercise set contains advanced, often innovative problems to challenge the student. The variety of these exercises allows the student to perceive the many creative applications of precalculus mathematics. In order not to discourage students from attempting complex exercises, there is no labeling to distinguish between these and the more routine exercises. **Scientific calculator exercises** are also not labeled, which allows students to decide when use of this tool is appropriate. **Graphing calculator exercises** appear in most of the sections and are designated.

Applications are taken from a variety of fields, such as the natural sciences, social sciences, business, and economics, and appear throughout the text. These range from a problem concerning the concentration of drugs in a patient's bloodstream to the application of oscillation as illustrated by a bungee jumper.

ANCILLARIES FOR INSTRUCTORS

Instructors Solutions Manual contains complete, worked-out solutions to each exercise found in the text. Over 500 graphs are included in the answers.

EXPTest is a computerized test bank for IBM and compatibles that allows users to view and edit all tests, enabling them to add to, delete from, and modify existing questions. A graphics importation feature permits display and printing of graphics, diagrams, and maps provided with the test banks. Demo disk available.

ExamBuilder is a testing program for the Macintosh offering all the features found in the EXPTest. Demo available.

Test Bank includes all the questions and answers found in the computerized tests as well as a sample chapter test for each chapter.

Transparencies 25 full-color acetates, appropriate for use with a precalculus course, are free to adopters.

ANCILLARIES FOR STUDENTS

Student Solutions Manual contains worked-out solutions for every fourth problem found in the text.

TrueBasic for Precalculus (Kemeny/Kurtz) is a software package ideal for classroom demonstrations, individual study, and problem solving.

Precalculus in Context: Functioning in the Real World (Davis/ Moran/Murphy) A lab manual consisting of 12 projects that encourage students to explore precalculus concepts. Graphics calculators and computer graphing software are used to solve each experiment and the corresponding exercises.

LIST OF SOURCES FOR HISTORICAL NOTES

Bell, E. T., *Men in Mathematics*, Simon & Schuster (1986).
Boyer, Carl B., *A History of Mathematics*, Wiley (1968).
Hollingdale, Stuart, *Makers of Mathematics*, The Penguin Group, London, England (1989).
Struik, Dirk, *A Concise History of Mathematics*, Dover Series (1987) and *A Sourcebook in Mathematics*, 1200–1800, Princeton University Press (1986).

ACKNOWLEDGMENTS

We would like to thank all of the people at PWS-KENT Publishing Company who worked with us in the development of this textbook. It was a pleasure to work with Developmental Editor Barbara Lovenvirth, who amicably guided us through this project. Special thanks also go to Pamela Rockwell, Senior Production Editor, for her sensitivity to our concerns during the production stage of this book.

Several other people also worked with us on this project. Jimmie L. Montgomery typed a portion of the text manuscript and Camilla Hightower typed some supplementary material for the textbook. Brett D. Shoelson proofread a portion of the manuscript and typed the Complete Solutions Manual and the Student Solutions Manual.

Finally, but not least, we are indebted to our families for their support and encouragement of this project. We especially wish to thank Ann and Diane Sharp.

R E V I E W E R S

We wish to thank the people who reviewed this text during the stages of its development. Their many helpful suggestions were invaluable to us.

Robert D. Baer
Miami University–Hamilton

Karen E. Barker
Indiana University–South Bend

Derek Bloomfield
Orange County Community College

Ben P. Bockstege
Broward Community College

John S. Cross
University of Northern Iowa

Susan M. Dimick
Spokane Community College

Margaret D. Dolgas
University of Delaware

Christopher Ennis
Western New England College

B. Merldene Friel
Humboldt State University

Vivian Heigl
University of Wisconsin–Parkside

Linda Kilgariff
University of North Carolina

Peter J. Livorsi
Oakton Community College

James E. Moran
Diablo Valley College

William Paul
Appalachian State University

David C. Vella
Skidmore College

Carroll G. Wells
Western Kentucky University

Elaine K. Whittlesy
Siena College

Phillip W. Bean
Jack C. Sharp
Thomas J. Sharp

CHAPTER 1

Essential Topics from Algebra

Following a brief introduction to set theory, this chapter will provide a careful development of the real numbers, the complex numbers, and the Cartesian plane. The final three sections present a detailed treatment of basic equations and inequalities. A thorough understanding of the material presented in this chapter is required for a complete mastery of the precalculus topics developed later in the text.

1.1   BRIEF INTRODUCTION TO SET THEORY

The concept of set theory is of fundamental importance in the study of mathematics. Set notation provides us with a convenient way to represent collections of numbers and is especially useful for representing solutions of equations and inequalities. Much of the terminology that is used to describe sets was first developed by the Danish mathematician Georg Cantor (1845–1918).

We can think of a **set** as a collection of distinct objects called **elements**. It is customary to use capital letters to represent sets and to enclose the elements in set braces { }. The following are examples of sets:

$$A = \{-2, 1, 3, 0\}, \qquad B = \{a, e, i, o, u\}, \qquad E = \{2, 4, 6, \ldots\}.$$

The ellipsis (three dots) used in the description of set E indicates that the pattern established by the first three elements continues indefinitely.

Instead of listing the elements in a set, we sometimes describe sets by using set-builder notation. In **set-builder notation**, the elements of a set are defined as the values of a given variable that satisfy an accompanying rule or condition. For example, the set S which consists of the letters that appear in the word *correct*, can be described as

$$S = \{x \mid x \text{ is a letter in the word } correct\},$$

where the symbol $\mid$ is read "such that," and the set of all isosceles triangles can be described as

$$T = \{t \mid t \text{ is a triangle in which exactly two sides have equal length}\}.$$

While the set S above can also be described as $S = \{c, o, r, e, t\}$, it would be impossible for us to list all the elements that belong to the set T.

The set that consists of all elements under consideration in a particular discussion is called the **universal set** and is denoted by the symbol U. On the other hand, the set that contains no elements is called the **empty set** and is denoted by the symbol $\varnothing$ or by { }.

We use the symbol $\in$ to represent the words "is an element of" or "is a member of." To illustrate using the set $S = \{-2, 1, 3, 0\}$, we see that $-2 \in S$, while $4 \notin S$. If every element of a set A is also an element of a set B, we say that A is a **subset** of B and we write $A \subseteq B$. Since the empty set $\varnothing$ contains no elements, it follows that $\varnothing \subseteq A$, for every set A. If $A \subseteq B$, and if the set B contains at least one element that is not in A, we say that A is a **proper subset** of B and we write $A \subset B$. For example, if

$$A = \{1, 3, 5, \ldots\}, B = \{3, 15, 7, 229\}, \text{ and } C = \{0, 5, 1813\},$$

you should verify that each of the following relationships holds:

$$B \subset A, \qquad C \nsubseteq A, \qquad A \subseteq A, \qquad \varnothing \subset A.$$

EXAMPLE 1

List all the subsets of each of the following sets:

a. $A = \{0, 5, -12\}$ **b.** $\varnothing$

Solution

a.   The subsets of $A = \{0, 5, -12\}$ are

$$\varnothing, \{0\}, \{5\}, \{-12\}, \{0, 5\}, \{0, -12\}, \{5, -12\}, \{0, 5, -12\}.$$

b.   The only subset of the empty set, $\varnothing$, is $\varnothing$ itself.

////// CAUTION: It is important to note that the empty set is not denoted by $\{\varnothing\}$. The set $\{\varnothing\}$ is *not* empty since it contains the element $\varnothing$.

If each of the sets A and B is a subset of the other, we say that A and B are **equal**, and we write $A = B$. For example, the sets $A = \{2, -7, \frac{1}{4}\}$ and $B = \{-7, \frac{1}{4} \sqrt[3]{8}\}$ are equal since each set is a subset of the other.

If U is the universal set and if $A \subseteq U$, the **complement of A**, denoted by A', is defined by

$$A' = \{x \mid x \in U \quad \text{and} \quad x \notin A\}.$$

For example, if $U = \{1, 2, 3, \ldots\}$, the complement of $A = \{11, 12, 13, \ldots\}$ is the set $A' = \{1, 2, 3, \ldots, 10\}$.

EXAMPLE 2

If $U = \{m \mid m$ is a month of the year$\}$ is the universal set, find the complement of each of the following sets:

a. $A = \{x \mid x$ is a 31-day month$\}$
b. $B = \{y \mid y$ is a month having exactly 30 days$\}$

Solution

a. The complement of A is the set consisting of all months not having 31 days. Thus,

$$A' = \{\text{February, April, June, September, November}\}.$$

b. The complement of B is the set whose elements are months that do not have exactly 30 days. Hence,

$$B' = \{\text{January, February, March, May, July,}$$
$$\text{August, October, December}\}.$$

Two sets can be combined to produce other sets using set operations called union and intersection. If A and B are sets, the **union** of A and B, denoted by $A \cup B$, is the set

$$A \cup B = \{x \mid x \in A \text{ or } x \in B\}.$$

Thus, $A \cup B$ is the set consisting of all elements that are in A or in B (or in both A and B).

The **intersection** of A and B, denoted by $A \cap B$, is the set

$$A \cap B = \{x \mid x \in A \text{ and } x \in B\}.$$

Hence, $A \cap B$ is the set consisting of all elements that are in *both* A and B.

EXAMPLE 3

Let $U = \{1, 2, 3, \ldots, 10\}$ be the universal set. If $A = \{1, 3, 4, 5\}$, $B = \{2, 3, 5, 7\}$, and $C = \{2, 8, 9, 10\}$, find each of the following sets:

a. $A \cap B$ **b.** $B \cup C$ **c.** $A \cup B'$ **d.** $(A \cup B')'$

Solution

a. $A \cap B = \{3, 5\}$
b. $B \cup C = \{2, 3, 5, 7, 8, 9, 10\}$

    c.   Since $B' = \{1, 4, 6, 8, 9, 10\}$, $A \cup B' = \{1, 3, 4, 5, 6, 8, 9, 10\}$.

    d.   From part (c), we know that $A \cup B' = \{1, 3, 4, 5, 6, 8, 9, 10\}$. Thus $(A \cup B')' = \{2, 7\}$.   _____

If A and B are sets such that $A \cap B = \varnothing$, we say that A and B are **disjoint**. For example, the sets $A = \{1, 3, 4, 5\}$ and $C = \{2, 8, 9, 10\}$ from Example 3 above have no elements in common and are therefore disjoint sets.

Venn Diagrams

Many of the set operations we have discussed so far can be displayed in pictures called Venn diagrams. **Venn diagrams** are named in honor of the English logician John Venn (1834–1923). Within such diagrams, sets appear as circles (or ovals) inside a rectangle that represents the universal set. Some examples of Venn diagrams are shown in Figure 1.1.

FIGURE 1.1
Venn diagrams

a. $A \subseteq B$

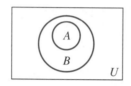

b. A' is in color

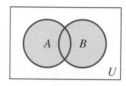

c.   Color denotes $A \cup B$ in the three diagrams below.

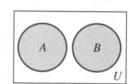
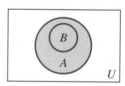
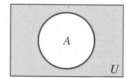

d.   Color denotes $A \cap B$ in the three diagrams below.

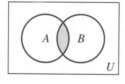
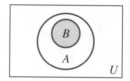
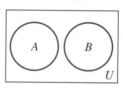

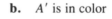

EXERCISES 1.1

1.   If $A = \{-3, 0, \frac{1}{2}, 7\}$ and $B = \{-1, 2, 7, -3\}$, find $A \cup B$ and $A \cap B$.

2.   If $S = \{a, e, i, o, u\}$ and $T = \{x \mid x$ is a letter of the English alphabet$\}$, find $S \cup T$ and $S \cap T$.

3.   Find all the subsets of $W = \{-1, 0, 1\}$.

4.   Describe the set $M = \{x \mid x$ is a letter in the word *Mississippi*$\}$ by listing its elements.

5.   If $U = \{1, 2, 3, \ldots\}$ is the universal set and $A = \{50, 51, 52, \ldots\}$, find A'.

6.   Let $U = \{1, 2, 3, \ldots, 10\}$ be the universal set. If $A = \{1, 2, 7, 9\}$, $B = \{5, 7, 9\}$, and $C = \{2, 3, 7\}$, find each of the following sets:

   a. $A \cup B$ **b.** $A \cap C$
   c. $A \cap B'$ **d.** $A \cup (B \cap C)$
   e. $(A \cup B) \cap (A \cup C)$ **f.** $A' \cap B$
   g. $A \cup C'$ **h.** $(B \cup C)'$
   i. $A \cap (B \cup C)'$ **j.** $(A \cap B') \cup (A \cap C')$

In Exercises 7–16, $U = \{0, 1, 2, 3, \ldots\}$ is the universal set and $A = \{0, 1\}$. Label each statement as True or False.

   7. $\{0\} \subseteq A$ **8.** $\varnothing \subset A$
   9. $0 \in \varnothing$ **10.** $A \cup U = U$
   11. $A' \cap U = U$ **12.** $A \cap \{0\} = 0$
   13. $\{\varnothing\} \subseteq A$ **14.** $A' \cap \varnothing = \{0\}$

15. $\varnothing \subset A'$ **16.** $A \cup \varnothing = \varnothing$

17. It can be shown that a set containing exactly n elements has exactly 2^n subsets. Determine the number of subsets each of the following sets has:

    a. $\{0, 1\}$ **b.** $\{0\}$ **c.** $\varnothing$

    d. $\{\varnothing\}$ **e.** $\{1, 2, 3, \ldots, 10\}$

    f. $\{10, 11, 12, \ldots, 50\}$

    g. $\{t \mid t$ is a letter in the word *Mississippi*$\}$

    h. $\{x \mid x$ is a letter in the English alphabet$\}$

In Exercises 18–25, shade each indicated set in the accompanying Venn diagram.

18. $(A \cap B)'$

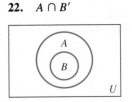

19. $(A \cap B) \cup C$

20. $(A \cup B)'$

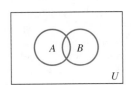

21. $A \cap (B \cup C)$

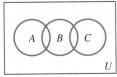

22. $A \cap B'$

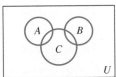

23. $A \cup (B \cap C)$

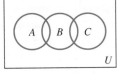

24. $A \cap (B \cap C)$

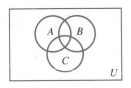

25. $(A \cap B) \cup (A \cap C)$

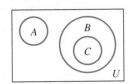

26. According to DeMorgan's laws [named for English mathematician Augustus DeMorgan (1806–1873)], the following relationships hold for any sets A and B:

$(A \cup B)' = A' \cap B'$;   The complement of the union is the intersection of the complements.

$(A \cap B)' = A' \cup B'$.   The complement of the intersection is the union of the complements.

Illustrate DeMorgan's laws by shading the appropriate sets in each of the following Venn diagrams:

   a. $(A \cup B)'$ $A' \cap B'$

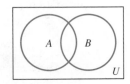
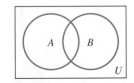

   b. $(A \cap B)'$ $A' \cup B'$

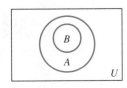
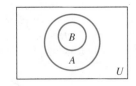

   c. $(A \cap B)'$ $A' \cup B'$

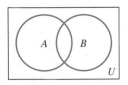
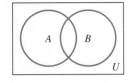

   d. $(A \cup B)'$ $A' \cap B'$

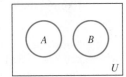
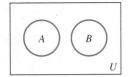

1.2 REAL NUMBERS

Much of our work in mathematics deals with the set of real numbers, which we will denote by the symbol $\mathbb{R}$. To help us describe the set $\mathbb{R}$, we first examine several of its subsets.

The set of **natural numbers** $\mathbb{N}$ consists of the numbers we use for counting:

$$\mathbb{N} = \{1, 2, 3, \ldots\}.$$

The natural numbers, together with their negatives and zero, make up the set of **integers** I:

$$I = \{\ldots, -3, -2, -1, 0, 1, 2, 3, \ldots\}.$$

The set of **rational numbers** Q consists of all numbers that can be expressed as a quotient of two integers:

$$Q = \{x \mid x = \frac{a}{b}, \text{ where } a \text{ and } b \text{ are integers and } b \neq 0\}.$$

We can also define rational numbers as numbers whose decimal representations either terminate or repeat (in "cycles" of one or more digits). The numbers $\frac{3}{4}$, -5, 2.19, $0.333\ldots$, 0, and $7.48215215215\ldots$ are examples of rational numbers.

The set of **irrational numbers** K consists of all numbers that cannot be expressed as ratios of integers. In contrast to rational numbers, the decimal representations for irrational numbers are *nonterminating* and *nonrepeating*. For example, π, $\sqrt{2}$, $\sqrt{3}/2$, and $0.1010010001\ldots$ (note the pattern of 0s and 1s) are all irrational numbers.

We see from their definitions that the set of rational numbers Q and the set of irrational numbers K are disjoint sets. The set of **real numbers** $\mathbb{R}$ is defined to be the union of these two sets. Thus

$$\mathbb{R} = Q \cup K.$$

There are several significant relationships among the various subsets of $\mathbb{R}$. Since every natural number is also an integer, it follows that $\mathbb{N} \subseteq I$. Also, since every integer a can be written as $a/1$, it follows that $I \subseteq Q$. Using $\mathbb{R}$ as the universal set, we can illustrate these relationships in a Venn diagram (see Figure 1.2 at left).

The following example illustrates a method for expressing a rational number, written as a repeating decimal, as the ratio of two integers:

FIGURE 1.2

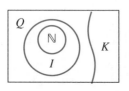

$U = \mathbb{R}$
$\mathbb{R}$ = set of real numbers
Q = set of rational numbers
K = set of irrational numbers
I = set of integers
$\mathbb{N}$ = set of natural numbers

E X A M P L E   4

Express the rational number $0.53161616\ldots$ as a ratio of two integers.

Solution   Noting that 16 is the repeating cycle of digits, we multiply the number $x = 0.53161616\ldots$ by 10^2, the power of ten that will move the decimal to the *beginning* of the repeating cycle.

$$10^2 \cdot x = 100x = 53.161616\ldots.$$

Next, we multiply x by 10^4, the power of ten that will move the decimal to the *end* of the first repeating cycle.

$$10^4 \cdot x = 10{,}000x = 5316.161616\ldots.$$

By subtracting $100x$ from $10{,}000x$, we can eliminate the repeating decimal and obtain

$$9900x = 5263, \quad \text{or}$$

$$x = \frac{5263}{9900}.$$

We should point out that the development of the real numbers was a long and difficult process. Early civilizations were probably content with only the natural numbers, since their fingers and toes were their only calculating devices. However, as time passed and new civilizations evolved, arithmetic calculations were developed that required additional types of numbers.

The ancient Greeks, particularly the Pythagoreans (500–400 B.C.), believed that every aspect of their lives could be explained in terms of natural numbers and their ratios. They were stunned when, using their own geometry, they encountered the number $\sqrt{2}$ (see accompanying figure).

Even after Pythagoras had shown that $\sqrt{2}$ was not a rational number, the Greeks refused to accept the irrationals as numbers. Using irrational numbers only for representing lengths, areas, or volumes, the Greeks continued to solve quadratic equations geometrically. The Hindus and Arabs were the first to fully accept irrational numbers and to use them in arithmetic calculations such as $\sqrt{2} \cdot \sqrt{3} = \sqrt{6}$.

Negative numbers had a similar history of difficult acceptance. Although introduced by the Hindus around A.D. 600, they did not gain full acceptance by the mathematical community for nearly one thousand years. Even as late as the mid-seventeenth century, French mathematician Blaise Pascal (1623–1662) referred to the subtraction of positive numbers from zero as "nonsense."

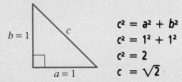

$$c^2 = a^2 + b^2$$
$$c^2 = 1^2 + 1^2$$
$$c^2 = 2$$
$$c = \sqrt{2}$$

Under the operations of addition and multiplication, the set of real numbers $\mathbb{R}$ satisfies the following properties:

Properties of Real Numbers

Let a, b, and c be real numbers.

(i) Closure properties
$a + b$ is a real number (Addition)
ab is a real number (Multiplication)

(ii) Commutative properties
$a + b = b + a$ (Addition)
$ab = ba$ (Multiplication)

(iii) Associative properties
$a + (b + c) = (a + b) + c$ (Addition)
$a(bc) = (ab)c$ (Multiplication)

(iv) Distributive property
$a(b + c) = ab + ac$

(v) Identity properties
$a + 0 = 0 + a = a$
(0 is called the **additive identity**.)
$a \cdot 1 = 1 \cdot a = a$
(1 is called the **multiplicative identity**.)

(vi) Inverse properties
$a + (-a) = (-a) + a = 0$
($-a$ is called the **additive inverse** of a.)
If $a \neq 0$, $a \cdot \dfrac{1}{a} = \dfrac{1}{a} \cdot a = 1$

$\left(\dfrac{1}{a} \text{ is called the } \textbf{multiplicative inverse} \text{ or } \textbf{recipro-} \right.$

cal of a. $\Big)$

These six properties are often called the **field properties**, and any collection of numbers that satisfies these properties is called a **field**.

Definitions of Subtraction and Division

If a and b are real numbers, $a - b = a + (-b)$. (Subtraction)

If a and b are real numbers and $b \neq 0$, then $a \div b = a(1/b)$. (Division)

Thus, to subtract b from a, we simply add the additive inverse (or negative) of b to a. Similarly, to divide a by b, where $b \neq 0$, we multiply a by the multiplicative inverse (or reciprocal) of b. By defining subtraction in terms of addition and division in terms of multiplication, we see that these two operations are also governed by the field properties.

EXAMPLE 5

Give the name of the real number property that is best illustrated by each of the following statements. Assume that all variables represent real numbers.

a. $3(x + y) = 3x + 3y$
b. $7 + 3m = 3m + 7$
c. $(\frac{1}{5})5 = 1$
d. $6 \cdot 1 = 1 \cdot 6$
e. $9(t + 4) = 9(4 + t)$
f. $8 \cdot 1 = 8$
g. $-\frac{1}{4} + \frac{1}{4} = 0$
h. $2(7w) = (2 \cdot 7)w$

Solution

a. Distributive property
b. Commutative property of addition
c. Inverse property of multiplication
d. Commutative property of multiplication
e. Commutative property of addition
f. Identity property of multiplication
g. Inverse property of addition
h. Associative property of multiplication

The Number Line

We can represent the set of real numbers geometrically by means of a **number line** (see Figure 1.3). In this configuration, each real number corresponds to a unique point on the number line, and conversely, each point on the number line

FIGURE 1.3
Number line

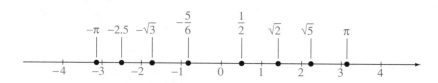

corresponds to a unique real number. In general, we will not distinguish between a given real number and the point to which it corresponds.

If a and b represent the same real number, we say that a and b are **equal** and we write $a = b$. Four properties of equality are described below.

Properties of Equality

Let a, b, and c be real numbers.
(i)   Reflexive property: $a = a$;
(ii) Symmetric property: If $a = b$, then $b = a$;
(iii) Transitive property: If $a = b$ and $b = c$, then $a = c$;
(iv) Substitution property: If $a = b$, then either can replace the other in any mathematical expression.

The correspondence between real numbers and points on the number line establishes an inherent ordering of the real numbers. As we can see from Figure 1.3, the real numbers are arranged in increasing order of size from left to right on the number line.

For real numbers a and b, we say that a **is less than** b and we write $a < b$ if a is located to the left of b on the number line. We note that whenever a is less than b, it is also true that b **is greater than** a, and we write $b > a$. From Figure 1.3, we make the following observations: $-2.5 < -2$, $3 < \pi$, $1 > 0$, and $\sqrt{2} < 2$.

We write $a \leq b$, read "a **is less than or equal to** b," if either $a < b$ or $a = b$ is true. The relationship $a \leq b$ also means that $b \geq a$, read "b **is greater than or equal to** a." For example, $1 \leq 5$, $4 \geq 4$, and $-\sqrt{3} \leq -1$.

Let a and b be real numbers such that $a < b$. We say that the real number x **is between a and b**, and we write $a < x < b$ when x satisfies both $a < x$ and $x < b$. Similarly, $a \leq x \leq b$ means that $a \leq x$ and $x \leq b$. For example, we see from Figure 1.3, that $-2 < 0 < 2$, $1 \leq \sqrt{2} \leq 2$, and $-\frac{5}{6} < 3 < \pi$.

From the fact that each real number corresponds to a unique point on the number line, we have the following law:

Law of Trichotomy

If a and b are real numbers, exactly one of the following is true:

(i) $a < b$;
(ii) $a = b$;
(iii) $b < a$.

For convenience, we often represent certain subsets of $\mathbb{R}$ using **interval notation**. For example, if a and b are real numbers such that $a < b$, the **open interval** (a, b) and the **closed interval** $[a, b]$ are defined as follows:

$$(a, b) = \{x \mid a < x < b\},$$
$$[a, b] = \{x \mid a \le x \le b\}.$$

In either case, the numbers a and b are called the **endpoints** of the interval. The brackets in the notation $[a, b]$ are used to indicate that the endpoints are included in the closed interval $[a, b]$, while the parentheses in the notation (a, b) indicate that the endpoints are not included in the open interval (a, b). In Figure 1.4 the intervals (a, b) and $[a, b]$ are displayed on the number line.

FIGURE 1.4

Open interval (a, b) Closed interval $[a, b]$

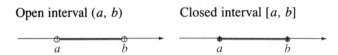

To represent the set of real numbers that are greater than the real number a, we use the interval (a, ∞), where the symbol ∞ (read "infinity") is used to indicate that the interval does not have an endpoint on the right. Similarly, the interval $(-\infty, a)$ represents the set of real numbers that are less than a. The intervals we have discussed thus far and several others are described in Table 1.1.

TABLE 1.1
Interval notation

Interval Notation	Set Description	Shown in Color on the Number Line
(a, b)	$\{x \mid a < x < b\}$	
$[a, b]$	$\{x \mid a \le x \le b\}$	
$[a, b)$	$\{x \mid a \le x < b\}$	
$(a, b]$	$\{x \mid a < x \le b\}$	
$(-\infty, a)$	$\{x \mid x < a\}$	
$(-\infty, a]$	$\{x \mid x \le a\}$	
(a, ∞)	$\{x \mid x > a\}$	
$[a, \infty)$	$\{x \mid x \ge a\}$	
$(-\infty, \infty)$	$\{x \mid x \in \mathbb{R}\}$	

EXAMPLE 6

Describe each of the following sets using set-builder notation. Display each set on the number line.

a. $[-1, 3]$ **b.** $(-\infty, 2)$ **c.** $(0, 4]$ **d.** $[\frac{1}{2}, \infty)$ **e.** $(-2, 1] \cup (4, \infty)$

Solution

a. $[-1, 3] = \{x \mid -1 \le x \le 3\}$

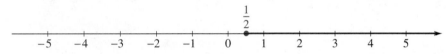

b. $(-\infty, 2) = \{x \mid x < 2\}$

c. $(0, 4] = \{x \mid 0 < x \le 4\}$

d. $[\frac{1}{2}, \infty) = \{x \mid x \ge \frac{1}{2}\}$

e. $(-2, 1] \cup (4, \infty) = \{x \mid -2 < x \le 1 \text{ or } x > 4\}$

Absolute Value and Distance

Associated with each real number a is a nonnegative real number that represents the distance between a and 0 on the number line. This nonnegative number is called the absolute value of a and is denoted by $|a|$.

Absolute Value of a Real Number

Let a be any real number. The **absolute value** of a is defined by

$$|a| = \begin{cases} a, & \text{if } a \ge 0 \\ -a, & \text{if } a < 0 \end{cases}.$$

From the definition of the absolute value of a real number a, we see that $|a|$ is always a nonnegative real number.

EXAMPLE 7

Write each of the following expressions as an equivalent expression that does not involve the absolute value symbol:

a. $|0|$
b. $|-2.5|$
c. $|1 - t|$, where $t > 1$
d. $|x|$, where x is a real number

Solution

a. $|0| = 0$, since $0 \ge 0$.
b. Since $-2.5 < 0, |-2.5| = -(-2.5) = 2.5$.

c. Since $1 - t < 0$ when $t > 1$, it follows that $|1 - t| = -(1 - t)$
$= t - 1$.

d. $|x| = \begin{cases} x, & \text{if } x \geq 0 \\ -x, & \text{if } x < 0 \end{cases}$.

Some useful properties of absolute values are summarized in Theorem 1.1.

THEOREM 1.1   Properties of Absolute Values

Let a and b be real numbers.

(i) $|a| \geq 0$;

(ii) $|a| = \sqrt{a^2}$;

(iii) $|-a| = |a|$;

(iv) $|ab| = |a| \cdot |b|$;

(v) $\left| \dfrac{a}{b} \right| = \dfrac{|a|}{|b|}$, if $b \neq 0$;

(vi) $|a| = |b|$, if and only if $a = b$ or $a = -b$;

(vii) $|a + b| \leq |a| + |b|$.   (Triangle inequality)

Part (i) of Theorem 1.1 follows at once from the definition of absolute value. The remaining parts of this theorem can be verified by considering cases determined by the signs of the numbers a and b (see Exercises 57–60).

Recall that $|a|$ measures the distance between a and 0 on the number line. We can extend this concept to define the distance between any two points on the number line.

Distance Between Points on the Number Line

Let a and b be real numbers. The **distance between a and b** on the number line is given by $d(a, b) = |b - a| = |a - b|$.

EXAMPLE 8

Find the distance between a and b in each case.

a. $a = 3$, $b = -\frac{1}{4}$

b. $a = -3.05$, $b = 2.336$

c. $a = \pi$, $b = \sqrt{2}$

Solution

a. $d(3, -\frac{1}{4}) = |-\frac{1}{4} - 3| = |-\frac{13}{4}| = \frac{13}{4}$

b. $d(-3.05, 2.336) = |2.336 - (-3.05)| = |5.386| = 5.386$

c. $d(\pi, \sqrt{2}) = |\sqrt{2} - \pi|$. Since $\sqrt{2} < \pi$ (see Figure 1.3), it follows that $|\sqrt{2} - \pi| = -(\sqrt{2} - \pi) = \pi - \sqrt{2}$. Thus $d(\pi, \sqrt{2}) = \pi - \sqrt{2}$.

EXERCISES 1.2

In Exercises 1–10, label each statement as True or False.

1. All integers are rational numbers.
2. All integers are natural numbers.
3. The product of two irrational numbers is always an irrational number.
4. The sum of two rational numbers is always a rational number.
5. The closure property of addition holds for the set I of integers.
6. The inverse property of addition holds for the set $\mathbb{N}$ of natural numbers.
7. The closure property holds for the operation of subtraction on the set I of integers.
8. The additive identity is a rational number.
9. The closure property holds for the operation of division on the set K of irrational numbers.
10. The multiplicative inverse of 1.5 is a rational number.

In Exercises 11–18, express each rational number as the ratio of two integers.

11. 2.09
12. -10.71
13. $0.818181\ldots$
14. $0.252525\ldots$
15. $7.444\ldots$
16. $-3.272727\ldots$
17. $0.8514514514\ldots$
18. $5.02373737\ldots$

In Exercises 19–28, give the name of the real number property that is best illustrated by each statement.

19. $5(r + s) = 5r + 5s$
20. $-\frac{2}{3} + \frac{2}{3} = 0$
21. $-6 + 0 = -6$
22. $(8 + m) + 5 = (m + 8) + 5$
23. $(\frac{7}{3})(\frac{3}{7}) = 1$
24. $(m + 2n) \cdot 7 = 7 \cdot (m + 2n)$
25. $-x + (x + 4) = (-x + x) + 4$
26. $-7(7y) = (-7 \cdot 7)y$
27. $(-2 + 2) + w = 0 + w$
28. $1 \cdot (-1) = -1$

In Exercises 29–36, give the name of the real number property that tells why each numbered statement follows from the previous statement.

$$4(x + 2) + 7 = 16$$
29. $(4x + 8) + 7 = 16$
30. $4x + (8 + 7) = 16$
$$4x + 15 = 16$$
$$(4x + 15) + (-15) = 16 + (-15)$$

31. $4x + [15 + (-15)] = 1$
32. $4x + 0 = 1$
33. $4x = 1$
$$(\tfrac{1}{4})(4x) = (\tfrac{1}{4})(1)$$
34. $(\tfrac{1}{4} \cdot 4)x = (\tfrac{1}{4} \cdot 1)$
35. $1 \cdot x = (\tfrac{1}{4} \cdot 1)$
36. $x = \tfrac{1}{4}$

In Exercises 37–42, write each expression as an equivalent expression that does not involve the absolute value symbol. Assume that all variables represent real numbers.

37. $|-2|$
38. $|\tfrac{1}{5} - 0.3|$
39. $|r - 1|$, if $r \geq 1$
40. $|3 - \pi|$
41. $|-y|$, if $y < 0$
42. $|k - 5|$, if $k < 5$

In Exercises 43–50, describe each set by using set-builder notation. Display each set on the number line.

43. $[-2, 1)$
44. $(3, \infty)$
45. $[-4, \infty)$
46. $(-\frac{5}{3}, \frac{1}{2})$
47. $(-\infty, -\sqrt{2}]$
48. $(-\pi, \pi)$
49. $(-\infty, 0) \cup (1, \frac{13}{8})$
50. $(-\infty, -3) \cup [-\frac{1}{2}, \infty)$

In Exercises 51–56, find the distance between a and b in each case.

51. $a = -3, b = 4$
52. $a = 2, b = -5$
53. $a = \frac{4}{9}, b = \frac{7}{3}$
54. $a = \sqrt{6}, b = \sqrt{3}$
55. $a = -\pi, b = \pi/4$
56. $a = 2.5, b = -1.8$

57. Show that $\sqrt{a^2} = |a|$, for all real numbers a. [*Hint*: Consider three cases: $a < 0$, $a = 0$, and $a > 0$.]
58. Show that $|-a| = |a|$, for all real numbers a.
59. Show that $|ab| = |a| \cdot |b|$, for all real numbers a and b. [*Hint*: Consider the following cases: (i) the numbers a and b agree in sign; (ii) the numbers a and b have opposite signs; and (iii) at least one of the two numbers is equal to 0.]
60. a. Find real numbers a and b such that $|a + b| = |a| + |b|$.
    b. Find real numbers a and b such that $|a + b| < |a| + |b|$.
    c. Is it possible to find real numbers a and b such that $|a + b| > |a| + |b|$? Why?

1.3 THE CARTESIAN PLANE

In a sense, our mathematical development to this point has been one-dimensional. We have described real numbers as points on a number line and have detailed many of their properties. We now extend our discussion to two dimensions by introducing ordered pairs of real numbers and their representation as points in a two-dimensional configuration known as the plane.

Ordered Pair of Real Numbers

If x and y are real numbers, (x, y) is called an **ordered pair** with **first component** x and **second component** y.

Two ordered pairs (a, b) and (c, d) are said to be **equal** if and only if $a = c$ and $b = d$. In other words, two ordered pairs are equal if and only if they have the same first component and the same second component. For example, $(x, y) = (2, -5)$ if and only if $x = 2$ and $y = -5$.

It is customary to denote the set of all ordered pairs of real numbers by $\mathbb{R}^2$. Thus

$$\mathbb{R}^2 = \{(x, y) \mid x \text{ and } y \text{ are real numbers}\}.$$

To display these ordered pairs of real numbers, we define a two-dimensional coordinate system called the **xy-plane**. We begin by constructing two number lines—one horizontal and one vertical—that intersect at the point called 0 (zero) on each. The horizontal number line, called the **x-axis**, is labeled in the usual manner with positive numbers to the right of 0. The vertical number line, called the **y-axis**, is labeled with positive numbers extending upward from 0. The point of intersection of these two axes is called the **origin** and is denoted by the capital letter O (see Figure 1.5).

We can now establish a one-to-one correspondence between the ordered pairs of real numbers and points in the plane. To assign a unique point in the plane to each ordered pair (a, b) of real numbers, we proceed as follows. Constructing a vertical line through the point a on the x-axis, and a horizontal line through the point b on the y-axis, we assign to (a, b) the point P where the two lines intersect (see Figure 1.6).

To assign a unique ordered pair of real numbers to each point in the plane, we essentially reverse the assignment process above. If P is a point in the plane,

FIGURE 1.5

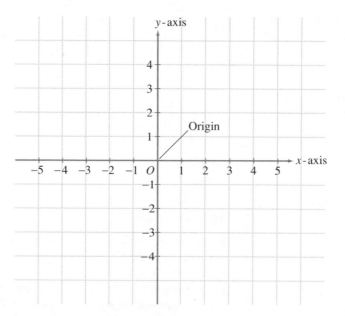

FIGURE 1.6
Associating ordered pairs of real
numbers with points in the plane

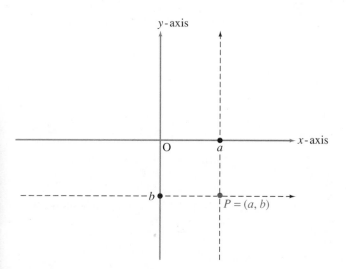

RENÉ DESCARTES
(1596–1650)

René Descartes, the French philosopher amd mathematician, has been called the father of modern philosophy. He sought a method of thinking that would give coherence to knowledge and lead to scientific truth. It was in *La Géometrie,* the 100-page appendix to his great treatise on method in science published in 1637, that Descartes linked two branches of mathematics: geometry and algebra. This work, together with that of fellow Frenchman Pierre Fermat (1601–1665), laid the foundation for analytic geometry. Descartes's discovery that there is a one-to-one correspondence between the points in a plane and the ordered pairs (x, y) of real numbers made a great impact on the development of mathematics.

we begin by constructing vertical and horizontal lines through P. If a denotes the point where the vertical line intersects the x-axis, and b denotes the point where the horizontal line intersects the y-axis, we assign the ordered pair (a, b) to the point P (see Figure 1.6).

Thus we see that every ordered pair (x, y) of real numbers can be represented by a unique point P in the plane, and conversely, every point P in the plane can be represented by a unique ordered pair (x, y) of real numbers. The first component of each ordered pair is called the **x-coordinate** or **abscissa**, and the second component is called the **y-coordinate** or **ordinate**. This two-dimensional system for representing ordered pairs of real numbers is called a **rectangular coordinate system** or **Cartesian coordinate system** to honor French mathematician René Descartes (1596–1650). We will refer to this system simply as the plane.

As shown in Figure 1.7, the plane is divided into four **quadrants**, which

FIGURE 1.7
The plane and its quadrants

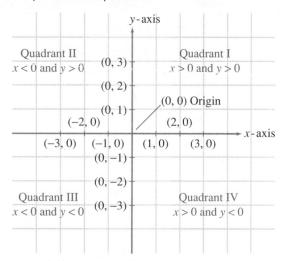

are numbered I–IV in a counterclockwise direction. We note that the x- and y-axes are not included in any of the quadrants.

EXAMPLE 9

Plot each of the following points in the plane:

$$A = (-2, -2), \qquad B = (\tfrac{7}{2}, -3), \qquad C = (-\pi, 3),$$
$$D = (0, -\sqrt{2}), \qquad E = (\sqrt{5}, 1.7)$$

Solution

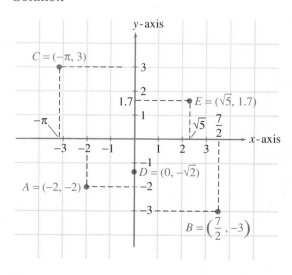

The Distance Formula

Unlike the set of real numbers $\mathbb{R}$, there is no order relation for the set $\mathbb{R}^2$. However, the concept of distance between points in the plane is extremely important. The following formula, called the distance formula, enables us to find the distance between any two points in the plane:

The Distance Formula

If $P_1 = (x_1, y_1)$ and $P_2 = (x_2, y_2)$ are points in the plane, the **distance between P_1 and P_2** is denoted by $d(P_1, P_2)$ and is given by the formula

$$d(P_1, P_2) = \sqrt{(x_2 - x_1)^2 + (y_2 - y_1)^2}.$$

We will verify the distance formula for the case where P_1 lies in quadrant III and P_2 lies in quadrant I (see Figure 1.8). A similar argument will hold in general.

Let Q denote the point (x_2, y_1). Constructing perpendicular line segments from P_1 to the x-axis and from P_2 to the y-axis, we see that

$$d(P_1, Q) = |x_2 - x_1| \qquad \text{and} \qquad d(Q, P_2) = |y_2 - y_1|.$$

FIGURE 1.8

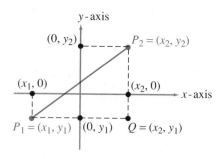

Since $x_2 - x_1$ and $y_2 - y_1$ are both positive, it follows that

$$d(P_1, Q) = x_2 - x_1 \quad \text{and} \quad d(Q, P_2) = y_2 - y_1.$$

By the Pythagorean theorem,

$$[d(P_1, P_2)]^2 = [d(P_1, Q)]^2 + [d(Q, P_2)]^2,$$

so that

$$[d(P_1, P_2)]^2 = (x_2 - x_1)^2 + (y_2 - y_1)^2.$$

Since the distance between points cannot be negative, taking the square root of both sides of the latter equation yields the desired result:

$$d(P_1, P_2) = \sqrt{(x_2 - x_1)^2 + (y_2 - y_1)^2}.$$

The Midpoint Formula

Using the distance formula, we can establish a formula for finding the midpoint of the line segment joining two points in the plane.

The Midpoint Formula

Let $A = (x_1, y_1)$ and $B = (x_2, y_2)$ be points in the plane. The midpoint M of the line segment $\overline{AB}$ is given by the formula

$$M = \left(\frac{x_1 + x_2}{2}, \frac{y_1 + y_2}{2} \right).$$

FIGURE 1.9

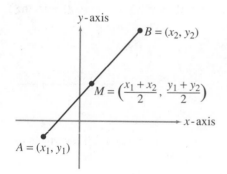

To verify that M is the midpoint of segment $\overline{AB}$, we must show that $d(A, M) = d(M, B)$, and that the points A, B, and M all lie on the same line. As shown in Figure 1.9, we will consider the case where A lies in quadrant III and B lies in quadrant I. A similar argument holds in general. Using the distance formula,

$$d(A, M) = \sqrt{\left(\frac{x_1 + x_2}{2} - x_1 \right)^2 + \left(\frac{y_1 + y_2}{2} - y_1 \right)^2}$$

$$= \sqrt{\left(\frac{x_1 + x_2 - 2x_1}{2} \right)^2 + \left(\frac{y_1 + y_2 - 2y_1}{2} \right)^2}$$

$$= \sqrt{\left(\frac{x_2 - x_1}{2} \right)^2 + \left(\frac{y_2 - y_1}{2} \right)^2}$$

$$= \sqrt{\frac{(x_2 - x_1)^2}{4} + \frac{(y_2 - y_1)^2}{4}}$$

$$= \frac{1}{2} \sqrt{(x_2 - x_1)^2 + (y_2 - y_1)^2}.$$

Similarly,

$$d(M, B) = \sqrt{\left(x_2 - \frac{x_1 + x_2}{2}\right)^2 + \left(y_2 - \frac{y_1 + y_2}{2}\right)^2}$$

$$= \sqrt{\left(\frac{2x_2 - x_1 - x_2}{2}\right)^2 + \left(\frac{2y_2 - y_1 - y_2}{2}\right)^2}$$

$$= \sqrt{\left(\frac{x_2 - x_1}{2}\right)^2 + \left(\frac{y_2 - y_1}{2}\right)^2}$$

$$= \sqrt{\frac{(x_2 - x_1)^2}{4} + \frac{(y_2 - y_1)^2}{4}}$$

$$= \frac{1}{2}\sqrt{(x_2 - x_1)^2 + (y_2 - y_1)^2}.$$

Thus, $d(A, M) = d(M, B)$. Since

$$d(A, M) + d(M, B) = \sqrt{(x_2 - x_1)^2 + (y_2 - y_1)^2} = d(A, B),$$

we know from plane geometry that the points A, B, and M lie on the same line.

 CAUTION: When calculating midpoints be sure to *add*—not subtract—the corresponding components of the endpoints of the line segment. To remember this, think of the midpoint of segment $\overline{AB}$ as the point whose coordinates are the "averages" of the corresponding components of A and B.

EXAMPLE 10

Find the midpoint M of the line segment $\overline{AB}$, where $A = (-1, 5)$ and $B = (2, -\frac{3}{2})$.

Solution   Substituting $(x_1, y_1) = (-1, 5)$ and $(x_2, y_2) = (2, -\frac{3}{2})$ into the midpoint formula, we get

$$M = \left(\frac{-1 + 2}{2}, \frac{5 + (-3/2)}{2}\right) = \left(\frac{1}{2}, \frac{7}{4}\right).$$

EXAMPLE 11

Show that $A = (-4, -3)$, $B = (6, 1)$, and $C = (5, -11)$ are the vertices of an isosceles triangle, and find the area of the triangle.

Solution   We use the distance formula to calculate the lengths of sides $\overline{AB}$, $\overline{BC}$, and $\overline{AC}$.

$$d(A, B) = \sqrt{(6 + 4)^2 + (1 + 3)^2} = \sqrt{116} = 2\sqrt{29},$$

$$d(B, C) = \sqrt{(5 - 6)^2 + (-11 - 1)^2} = \sqrt{145},$$

$$d(A, C) = \sqrt{(5 + 4)^2 + (-11 + 3)^2} = \sqrt{145}.$$

Since the sides $\overline{AC}$ and $\overline{BC}$ have the same length, triangle ABC is an isosceles triangle.

FIGURE 1.10

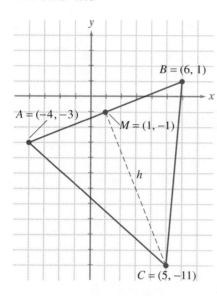

We can now calculate the area of the triangle. Since sides $\overline{AC}$ and $\overline{BC}$ have equal length, the line segment from vertex C to the midpoint M of side $\overline{AB}$ is an altitude of the triangle (see Figure 1.10).

Using the midpoint formula, we find that

$$M = \left(\frac{-4 + 6}{2}, \frac{-3 + 1}{2} \right) = (1, -1).$$

The length of altitude $\overline{MC}$ is

$$d(M, C) = \sqrt{(5 - 1)^2 + (-11 + 1)^2} = \sqrt{116} = 2\sqrt{29},$$

and, since the length of the base $\overline{AB}$ is also $2\sqrt{29}$, the area $\mathcal{A}$ of triangle ABC is

$$\mathcal{A} = \frac{1}{2} bh = \frac{1}{2} (2\sqrt{29})(2\sqrt{29}) = \frac{1}{2} (4 \cdot 29) = 58 \text{ square units.}$$

EXERCISES 1.3

In Exercises 1–6, plot each point in the plane.

1. $(-2, 5)$
2. $(1.5, -\pi)$
3. $(-\sqrt{3}/2, \frac{1}{2})$
4. $(-\sqrt{2}, \frac{3}{4})$
5. $(-\pi/2, -2.5)$
6. $(-1.8, \sqrt{2}/2)$

In Exercises 7–14, find the distance between each pair of points.

7. $(1, -2), (-1, 2)$
8. $(0, 3), (-1, 5)$
9. $(\frac{1}{2}, -2), (\frac{3}{2}, 0)$
10. $(3, -\sqrt{2}), (1, \sqrt{2})$
11. $(\sqrt{3}/2, \frac{1}{2}), (0, \frac{5}{4})$
12. $(-1, -\pi), (-1, 3\pi)$
13. $(-a, 2a), (a, 0),$ if $a > 0$
14. $(a/4, a), (a, 2a),$ if $a > 0$

In Exercises 15–22, find the midpoint M of the line segment joining each pair of points.

15. $(-2, 4), (2, -6)$
16. $(3, -1), (-2, 7)$
17. $(1.5, -3), (3.7, 1)$
18. $(\frac{7}{8}, -\frac{3}{4}), (-\frac{1}{8}, 0)$
19. $(\sqrt{3}/2, \frac{1}{2}), (\sqrt{3}/2, -\frac{1}{2})$
20. $(\pi/2, -1), (\pi/3, 0)$
21. $(-a, a), (3a, -a)$
22. $(a/8, a/2), (a/4, a)$

23. Find the point C that is three-fourths of the way from $A = (-1, 3)$ to $B = (5, 17)$.
24. Find the point C that is one-fourth of the way from $A = (2, -3)$ to $B = (-\frac{1}{2}, 7)$.
25. Find the perimeter of the triangle whose vertices are $R = (-3, 2), S = (2, -3),$ and $T = (1, 1)$.
26. Find all values of x for which the distance between the points $(1, 2)$ and $(x, x + 3)$ is equal to 2.
27. Find all points on the x-axis that are 5 units away from the point $Q = (0, 4)$.

28. Find the value of a for which the point $P = (a, a)$ is equidistant from the points $Q = (-1, 2)$ and $R = (10, -1)$.
29. Find all values of y for which the point $(y - 1, y)$ is twice as far from the point $(4, -5)$ as it is from the point $(-3, 2)$.
30. If $A = (2, 4)$ and $B = (0, -4)$ are the endpoints of the diameter of a circle, find the center and radius of the circle.

In Exercises 31–34, determine whether the given point P lies inside, outside, or on the circle with center $C = (-2, 3)$ and radius 4.

31. $P = (-1, 1)$
32. $P = (2, 3)$
33. $P = (-2, 7)$
34. $P = (-\frac{1}{2}, -3)$

35. Show that $A = (1, 0), B = (9, 0),$ and $C = (5, 4\sqrt{3})$ are the vertices of an equilateral triangle and find the area of the triangle.
36. Show that $P = (5, -2), Q = (6, -6),$ and $R = (2, -5)$ are the vertices of an isosceles triangle and find the area of the triangle.

In Exercises 37–42, the vertices of triangle ABC are given. In each case, determine whether the triangle is isosceles, equilateral, or neither.

37. $A = (1, -1), B = (-1, 1), C = (-\sqrt{3}, -\sqrt{3})$
38. $A = (2, -5), B = (6, -6), C = (5, -2)$
39. $A = (-5, 1), B = (10, 7), C = (18, -8)$
40. $A = (-9, 1), B = (6, 9), C = (14, -6)$
41. $A = (2, -3), B = (5, 0), C = (6, -4)$
42. $A = (10, -1), B = (6, -6), C = (-1, 2)$

43. Show that the point $P = (-7, 5)$ lies on the perpendicular bisector of line segment $\overline{AB}$, where $A = (-3, -5)$ and

$B = (3, 1)$. [*Hint*: The **perpendicular bisector** of line segment $\overline{AB}$ contains all points that are equidistant from the points A and B.]

44. Two or more points that lie on the same straight line are said to be **collinear**. Show that the points $P = (3, 10)$, $Q = (2, 3)$, and $R = (1, -4)$ are collinear.

In Exercises 45–48, determine whether or not the given points are collinear.

45. $(3, 2), (4, 6), (0, -8)$

46. $(-2, 1), (-4, 0), (2, 3)$

47. $(0, -1), (-2, 4), (1, -\frac{7}{2})$

48. $(3, 2), (6, 0), (-\frac{3}{2}, 5)$

49. Show that the points $T = (-5, 1)$, $U = (7, -4)$, $V = (12, 8)$, and $W = (0, 13)$ are the vertices of a square. [*Hint*: A square has four equal sides and equal diagonals.]

50. Show that the points $A = (-1, 1)$, $B = (2, -2)$, $C = (5, 1)$, and $D = (2, 4)$ are the vertices of a square (see Exercise 49).

51. Show that the points $A = (5, 1)$, $B = (1, 3)$, and $C = (-1, -11)$ are the vertices of a right triangle and find the area of the triangle.

52. The Indian mathematician Bhaskara (1114–1158) published a one-word proof of the Pythagorean theorem. Show that the "proof," which is shown here, is indeed a proof.

BEHOLD!

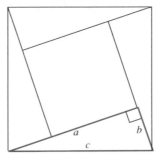

1.4 COMPLEX NUMBERS

In the set of real numbers, $\sqrt{x}$ is undefined when $x < 0$. For example, $\sqrt{-1}$ cannot represent a real number since no real number can be squared to obtain -1. However, we can use the plane to describe a set of numbers in which such square roots exist. We begin by representing the (nonreal) number $\sqrt{-1}$ by the symbol i. We call i the **imaginary unit** and use it to define a set of numbers called the complex numbers.

Set of Complex Numbers

The set of **complex numbers** $\mathscr{C}$ is defined as follows:

$$\mathscr{C} = \{a + bi \mid a \text{ and } b \text{ are real numbers and } i^2 = -1\}$$

The real number a is called the **real part** of $a + bi$, and the real number b is called the **imaginary part**.

Some examples of complex numbers are $-4 + 7i$, $\frac{1}{3} - 2i$, $\sqrt{5}\, i$, and 12. We note that the real part of the complex number $\sqrt{5}\, i$ is 0, while the imaginary part is $\sqrt{5}$. You should identify the real part and the imaginary part of each of the other three complex numbers.

Just as real numbers can be represented as points on the number line, complex numbers can be represented as points in the plane. By referring to the x-axis as the **real axis** and to the y-axis as the **imaginary axis**, there is a natural association between the complex number $a + bi$ and the point (a, b) in the plane. In light of this arrangement, the xy-plane is also known as the **complex plane**.

EXAMPLE 12

Plot each of the following complex numbers in the complex plane:

$$z_1 = -1 + 4i, \qquad z_2 = 2 - 3i, \qquad z_3 = -3 - 3i, \qquad z_4 = \sqrt{5} + 2i.$$

Solution

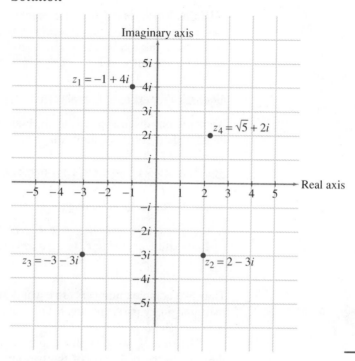

FIGURE 1.11

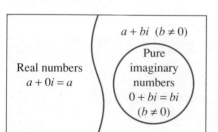

$U = \mathscr{C}$

Since every real number a can be written as $a + 0i$, it follows that the set of real numbers $\mathbb{R}$ is a subset of the set of complex numbers $\mathscr{C}$. If $a = 0$ and $b \neq 0$, the complex number $a + bi = 0 + bi = bi$ is called a **pure imaginary number** (see Figure 1.11).

Now that we have defined complex numbers, we can define the square root of a negative number.

Square Root of a Negative Number

If a is a positive real number,

$$\sqrt{-a} = \sqrt{a}\, i.$$

We see from this definition that the square root of a negative number is always a pure imaginary number. For example, $\sqrt{-25} = \sqrt{25}\, i = 5i$, and $\sqrt{-7} = \sqrt{7}\, i$.

EXAMPLE 13

Evaluate each of the following:

a. $\sqrt{-12}$ **b.** $\sqrt{-\pi}$ **c.** $\sqrt{-4}\sqrt{-9}$

Solution

a. $\sqrt{-12} = \sqrt{12}\, i = 2\sqrt{3}\, i$ **b.** $\sqrt{-\pi} = \sqrt{\pi}\, i$

c. $\sqrt{-4}\,\sqrt{-9} = (\sqrt{4}\, i)(\sqrt{9}\, i) = (2i)(3i) = 6i^2 = 6(-1) = -6.$

CAUTION: When simplifying an expression involving square roots of negative numbers, always begin by using the definition to express the square roots in terms of i. Failure to do this can lead to incorrect results. For example, $\sqrt{-4}\,\sqrt{-4} = (2i)(2i) = 4i^2 = -4$, but $\sqrt{-4}\,\sqrt{-4} \neq \sqrt{(-4)(-4)} = \sqrt{16} = 4.$

Algebraic Properties of Complex Numbers

We begin our discussion of the algebraic properties of complex numbers by defining equality of two complex numbers.

Equality of Complex Numbers

Let $z_1 = a + bi$ and let $z_2 = c + di$ be complex numbers. We say that z_1 and z_2 are **equal**, and write $z_1 = z_2$, if and only if $a = c$ and $b = d$.

Thus two complex numbers are equal if and only if their real parts are equal and their imaginary parts are equal. For example, $a + bi = -3 + 5i$ if and only if $a = -3$ and $b = 5$.

The sum of two complex numbers is a complex number whose real part is the sum of the real parts of the two numbers and whose imaginary part is the sum of their imaginary parts. The subtraction of complex numbers is defined in a similar manner.

Sum and Difference of Complex Numbers

Let $z_1 = a + bi$ and $z_2 = c + di$ be complex numbers. Then

$$z_1 + z_2 = (a + c) + (b + d)i,$$
$$z_1 - z_2 = (a - c) + (b - d)i.$$

EXAMPLE 14

If $z_1 = -3 + 4i$, $z_2 = 5 - \frac{1}{2}i$, and $z_3 = \sqrt{3}\, i$, find each of the following:

a. $z_1 + z_2$ **b.** $z_2 + z_3$ **c.** $z_3 - z_1$

Solution

a. $z_1 + z_2 = (-3 + 4i) + (5 - \frac{1}{2}i) = (-3 + 5) + (4 - \frac{1}{2})i = 2 + \frac{7}{2}i$

b. $z_2 + z_3 = (5 - \frac{1}{2}i) + \sqrt{3}i = 5 + (-\frac{1}{2} + \sqrt{3})i = 5 + \dfrac{2\sqrt{3} - 1}{2}i$

c. $z_3 - z_1 = \sqrt{3}\, i - (-3 + 4i) = 3 + (\sqrt{3} - 4)i$

We can multiply complex numbers as if they were binomial expressions in i, replacing i^2 by -1 whenever it appears.

$$\begin{aligned}(a + bi)(c + di) &= ac + bci + adi + bdi^2 \\ &= ac + bd(-1) + (bc + ad)i \\ &= (ac - bd) + (ad + bc)i.\end{aligned}$$

This leads us to the following definition:

Product of Complex Numbers

Let $z_1 = a + bi$ and $z_2 = c + di$ be complex numbers. Then

$$z_1 \cdot z_2 = (a + bi)(c + di) = (ac - bd) + (ad + bc)i.$$

EXAMPLE 15

Let $z_1 = -2 + 3i$, $z_2 = 5 - i$, and $z_3 = -4i$. Express each of the following products in the form $a + bi$:

a. $z_1 \cdot z_2$ **b.** $z_2 \cdot z_3$ **c.** z_1^2

Solution

a.
$$\begin{aligned}z_1 \cdot z_2 &= (-2 + 3i)(5 - i) \\ &= -10 + 15i + 2i - 3i^2 \\ &= -10 - 3(-1) + 17i \\ &= -7 + 17i.\end{aligned}$$

b.
$$\begin{aligned}z_2 \cdot z_3 &= (5 - i)(-4i) = -20i + 4i^2 = -20i + 4(-1) \\ &= -4 - 20i.\end{aligned}$$

c.
$$\begin{aligned}z_1^2 = z_1 \cdot z_1 &= (-2 + 3i)(-2 + 3i) \\ &= 4 - 6i - 6i + 9i^2 \\ &= 4 + 9(-1) - 12i \\ &= -5 - 12i.\end{aligned}$$

Continuing our discussion of the algebra of complex numbers, we define the **conjugate** of the complex number $z = a + bi$ to be the complex number $\bar{z} = a - bi$. For example, the conjugate of $z_1 = -5 + 2i$ is $\bar{z}_1 = -5 - 2i$, and the conjugate of $z_2 = \frac{1}{2} - (\sqrt{3}/2)i$ is $\bar{z}_2 = \frac{1}{2} + (\sqrt{3}/2)i$. The product of a complex number and its conjugate is always a real number, for if $z = a + bi$, then $\bar{z} = a - bi$ and

$$\begin{aligned}z \cdot \bar{z} = (a + bi)(a - bi) &= a^2 + abi - abi - b^2i^2 \\ &= a^2 - b^2(-1) = a^2 + b^2.\end{aligned}$$

It is this property of conjugates that enables us to find the quotient of two complex numbers.

Consider the quotient $(4 - 3i)/(7 + 2i)$. We can express this quotient in the form $a + bi$ by multiplying both numerator and denominator by the conjugate of the denominator.

$$\frac{4 - 3i}{7 + 2i} = \frac{4 - 3i}{7 + 2i} \cdot \frac{7 - 2i}{7 - 2i}$$

$$= \frac{28 - 21i - 8i + 6i^2}{7^2 + 2^2}$$

$$= \frac{22 - 29i}{53}$$

$$= \frac{22}{53} - \frac{29}{53}i.$$

In general if $c + di \neq 0$,

$$\frac{a + bi}{c + di} = \frac{a + bi}{c + di} \cdot \frac{c - di}{c - di}$$

$$= \frac{(ac + bd) + (bc - ad)i}{c^2 + d^2}$$

$$= \frac{ac + bd}{c^2 + d^2} + \frac{bc - ad}{c^2 + d^2} i,$$

which leads us to the following definition:

Quotient of Complex Numbers

Let $z_1 = a + bi$ and $z_2 = c + di$ be complex numbers. If $z_2 \neq 0$, then

$$\frac{z_1}{z_2} = \frac{a + bi}{c + di} = \left(\frac{ac + bd}{c^2 + d^2} \right) + \left(\frac{bc - ad}{c^2 + d^2} \right) i.$$

Since the quotient z_1/z_2 ($z_2 \neq 0$) can be obtained by multiplying and dividing by $\bar{z}_2$, there is no need to memorize this definition. The following example illustrates this fact:

EXAMPLE 16

Express each of the following quotients in the form $a + bi$:

a. $\dfrac{1 - 4i}{6 + i}$ **b.** $\dfrac{2 - \frac{1}{2}i}{\frac{3}{4}i}$

Solution

a. $\dfrac{1 - 4i}{6 + i} = \dfrac{1 - 4i}{6 + i} \cdot \dfrac{6 - i}{6 - i}$

$$= \frac{6 - 24i - i + 4i^2}{36 - i^2}$$

$$= \frac{2 - 25i}{37}$$

$$= \frac{2}{37} - \frac{25}{37}i.$$

b. $\dfrac{2 - \frac{1}{2}i}{\frac{3}{4}i} = \dfrac{2 - \frac{1}{2}i}{\frac{3}{4}i} \cdot \dfrac{-\frac{3}{4}i}{-\frac{3}{4}i}$

$$= \frac{-\frac{3}{2}i + \frac{3}{8}i^2}{-\frac{9}{16}i^2}$$

$$= \frac{-\frac{3}{8} - \frac{3}{2}i}{\frac{9}{16}}$$

$$= -\frac{2}{3} - \frac{8}{3}i.$$

The set of complex numbers $\mathscr{C}$, under the operations of addition and multiplication defined earlier in this section, satisfies the same field properties that real numbers do under ordinary addition and multiplication.

Field Properties of Complex Numbers

If z_1, z_2, and z_3 are complex numbers, then

(i)   Closure properties $z_1 + z_2$ and $z_1 z_2$ are complex numbers
(ii) Commutative properties $z_1 + z_2 = z_2 + z_1$ and $z_1 z_2 = z_2 z_1$
(iii) Associative properties $z_1 + (z_2 + z_3) = (z_1 + z_2) + z_3$
                                  and $z_1(z_2 z_3) = (z_1 z_2)z_3$
(iv) Distributive property $z_1(z_2 + z_3) = z_1 z_2 + z_1 z_3$
(v)   Identity properties   The identity elements for addition and multiplication are $0 = 0 + 0i$ and $1 = 1 + 0i$, respectively.
(vi) Inverse properties   The additive inverse of $z = a + bi$ is $-z = -a - bi$, and if $z \neq 0$, the multiplicative inverse or reciprocal of z is $1/z = 1/(a + bi)$.

EXAMPLE 17

Express each of the following in the form $a + bi$:

a.   the additive inverse of $-5 + 13i$ **b.**   the reciprocal of $6 - 2i$

Solution

a.   The additive inverse of $-5 + 13i$ is $-(-5 + 13i) = 5 - 13i$.
b.   The reciprocal of $6 - 2i$ is

$$\frac{1}{6 - 2i} = \frac{1}{6 - 2i} \cdot \frac{6 + 2i}{6 + 2i}$$

$$= \frac{6 + 2i}{40}$$

$$= \frac{3}{20} + \frac{1}{20}i.$$

The Absolute Value of a Complex Number

The concept of absolute value provides yet another parallel between real and complex numbers.

FIGURE 1.12

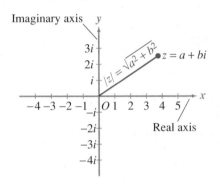

Absolute Value of a Complex Number

If $z = a + bi$ is a complex number, the **absolute value** or **modulus** of z is defined by

$$|z| = \sqrt{a^2 + b^2}.$$

Just as $|a|$ denotes the distance between the point a and 0 on the number line, $|z|$ represents the distance between the point z and the origin O in the complex plane (see Figure 1.12). To see that the definition of the absolute value of a complex number is consistent with the definition of the absolute value of a real number, let $z = a$, where a is a real number. Then

$$|z| = |a + 0i| = \sqrt{a^2 + 0^2} = \sqrt{a^2} = |a|.$$

EXAMPLE 18

Find the absolute value of each of the following complex numbers:

a. $-5 + 2i$ **b.** $-13i$

Solution

a. $|-5 + 2i| = \sqrt{(-5)^2 + 2^2} = \sqrt{29}$

b. $|-13i| = \sqrt{0^2 + (-13)^2} = \sqrt{169} = 13$

The following theorem, stated without proof, gives several properties of absolute values of complex numbers:

THEOREM 1.2   **Properties of Absolute Values of Complex Numbers**

Let z, z_1, and z_2 be complex numbers. Then

(i) $|z_1 z_2| = |z_1|\,|z_2|$;

(ii) $\left|\dfrac{z_1}{z_2}\right| = \dfrac{|z_1|}{|z_2|}$,   if $z_2 \neq 0$;

(iii) $|\bar{z}| = |z|$;

(iv) $|z^n| = |z|^n$, n a positive integer.

EXAMPLE 19

Find the absolute value of each of the following complex numbers:

a. $\dfrac{3 + 7i}{-5i}$ **b.** $\left(\dfrac{1}{2} - \dfrac{\sqrt{3}}{2}i\right)^{10}$

Solution

a. $\left|\dfrac{3 + 7i}{-5i}\right| = \dfrac{|3 + 7i|}{|-5i|}$ by part (ii) of Theorem 1.2

$$= \frac{\sqrt{3^2 + 7^2}}{\sqrt{(-5)^2}}$$

$$= \frac{\sqrt{58}}{5}.$$

b. $\left| \left(\frac{1}{2} - \frac{\sqrt{3}}{2}i \right)^{10} \right| = \left| \frac{1}{2} - \frac{\sqrt{3}}{2}i \right|^{10}$ **by part (iv) of Theorem 1.2**

$$= \left(\sqrt{\left(\frac{1}{2}\right)^2 + \left(-\frac{\sqrt{3}}{2}\right)^2} \right)^{10}$$

$$= \left(\sqrt{\frac{1}{4} + \frac{3}{4}} \right)^{10}$$

$$= 1^{10}$$

$$= 1.$$

Powers of i

We conclude this section by illustrating a method for evaluating i^n, where n is a positive integer greater than 2. The method uses the fact that $i^2 = -1$ and the fact that any even power of -1 is equal to 1, while any odd power of -1 is equal to -1.

EXAMPLE 20

Express each of the following in the form $a + bi$:

a. i^{28} **b.** i^{315}

Solution

a. $i^{28} = (i^2)^{14}$

$\quad\quad = (-1)^{14}$

$\quad\quad = 1.$ any even power of -1 is equal to 1

b. $i^{315} = i^{314} \cdot i$ factoring out the maximum even power of i

$\quad\quad = (i^2)^{157} \cdot i$

$\quad\quad = (-1)^{157} \cdot i$

$\quad\quad = -1 \cdot i$ any odd power of -1 is equal to -1

$\quad\quad = -i.$

EXERCISES 1.4

In Exercises 1–8, plot each point in the complex plane.

1. $-2 + 3i$ **2.** $\frac{1}{2} - \frac{3}{4}i$

3. $\frac{7}{2}i$ **4.** $-1 - i$

5. $-4 + 3.2i$ **6.** $\frac{1}{2} + (\sqrt{3}/2)i$

7. $\sqrt{3} - 2i$ **8.** $\pi - i$

In Exercises 9–18, evaluate each expression.

9. $\sqrt{-81}$ **10.** $\sqrt{-18}$

11. $-\sqrt{-32}$ **12.** $-\sqrt{-14}$

13. $\sqrt{-12}\,\sqrt{-12}$ **14.** $(\sqrt{-9})(-\sqrt{9})$

15. $(-\sqrt{-3})(\sqrt{-12})$ **16.** $(\sqrt{-6})(\sqrt{-8})$

17. $\sqrt{-27}/\sqrt{-3}$ **18.** $\sqrt{-56}/\sqrt{-7}$

In Exercises 19–36, perform the indicated operations. Express answers in the form $a + bi$.

19. $(2 + 3i) + (5 - 7i)$ **20.** $(-1 + i) - (4 - i)$

21. $(\frac{1}{4} - 2i) - (-8 + \frac{3}{2}i)$ **22.** $(-\frac{7}{2} - 3i) + \frac{3}{2}i$

23. $(-7 - i)(3 + 5i)$ **24.** $(2 - 4i)(-3 + 2i)$

25. $(1 - 2i)^2$

26. $(-5 + 8i)^2$

27. $(-5 + i)^3$

28. $(\sqrt{3} - 2i)^3$

29. $(3 - i)(3 + i)(-i)$

30. $(2 + \sqrt{2}\,i)(2 - \sqrt{2}\,i)(\sqrt{2}\,i)$

31. $\dfrac{1}{1 - 4i}$

32. $\dfrac{1}{-3 + i}$

33. $\dfrac{7 + 6i}{-i}$

34. $\dfrac{2i}{-1 + 4i}$

35. $\dfrac{1 - \pi i}{1 + \pi i}$

36. $\dfrac{1}{i} - \dfrac{i}{2}$

In Exercises 37–40, find the additive inverse of each complex number.

37. $-2 + 4i$

38. $1 - i$

39. $-\frac{1}{2}i$

40. $\frac{1}{2} + (\sqrt{3}/2)i$

In Exercises 41–44, find the reciprocal of each complex number. Express answers in the form $a + bi$.

41. $1 - 3i$

42. $-6i$

43. $(\sqrt{2}/2) + 2i$

44. $-5 + (\sqrt{3}/2)i$

In Exercises 45–56, evaluate each expression.

45. i^{12}

46. i^{37}

47. i^{239}

48. i^{1388}

49. $(-i)^{42}$

50. $(-2i)^9$

51. $-i^{223}$

52. $-i^{80}$

53. i^{-28}

54. i^{-119}

55. i^{31}/i^{19}

56. $i^{26} \cdot i^{-37}$

In Exercises 57–70, find the absolute value of each complex number.

57. $-3 + 7i$

58. $12 - 8i$

59. $\frac{1}{2} - \frac{3}{4}i$

60. $-(\sqrt{3}/2) + \frac{1}{2}i$

61. $-16i$

62. -9

63. $(-3 + 4i)^2$

64. $(1 - i)^3$

65. i^{51}

66. i^{104}

67. $-1/3i$

68. $(-2 + 3i)/-4i$

69. $\overline{12 - 6i}$

70. $\overline{3 + 4i}$

71. If $z_1 = z_2$, where $z_1 = -2 + 5i$ and $z_2 = x - yi$, find x and y.

72. Express $(1 + i)^2$ in the form $a + bi$, and use the result to express each of the following in the form $a + bi$:
  a. $(1 + i)^6$ **b.** $(1 + i)^9$ **c.** $(1 + i)^{10}$

73. A complex number z is called an **nth root of unity** if and only if $z^n = 1$.
  a. Show that $-\frac{1}{2} + (\sqrt{3}/2)i$ is a cube root of unity.
  b. Show that i is an eighth root of unity.

74. Verify each of the following properties of the complex number field:
  a. the commutative property of addition
  b. the associative property of multiplication
  c. the distributive law

75. If $z = a + bi$ is a complex number, verify that $|\overline{z}| = |z|$.

76. Let $z_1 = a + bi$ and $z_2 = c + di$ be complex numbers. Verify each of the following:
  a. $\overline{z_1 + z_2} = \overline{z_1} + \overline{z_2}$ **b.** $\overline{z_1 \cdot z_2} = \overline{z_1} \cdot \overline{z_2}$
  c. $\overline{z_1^n} = \overline{z_1}^n$, n any positive integer
  d. If z_1 is a real number, $\overline{z_1} = z_1$.

1.5 LINEAR AND QUADRATIC EQUATIONS

An **equation** is a statement that two mathematical expressions are equal. The following are examples of equations in one variable:

$$2(x + 3) - (x - 5) = 0$$
$$t = 6 - 2t$$
$$3y^4 = 25 - y^2$$
$$2x^2 - 1 = 17.$$

The values that are assigned to the variable determine whether or not an equation is true. For example, consider the equation

$$2x^2 - 1 = 17. \qquad (1)$$

If $x = 2$, equation (1) is false since $2(2)^2 - 1 = 7 \neq 17$. On the other hand, if $x = 3$, equation (1) is true since $2(3)^2 - 1 = 17$. The values of the variable for which an equation is true are called **solutions** or **roots** of the equation. For example, 3 is a solution of equation (1). The set of all solutions of an equation is called the **solution set**. It can be shown that the solution set for equation (1) above is $\{-3, 3\}$.

We **solve** an equation by finding its solution set. Two equations that have the same solution set are called **equivalent equations**. The general procedure for solving an equation is to replace the equation with equivalent equations until an equation is obtained whose solution set is apparent. Recall that we can convert an equation to an equivalent equation by adding or subtracting the same expression on both sides or by multiplying or dividing both sides by the same nonzero expression. For example, the following equations are all equivalent:

$$2x + 1 = 5, \qquad 2x = 4, \qquad x = 2. \tag{2}$$

Each of the equations in (2) above is a linear equation. In general, a **linear equation** in one variable x is an equation that can be written in the form

$$ax + b = 0,$$

where a and b are real numbers and $a \neq 0$. In the following example, we solve some linear equations:

EXAMPLE 21

Find the solution set of each equation.

a. $2(x - 5) - 3(1 - x) = 12x - 2(x - 3)$

b. $\dfrac{x + 1}{5} - \dfrac{x}{10} = \dfrac{7}{6} + 3x$

Solution

a. The following equations are equivalent:

$$
\begin{aligned}
2(x - 5) - 3(1 - x) &= 12x - 2(x - 3) \\
2x - 10 - 3 + 3x &= 12x - 2x + 6 \\
5x - 13 &= 10x + 6 \\
-5x &= 19 \\
x &= -\frac{19}{5}.
\end{aligned}
$$

Therefore, the solution set is $\{-\frac{19}{5}\}$.

b. We can simplify the equation

$$\frac{x + 1}{5} - \frac{x}{10} = \frac{7}{6} + 3x \tag{3}$$

by multiplying both sides by 30, the least common denominator (LCD) of all the fractions in the equation. Thus, equation (3) is equivalent to

$$30\left(\frac{x + 1}{5}\right) - 30\left(\frac{x}{10}\right) = 30\left(\frac{7}{6}\right) + 30(3x), \qquad \text{or}$$

$$6(x + 1) - 3x = 35 + 90x.$$

The latter equation is equivalent to each of the following:

$$
\begin{aligned}
6x + 6 - 3x &= 35 + 90x \\
3x + 6 &= 35 + 90x \\
-87x &= 29 \\
x &= -\frac{29}{87} = -\frac{1}{3}.
\end{aligned}
$$

Hence, the solution set is $\{-\frac{1}{3}\}$.

Linear equations often occur in applications, as the following example illustrates:

E X A M P L E 22

A movie theater charges \$4.50 admission for adults and \$3.00 for children. The total income from ticket sales for a recent weekend was \$4200.00. There were four times as many adult tickets sold as children's tickets. How many tickets of each kind were sold?

Solution If x represents the number of children's tickets sold, then $4x$ represents the number of adult tickets sold. It follows that the income from the children's tickets is given by $(3.00)x$ and the income from the adult tickets is given by $(4.50)(4x)$. Since the total income from ticket sales is \$4,200.00, we obtain the equation

$$(3.00)x + (4.50)(4x) = 4200.00.$$

Multiplying both sides by 100 to eliminate the decimals and combining like terms, we obtain

$$300x + 450(4x) = 420,000$$
$$2,100x = 420,000$$
$$x = 200.$$

Therefore, 200 children's tickets were sold and $4(200) = 800$ adult tickets were sold. _____

We are sometimes interested in solving a given mathematical formula for a particular variable in terms of the other variables.

E X A M P L E 23

Solve each formula for the indicated variable.

a. $P = 2L + 2W$, for W **b.** $A = P(1 + rt)$, for t.

Solution

a. Solving for W, we have

$$P = 2L + 2W$$
$$P - 2L = 2W$$
$$W = \frac{P - 2L}{2}.$$

b. Solving for t, we get

$$A = P(1 + rt)$$
$$A = P + Prt$$
$$A - P = Prt$$
$$t = \frac{A - P}{Pr}. \qquad \text{provided } P \neq 0 \text{ and } r \neq 0$$

_____

Quadratic Equations

We now turn our attention to quadratic equations. A **quadratic equation** in the variable x is an equation that can be written in the form

$$ax^2 + bx + c = 0,$$

where a, b, and c are real numbers and $a \neq 0$. For example, the equation

$$2x^2 = x - 3$$

is a quadratic equation since it can be written in the form

$$2x^2 - x + 3 = 0.$$

A quadratic equation, written in the form $ax^2 + bx + c = 0$, is said to be in **standard form**.

There are several methods which can be used to solve quadratic equations. We illustrate three of these methods, beginning with the **factoring method**. When the expression $ax^2 + bx + c$ can be readily factored, the roots of the equation

$$ax^2 + bx + c = 0$$

can be obtained by using the following property of real numbers:

Zero-Product Property

Let a and b be real numbers. Then

$$ab = 0 \quad \text{if and only if} \quad a = 0 \text{ or } b = 0.$$

We illustrate the factoring method in an example.

E X A M P L E 24

Use the factoring method to find the solution set of each equation.

a. $6x^2 + x - 2 = 0$ **b.** $4x^2 = 4x - 1$

Solution

a. Factoring and using the zero-product property, we have

$$6x^2 + x - 2 = 0$$
$$(3x + 2)(2x - 1) = 0$$
$$3x + 2 = 0 \quad \text{or} \quad 2x - 1 = 0$$
$$3x = -2 \quad \text{or} \quad 2x = 1$$
$$x = -\frac{2}{3} \quad \text{or} \quad x = \frac{1}{2}.$$

Thus, the solution set is $\{-\frac{2}{3}, \frac{1}{2}\}$.

b. In standard form, the equation

$$4x^2 = 4x - 1$$

becomes $$4x^2 - 4x + 1 = 0.$$

Proceeding as in part (a), we have

$$(2x - 1)(2x - 1) = 0.$$

By the zero-product property, the latter equation is equivalent to

$$2x - 1 = 0, \qquad \text{or}$$

$$x = \frac{1}{2}.$$

Hence, the solution set is $\{\frac{1}{2}\}$. _____

Notice that the quadratic equation $4x^2 - 4x + 1 = 0$ from part (b) of the previous example has only one solution. In general, when a quadratic equation has only one solution, that solution is called a **double root** or a **root of multiplicity two**.

As we see from the previous example, the factoring method (when applicable) is an efficient method for solving quadratic equations. Sometimes, however, quadratic equations cannot readily be solved by this method. For example, consider the equations

$$x^2 - 2x - 4 = 0 \qquad \text{and} \qquad 3x^2 + 5x - 1 = 0. \qquad \textbf{(4)}$$

Although neither of the equations in (4) can readily be solved by the factoring method, both can be solved by using either of the next two methods we discuss.

Observe that in the solution of part (b) of Example 24 above, the quadratic polynomial $4x^2 - 4x + 1$ has the factored form

$$(2x - 1)(2x - 1) = (2x - 1)^2.$$

In general, any quadratic polynomial that has the factored form $(ax + b)^2$ is called a **perfect square polynomial**, since it is the square of a linear polynomial. This concept of perfect square quadratic polynomials is an integral part of our second method for solving quadratic equations. This method, called completing the square, makes use of the following property:

Square Root Property

For any real number d,

$$x^2 = d \quad \text{if and only if} \quad x = \pm\sqrt{d}.$$

From any quadratic polynomial of the form $x^2 + kx$, where k is a real number, we can obtain a perfect square by adding $(k/2)^2 = k^2/4$, which is the square of half the coefficient of x. For example, from $x^2 + 3x$, we can obtain a perfect square by adding $(\frac{3}{2})^2 = \frac{9}{4}$ as follows:

$$x^2 + 3x + \frac{9}{4} = \left(x + \frac{3}{2}\right)\left(x + \frac{3}{2}\right) = \left(x + \frac{3}{2}\right)^2.$$

The process of adding $(k/2)^2 = k^2/4$ to $x^2 + kx$ to obtain the perfect square polynomial

$$x^2 + kx + \frac{k^2}{4} = \left(x + \frac{k}{2}\right)\left(x + \frac{k}{2}\right) = \left(x + \frac{k}{2}\right)^2$$

is called **completing the square**.

To see how completing the square can be used to solve a quadratic equation, consider the equation

$$5x^2 + 30x - 15 = 0.$$

Since the coefficient of x^2 is $5 \neq 1$, we divide both sides by 5 to obtain

$$x^2 + 6x - 3 = 0.$$

Adding 3 to both sides yields

$$x^2 + 6x = 3.$$

Since the quadratic polynomial $x^2 + 6x$ has the form $x^2 + kx$, we can complete the square by adding $(k/2)^2 = (\frac{6}{2})^2 = 9$ to both sides to obtain

$$x^2 + 6x + 9 = 12, \quad \text{or}$$
$$(x + 3)^2 = 12.$$

By the square root property,

$$x + 3 = \pm\sqrt{12} = \pm 2\sqrt{3}.$$

Thus, $\qquad x = -3 + 2\sqrt{3} \qquad \text{or} \qquad x = -3 - 2\sqrt{3}.$

Therefore, there are two real solutions, and the solution set is

$$\{-3 - 2\sqrt{3}, \ -3 + 2\sqrt{3}\}.$$

EXAMPLE 25

Use the method of completing the square to find the solution set of each equation.

a. $4x^2 - 4x + 2 = 0$ **b.** $5x^2 + 4 = -12x - 4x^2$

Solution

a. We begin by dividing both sides of the equation $4x^2 - 4x + 2 = 0$ by 4 to obtain

$$x^2 - x + \frac{1}{2} = 0.$$

Completing the square, we have

$$x^2 - x = -\frac{1}{2}$$

$$x^2 - x + \frac{1}{4} = -\frac{1}{2} + \frac{1}{4}$$

$$\left(x - \frac{1}{2}\right)^2 = -\frac{1}{4}.$$

By the square root property,

$$x - \frac{1}{2} = \pm\sqrt{-\frac{1}{4}} = \pm\frac{1}{2}i.$$

Hence, $\qquad x = \frac{1}{2} + \frac{1}{2}i \qquad \text{or} \qquad x = \frac{1}{2} - \frac{1}{2}i.$

Thus, there are two complex solutions, which are conjugates, and the solution set is $\{\frac{1}{2} - \frac{1}{2}i, \frac{1}{2} + \frac{1}{2}i\}$.

b. In standard form, the equation $5x^2 + 4 = -12x - 4x^2$ becomes

$$9x^2 + 12x + 4 = 0.$$

Completing the square, we get

$$x^2 + \frac{4}{3}x + \frac{4}{9} = 0$$

$$x^2 + \frac{4}{3}x = -\frac{4}{9}$$

$$x^2 + \frac{4}{3}x + \frac{4}{9} = -\frac{4}{9} + \frac{4}{9}$$

$$\left(x + \frac{2}{3}\right)^2 = 0.$$

Hence,

$$x + \frac{2}{3} = 0, \qquad \text{or}$$

$$x = -\frac{2}{3}.$$

Therefore, $-\frac{2}{3}$ is a double root, and the solution set is $\left\{-\frac{2}{3}\right\}$.

The Quadratic Formula

Although the method of completing the square can be used to solve any quadratic equation, there is another method that is often easier to use. This third method, which also applies to any quadratic equation, is based upon the use of a formula called the **quadratic formula**. To derive the quadratic formula, we solve the quadratic equation

$$ax^2 + bx + c = 0$$

by completing the square as follows:

Subtracting the constant term from both sides and dividing both sides by $a \neq 0$, we have

$$x^2 + \frac{b}{a}x = -\frac{c}{a}. \tag{5}$$

Now, to complete the square, we add

$$\left(\frac{b}{2a}\right)^2 = \frac{b^2}{4a^2}$$

to both sides of equation (5) to obtain

$$x^2 + \frac{b}{a}x + \frac{b^2}{4a^2} = -\frac{c}{a} + \frac{b^2}{4a^2}, \qquad \text{or}$$

$$\left(x + \frac{b}{2a}\right)^2 = \frac{b^2 - 4ac}{4a^2}.$$

By the square root property,

$$x + \frac{b}{2a} = \pm\sqrt{\frac{b^2 - 4ac}{4a^2}}.$$

It follows that

$$x + \frac{b}{2a} = \pm\frac{\sqrt{b^2 - 4ac}}{2a}.$$ **(See Exercise 68.)**

Hence,

$$x + \frac{b}{2a} = \frac{\sqrt{b^2 - 4ac}}{2a} \quad \text{or} \quad x + \frac{b}{2a} = -\frac{\sqrt{b^2 - 4ac}}{2a}$$

$$x = \frac{-b + \sqrt{b^2 - 4ac}}{2a} \quad \text{or} \quad x = \frac{-b - \sqrt{b^2 - 4ac}}{2a}.$$

It is customary to express the solutions in the form

$$x = \frac{-b \pm \sqrt{b^2 - 4ac}}{2a}.$$

To summarize, we have

Quadratic formula

The solutions of the quadratic equation

$$ax^2 + bx + c = 0, \quad a \neq 0$$

are given by the formula

$$x = \frac{-b \pm \sqrt{b^2 - 4ac}}{2a}.$$

E X A M P L E 26

Use the quadratic formula to find the solution set of each equation.

a. $2x^2 - 3x - 5 = 0$ **b.** $4x(x + 3) = -9$
c. $2x^2 - 3x = -4$

Solution

a. For the equation $2x^2 - 3x - 5 = 0$, we see that $a = 2$, $b = -3$, and $c = -5$. Thus, we have

$$x = \frac{-(-3) \pm \sqrt{(-3)^2 - 4(2)(-5)}}{2(2)}$$

$$= \frac{3 \pm \sqrt{49}}{4}$$

$$= \frac{3 \pm 7}{4}.$$

Hence, $x = \dfrac{3 + 7}{4} = \dfrac{10}{4} = \dfrac{5}{2}$ or $x = \dfrac{3 - 7}{4} = -\dfrac{4}{4} = -1$.

Thus, there are two real solutions, and the solution set is $\{-1, \frac{5}{2}\}$.

b. In standard form, the equation $4x(x + 3) = -9$ becomes $4x^2 + 12x + 9 = 0$. Hence, $a = 4$, $b = 12$, and $c = 9$, and we have

$$x = \frac{-12 \pm \sqrt{(12)^2 - 4(4)(9)}}{2(4)}$$

$$= \frac{-12 \pm \sqrt{0}}{8}$$

$$= -\frac{3}{2}.$$

Therefore, $-\frac{3}{2}$ is a double root, and the solution set is $\{-\frac{3}{2}\}$.

c. Since the standard form of the equation $2x^2 - 3x = -4$ is $2x^2 - 3x + 4 = 0$, it follows that

$$x = \frac{-(-3) \pm \sqrt{(-3)^2 - 4(2)(4)}}{2(2)}$$

$$= \frac{3 \pm \sqrt{-23}}{4}$$

$$= \frac{3}{4} \pm \frac{\sqrt{23}}{4}i.$$

Hence, there are two complex solutions, which are conjugates, and the solution set is

$$\left\{ \frac{3}{4} - \frac{\sqrt{23}}{4}i, \frac{3}{4} + \frac{\sqrt{23}}{4}i \right\}.$$

The quantity $b^2 - 4ac$, which appears under the radical sign in the quadratic formula, is called the **discriminant**. The discriminant determines the nature of the solutions of the equation $ax^2 + bx + c = 0$. In part (a) of Example 26, the discriminant is positive, and there are two real solutions. In part (b) of Example 26, the discriminant is zero, and there is one real solution of multiplicity two (a double root). In part (c) of Example 26, where the discriminant is negative, there are two complex solutions which are conjugates. In general, we have the facts listed in Table 1.2.

TABLE 1.2

Discriminant	Nature of the solutions of $ax^2 + bx + c = 0$, where a, b, and c are real numbers and $a \neq 0$
$b^2 - 4ac > 0$	There are two distinct real solutions.
$b^2 - 4ac = 0$	There is one real solution of multiplicity two (a double root).
$b^2 - 4ac < 0$	There are two complex solutions, which are conjugates.

In the following examples, we see how quadratic equations are used in applications:

EXAMPLE 27

One leg of a right triangle is 14 meters shorter than the other. If the length of the hypotenuse is 26 meters, find the length of each leg.

Solution   If x represents the length of the longer leg, then $x - 14$ represents the length of the shorter leg. By the Pythagorean theorem, we have

$$x^2 + (x - 14)^2 = (26)^2.$$

Hence,
$$x^2 + (x^2 - 28x + 196) = 676$$
$$2x^2 - 28x - 480 = 0$$
$$x^2 - 14x - 240 = 0$$
$$(x - 24)(x + 10) = 0$$
$$x - 24 = 0 \quad \text{or} \quad x + 10 = 0$$
$$x = 24 \quad \text{or} \quad x = -10.$$

Since x represents length, we discard -10. Therefore, the lengths of the legs are 24 meters and $24 - 14 = 10$ meters.

EXAMPLE 28

Solve the equation $A = 3r^2 + 4r - 3$, for r.

Solution   We first note that the equation

$$A = 3r^2 + 4r - 3 \tag{6}$$

is a quadratic equation in the variable r. Expressing equation (6) in standard form, we have

$$3r^2 + 4r - (A + 3) = 0.$$

Hence, $a = 3$, $b = 4$, and $c = -(A + 3)$. By the quadratic formula, we have

$$r = \frac{-4 \pm \sqrt{4^2 - 4(3)[-(A + 3)]}}{2(3)}$$

$$= \frac{-4 \pm \sqrt{16 + 12(A + 3)}}{6}$$

$$= \frac{-4 \pm \sqrt{12A + 52}}{6}$$

$$= \frac{-4 \pm 2\sqrt{3A + 13}}{6}$$

$$= \frac{-2 \pm \sqrt{3A + 13}}{3}.$$

Therefore, $r = \dfrac{-2 - \sqrt{3A + 13}}{3}$ or $r = \dfrac{-2 + \sqrt{3A + 13}}{3}.$

EXERCISES 1.5

In Exercises 1–8, find the solution set of each equation.

1. $3x + 15 = 0$

2. $\dfrac{x - 2}{3} - \dfrac{x}{6} = \dfrac{7}{5} + 2x$

3. $2(3x - 1) - 4 = 5 + 7(x - 1)$

4. $\dfrac{3x - 1}{5} - \dfrac{2x + 1}{10} = \dfrac{1 - x}{4} - 3$

5. $\dfrac{3x}{4} - 6 = 2 + \dfrac{x}{3}$

6. $2(x - 3) + 6x = 2(x + 2)$

7. $0.01x - 0.1(x - 10) = 0.02x + 1$
8. $0.03x + 0.5(2x - 3) = 10$

In Exercises 9–20, solve each formula for the indicated variable.

9. $P = 2L + 2W$, for L (Perimeter of a rectangle)
10. $A = P(1 + rt)$, for r (Amount at simple interest)
11. $A = LW$, for W (Area of a rectangle)
12. $A = \frac{1}{2} bh$, for h (Area of a triangle)
13. $C = 2\pi r$, for r (Circumference of a circle)
14. $F = \frac{9}{5} C + 32$, for C (Celsius to Fahrenheit)
15. $C = \frac{5}{9} (F - 32)$, for F (Fahrenheit to Celsius)
16. $S = 2\pi rh + 2\pi r^2$, for h (Surface area of a cylinder)
17. $V = \pi r^2 h$, for h (Volume of a cylinder)
18. $I = Prt$, for t (Simple interest)
19. $Ax + By + C = 0$, for y (Equation of a line)
20. $S = B + Bxy$, for B

21. At a recent performance of the Community Theater, adult tickets were $5.00 and student tickets were $2.75. The total income from ticket sales was $798.75. There were three times as many adult tickets sold as student tickets. How many tickets of each kind were sold?

Photo R. Laird/FPG International

22. The length of a rectangular lot is 5 meters less than two times its width. The perimeter of the lot is 440 meters. Find the dimensions of the lot and its area.
23. A mechanic is paid time-and-one-half for each hour worked over 40 hours in a week. In a recent week, she worked 47 hours and earned $454.50. What is her standard hourly salary?

24. Tommy bought a pair of boots at a 40% off sale for $48. What was the original price of the boots?
25. A salesman is paid a weekly salary of $250 plus a 10% commission on the dollar amount of sales he makes each week. Find the dollar amount of sales in a week when his total salary was $302.
26. The campus bookstore has a 20% off sale on calculators. What was the original price of a calculator that is on sale for $30?
27. The cost of renting a lawn mower is $12 per day. A new lawn mower is on sale for $168. After how many days of renting will the rental fee be the same as the sale price of the new mower?
28. The population of a town is 25,000. If the population is growing at the rate of 1500 people per year, in how many years will the population be 31,750?

In Exercises 29–34, use the factoring method to find the solution set of each equation.

29. $6x^2 + 7x - 3 = 0$
30. $x^2 = -3(15 + 6x)$
31. $x^2 - 42 = -19x$
32. $2(x + 3) + 3x = 2x^2 - 6$
33. $x(6x - 5) = 4$
34. $x(3x + 1) = 2$

In Exercises 35–40, find the solution set of each equation by completing the square.

35. $2x^2 - 4x - 3 = 0$
36. $3x + 2 = 2x^2$
37. $3x(x - 4) = -13$
38. $x^2 - 2x = 1$
39. $4x^2 = 3(4x - 3)$
40. $x(3x + 6) = 9$

In Exercises 41–46, use the quadratic formula to find the solution set of each equation.

41. $2x^2 = 4x - 1$
42. $x^2 = 4x - 13$
43. $x^2 = 2(x - 1)$
44. $4x^2 = 4x - 1$
45. $9x^2 = -8(3x + 2)$
46. $x^2 = 4(x + 2)$

In Exercises 47–52, find the discriminant and, without solving the equation, discuss the nature of its roots.

47. $x^2 - 2x - 4 = 0$
48. $3x^2 + 3x + 1 = 0$
49. $9x^2 + 6x + 1 = 0$
50. $4x^2 + 3x - 10 = 0$
51. $5x^2 - 2x + 1 = 0$
52. $x^2 + 5x + 4 = 0$

In Exercises 53–60, find the solution set of each equation.

53. $3x^2 + x = 4(3x + 1)$
54. $16x(2 - x) = 8$
55. $7x = 2x^2 - 1$
56. $3x^2 = 4x + 5$
57. $5x^2 = 4x$
58. $x(5x - 2) = 4(2 - 5x)$
59. $9x^2 = 5(6x - 5)$
60. $x(6 - x) = 18$

In Exercises 61–66, solve each formula for the indicated variable.

61. $A = \pi r^2$, for r (Area of a circle)
62. $x^2 + y^2 = r^2$, for y (Equation of a circle)
63. $D = b^2 - 4ac$, for b (Discriminant)
64. $V = \pi r^2 h$, for r (Volume of a cylinder)
65. $A = 4h^2 - 3h - 5$, for h
66. $S = 2\pi rh + 2\pi r^2$, for r (Surface area of a cylinder)
67. Consider the solutions of the quadratic equation $ax^2 + bx + c = 0$:

$$r_1 = \frac{-b + \sqrt{b^2 - 4ac}}{2a} \quad \text{and}$$

$$r_2 = \frac{-b - \sqrt{b^2 - 4ac}}{2a}.$$

Show that $r_1 + r_2 = -b/a$ and that $r_1 \cdot r_2 = c/a$.

68. Explain why the expressions $\pm 2|a|$ and $\pm 2a$, in the derivation of the quadratic formula, are identical for all real values of a.

69. One leg of a right triangle is 21 meters shorter than the other, and the hypotenuse is 39 meters long. Find the lengths of the legs.

70. Two times the square of an integer is 15 more than the integer. Find the integer.

71. The length of a rectangular lot is 6 feet more than twice its width. The area of the lot is 360 square feet. Find the dimensions of the lot and its perimeter.

72. An open rectangular box with a volume of 338 cubic inches is formed from a square piece of cardboard by cutting out 2-inch squares from the corners and then folding up the flaps. Find the dimensions and the area of the original piece of cardboard.

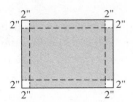

73. The product of two positive numbers is 35. Find the two numbers if one number is 9 less than twice the other.

74. The sum of the squares of two consecutive positive even integers is 340. Find the integers.

75. A child, who is sitting on a Ferris wheel, throws a ball upward. The height h of the ball above the ground t seconds after it is thrown is given by the equation

$$h = -16t^2 + 32t + 240,$$

where h is in feet. When does the ball strike the ground?

76. A rectangular sheet of cardboard is 46 centimeters wide and 32 centimeters long. How many centimeters must be subtracted from both its length and its width (see the accompanying figure) so that the area decreases by 432 square centimeters?

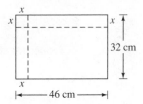

77. The perimeter of a rectangle is 40 meters and its area is 91 square meters. Find the dimensions of the rectangle.

78. The area of a triangle is 120 square inches. The height is 14 inches shorter than the base. Find the base and height of the triangle.

79. Three times the square of a certain positive integer is 45 more than 6 times the integer. Find the integer.

80. A rancher wants to use 600 yards of fencing to enclose two identical rectangular grazing pens (see the accompanying figure). If the total enclosed area is to be 15,000 square yards, find the dimensions of each pen.

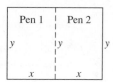

1.6   LINEAR AND QUADRATIC INEQUALITIES

If the equality symbol in an equation is replaced by one of the symbols $<$, $\leq$, $>$, or $\geq$, we obtain an **inequality**. The following are examples of inequalities in one variable:

$$7x > 5 + 2x$$
$$3t - 10 \leq 2t + 9$$
$$y^2 \geq 10 - 3y^3$$
$$2x - 1 < 5.$$

The values that are assigned to the variable determine whether or not the inequality is true. For example, consider the inequality

$$2x - 1 < 5. \qquad \qquad (1)$$

If $x = 4$, inequality (1) is false since $2(4) - 1 = 7 \not< 5$. On the other hand, if $x = 2$, inequality (1) is true since $2(2) - 1 = 3 < 5$. The values of the variables for which an inequality is true are called **solutions** of the inequality. For example, 2 is a solution of inequality (1). The set of all solutions of an inequality is called the **solution set**. It can be shown that the solution set of inequality (1) is $\{x \mid x < 3\}$.

We **solve** an inequality by finding its solution set. Two inequalities that have the same solution set are called **equivalent inequalities**. The general procedure for solving an inequality is to replace the inequality with equivalent inequalities until an inequality is obtained whose solution set is apparent. Some important properties of inequalities are summarized below.

Properties of Inequalities

Let a, b, and c be real numbers. Then

(i) $a < b$ if and only if $a + c < b + c$;   (Additive law of order)

(ii) If $c > 0$, then $a < b$ if and only if $ac < bc$;   (Multiplicative law of order)

(iii) If $c < 0$, then $a < b$ if and only if $ac > bc$;   (Reversal law of order)

(iv) If $a < b$ and $b < c$, then $a < c$.   (Transitive law of order)

From properties (i) and (ii), we see that inequalities behave like equations whenever we add or subtract the same expression on both sides or multiply or divide both sides by the same *positive* expression. However, according to property (iii), whenever we multiply or divide both sides by the same *negative* expression, we *must reverse the inequality symbol*.

Linear Inequalities

Perhaps the simplest type of inequality is a linear inequality. A **linear inequality** in one variable x is an inequality that can be written in the form

$$ax + b < 0 \qquad \text{or} \qquad ax + b \le 0,$$

where a and b are real numbers and $a \ne 0$. Our first example involves linear inequalities.

E X A M P L E   29

Find the solution set of each inequality. Display each solution set on the number line.

a. $\dfrac{x - 1}{10} - \dfrac{x - 3}{4} > \dfrac{x - 7}{2} + \dfrac{1}{5}$ **b.** $10 \le 1 - 3x < 22$

Solution

a. To simplify the inequality

$$\frac{x - 1}{10} - \frac{x - 3}{4} > \frac{x - 7}{2} + \frac{1}{5} \tag{2}$$

we first multiply both sides by 20, the LCD. Therefore, inequality (2) is equivalent to

$$20\left(\frac{x-1}{10}\right) - 20\left(\frac{x-3}{4}\right) > 20\left(\frac{x-7}{2}\right) + 20\left(\frac{1}{5}\right)$$

or $2(x-1) - 5(x-3) > 10(x-7) + 4.$ **(3)**

Inequality (3) is equivalent to each of the following:

$$2x - 2 - 5x + 15 > 10x - 70 + 4$$
$$-3x + 13 > 10x - 66$$
$$-13x > -79$$
$$x < \frac{79}{13}. \qquad \text{Reversal law of order}$$

FIGURE 1.13

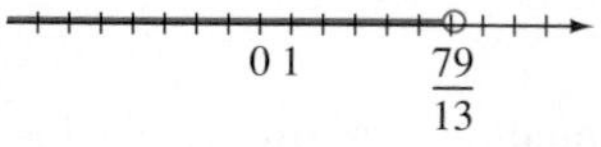

Thus, the solution set is $\{x \mid x < \frac{79}{13}\} = (-\infty, \frac{79}{13})$, which is shown in Figure 1.13.

b. We note that a real number x is a solution of the inequality

$$10 \le 1 - 3x < 22 \qquad \qquad \textbf{(4)}$$

if and only if it is a solution of both of the inequalities

$$10 \le 1 - 3x \qquad \text{and} \qquad 1 - 3x < 22. \qquad \textbf{(5)}$$

We could solve each of the inequalities in (5) separately. The solution set for inequality (4) could then be found by forming the intersection of the two solution sets thus obtained. However, it is more efficient to solve inequality (4) by working with both inequalities simultaneously, as we now illustrate. Subtracting 1 from each part of inequality (4) yields

$$9 \le -3x < 21.$$

Dividing each part of the latter inequality by -3 and reversing the inequality symbols, we have

$$-3 \ge x > -7 \qquad \text{or} \qquad -7 < x \le -3.$$

FIGURE 1.14

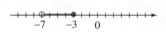

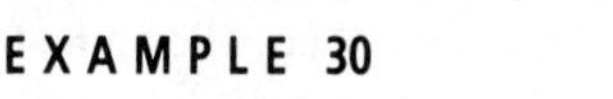

It follows that the solution set is $\{x \mid -7 < x \le -3\} = (-7, -3]$, which is shown in Figure 1.14.

In our next example, we see how linear inequalities can be used in applications.

EXAMPLE 30

In a certain city, the temperature in degrees Fahrenheit has the average annual range: $24.8 \le F \le 95$. Use the formula $F = (9C/5) + 32$, which relates F (degrees Fahrenheit) and C (degrees Celsius), to find the average annual range in degrees Celsius.

Solution   Substituting $(9C/5) + 32$ for F in the inequality

$$24.8 \le F \le 95$$

we obtain

$$24.8 \le \frac{9C}{5} + 32 \le 95.$$

The latter inequality is equivalent to each of the following:

$$-7.2 \le \frac{9C}{5} \le 63$$

$$\frac{5}{9}(-7.2) \le \frac{5}{9}\left(\frac{9C}{5}\right) \le \frac{5}{9}(63)$$

$$-4 \le C \le 35.$$

It follows that the temperature C, in degrees Celsius, has the average annual range: $-4 \le C \le 35$.

Quadratic Inequalities

We now turn our attention to solving quadratic inequalities. A **quadratic inequality** is an inequality that can be written in either of the forms

$$ax^2 + bx + c < 0 \qquad (6)$$

or

$$ax^2 + bx + c \le 0, \qquad (7)$$

where a, b, and c are real numbers and $a \ne 0$. Inequality (6) and inequality (7) are both said to be in **standard form**. The equation

$$ax^2 + bx + c = 0$$

is called the **associated quadratic equation** for both inequalities.

The solution set of a quadratic inequality of the form $ax^2 + bx + c \le 0$ is the union of the solution set of the quadratic inequality $ax^2 + bx + c < 0$ and the solution set of the associated quadratic equation $ax^2 + bx + c = 0$. Thus, we will first focus our attention on solving quadratic inequalities of the form $ax^2 + bx + c < 0$.

One technique for solving quadratic inequalities of the form $ax^2 + bx + c < 0$ is called the **cut-point method**. To use this method, we first find the real roots (if any) of the associated quadratic equation $ax^2 + bx + c = 0$.

If the equation $ax^2 + bx + c = 0$ has real roots, these real roots are called **cut-points**. We use the cut-points to divide the number line into open intervals. For each of these intervals, it can be shown that if *any* number in the interval, when substituted for x in the polynomial $ax^2 + bx + c$, yields a negative number, then *every* number in the interval yields a negative number. Similarly, if *any* number in the interval, when substituted for x in the polynomial $ax^2 + bx + c$, yields a positive number, then *every* number in the interval yields a positive number. Thus, in each open interval, we select a convenient number called a **test point** and use this value of x to determine the sign of the polynomial $ax^2 + bx + c$ on the entire interval.

If the equation $ax^2 + bx + c = 0$ has no real roots, there are no cut-points. In this case, it can be shown that the quadratic polynomial $ax^2 + bx + c$ is either always positive or always negative.

We illustrate this procedure with several examples.

EXAMPLE 31

Use the cut-point method to find the solution set of the quadratic inequality $2x^2 - 5x - 3 < 0$. Display the solution set on the number line.

Solution   We first consider the associated quadratic equation $2x^2 - 5x - 3 = 0$ to find the cut-points (if any). Factoring and using the zero-product property, we have

$$(2x + 1)(x - 3) = 0$$
$$2x + 1 = 0 \quad \text{or} \quad x - 3 = 0$$
$$x = -\frac{1}{2} \quad \text{or} \quad x = 3.$$

Hence, there are two cut-points, $-\frac{1}{2}$ and 3.

Using these two cut-points, we divide the number line into the three open intervals:

$$\left(-\infty, -\frac{1}{2}\right), \left(-\frac{1}{2}, 3\right), \text{ and } (3, \infty). \tag{8}$$

From each of the open intervals in (8), we select a convenient number as a test point and use this value of x to determine the sign of the polynomial $2x^2 - 5x - 3$ on that interval. Choosing $x = -1$ in the interval $(-\infty, -\frac{1}{2})$, we have

$$2x^2 - 5x - 3 = 2(-1)^2 - 5(-1) - 3 = 2 + 5 - 3 = 4 > 0.$$

Since the test point yields a positive number, it follows that the polynomial $2x^2 - 5x - 3$ is positive on the entire interval $(-\infty, -\frac{1}{2})$.

In the interval $(-\frac{1}{2}, 3)$ we test $x = 0$:

$$2x^2 - 5x - 3 = 2(0)^2 - 5(0) - 3 = -3 < 0.$$

Since the test point yields a negative number, the polynomial $2x^2 - 5x - 3$ is negative on the entire interval $(-\frac{1}{2}, 3)$.

Finally, in the interval $(3, \infty)$, we test $x = 4$:

$$2x^2 - 5x - 3 = 2(4)^2 - 5(4) - 3 = 9 > 0.$$

It follows that the polynomial $2x^2 - 5x - 3$ is positive on the interval $(3, \infty)$. We summarize our work in Table 1.3.

TABLE 1.3

Interval	Test Point	Value of $2x^2 - 5x - 3$ at Test Point	Sign of $2x^2 - 5x - 3$ at Test Point
$(-\infty, -\frac{1}{2})$	-1	4	$+$
$(-\frac{1}{2}, 3)$	0	-3	$-$
$(3, \infty)$	4	9	$+$

From Table 1.3, we see that the solution set of the inequality $2x^2 - 5x - 3 < 0$ is the open interval $(-\frac{1}{2}, 3)$, which is shown in Figure 1.15.

FIGURE 1.15

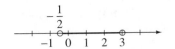

Notice that in Table 1.3, we found the value of the polynomial $2x^2 - 5x - 3$ at each test point. However, the essential information is simply the sign of $2x^2 - 5x - 3$ at each test point. If the sign can be readily determined, it is unnecessary to compute the actual value. However, we will continue to give both the value and the sign in each case.

E X A M P L E 32

Use the cut-point method to find the solution set of the inequality $3x^2 \leq -1 - 10x$. Display the solution set on the number line.

Solution   In standard form, the inequality $3x^2 \leq -1 - 10x$ becomes

$$3x^2 + 10x + 1 \leq 0.$$

We thus consider the associated quadratic equation

$$3x^2 + 10x + 1 = 0 \tag{9}$$

to find the cut-points (if any). Using the quadratic formula, we find that the solutions of equation (9) are

$$x = \frac{-5 - \sqrt{22}}{3} \approx -3.23 \quad \text{and} \quad x = \frac{-5 + \sqrt{22}}{3} \approx -0.10.$$

These two cut-points divide the number line into three open intervals:

$$\left(-\infty, \frac{-5 - \sqrt{22}}{3}\right), \left(\frac{-5 - \sqrt{22}}{3}, \frac{-5 + \sqrt{22}}{3}\right),$$

$$\text{and} \left(\frac{-5 + \sqrt{22}}{3}, \infty\right). \tag{10}$$

Proceeding as we did in Example 31, we choose a test point from each interval given in (10) and use it to determine the sign of the polynomial $3x^2 + 10x + 1$ on that interval. Our results are summarized in Table 1.4.

T A B L E 1.4

Interval	Test Point	Value of $3x^2 + 10x + 1$ at Test Point	Sign of $3x^2 + 10x + 1$ at Test Point
$\left(-\infty, \dfrac{-5 - \sqrt{22}}{3}\right)$	-4	9	$+$
$\left(\dfrac{-5 - \sqrt{22}}{3}, \dfrac{-5 + \sqrt{22}}{3}\right)$	-1	-6	$-$
$\left(\dfrac{-5 + \sqrt{22}}{3}, \infty\right)$	0	1	$+$

It follows from Table 1.4, that the solution set of the inequality $3x^2 + 10x + 1 < 0$ is the open interval $((-5 - \sqrt{22})/3, (-5 + \sqrt{22})/3)$. The solution set of the inequality $3x^2 + 10x + 1 \leq 0$ must also include the cut-points

$$\frac{-5 - \sqrt{22}}{3} \quad \text{and} \quad \frac{-5 + \sqrt{22}}{3}$$

since they are solutions of $3x^2 + 10x + 1 = 0$. Hence the solution set of the original inequality is the closed interval $[(-5 - \sqrt{22})/3, (-5 + \sqrt{22})/3]$, which is shown in Figure 1.16.

F I G U R E 1.16

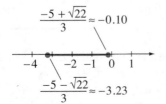

$\dfrac{-5 + \sqrt{22}}{3} \approx -0.10$

$\dfrac{-5 - \sqrt{22}}{3} \approx -3.23$

EXAMPLE 33

Use the cut-point method to find the solution set of the inequality $x^2 + 2x \geq -5$. Display the solution set on the number line.

Solution   Since $x^2 + 2x \geq -5$ is equivalent to

$$x^2 + 2x + 5 \geq 0,$$

we consider the associated quadratic equation

$$x^2 + 2x + 5 = 0 \qquad (11)$$

to find the cut-points (if any). It can be shown, using the discriminant, that equation (11) has no real roots. Since there are no cut-points, the polynomial $x^2 + 2x + 5$ is either always positive or always negative. For convenience, we chose $x = 0$ as a test point and find that

$$x^2 + 2x + 5 = (0)^2 + 2(0) + 5 = 5 > 0.$$

Therefore, the polynomial $x^2 + 2x + 5$ is always positive. It follows that the solution set for the original inequality is $\mathbb{R}$, which is shown in Figure 1.17.

FIGURE 1.17

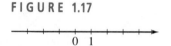

EXAMPLE 34

Use the cut-point method to find the solution set of the inequality $-x^2 + 6x - 9 > 0$. Display the solution set on the number line.

Solution   For the inequality

$$-x^2 + 6x - 9 > 0$$

we must consider the associated quadratic equation

$$-x^2 + 6x - 9 = 0, \qquad \text{or} \qquad x^2 - 6x + 9 = 0$$

to determine the cut-points (if any). Factoring and using the zero-product property, we have

$$(x - 3)(x - 3) = 0$$
$$x - 3 = 0$$
$$x = 3.$$

Hence, 3 is a double root, and there is only one cut-point. The cut-point 3 determines the two open intervals: $(-\infty, 3)$ and $(3, \infty)$. Proceeding as we did in the three previous examples, we use a test point in each of these intervals to determine the sign of $-x^2 + 6x - 9$ on that interval. Our results are summarized in Table 1.5. It follows from Table 1.5, that the solution set for $-x^2 + 6x - 9 > 0$ is $\varnothing$.

TABLE 1.5

Interval	Test Point	Value of $-x^2 + 6x - 9$ at Test Point	Sign of $-x^2 + 6x - 9$ at Test Point
$(-\infty, 3)$	0	-9	$-$
$(3, \infty)$	4	-1	$-$

E X E R C I S E S **1.6**

In Exercises 1–20, find the solution set of each inequality. Show each solution set on the number line.

1. $-4x - 16 \geq 0$

2. $-4(x - 2) + 3(x - 4) \leq 6(x - 7) + 3$

3. $\dfrac{2x - 1}{2} - \dfrac{1 - x}{6} > \dfrac{x}{3} + 2$

4. $-2x + 1 \leq 4 + x$

5. $-5 \leq 1 - 2x < 9$

6. $0 < 2 - 3x \leq 4$

7. $4(x - 6) - 5x > 11$

8. $2(4x - 1) - 6(2x - 5) < 0$

9. $\dfrac{x - 2}{3} + \dfrac{x + 1}{8} > \dfrac{5}{6}$

10. $\dfrac{2x - 3}{10} - \dfrac{3x - 1}{4} \leq x - 1$

11. $-1 \leq 3x + 4 < 10$

12. $-6 < -4x - 5 \leq -2$

13. $\dfrac{4x - 3}{-2} \leq 12$

14. $-30x - 2 \geq 5(2 - 3x)$

15. $6 - 3x \geq -14 + 2x$

16. $0 \leq 1 - 5x < 7$

17. $-7 \leq 2x - 3 \leq -4$

18. $\dfrac{2x}{3} + \dfrac{x - 3}{2} < \dfrac{2x}{5} + \dfrac{3x + 6}{10}$

19. $\dfrac{3}{8} + \dfrac{x - 4}{2} \geq \dfrac{x + 3}{4}$

20. $-3 < \dfrac{2x}{3} + 3 < 9$

In Exercises 21–40, use the cut-point method to find the solution set of each inequality. Show each solution set on the number line.

21. $x^2 + x - 12 \leq 0$ 22. $x^2 - 6x - 5 > 0$

23. $4x^2 - 4x \geq 3$ 24. $x^2 < 3 - 2x$

25. $x^2 \leq 4x + 2$ 26. $-2x^2 > 20x + 52$

27. $3x^2 + 2x < \frac{1}{3}$ 28. $9 - x^2 < 0$

29. $-x^2 - 4 > 0$ 30. $x^2 + 4x \leq 5$

31. $12 < x(x + 1)$ 32. $x(3x + 4) < 7$

33. $2x^2 \leq -3x - 4$ 34. $1 - 2x^2 \geq -5 + 4x$

35. $-(x^2 - 6x) < -16$ 36. $-x^2 \geq 1 - 3x$

37. $3x^2 \geq 7x + 20$ 38. $x^2 \leq -7x - 6$

39. $2x^2 > -2x - 1$ 40. $x^2 < -x - 5$

41. In a certain city, the temperature in degrees Celsius has the average annual range: $-20 \leq C \leq 30$. Use the formula $C = \frac{5}{9}(F - 32)$, which relates C (degrees Celsius) and F (degrees Fahrenheit), to find the average annual range in degrees Fahrenheit.

42. For a collection of triangles all having a base of 12 centimeters, it was found that the areas of the triangles had a range that satisfies $8 \text{ cm}^2 \leq A \leq 150 \text{ cm}^2$. Use the formula $A = \frac{1}{2}bh$, for the area of a triangle, to find the range for the heights of the triangles.

43. A person's IQ (Intelligence Quotient) is found by the formula

$$IQ = \dfrac{100 \, M}{C},$$

where M is the mental age and C is the chronological age. If the IQ range for a group of ten-year-olds satisfies $90 \leq IQ \leq 150$, find the mental age range of this group.

44. A company that makes bird houses finds that for each week its cost equation is

$$C = 30 + 3.5x \text{ (dollars)}$$

and its revenue equation is

$$R = 5x \text{ (dollars)},$$

where x is the number of bird houses it makes and sells per week. The company will make a profit when the revenue is greater than the cost; that is, when

$$R > C.$$

How many bird houses must be made and sold per week for the company to make a profit?

45. A potter finds that for each week his cost equation is

$$C = 25 + 2.5x \text{ (dollars)}$$

Photo Peter Vandermark/Stock, Boston

and his revenue equation is

$$R = 3x \text{ (dollars)},$$

where x is the number of pots he makes and sells per week. How many pots must he make and sell per week to make a profit? (Refer to exercise 44.)

46. The height of a projectile is given by the equation

$$h = -16t^2 + 96t$$

where h is in feet and t is time in seconds.

a. Find the time interval during which the projectile is above the ground.

b. Find the time interval during which the projectile is at least 128 feet high.

47. The height of a model rocket is given by the equation

$$h = -16t^2 + 88t,$$

where h is in feet and t is time in seconds.

a. Find the time interval during which the rocket is above the ground.

b. Find the time interval during which the rocket is at least 120 feet high.

48. A frog can jump to a height of $h = -16t^2 + 8t$ feet in t seconds.

a. Find the time interval during which the frog is above the ground.

b. Find the time interval during which the frog is at most 1 foot high.

1.7   MISCELLANEOUS EQUATIONS AND INEQUALITIES

Our work with linear and quadratic equations and inequalities in the last two sections provides us with a basis for solving a variety of other types of equations and inequalities. In this section, our study will include equations and inequalities that are quadratic in form as well as equations and inequalities involving absolute values, radicals, and rational expressions. We begin by considering equations that involve radicals.

When we solve an equation that contains radicals, it is sometimes necessary to raise both sides of the equation to the same positive integral power. The following property is especially useful in such cases:

///

Power Property of Equations

Let P and Q be algebraic expressions and let n be a positive integer. Then the solution set of the equation

$$P = Q$$

is a subset of the solution set of the equation

$$P^n = Q^n.$$

We see from the power property of equations that the equations

$$P = Q \quad \text{and} \quad P^n = Q^n$$

are *not* necessarily equivalent. In particular, it is possible that some of the solutions of the equation $P^n = Q^n$ are not solutions of the equation $P = Q$. In general, a solution of the equation $P^n = Q^n$ is called a **potential solution** of the equation $P = Q$. It is important to realize that each potential solution must be checked by substitution into the equation $P = Q$. A potential solution that does not satisfy the equation $P = Q$ is called an **extraneous solution**, and is *not* included in the solution set.

For example, consider the equations

$$x = -4 \hspace{4em} \textit{(1)}$$

and
$$x^2 = (-4)^2 = 16. \tag{2}$$

Notice that the solution set for equation (2) is $\{-4, 4\}$, while the solution set for equation (1) is $\{-4\}$. Since 4 is a solution of equation (2), but is not a solution of equation (1), 4 is an extraneous solution of equation (1).

EXAMPLE 35

Find the solution set of each equation.

a. $\sqrt[3]{x^2 - 2} = -3$ **b.** $\sqrt{x} + \sqrt{x - 16} = 2$
c. $x = \sqrt{3 - 2x}$

Solution

a. The equation
$$\sqrt[3]{x^2 - 2} = -3 \tag{3}$$

can be solved by first cubing both sides as follows:
$$(\sqrt[3]{x^2 - 2})^3 = (-3)^3$$
$$x^2 - 2 = -27$$
$$x^2 = -25$$
$$x = \pm 5i.$$

However, since we cubed both sides, we must check both potential solutions in equation (3). If $x = -5i$, then
$$\sqrt[3]{x^2 - 2} = \sqrt[3]{(-5i)^2 - 2} = \sqrt[3]{-27} = -3.$$

Thus, $-5i$ is a solution of equation (3). Similarly, it can be shown that $5i$ is a solution of equation (3), and thus the solution set is $\{-5i, 5i\}$.

b. Since the equation
$$\sqrt{x} + \sqrt{x - 16} = 2 \tag{4}$$

involves two terms that contain radicals, we first separate the radicals obtaining
$$\sqrt{x - 16} = 2 - \sqrt{x}.$$

ı sides yields
$$x - 16 = 4 - 4\sqrt{x} + x.$$

Isolating the radical, we have
$$-20 = -4\sqrt{x}, \qquad \text{or}$$
$$\sqrt{x} = 5.$$

Squaring both sides, we get
$$x = 25.$$

Since we squared both sides, we must check this potential solution in equation (4). If $x = 25$, then
$$\sqrt{x} + \sqrt{x - 16} = \sqrt{25} + \sqrt{25 - 16} = 5 + 3 = 8 \neq 2.$$

Thus, 25 is an extraneous solution, and the solution set is $\varnothing$.

c. To eliminate the radical in the equation
$$x = \sqrt{3 - 2x}, \tag{5}$$

we square both sides, obtaining

$$x^2 = 3 - 2x, \quad \text{or}$$
$$x^2 + 2x - 3 = 0.$$

Hence,
$$(x + 3)(x - 1) = 0$$
$$x + 3 = 0 \quad \text{or} \quad x - 1 = 0$$
$$x = -3 \quad \text{or} \quad x = 1.$$

We must check both potential solutions in equation (5). If $x = -3$, then

$$\sqrt{3 - 2x} = \sqrt{3 - 2(-3)} = \sqrt{9} = 3 \neq -3 \,(= x).$$

Thus, -3 is an extraneous solution. If $x = 1$, then

$$\sqrt{3 - 2x} = \sqrt{3 - 2(1)} = \sqrt{1} = 1(=x).$$

Therefore, 1 is the only solution and the solution set is $\{1\}$.

Equations That Are Quadratic in Form

We now turn our attention to equations that are similar in form to quadratic equations. For example, consider the equation

$$x^4 - 5x^2 - 36 = 0. \tag{6}$$

Although equation (6) is not a quadratic equation, if we let $u = x^2$ (so that $u^2 = x^4$), equation (6) can be written in the form

$$u^2 - 5u - 36 = 0. \tag{7}$$

Equation (7), which is a quadratic equation, can be solved for u. Then, since $u = x^2$, the solutions of equation (6) can be found. We say that equation (6) is **quadratic in form** since it can be written as

$$au^2 + bu + c = 0, \quad a \neq 0,$$

for some algebraic expression u.

E X A M P L E   36

Find the solution set of each equation.

a. $x^4 - 5x^2 - 36 = 0$ **b.** $2x^{2/3} - x^{1/3} - 3 = 0$

Solution

a. If we let $u = x^2$, then $u^2 = x^4$, and the equation

$$x^4 - 5x^2 - 36 = 0 \tag{8}$$

becomes

$$u^2 - 5u - 36 = 0.$$

Factoring and using the zero-product property, we have

$$(u - 9)(u + 4) = 0$$
$$u = 9 \quad \text{or} \quad u = -4.$$

Since $u = x^2$,
$$x^2 = 9 \quad \text{or} \quad x^2 = -4$$
$$x = \pm 3 \quad \text{or} \quad x = \pm 2i. \tag{9}$$

You should verify that each of the solutions given in (9) satisfies equation (8). Therefore, the solution set is $\{-3, 3, -2i, 2i\}$.

b. Since the equation

$$2x^{2/3} - x^{1/3} - 3 = 0 \qquad\qquad \textit{(10)}$$

can be written as

$$2(x^{1/3})^2 - x^{1/3} - 3 = 0,$$

we see that equation (10) is quadratic in form. Letting $u = x^{1/3}$ (so that $u^2 = x^{2/3}$), we obtain the quadratic equation

$$2u^2 - u - 3 = 0.$$

Factoring and using the zero-product property, we have

$$(2u - 3)(u + 1) = 0$$
$$2u - 3 = 0 \qquad \text{or} \qquad u + 1 = 0$$
$$u = \frac{3}{2} \qquad \text{or} \qquad u = -1.$$

Since $u = x^{1/3}$,

$$x^{1/3} = \frac{3}{2} \qquad \text{or} \qquad x^{1/3} = -1$$
$$x = \left(\frac{3}{2}\right)^3 \qquad \text{or} \qquad x = (-1)^3$$
$$x = \frac{27}{8} \qquad \text{or} \qquad x = -1.$$

You should verify that $x = \frac{27}{8}$ and $x = -1$ both satisfy equation (10). Thus, the solution set is $\{-1, \frac{27}{8}\}$.

Equations Involving Rational Expressions

We will now consider equations that involve rational expressions. When solving such equations, it is important to exclude from consideration all values of the variable that would cause any denominator to become zero. Such values are called **excluded values** for the equation.

E X A M P L E 37

Find the solution set of each equation.

a. $\dfrac{3x}{x - 2} - 4 = \dfrac{14 - 4x}{x - 2}$ **b.** $\dfrac{x}{x - 2} - \dfrac{4}{x + 3} = \dfrac{10}{x^2 + x - 6}$

Solution

a. For the equation

$$\frac{3x}{x - 2} - 4 = \frac{14 - 4x}{x - 2}, \qquad\qquad \textit{(11)}$$

we see that 2 is the only excluded value. Hence, we will assume that $x \neq 2$ and multiply both sides of equation (11) by $x - 2$ (the LCD) to get

$$(x - 2)\left(\frac{3x}{x - 2}\right) - (x - 2)(4) = (x - 2)\left(\frac{14 - 4x}{x - 2}\right)$$

$$3x - 4x + 8 = 14 - 4x$$

$$3x = 6$$

$$x = 2.$$

Since 2 is an excluded value, it follows that the solution set for equation (11) is $\emptyset$.

b. For the equation

$$\frac{x}{x - 2} - \frac{4}{x + 3} = \frac{10}{x^2 + x - 6},$$

we note that $x^2 + x - 6 = (x - 2)(x + 3)$. Therefore, the excluded values are 2 and -3. Assuming that $x \neq 2$ and that $x \neq -3$, it follows that $(x - 2)(x + 3)$, which is the LCD, is nonzero. Multiplying both sides of equation (12) by $(x - 2)(x + 3)$, we have

$$(x - 2)(x + 3)\left(\frac{x}{x - 2}\right) - (x - 2)(x + 3)\left(\frac{4}{x + 3}\right)$$

$$= (x - 2)(x + 3)\left[\frac{10}{(x - 2)(x + 3)}\right], \quad \text{or}$$

$$x(x + 3) - 4(x - 2) = 10$$

$$x^2 + 3x - 4x + 8 = 10$$

$$x^2 - x - 2 = 0.$$

Factoring and using the zero-product property, we have

$$(x - 2)(x + 1) = 0$$

$$x - 2 = 0 \quad \text{or} \quad x + 1 = 0$$

$$x = 2 \quad \text{or} \quad x = -1.$$

Since 2 is an excluded value while -1 is not, the solution set is $\{-1\}$.

Equations and Inequalities Involving Absolute Value

Next, we consider equations and inequalities that involve absolute values. The following theorem will be of fundamental importance in our work.

THEOREM 1.3

Let d be a positive real number and let x be any real number. Then

(i) $|x| = d$ if and only if $x = \pm d$;

(ii) $|x| < d$ if and only if $-d < x < d$;

(iii) $|x| > d$ if and only if $x < -d$ or $x > d$.

FIGURE 1.18

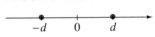

FIGURE 1.19

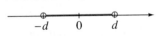

FIGURE 1.20

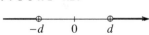

In lieu of a proof, we give the following geometric interpretation. (Recall that $|x|$ represents the distance between x and 0 on the number line.)

(i) The statement $|x| = d$ holds if and only if the distance between x and 0 is exactly d units; that is, if and only if $x = \pm d$ (see Figure 1.18).

(ii) The statement $|x| < d$ means that the distance between x and 0 is less than d, which, in turn, means that $-d < x < d$ (see Figure 1.19).

(iii) The statement $|x| > d$ holds if and only if the distance between x and 0 is greater than d; that is, if and only if $x < -d$ or $x > d$ (see Figure 1.20).

E X A M P L E 38

Find the solution set of each equation or inequality. Display each solution set on the number line.

a. $|2x - 1| - |-3| < 4$ **b.** $|1 - 3x| \geq 5$ **c.** $|4x + 3| = |5x - 6|$

Solution

a. Since $|-3| = 3$, the inequality

$$|2x - 1| - |-3| < 4$$

is equivalent to

$$|2x - 1| - 3 < 4,$$

or $$|2x - 1| < 7. \qquad (13)$$

By part (ii) of Theorem 1.3, inequality (13) is equivalent to $-7 < 2x - 1 < 7$.

Hence, $$-6 < 2x < 8$$
$$-3 < x < 4.$$

FIGURE 1.21

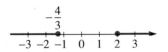

Therefore, the solution set is $\{x \mid -3 < x < 4\} = (-3, 4)$ (see Figure 1.21).

b. By parts (i) and (iii) of Theorem 1.3, the inequality

$$|1 - 3x| \geq 5$$

is equivalent to

$$1 - 3x \leq -5 \quad \text{or} \quad 1 - 3x \geq 5.$$

Therefore, $$-3x \leq -6 \quad \text{or} \quad -3x \geq 4$$

$$x \geq 2 \quad \text{or} \quad x \leq -\frac{4}{3}.$$

It follows that the solution set is $\{x \mid x \leq -\frac{4}{3} \text{ or } x \geq 2\} = (-\infty, -\frac{3}{4}] \cup [2, \infty)$, which is displayed in Figure 1.22.

FIGURE 1.22

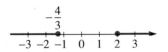

$-\dfrac{4}{3}$

c. By part (vi) of Theorem 1.1, the equation

$$|4x + 3| = |5x - 6|$$

is equivalent to

$$4x + 3 = \pm(5x - 6).$$

FIGURE 1.23

Therefore, $4x + 3 = 5x - 6$ or $4x + 3 = -(5x - 6)$
$-x = -9$ or $9x = 3$
$x = 9$ or $x = \dfrac{1}{3}.$

Hence, the solution set is $\{\frac{1}{3}, 9\}$ (see Figure 1.23).   ⎯⎯⎯⎯⎯⎯

Rational Inequalities

We now turn our attention to rational inequalities. A **rational inequality** is an inequality that can be written in either of the forms

$$\frac{P}{Q} < 0 \quad \text{or} \quad \frac{P}{Q} \le 0, \tag{14}$$

where P and Q are polynomials that have no common factors. We will solve rational inequalities by using a modified version of the cut-point method discussed in Section 1.6. The cut-points for either of the inequalities given in (14) are the real numbers (if any) for which $P = 0$ or $Q = 0$. It can be shown that the algebraic sign of the expression P/Q remains the same over each of the open intervals determined by the cut-points.

EXAMPLE 39

Find the solution set of each inequality. Display each solution set on the number line.

a. $\dfrac{x^2 + 2x - 3}{x^2 - 2x - 8} \le 0$ **b.** $\dfrac{x + 3}{2 - x} > 1$

Solution

a. The inequality

$$\frac{x^2 + 2x - 3}{x^2 - 2x - 8} \le 0 \tag{15}$$

is equivalent to

$$\frac{(x + 3)(x - 1)}{(x - 4)(x + 2)} \le 0.$$

Hence, there are four cut-points: -3, 1, 4, and -2. We note that the cut-points 4 and -2 are excluded values, since they cause the denominator of the rational expression on the left side of inequality (15) to be zero. The four cut-points divide the number line into the five open intervals:

$$(-\infty, -3), \quad (-3, -2), \quad (-2, 1), \quad (1, 4), \quad \text{and} \quad (4, \infty). \tag{16}$$

From each of the open intervals given in (16), we select a convenient number to use as a test point for determining the sign of the quotient

$$\frac{x^2 + 2x - 3}{x^2 - 2x - 8}$$

TABLE 1.6

on that interval. Our work is summarized in Table 1.6.

Interval	Test Point	Value of $\dfrac{x^2 + 2x - 3}{x^2 - 2x - 8}$ at Test Point	Sign of $\dfrac{x^2 + 2x - 3}{x^2 - 2x - 8}$ at Test Point
$(-\infty, -3)$	-4	$\frac{5}{16}$	$+$
$(-3, -2)$	$-\frac{5}{2}$	$-\frac{7}{13}$	$-$
$(-2, 1)$	0	$\frac{3}{8}$	$+$
$(1, 4)$	2	$-\frac{5}{8}$	$-$
$(4, \infty)$	5	$\frac{32}{7}$	$+$

From Table 1.6, we see that the quotient

$$\frac{x^2 + 2x - 3}{x^2 - 2x - 8}$$

is negative over each of the open intervals $(-3, -2)$ and $(1, 4)$. Since the cut-points -3 and 1 are solutions of the equation

$$\frac{x^2 + 2x - 3}{x^2 - 2x - 8} = 0$$

they are also solutions of the inequality

$$\frac{x^2 + 2x - 3}{x^2 - 2x - 8} \le 0.$$

FIGURE 1.24

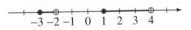

However, as noted above, the cut-points 4 and -2 are excluded values. Therefore, the solution set for the original inequality is $[-3, -2) \cup [1, 4)$, which is shown in Figure 1.24.

b. We first replace the inequality

$$\frac{x + 3}{2 - x} > 1$$

with an equivalent inequality that has zero on the right side:

$$\frac{x + 3}{2 - x} - 1 > 0. \qquad (17)$$

Since
$$\frac{x + 3}{2 - x} - 1 = \frac{x + 3 - (2 - x)}{2 - x}$$
$$= \frac{2x + 1}{2 - x},$$

inequality (17) is equivalent to

$$\frac{2x + 1}{2 - x} > 0. \qquad (18)$$

Therefore, the cut-points are $-\frac{1}{2}$ and 2, where 2 is an excluded value. The cut-points divide the number line into the three open intervals:

$$\left(-\infty, -\frac{1}{2}\right), \left(-\frac{1}{2}, 2\right), \text{ and } (2, \infty).$$

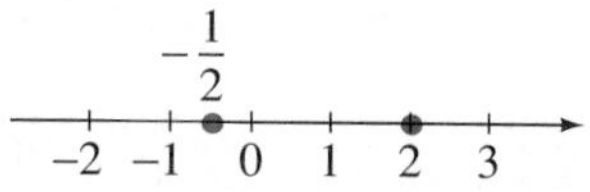

Proceeding as in part (a), we obtain the results shown in Table 1.7.

TABLE 1.7

Interval	Test Point	Value of $\dfrac{2x + 1}{2 - x}$ at Test Point	Sign of $\dfrac{2x + 1}{2 - x}$ at Test Point
$(-\infty, -\frac{1}{2})$	-1	$-\frac{1}{3}$	$-$
$(-\frac{1}{2}, 2)$	0	$\frac{1}{2}$	$+$
$(2, \infty)$	3	-7	$-$

From Table 1.7, we see that the quotient

$$\frac{2x + 1}{2 - x} \qquad (19)$$

is positive on the open interval $(-\frac{1}{2}, 2)$. The cut-point $-\frac{1}{2}$ does not satisfy inequality (18), since it causes the quotient given in (19) to be zero. As mentioned above, the cut-point 2 is an excluded value. Therefore, the solution set of inequality (18) and, hence, of the original equivalent inequality is the open interval $(-\frac{1}{2}, 2)$, which is shown in Figure 1.25.

FIGURE 1.25

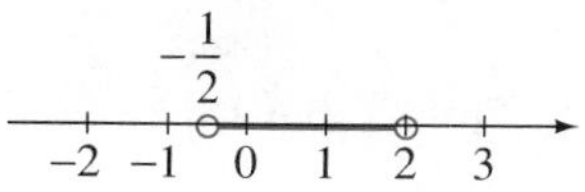

We conclude this section with an example involving an inequality that is quadratic in form.

EXAMPLE 40

Find the solution set of the inequality $4x^4 \le 45 - 11x^2$. Display the solution set on the number line.

Solution   We begin by replacing the inequality

$$4x^4 \le 45 - 11x^2$$

with an equivalent inequality that has zero on the right side:

$$4x^4 + 11x^2 - 45 \le 0.$$

To find the cut-points (if any), we consider the equation

$$4x^4 + 11x^2 - 45 = 0. \qquad (20)$$

If we let $u = x^2$ (so that $u^2 = x^4$), equation (20) becomes the quadratic equation

$$4u^2 + 11u - 45 = 0.$$

Factoring and using the zero-product property, we have

$$(u + 5)(4u - 9) = 0$$

$$u = -5 \quad \text{or} \quad u = \frac{9}{4}.$$

Since $u = x^2$,

$$x^2 = -5 \quad \text{or} \quad x^2 = \frac{9}{4}.$$

Since cut-points must be real numbers, and since the equation

$$x^2 = -5$$

has no real roots, we need only consider the equation

$$x^2 = \frac{9}{4}, \quad \text{or}$$

$$x = \pm\frac{3}{2}.$$

Thus, $-\frac{3}{2}$ and $\frac{3}{2}$ are the only cut-points. These two cut-points divide the number line into the three open intervals:

$$\left(-\infty, -\frac{3}{2}\right), \left(-\frac{3}{2}, \frac{3}{2}\right), \text{ and } \left(\frac{3}{2}, \infty\right). \tag{21}$$

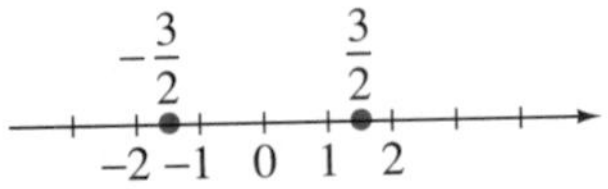

We use test points from each of the intervals given in (21) to determine the sign of

$$4x^4 + 11x^2 - 45$$

on that interval. Our work is summarized in Table 1.8. From Table 1.8, we see that the polynomial

$$4x^4 + 11x^2 - 45$$

is negative on the open interval $(-\frac{3}{2}, \frac{3}{2})$. Since the cut-points $-\frac{3}{2}$ and $\frac{3}{2}$ are solutions of the equation

$$4x^4 + 11x^2 - 45 = 0,$$

they are also solutions of the inequality

$$4x^4 + 11x^2 - 45 \leq 0.$$

Therefore, the solution set of the original inequality is the closed interval $[-\frac{3}{2}, \frac{3}{2}]$, which is shown in Figure 1.26.

FIGURE 1.26

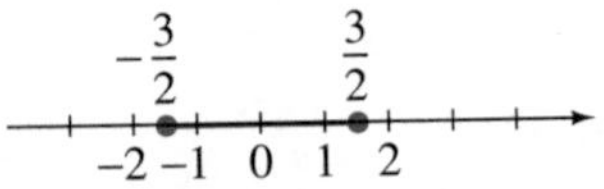

TABLE 1.8

Interval	Test Point	Value of $4x^4 + 11x^2 - 45$ at Test Point	Sign of $4x^4 + 11x^2 - 45$ at Test Point
$(-\infty, -\frac{3}{2})$	-2	63	$+$
$(-\frac{3}{2}, \frac{3}{2})$	0	-45	$-$
$(\frac{3}{2}, \infty)$	2	63	$+$

EXERCISES 1.7

In Exercises 1–18, find the solution set of each equation.

1. $\sqrt{x-3} = 5$

2. $\sqrt{2x-4} = 4$

3. $\sqrt[3]{5x-2} = -2$

4. $\sqrt[3]{3x-4} = -4$

5. $\sqrt{x-1} = 3 + \sqrt{x+1}$

6. $\sqrt{9x^2-2} = 3x - 1$

7. $\sqrt{x-4} = 4 - x$

8. $\sqrt{x^2+5} = x - 2$

9. $\sqrt{x^2+3x+4} = x + 1$

10. $\sqrt[3]{2x+1} = -1$

11. $\sqrt{x^2-5x+10} = 2$

12. $\sqrt{x+25} = \sqrt{x} + 1$

13. $\sqrt{x^2+9} = x + 3$

14. $\sqrt[3]{x^2-91} = -3$

15. $\sqrt{x+1} = \sqrt{2x+2}$

16. $1 - \sqrt{x-1} = \sqrt{x+1}$

17. $\sqrt{x+5} = x - 1$

18. $\sqrt{x+1} - \sqrt{x} = 3$

In Exercises 19–30, find the solution set of each equation.

19. $2x^{2/3} + x^{1/3} - 3 = 0$

20. $5(x^2-9)^2 + 44(x^2-9) + 32 = 0$

21. $x^4 + 5x^2 - 36 = 0$

22. $\dfrac{x-1}{x+1} - 2\left(\dfrac{x+1}{x-1}\right) + 1 = 0$

23. $x^6 + 9x^3 + 8 = 0$

24. $x - \sqrt{x} - 2 = 0$

25. $2\left(\dfrac{1-x}{x+2}\right) - 3\left(\dfrac{x+2}{1-x}\right) - 1 = 0$

26. $15(x+1) + \dfrac{4}{x+1} - 23 = 0$

27. $(x^2+2)^2 - 4(x^2+2) - 5 = 0$

28. $x^{2/5} - x^{1/5} - 2 = 0$

29. $x^{2/3} - x^{1/3} - 12 = 0$

30. $2x^4 + 17x^2 - 9 = 0$

In Exercises 31–40, find the solution set of each equation.

31. $\dfrac{x}{x-2} + \dfrac{2}{2-x} = -\dfrac{2}{3}$

32. $\dfrac{3}{2x-1} + 4 = \dfrac{6x}{2x-1}$

33. $2 - \dfrac{4}{x} = \dfrac{9}{x+5}$

34. $\dfrac{2x}{x-1} - 3 = \dfrac{7-3x}{x-1}$

35. $\dfrac{2}{3x} - \dfrac{5}{x+1} = \dfrac{1}{2x}$

36. $\dfrac{3x}{x-2} - 4 = \dfrac{14-4x}{x-2}$

37. $4 + \dfrac{17}{x} = \dfrac{15}{x^2}$

38. $15x - \dfrac{1}{x} = -2$

39. $\dfrac{2x}{3x-1} - \dfrac{1}{x+2} = \dfrac{3-2x}{3x^2+5x-2}$

40. $\dfrac{2x}{x-2} + \dfrac{5}{x+4} = \dfrac{6x-6}{x^2+2x-8}$

In Exercises 41–58, find the solution set of each equation or inequality. Display each solution set on the number line.

41. $|x-4| = 6$

42. $|2x+5| = 13$

43. $|2-5x| \geq 3$

44. $|5x+2| \leq 3$

45. $|6-7x| < 1$

46. $|5-2x| > 7$

47. $|5x+2| = |3x-4|$

48. $|3x+15| = |1-6x|$

49. $|7-6x| \geq 5$

50. $|2-x| \geq 10$

51. $|2-3x| - |-7| < 1$

52. $|3x| - |-4| > 5$

53. $|2x| - |-3| \geq -|-1|$

54. $|3x| - |-5| < -|-2|$

55. $\left|\dfrac{x}{2} - 3\right| < 5$

56. $|1-10x| - |-6| > -|-3|$

57. $\left|3 - \dfrac{x}{4}\right| \geq 10$

58. $\left|1 - \dfrac{x}{2}\right| - |-10| \leq -|-5|$

In Exercises 59–68, find the solution set of each inequality. Show each solution set on the number line.

59. $\dfrac{2x+3}{6-x} \geq 1$

60. $\dfrac{x+4}{3-x} < 1$

61. $\dfrac{x^2+x-2}{2-x} \geq 0$

62. $\dfrac{x^2+x-12}{5-x} \leq 0$

63. $\dfrac{x^2-2x-3}{x^2+2x-8} \leq 0$

64. $\dfrac{x^2-x-12}{x^2+x-2} \geq 0$

65. $x^4 - 5x^2 > -4$

66. $x^4 - 41x^2 < -400$

67. $\dfrac{x^2-4}{1-x} \leq 0$

68. $16x^4 \geq 9 - 7x^2$

CHAPTER 1 REVIEW EXERCISES

1. If $H = \{1, 2, 3, 4, 5, 6\}$, $J = \{1, 3, 5\}$, and $K = \{2, 4, 6\}$, find each of the following sets:

  a. $H \cap K$ **b.** $J \cap K$

  c. $H \cup J$ **d.** $J \cup K$

  e. $H \cap (J \cup K)$ **f.** $(H \cap J) \cup (H \cap K)$

2. Describe the set $V = \{x \mid x \text{ is a vowel in the word } \textit{caribou}\}$ by listing its elements.

3. Determine all subsets of the set $S = \{-2, 0, 2\}$.

4. If $U = \{0, 1, 2, 3, \ldots\}$ is the universal set, find the complement of each of the following sets:

  a. $A = \{0\}$ **b.** $B = \{11, 12, 13, \ldots\}$

  c. U **d.** $\varnothing$

In Exercises 5–10, shade the indicated set in the accompanying Venn diagram.

5. $A \cup B$

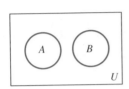

6. $A \cup B'$

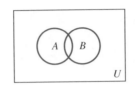

7. $A \cap (B' \cap C)$

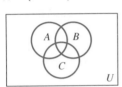

8. $A \cup (B \cap C)$

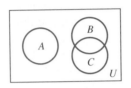

9. $(A \cap B) \cup (B \cap C)$

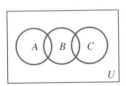

10. $(A \cap B') \cup C$

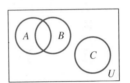

In Exercises 11 and 12, express each number as the quotient of two integers.

11. $2.137137137\ldots$

12. $0.0484848\ldots$

In Exercises 13–18, give the name of the real number property that is best illustrated by each statement. Assume all variables represent real numbers.

13. $(x + 3) + 5 = x + (3 + 5)$

14. $6x + 6 \cdot 17 = 6(x + 17)$

15. $8(y + 0) = 8y$

16. $-\sqrt{2} + \sqrt{2} = 0$

17. $1 \cdot 0 = 0$

18. $\pi(1/\pi) = 1$

In Exercises 19 and 20, write each expression as an equivalent expression that does not involve the absolute value symbol.

19. $|\pi - 3.14|$

20. $|3 - y|$, where $y > 3$

21. Describe each of the following sets using set-builder notation. Display each set on the number line.
  a. $(-2, 4]$
  b. $[-1, \infty)$
  c. $(-3, -1) \cup [1, \infty)$
  d. $(-\infty, -\pi] \cup (0, \frac{5}{4}]$

22. In each case, find the distance between the points a and b on the number line.
  a. $a = -8, b = -1$
  b. $a = -\frac{3}{4}, b = \frac{7}{8}$

23. Plot each of the following points in the plane.
  a. $(-2, 3)$
  b. $(\frac{1}{2}, -4)$
  c. $(-\sqrt{2}, \sqrt{2})$
  d. $(\pi/2, -\frac{3}{4})$

In Exercises 24–25, find the distance between the points A and B, and the midpoint of the line segment $\overline{AB}$.

24. $A = (-3, 4), B = (2, -1)$

25. $A = (\frac{5}{4}, 3), B = (\frac{1}{2}, -1)$

26. Show that $Q = (6, -6)$ is a point on the perpendicular bisector of the line segment that joins $P_1 = (5, -2)$ and $P_2 = (2, -5)$.

27. Find the point P that lies one-fourth of the way from $J = (-3, 1)$ to $K = (5, -11)$.

28. Show that $A = (4, 1)$, $B = (3, 4)$, and $C = (-5, -2)$ are the vertices of a right triangle and find its area.

29. Display the portion of the plane that contains all ordered pairs (x, y) such that $xy > 0$.

30. Plot each of the following points in the complex plane.
  a. $-3i$
  b. $-4 + i$
  c. $2 + \pi i$
  d. $-\frac{1}{2} - \frac{3}{4}i$

In Exercises 31–42, evaluate each expression. Leave your answers in the form $a + bi$.

31. $(2 + 3i) - (5 - i)$

32. $(-2 + i) + (7 - 4i)$

33. $(1 - 2i)(1 + 3i)$

34. $i(5 - 3i)$

35. $\dfrac{2 + 3i}{1 - i}$

36. $\dfrac{1}{-9i}$

37. $|2 - 3i|$

38. $|\frac{3}{2} - \frac{1}{2}i|$

39. i^{23}

40. i^{253}

41. $|i^{17}|$

42. $\dfrac{1}{i^{39}}$

43. Find the additive inverse of the complex number $z = -2 + 6i$.

44. If $z_1 = a + bi$ and $z_2 = c + di$ are complex numbers, show that $\overline{z_1 \cdot z_2} = \overline{z_1} \cdot \overline{z_2}$.

45. Express the reciprocal of the complex number $z = -2 + 3i$ in the form $a + bi$.

46. Verify that $\frac{1}{2} + (\sqrt{3}/2)i$ is a cube root of -1 by showing that $[\frac{1}{2} + (\sqrt{3}/2)i]^3 = -1$.

In Exercises 47–52, find the solution set of each equation.

47. $3(x - 1) - 2(1 - x) = 2x - 3(x + 2)$

48. $\dfrac{2x + 1}{5} - \dfrac{x}{2} = \dfrac{1}{10} - x$

49. $\dfrac{3}{x + 2} - \dfrac{5}{x - 1} = \dfrac{2}{x + 2}$

50. $\dfrac{x}{x - 1} - \dfrac{1}{1 - x} = \dfrac{2}{x - 1}$

51. $\dfrac{x}{x - 3} + \dfrac{3}{3 - x} = -1$

52. $0.03x - 0.4(2x - 20) = 30$

In Exercises 53–58, solve each formula for the indicated variable.

53. $Ax + By + C = 0$, for B

54. $S = 4\pi r^2$, for r

55. $A = \frac{1}{2}(b_1 + b_2)h$, for b_1

56. $D = b^2 - 4ac$, for a

57. $y = 2x^2 - 4x + 1$, for x

58. $\dfrac{1}{x} - \dfrac{2}{y} = \dfrac{3}{w}$, for w

In Exercises 59–62, find the solution set for each inequality. Display each solution set on the number line.

59. $-2(x - 1) - 3(x - 4) \leq 4(x - 1) + 2$

60. $-4 \le 2 - 3x < 5$

61. $-1 \le 5 - 2x < 3$ **62.** $-4 \le -x/2 < -1$

In Exercises 63–66, find the discriminant and, without solving the equation, discuss the nature of the roots.

63. $x^2 + 3x + 5 = 0$ **64.** $4x^2 + 6x - 1 = 0$

65. $2x^2 - x - 3 = 0$ **66.** $25x^2 - 30x + 9 = 0$

In Exercises 67–80, find the solution set of each equation.

67. $6x^2 - x - 12 = 0$ **68.** $x^2 - 4x - 4 = 0$

69. $x(x - 2) = -1$ **70.** $\dfrac{5x^2}{x + 1} = 3$

71. $\dfrac{x^2}{2} - \dfrac{x}{3} + 1 = 0$ **72.** $3x(x + 1) = 20$

73. $1 - \dfrac{1}{x} = \dfrac{2}{3x^2}$ **74.** $x^2 = -1(x + 3)$

75. $2x^2 - 2x + 1 = 0$ **76.** $2x^2 - 3x - 5 = 0$

77. $x(4x + 2) = 5$ **78.** $x^2 = 3(3x - 1)$

79. $x^2 = 4x + 1$ **80.** $2(x^2 - 2) = -7x$

In Exercises 81–88, use the cut-point method to find the solution set of each inequality. Display each solution set on the number line.

81. $x^2 < x + 6$ **82.** $2x^2 \ge 6 - 4x$

83. $6x^2 > 2(5x + 2)$ **84.** $-2x^2 < -1 - x$

85. $x(x - 4) < 1$ **86.** $x^2 < -2(2x + 1)$

87. $x^2 \ge 4(2x - 5)$ **88.** $x(x + 4) > -1$

In Exercises 89–96, find the solution set of each equation or inequality. Display each solution set on the number line.

89. $|1 - 2x| = 3$ **90.** $|2x - 7| = |1 - x|$

91. $|3 - 2x| \ge 1$ **92.** $|1 - x| > 15$

93. $|3 - 5x| \le 3$ **94.** $|2x + 10| - |-12| < -|-3|$

95. $\left| \dfrac{1 - x}{2} \right| > 5$ **96.** $\left| \dfrac{x}{2} - 1 \right| - |-2| > |-3|$

In Exercises 97–108, find the solution set of each equation.

97. $\sqrt{x - 4} = \sqrt{x} - 4$

98. $\sqrt{2x} - 4 = \sqrt{2x + 4}$

99. $\sqrt{x - 5} = 5 - \sqrt{x}$ **100.** $\sqrt{x^2 - 4} = x - 2$

101. $\sqrt{x^2 + 25} = x + 5$ **102.** $\sqrt{x + 4} = \sqrt{x} + 2$

103. $x^4 + 5x^2 + 4 = 0$ **104.** $x^{2/3} - x^{1/3} - 6 = 0$

105. $x - 6\sqrt{x} + 8 = 0$ **106.** $2x^4 - x^2 - 3 = 0$

107. $(x + 1)^2 - 8(x + 1) + 15 = 0$

108. $\left(\dfrac{2}{x - 1} \right)^2 + 2\left(\dfrac{2}{x - 1} \right) - 3 = 0$

In Exercises 109–114, find the solution set of each inequality. Display each solution set on the number line.

109. $\dfrac{x - 1}{2 - x} \le 0$ **110.** $\dfrac{x^2 + 4x + 3}{3 - x} \ge 0$

111. $\dfrac{x^2 + x - 6}{x^2 - x - 6} > 0$ **112.** $\dfrac{x + 3}{5 - x} < 1$

113. $x^4 < -13x^2 - 36$ **114.** $x^4 - 34x^2 \le -225$

115. At a recent football game, adult tickets were \$4.00 and student tickets were \$2.50. The total income from the ticket sales was \$4400.00. The number of student tickets sold was four-thirds of the number of adult tickets sold. How many tickets of each kind were sold?

116. The length of a rectangular lot is 10 meters more than two times its width. The perimeter of the lot is 380 meters. Find the dimensions of the lot and its area.

117. A plumber is paid double-time for each hour worked over 40 hours in a week. In a recent week, he worked 52 hours and earned \$608.00. What is his standard hourly salary?

118. The sum of a number and its reciprocal is $\frac{58}{21}$. Find the two numbers.

119. The length of a rectangular lot is 10 feet more than twice its width. The area of the lot is 1500 square feet. Find the dimensions of the lot.

120. Twice the square of an integer is 45 more than the integer. Find the integer.

CHAPTER 2

Relations,
Functions,
and Graphs

The concept of a relation is playing an increasingly important role in modern mathematics. In this chapter, we study relations; in particular, we study an important type of relation called a function. Functions are crucial to the study and applications of mathematics. Therefore, an understanding of functions and their graphs is essential. We begin by considering linear and quadratic functions and their graphs. We also investigate methods for graphing functions that will be used in our study of functions in subsequent chapters. This chapter ends with a discussion of one-to-one functions and their inverses.

2.1 RELATIONS AND FUNCTIONS

The concept of a relation plays an important role in the study of mathematics. Simply put, a **relation** is a correspondence between the elements of a set A and the elements of a set B. Each of us is familiar with the relations defined by the following correspondences:

1. To each person there corresponds a weight and to each there corresponds a height.
2. To each house there corresponds a house number.
3. To each rectangle there corresponds a perimeter.

Each student of geometry knows that the relation between the area A of a circle and its radius r is given by the equation $A = \pi r^2$. Likewise, students of economics generally agree that a relationship exists between the demand d for a product and its price p.

Suppose John, Allison, Kate, Kevin, and Marcy are five college students whose grade point averages (GPA's) are, respectively, 2.71, 3.45, 2.67, 3.56, and 3.21. Then we can pair these students with their grade point averages as follows:

(John, 2.71)

(Allison, 3.45)

(Kate, 2.67)

(Kevin, 3.56)

(Marcy, 3.21)

As you can see, we have formed ordered pairs with a student as the first component and his or her corresponding GPA as the second.

The importance of establishing relationships among various types of phenomena underscores the necessity for describing a relation in precise mathematical language.

DEFINITION 2.1

A **relation** is a set of ordered pairs.

Let us consider the relations defined below:

$R_1 = \{$(Tom, 192 lb), (Tom, 6′1″), (Julie, 125 lb), (Julie, 5′7″), (Paul, 182 lb), (Paul, 5′11″), (George, 177 lb), (George, 5′9″), (Carol, 142 lb), (Carol, 5′8″)$\}$;

$R_2 = \{$(Tom's house, 879), (Julie's house, 4627), (Paul's house, 133), (George's house, 5), (Carol's house, 1961)$\}$;

$R_3 = \{$(4 by 3, 14), (2 by 6, 16), (6 by 8, 28), (3 by 5, 16)$\}$;

$R_4 = \{(x, y) \mid x^2 + y^2 = 1, x, y \in \mathbb{R}\}$.

In relation R_1, Tom, Julie, Paul, George, and Carol are paired with their weights and with their heights. One interpretation of relation R_2 is that houses are paired with their house numbers. In relation R_3, if one interprets the first component of each ordered pair as the dimensions of a rectangle, then the second

North Wind Picture Archives

LEONHARD EULER
(1707–1783)

Born in Basel, Switzerland in 1707, Leonhard Euler was the leading mathematician of the eighteenth century and, without question, the most prolific writer of mathematics in history. Euler published an estimated 530 books and papers during his lifetime and wrote, in addition, a wealth of unpublished manuscripts. Known as a successful notation builder, Euler is credited with introducing the functional notation $f(x)$ and is thought to be the first to use the Greek letter π for 3.14159..., i for $\sqrt{-1}$, and e for 2.71828.... The use of circles to represent sets originated with Euler—we might more accurately refer to Venn diagrams, therefore, as Euler diagrams. Euler's accomplishments are all the more remarkable in view of the fact that he was totally blind during the last seventeen years of his life.

component gives the perimeter of that rectangle. We see that relation R_4 consists of all ordered pairs (x, y) of real numbers whose components satisfy the equation $x^2 + y^2 = 1$. Thus, R_4 contains such ordered pairs as $(1, 0)$, $(0, 1)$, $(\sqrt{2}/2, -\sqrt{2}/2)$, $(\sqrt{3}/2, -\frac{1}{2})$, etc.

The set consisting of the first components of the ordered pairs in a relation and the set consisting of the second components of the ordered pairs in a relation are extremely important.

DEFINITION 2.2

Let R be a relation. The **domain** of R, denoted by dom (R), is

$$\{x \mid x \text{ is the first component of an ordered pair in } R\}.$$

The **range** of R, denoted by ran (R), is

$$\{y \mid y \text{ is the second component of an ordered pair in } R\}.$$

EXAMPLE 1

Consider the relations R_1, R_2, and R_3 defined on page 61. Give the domain and range of each.

Solution   For the relations R_1, R_2, and R_3,

dom (R_1) = {Tom, Julie, Paul, George, Carol};
ran (R_1) = {192 lb, 6'1", 125 lb, 5'7", 182 lb, 5'11", 177 lb,
5'9", 142 lb, 5'8"};
dom (R_2) = {Tom's house, Julie's house, Paul's house, George's house,
Carol's house};
ran (R_2) = {879, 4627, 133, 5, 1961}.
dom (R_3) = {4 by 3, 2 by 6, 6 by 8, 3 by 5};
ran (R_3) = {14, 16, 28}.

Graphs of Relations

The ordered pairs that belong to a relation whose domain and range are subsets of $\mathbb{R}$ can be interpreted geometrically as points in the plane. The collection of all such points is called the graph of the relation. More specifically, we have

---

Graph of a Relation

Let R be a relation whose domain and range are subsets of $\mathbb{R}$. The **graph** of R consists of all those points in the plane corresponding to the ordered pairs in R.

FIGURE 2.1

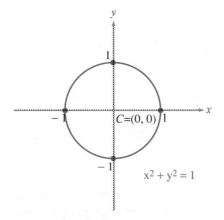

The phrase "sketch the graph of R" means to plot enough points on the graph to illustrate the important features of the graph. When a relation is defined by an equation, we say that the graph of the relation is the graph of the equation. For example, the graph of relation R_4, defined above, is the graph of the equation $x^2 + y^2 = 1$, which is the circle with radius 1 and center $(0, 0)$ (see Figure 2.1).

From Figure 2.1, we see that if $P = (x, y)$ is a point on the circle $x^2 + y^2 = 1$, then $-1 \le x \le 1$ and $-1 \le y \le 1$. On the other hand, if $-1 \le x \le 1$, there exists at least one point P on the circle whose abscissa is x. Similarly, if $-1 \le y \le 1$, there exists at least one point Q on the circle whose ordinate is y. Thus, $\text{dom}\,(R_4) = [-1, 1]$ and $\text{ran}\,(R_4) = [-1, 1]$.

EXAMPLE 2

Sketch the graph of the relation R_5 defined by $R_5 = \{(x, y) \mid |x| < 3 \text{ and } |y| \le 1\}$. Give the domain and range of this relation.

FIGURE 2.2

a.

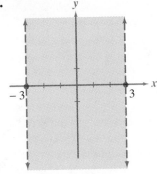

Solution To find the collection of ordered pairs whose components satisfy the inequalities $|x| < 3$ and $|y| \le 1$, we replace $|x| < 3$ by the equivalent inequality $-3 < x < 3$ and replace $|y| \le 1$ by the equivalent inequality $-1 \le y \le 1$. Recall that $-3 < x < 3$ if and only if x is in the open interval $(-3, 3)$. Thus, the set of ordered pairs (x, y) satisfying the inequality $-3 < x < 3$ is the infinite vertical strip shown in Figure 2.2(a). The sides of this infinite vertical strip are dashed to indicate that -3 and 3 are not in the open interval $(-3, 3)$. Since $-1 \le y \le 1$ if and only if y is the closed interval $[-1, 1]$, the set of ordered pairs (x, y) for which $-1 \le y \le 1$ is the set of points in the infinite horizontal strip shown in Figure 2.2(b). The sides of this infinite horizontal strip are solid to indicate that -1 and 1 are in the closed interval $[-1, 1]$. Since each ordered pair (x, y) in R_5 must satisfy *both* of the inequalities

$$-3 < x < 3 \qquad \text{and} \qquad -1 \le y \le 1,$$

b.

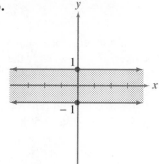

the graph of R_5 is the intersection of the two strips, as shown in Figure 2.2(c). Hence, $\text{dom}\,(R_5) = (-3, 3)$ and $\text{ran}\,(R_5) = [-1, 1]$.

Sometimes, relations are defined by tables or graphs. Table 2.1 below gives the amount of rainfall, measured in inches, for Macon, Georgia in 1987. The relation determined by Table 2.1 is

$R_6 = \{$(Jan, 6.92), (Feb, 7.01), (Mar, 4.25), (Apr, 0.68), (May, 2.78), (Jun, 5.10), (Jul, 1.38), (Aug, 3.21), (Sep, 1.50), (Oct, 0.05), (Nov, 3.36), (Dec, 1.88)$\}$.

c.

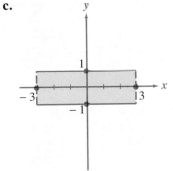

Hence, $\text{dom}\,(R_6) = \{$Jan, Feb, Mar, Apr, May, Jun, Jul, Aug, Sept, Oct, Nov, Dec$\}$ and $\text{ran}\,(R_6) = \{6.92, 7.01, 4.25, 0.68, 2.78, 5.10, 1.38, 3.21, 1.50, 0.05, 3.36, 1.88\}$.

Let R_7 be the relation whose graph is given in Figure 2.3 (see page 64). Assume that the cost of mailing a letter is 29 cents if its weight does not exceed one ounce, and that the cost increases by 23 cents per ounce or fraction of an ounce thereafter. The graph shown in Figure 2.3 gives the cost of mailing a first-class letter whose weight is not greater than 11 ounces. Note that the points $(1, 0.29), (2, 0.52), (3, 0.75), (4, 0.98)$, etc. are represented by solid circles to indicate they are points on the graph, while the points $(0, 0.29), (1, 0.52), (2, 0.75)$, etc. are represented by open circles to indicate that they are *not* points on the graph. Thus, for example, the cost of mailing a first-class letter weighing 2 ounces is 52 cents, not 75 cents.

TABLE 2.1

Jan	Feb	Mar	Apr	May	Jun	Jul	Aug	Sep	Oct	Nov	Dec
6.92	7.01	4.25	0.68	2.78	5.10	1.38	3.21	1.50	0.05	3.36	1.88

FIGURE 2.3

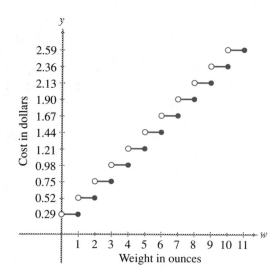

EXAMPLE 3

Consider the relation R_7 defined by the graph shown in Figure 2.3.

a. Give dom (R_7) and ran (R_7).
b. Give the cost of mailing a first-class letter whose weight is
    (i)   1.4 oz   **(ii)**   2.3 oz   **(iii)**   2.7 oz

Solution

a. Let us interpret the graph of R_7. If the weight w of a letter is one ounce or less, the cost of mailing it is 29 cents. Moreover, if $1 < w \leq 2$, the cost is 52 cents. Continuing, we see that if $2 < w \leq 3$, the cost of mailing the letter is 75 cents, etc.

Hence, from the graph, we see that dom $(R_7) = (0, 11]$ and ran (R_7) = {0.29, 0.52, 0.75, 0.98, 1.21, 1.44, 1.67, 1.90, 2.13, 2.36, 2.59}.

b. **(i)**   Since $1 < 1.4 \leq 2$, the cost is 52 cents.
    (ii)   Since $2 < 2.3 \leq 3$, the cost is 75 cents.
    (iii)   As in part (ii), the cost of mailing the letter is 75 cents since $2 < 2.7 \leq 3$.

Circles

Earlier in this section, we noted that the graph of the equation

$$x^2 + y^2 = 1$$

is a circle with center $(0, 0)$ and radius 1. Thus, a point $P = (x, y)$ lies on this circle if and only if the distance from P to the origin $(0, 0)$ is 1. More generally, a circle is defined as the set of all points in a plane that lie a given distance from a given point. The given point is called the **center** of the circle, and the common distance from the center is called the **radius**. As shown in Figure 2.4, a point $P = (x, y)$ is on the circle with center $C = (h, k)$ and radius $r > 0$ if and only if

$$d(C, P) = r.$$

Thus, using the distance formula, a point $P = (x, y)$ is on this circle if and only if

FIGURE 2.4

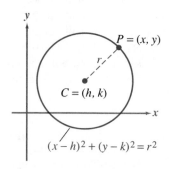

$$\sqrt{(x - h)^2 + (y - k)^2} = r.$$

The latter equation is equivalent to the equation

$$(x - h)^2 + (y - k)^2 = r^2, \quad \text{where } r > 0.$$

Summarizing, we have

Standard Form of the Equation of a Circle

The standard form of the equation of a circle with radius $r > 0$ and center $C = (h, k)$ is

$$(x - h)^2 + (y - k)^2 = r^2. \tag{1}$$

From equation (1), it follows that the standard form for the circle of radius 1 and center $(0, 0)$ is

$$(x - 0)^2 + (y - 0)^2 = 1^2, \quad \text{or}$$
$$x^2 + y^2 = 1.$$

This circle is called a **unit circle** since the radius $r = 1$. The unit circle $x^2 + y^2 = 1$ plays a central role in the development of the trigonometric functions that we will cover in Chapter 5.

EXAMPLE 4

FIGURE 2.5
$(x - 4)^2 + (y + 2)^2 = 18$

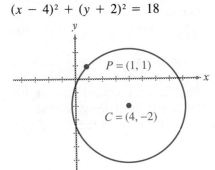

Find the equation, in standard form, for the circle that has center $C = (4, -2)$ and contains the point $P = (1, 1)$. Sketch the graph of this circle.

Solution Since the point $P = (1, 1)$ is on the circle, the radius r is $d(C, P)$. By the distance formula, we have

$$\begin{aligned} r &= \sqrt{(1 - 4)^2 + (1 - (-2))^2} \\ &= \sqrt{(-3)^2 + 3^2} \\ &= \sqrt{9 + 9} \\ &= \sqrt{18} \\ &= 3\sqrt{2}. \end{aligned}$$

Using the standard form of the equation of a circle with $h = 4$, $k = -2$, and $r = 3\sqrt{2}$, we obtain the equation

$$(x - 4)^2 + (y - (-2))^2 = (3\sqrt{2})^2, \quad \text{or}$$
$$(x - 4)^2 + (y + 2)^2 = 18.$$

The graph of this circle is shown in Figure 2.5.

By expanding the terms and simplifying, the equation

$$(x - h)^2 + (y - k)^2 = r^2$$

can be written in the form

$$x^2 + y^2 + Dx + Ey + F = 0 \tag{2}$$

for some real numbers D, E, and F (see Exercise 6).

Conversely, if we begin with equation (2), it is always possible, by completing the squares in x and y, to obtain an equation of the form

$$(x - h)^2 + (y - k)^2 = c.$$

If $c > 0$, the graph is a circle with center $C = (h, k)$ and radius $r = \sqrt{c}$. If $c = 0$, the graph consists of the single point (h, k). Finally, if $c < 0$, the equation has no real solutions, and, therefore, there is no graph.

EXAMPLE 5

Show that the equation

$$x^2 + y^2 + 2x + 6y + 6 = 0$$

is the equation of a circle. Find the center and radius, and sketch the graph.

Solution   First, we write the given equation in the form

$$(x^2 + 2x \quad) + (y^2 + 6y \quad) = -6. \tag{3}$$

FIGURE 2.6

$(x + 1)^2 + (y + 3)^2 = 4$

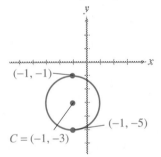

$(-1, -1)$

$(-1, -5)$

$C = (-1, -3)$

To complete the square in x, we add $(\frac{2}{2})^2$, or 1, to each side of equation (3). To complete the square in y, we add $(\frac{6}{2})^2$, or 9, to each side of this equation. Thus, we have

$$(x^2 + 2x + 1) + (y^2 + 6y + 9) = -6 + 1 + 9. \tag{4}$$

Simplifying equation (4), we get

$$(x + 1)^2 + (y + 3)^2 = 4. \tag{5}$$

We recognize equation (5) as the standard form of the equation for the circle having center $C = (-1, -3)$ and radius $r = 2$. The graph of this circle is shown in Figure 2.6.

Functions

Every circle is a relation. Notice that the relation defined by the equation

$$(x + 1)^2 + (y + 3)^2 = 4$$

contains distinct ordered pairs having the same first component such as $(-1, -1)$ and $(-1, -5)$ (see Figure 2.6). However, a close inspection of relation R_2, given on page 61, reveals that no two distinct ordered pairs in this relation have the same first component. A relation with the property that no two distinct ordered pairs have the same first component is called a function. Functions play a vital role in the study of mathematics.

DEFINITION 2.3

A **function** is a relation with the property that no two distinct ordered pairs have the same first component.

From the discussion above, we see that not every relation is a function. Using the graph of a relation R, we can determine whether or not R is a function.

Recall that a vertical line is characterized by the fact that all the points on the line have the same x-coordinate. Thus, if a vertical line intersects the graph of a relation more than once, then there are at least two ordered pairs in the relation with the same first component. Hence, the relation is not a function. Summarizing the preceding remarks, we have

Vertical Line Test

A graph in the xy-plane is the graph of a function if and only if any vertical line intersects the graph in at most one point.

Note that the phrase "in at most one point" means once or not at all.

Let us investigate the graphs of two of the relations we have defined in this section. First, consider the graph of relation R_5, which was defined on page 63. As shown in Figure 2.7(a), there exists a vertical line that intersects the graph of R_5 more than once. Hence, relation R_5 is not a function. However, relation R_7, as defined on page 64 and graphed in Figure 2.7(b), is a function, since no vertical line intersects its graph more than once.

FIGURE 2.7

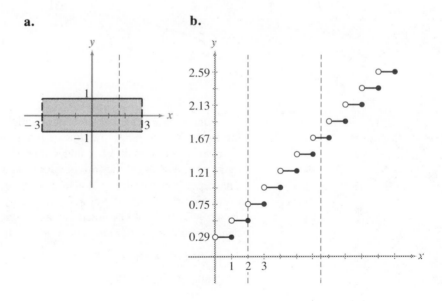

a.

b.

EXAMPLE 6

Consider the relations whose graphs are shown in Figure 2.8 (page 68). Which of the relations are functions?

Solution   The graphs of the relations shown in parts (a) and (d) of Figure 2.8 pass the vertical line test, and, therefore, these relations are functions. However, the graphs of the relations sketched in parts (b) and (c) of Figure 2.8 *do not* pass the vertical line test, and hence, these relations are not functions.

FIGURE 2.8

a.

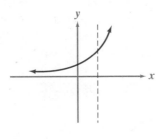

b.

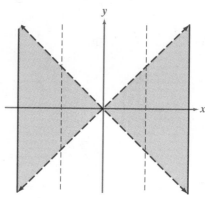

c.

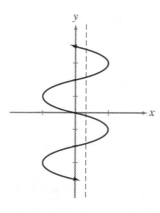

d.

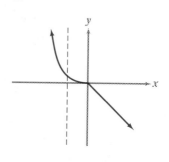

Functions are often defined by equations. For example, it can be shown that the equation $y = 4x + 3$ defines a function f. In this equation, x represents an element in the domain of f, and y represents the corresponding element in the range of f. We call x the **independent variable** and y the **dependent variable**. The dependent variable y is also denoted by the symbol $f(x)$, which is read "f of x" or "the value of f at x." Therefore, $(x, y) \in f$ if and only if $y = f(x) = 4x + 3$. It should be noted that any letter may be used to denote the independent variable in the defining equation of a function. For example, the equations

$$f(x) = 4x + 3, \qquad f(t) = 4t + 3, \qquad \text{and} \qquad f(z) = 4z + 3$$

all define the same function.

EXAMPLE 7

Let $f(x) = x^2 + x - 1$. Evaluate each of the following:

a. $f(-2)$ **b.** $f(\frac{1}{4})$
c. $f(-3 + h)$ **d.** $f(a + h)$

Solution

a. Substituting -2 for x, we get

$$f(-2) = (-2)^2 + (-2) - 1 = 4 - 2 - 1 = 1.$$

b. For $x = \frac{1}{4}$, we get

$$f\left(\frac{1}{4}\right) = \left(\frac{1}{4}\right)^2 + \frac{1}{4} - 1 = \frac{1}{16} + \frac{1}{4} - 1 = -\frac{11}{16}.$$

c. Letting $x = -3 + h$, we have

$$\begin{aligned} f(-3 + h) &= (-3 + h)^2 + (-3 + h) - 1 \\ &= 9 - 6h + h^2 - 3 + h - 1 \\ &= h^2 - 5h + 5. \end{aligned}$$

d. For $x = a + h$, we get

$$\begin{aligned} f(a + h) &= (a + h)^2 + (a + h) - 1 \\ &= a^2 + 2ah + h^2 + a + h - 1. \end{aligned}$$

Notice that if $a = -3$, $f(a + h) = f(-3 + h) = h^2 - 5h + 5$ as we found in part (c).

While an equation can be used to define a function, the function is not determined until the domain is specified. For example, the function f_1 defined by $f_1(x) = 6x + 3$ for $x \in [1, 2]$ is not the same as the function f_2 defined by $f_2(x) = 6x + 3$ for $x \in [1, 6]$. The following rule applies when the domain of a function is not specified:

Agreement on the Domain of a Function

If a function f is defined by an equation and the domain is not given, then the domain is assumed to be the set of all real numbers x for which $f(x)$ is a real number.

EXAMPLE 8

Find the domain of each function.

a. $f_1(x) = x^2 - x - 2$ **b.** $f_2(x) = \dfrac{x}{x^2 - x - 2}$

c. $f_3(x) = \sqrt{x^2 - x - 2}$ **d.** $f_4(x) = \dfrac{1}{\sqrt{x^2 - x - 2}}$

Solution

a. In Chapter 1 we learned that sums and products of real numbers are also real numbers. Hence, if x is a real number, then $x^2 - x - 2$ is also a real number, and thus $\text{dom}(f_1) = \mathbb{R}$.

b. Since division by zero is undefined, $\text{dom}(f_2)$ contains no numbers x such that $x^2 - x - 2 = 0$. Solving the quadratic equation $x^2 - x - 2 = 0$, we find the solutions are -1 and 2. Consequently, $\text{dom}(f_2) = \{x \mid x \in \mathbb{R},\ x \neq -1,\ \text{and}\ x \neq 2\}$.

c. Since the square root of a negative number is not a real number, we must require that $x^2 - x - 2 \geq 0$. Using methods from Chapter 1, we find that the solution set of this inequality is $(-\infty, -1] \cup [2, \infty)$. Thus, by our agreement on domains, $\text{dom}(f_3) = (-\infty, -1] \cup [2, \infty)$.

d. As noted in part (c), $x^2 - x - 2$ must be greater than or equal to zero in order for $\sqrt{x^2 - x - 2}$ to be a real number. However, since we cannot divide by zero, we require that $x^2 - x - 2 > 0$. The solution set for this inequality is $(-\infty, -1) \cup (2, \infty)$, and therefore, $\text{dom}(f_4) = (-\infty, -1) \cup (2, \infty)$. _____

In calculus, the **difference quotient**

$$\frac{f(x + h) - f(x)}{h}, \quad \text{for } h \neq 0,$$

is used to determine some basic properties of the function f. In Example 9, we illustrate how to compute and simplify difference quotients.

EXAMPLE 9

a. If $f(x) = x^2 + 2x - 3$, compute and simplify the difference quotient

$$\frac{f(2 + h) - f(2)}{h}.$$

b. If $g(x) = 2/x$, compute and simplify the difference quotient

$$\frac{g(x + h) - g(x)}{h}.$$

Solution

a. First, we compute $f(2 + h)$.

$$f(x) = x^2 + 2x - 3$$
$$f(2 + h) = (2 + h)^2 + 2(2 + h) - 3$$
$$= 4 + 4h + h^2 + 4 + 2h - 3$$
$$= h^2 + 6h + 5.$$

Next, we subtract $f(2)$ from $f(2 + h)$.

$$f(2 + h) - f(2) = h^2 + 6h + 5 - [(2)^2 + 2(2) - 3]$$
$$= h^2 + 6h + 5 - 5$$
$$= h^2 + 6h.$$

Dividing $f(2 + h) - f(2)$ by h and simplifying, we have

$$\frac{f(2 + h) - f(2)}{h} = \frac{h^2 + 6h}{h}$$
$$= h + 6.$$

b. Since $g(x) = 2/x$, $g(x + h) = 2/(x + h)$. Therefore,

$$g(x + h) - g(x) = \frac{2}{x + h} - \frac{2}{x}$$
$$= \frac{2x - 2(x + h)}{x(x + h)}$$
$$= \frac{2x - 2x - 2h}{x^2 + xh}$$
$$= \frac{-2h}{x^2 + xh}.$$

Hence,

$$\frac{g(x + h) - g(x)}{h} = \frac{-2h/(x^2 + xh)}{h}$$

$$= \frac{-2}{x^2 + xh}.$$

Applications

Functions play an important role in solving applied problems, as our next three examples illustrate.

E X A M P L E 10

An open box with a rectangular base is to be constructed from a rectangular piece of cardboard 10 inches wide and 15 inches long, by cutting out equal squares of length x from each corner and then bending up the sides. Express the volume of the box as a function of x.

Solution

FIGURE 2.9

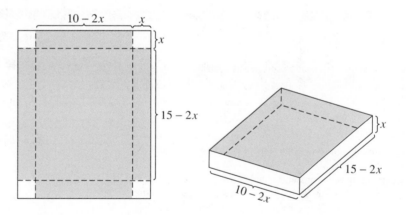

The volume of the rectangular box is given by the equation

volume = (length of base)(width of base)(height).

From Figure 2.9 we see that

volume = $V(x)$ = $(15 - 2x)(10 - 2x)x$
= $150x - 50x^2 + 4x^3$.

E X A M P L E 11

FIGURE 2.10

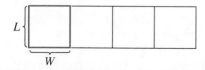

A farmer wishes to make each of four adjoining rectangular pens 1000 square feet in area, as shown in Figure 2.10. Express the amount of fencing required as a function of W, the width of each pen.

Solution   Since the area of a rectangle is the product of its length and its width, we have $1000 = L \cdot W$. Solving for L, we get

$$L = \frac{1000}{W}.$$

From Figure 2.10, we see that F, the total amount of fencing needed, is

$$F(W) = 8W + 5L$$
$$= 8W + 5\left(\frac{1000}{W}\right)$$
$$= 8W + \frac{5000}{W}.$$

E X A M P L E 12

The sum of two numbers is 40. Express the product of the two numbers as a function of one variable.

Solution   Let x and z represent the two numbers. Since $x + z = 40$, it follows that $z = 40 - x$. Thus the product P of the two numbers is given by

$$P = xz = x(40 - x).$$

We write $P(x) = x(40 - x) = 40x - x^2$ to indicate that P is a function of the variable x.

EXERCISES 2.1

In Exercises 1–4, graph each relation and give its domain and range.

1. $A = \{(x, y) \mid 0 \le x \le 3, 0 \le y \le 2, x \text{ and } y \text{ are integers}\}$.
2. $B = \{(x, y) \mid |y| \le x, 0 \le x < 2, x \text{ and } y \text{ are integers}\}$.
3. $C = \{(x, y) \mid y = x/2, x \in \{-3, -2, 0, 1, 4\}\}$.
4. $D = \{(x, y) \mid |y| < |x|, -3 \le x \le 3\}$.
5. Determine which of the relations in Exercises 1–4 are functions.
6. By expanding the terms and simplifying, show that the equation $(x - h)^2 + (y - k)^2 = r^2$ can be written in the form $x^2 + y^2 + Dx + Ey + F = 0$ for some real numbers D, E, and F.

In Exercises 7–14, give the equation, in standard form, of the circle satisfying the given conditions.

7. Center: $(3, -4)$
    radius: 10
8. Center: $(-\sqrt{3}, 1)$
    radius: $\sqrt{6}$
9. Center: $(2, -1)$
    containing the point
    $P = (-1, 5)$
10. Center: $(\frac{3}{4}, -\frac{1}{10})$
    containing the point
    $P = (\frac{5}{4}, 2)$
11. Center: $(1, -2)$
    tangent to the y-axis
12. Center: $(1, -2)$
    tangent to the x-axis
13. Endpoints of a diameter: $P = (-1, 1)$ and $Q = (5, 3)$
14. Endpoints of a diameter: $P = (-3, 6)$ and $Q = (-1, 1)$

In Exercises 15–22, find the center and radius of the circle defined by the given equation.

15. $x^2 + y^2 = 36$
16. $x^2 = 64 - y^2$
17. $x^2 + y^2 - 2x - 2y + 1 = 0$
18. $x^2 + y^2 + 6y = 25$
19. $9x^2 + 9y^2 - 6x - 225 = 0$
20. $16x^2 + 16y^2 + 16x - 8y + 3 = 0$
21. $4x^2 + 4y^2 - 4x - 8y - 11 = 0$
22. $\frac{1}{2}x^2 + \frac{1}{2}y^2 - 6x - 4y + 8 = 0$

Give the domain and range of each of the relations whose graphs are given in Exercises 23–28. Determine which relations are functions.

23.

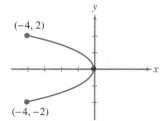

24.

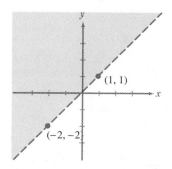

25.

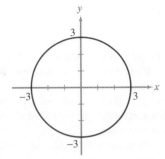

26.

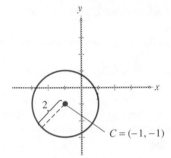

$C = (-1, -1)$

27.

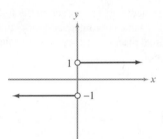

28.

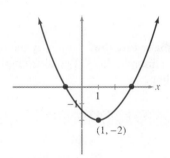

$(1, -2)$

29. If $f(x) = 3x^2 - x - 1$, find $f(0), f(-1)$, and $f(2)$.

30. If $f(x) = \sqrt{x - 2} - 3x$, find $f(6), f(4)$, and $f(2)$.

31. Given that $f(x) = x/(x^2 - x - 6)$, find (if possible) $f(0)$, $f(2)$, and $f(3)$.

32. If $f(x) = |x - 2| + 3$, find $f(-1), f(0)$, and $f(1)$.

In Exercises 33–36, find each of the following, where a and h are real numbers and $h \neq 0$.

a. $f(a)$

b. $f(a + h)$

c. $\dfrac{f(a + h) - f(a)}{h}$

d. $f\left(\dfrac{1}{a}\right)$

e. $\dfrac{1}{f(a)}$

f. $f(a^2)$

33. $f(x) = \sqrt{6x - 5}$

34. $f(x) = \dfrac{2x + 3}{x^2 + x - 1}$

35. $f(x) = \dfrac{x + 1}{x^3 + 4x}$

36. $f(x) = \dfrac{\sqrt{x}}{2x^2 - 10x + 5}$

In Exercises 37–42, find and simplify the difference quotient

$$\frac{f(x + h) - f(x)}{h}, \quad h \neq 0.$$

37. $f(x) = 2x^2 - x + 1$

38. $f(x) = 5x^2 + 2x - 1$

39. $f(x) = \sqrt{x}$

40. $f(x) = 1/\sqrt{2x}$

41. $f(x) = 10/x$

42. $f(x) = 5/(2x - 3)$

In Exercises 43–50, determine the domain of each function.

43. $f(x) = x^3 - x^2 + 2x - 1$

44. $f(x) = (x^2 - 2x)^4$

45. $f(x) = \sqrt{2x + 3}$

46. $f(x) = \sqrt{x}/(x - 5)$

47. $f(x) = 1/(2x + 1)$

48. $f(x) = |x|/x$

49. $f(x) = |x| - 4$

50. $f(x) = (x - 3)/(x^2 - x - 6)$

51. A man wishes to have an enclosed rectangular garden in his back yard. He has 100 feet of fencing with which to enclose the garden. Express the area A of the garden as a function of the width x.

52. The management of SAV Department Store wishes to enclose a 1000-square-foot area outside the building. This rectangular area will display outdoor furniture. One side is provided by the external wall of the store. Cedar boards will be used for two sides of the enclosure, and the fourth side will be made of steel fencing. If cedar boards cost \$5 per linear foot and steel fencing costs \$2 per linear foot, express the cost, C, of enclosing the rectangular plot as a function of the plot's length x.

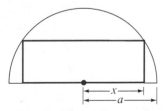

53. A rectangle is to be inscribed in a semicircle of radius a as shown in the accompanying figure. Express the area A of the rectangle as a function of x.

54. A man is at point A on a bank of a straight river 2 kilometers wide and wants to reach point B 6 kilometers downstream on the opposite bank as quickly as possible (see

Figure 2.11). Suppose he rows his boat to some point, D, between C and B and then runs to B. If he can row at 4 kilometers per hour and run at 6 kilometers per hour, express T, the total time to make the trip, as a function of x, the distance from C to D.

FIGURE 2.11

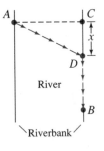

River

Riverbank

Exercises 55–58 refer to the following relations:

$A = \{(x, y) \mid x = y; x, y \in \mathbb{R}\}$;
$B = \{(x, y) \mid y < x; x, y \in \mathbb{R}\}$;
$C = \{(x, y) \mid x \text{ is the sister of } y; x \text{ and } y \text{ are people}\}$;
$D = \{(x, y) \mid x \text{ is the mother of } y; x \text{ and } y \text{ are people}\}$;
$E = \{(x, y) \mid x - y \text{ is even}; x \text{ and } y \text{ are integers}\}$.

55. A relation R is said to be **reflexive** if $(x, x) \in R$ for every x in dom(R). Which of the above relations are reflexive?

56. A relation R is said to be **symmetric** if whenever $(x, y) \in R$, then $(y, x) \in R$. Which of the above relations are symmetric?

57. A relation R is said to be **transitive** if whenever $(x, y) \in R$ and $(y, z) \in R$, then $(x, z) \in R$. Which of the above relations are transitive?

58. If a relation R is reflexive, symmetric, and transitive, then R is called an **equivalence relation**. Which of the above relations are equivalence relations?

59. Give an example of a relation that is reflexive and symmetric but not transitive.

60. Give an example of a relation that is reflexive and transitive but not symmetric.

61. Give an example of a relation that is symmetric and transitive but not reflexive.

62. Is the relation $R = \{(0, 0)\}$ an equivalence relation? Explain.

63. Using a graphing calculator, graph each of the following equations in the same coordinate system. Approximate, to two decimal places, the coordinates of the points of intersection.
  a. $x^2 + y^2 = 4$ **b.** $(x - 1)^2 + y^2 = 5$

64. Using a graphing calculator, graph each of the following equations in the same coordinate system. Approximate, to two decimal places, the coordinates of the points of intersection.
  a. $y^2 = 1 - (x + 1)^2$ **b.** $x^2 + (y + 1)^2 = 4$

FIGURE 2.12

a.

Positive slope

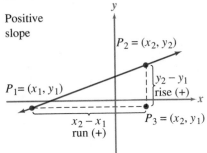

b.

Negative slope

2.2   LINEAR FUNCTIONS AND THEIR GRAPHS

A function whose defining equation has the form

$$f(x) = mx + b$$

where m and b are real numbers, is called a **linear function**. To justify the term *linear*, we must introduce several concepts pertaining to lines. Recall from plane geometry that two points determine a unique line.

DEFINITION 2.4

Let l be the line determined by the points $P_1 = (x_1, y_1)$ and $P_2 = (x_2, y_2)$, where $x_1 \neq x_2$. The **slope m** of l is

$$m = \frac{y_2 - y_1}{x_2 - x_1}.$$

The slope can be interpreted geometrically as the ratio of the change in y (**rise**) to the change in x (**run**) as we move from P_1 to P_2 (see Figure 2.12). It should also be noted that the slope of a line does not depend on the two points chosen, nor on which point we call P_1.

EXAMPLE 13

Find the slope (if it exists) of the line through each pair of points. Sketch the graph of each line.

a. $P_1 = (-1, 6)$, $P_2 = (2, -3)$ **b.** $P_1 = (-5, 2)$, $P_2 = (1, 2)$

c. $P_1 = (-1, 5)$, $P_2 = (-1, -4)$ **d.** $P_1 = (3, -2)$ $P_2 = (-5, -6)$

Solution

a. $m = \dfrac{-3 - 6}{2 - (-1)} = \dfrac{-9}{3} = -3$ **b.** $m = \dfrac{2 - 2}{1 - (-5)} = \dfrac{0}{6} = 0$

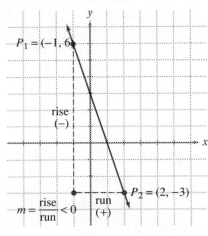
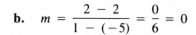
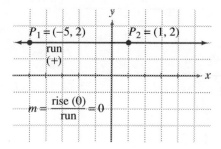

c. $m = \dfrac{-4 - 5}{-1 - (-1)}$

Since $-1 - (-1) = 0$, and since division by 0 is not defined, the slope of this line is undefined.

d. $m = \dfrac{(-6) - (-2)}{-5 - 3} = \dfrac{-4}{-8} = \dfrac{1}{2}$

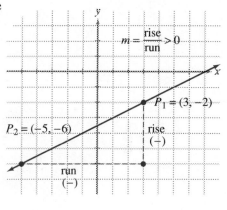
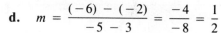
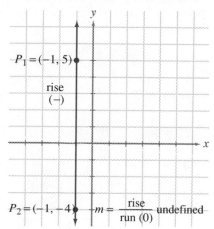

A line that has slope 0, as in part (b) of Example 13, is called *horizontal*, while a line whose slope is undefined, as in part (c) of Example 13, is called *vertical*. Every line parallel to the *x*-axis is horizontal, and every line parallel to the *y*-axis is vertical.

FIGURE 2.13

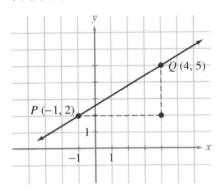

If we are given a point $P = (x_0, y_0)$ on a nonvertical line l with slope m, we can find a second point $Q = (x, y)$ that lies on line l by making the following observation:

> If a line has slope $m = r/s$ where $s > 0$, then for every change of s units in the horizontal direction, the line rises r units if r is positive and falls $|r|$ units if r is negative. (Note that if $r = 0$, the line is horizontal).

Suppose, for example, that line l has slope $m = \frac{3}{5}$ and passes through the point $P = (-1, 2)$. Based on the above discussion, we can find a second point Q on l by moving horizontally 5 units to the right of P and then moving vertically 3 units upward. Hence $Q = (-1 + 5, 2 + 3) = (4, 5)$ is a second point on the line l. Using points P and Q, we sketch the graph of the line l in Figure 2.13.

E X A M P L E 14

Sketch the graph of the line l containing the point $P = (1, -1)$ and having slope $-\frac{3}{2}$.

FIGURE 2.14

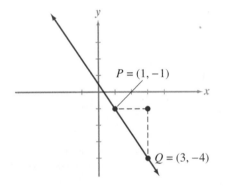

Solution   Starting at the point $P = (1, -1)$, we move horizontally 2 units to the right and vertically 3 units downward, arriving at the point

$$Q = (1 + 2, -1 - 3) = (3, -4),$$

which lies on line l (see Figure 2.14).

Equations of Lines

We now turn our attention to finding equations of lines. First, assume l is the nonvertical line passing through $P_0 = (x_0, y_0)$ having slope m. A point $P = (x, y)$ lies on line l if and only if the line determined by $P = (x, y)$ and $P_0 = (x_0, y_0)$ is line l. Therefore, the slope of the line determined by P and P_0, which is $(y - y_0)/(x - x_0)$, must be equal to m. Equating the two slopes, we get the equation

$$\frac{y - y_0}{x - x_0} = m, \qquad \text{or} \qquad y - y_0 = m(x - x_0).$$

Thus, we have

Point-Slope Form of the Equation of a Line

An equation of the nonvertical line that contains the point $P_0 = (x_0, y_0)$ and has slope m is

$$y - y_0 = m(x - x_0).$$

E X A M P L E 15

Find an equation, in point-slope form, of the line that contains the point $P = (4, -1)$ and has slope 3.

Solution   Since $(x_0, y_0) = (4, -1)$ and $m = 3$, an equation for this line, in point-slope form, is

$$y + 1 = 3(x - 4).$$ _____

If a line l intersects the x-axis at the point $(a, 0)$, then $(a, 0)$ is called the **x-intercept** of l. Similarly, if a line l intersects the y-axis at $(0, b)$, the point $(0, b)$ is called the **y-intercept** of l. An equation of the line l having slope m and y-intercept $(0, b)$ is

$$y - b = m(x - 0), \quad \text{or}$$
$$y = mx + b.$$

The latter equation is called the **slope-intercept form** of the equation of a line.

Slope-Intercept Form of the Equation of a Line

An equation of the line with slope m and y-intercept $(0, b)$ is

$$y = mx + b.$$

EXAMPLE 16

Find the slope-intercept form of the equation of each of the following lines:

a.   the line passing through $P_1 = (3, 4)$ and $P_2 = (-5, 6)$;
b.   the horizontal line passing through the point $(5, 6)$.

Solution

a.   Since the slope of the line passing through $P_1 = (3, 4)$ and $P_2(-5, 6)$ is

$$m = \frac{6 - 4}{-5 - 3} = \frac{2}{-8} = -\frac{1}{4},$$ an equation of this line is $y - 4 = -\frac{1}{4}(x - 3)$. Solving for y, we obtain the slope-intercept form

$$y = -\frac{1}{4}x + \frac{19}{4}.$$

b.   Since the line is horizontal and contains the point $(5, 6)$, every point on the line has 6 as its y-coordinate. Hence, $(0, 6)$ is the y-intercept. Since every horizontal line has slope 0, the slope-intercept form of the equation of the line is $y = 0x + 6$, or $y = 6$.   _____

FIGURE 2.15 $x = \sqrt{6}$

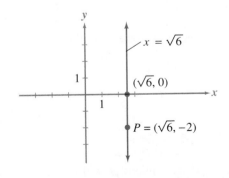

Neither the point-slope form nor the slope-intercept form is appropriate when finding an equation of a vertical line since the slope of a vertical line is undefined. Recall that all the points on a vertical line have the same x-coordinate. Thus, if $P = (x_1, y_1)$ is any point on the vertical line l, then $x = x_1$ is an equation for l. For example, $x = \sqrt{6}$ is an equation of the vertical line passing through the point $P = (\sqrt{6}, -2)$. Its graph is given in Figure 2.15.

Any equation of a line can be written in the form $ax + by + c = 0$ where a, b, and c are real numbers and a and b are not both zero. For example, the equation $y = -3x + 2$ can be written in the equivalent form $3x + y - 2 = 0$, and the equation $y = 6$ has the equivalent form $0x + y - 6 = 0$.

General Form of the Equation of a Line

Any equation of a line can be written in the form $ax + by + c = 0$ where a, b, and c are real numbers and a and b are not both zero. This equation is called the **general form** of the equation of a line.

Thus, we see that every line is the graph of an equation of the form $ax + by + c = 0$, where a, b, and c are real numbers and a and b are not both zero. Conversely, as we shall now show, the graph of the equation $ax + by + c = 0$ is a line, provided a and b are not both zero.

First, let us assume that $b \neq 0$. Solving the equation $ax + by + c = 0$ for y, we obtain

$$y = \left(-\frac{a}{b}\right)x - \frac{c}{b},$$

which is an equation of the line with slope $-\dfrac{a}{b}$ and y-intercept $\left(0, -\dfrac{c}{b}\right)$. However, if $b = 0$, then $a \neq 0$, and the equation $ax + by + c = 0$ becomes

$$ax + c = 0, \quad \text{or}$$

$$x = -\frac{c}{a},$$

which is an equation of the vertical line with x-intercept $(-c/a, 0)$. The following theorem summarizes these results:

THEOREM 2.1

Every line is the graph of an equation of the form $ax + by + c = 0$, where a and b are not both 0. Conversely, the graph of an equation of the form $ax + by + c = 0$ is a line provided that a and b are not both zero.

EXAMPLE 17

Find the slope, x-intercept, and y-intercept, if they exist, for each of the following lines:

a. $2x - y + 4 = 0$ **b.** $y + \pi = 0$ **c.** $x - \dfrac{\sqrt{3}}{2} = 0$

Solution

a. The slope-intercept form of the equation $2x - y + 4 = 0$ is $y = 2x + 4$. Thus, the slope of this line is 2 and its y-intercept is the point $(0, 4)$. To find the x-intercept, we set $y = 0$ and solve for x, getting

$$0 = 2x + 4, \qquad \text{or}$$
$$x = -2.$$

Hence, the point $(-2, 0)$ is the x-intercept.

b. Solving the equation $y + \pi = 0$ for y, we get $y = -\pi$. We recognize this equation as an equation of the horizontal line passing through the point $(0, -\pi)$. Therefore, the y-intercept is $(0, -\pi)$. Since the line is horizontal, it has slope zero. There is no x-intercept for this line, since it is parallel to the x-axis.

c. Solving this equation for x, we get $x = \sqrt{3}/2$. We recognize this as an equation of the vertical line passing through the point $(\sqrt{3}/2, 0)$. Thus the x-intercept is $(\sqrt{3}/2, 0)$. Since the line is parallel to the y-axis, it has no y-intercept, and its slope is undefined.

FIGURE 2.16

Graph of the linear function f defined by $f(x) = 6x + 3$

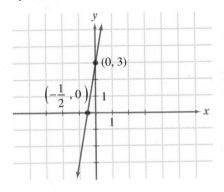

Returning to the opening remarks of this section, we now see that functions whose defining equations have the form $y = f(x) = mx + b$ are justifiably called linear functions. For example, we see in Figure 2.16 that the graph of the linear function f defined by $f(x) = 6x + 3$ is the line with equation $y = 6x + 3$.

Parallel and Perpendicular Lines

Let us now observe the relationship between the slopes of lines that are either parallel or perpendicular. Recalling that the slope of a vertical line is undefined, we focus our attention on nonvertical lines. It can be shown that two nonvertical lines are parallel if and only if they have the same slope. However, for perpendicular lines, the relationship between the slopes is not as clear. If l_1 with slope m_1, and l_2 with slope m_2 are nonvertical perpendicular lines, it can be shown that $m_1 m_2 = -1$ or $m_1 = (-1/m_2)$; that is, the slopes are negative reciprocals of one another. These facts are summarized below.

THEOREM 2.2   Slopes of Parallel and Perpendicular Lines

Let m_1 and m_2 be the slopes of two nonvertical lines. Then

(i)   the lines are parallel if and only if $m_1 = m_2$;

(ii) the lines are perpendicular if and only if $m_1 m_2 = -1$ [or $m_1 = (-1/m_2)$].

EXAMPLE 18

Find $f(x)$ given that f is the linear function containing the point $(2, -1)$ whose graph is

a. parallel to the line $2x - y = 4$;

b. perpendicular to the line $2x - y = 4$.

Solution

a. First, we determine the slope of the line $2x - y = 4$ by writing the equation in slope-intercept form. Solving for y, we get $y = 2x - 4$, and therefore, the slope is 2. Since parallel lines have equal slope by part (i) of Theorem 2.2, an equation of the line passing through $P = (2, -1)$ parallel to the line $2x - y = 4$ is

$$y - (-1) = 2(x - 2)$$
$$y + 1 = 2x - 4$$
$$y = 2x - 5.$$

Hence, $f(x) = 2x - 5.$

b. Since the line $2x - y = 4$ has slope 2, we know by part (ii) of Theorem 2.2 that a line perpendicular to it has slope $-\frac{1}{2}$. Therefore, an equation of such a line through $P = (2, -1)$ is

$$y - (-1) = -\frac{1}{2}(x - 2)$$

$$y + 1 = -\frac{1}{2}x + 1$$

$$y = -\frac{1}{2}x.$$

Thus, $f(x) = -\frac{1}{2}x.$

Applications

Linear functions often arise in applications. We end this section with several examples.

EXAMPLE 19

On an average summer day in a large city, the air pollution index is 10 parts per million at 10:00 A.M., and it increases linearly by 5 parts per million each hour thereafter until 5:00 P.M. Let $P(x)$ be the amount of pollutants in the air x hours after 10:00 A.M.

a. Express P as a linear function of x.
b. What is the air pollution index at 2:00 P.M.?
c. Sketch the graph of P for x in the interval $[0, 7]$.

Solution

a. Since we want $P(x)$ to be the amount of pollutants in the air x hours after 10:00 A.M., $x = 0$ at 10:00 A.M. and $x = 7$ at 5:00 P.M. We know that at $x = 0$ the pollution index is 10 parts per million and at $x = 7$ the pollution index is $10 + 5(7) = 45$. Thus, the graph of P contains the points $(0, 10)$ and $(7, 45)$. Since the slope of the line containing these two points is $m = \dfrac{45 - 10}{7} = \dfrac{35}{7} = 5$, the defining equation of P (in slope-intercept form) is $y = P(x) = 5x + 10$.
b. At 2:00 P.M., $x = 4$. Hence, the air pollution index is $P(4) = 5(4) + 10 = 30$ parts per million.
c. Using the points $(0, 10)$ and $(7, 45)$ from part (a), we graph the linear function P for x in the interval $[0, 7]$ in the accompanying figure.

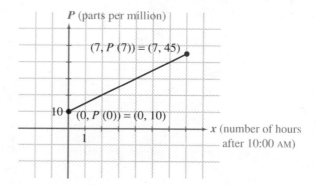

EXAMPLE 20

For tax purposes, a computer purchased by a company for $50,000 is assumed to have a salvage value of $10,000 after ten years. The company uses the method of linear depreciation to depreciate the value of the computer over a 10-year period.

a. Give an equation that defines V, the value of the computer in dollars, as a function of time t.
b. Find the value of the computer after six years.
c. At what time is the value of the computer (theoretically) $15,000?
d. Sketch the graph of V for t in the interval $[0, 10]$.

Solution

a. We begin by letting $V(t)$ denote the value, in dollars, of the computer t years after it is purchased. At the time of purchase, $t = 0$ and $V(0) = 50,000$. Also $V(10) = 10,000$. It follows that $(0, 50,000)$ and $(10, 10,000)$ are points on the graph of V. Thus, the slope of the line defined by V is $m = (10,000 - 50,000)/(10 - 0) = -4000$. Since $(0, 50,000)$ is the y-intercept of this line, its equation, in slope-intercept form, is $y = V(t) = -4000t + 50,000$.

b. Substituting 6 for t in the equation $V(t) = -4000t + 50,000$, we find that

$$V(6) = -4000(6) + 50,000$$
$$= -24,000 + 50,000$$
$$= 26,000.$$

Therefore, the value of the computer after six years is (theoretically) $26,000.

c. To determine the time at which the computer is valued at $15,000 by the company, we set $V(t) = -4000t + 50,000 = 15,000$ and solve for t. Thus, we have

$$-4000t + 50,000 = 15,000$$
$$-4000t = -35,000, \quad \text{or}$$
$$t = 8.75.$$

Therefore, eight years and 9 months after the computer is purchased, it is valued at $15,000 by the company.

d. The graph of V over the interval $[0, 10]$ is shown below.

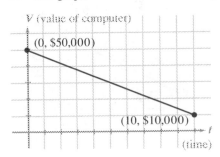

V (value of computer)

(0, $50,000)

(10, $10,000)

t
(time)

EXAMPLE 21

According to Hooke's Law, the force f required to stretch a spring x units beyond its natural length is given by $f(x) = kx$, where k is a constant called the **spring constant**, and f is called the **force function**. If a force of 9 pounds is required to stretch a spring from its natural length of 6 inches to a length of 8 inches, find the force required to stretch the spring from its natural length to a length of 10 inches.

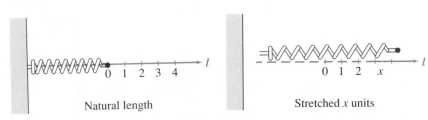

Natural length   Stretched x units

Solution   Since a force of 9 pounds is required to stretch the spring 2 inches, $9 = k(2)$, and hence $k = \frac{9}{2}$. Thus, the force function is defined by $f(x) = \frac{9}{2}x$ for $x \geq 0$. Since 10 inches is 4 inches greater than the natural length of the spring, the amount of force required to stretch the spring from its natural length of 6 inches to a length of 10 inches is $f(4) = \frac{9}{2}(4) = 18$ pounds.

EXERCISES 2.2

In Exercises 1–8, find the slope, x-intercept, and y-intercept (if they exist) for each line. Sketch the graph of each line.

1. $y = 3x + 6$
2. $2x - y = 4$
3. $y = 5 - 6x$
4. $2x - 3y + 4 = 0$
5. $3y - 4 = 2x + 3$
6. $5x - 4 = 2y - 3$
7. $x = 8$
8. $-y = 10$

In Exercises 9–16, find f(x) given that f is a linear function whose graph satisfies the given conditions.

9. Passes through $P = (-1, 2)$ and $Q = (4, -3)$.
10. Has slope -2 and y-intercept $(0, -1)$.
11. Passes through $P = (-3, -1)$ and is perpendicular to the line $3x - 2y + 1 = 0$.

12. Passes through $P = (2, -5)$ and is parallel to the x-axis.
13. $f(5) = 3$ and $f(-2) = 7$.
14. Passes through $P = (3, -4)$ with x-intercept $(-1, 0)$.
15. Passes through $P = (-\sqrt{2}, 1)$ and is parallel to the line $x - y + 4 = 0$.
16. Is the perpendicular bisector of the line segment determined by $P_1 = (-3, 2)$ and $P_2 = (4, -1)$.

In Exercises 17–20, find the slope-intercept form of the equation of the line with slope m and y-intercept (0, b). Find the x-intercept of the line and sketch the graph.

17. $m = -2, b = 3$
18. $m = \sqrt{2}, b = -2$
19. $m = 1/\sqrt{3}, b = 0$
20. $m = -\frac{1}{2}, b = \sqrt{5}$

In Exercises 21–26, express each linear equation in slope-intercept form and give the slope and y-intercept of each line. Also, sketch the graph of each line.

21. $6x - 4y + 10 = 0$ **22.** $6x = 4 - 3y$
23. $y = -2$ **24.** $y = 0$
25. $x = -3y + 5$ **26.** $\frac{1}{2}x = 3y + 2$

27. Find a real number c such that $P = (-2, -3)$ lies on the line $cx + y + 9 = 0$.

28. Find a real number c such that the line $2x + cy + 4 = 0$ has x-intercept $(-2, 0)$.

29. Prove that an equation of the nonvertical line through $P_1 = (x_1, y_1)$ and $P_2 = (x_2, y_2)$ is

$$(y - y_1)(x_2 - x_1) = (y_2 - y_1)(x - x_1).$$

This is called the **two-point form** of the equation of a line.

30. Give the defining equation of the linear function f satisfying each of the following conditions.
   a. Its graph is the line passing through $P = (-1, 3)$ with slope $-\frac{1}{4}$.
   b. Its graph is the line with intercepts $(\sqrt{2}, 0)$ and $(0, -\frac{1}{3})$.
   c. $f(1) = -6$ and its graph is parallel to the line $2x - 3y + 4 = 0$.

31. An electronics store sells a television that costs $250 for $400 and a stereo that costs $550 for $800.
   a. If the markup policy of the store for items that cost over $100 is assumed to be linear and is reflected in these prices, write an equation that expresses the retail price R as a function of the cost C.
   b. What would be the retail price of a television whose cost is $300?
   c. What was the cost of a stereo whose retail price is $1000?

32. There is a linear relationship between temperature measured in degrees Fahrenheit (F) and temperature measured in degrees Celsius (C). The freezing point of water is 0° C and 32° F, while the boiling point is 100° C and 212° F.

   a. Express C as a function of F.
   b. Calculate C when F = 100°.
   c. Calculate F when C = -40°.

33. An investor paid $100,000 for rental property that he depreciates over a 20-year period using the method of linear depreciation. This means that at the end of 20 years the investor is allowed by the IRS to set the value of the property at $0.00.
   a. Give an equation that defines V, the value of the rental property, as a function of time t.
   b. How many years after the property was purchased is it (theoretically) worth $42,500?

34. A teacher has given a test on which the highest grade is 80 and the lowest grade is 50. Determine a linear function f that can be used to distribute the test grades so that the grade of 50 becomes 60, and 80 becomes 100.

35. A force of 15 pounds is required to stretch a spring from its natural length of 10 inches to a length of 15 inches. Use Hooke's Law (see Example 21) to calculate the force needed to stretch the spring from its natural length to a length of 12 inches.

36. An automobile salesperson receives a monthly salary of $500 plus $200 for each automobile sold.
   a. Give an equation that expresses her monthly salary S in terms of n, the number of cars she sells.
   b. Give the domain of the function S.
   c. Sketch the graph of S.

37. Using a graphing calculator, graph each of the following equations in the same coordinate system. Approximate, to two decimal places, the coordinates of the points of intersection.
   a. $\frac{\sqrt{3}}{3}x + y = 4$ **b.** $\sqrt{3}x - y = 2$

38. Using a graphing calculator, graph each of the following equations in the same coordinate system. Approximate, to two decimal places, the coordinates of the points of intersection.
   a. $(x - 2)^2 + y^2 = 9$ **b.** $\sqrt{2}x - y + 3 - 2\sqrt{2} = 0$.

2.3 QUADRATIC FUNCTIONS AND THEIR GRAPHS

A function whose defining equation has the form

$$f(x) = ax^2 + bx + c, \tag{1}$$

where a, b, and c are real numbers and $a \neq 0$, is called a **quadratic function**. To determine some useful properties of quadratic functions, we will transform equation (1) into the form

$$f(x) = a(x + h)^2 + k, \quad \text{where } a, h, \text{ and } k \text{ are constants,}$$

by using the method of completing the square.

First, we factor a, the coefficient of x^2, from the first two terms on the right side of equation (1) to get

$$f(x) = a\left(x^2 + \frac{b}{a}x\right) + c. \tag{2}$$

To complete the square for the expression

$$x^2 + \frac{b}{a}x,$$

we add $b^2/4a^2$ to get

$$x^2 + \frac{b}{a}x + \frac{b^2}{4a^2}.$$

We can now rewrite equation (2) in the form

$$f(x) = a\left(x^2 + \frac{b}{a}x + \frac{b^2}{4a^2}\right) + c - \frac{b^2}{4a}. \tag{3}$$

To see why we subtract $b^2/4a$ instead of $b^2/4a^2$ on the right side, observe that every term of the expression $x^2 + (b/a)x + (b^2/4a^2)$ in equation (3) is multiplied by the number a. Since the expression $x^2 + (b/a)x + (b^2/4a^2)$ is a perfect square trinomial that can be written as $[x + (b/2a)]^2$, and since $c - (b^2/4a) = (4ac - b^2)/4a$, equation (3) can be expressed as

$$f(x) = a\left(x + \frac{b}{2a}\right)^2 + \frac{4ac - b^2}{4a}.$$

Letting $h = b/2a$ and $k = (4ac - b^2)/4a$, the latter equation becomes

$$f(x) = a(x + h)^2 + k, \tag{4}$$

as desired.

The graph of a quadratic function is a U-shaped curve called a **parabola**. If $a > 0$, the parabola opens upward, and from equation (4), we see that the **minimum** value of f occurs when

$$x + h = 0,$$

or when

$$x = -h = -\frac{b}{2a}.$$

On the other hand, if $a < 0$, the parabola opens downward, and the **maximum** value of f occurs when

$$x + h = 0,$$

or when

$$x = -h = -\frac{b}{2a}.$$

(From equation (4), notice that if $x = -h = -(b/2a)$, then $f(-h) = f(-b/2a) = k = (4ac - b^2)/4a$.)

In either case, the point

$$V = \left(-\frac{b}{2a}, f\left(-\frac{b}{2a}\right)\right) \tag{5}$$

is called the **vertex** of the parabola. The vertical line $x = -b/2a$, which passes through the vertex, is called the **axis of symmetry**. The axis of symmetry divides the graph into halves that would coincide if the graph were folded along this line.

The domain of any quadratic function is $\mathbb{R}$, and the range is always an interval. From Figure 2.17, we see that ran (f) is the interval $[f(-b/2a), \infty)$ if $a > 0$, while ran $(f) = (-\infty, f(-b/2a)]$ if $a < 0$.

FIGURE 2.17
Graph of $f(x) = ax^2 + bc + c$
a. $a > 0$

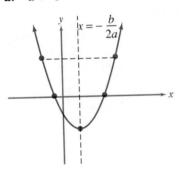

b. $a < 0$

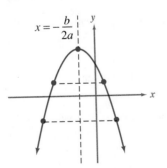

When sketching the graph of a quadratic function, we usually label the vertex, the axis of symmetry, and the x- and y-intercepts of the graph. The procedure for graphing a quadratic function is outlined below.

Procedure for Graphing Quadratic Functions

$$y = f(x) = ax^2 + bx + c, \quad a \neq 0.$$

(i)   Plot the vertex, sketch the axis of symmetry, and determine whether the parabola opens upward or downward.

(ii) Determine the y-intercept, which is $(0, f(0))$, and find the x-intercepts, which, if they exist, are the real solutions of the equation $ax^2 + bx + c = 0$. Plot these points on the coordinate axes.

(iii) Draw the graph as a smooth U-shaped curve that passes through the intercepts and the vertex. If necessary, plot some additional points on either side of the axis of symmetry.

E X A M P L E   22

Sketch the graph of the quadratic function defined by $f(x) = x^2$. Find ran (f).

Solution   The equation $f(x) = x^2$ may be written in the form $f(x) = x^2 + 0x + 0$. Referring to equation (1) on page 83, we see that $a = 1$, $b = 0$, and $c = 0$. Since $a > 0$, the parabola opens upward. Noting that $-b/2a = 0$, we see that the vertex is $V = (0, f(0)) = (0, 0)$. It follows that the line $x = 0$ (the y-axis) is the axis of symmetry and the origin $(0, 0)$ is both the y-intercept and the only x-intercept.

After plotting some additional points, we obtain the graph of f shown in Figure 2.18. From the graph, we see that ran $(f) = [0, \infty)$.

F I G U R E   2.18

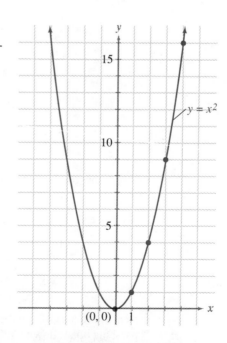

x	y
1	1
2	4
3	9
4	16

EXAMPLE 23

Sketch the graph of the quadratic function defined by $f(x) = 2x^2 + 8x - 1$, and find ran (f).

Solution   In this equation, $a = 2$, $b = 8$, and $c = -1$. Since $a > 0$, the parabola opens upward. Since $-b/2a = -8/[2(2)] = -2$, it follows that the vertex is

$$V = (-2, f(-2))$$
$$= (-2, -9).$$

Hence, the vertical line $x = -2$ is the axis of symmetry for the graph. Solving the equation

$$2x^2 + 8x - 1 = 0$$

for x, we find that the x-intercepts are $((-4 + 3\sqrt{2})/2, 0) \approx (0.12, 0)$ and $((-4 - 3\sqrt{2})/2, 0) \approx (-4.12, 0)$. In addition, the y-intercept is $(0, -1)$. We plot several more points to determine the graph (see Figure 2.19). From the graph, we see that ran $(f) = [-9, \infty)$.

FIGURE 2.19

x	y
-1	-7
0	-1
1	9

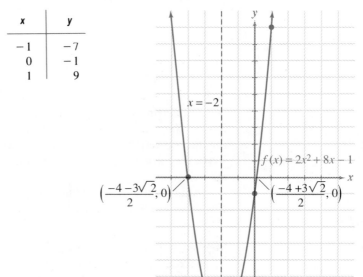

EXAMPLE 24

Sketch the graph of the function defined by $f(x) = -3x^2 + x$ and find ran (f).

Solution   In this case, we see that $a = -3$, $b = 1$, and $c = 0$. Since $-b/2a = -[1/2(-3)] = \frac{1}{6}$, the vertex is

$$V = \left(\frac{1}{6}, f\left(\frac{1}{6}\right)\right)$$
$$= \left(\frac{1}{6}, \frac{1}{12}\right).$$

Hence, the axis of symmetry is the vertical line $x = \frac{1}{6}$. Since $a = -3 < 0$, the parabola opens downward. The y-intercept $(0, f(0))$ is the point $(0, 0)$. Since the

FIGURE 2.20

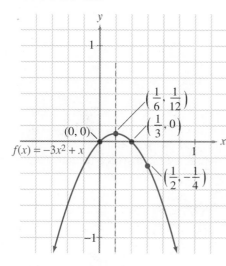

solutions of the equation $-3x^2 + x = 0$ are 0 and $\frac{1}{3}$, the x-intercepts are $(0, 0)$ and $(\frac{1}{3}, 0)$. We plot some additional points to obtain the graph shown in Figure 2.20, from which we see that ran $(f) = (-\infty, \frac{1}{12}]$.

The Graphical Method for Solving Quadratic Inequalities

In Chapter 1, we used the cut-point method to solve quadratic inequalities of the form

$$ax^2 + bx + c > 0, \quad \text{(or with } > \text{ replaced by } \geq, <, \text{ or } \leq)$$

where a, b, and c are real numbers and $a \neq 0$. As you recall, the cut-point method consists of using the real solutions, if any, of the associated quadratic equation $ax^2 + bx + c = 0$ to divide the number line into open intervals. A test value from each of these open intervals is chosen and then used to determine whether that interval is part of the solution set of the inequality. We now illustrate a second method, the **graphical method**, for solving quadratic inequalities.

EXAMPLE 25

Find the solution set of the quadratic inequality $x^2 + x - 2 > 0$.

Solution   We first graph the equation $y = x^2 + x - 2$. Using the method outlined earlier in this section, we obtain the graph shown in Figure 2.21. The solutions of the given inequality are the values of x for which the corresponding values of y are positive. From the graph, we see that the values of y are positive when x is in either of the open intervals $(-\infty, -2)$ or $(1, \infty)$. Therefore, the solution set of the given inequality is $(-\infty, -2) \cup (1, \infty)$.

FIGURE 2.21

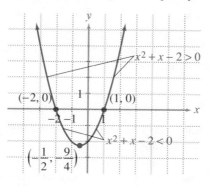

EXAMPLE 26

Find the solution set of the quadratic inequality $7x \leq 20 - 6x^2$.

FIGURE 2.22

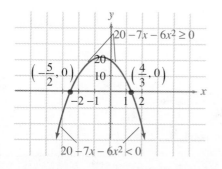

Solution   Since the inequality $7x \leq 20 - 6x^2$ is equivalent to the inequality $20 - 7x - 6x^2 \geq 0$, we begin by graphing the equation $y = 20 - 7x - 6x^2$ as shown in Figure 2.22. The solution set for the given inequality consists of all values of x for which the corresponding value of y is nonnegative. Thus, from the graph, we see that the solution set is $[-\frac{5}{2}, \frac{4}{3}]$.

Applications

The importance of quadratic functions is underscored by their occurrence in a wide range of applications. We now discuss several examples whose solutions require the investigation of quadratic functions.

EXAMPLE 27

Find the point on the graph of $f(x) = \sqrt{x + 1}$ that is closest to the point $(2, 0)$.

Solution First, we sketch the graph of $f(x) = \sqrt{x + 1}$ (see Figure 2.23). We must choose $x \geq -1$ for $\sqrt{x + 1}$ to be a real number. Thus, the domain of f is $[-1, \infty)$. Any point P on the graph of $f(x) = \sqrt{x + 1}$ has the form $(x, \sqrt{x + 1})$ for some value of $x \geq -1$. Hence, by the distance formula, the distance from $P_0 = (2, 0)$ to $P = (x, \sqrt{x + 1})$ is given by

$$d(P_0, P) = \sqrt{(x - 2)^2 + (\sqrt{x + 1} - 0)^2}$$
$$= \sqrt{x^2 - 4x + 4 + x + 1}$$
$$= \sqrt{x^2 - 3x + 5}.$$

Notice that we have expressed the distance $d(P_0, P)$ in terms of the x-coordinate of P. Also notice that the value of x that minimizes $d(P_0, P) = \sqrt{x^2 - 3x + 5}$ also minimizes the square of this distance, which is $x^2 - 3x + 5$. Therefore, we need only find the value of x that minimizes $x^2 - 3x + 5$. Since the graph of $y = x^2 - 3x + 5$ is a parabola opening upward, it follows that $x = -b/2a = \frac{3}{2}$, the x-coordinate of the vertex, is the value of x that minimizes $d(P_0, P)$. Because the y-coordinate of P is given by

$$y = f(x) = \sqrt{x + 1},$$

we have

$$y = \sqrt{\frac{3}{2} + 1} = \frac{\sqrt{10}}{2}.$$

Hence, $(\frac{3}{2}, \sqrt{10}/2)$ is the point on the graph of $f(x) = \sqrt{x + 1}$ which is closest to the point $(2, 0)$.

FIGURE 2.23

x	y
-1	0
0	1
3	2
8	3

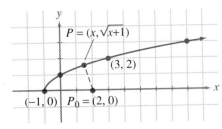

EXAMPLE 28

The sum of two numbers is 64. Find the two numbers if their product is to be as large as possible.

Solution Let x represent one of the numbers. Since the sum of the two numbers is 64, the other number can be expressed as $64 - x$. Hence, the product P of the two numbers is a function of x whose defining equation is

$$P(x) = x(64 - x)$$
$$= 64x - x^2.$$

We see that the product P is a quadratic function whose graph is a parabola opening downward. Hence, the product P has its largest value at the vertex of its graph, and the x-coordinate of the vertex is one of the numbers we are seeking. Since the x-coordinate of the vertex of this parabola is

$$x = -\frac{b}{2a} = -\frac{64}{2(-1)} = 32,$$

one of the numbers is 32. The other number is $64 - 32 = 32$.

EXAMPLE 29

A shot-putter heaves a 10-pound shot. Its height in feet after t seconds is given by $h(t) = -16t^2 + 32t + 6$.

a. What is the maximum height attained by the shot?
b. When does the shot hit the ground?

Solution

a. The height function h is a quadratic function whose graph is a parabola opening downward. Thus, the maximum height attained is the y-coordinate of the vertex. Since $-b/2a = -32/[2(-16)] = 1$,

$$y = h(1) = -16(1)^2 + 32(1) + 6 = 22 \text{ feet.}$$

b. At the time the shot hits the ground, its height above ground is zero. Setting $h(t) = 0$, we get $-16t^2 + 32t + 6 = 0$ or $8t^2 - 16t - 3 = 0$. Solving for t using the quadratic formula, we find that $t = (16 \pm \sqrt{352})/16$. We disregard the negative number $t = (16 - \sqrt{352})/16$ and conclude that the shot hits the ground $(16 + \sqrt{352})/16 \approx 2.17$ seconds after it is put.

Today, mathematical methods are incorporated in the economic decision-making processes of business firms. Economists and business analysts rely on mathematical tools to help them predict the effects of various policy alternatives and to choose reasonable courses of action from a myriad of possibilities. We now show how linear and quadratic functions play a part in this process. For example, when deciding whether to produce a new product, management often considers the following functions:

Cost function C: $C(x) = $ the cost of producing x units of the product;

Revenue function R: $R(x) = $ the revenue generated by producing x units of the product;

Profit function P: $P(x) = R(x) - C(x) = $ the profit (or loss) generated by producing x units of the product

We note that, theoretically, the functions C, R, and P are defined only for nonnegative integers. This follows from the fact that it makes no sense to discuss the cost of producing 6.3 computers or the revenue generated by producing -3 airplane engines. However, in practice, it is customary to treat these functions as though the domain of each is an interval subset of the set of nonnegative real numbers. We illustrate these ideas in the following example:

EXAMPLE 30

After extensive surveys, the research department of a company recommended to management that a certain new product be manufactured and marketed. The research department, in conjunction with the financial department of the firm, obtained cost and revenue functions as follows.

The cost C (in dollars) for producing x units was found to be

$$C(x) = 65,000 + 60x$$

where $65,000 represents fixed costs (tooling and overhead) and $60 is the variable cost per unit. In turn, it was found that there will be a **demand** for $x = 10,000 - 50p$ units if p is the price per unit. Expressing the cost function as a function of p, we get

$$C(p) = 65,000 + 60(10,000 - 50p),$$

or $\qquad C(p) = 665,000 - 3000p.$ **a linear function**   *(6)*

The revenue R generated is defined by $R = xp$, where x is the number of units sold if the price for each unit is p. Since $x = 10,000 - 50p$, the revenue R can be expressed as a function of p as follows:

$$R(p) = (10,000 - 50p)p,$$

or $\qquad R(p) = 10,000p - 50p^2.$ **a quadratic function**   *(7)*

a. At what price will the maximum revenue occur?
b. Find the **break-even** point(s); that is, find the price at which the cost equals the revenue.

Solution   We first graph equations (6) and (7) in the same coordinate system.

FIGURE 2.24

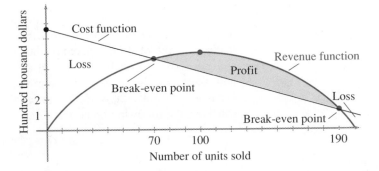

a. Since the revenue function is a parabola opening downward, the maximum revenue is the y-coordinate of the vertex. Inspection of the graph of the revenue function in Figure 2.24 reveals that the maximum revenue occurs when the price of a unit is $100.
b. To find the break-even points, we find the values of p at which the cost equals the revenue. Setting $C(p) = R(p)$, we get

$$665,000 - 3000p = 10,000p - 50p^2.$$

Solving for p, we have

$$665,000 - 13,000p + 50p^2 = 0$$
$$13,000 - 260p + p^2 = 0 \qquad \text{dividing each side by 50}$$
$$(p - 190)(p - 70) = 0.$$

Hence $p = 190 or $p = 70, and the company will break even if the price per unit is either $70 or $190. Moreover, if the price is between $70 and $190, the company makes a profit, since the profit function P defined by $y = P(p) = R(p) - C(p)$ is positive for $p \in (70, 190)$. On the other hand, it loses money if $p < 70 or $p > 190, as shown in Figure 2.24.

EXERCISES 2.3

For each of the quadratic functions defined in Exercises 1–12,
a. find the maximum or minimum value;
b. sketch the graph and give the vertex, axis of symmetry, and the x- and y-intercepts (if they exist);
c. find the range.

1. $f(x) = 2(x + 1)^2 - 3$
2. $f(x) = -3(x - 1)^2 + 2$
3. $f(x) = 4x^2 - x + 10$
4. $f(x) = 3x^2 - x - 10$
5. $f(x) - 5 = -(x + 4)^2 + 6$
6. $f(x) = -(x - 3)^2 + 5\pi$
7. $f(x) = \frac{1}{3}x^2 + \frac{1}{3}x - \frac{1}{3}$
8. $f(x) = \sqrt{2}x^2 + 5$
9. $f(x) = -9x^2 + 12x - 4$
10. $f(x) = 2x^2 + \sqrt{2}x$
11. $3[f(x) + 1] = (x - 8)^2$
12. $\frac{1}{2}[f(x) - 4] = (x + 3)^2$

In Exercises 13–20, use the graphical method to solve the given quadratic inequalities.

13. $x^2 - 4x < 0$
14. $x^2 + 9x > 0$
15. $x^2 + 2x \geq 3$
16. $x^2 + 16 \geq 0$
17. $x^2 + 10x + 25 > 0$
18. $14x > x^2 + 49$
19. $-10 \geq 3x - x^2$
20. $x^2 + 2x \leq 24$
21. Find the points of intersection of the graphs of $f(x) = x^2 - 2x + 5$ and $g(x) = 11 + 2x - x^2$.
22. How far from the origin is the vertex of the parabola $y = x^2 - 4x + 5$?

In Exercises 23–25, find a quadratic function satisfying the given conditions.

23. The graph passes through the origin, and the vertex is (4, 4).
24. The axis of symmetry is the line $x = -2$. The y-intercept is (0, 4), and there is only one x-intercept.
25. The vertex is $(-1, 4)$ and one x-intercept is (2, 0).
26. Find the maximum area of a rectangle with the property that the sum of any two of its adjacent sides is 4.
27. A real estate company manages an apartment building containing 100 units. When the rent for each unit is $350 per month, all apartments are occupied. However, for each $25 increase in monthly rent per unit, one of the units becomes vacant. Each vacant unit costs the management $15 per month for taxes and upkeep, and each occupied unit costs the management $50 per month for taxes, service, upkeep, and water. What rent should be charged for a maximum profit?
28. Prove that if x and y are numbers whose sum is S, then the product $P = xy$ will be maximum when $x = y = S/2$.
29. A shot-putter heaves a 12-pound shot. The height in feet of the shot after t seconds is given by $h(t) = -16t^2 + 30t + 8$.
   a. What is the maximum height attained by the shot?
   b. How long is the shot airborne?

Photo Daley Thompson/FPG International

30. A suspension cable that helps support a small footbridge between two sheer cliffs of equal height is approximately the shape of a parabola. Assume that the height, in meters, of the cable above the bridge is given by $h(x) = \frac{1}{4}x^2 - \frac{4}{5}x + 10$ where x is the distance in meters from one end of the bridge.
   a. What is the shortest distance from the bridge to the cable?
   b. How long is the bridge?
31. Suppose in Example 30, the demand function is changed to

$$x = 12,000 - 40p,$$

and the cost function is changed to

$$C(x) = 200,000 + 50x.$$

   a. Express the cost C as a linear function of the price p.
   b. Express the revenue R as a quadratic function of the price p.
   c. Graph the cost and revenue functions found in parts (a) and (b) in the same coordinate system. Identify the regions of profit and loss.
   d. Find the break-even points to the nearest dollar.

32. The cables of an evenly loaded suspension bridge assume the form of a parabola. The cables for the control span of such a bridge are supported by towers that are 400 feet apart. If the cables are attached to the towers at points 200 feet above the floor of the bridge, how long must a vertical strut be that is 50 feet from one of the towers? Assume that the cable touches the base of the bridge at the midpoint (see Figure 2.25).

FIGURE 2.25

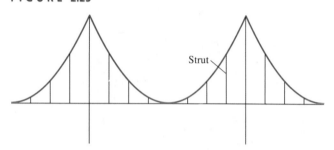

33. Using a graphing calculator, graph each of the following equations in the same coordinate system. Approximate, to two decimal places, the coordinates of the points of intersection.
  a. $y = \sqrt{3}x^2 - 4x + 2$ **b.** $y = 6 - 2\sqrt{3}x^2$

34. Using a graphing calculator, graph each of the following equations in the same coordinate system. Approximate, to two decimal places, the coordinates of the points of intersection.
  a. $y = x^2 + 0.4x$ **b.** $y = x^2 + 0.9x - 8$

35. **a.** Using a graphing calculator, graph the equation $y = 0.4x^2 - \frac{3}{4}x - 1$.
   b. Approximate, to two decimal places, the coordinates of the vertex of this parabola.
   c. Determine the values of x for which $0.4x^2 - \frac{3}{4}x - 1 \leq 0$.

36. **a.** Using a graphing calculator, graph the equation $y = 4 + 0.2x - 0.7x^2$.
   b. Approximate, to two decimal places, the coordinates of the vertex of this parabola.
   c. Determine the values of x for which $4 + 0.2x - 0.7x^2 \leq 0$.

2.4 TECHNIQUES OF GRAPHING

Graphs are valuable tools for assisting in the introduction and development of concepts in calculus. For this reason, it is essential that we learn to sketch accurately the graphs of certain equations. In this section we will discuss techniques that can be helpful to us when we graph equations.

Symmetry

Let us consider the graph of the equation $y = x^2 + 2$. From our work in Section 2.3, we know that the graph of this equation is a parabola that opens upward. Its axis of symmetry is the y-axis, and its vertex (and y-intercept) is the point $(0, 2)$. The graph, which has no x-intercept, is shown in Figure 2.26.

As we learned in the previous section, if we "fold" the graph along the axis of symmetry, which is the y-axis, the left and right halves will coincide. Stated another way, the left half of the graph is the mirror image of the right half across the y-axis, and we say that the graph is **symmetric with respect to the y-axis**. As we see in Figure 2.26, a graph is symmetric with respect to the y-axis if and only if the point $(-x, y)$ is on the graph whenever the point (x, y) lies on the graph. The points (x, y) and $(-x, y)$ are called **reflections of one another through the y-axis**. Moreover, in Figure 2.26, the portion of the graph to the right of the y-axis is said to be the **reflection, through the y-axis**, of that portion of the graph to the left of the y-axis and vice versa.

Consider the graph of $x = |y|$ given in Figure 2.27. Observe that if the graph were folded along the x-axis, then the upper half would coincide with the lower half. In this case, the point $(x, -y)$ is on the graph whenever the point (x, y) lies on the graph, and we say that the graph is **symmetric with respect to the x-axis**. The points (x, y) and $(x, -y)$ are called **reflections of one another through the x-axis**. In Figure 2.27, the portions of the graph above and below the x-axis are said to be reflections of one another through the x-axis.

FIGURE 2.26
$y = x^2 + 2$

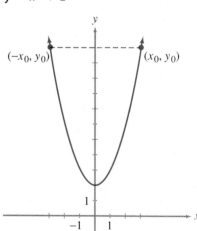

FIGURE 2.27

$x = |y|$	y
2	-2
1	-1
0	0
1	1
2	2

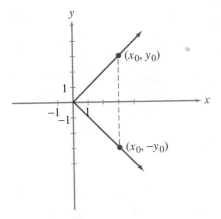

FIGURE 2.28

x	$y = x^3$
-2	-8
-1	-1
0	0
1	1
2	8

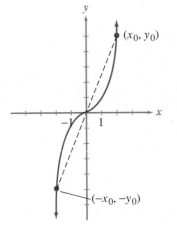

Finally, consider the graph of $y = x^3$ shown in Figure 2.28. For each point (x, y) on the graph, the point $(-x, -y)$ is also on the graph. In this case, we say that the graph is **symmetric with respect to the origin**. The points (x, y) and $(-x, -y)$ are **reflections of one another through the origin**. Notice that the graph in Figure 2.28 can be divided into halves, one in quadrant I and the other in quadrant III. The portion of the graph in quadrant III is said to be the **reflection, through the origin**, of the portion in quadrant I and vice versa.

In summary, we have

Symmetry of Graphs

A graph is symmetric with respect to

(i) **the y-axis** if and only if $(-x, y)$ is on the graph whenever (x, y) is on the graph;

(ii) **the x-axis** if and only if $(x, -y)$ is on the graph whenever (x, y) is on the graph;

(iii) **the origin** if and only if $(-x, -y)$ is on the graph whenever (x, y) is on the graph.

In general, the graph of a function is not symmetric with respect to the x-axis (see Exercise 48).

When graphing an equation, it is often helpful to determine which, if any, symmetries are present. For example, if we know that the graph of $y = x^3$ is symmetric with respect to the origin, we can sketch the right side of the graph, and reflect the resulting curve through the origin to obtain the entire graph. Also, it is helpful to know that when any two of the symmetries are present in a graph, so is the third (see Exercises 49 and 50). The tests for symmetry given in Table 2.2 are easily applied and should be used when graphing equations.

Applying the test for symmetry with respect to the origin to the equation $y = x^3$, we replace x with $-x$ and y with $-y$ and observe that an equivalent equation is obtained.

$$-y = (-x)^3$$
$$-y = -x^3, \quad \text{or}$$
$$y = x^3.$$

Thus, the graph is symmetric with respect to the origin. If we replace x with $-x$, the equation becomes

$$y = (-x)^3, \quad \text{or}$$
$$y = -x^3.$$

TABLE 2.2
Symmetry tests for graphing equations

The graph of an equation is symmetric with respect to the	Provided that an equivalent equation is obtained when
y-axis	x is replaced with $-x$
x-axis	y is replaced with $-y$
origin	x and y are replaced with $-x$ and $-y$, respectively.

Since the latter equation is not equivalent to the equation $y = x^3$, the graph is not symmetric with respect to the y-axis. You should verify that the graph is not symmetric with respect to the x-axis.

EXAMPLE 31

Test the graphs of each of the following equations for symmetry and sketch the graph of each equation:

a. $y = x^4$ **b.** $x^2 + y^2 = 25$ **c.** $y = x^3 + x^2$

Solution

a. First, observe that in the equation $y = x^4$ the variable x occurs only to the even power four. Hence, an equivalent equation is obtained when x is replaced by $-x$, as we see below.

$$y = (-x)^4, \quad \text{or}$$
$$y = x^4.$$

Therefore, the graph is symmetric with respect to the y-axis. You should verify that the graph is not symmetric with respect to the x-axis nor with respect to the origin. Since $x^4 \geq 0$ for all values of x, no point on the graph lies below the x-axis. We sketch the portion of the graph in the first quadrant and then reflect this portion through the y-axis. The graph is shown in Figure 2.29(a).

b. Since both x and y occur only to even powers in the equation $x^2 + y^2 = 25$, we obtain an equivalent equation if x is replaced with $-x$, and y with $-y$, as we show below.

$$(-x)^2 + (-y)^2 = 25, \quad \text{or}$$
$$x^2 + y^2 = 25.$$

Thus, the graph is symmetric with respect to the origin. You should verify that the graph, which is the circle shown in Figure 2.29(b), is also symmetric with respect to the x-axis and the y-axis.

c. Observe that the equation $y = x^3 + x^2$ contains both even and odd powers of x. To test for symmetry with respect to the y-axis, we replace x by $-x$ in this equation to get

$$y = (-x)^3 + (-x)^2, \quad \text{or}$$
$$y = -x^3 + x^2.$$

Since the last equation is not equivalent to the original equation, the graph is not symmetric with respect to the y-axis. You should verify that the graph also fails to be symmetric with respect to the x-axis and to the origin. The graph of the equation is shown in Figure 2.29(c).

FIGURE 2.29

a. $y = x^4$

x	0	1	2
y	0	1	16

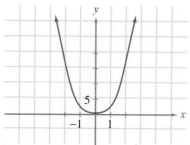

b. $x^2 + y^2 = 25$

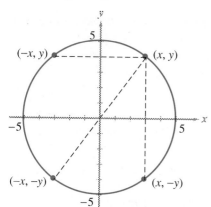

Even and Odd Functions

We observed earlier that the graph of an equation is symmetric with respect to the y-axis if and only if whenever the point $(-x, y)$ is on the graph so is the

c. $y = x^3 + x^2$

x	0	1	2	$-\frac{2}{3}$	-1	-2
y	0	2	12	$\frac{4}{27}$	0	-4

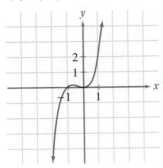

point (x, y). It follows that the graph of a function f is symmetric with respect to the y-axis if and only if $f(-x) = f(x)$ for all x in the domain of f. Such a function is called an *even* function. In a similar fashion, the graph of a function is symmetric with respect to the origin if and only if $f(-x) = -f(x)$ for all x in the domain of f. Such a function is called an *odd* function.

In summary, we have

DEFINITION 2.5

A function f is **even** if and only if $f(-x) = f(x)$ for all x in the domain of f.

A function f is **odd** if and only if $f(-x) = -f(x)$ for all x in the domain of f.

EXAMPLE 32

Determine which of the functions defined below are even, which are odd, and which are neither. Discuss any resulting symmetries and sketch the graph of each function.

a. $f(x) = 4/x$ b. $f(x) = x^2 + 4x + 4$ c. $f(x) = x^4 - 2x^2$

Solution

a. First, we compute $f(-x)$.

$$f(-x) = \frac{4}{-x}$$
$$= -\frac{4}{x}.$$

Since $f(-x) = -4/x = -f(x)$, f is an odd function. Hence, the graph of f is symmetric with respect to the origin. The graph of f is shown in Figure 2.30(a) on page 96.

b. Computing $f(-x)$, we obtain

$$f(-x) = (-x)^2 + 4(-x) + 4$$
$$= x^2 - 4x + 4.$$

Since $f(-x) \neq f(x)$, f is not even. Comparing $f(-x)$ with $-f(x) = -x^2 - 4x - 4$, we see also that f is not odd since $f(-x) \neq -f(x)$. Hence, the graph of f is not symmetric with respect to the y-axis nor to the origin. The graph of f is shown in Figure 2.30(b).

c. Finding $f(-x)$, we have

$$f(-x) = (-x)^4 - 2(-x)^2$$
$$= x^4 - 2x^2.$$

Since $f(-x) = f(x)$, f is even, and therefore its graph is symmetric with respect to the y-axis. The graph of f is shown in Figure 2.30(c).

FIGURE 2.30   **a.** $f(x) = \dfrac{4}{x}$ **b.** $f(x) = x^2 + 4x + 4$ **c.** $f(x) = x^4 - 2x^2$

x	$f(x)$
1	4
2	2
4	1
$\frac{1}{2}$	8
$\frac{1}{3}$	12

x	$f(x)$
0	4
-1	1
-2	0
-3	1
-4	4

x	$f(x)$
0	0
1	-1
2	8

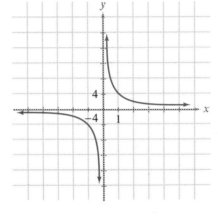

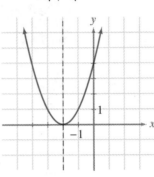

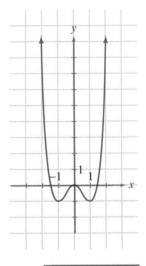

Graphs of Some Basic Functions

We can often obtain the graph of one function by using the graph of another. For this reason, it is convenient for us to be familiar with the graphs of a collection of basic functions. We have previously encountered three of these functions: the identity function, defined by $f(x) = x$; the squaring function, defined by $f(x) = x^2$; and the cubing function, defined by $f(x) = x^3$. In our next example, we investigate three additional basic functions.

EXAMPLE 33

Find the domain and sketch the graph of each of the following functions. Use the graph to find the range of each function.

a. $f(x) = |x|$ the absolute value function
b. $g(x) = \sqrt{x}$ the square root function
c. $h(x) = 1/x$ the reciprocal function

FIGURE 2.31 $f(x) = |x|$

x	0	1	2
$f(x)$	0	1	2

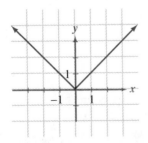

Solution

a. Since $|x|$ is a real number for each real number x, dom $(f) = \mathbb{R}$. Checking for symmetry, we see that

$$f(-x) = |-x|$$
$$= |x|$$
$$= f(x).$$

Thus f is even, and the graph of f is symmetric with respect to the y-axis. Making a table of values for the portion of the graph in quadrant I and using symmetry, we obtain the graph shown in Figure 2.31. From the graph, we see that ran $(f) = [0, \infty)$.

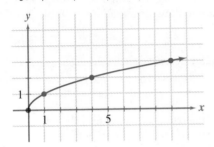

$$g(x) = \sqrt{x}$$

x	0	1	4	9
$g(x)$	0	1	2	3

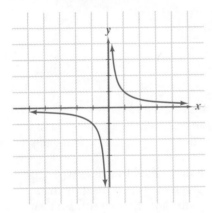

$$h(x) = 1/x$$

x	1	2	3	$\frac{1}{2}$	$\frac{1}{3}$	$\frac{1}{4}$
y	1	$\frac{1}{2}$	$\frac{1}{3}$	2	3	4

FIGURE 2.34

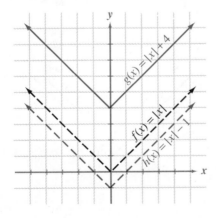

b. We know that $\text{dom}(g) = [0, \infty)$ since $\sqrt{x}$ is a real number if and only if $x \geq 0$. Using the tests for symmetry, it can be shown that the graph of g has no symmetry. Forming a table of values, we get the graph of g, shown in Figure 2.32, and we see that $\text{ran}(g) = [0, \infty)$.

c. Since division by zero is undefined, $\text{dom}(h) = (-\infty, 0) \cup (0, \infty)$. Now

$$h(-x) = \frac{1}{-x} = -\frac{1}{x} = -h(x).$$

Hence h is odd, and the graph of h is symmetric with respect to the origin. From the graph of h, shown in Figure 2.33, we see than $\text{ran}(h) = (-\infty, 0) \cup (0, \infty)$.

Translations and Reflections

We now use this collection of basic functions to obtain the graphs of many additional functions without having to construct tables of values. As our first example, let us consider the graph of $f(x) = |x|$, with which we are familiar (see Figure 2.31). How do the graphs of $g(x) = |x| + 4$ and $h(x) = |x| - 1$ compare to the graph of f? First, we observe that the domain of each of these functions is $\mathbb{R}$. Moreover, for each $x \in \mathbb{R}$, $g(x) = |x| + 4 = f(x) + 4$. This means that the graph of g can be obtained by moving the graph of f upward four units. We call such a movement a **translation**. Similarly, $h(x) = |x| - 1 = f(x) - 1$, and the graph of h can be obtained by translating the graph of f one unit downward. From Figure 2.34, we see that $\text{ran}(f) = [0, \infty)$, while $\text{ran}(g) = [4, \infty)$ and $\text{ran}(h) = [-1, \infty)$.

To obtain the graph of $g(x) = |x + 3|$ and $h(x) = |x - 2|$ from the graph of $f(x) = |x|$ (each of the three functions having $\mathbb{R}$ as its domain), observe that for each x, $g(x) = |x + 3| = f(x + 3)$. For example,

$$g(-4) = f(-4 + 3) = f(-1), \quad g(-3) = f(-3 + 3) = f(0),$$
$$g(-2) = f(-2 + 3) = f(1), \quad \text{etc.}$$

Hence, if $g(x_1) = f(x_2)$, then $x_2 = x_1 + 3$ or $x_1 = x_2 - 3$. It follows that the graph of g can be obtained by translating the graph of f three units to the left, as shown in Figure 2.35. Since $h(x) = |x - 2| = f(x - 2)$, we see that

$$h(0) = f(0 - 2) = f(-2), \quad h(1) = f(1 - 2) = f(-1),$$
$$h(2) = f(2 - 2) = f(0), \quad \text{etc.}$$

It follows that if $h(x_1) = f(x_2)$, then $x_2 = x_1 - 2$ or $x_1 = x_2 + 2$. Therefore, the graph of h can be obtained by translating the graph of f two units to the right.

FIGURE 2.35

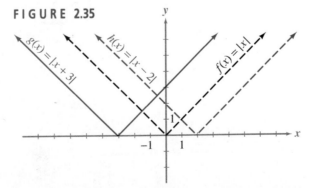

FIGURE 2.36

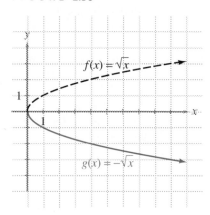

We now illustrate how the graph of $g(x) = -f(x)$ can be readily obtained from the graph of $y = f(x)$ by considering $f(x) = \sqrt{x}$ and $g(x) = -f(x) = -\sqrt{x}$. We note that dom $(f) = $ dom $(g) = [0, \infty)$, and observe that the point (x, y) is on the graph of f if and only if the point $(x, -y)$ is on the graph of g. Hence, the graph of g can be obtained by reflecting the graph of f through the x-axis, as shown in Figure 2.36. We summarize these observations in Table 2.3.

TABLE 2.3

Relationship Between f and g	Method of Obtaining the Graph of g from the Graph of f
(i) $g(x) = f(x) + c$ $\left.\right\}$ $(c > 0)$	Translate c units upward
(ii) $g(x) = f(x) - c$	Translate c units downward
(iii) $g(x) = f(x + c)$ $\left.\right\}$ $(c > 0)$	Translate c units to the left
(iv) $g(x) = f(x - c)$	Translate c units to the right
(v) $g(x) = -f(x)$	Reflect the graph through the x-axis

EXAMPLE 34

Graph the functions defined below.

a. $g(x) = x^2 + 4$ **b.** $g(x) = |x + 2| - 1$

c. $g(x) = \dfrac{1}{x - 3}$ **d.** $g(x) = -|x| + 1$

Solution

a. If we let $f(x) = x^2$, then $g(x) = x^2 + 4$ is of the form $g(x) = f(x) + c$ with $c = 4$. According to rule (i) in Table 2.3, the graph of g can be obtained from the graph of f by translating the graph of f four units upward as shown in Figure 2.37.

b. To sketch the graph of g, we first sketch the graph of $h(x) = |x + 2|$. According to rule (iii) in Table 2.3, the graph of h is obtained by translating the graph of $f(x) = |x|$ two units to the left. By rule (ii), we then translate the graph of h one unit downward to obtain the graph of g as pictured in Figure 2.38.

FIGURE 2.37

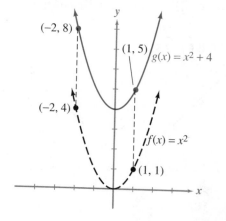

FIGURE 2.38

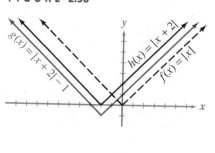

c. Let $f(x) = 1/x$. By rule (iv) in Table 2.3, we obtain the graph of g by translating the graph of f three units to the right, as shown in Figure 2.39.

d. Let $f(x) = |x|$ and $h(x) = -|x|$. By rule (v), we obtain the graph of h by reflecting the graph of f through the x-axis. Translating the graph of h one unit upward, we obtain the graph of g, as shown in Figure 2.40.

FIGURE 2.39

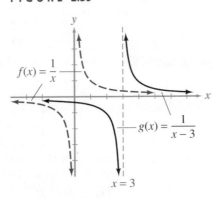

FIGURE 2.40

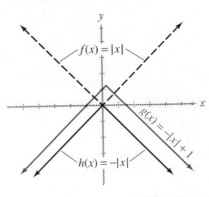

Piecewise-Defined Functions

Sometimes functions are described in terms of more than one expression. Such functions, which are called **piecewise-defined functions**, are often encountered in calculus. Consider, for example, the function f defined by

$$f(x) = \begin{cases} x + 3, & \text{if } x \geq 1 \\ x^2, & \text{if } x < 1 \end{cases}.$$

FIGURE 2.41

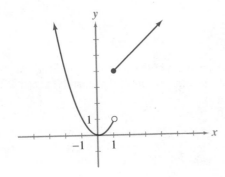

To graph f, we first determine its domain. Since every real number satisfies exactly one of the inequalities, $x \geq 1$ or $x < 1$, $\text{dom}(f) = \mathbb{R}$. The graph of f coincides with the graph of the line $y = x + 3$ on $[1, \infty)$ and coincides with the graph of the parabola $y = x^2$ on $(-\infty, 1)$ (see Figure 2.41). In Figure 2.41, note that the point $(1, 1)$ is indicated with an open circle to show that it is *not* a point on the graph of f. From the graph, we see that $\text{ran}(f) = [0, \infty)$.

The following is an example of another type of function often encountered in calculus:

EXAMPLE 35

Let f be the function defined by $f(x) = |x - 2|/(x - 2)$. Find $\text{dom}(f)$ and sketch the graph of f.

Solution The only value of x for which the denominator is zero is 2. Thus, $\text{dom}(f) = (-\infty, 2) \cup (2, \infty)$. Notice that if $x > 2$, $|x - 2| = x - 2$, while, if $x < 2$, $|x - 2| = -(x - 2)$. It follows that if $x > 2$,

$$f(x) = \frac{|x - 2|}{x - 2}$$

$$= \frac{x - 2}{x - 2}$$

$$= 1.$$

FIGURE 2.42

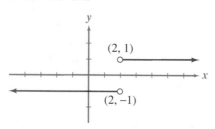

$$f(x) = \begin{cases} 1, & \text{if } x > 2 \\ -1, & \text{if } x < 2 \end{cases}$$

On the other hand, if $x < 2$,

$$f(x) = \frac{|x - 2|}{x - 2}$$

$$= \frac{-(x - 2)}{x - 2}$$

$$= -1.$$

Thus, we can write $f(x)$ as follows:

$$f(x) = \begin{cases} 1, & \text{if } x > 2 \\ -1, & \text{if } x < 2 \end{cases}.$$

From the graph of f, shown in Figure 2.42, we see that ran $(f) = \{-1, 1\}$.

EXERCISES 2.4

In Exercises 1–10, determine if the function is odd, even, or neither. Discuss the symmetry (if any) of each graph.

1. $f(x) = 4x$
2. $f(x) = 3x - 4$
3. $f(x) = x^2 - 5x + 2$
4. $f(x) = 3x^5 - 2x^4$
5. $f(x) = 1/x^6$
6. $f(x) = x^2/(x^2 + 4)$
7. $f(x) = \sqrt{x + 2}$
8. $f(x) = x + x^{-2}$
9. $f(x) = |x - 3|$
10. $f(x) = x^5 - 2x^{1/5}$

In Exercises 11–16, determine if the given relations are functions. Give the domain and range of each. Identify any odd or even functions.

11.

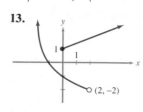

12.

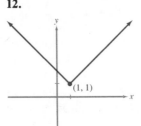

13.

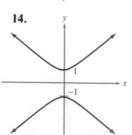

14.

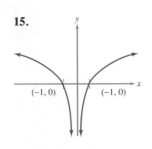

15.

16.

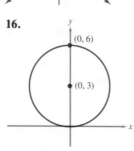

In Exercises 17–20, use the graph of $g(x) = x^2$ to obtain the graphs of the given functions. Sketch both graphs in the same coordinate system.

17. $f(x) = x^2 - 4$
18. $f(x) = (x - 3)^2$
19. $f(x) = (x + 4)^2$
20. $f(x) = (-x + 2)^2$

In Exercises 21–24, use the graph of $g(x) = 1/x$ to obtain the graphs of the given functions. In each case, sketch both graphs in the same coordinate system.

21. $f(x) = (1/x) + 5$
22. $f(x) = 1/(x - 6)$
23. $f(x) = -(1/x) - 2$
24. $f(x) = (1 - 4x)/x$

In Exercises 25–40, sketch the graph of the given functions using the techniques discussed in this section. Give dom (f) and ran (f) in each case.

25. $f(x) = \sqrt{x + 10}$
26. $f(x) = \sqrt{x - \sqrt{2}}$
27. $f(x) = x^3 + x$
28. $f(x) = x^4 - x^2 - 8$
29. $f(x) = 4 - |x|$
30. $f(x) = x + |x|$
31. $f(x) = |x + 2| - 3$
32. $f(x) = |x - \sqrt{3}|/(x - \sqrt{3})$
33. $f(x) = (x + 1)^3 - 2$
34. $f(x) = 2(x + 1)^3 - 1$
35. $f(x) = (1/x^6) - 2$
36. $f(x) = x^2/(x^2 + 4)$

37. $f(x) = \begin{cases} 1, & \text{if } x \geq 0 \\ -2, & \text{if } x < 0 \end{cases}$

38. $f(x) = \begin{cases} x + 1, & \text{if } x \geq -1 \\ x, & \text{if } -2 < x < -1 \\ 1, & \text{if } x \leq -2 \end{cases}$

39. $f(x) = \begin{cases} x^2, & \text{if } x \text{ is an integer} \\ 0, & \text{otherwise} \end{cases}$

40. $f(x) = \begin{cases} (x^2 - 9)/(x - 3), & \text{if } x \neq 3 \\ 2, & \text{if } x = 3 \end{cases}$

41. The symbol $[\![x]\!]$ denotes the greatest integer n such that $n \leq x$. Thus, the function f defined by $f(x) = [\![x]\!]$ is referred to as the **greatest integer function**. Find the domain and range of the greatest integer function and sketch its graph.

42. Determine whether the greatest integer function is odd, even, or neither.

In Exercises 43–45, use the graph of $g(x) = [\![x]\!]$ to find the graph of the given functions.

43. $f(x) = [\![x - 2]\!]$

44. $f(x) = [\![x]\!] - 4$ **45.** $f(x) = -2[\![x]\!]$

46. Under what conditions on the integer n is the function defined by $f(x) = x^n$ even? Under what conditions is it odd?

47. Show that the only function that is both even and odd is the **zero function** defined by $f(x) = 0$ for each real number x.

48. Are there any functions whose graphs are symmetric with respect to the x-axis? Explain your answer.

49. Prove that a graph that is symmetric with respect to the

x-axis and to the y-axis is also symmetric with respect to the origin.

50. Prove that a graph that possesses any two of the types of symmetry necessarily possesses the third type.

51. Using a graphing calculator, sketch the graph of each equation in the same coordinate system.
  a. $y = 0.9x^2$ **b.** $y = 0.9(x + \sqrt{3})^2$
  c. $y = 0.9(x + \sqrt{3})^2 - 0.2$

52. Using a graphing calculator, sketch the graph of each equation in the same coordinate system.
  a. $y = \sqrt{5}x^2 + 1$ **b.** $y = -(\sqrt{5}x^2 + 1)$
  c. $y = -[\sqrt{5}(x - 0.9^2] + 1$

2.5 THE ALGEBRA AND COMPOSITION OF FUNCTIONS

Just as two real numbers can be added, subtracted, multiplied, or divided (provided we do not divide by zero) to obtain another real number, two functions can be added, subtracted, multiplied, or divided (avoiding division by zero) to obtain another function. For instance, in Section 2.3 of this chapter, we found that the profit function P is equal to the revenue function R minus the cost function C.

DEFINITION 2.6

Let f and g be functions, and let $D = \text{dom}(f) \cap \text{dom}(g)$ where $D \neq \emptyset$. Then

(i)   the **sum** of f and g is the function $f + g$ defined by
$$(f + g)(x) = f(x) + g(x), \text{ for all } x \in D;$$

(ii) the **difference** of f and g is the function $f - g$ defined by
$$(f - g)(x) = f(x) - g(x), \text{ for all } x \in D;$$

(iii) the **product** of f and g is the function fg defined by
$$(fg)(x) = f(x) \cdot g(x), \text{ for all } x \in D;$$

(iv) the **quotient** of f and g is the function f/g defined by
$$\left(\frac{f}{g}\right)(x) = \frac{f(x)}{g(x)}, \text{ for all } x \in D \text{ where } g(x) \neq 0.$$

We illustrate these definitions in the following example:

EXAMPLE 36

Let $f(x) = |x - 1|$ and $g(x) = x - 1$. Find

a. $(f + g)(x)$ **b.** $(f - g)(x)$ **c.** $(fg)(x)$ **d.** $(f/g)(x)$

Give the domain of each function.

Solution   First, observe that $\text{dom}(f) = \text{dom}(g) = \mathbb{R}$. Hence, $D = \text{dom}(f) \cap \text{dom}(g) = \mathbb{R}$.

a. Using the definition of the sum function $f + g$, we get

$$(f + g)(x) = f(x) + g(x)$$
$$= |x - 1| + x - 1, \qquad x \in \mathbb{R}.$$

b. For the difference $f - g$, we get

$$(f - g)(x) = f(x) - g(x)$$
$$= |x - 1| - (x - 1)$$
$$= |x - 1| - x + 1, \qquad x \in \mathbb{R}.$$

c. The definition of the product function fg gives us

$$(fg)(x) = f(x) \cdot g(x)$$
$$= |x - 1|(x - 1), \qquad x \in \mathbb{R}.$$

d. From the definition of the quotient function $\dfrac{f}{g}$, we have

$$\left(\frac{f}{g}\right)(x) = \frac{f(x)}{g(x)}$$

$$= \frac{|x - 1|}{x - 1}, \qquad x \in \mathbb{R} \text{ and } x \neq 1.$$

Note that 1 is the only value of x for which $g(x) = 0$.

We can express each function found in the solution of Example 36 in a form that is free of absolute values. For example, using a procedure similar to the one used in the solution of Example 35 in Section 2.4, we can express $\left(\dfrac{f}{g}\right)(x)$, found in part (d) of Example 36, as follows:

$$\left(\frac{f}{g}\right)(x) = \begin{cases} 1, & \text{if } x > 1 \\ -1, & \text{if } x < 1 \end{cases}.$$

We graph $\dfrac{f}{g}$ in Figure 2.43.

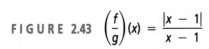

FIGURE 2.43 $\left(\dfrac{f}{g}\right)(x) = \dfrac{|x - 1|}{x - 1}$

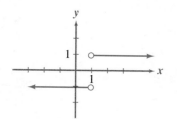

Composition of Functions

We can also obtain another function using two given functions by performing an operation called **composition**.

DEFINITION 2.7   Composition of Functions

Suppose f and g are functions such that dom $(f) \cap$ ran $(g) \neq \varnothing$. Then the **composition of f with g**, denoted by $f \circ g$, is the function defined by

$$(f \circ g)(x) = f[g(x)],$$

where dom $(f \circ g)$ is the set of all x in dom (g) for which $g(x)$ is in dom (f).

The function $f \circ g$ is called a **composite function**. Let us outline the process of evaluating the function $f \circ g$ at the number x. First, we must evaluate g at x, which implies that dom $(f \circ g) \subseteq$ dom (g). (Why?) We then evaluate f at $g(x)$, which means that $g(x)$ must be in dom (f). Therefore, dom $(f \circ g)$ is the set of all real numbers in dom (g) for which $g(x)$ is in dom (f). The composition of f with g is illustrated in Figure 2.44.

FIGURE 2.44

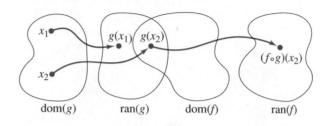

EXAMPLE 37

Let f and g be defined by $f(x) = x^2 - x$ and $g(x) = 6x - 4$. Find each of the following composite functions. Give the domain of each.

a. $(f \circ g)(x)$ **b.** $(g \circ f)(x)$

Solution

a.
$$
\begin{aligned}
(f \circ g)(x) &= f[g(x)] \\
&= [g(x)]^2 - g(x) \\
&= (6x - 4)^2 - (6x - 4) \\
&= 36x^2 - 48x + 16 - 6x + 4 \\
&= 36x^2 - 54x + 20.
\end{aligned}
$$

Now dom $(g) = \mathbb{R}$ and since the graph of g is a line with slope 6, ran $(g) = \mathbb{R}$. Also dom $(f) = \mathbb{R}$. Hence, if x is in dom (g), $g(x)$ is in dom (f). Therefore, dom $(f \circ g) = \mathbb{R}$.

b.
$$
\begin{aligned}
(g \circ f)(x) &= g[f(x)] \\
&= 6[f(x)] - 4 \\
&= 6(x^2 - x) - 4 \\
&= 6x^2 - 6x - 4.
\end{aligned}
$$

Since dom $(g) = \mathbb{R}$, we know that ran (f) is a subset of dom (g). (We could have used this approach in part (a) also.) Therefore, dom $(g \circ f) =$ dom $(f) = \mathbb{R}$.

Example 37 illustrates the fact that *in general*, $f \circ g \neq g \circ f$.

EXAMPLE 38

Let f and g be functions defined by $f(x) = x^2$ and $g(x) = \sqrt{x - 1}$. Find each of the following composite functions. Give the domain of each.

a. $(f \circ g)(x)$ **b.** $(g \circ f)(x)$

Solution

a.
$$
\begin{aligned}
(f \circ g)(x) &= f[g(x)] \\
&= [g(x)]^2 \\
&= (\sqrt{x - 1})^2 \\
&= x - 1.
\end{aligned}
$$

Here, it is tempting to say that dom $(f \circ g) = \mathbb{R}$ since $(f \circ g)(x) = x - 1$. However, this is incorrect! Observe that if we choose $x = 0$, $(f \circ g)(0)$ must be equal to $f[g(0)]$. But $g(0)$ is not defined since $\sqrt{0 - 1} = \sqrt{-1}$, which is not a real number. In determining dom $(f \circ g)$, we employ the methods used in the solution of Example 37. In finding dom (g), we require that $x - 1 \geq 0$ or $x \geq 1$, since the square root of a negative number is not a real number. Hence, dom $(g) = [1, \infty)$. Since dom $(f) = \mathbb{R}$, ran $(g) \subseteq$ dom (f), and therefore, dom $(f \circ g) =$ dom $(g) = [1, \infty)$.

b. $(g \circ f)(x) = g[f(x)]$

$$= \sqrt{f(x) - 1}$$

$$= \sqrt{x^2 - 1}.$$

Since dom $(f) = \mathbb{R}$, dom $(g \circ f) = \{x \mid x^2 - 1 \geq 0\}$. Using methods outlined in Section 1.6 of Chapter 1, we find that $x^2 - 1 \geq 0$ if and only if $x \geq 1$ or $x \leq -1$. Hence, dom $(g \circ f) = (-\infty, -1] \cup [1, \infty)$.

EXAMPLE 39

Let f and g be functions defined by $f(x) = 1/(x - 4)$ and $g(x) = \sqrt{x}$. Find each of the following. Give the domain of each composite function.

a. $(f \circ g)(x)$ **b.** $(g \circ f)(x)$

Solution

a. $(f \circ g)(x) = f[g(x)]$

$$= \frac{1}{g(x) - 4}$$

$$= \frac{1}{\sqrt{x} - 4}.$$

We know that dom $(g) = [0, \infty)$. However, we must not give x a value that will make $\sqrt{x} - 4 = 0$. Since 16 is the only number in the domain of g that makes $\sqrt{x} - 4 = 0$, dom $(f \circ g) = [0, 16) \cup (16, \infty)$.

b. $(g \circ f)(x) = g[f(x)]$

$$= \sqrt{f(x)}$$

$$= \sqrt{\frac{1}{x - 4}}$$

$$= \frac{1}{\sqrt{x - 4}}.$$

Now dom $(f) = (-\infty, 4) \cup (4, \infty)$. However, for x to be in dom $(g \circ f)$, $x - 4$ must be positive. (Why?) Therefore, x must be greater than 4, and thus, dom $(g \circ f) = (4, \infty)$.

EXAMPLE 40

The radius r, in centimeters, of a spherical balloon being inflated is a function of time t, in minutes, given by $r(t) = 3\sqrt[3]{t + 8}$ where $0 < t < 5$. Express the volume and the surface area of the balloon as functions of time t.

Solution   We know that the volume V of a sphere is a function of its radius given by the formula

$$V = V(r) = \frac{4}{3}\pi r^3. \tag{1}$$

Since the radius is a function of time t defined by

$$r(t) = 3\sqrt[3]{t + 8}, \tag{2}$$

we substitute $3\sqrt[3]{t + 8}$ for r in equation (1), getting

$$V = V(r) = V[r(t)] = \frac{4}{3}\pi(3\sqrt[3]{t + 8})^3$$

$$= \frac{4}{3}\pi[27(t + 8)]$$

$$= 36\pi(t + 8)$$

$$= 36\pi t + 288\pi.$$

In a similar fashion, since the surface area S of a sphere is given by the formula

$$S = S(r) = 4\pi r^2,$$

we have $$S = S(r) = S[r(t)] = 4\pi(3\sqrt[3]{t + 8})^2$$

$$= 36\pi(\sqrt[3]{t + 8})^2.$$

Expressing a given function as the composition of two functions (which are not unique) is illustrated in the following examples:

EXAMPLE 41

Find functions f and g such that the function h defined by $h(x) = \sqrt{x^2 + 2x - 1}$ can be expressed as $(f \circ g)(x)$.

Solution   Observe that $h(x)$ is the square root of $x^2 + 2x - 1$. Let f be the function $f(x) = \sqrt{x}$ and g be the function defined by $g(x) = x^2 + 2x - 1$. Then

$$(f \circ g)(x) = f[g(x)]$$

$$= \sqrt{g(x)}$$

$$= \sqrt{x^2 + 2x - 1}$$

$$= h(x).$$

EXAMPLE 42

Find functions f and g such that the function h defined by $h(x) = [x/(x + 1)]^4$ can be expressed as $(f \circ g)(x)$.

Solution   For each x in dom (h), $h(x)$ is the fourth power of $x/(x + 1)$. Thus, if we let $f(x) = x^4$ and $g(x) = x/(x + 1)$, we get

$$(f \circ g)(x) = f[g(x)]$$

$$= [g(x)]^4$$

$$= \left(\frac{x}{x + 1}\right)^4$$

$$= h(x).$$

EXERCISES 2.5

In Exercises 1–12, compute

a. $f + g$ **b.** $f - g$ **c.** fg **d.** $\dfrac{f}{g}$

Simplify and give the domain of each function.

1. $f(x) = 2 + x$,
   $g(x) = 2 - x$

2. $f(x) = 4 - x$,
   $g(x) = 3x + 2$

3. $f(x) = x^2 - x + 15$,
   $g(x) = 6 - x^2$

4. $f(x) = 1 - x^2 - x^3$,
   $g(x) = x^3 - 8$

5. $f(x) = (x + 4)/5$,
   $g(x) = 2/x$

6. $f(x) = 4x^6 - 5x^3$,
   $g(x) = |x|$

7. $f(x) = x^5$, $g(x) = x^{1/5}$

8. $f(x) = 1/(x - 1)$,
   $g(x) = (x^2 - 1)/x$

9. $f(x) = \sqrt{x + 3}$,
   $g(x) = |x - 2|$

10. $f(x) = |x - 5|$,
   $g(x) = x^2$

11. $f(x) = (x + 3)/(x - 2)$,
   $g(x) = 1/(x + 2)$

12. $f(x) = 1/\sqrt{x^2 - 5}$,
   $g(x) = \sqrt{x^2 - 5}$

13. Let $f(1) = 4$ and $g(1) = -3$. Compute
   a. $(f + g)(1)$ **b.** $(f - g)(1)$

   c. $(fg)(1)$ **d.** $\left(\dfrac{f}{g}\right)(1)$

14. Let $f(-2) = 3$, $g(-2) = -5$, and $h(-2) = -2$. Compute each.

a. $(f + g)(-2)$ **b.** $\left(\dfrac{f - h}{g}\right)(-2)$

c. $\left(\dfrac{fg}{h - g}\right)(-2)$ **d.** $\left(\dfrac{f - h}{g - h}\right)(-2)$

In Exercises 15–24,

   a. Compute $(f \circ g)(x)$ and $(g \circ f)(x)$
   b. Give the domain of each composite function.

15. $f(x) = 4 - x$,
   $g(x) = 5 + 2x$

16. $f(x) = \sqrt{5} + x$,
   $g(x) = 10 - \pi x$

17. $f(x) = 4x - 3$,
   $g(x) = 1 - x^2$

18. $f(x) = \dfrac{x}{x^2 - 1}$,

   $g(x) = \dfrac{x - 5}{2}$

19. $f(x) = x/(x + 1)$,
   $g(x) = 1/x$

20. $f(x) = |x - 1|/2$,
   $g(x) = 3x^2 - 4x + 5$

21. $f(x) = 6 - |x|$,
   $g(x) = (x^2 - 1)/x^2$

22. $f(x) = \sqrt{6x - 2}$,
   $g(x) = 1/(x + 1)$

23. $f(x) = \sqrt{x}$,
   $g(x) = x^4 + 3x^2$

24. $f(x) = (x + 1)/x^2$,
   $g(x) = 1/(x + 1)$

25. Suppose f, g, and h are functions. We define $f \circ g \circ h$ as follows:

$$(f \circ g \circ h)(x) = f\{g[h(x)]\}.$$

For example, if $f(x) = 3x + 2$, $g(x) = 6x^2$, and $h(x) = x/3$,

$$(f \circ g \circ h)(x) = f\{g[h(x)]\}$$

$$= f\left[g\left(\dfrac{x}{3}\right)\right]$$

$$= f\left[6\left(\dfrac{x}{3}\right)^2\right]$$

$$= f\left(\dfrac{2x^2}{3}\right)$$

$$= 3\left(\dfrac{2x^2}{3}\right) + 2$$

$$= 2x^2 + 2.$$

For $f(x) = x^2 - 1$, $g(x) = 2 - x$, and $h(x) = x/3$, compute

a. $(f \circ g \circ h)(x)$ **b.** $(g \circ f \circ h)(x)$
c. $(g \circ h \circ f)(x)$ **d.** $(h \circ f \circ g)(x)$
e. $(f \circ h \circ h)(x)$ **f.** $(g \circ g \circ f)(x)$

26. For $f(x) = 1 + x^2$, $g(x) = (x - 1)/3$, and $h(x) = 2/x$, compute

a. $(f \circ g \circ h)(x)$ **b.** $(g \circ f \circ h)(x)$
c. $(g \circ h \circ f)(x)$ **d.** $(h \circ f \circ g)(x)$

27. Given that $f(-1) = 5$, $f(\sqrt{7}) = -2$, $g(2) = -1$, $g(6) = \pi$, $h(-2) = 6$, and $h(5) = -4$, compute

a. $(h \circ f \circ g)(2)$ **b.** $(g \circ h \circ f)(\sqrt{7})$

28. Given that f, g, and h are the functions with the accompanying graphs, compute each of the following:

a. $(f \circ g)(2)$ **b.** $(g \circ h)(-1)$
c. $h[g(4)]$ **d.** $(f \circ h \circ g)(2)$
e. $g\{f[h(0)]\}$

a. $y = f(x)$

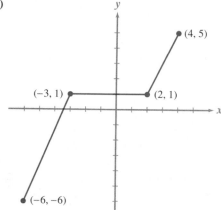

b. $y = g(x)$

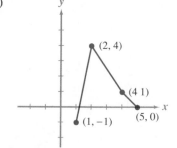

c. $y = h(x)$

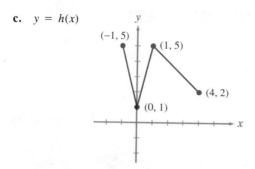

29. Compute each of the following, using the accompanying graphs of f and g:

a. $(f \circ g)(4)$ **b.** $(g \circ f)(4)$

c. $(f \circ g)(3)$ **d.** $(g \circ f)(3)$

e. $(f \circ g)(16)$ **f.** $(g \circ f)(-6)$

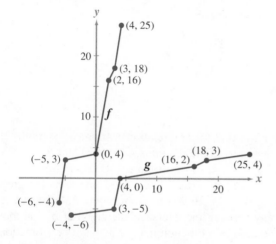

In Exercises 30–37, find functions f and g such that $h = f \circ g$. Verify your answer by forming the composition. The functions f and g are not unique.

30. $h(x) = \sqrt{2x^2 - x + 1}$

31. $h(x) = \sqrt{x + 3}$

32. $h(x) = |10 - x^2|$ **33.** $h(x) = 5/(x - 3)^3$

34. $h(x) = |x| + \sqrt{x}$ **35.** $h(x) = \sqrt[5]{(x - 2)^4}$

36. $h(x) = \dfrac{5|1 - x^2| + 6}{|1 - x^2|}$ **37.** $h(x) = \left(\dfrac{10x}{8 - x}\right)^{-5}$

38. Express the function $f + g$ found in the solution of part (a) of Example 36 as a piecewise-defined function that is free of absolute values. [*Hint:* Determine the value of $f(x)$ for $x \in (-\infty, 1)$ and for $x \in [1, \infty)$.]

39. Express the function fg found in part (c) of Example 36 as a piecewise-defined function that is free of absolute values. [*Hint:* See Problem 38.]

40. A balloon in the shape of a soft drink can is being inflated. If the radius of its base is a function of time t given by $r(t) = 10 + 3t$ and the volume $V(t) = t^3 - t^2 + t + 5$, express the height of the balloon as a function of time t.

41. A large spherical balloon is being inflated with gas. If the radius is increasing at the rate of 2 feet per minute, express the volume and surface area of the balloon as functions of time.

Photo © Craig Blouin/Offshoot Stock

42. Suppose that in an experiment concerning bacteria it is known that the number of bacteria present, N, is a function of temperature $T(°C)$ given by

$$N(T) = T^3 + 10,000.$$

In addition, suppose that t hours after the experiment begins the temperature is given by

$$T(t) = 5t + 20, \qquad 0 \le t \le 2.$$

Compute

a. $(N \circ T)(t)$

b. How many bacteria are present when $t = 0$? $t = 1$? $t = 2$?

43. Prove that the product of two even functions is even.

44. Prove that the product of two odd functions is even.

45. Prove that the product of an odd function and an even function is odd.

46. Prove that every function f can be written as the sum of an odd function with an even function. [*Hint:* Consider the functions g and h defined by

$$g(x) = \frac{1}{2}[f(x) + f(-x)] \qquad \text{and}$$

$$h(x) = \frac{1}{2}[f(x) - f(-x)].]$$

2.6 ONE-TO-ONE FUNCTIONS AND THEIR INVERSES

At the beginning of this chapter, we found that functions are special types of relations. Let us consider the functions A and B defined as follows:

$$A = \{(0, 2), (1, 4), (-1, 4), (2, 5), (-2, -4)\};$$
$$B = \{(2, 1), (1, 3), (4, 8), (-1, -2), (-2, -8)\}.$$

We now form two new relations by interchanging the components in the ordered pairs of A and B to get

$$A_1 = \{(2, 0), (4, 1), (4, -1), (5, 2), (-4, 2)\} \quad \text{and}$$
$$B_1 = \{(1, 2), (3, 1), (8, 4), (-2, -1), (-8, -2)\}.$$

The relations A_1 and B_1 are called the inverse relations of A and B, respectively.

DEFINITION 2.8   Inverse Relation

Let A and B be relations. Then B is called the **inverse relation** of A provided

$$B = \{(x, y) \mid (y, x) \in A\}.$$

The relation B is denoted by A^{-1}.

It follows from Definition 2.8 that for any relation A, $\text{dom}(A) = \text{ran}(A^{-1})$ and $\text{ran}(A) = \text{dom}(A^{-1})$. The graphs of the relations $A_1 = A^{-1}$ and $B_1 = B^{-1}$, defined above, are shown in Figure 2.45. From Figure 2.45(a), we see that the relation A^{-1} is not a function since the vertical line $x = 4$ intersects the graph of A^{-1} at the points $(4, 1)$ and $(4, -1)$. However, from Figure 2.45(b), we see that B^{-1} is a function since any vertical line intersects the graph of B^{-1} at most one time. Thus, the inverse relation of a function may or may not be a function.

FIGURE 2.45

a.

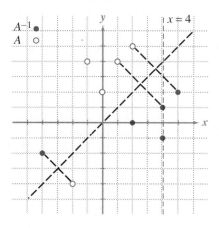

b.

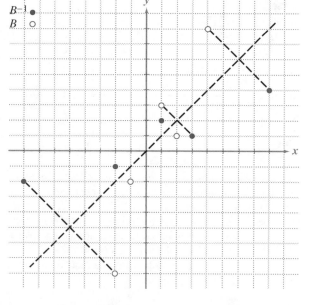

One-to-One Functions

What condition, or conditions, must be satisfied by a function to insure that its inverse relation is a function? To answer this question, we compare functions A and B, as previously defined. We see that A contains the ordered pairs $(1, 4)$ and $(-1, 4)$, which have the same second component. The ordered pairs in $A_1 = A^{-1}$ formed by interchanging the components in $(1, 4)$ and $(-1, 4)$ are $(4, 1)$ and $(4, -1)$, respectively. Hence, relation A^{-1} is not a function, since it contains two distinct ordered pairs with the same first component. On the other hand, no two ordered pairs in B have the same second component. Hence, B^{-1} contains no two ordered pairs with the same first component and therefore is a function. The function B is called a one-to-one-function.

D E F I N I T I O N 2.9   One-to-One Function

A function f is **one-to-one**, provided no two ordered pairs in f have the same second component.

It follows from Definition 2.9 that the inverse relation of a one-to-one function is a function. Also, we see that f is one-to-one provided that for all x_1, x_2 in dom (f), $f(x_1) = f(x_2)$ implies that $x_1 = x_2$. We can determine whether or not a function f is one-to-one by investigating its graph.

Horizontal Line Test

A function f is **one-to-one** if and only if *any* horizontal line intersects the graph of f at *most one time*.

Remember that "at most one time" means exactly one time or not at all. In Figure 2.46(a) on page 110, we see that the horizontal line $y = 2$ intersects the graph of f at the points $(-\frac{7}{2}, 2)$ and $(3, 2)$. Hence, the function f contains the ordered pairs $(-\frac{7}{2}, 2)$ and $(3, 2)$, which have the same second component. Therefore, f is not one-to-one. However, from Figure 2.46(b), we see that no horizontal line intersects the graph of g more than one time. Thus, g does not contain two distinct ordered pairs with the same second component and therefore is one-to-one.

Increasing, Decreasing, and Constant Functions

Other properties of the functions f and g are revealed by their graphs in Figure 2.46. For example, if you look from left to right along the graph of f, you will notice that the graph is falling for values of x in the interval $[-6, -3]$; is horizontal for values of x in the interval $[-3, 1]$; and is rising for values of x in the interval $[1, 6]$. We say that the function f is decreasing on $[-6, -3]$, is constant on $[-3, 1]$, and is increasing on $[1, 6]$. More generally we have

$$f(x) = \begin{cases} -2x - 5, & \text{if } -6 \le x < -3 \\ 1, & \text{if } -3 \le x \le 1 \\ \frac{1}{2}(x + 1), & \text{if } 1 < x \le 6 \end{cases}$$

FIGURE 2.46

a. Not a one-to-one function

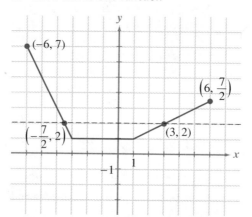

b. A one-to-one function

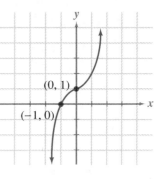

$g(x) = x^3 + 1$

Increasing, Decreasing, and Constant Functions

Suppose the interval $I \subseteq \text{dom}(f)$ and x_1 and x_2 are in I. Then

(i) f is **increasing** on I provided that $f(x_1) < f(x_2)$ whenever $x_1 < x_2$;
(ii) f is **decreasing** on I provided that $f(x_1) > f(x_2)$ whenever $x_1 < x_2$;
(iii) f is **constant** on I provided that $f(x_1) = f(x_2)$ for every x_1 and x_2 in I.

From Figure 2.46(b), we see that the function g defined by $g(x) = x^3 + 1$ is increasing on $(-\infty, \infty)$.

EXAMPLE 43

Find the domain of each function whose graph is shown in Figure 2.47. Determine the intervals on which the function is increasing, the intervals on which it is decreasing, and those on which it is constant.

Solution

a.   For each $x \in \mathbb{R}$, $3x - x^3$ is a real number. Hence, if $f(x) = 3x - x^3$, $\text{dom}(f) = (-\infty, \infty)$. By inspecting the graph of f, we see that f is increasing on the interval $[-1, 1]$ and is decreasing on the intervals $(-\infty, -1]$ and $[1, \infty)$.

b.   Since $g(x) = (x - 1)^3$, $\text{dom}(g) = (-\infty, \infty)$. From the graph of g, we see that g is increasing on $(-\infty, \infty)$.

c.   The domain of the function h defined by $h(x) = |x + 1|/(x + 1)$ is $(-\infty, -1) \cup (-1, \infty)$. From Figure 2.47(c), we conclude that h is constant on the interval $(-\infty, -1)$ and is also constant on the interval $(-1, \infty)$. (It is a mistake to say that h is constant on $(-\infty, -1) \cup (-1, \infty)$. Why?)

FIGURE 2.47

a. $f(x) = 3x - x^3$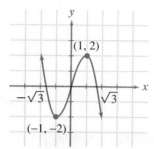

b. $g(x) = (x - 1)^3$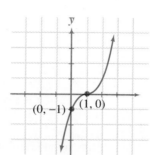

c. $h(x) = \dfrac{|x + 1|}{x + 1}$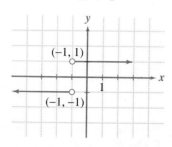

In Example 43, we found that the function defined by $g(x) = (x - 1)^3$ is increasing on the interval $(-\infty, \infty)$. Moreover, from the graph of g shown in Figure 2.47(b), we see that g is one-to-one. In general, if a function is increasing on its domain, then the function is one-to-one.

THEOREM 2.3

(i)   If f is increasing on its domain, then f is a one-to-one function;
(ii)   If f is decreasing on its domain, then f is a one-to-one function.

Proof

(i)   To prove that a function f is one-to-one, it is sufficient to show that if $x_1 \neq x_2$, then $f(x_1) \neq f(x_2)$. Suppose x_1 and x_2 $(x_1 \neq x_2)$ are in dom (f). We can assume without loss of generality that $x_1 < x_2$. Since f is increasing, $f(x_1) < f(x_2)$. Hence $f(x_1) \neq f(x_2)$, and therefore f is one-to-one.

(ii)   The proof is similar to that of part (i).

In the absence of the graph of a function f, we can use algebraic techniques to determine if f is one-to-one, as our next two examples illustrate.

EXAMPLE 44

Determine if the function f defined by $f(x) = x/(x - 2)$ is one-to-one.

Solution   Let x_1 and x_2 be in dom (f), and assume that $f(x_1) = f(x_2)$. If we can show that x_1 must be equal to x_2, it will follow that f is one-to-one. However, if we can find values of x_1 and x_2 such that $f(x_1) = f(x_2)$ and $x_1 \neq x_2$, we will have shown that f is not one-to-one. We proceed as follows: Suppose $f(x_1) = f(x_2)$. Then

$$\frac{x_1}{x_1 - 2} = \frac{x_2}{x_2 - 2}.$$

Multiplying by the LCD, we get

$$\frac{x_1}{x_1 - 2}(x_1 - 2)(x_2 - 2) = \frac{x_2}{x_2 - 2}(x_1 - 2)(x_2 - 2)$$
$$x_1(x_2 - 2) = x_2(x_1 - 2)$$
$$x_1 x_2 - 2x_1 = x_2 x_1 - 2x_2$$
$$-2x_1 = -2x_2$$
$$x_1 = x_2.$$

Therefore, f is a one-to-one function.

EXAMPLE 45

Determine if the function f defined by $f(x) = x + |x|$ is one-to-one.

Solution   Let x_1 and x_2 be in dom (f), and assume $f(x_1) = f(x_2)$. Then $x_1 + |x_1| = x_2 + |x_2|$. Does this imply that $x_1 = x_2$? No! Let $x_1 = -1$ and $x_2 = -2$. Then $x_1 + |x_1| = -1 + |-1| = -1 + 1 = 0$. Similarly, $x_2 + |x_2| = -2 + |-2| = -2 + 2 = 0$. Therefore, f is not a one-to-one function.

Finding the Inverse of a One-to-One Function

Although all relations have inverses, only the inverses of one-to-one functions are functions. If a one-to-one function f is defined by a single equation, we can use the following procedure to find its inverse function f^{-1}:

Procedure for Finding $f^{-1}(x)$

Let the one-to-one function f be defined by $y = f(x)$. To find $f^{-1}(x)$,

(i)   interchange x and y in the equation $y = f(x)$; and

(ii)   solve the new equation for y in terms of x. Then $y = f^{-1}(x)$.

EXAMPLE 46

Show that the function f defined by $f(x) = 2x + 3$ is one-to-one and find $f^{-1}(x)$. Sketch the graphs of f and f^{-1} in the same coordinate system.

Solution   Since f is a linear function, dom $(f) = \mathbb{R}$. The graph of f is the line with slope 2 and y-intercept $(0, 3)$ shown in Figure 2.48. Inspecting the graph, we see that f is increasing on $\mathbb{R}$ and therefore is one-to-one. From the graph of f, we see that ran $(f) = \mathbb{R}$. Thus, dom $(f^{-1}) = $ ran $(f) = \mathbb{R}$. To find $f^{-1}(x)$, we proceed as follows. First we write the equation $f(x) = 2x + 3$ as

$$y = 2x + 3.$$

ing x and y and solving for y, we have

$$x = 2y + 3$$
$$2y = x - 3$$
$$y = \frac{x - 3}{2}.$$

Hence $f^{-1}(x) = (x - 3)/2$. The graphs of f and f^{-1} are shown in Figure 2.49.

FIGURE 2.48

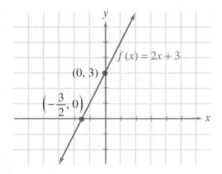

FIGURE 2.49

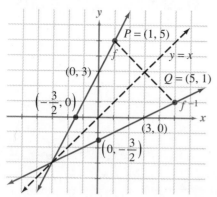

In Figure 2.49, we see that a certain symmetry exists between the graphs of $f(x) = 2x + 3$ and $f^{-1}(x) = (x - 3)/2$. Since the point $P = (1, 5)$ is on the graph of f, the point $Q = (5, 1)$ is on the graph of f^{-1}. The line $y = x$ is the perpendicular bisector of the line segment $\overline{PQ}$. We say that P and Q are **reflections of one another through the line $y = x$.** Moreover, for any point (a, b) on the graph of f, the point (b, a) is on the graph of f^{-1}, and these points are reflections of one another through the line $y = x$. We say that the graph of f and f^{-1} are reflections of one another through the line $y = x$. In general, the graph of a one-to-one function f and the graph of its inverse f^{-1} are reflections of one another through the line $y = x$.

Let us see what happens when we form the composite $f \circ f^{-1}$ and the composite $f^{-1} \circ f$ using the functions $f(x) = 2x + 3$ and $f^{-1}(x) = (x - 3)/2$. Since $\operatorname{ran}(f^{-1}) = \operatorname{dom}(f)$, $\operatorname{dom}(f \circ f^{-1}) = \operatorname{dom}(f^{-1}) = \mathbb{R}$. Thus for $x \in \mathbb{R}$,

$$
\begin{aligned}
(f \circ f^{-1})(x) &= f[f^{-1}(x)] \\
&= 2[f^{-1}(x)] + 3 \\
&= 2\left(\frac{x - 3}{2}\right) + 3 \\
&= x - 3 + 3 \\
&= x.
\end{aligned}
$$

Since $\operatorname{ran}(f) = \operatorname{dom}(f^{-1})$, $\operatorname{dom}(f^{-1} \circ f) = \operatorname{dom}(f) = \mathbb{R}$. Hence, for $x \in \mathbb{R}$,

$$
\begin{aligned}
(f^{-1} \circ f)(x) &= f^{-1}[f(x)] \\
&= \frac{f(x) - 3}{2} \\
&= \frac{(2x + 3) - 3}{2} \\
&= \frac{2x}{2} \\
&= x.
\end{aligned}
$$

Hence $(f \circ f^{-1})(x) = x$ for all $x \in \operatorname{dom}(f^{-1}) = \mathbb{R}$, and $(f^{-1} \circ f)(x) = x$ for all $x \in \operatorname{dom}(f) = \mathbb{R}$.

In general, we have

THEOREM 2.4

If f is a one-to-one function, then

(i) $(f \circ f^{-1})(x) = x$ for each x in dom (f^{-1}); and

(ii) $(f^{-1} \circ f)(x) = x$ for each x in dom (f).

This theorem is illustrated in the following example:

EXAMPLE 47

Show that the function f defined by $f(x) = 1/(x - 2)$ is one-to-one. Find $f^{-1}(x)$, and verify Theorem 2.4 in this case.

Solution   The domain of f, whose graph is shown in Figure 2.50(a), is $(-\infty, 2) \cup (2, \infty)$. From Figure 2.50(a), we see that the graph of f satisfies the horizontal line test. Hence f is one-to-one.

FIGURE 2.50

a. $f(x) = \dfrac{1}{x - 2}$

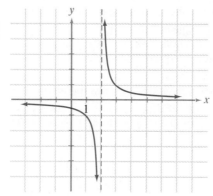

b. $f^{-1}(x) = \dfrac{1 + 2x}{x}$

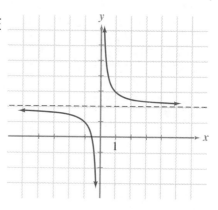

From the graph of f, we see that $\operatorname{ran}(f) = (-\infty, 0) \cup (0, \infty)$. Thus, $\operatorname{dom}(f^{-1}) = (-\infty, 0) \cup (0, \infty)$. To find $f^{-1}(x)$, we write the equation $f(x) = 1/(x - 2)$ in the form

$$y = \frac{1}{x - 2}.$$

Interchanging x and y in the latter equation and solving for y, we have

$$x = \frac{1}{y - 2}$$
$$x(y - 2) = 1$$
$$xy - 2x = 1$$
$$xy = 1 + 2x$$
$$y = \frac{1 + 2x}{x}, \qquad x \neq 0.$$

Therefore, $f^{-1}(x) = (1 + 2x)/x$ for $x \neq 0$. (See Figure 2.50(b).)

Let us verify Theorem 2.4. For $x \in (-\infty, 0) \cup (0, \infty)$,

$$(f \circ f^{-1})(x) = f[f^{-1}(x)]$$
$$= \frac{1}{f^{-1}(x) - 2}$$
$$= \frac{1}{\dfrac{1 + 2x}{x} - 2}$$
$$= \frac{1}{\dfrac{1 + 2x - 2x}{x}}$$
$$= \frac{1}{\dfrac{1}{x}}$$
$$= x.$$

FIGURE 2.51

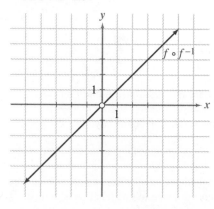

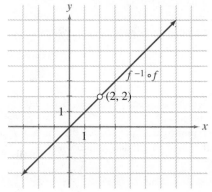

If $x \in (-\infty, 2) \cup (2, \infty)$,

$$(f^{-1} \circ f)(x) = f^{-1}[f(x)]$$
$$= \frac{1 + 2f(x)}{f(x)}$$
$$= \frac{1 + 2\left(\dfrac{1}{x-2}\right)}{\dfrac{1}{x-2}}$$
$$= \left[1 + 2\left(\frac{1}{x-2}\right)\right](x-2)$$
$$= x - 2 + 2$$
$$= x.$$

The graphs of $f \circ f^{-1}$ and $f^{-1} \circ f$ are shown in Figure 2.51.

Given a one-to-one function f, the following theorem can be used to determine if a given function g is f^{-1}:

THEOREM 2.5

Let f be a one-to-one function with domain A and range B. If g is a function with domain B and range A such that

(i) $(f \circ g)(x) = x$ for all x in B; and   (ii) $(g \circ f)(x) = x$ for all x in A.

then $g = f^{-1}$.

This theorem is illustrated in the next example.

EXAMPLE 48

Let f be the function defined by $f(x) = x^2 + 2$ for $x \le 0$. Verify that the function g defined by $g(x) = -\sqrt{x - 2}$ for $x \ge 2$ is f^{-1}. Sketch the graphs of f and g in the same coordinate system (see Figure 2.52).

Solution   If $x \in \text{dom}(g) = [2, \infty)$, then

$$(f \circ g)(x) = f[g(x)]$$
$$= [g(x)]^2 + 2$$
$$= (-\sqrt{x-2})^2 + 2$$
$$= x - 2 + 2$$
$$= x, \qquad x \ge 2.$$

Also, if $x \in \text{dom}(f) = (-\infty, 0]$, then

$$(g \circ f)(x) = g[f(x)]$$
$$= -\sqrt{f(x) - 2}$$
$$= -\sqrt{(x^2 + 2) - 2}$$
$$= -\sqrt{x^2}$$
$$= -|x|$$
$$= -(-x) \qquad \text{(since } x \le 0\text{)}$$
$$= x.$$

FIGURE 2.52

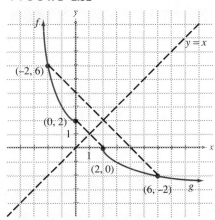

Therefore, by Theorem 2.5, $g = f^{-1}$.

EXERCISES 2.6

In Exercises 1–10,
a. Sketch the graph of each function.
b. Determine the intervals on which the function is increasing and those on which it is decreasing.
c. Determine which of the functions are one-to-one.

1. $f(x) = 4 - 3x$

2. $f(x) = 1/(x + 3)$

3. $f(x) = x^2 + 4x - 1$

4. $f(x) = 6 - x - 8x^2$

5. $f(x) = |x - 1|$

6. $f(x) = |2x + 3|$

7. $f(x) = \sqrt[3]{x}$

8. $f(x) = 4 - \sqrt{x - 1}$

9. $f(x) = \begin{cases} x^2, & \text{if } x \geq 1 \\ 3x - 1, & \text{if } x < 1 \end{cases}$

10. $f(x) \begin{cases} -x, & \text{if } x \geq 2 \\ x, & \text{if } -2 < x < 2 \\ x + 4, & \text{if } x \leq -2 \end{cases}$

In Exercises 11–16, prove that f and g are inverses of one another, and sketch the graphs of f and g in the same coordinate system.

11. $f(x) = 8 - 3x$,
   $g(x) = (8 - x)/3$

12. $f(x) = \pi + 5x$,
   $g(x) = (x - \pi)/5$

13. $f(x) = \sqrt[3]{x + 4}$,
   $g(x) = x^3 - 4$.

14. $f(x) = x^2 - 3, x \geq 0$,
   $g(x) = \sqrt{x + 3}$

15. $f(x) = 1/x; g(x) = 1/x$

16. $f(x) = -3/x$,
   $g(x) = -3/x$

In Exercises 17–24, show that f is one-to-one and use the algebraic method to find $f^{-1}(x)$. Sketch the graph of f and f^{-1} on the same coordinate system.

17. $f(x) = 9x - 10$

18. $f(x) = -\frac{1}{3}x + 10$

19. $f(x) = x^{3/5}$

20. $f(x) = -\sqrt{3x - 2}$

21. $f(x) = 2 + \sqrt{x}$

22. $f(x) = -x^2 - 2, \quad x \geq 0$

23. $f(x) = (x + 2)^2 - 3, \quad x \leq -2$

24. $f(x) = (2x - 7)/(x + 1)$

25. Let f be defined by $f(x) = (2 - x)/5$.
   a. Find $f^{-1}(x)$.
   b. Find $f^{-1}(1)$ and $1/f(1)$. Comment.

26. Let f be defined by $f(x) = x^3 + x - 1$.
   a. Find $(f \circ f^{-1})(1)$
   b. Find $f^{-1}[f(2)]$
   c. Find $f(-1)$ and use your answer to find $f^{-1}(-3)$.
   d. Find $f(0)$ and use your answer to find $f^{-1}(-1)$.

27. Suppose f is the linear function defined by $f(x) = ax + b$ where $a \neq 0$. Find $f^{-1}(x)$.

28. Suppose f is the function defined by $f(x) = x^2 - 2x + 4$ for $x \leq 1$. Find $f^{-1}(x)$.

29. Can a one-to-one function be even? Explain.

30. Prove that the inverse of a one-to-one function is unique.

31. If R is a relation, prove that $(R^{-1})^{-1} = R$.

32. a. Using a graphing calculator, sketch the graph of f and g in the same coordinate system, where

$$f(x) = \frac{2}{x + \sqrt{5}} \quad \text{and} \quad g(x) = \frac{2 - \sqrt{5}x}{x}$$

   b. Using the graphs obtained in part (a), determine if g is the inverse of f.

33. a. Using a graphing calculator, sketch the graph of f and g in the same coordinate system where

$$f(x) = 2x + \sqrt{2} \quad \text{and} \quad g(x) = \frac{1}{2x + \sqrt{2}}$$

   b. Using the graphs obtained in part (a), determine if g is the inverse of f.

34. a. Using a graphing calculator, sketch the graph of f, g, and h in the same coordinate system where

$$f(x) = 0.9x^3 - 0.6$$
$$g(x) = 0.9\sqrt[3]{x + 0.6}, \quad \text{and}$$
$$h(x) = \sqrt[3]{\tfrac{10}{9}x + \tfrac{2}{3}}$$

   b. Using the graphs obtained in part (a), determine the inverse of f.

CHAPTER 2 REVIEW EXERCISES

In Exercises 1–8, determine which of the given relations are functions. Given the domain and range of each relation.

1.

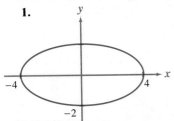

2.

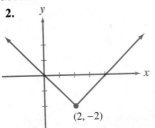

(2, −2)

3.

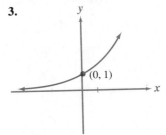

(0, 1)

4.

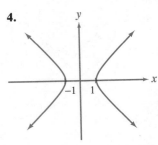

5.

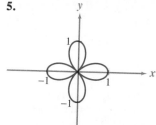

6.

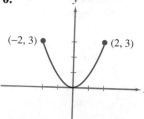

(−2, 3) • • (2, 3)

7.

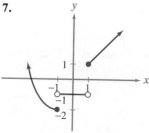

8.

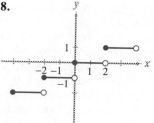

In Exercises 9–16, determine the center and radius of the circle with the given equation.

9. $x^2 + y^2 = 49$

10. $x^2 + y^2 = 15$

11. $(x − 5)^2/(12) + (y + 2)^2/(12) = 1$

12. $(x + \sqrt{7})^2/(5) + (y − 2\sqrt{3})^2/(5) = 1$

13. $x^2 + y^2 − 10x − 10y + 35 = 0$

14. $x^2 + y^2 − 2\sqrt{3}x + 2\sqrt{2}y + 4 = 0$

15. $x^2 + y^2 − 1.4x − 0.2y − 3.5 = 0$

16. $x^2 + y^2 − 2\pi x + 4\pi y + 5\pi^2 − 1 = 0$

In Exercises 17–30, give the domain of the functions defined by the given equations and sketch the graph of each. Discuss symmetry.

17. $f(x) = 5x − 6$

18. $f(x) = 9 − 7x$

19. $f(x) = 10 − x^2$

20. $f(x) = 3x^2 + x + 8$

21. $f(x) = 4/(x + 3)$

22. $f(x) = 1/(x^2 + 7x + 12)$

23. $f(x) = |x| − 4$

24. $f(x) = |x − 5| + 10$

25. $f(x) = \sqrt{x + 5}$

26. $f(x) = \sqrt{x − 3} − 5$

27. $f(x) = \begin{cases} x^2, & \text{if } x \geq 3 \\ x − 1, & \text{if } −1 < x < 3 \\ 5 − x, & \text{if } x \leq −1 \end{cases}$

28. $f(x) = \begin{cases} 1/(x − 2), & \text{if } x > 2 \\ |x − 2|, & \text{if } x \leq 2 \end{cases}$

29. $f(x) = [[2x]]$

30. $f(x) = [[\frac{1}{2}x + 1]]$

In Exercises 31–34, compute

a. $f(2)$ **b.** $f(−1)$ **c.** $f(a + 2)$

d. $\dfrac{f(a + h) − f(a)}{h}$ **e.** $f\left(\dfrac{1}{a}\right)$

31. $f(x) = 11 − 3x − x^2$

32. $f(x) = \sqrt{6x + 1}$

33. $f(x) = 1/2x$

34. $f(x) = 2|x + 3|$

In Exercises 35–40, find the slope and the x- and y-intercepts and sketch the graph of each line.

35. $3x − y = 4$

36. $4x + 2y − 15 = 0$

37. $x = −3$

38. $4x = 3y − 5$

39. $\frac{2}{5}y = −x + 1$

40. $4y = 9$

In Exercises 41–44, find an equation in general form for the line satisfying the following conditions:

41. Passing through $P = (−1, 5)$ and $Q = (3, 2)$.

42. Passing through $P = (−2, 5)$ and $Q = (−2, 1)$.

43. Perpendicular to the line $3x − 2y + 4 = 0$ and passing through $P = (0, 5)$.

44. Tangent to the circle $x^2 + y^2 = 16$ at the point $P = (0, −4)$.

In Exercises 45–50, sketch the graph and give the maximum or minimum value of each function.

45. $f(x) = x^2 − 2x − 8$

46. $f(x) = 2 − x − 3x^2$

47. $f(x) + 2 = 6 − x^2$

48. $f(x) = −(x + 1)^2 + 4$

49. $2x^2 − 1 = 4f(x) + 3$

50. $f(x) = 5x^2 + \pi$

51. Extend the graph of the function given below to the interval $[−5, 0]$ so that the function is

a. even **b.** odd

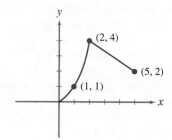

(2, 4)

(5, 2)

(1, 1)

52. Find an equation for the quadratic function whose graph passes through the points $P = (−1, 9)$, $Q = (0, 3)$, and $R = (2, 3)$.

In Exercises 53–56, determine the intervals on which the given function is increasing and those on which it is decreasing.

53.

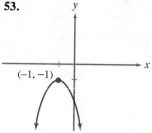

(−1, −1)

54.

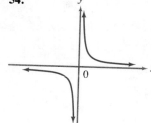

0

55.

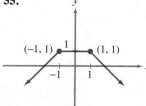

(−1, 1) • • (1, 1)

−1 1

56.

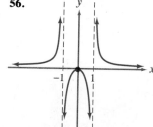

−1 1

In Exercises 57–60, given that $f(0) = -1$, $g(0) = 2$, and $h(0) = 4$, compute the indicated functional value.

57. $\left(\dfrac{f + g}{h - g}\right)(0)$

58. $[(f - g) + h](0)$

59. $\left(\dfrac{f}{g} + h\right)(0)$

60. $(4f + 3g)(0)$

In Exercises 61–64, compute $(f \circ g)(x)$ and state the domain of each composition.

61. $f(x) = (x - 4)/3$,
     $g(x) = x^2 - x$

62. $f(x) = 1/x$, $g(x) = 1/x$

63. $f(x) = 1/(\sqrt{x} - 4)$,
     $g(x) = x^2 - 6x$

64. $f(x) = \sqrt{x + 1}$,
     $g(x) = x^2 - 2x - 9$

In Exercises 65–68, determine which of the functions are one-to-one functions.

65. $f(x) = x^3 + 2$
66. $f(x) = (x - 2)^2 + 3$, $x \geq 2$
67. $f(x) = |x + 3| + 1$, $x \leq -3$
68. $f(x) = 2/(3x - 1)$

In Exercises 69–72, show that f is a one-to-one function, find $f^{-1}(x)$, and sketch the graph of f and f^{-1} on the same coordinate system.

69. $f(x) = \sqrt[5]{x}$

70. $f(x) = (x + 4)/9$

71. $f(x) = \sqrt{9 - x^2}$,
     $-3 \leq x \leq 0$

72. $f(x) = \sqrt[3]{x + 1}$

73. A wire 20 inches long is to be cut into four pieces to form a rectangle whose shortest side has length x. Express the area of the rectangle as a function of x. Find the maximum possible value of the area of the rectangle.

74. An arch has the shape of a parabola with its vertex at the top and its axis vertical. If the arch is 30 feet wide at the base and 15 feet high in the center, how wide is it 5 feet above the base?

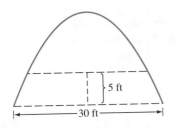

75. A ball thrown directly upward with a speed of 64 feet per second moves according to the law

$$y = 64t - 16t^2$$

   a.   How high does the ball travel?
   b.   When does it attain its maximum height?
   c.   At what time is the ball 30 feet above ground?

76. The perimeter of a rectangle is 32 feet. Find the dimensions that will maximize the area. What is the maximum area possible?

77. Use slopes to show that the triangle with vertices $(0, -1)$, $(1, 3)$, and $(4, -2)$ is a right triangle.

78. A car rental company leases luxury automobiles for $50 a day plus $0.30 per mile. Write an equation for C, the cost in dollars, in terms of x, the number of miles driven, if the car is rented for a. 5 days. b. n days.

79. A manufacturer finds that 50,000 VCRs are sold per month at a price of $300 each but that only 30,000 are sold each month if the price is $400 each. Suppose that the demand function f for these VCRs is linear (see Example 25).
   a.   Find a formula for $f(p)$ where p is the price for a VCR.
   b.   Find the selling price per VCR if the monthly demand is 20,000.
   c.   Should the company price each unit as high as $600? Why?

80. In a certain state income taxes are levied on taxable income as indicated below:

Income of $0 up to $999 taxed at 0 percent;

Income of $1,000 up to $4,999, $10 plus 5% of the excess over $1,000;

Income of $5,000 up to $9,999, $210 plus 6% of the excess over $5,000;

Income of at least $10,000, $510 plus 3% of the excess over $10,000.

   a.   Find an equation for the line determined by each of the tax intervals.
   b.   Sketch the graph of this "income tax curve."

CHAPTER 3

Polynomial Functions and Their Zeros

Linear and quadratic functions, studied in the previous chapter, are part of a much larger class of functions known as polynomial functions. In this chapter, we investigate polynomial functions and discuss techniques that can be used to obtain their graphs. Since the zeros of a polynomial play a central role in the determination of its graph, much of our effort is concentrated on the development of step-by-step procedures for finding the zeros of a polynomial function. In the final section of this chapter, we study the graphs of functions whose defining equations are expressed as the quotient of two polynomials. Functions of this type are called rational functions.

3.1 POLYNOMIAL EQUATIONS: AN OVERVIEW

In Chapter 1, methods for solving linear and quadratic equations were discussed. These equations are special cases of the general polynomial equation, which is defined as follows:

///

DEFINITION 3.1

Let n be a nonnegative integer and let $a_0, a_1, a_2, \ldots, a_n$ be complex* numbers. If $a_n \neq 0$, the function P defined by

$$P(x) = a_n x^n + a_{n-1} x^{n-1} + \cdots + a_1 x + a_0$$

is called a **polynomial function of degree n**. The equation

$$a_n x^n + a_{n-1} x^{n-1} + \cdots + a_1 x + a_0 = 0$$

is called a **polynomial equation of degree n**. We say that a complex number c is a **zero** of the function P and a **root** of the equation $P(x) = 0$ if and only if $P(c) = 0$.

For example, -2 is a zero of the polynomial function P defined by $P(x) = x^3 + 3x^2 + x - 2$ and a root of the polynomial equation $x^3 + 3x^2 + x - 2 = 0$ since $(-2)^3 + 3(-2)^2 + (-2) - 2 = 0$.

EXAMPLE 1

Find the set of all zeros of the polynomial function P defined by $P(x) = x^3 + x^2 + 4x + 4$.

Solution   To find the zeros of this polynomial function, we will solve the associated polynomial equation

$$x^3 + x^2 + 4x + 4 = 0.$$

We factor by grouping and set each of the resulting factors equal to zero.

$$(x^3 + x^2) + (4x + 4) = 0$$
$$x^2(x + 1) + 4(x + 1) = 0$$
$$(x + 1)(x^2 + 4) = 0$$
$$x + 1 = 0 \quad \text{or} \quad x^2 + 4 = 0.$$

Since the solutions of the latter equations are -1 and $\pm 2i$, respectively, the set of all zeros is $\{-1, -2i, 2i\}$.

We should not be deceived by the simplicity of the solution of the previous example. Finding roots of polynomial equations is often a very difficult task, and the factoring method used here is rarely successful. Although solutions can easily be found for polynomial equations of degrees one and two, this is not the case for equations of degree three or higher.

_____

*Since every real number is also a complex number, some or all of these coefficients may be real numbers.

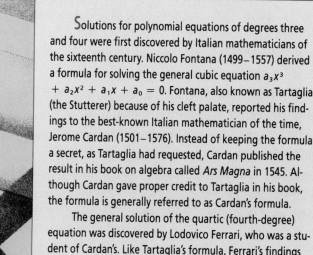

Solutions for polynomial equations of degrees three and four were first discovered by Italian mathematicians of the sixteenth century. Niccolo Fontana (1499–1557) derived a formula for solving the general cubic equation $a_3x^3 + a_2x^2 + a_1x + a_0 = 0$. Fontana, also known as Tartaglia (the Stutterer) because of his cleft palate, reported his findings to the best-known Italian mathematician of the time, Jerome Cardan (1501–1576). Instead of keeping the formula a secret, as Tartaglia had requested, Cardan published the result in his book on algebra called *Ars Magna* in 1545. Although Cardan gave proper credit to Tartaglia in his book, the formula is generally referred to as Cardan's formula.

The general solution of the quartic (fourth-degree) equation was discovered by Lodovico Ferrari, who was a student of Cardan's. Like Tartaglia's formula, Ferrari's findings first appeared in the book *Ars Magna*.

Inspired by the discoveries of Tartaglia and Ferrari, mathematicians turned their attention to polynomial equations of degrees five, six, and higher. For almost three centuries their laborious efforts produced no new results. Finally, in 1832, a brilliant young French mathematician named Evariste Galois (1811–1832) showed that it is impossible to find general algebraic formulas for the solutions of polynomial equations of degree five or higher.

Although Galois' discovery ended the search for formula solutions of polynomial equations, it by no means curtailed the mathematical investigation of other methods for finding zeros of polynomials. In fact, a great deal of current research in mathematics is involved with finding, or approximating, zeros of polynomials.

As mentioned earlier, algebraic formulas are available for solving polynomial equations of degrees three and four. However, these formulas are quite complicated, and they will not be discussed here. Instead, we turn our attention to other methods which can be used to find the zeros of a polynomial function.

3.2 DIVISION OF POLYNOMIALS

We recall that a quotient of two polynomials is called a rational expression. Instead of dividing the numerator by the denominator, we often simplify rational expressions by canceling common factors of the numerator and denominator. However, as we develop methods for finding zeros of polynomials later in this chapter, we will often find it necessary to divide one polynomial by another. The essential features of the division process are summarized by the division algorithm for polynomials.

Division Algorithm for Polynomials

If $F(x)$ and $G(x)$ are polynomials such that $G(x) \neq 0$, then there exist unique polynomials $Q(x)$ and $R(x)$ such that

$$F(x) = G(x)Q(x) + R(x),$$

where either $R(x) = 0$ or the degree of $R(x)$ is less than the degree of $G(x)$.

The polynomials $F(x)$, $G(x)$, $Q(x)$, and $R(x)$, which appear in the division algorithm above, are referred to, respectively, as the **dividend**, **divisor**, **quotient**, and **remainder**. Since a proof of the division algorithm requires mathematics beyond the scope of this text, it will not be included here. Instead, we illustrate this theorem by dividing the polynomial $F(x) = 6x^3 - x^2 - 15$ by the polynomial $G(x) = 2x - 3$.

Polynomial division is similar to the long division process used for integers. However, before we divide we must be certain that both the divisor and the dividend are arranged in descending powers of the variable. Since the x term is "missing" from the dividend, we supply $0 \cdot x$ to insure the proper alignment of terms and write

$$2x - 3 \overline{)\, 6x^3 - x^2 + 0x - 15} \qquad (1)$$

The steps of the division process are outlined below.

Step 1. Divide the first term of the divisor into the first term of the dividend and place the result over the first term in the dividend.

Step 2. Multiply each term of the divisor by the result obtained in step 1 and align the terms under like terms in the dividend.

Step 3. Subtract the product obtained in step 2 from the dividend.

Step 4. Using the result from step 3 as the "new" dividend, repeat steps 1–3 above until step 3 produces either 0 or a polynomial whose degree is less than the degree of the divisor.

Following steps 1, 2, and 3, quotient (1) becomes

$$
\begin{array}{l}
\text{Step 1} \longrightarrow \quad 3x^2 \\
2x - 3 \overline{)\, 6x^3 - \ x^2 + 0x - 15} \\
\text{Step 2} \longrightarrow \ 6x^3 - 9x^2 \\
\hline
\text{Step 3} \longrightarrow \qquad\qquad 8x^2 + 0x - 15
\end{array}
$$

Repeating steps 1–3 using $8x^2 + 0x - 15$ as the dividend, we have

$$
\begin{array}{l}
\qquad\qquad 3x^2 + 4x \\
2x - 3 \overline{)\, 6x^3 - \ x^2 + 0x \ - 15} \\
\qquad\quad\ 6x^3 - 9x^2 \\
\hline
\qquad\qquad\quad 8x^2 + 0x \ - 15 \\
\qquad\qquad\quad 8x^2 - 12x \\
\hline
\qquad\qquad\qquad\qquad 12x - 15
\end{array}
$$

Using $12x - 15$ as the dividend, we repeat steps 1–3.

$$
\begin{array}{l}
\qquad\qquad 3x^2 + 4x \ + 6 \\
2x - 3 \overline{)\, 6x^3 - \ x^2 + 0x \ - 15} \\
\qquad\quad\ 6x^3 - 9x^2 \\
\hline
\qquad\qquad\quad 8x^2 + 0x \ - 15 \\
\qquad\qquad\quad 8x^2 - 12x \\
\hline
\qquad\qquad\qquad\quad 12x - 15 \\
\qquad\qquad\qquad\quad 12x - 18 \\
\hline
\qquad\qquad\qquad\qquad\qquad 3
\end{array}
$$

Since 3 has degree 0 and the divisor $2x - 3$ has degree 1, the division process is concluded. Thus $Q(x) = 3x^2 + 4x + 6$ is the quotient and $R(x) = 3$ is the remainder. From the division algorithm for polynomials, we see that

$$
\begin{aligned}
Q(x) \cdot G(x) + R(x) &= (3x^2 + 4x + 6)(2x - 3) + 3 \\
&= 6x^3 - x^2 - 18 + 3 \\
&= 6x^3 - x^2 - 15 \\
&= F(x).
\end{aligned}
$$

EXAMPLE 2

Find the quotient and remainder when $3x^4 - 8x^2 - 16$ is divided by $x - 2$.

Solution Proceeding as above, we have

$$
\begin{array}{r}
3x^3 + 6x^2 + 4x + 8 \\
x - 2 \overline{)\; 3x^4 + 0x^3 - 8x^2 + 0x - 16} \\
\underline{3x^4 - 6x^3} \\
6x^3 - 8x^2 + 0x - 16 \\
\underline{6x^3 - 12x^2} \\
4x^2 + 0x - 16 \\
\underline{4x^2 - 8x} \\
8x - 16 \\
\underline{8x - 16} \\
0
\end{array}
$$

Thus $Q(x) = 3x^3 + 6x^2 + 4x + 8$ is the quotient and $R(x) = 0$ is the remainder. By the division algorithm,

$$3x^4 - 8x^2 - 16 = (3x^3 + 6x^2 + 4x + 8)(x - 2),$$

which means that $x - 2$ is a factor of $3x^4 - 8x^2 - 16$.

EXAMPLE 3

Find the quotient and the remainder when $3x^4 - 2x^2 + 7x - 5$ is divided by $2x^2 - 4$.

Solution As in Example 2,

$$
\begin{array}{r}
\frac{3}{2}x^2 + 2 \\
2x^2 + 0x - 4 \overline{)\; 3x^4 + 0x^3 - 2x^2 + 7x - 5} \\
\underline{3x^4 + 0x^3 - 6x^2} \\
4x^2 + 7x - 5 \\
\underline{4x^2 + 0x - 8} \\
7x + 3
\end{array}
$$

Hence $Q(x) = \frac{3}{2}x^2 + 2$ is the quotient and $R(x) = 7x + 3$ is the remainder. You should illustrate the division algorithm by showing that

$$\left(\frac{3}{2}x^2 + 2\right)(2x^2 - 4) + (7x + 3) = 3x^4 - 2x^2 + 7x - 5.$$

Synthetic Division

In the search for zeros of polynomials we often must divide a polynomial by a binomial of the form $x - c$. When the divisor has this form, we can streamline the long division process by a procedure known as **synthetic division**. The basic idea behind synthetic division is that the essential information of the long division process is carried by the coefficients of the various polynomials. To illustrate, we outline the synthetic division procedure using the polynomials $F(x) = 3x^4 - 8x^2 - 16$ and $G(x) = x - 2$ from Example 2.

Step 1. As in long division, we write the powers of the variable in descending order and supply missing powers with zero coefficients. We then align the coefficients of the dividend in a horizontal row and place $c = 2$ in a ⌐ symbol to the left.

$$\underline{2}\rfloor \quad 3 \quad 0 \quad -8 \quad 0 \quad -16$$

Step 2. Skipping a line, draw a horizontal line under the dividend coefficients and bring down the first coefficient.

$$\begin{array}{c|ccccc} \underline{2}\rfloor & 3 & 0 & -8 & 0 & -16 \\ \hline & 3 \end{array}$$

Step 3. Multiply the number under the line by 2 and place the result under the next dividend coefficient.

$$\begin{array}{c|ccccc} \underline{2}\rfloor & 3 & 0 & -8 & 0 & -16 \\ & & 6 \\ \hline & 3 \end{array}$$

Step 4. Add the numbers in the second column and place the sum under the line.

$$\begin{array}{c|ccccc} \underline{2}\rfloor & 3 & 0 & -8 & 0 & -16 \\ & & 6 \\ \hline & 3 & 6 \end{array}$$

Step 5. Repeat steps 3 and 4 for the remaining dividend coefficients.

$$\begin{array}{c|ccccc} \underline{2}\rfloor & 3 & 0 & -8 & 0 & -16 \\ & & 6 & 12 & 8 & 16 \\ \hline & 3 & 6 & 4 & 8 & 0 \end{array}$$

$$\underbrace{}_{\text{Coefficients of } Q(x)} \quad \underbrace{}_{R(x)}$$

The division process is now complete. The coefficients of the quotient and the remainder appear in the bottom row. Since the divisor has degree 1, the degree of the quotient will be one less than the degree of the dividend and the remainder will be a constant. Hence $Q(x) = 3x^3 + 6x^2 + 4x + 8$ and $R(x) = 0$, which agree with the results of Example 2.

EXAMPLE 4

Use synthetic division to show that $x + 3$ is a factor of the polynomial $x^5 + 3x^4 + 2x^2 + 4x - 6$.

Solution   Using synthetic division, we divide $x^5 + 3x^4 + 2x^2 + 4x - 6$ by $x + 3$.

$$\begin{array}{c|cccccc} \underline{-3}\rfloor & 1 & 3 & 0 & 2 & 4 & -6 \\ & & -3 & 0 & 0 & -6 & 6 \\ \hline & 1 & 0 & 0 & 2 & -2 & 0 \end{array}$$

Thus the quotient is $x^4 + 2x - 2$ and the remainder is 0. Since $R(x) = 0$, the division algorithm for polynomials tells us that

$$x^5 + 3x^4 + 2x^2 + 4x - 6 = (x^4 + 2x - 2)(x + 3),$$

and it follows that $x + 3$ is a factor of $x^5 + 3x^4 + 2x^2 + 4x - 6$.

EXAMPLE 5

Using synthetic division, find the value(s) of k for which $x - 2$ is a factor of $x^3 + k^2x^2 - 3kx + k^2 - 7$.

Solution   We must find the value(s) of k for which the remainder is 0 when $x^3 + k^2x^2 - 3kx + k^2 - 7$ is divided by $x - 2$. Using synthetic division, we have

$$
\begin{array}{r|ccccc}
2 & 1 & k^2 & -3k & k^2 - 7 \\
  &   & 2   & 2k^2 + 4 & 4k^2 - 6k + 8 \\
\hline
  & 1 & k^2 + 2 & 2k^2 - 3k + 4 & 5k^2 - 6k + 1
\end{array}
$$

Setting the remainder equal to 0 gives us the quadratic equation $5k^2 - 6k + 1 = 0$. Solving this equation, we find that the values of k are $\frac{1}{5}$ and 1.

Until now, we have only considered polynomials whose coefficients were real numbers. The division algorithm also applies to polynomials having nonreal complex coefficients, as we now illustrate.

EXAMPLE 6

Use synthetic division to divide $x^4 + ix^3 + 2x^2 + (1 - i)x - 2 + 3i$ by $x + 2i$.

Solution   Since $c = -2i$, we have

$$
\begin{array}{r|ccccc}
-2i & 1 & i & 2 & 1 - i & -2 + 3i \\
    &   & -2i & -2 & 0 & -2 - 2i \\
\hline
    & 1 & -i & 0 & 1 - i & -4 + i
\end{array}
$$

Thus $Q(x) = x^3 - ix^2 + 1 - i$ is the quotient, and $R(x) = -4 + i$ is the remainder. You should verify the division algorithm for this example.

EXAMPLE 7

Use synthetic division to show that $x - i$ is a factor of the polynomial $x^4 - 1$.

Solution   Synthetic division yields

$$
\begin{array}{r|ccccc}
i & 1 & 0 & 0 & 0 & -1 \\
  &   & i & -1 & -i & 1 \\
\hline
  & 1 & i & -1 & -i & 0
\end{array}
$$

Since the remainder is 0, $x^4 - 1 = (x^3 + ix^2 - x - i)(x - i)$, and $x - i$ is a factor of $x^4 - 1$.

EXERCISES 3.2

In Exercises 1–10, use long division to find the quotient and the remainder when the first polynomial is divided by the second.

1. $6m^2 + 5m - 1; 2m + 3$
2. $x^2 + 4x - 2; x - 3$
3. $y^4 + 1; y - 2$
4. $-12t^3 + 4t^2 - 1; 4t^2 + 1$
5. $a^5 - 3a^4 + 3a - 1; a^2 - 1$
6. $2y^3 + 4y^2 - y + 2; y + \frac{1}{2}$
7. $-x^4 - 6x^3 + 2x^2 + x - 1; 2x^2 + x - 2$
8. $t^4 - 1; t^2 + \sqrt{2}t - 1$
9. $x^5 + x^3; x - i$
10. $r^5 - k; r^2 - k$,   where k is a constant.

In Exercises 11–38, use synthetic division to find the quotient and the remainder when the first polynomial is divided by the second.

11. $3x^3 - 8x + 1; x + 2$
12. $5m^3 - 6m^2 - 28m - 2; m - 2$
13. $y^4 - 3y^3 - 5y^2 + 2y - 16; y - 3$
14. $p^5 - 6p^2 + 4p^3 - 9; p + 1$
15. $x^3 - 8x^2 + 17x - 10; x - 5$
16. $t^5 + 2t^4 - t^3 - 2t^2 + 2t + 4; t + 2$
17. $x^7 - x^3 + x; x - 1$
18. $2x^3 + 9x^2 - 17x + 6; x + 6$
19. $x^6 - 1; x - 1$
20. $x^4 + 1; x + 1$
21. $1 - x^5; x - 1$
22. $3y^5 - y^2 + 1; y - 2$
23. $x^4 + a^3x^3 + a^2x^2 + a^3x + a^4; x - a$
    (a is a constant)
24. $1 - a^2y^2 + 2y^4; y + a$ (a is a constant)
25. $8x^3 - 4x^2 - 2x + 1; x - \frac{1}{2}$
26. $2y^3 - y^2 + 3y - 1; y + \frac{1}{2}$
27. $x^3 + x^2 + \frac{1}{2}x + \frac{1}{8}; x + \frac{1}{2}$
28. $\frac{1}{3}x^3 - \frac{2}{9}x^2 + \frac{1}{27}x + 1; x - \frac{1}{3}$
29. $x^3 - x^2 + 4x - 4; x + 2i$
30. $y^4 + 2iy^3 - iy + 5; y + i$
31. $x^4 - ix^2 + 1; x - 1 + i$
32. $x^3 + 2x^2 - 1; x + 1 - i$
33. $x^5 - x^3; x - i$
34. $x^4 + 3x^2 - 1; x + i$
35. $2y^4 - 7iy^3 - 7y^2 + 2iy; y - \frac{1}{2}i$
36. $4x^4 + 5ix^3 - x^2 + 8ix; x + \frac{1}{4}i$
37. $m^5 + (1 - i)m^3 - 2 + i; m + 2i$
38. $x^3 + (1 - i)x^2 + (2 + i)x + 2; x - i$

In Exercises 39–48, use synthetic division to show that the second expression is a factor of the first.

39. $y^4 - 2y^3 + y^2 - 10y - 6; y - 3$
40. $2x^3 - 4x^2 - 3x + 3; x + 1$
41. $x^3 + x^2 + x + \frac{3}{8}; x + \frac{1}{2}$
42. $2y^3 + 3y^2 - 8y + 3; y - \frac{1}{2}$
43. $2x^4 - 7ix^3 - 7x^2 + 2ix; x - \frac{1}{2}i$
44. $x^3 + 3ix^2 - 3x - i; x + i$
45. $2m^3 + 3m^2 - 2im + 2 - 8i; m - 1 - i$
46. $x^3 + ix^2 + (-3 + i)x + 2 - 2i; x - 1 + i$
47. $\sqrt{2}x^5 - (\sqrt{2}/2)x^3 - x^2 - 4; x - \sqrt{2}$
48. $-x^4 + x^2 + 2; x + \sqrt{2}$

In Exercises 49–52, use synthetic division to find all values of k (if any) for which the second polynomial is a factor of the first.

49. $x^4 + k^2x + k - 10; x + 2$
50. $2x^3 + kx^2 + x - 2k; x - 1$
51. $k^2x^4 + 3kx^2 + 2; x + 1$
52. $x^3 + 2kx^2 - 27x + 11k; x + 1$
53. Find all values of c such that $x - 1$ is a factor of the polynomial $c^2x^4 - 4c^2 + 9$.
54. Use synthetic division to find the remainder when $ax^2 + bx + c$ is divided by $x - r$, where a, b, c, and r are constants.
55. Let a and b be constants. If $x - a$ is a factor of $x^2 + 2ax - 3b^2$, show that $a^2 = b^2$.
56. Show that $x - a$ is not a factor of $x^4 + 2x^2 + 1$ for any real number a. [*Hint*: Use synthetic division and the division algorithm for polynomials.]
57. When $x^3 + kx + 2$ is divided by $x - 1$ the remainder is -5. Find the value of k.
58. Determine a value of k such that $x + 2$ will be a factor of the polynomial $x^3 - kx + 6$.

3.3 REMAINDER THEOREM AND FACTOR THEOREM

In this section we prove two theorems that will form the basis for much of our subsequent work with polynomials.

Consider the polynomial $P(x) = 2x^3 - 5x^2 + 8x - 3$. Dividing $P(x)$ by $x - 2$ using synthetic division,

$$
\begin{array}{r|rrrr}
2 & 2 & -5 & 8 & -3 \\
  &   & 4 & -2 & 12 \\
\hline
  & 2 & -1 & 6 & 9
\end{array}
$$

we obtain the quotient $2x^2 - x + 6$ and a remainder of 9. If we evaluate $P(x)$ at $x = 2$, we find that

$$P(2) = 2(2)^3 - 5(2)^2 + 8(2) - 3$$
$$= 9.$$

Thus the remainder, when $P(x)$ is divided by $x - 2$, is the same as the value of $P(x)$ at $x = 2$. The theorem that follows affirms that this result is more than a coincidence:

Remainder Theorem

When a polynomial $P(x)$ is divided by $x - c$, where c is a constant, the remainder is $P(c)$.

Proof   When $P(x)$ is divided by $x - c$, we know by the division algorithm for polynomials that there exist polynomials $Q(x)$ and $R(x)$ such that

$$P(x) = Q(x)(x - c) + R(x) \tag{1}$$

where either $R(x) = 0$ or $R(x)$ has degree 0. In either case, $R(x)$ is a constant k, and we can write equation (1) above as

$$P(x) = Q(x)(x - c) + k.$$

Thus if $x = c$, we have

$$P(c) = Q(c)(c - c) + k$$
$$= 0 + k$$
$$= k.$$

EXAMPLE 8

If $P(x) = x^4 - 3x^2 + 4$, use the remainder theorem to evaluate each of the following:

a. $P(-3)$ **b.** $P(-i)$

Solution

a. Dividing $P(x)$ by $x + 3$ using synthetic division yields

$$
\begin{array}{r|rrrrr}
-3 & 1 & 0 & -3 & 0 & 4 \\
   &   & -3 & 9 & -18 & 54 \\
\hline
   & 1 & -3 & 6 & -18 & 58 \\
\end{array}
$$

Hence, by the remainder theorem, $P(-3) = 58$.

b. Dividing $P(x)$ by $x + i$, we have

$$
\begin{array}{r|rrrrr}
-i & 1 & 0 & -3 & 0 & 4 \\
   &   & -i & -1 & 4i & 4 \\
\hline
   & 1 & -i & -4 & 4i & 8 \\
\end{array}
$$

Hence, by the remainder theorem, $P(-i) = 8$.

Although the remainder theorem can be used to evaluate a polynomial $P(x)$ at a particular value of x, perhaps its most useful application is to help us prove the following theorem:

Factor Theorem

Let $P(x)$ be a polynomial and let c be a constant. If $P(c) = 0$, then $x - c$ is a factor of $P(x)$. Conversely, if $x - c$ is a factor of $P(x)$, then $P(c) = 0$.

Proof   Assume first that $P(c) = 0$. By the remainder theorem, $P(c)$ is the remainder when $P(x)$ is divided by $x - c$. Since $P(c) = 0$, the division algorithm for polynomials tells us that

$$P(x) = Q(x)(x - c),$$

for some polynomial $Q(x)$. Thus $x - c$ is a factor of $P(x)$. Conversely, if $x - c$ is a factor of $P(x)$, we know that the remainder is 0 when $P(x)$ is divided by $x - c$. Hence, by the remainder theorem, $P(c) = 0$.   _____

EXAMPLE 9

Use the factor theorem to show that the second polynomial is a factor of the first in each case.

a. $x^3 + 4x^2 - 9$; $x + 3$ **b.** $x^3 + 5x^2 + 10x + 12$; $x + 1 + \sqrt{3}i$

Solution

a. Let $P(x) = x^3 + 4x^2 - 9$. Then $P(-3) = (-3)^3 + 4(-3)^2 - 9 = 0$. Thus, by the factor theorem, $x + 3$ is a factor of $P(x)$.

b. We use synthetic division to divide $P(x) = x^3 + 5x^2 + 10x + 12$ by $x + 1 + \sqrt{3}i$.

$-1 - \sqrt{3}i$	1	5	10	12
		$-1 - \sqrt{3}i$	$-7 - 3\sqrt{3}i$	-12
	1	$4 - \sqrt{3}i$	$3 - 3\sqrt{3}i$	0

Hence $P(-1 - \sqrt{3}i) = 0$, and $x + 1 + \sqrt{3}i$ is a factor of $P(x)$ by the factor theorem.   _____

EXAMPLE 10

Find all values of k for which $x - i$ is a factor of $P(x) = x^4 + 2kx^2 + k^2 - 4$.

Solution   For $x - i$ to be a factor of $P(x) = x^4 + 2kx^2 + k^2 - 4$, the factor theorem tells us that $P(i)$ must equal 0. Since

$$\begin{aligned}
P(i) &= i^4 + 2ki^2 + k^2 - 4 \\
&= 1 - 2k + k^2 - 4 \\
&= k^2 - 2k - 3,
\end{aligned}$$

the values of k we seek are the solutions of the quadratic equation

$$k^2 - 2k - 3 = 0.$$

Solving this equation by factoring, we find that $k = -1$ or $k = 3$.

The following example shows how we can use the factor theorem to find a polynomial of specified degree whose zeros are given:

EXAMPLE 11

Find a polynomial of degree 4 that has -3, 5, $-2i$, and $2i$ as zeros.

Solution   We know by the factor theorem that $x + 3$, $x - 5$, $x + 2i$, and $x - 2i$ must be factors of a polynomial having -3, 5, $-2i$, and $2i$ as zeros.

Since
$$\begin{aligned}
P(x) &= (x + 3)(x - 5)(x + 2i)(x - 2i) \\
&= (x^2 - 2x - 15)(x^2 + 4) \\
&= x^4 - 2x^3 - 11x^2 - 8x - 60
\end{aligned}$$

has degree 4, $P(x) = x^4 - 2x^3 - 11x^2 - 8x - 60$ is a polynomial that has the desired properties. We note that this solution is *not* unique, since $k \cdot P(x)$ is also a solution for any nonzero constant k.

EXAMPLE 12

Find a fourth-degree polynomial equation with integer coefficients that has the following roots: $-\frac{3}{5}$, $\frac{1}{2}$ and 1 (a double root).

Solution   By the factor theorem, $x + \frac{3}{5}$, $x - \frac{1}{2}$ and $(x - 1)^2$ are factors of any polynomial having the given zeros. Thus

$$\left(x + \frac{3}{5} \right)\left(x - \frac{1}{2} \right)(x - 1)^2 = 0 \tag{2}$$

is a fourth-degree polynomial equation that has the desired roots. To express equation (2) with integer coefficients, we note that the equations

$$x + \frac{3}{5} = 0 \qquad \text{and} \qquad x - \frac{1}{2} = 0$$

are equivalent to $5x + 3 = 0$ and $2x - 1 = 0$, respectively. Hence

$$(5x + 3)(2x - 1)(x - 1)^2 = 0, \qquad \text{or}$$
$$10x^4 - 19x^3 + 5x^2 + 7x - 3 = 0$$

is a polynomial equation having all the desired properties.

The preceding examples have illustrated several applications of the factor theorem. This powerful theorem can also help us find the solutions of the general polynomial equation $P(x) = 0$, where $P(x)$ is a polynomial of degree $n > 0$.

For example, if we know that a is one of the solutions of $P(x) = 0$, the factor theorem permits us to write

$$P(x) = Q(x)(x - a)$$

where $Q(x)$ is a polynomial of degree $n - 1$. This means that we can replace the equation $P(x) = 0$ by the equivalent equation

$$Q(x)(x - a) = 0.$$

The latter equation leads us to two equations:

$$Q(x) = 0 \quad \text{or} \quad x - a = 0$$

Since $x - a = 0$ simply reaffirms that $x = a$ is a root, we need only solve

$$Q(x) = 0$$

in order to find the remaining roots of $P(x) = 0$. The equation $Q(x) = 0$ is usually referred to as the **reduced equation**.

E X A M P L E 13

If -3 is a double root of the equation $2x^4 + 11x^3 + 9x^2 - 27x - 27 = 0$, find the remaining solutions.

Solution   Let $P(x) = 2x^4 + 11x^3 + 9x^2 - 27x - 27$. Since -3 is a root of $P(x) = 0$, it follows from the factor theorem that $x + 3$ is a factor of $P(x)$. Using synthetic division, we divide $P(x)$ by $x + 3$.

$$
\begin{array}{r|rrrrr}
-3 & 2 & 11 & 9 & -27 & -27 \\
   &   & -6 & -15 & 18 & 27 \\
\hline
   & 2 & 5 & -6 & -9 & 0
\end{array}
$$

Thus $P(x) = (x + 3)(2x^3 + 5x^2 - 6x - 9)$ and the remaining roots of $P(x) = 0$ must be solutions of the reduced equation

$$2x^3 + 5x^2 - 6x - 9 = 0.$$

Since -3 is a double root of $P(x) = 0$, $x + 3$ must be a factor of $2x^3 + 5x^2 - 6x - 9$. Synthetic division yields

$$
\begin{array}{r|rrrr}
-3 & 2 & 5 & -6 & -9 \\
   &   & -6 & 3 & 9 \\
\hline
   & 2 & -1 & -3 & 0
\end{array}
$$

Thus $2x^3 + 5x^2 - 6x - 9 = (x + 3)(2x^2 - x - 3)$ and we can write

$$P(x) = (x + 3)^2(2x^2 - x - 3) = 0.$$

By factoring, we find that -1 and $\frac{3}{2}$ are the solutions of the "new" reduced equation $2x^2 - x - 3 = 0$. Hence the solution set is

$$\left\{ -3 \text{ (double root)}, -1, \frac{3}{2} \right\}.$$

E X A M P L E 14

Find all solutions of the equation $12x^4 + 20x^3 - 5x^2 + 5x - 2 = 0$, given that $-i/2$ and $i/2$ are two of the solutions.

Solution   Let $P(x) = 12x^4 + 20x^3 - 5x^2 + 5x - 2$. Since $-i/2$ and $i/2$ are solutions of $P(x) = 0$, the factor theorem tells us that $x + (i/2)$ and $x - (i/2)$ are factors of $P(x)$. The equation $P(x) = 0$ can thus be written as

$$\left(x + \frac{i}{2}\right)\left(x - \frac{i}{2}\right)Q(x) = 0 \qquad (3)$$

for some polynomial $Q(x)$.

We could now proceed as in the previous example and divide $P(x)$ by $x + (i/2)$ using synthetic division. However, this synthetic division is rather involved, so we take an alternate approach using long division. Since the equations $x + (i/2) = 0$ and $x - (i/2) = 0$ are equivalent to $2x + i = 0$ and $2x - i = 0$, respectively, we can express equation (3) above as

$$(2x + i)(2x - i)Q(x) = 0,$$

or
$$(4x^2 + 1)Q(x) = 0. \qquad (4)$$

To find $Q(x)$, we use long division to divide $P(x)$ by $4x^2 + 1$.

$$
\begin{array}{r}
3x^2 + 5x - 2 \\
4x^2 + 0x + 1 \overline{\smash{\big)}\, 12x^4 + 20x^3 - 5x^2 + 5x - 2} \\
\underline{12x^4 + 0x^3 + 3x^2} \\
20x^3 - 8x^2 + 5x - 2 \\
\underline{20x^3 + 0x^2 + 5x} \\
- 8x^2 + 0x - 2 \\
\underline{- 8x^2 + 0x - 2} \\
0
\end{array}
$$

Thus $Q(x) = 3x^2 + 5x - 2$ and we can write equation (4) as

$$(4x^2 + 1)(3x^2 + 5x - 2) = 0.$$

Since -2 and $\frac{1}{3}$ are the roots of the reduced equation $3x^2 + 5x - 2 = 0$, the solution set of $P(x) = 0$ is $\{-i/2, i/2, -2, \frac{1}{3}\}$.

Examples 13 and 14 provide a comparison of repeated synthetic division and long division as methods for finding roots of polynomial equations. In a given situation, you should choose the method that is easier to apply.

EXERCISES 3.3

In Exercises 1–10, use the remainder theorem to find the remainder when the first polynomial is divided by the second.

1. $x^3 + 3x^2 - 9x + 5; x + 5$
2. $-x^2 + 3x - 10; x - 20$
3. $2t^4 - 3t^3 + 3t + 4; t + 1$
4. $8x^3 - 16x + 7; x - \frac{1}{2}$
5. $x^5 - ix + 1; x - i$
6. $3x^3 - 7x^2 + 27x - 63; x + 2i$
7. $1 + 2y + y^2 - y^4; y - \sqrt{2}$
8. $x^4 - 2x^3 + 6x^2 - 2x + 5; x - 1 + 2i$
9. $m^4 - m^2 + 1; m + 1 - i$
10. $x^4 - 1; x - (i/2)$

In Exercises 11–22, use the factor theorem to determine whether or not the second polynomial is a factor of the first.

11. $x^3 - 8x^2 + 17x - 10; x - 5$

12. $2y^3 + 9y^2 - 17y + 6; y + 6$
13. $6x^4 - 19x^3 - 25x^2 + 18x + 8; x + \frac{1}{3}$
14. $2t^4 - 3t^2 + 1; t - \frac{1}{2}$
15. $x^4 + 3x^2 - 4; x + \sqrt{2}$
16. $\sqrt{2}x^5 - (\sqrt{2}/2)x^3 - x^2 - 4; x - \sqrt{2}$
17. $x^4 + 3ix^3 - 2x^2 - 6ix; x + 3i$
18. $x^3 + (1 - i)x^2 + (2 + i)x + 2; x - i$
19. $m^3 + 2m^2 + 1; m - 1 + i$
20. $x^3 + 2x^2 + 5x; x + 1 - 2i$
21. $3x^4 + x^3 - 5x^2 - x - 2; 3x - 2$
    [Hint: $3x - 2 = 3(x - \frac{2}{3})$]
22. $4y^5 - 17y^3 - 2; 2y + 1$

In Exercises 23–32, find a polynomial having integer coefficients that has the given degree and the given zeros.

23. degree 3; 2, -1, and 3

24. degree 3; 5, $-\sqrt{2}$, and $\sqrt{2}$
25. degree 3; -5, 0, and 2
26. degree 3; -1, 1, and $\frac{1}{2}$
27. degree 4; $-i$, i, -2, and 1
28. degree 4; $1 - 2i$, $1 + 2i$, and -3 (a double root)
29. degree 4; $-\frac{1}{2}$, $\frac{3}{4}$, 0, and 2
30. degree 4; $-\frac{2}{3}$, $\frac{1}{8}$, $-i$, and i
31. degree 5; $-2i$, $2i$, 4, and -1 (a double root)
32. degree 5; $-i/3$, $i/3$, -2, 2, and 0

In Exercises 33–42, a polynomial equation is followed by one or more of its solutions. In each case, find all remaining solutions.

33. $x^3 - 6x^2 + 12x - 8 = 0$; 2
34. $t^3 + 7t^2 + 11t + 5 = 0$; -1
35. $2y^3 + 5y^2 - y - 6 = 0$; $-\frac{3}{2}$
36. $5x^3 - 2x^2 + 5x - 2 = 0$; $\frac{2}{5}$
37. $m^5 + m^3 - m^2 = 1$; $-i$ and i
38. $x^5 + 9x^3 + x^2 + 9 = 0$; $-3i$ and $3i$
39. $4x^4 - 13x^2 + 3 = 0$; $-\sqrt{3}$ and $\sqrt{3}$
40. $6x^5 + 5x^4 - 29x^3 - 25x^2 - 5x = 0$; $-\frac{1}{2}$ and $-\frac{1}{3}$
41. $x^4 - 2x^3 + 6x^2 - 2x + 5 = 0$; $1 - 2i$ and $1 + 2i$
42. $2y^4 - 7y^3 + 11y^2 + 3y = -7$; $2 - \sqrt{3}i$ and $2 + \sqrt{3}i$
43. Show that $x^4 + 3x^2 + 4$ has no factor of the form $x - c$ where c is a real number.
44. If $P(x) = k^2 x^4 + 3kx^2 + 2$, find all values of k for which $x + 1$ is a factor of $P(x)$.

45. For which value(s) of k is $x - 2$ a factor of $kx^3 + 3x^2 - 4k^2$?
46. Determine values of a and b such that $x - 1$ is a factor of both $x^3 + x^2 - ax + b$ and $x^3 - x^2 + ax + b$.
47. If n is an odd positive integer, show that $x + a$ is a factor of $x^n + a^n$.
48. Suppose c is a zero of the polynomial $x^3 - x + 1$. Show that c is also a zero of the polynomial $-x^6 + 2x^4 - x^2 + 1$. [*Hint*: Use $c^3 = c - 1$ to find expressions for x^6 and x^4.]
49. Find a polynomial $P(x)$ that satisfies the following conditions:

    $P(x)$ has degree 3;
    -1 and 2 are zeros of $P(x)$;
    $P(1) = 6$.

50. Find a polynomial $P(x)$ that satisfies the following conditions:

    $P(x)$ has degree 4;
    $-\frac{1}{2}$, $-2i$ and $2i$ are roots of $P(x) = 0$;
    $P(0) = 4$.

51. Show that $x + 1$ is a factor of $3x^{101} + 9x^{78} + 2x^{15} - 4$.
52. Find the remainder when $P(x) = x^{84} - 12x^{43} + 7x - 5$ is divided by $x + 1$.

3.4   FUNDAMENTAL THEOREM OF ALGEBRA AND RATIONAL ROOT THEOREM

In the previous section, we found that knowing one root of a polynomial equation often made it possible to find the remaining roots. But how do we find a root of $P(x)$, or more importantly, how do we know that such a root even exists? Moreover, how do we know when we have found all the roots of the equation $P(x) = 0$? Each of these important, fundamental questions will be addressed in this section.

An answer to the last of our questions is provided by the theorem that follows. This theorem, often referred to as the "fundamental theorem of algebra," was first proved by the brilliant German mathematician Karl Friedrich Gauss (1777–1855) when he was only 22 years old. Since the proof involves mathematical concepts beyond the scope of this text, it will not be given here.

Fundamental Theorem of Algebra

If the polynomial equation $P(x) = 0$ has degree $n > 0$, then there exists a complex number r such that $P(r) = 0$.

As an immediate consequence of the fundamental theorem of algebra, we see that every nonconstant polynomial $P(x)$ has at least one complex zero. As we shall see, this theorem in conjunction with the division algorithm for polynomials and the factor theorem, will enable us to determine all the zeros of nonconstant polynomials with real coefficients.

Let $P(x) = 0$ be a polynomial equation of degree $n > 0$. By the fundamental theorem of algebra, we know that $P(x) = 0$ has at least one root, say r_1. Since $x - r_1$ must be a factor of $P(x)$, the division algorithm guarantees the existence of a polynomial $Q_1(x)$ such that

$$P(x) = Q_1(x)(x - r_1). \qquad (1)$$

Since $P(x)$ has degree $n > 0$ and $x - r_1$ has degree 1, it follows that the degree of $Q_1(x)$ is $n - 1$. If $n = 1$, $Q_1(x)$ has degree 0 and is thus a nonzero constant k_1. In this case, equation (1) above becomes

$$P(x) = k_1(x - r_1).$$

If $n > 1$, $Q_1(x)$ has degree $n - 1 > 0$ and the fundamental theorem can be applied to the reduced equation $Q_1(x) = 0$. Hence $Q_1(x)$ has a complex zero r_2 and we can write $Q_1(x) = Q_2(x)(x - r_2)$, for some polynomial $Q_2(x)$. We can now express equation (1) as

$$P(x) = Q_2(x)(x - r_1)(x - r_2) \qquad (2)$$

where the polynomial $Q_2(x)$ has degree $n - 2 \geq 0$. If $n = 2$, $Q_2(x)$ is a nonzero constant k_2 and equation (2) becomes

$$P(x) = k_2(x - r_1)(x - r_2).$$

In $n > 2$, we again apply the fundamental theorem and express equation (2) in the form

$$P(x) = Q_3(x)(x - r_1)(x - r_2)(x - r_3).$$

We continue in like manner until the polynomial $Q_n(x)$ is a nonzero constant k_n. In the process, we have found complex numbers $r_1, r_2, \ldots, r_n$ such that $P(x)$ can be expressed as

$$P(x) = k_n(x - r_1)(x - r_2) \cdots (x - r_n).$$

We have proved the following theorem:

THEOREM 3.1

If $P(x)$ is a polynomial of degree $n > 0$, then there exist complex numbers $r_1, r_2, \ldots, r_n$ (not necessarily all distinct) and a nonzero constant k such that

$$P(x) = k(x - r_1)(x - r_2) \cdots (x - r_n).$$

Theorem 3.1 tells us that a polynomial of degree $n > 0$ can be expressed as a product of n linear factors. We now illustrate this theorem with two polynomials whose factors are readily obtainable.

North Wind Picture Archives

KARL FRIEDRICH GAUSS
(1777–1855)

Known as the prince of mathematics, Karl Gauss's genius was recognized at a very early age. Before he was three years old he corrected an error in his father's payroll report, and by the age of ten he was confounding his teachers with his mathematical prowess. Gauss completed college at the age of fifteen and proved the Fundamental Theorem of Algebra as part of the doctoral thesis submitted to the University of Helmstedt when he was twenty-two. Many of Gauss's important mathematical results were discovered in a diary he began as early as 1796 but that was not found until forty years after his death. In this diary he settled a 2,000-year-old question by demonstrating how to construct a seventeen-sided polygon using a compass and a straightedge. In 1807 Gauss turned down a professorship in mathematics to become director of the observatory in Göttingen, Germany, where he spent the remaining years of his life totally devoted to the study of mathematics, physics, and astronomy.

EXAMPLE 15

Express in the form $P(x) = k(x - r_1)(x - r_2) \cdots (x - r_n)$ each of the following polynomials:

a. $P(x) = 2x^3 - x^2 - 8x + 4$ **b.** $P(x) = x^3 + 1$

Solution

a. We can factor $P(x)$ as follows:

$$2x^3 - x^2 - 8x + 4 = (2x^3 - x^2) - (8x - 4)$$
$$= x^2(2x - 1) - 4(2x - 1)$$
$$= (2x - 1)(x^2 - 4)$$
$$= (2x - 1)(x + 2)(x - 2).$$

Since $2x - 1 = 2(x - \frac{1}{2})$, we can express $P(x)$ in the desired form as

$$P(x) = 2\left(x - \frac{1}{2}\right)(x - (-2))(x - 2).$$

b. By the sum-of-cubes formula,

$$x^3 + 1 = (x + 1)(x^2 - x + 1).$$

To find the factors of $x^2 - x + 1$, we first use the quadratic formula to solve $x^2 - x + 1 = 0$. Since the roots are $(1 \pm \sqrt{3}i)/2$, we can factor $P(x)$ as

$$P(x) = (x + 1)\left(x - \frac{1 + \sqrt{3}i}{2}\right)\left(x - \frac{1 - \sqrt{3}i}{2}\right).$$

Using Theorem 3.1, we can factor a polynomial $P(x)$ as follows:

$$P(x) = k(x - r_1)(x - r_2) \cdots (x - r_n).$$

By the factor theorem, each of the numbers $r_1, r_2, \ldots, r_n$ is a root of the equation $P(x) = 0$. However, as Theorem 3.1 states, the numbers $r_1, r_2, \ldots, r_n$ are not necessarily all distinct. For example, consider the equation

$$(x + 1)(x - 2)(x - 2)(x - 2) = 0 \qquad (3)$$

where $r_1 = -1$, and $r_2 = r_3 = r_4 = 2$. Although equation (3) has only two roots, namely -1 and 2, the root 2, which appears three times, is called a **triple root** or a **root of multiplicity 3**. The concept of multiple roots is formalized by the following definition:

DEFINITION 3.2

Let $P(x)$ be a polynomial of degree $n > 0$ and let r be a root of $P(x) = 0$. If m is the largest positive integer such that $(x - r)^m$ is a factor of $P(x)$, then r is said to be a **root of multiplicity** m.

We are now able to answer another of the questions posed earlier in this section: How do we know when we have found all the roots of a polynomial equation $P(x) = 0$?

THEOREM 3.2

A polynomial equation of degree $n > 0$ has exactly n roots, where a root of multiplicity m is counted m times.

Proof   Let $P(x) = 0$ be a polynomial equation of degree $n > 0$. By Theorem 3.1 we know that $P(x) = 0$ has at least n roots, counting multiplicities, namely $r_1, r_2, \ldots, r_n$. Let us assume that r is a number which is different from $r_1, r_2, \ldots, r_n$. Theorem 3.1 permits us to write

$$P(r) = k(r - r_1)(r - r_2) \cdots (r - r_n), \qquad (4)$$

where k is a nonzero constant.

Since r is different from $r_1, r_2, \ldots, r_n$, and since $k \neq 0$, each factor on the right side of equation (4) is different from 0. Thus $P(r)$ cannot equal 0, which implies that r is not a solution of $P(x) = 0$. Hence $P(x) = 0$ has no roots other than $r_1, r_2, \ldots, r_n$.

The Rational Root Theorem

One question posed at the beginning of this section remains unanswered: How can we find a root of the polynomial equation $P(x) = 0$? We have indeed saved the most difficult question for last! In fact, we recall from Section 3.1 that there is no general answer to this question. There are, however, several techniques for finding the roots of the equation $P(x) = 0$ when $P(x)$ satisfies certain conditions. For example, if $P(x)$ has integer coefficients and if $P(x) = 0$ has a rational root r, we can find this root by applying the following theorem.

Rational Root Theorem

Let $P(x) = a_n x^n + a_{n-1} x^{n-1} + \cdots + a_1 x + a_0$ be a polynomial of degree $n > 0$, where $a_0, a_1, \ldots, a_n$ are integers. If c/d is a nonzero rational number, expressed in lowest terms, and if c/d is a root of $P(x) = 0$, then c is a factor of a_0 and d is a factor of a_n.

Proof   Let us assume that c/d, in lowest terms, is a root of $P(x) = 0$. We know by the factor theorem that $x - c/d$ is a factor of $P(x)$. Thus $d[x - (c/d)] = dx - c$ is a factor of $P(x)$, and we can use the division algorithm for polynomials to write

$$P(x) = (dx - c)Q(x), \qquad (5)$$

where $Q(x)$ is a polynomial of degree $n - 1$.

If $n = 1$, $Q(x)$ is a nonzero constant k and equation (5) becomes

$$P(x) = k(dx - c) = kdx - kc. \qquad (6)$$

We see at once from equation (6) that c is a factor of the constant term kc, and d is a factor of the leading coefficient kd.

If $n > 1$, we can write $Q(x) = b_{n-1}x^{n-1} + \cdots + b_1x + b_0$ and equation (5) becomes

$$a_nx^n + a_{n-1}x^{n-1} + \cdots + a_1x + a_0$$
$$= (dx - c)(b_{n-1}x^{n-1} + \cdots + b_1x + b_0).$$

Equating the coefficients of x^n on each side of this equation gives us

$$a_n = db_{n-1},$$

which means that d is a factor of the leading coefficient a_n. When the coefficients of the constant terms are equated, we obtain

$$a_0 = -cb_0.$$

Hence c is a factor of a_0 and the proof is complete.   ──────────

We conclude this section with two examples that show how the rational root theorem can be used to solve polynomial equations.

EXAMPLE 16

Find all rational roots (if any) of the equation $2x^4 - 7x^3 - 6x^2 + 44x - 40 = 0$. If possible, find all the roots of this equation.

Solution   Suppose that c/d is a rational number, expressed in lowest terms, that is a solution of the equation

$$2x^4 - 7x^3 - 6x^2 + 44x - 40 = 0. \qquad (7)$$

By the rational root theorem, c must be a factor of the constant term 40, and d must be a factor of the leading coefficient 2. Hence the possible values of c, d, and c/d are

$$c: \quad \pm1, \ \pm2, \ \pm4, \ \pm5, \ \pm8, \ \pm10, \ \pm20, \ \pm40$$
$$d: \quad \pm1, \ \pm2$$
$$\frac{c}{d}: \quad \pm1, \ \pm2, \ \pm4, \ \pm5, \ \pm8, \ \pm10, \ \pm20, \ \pm40, \ \pm\frac{1}{2}, \ \pm\frac{5}{2}.$$

Using synthetic division, we test the values of c/d in their listed order until we find that 2 is a root of equation (7).

$$\underline{2|} \quad \begin{array}{rrrrr} 2 & -7 & -6 & 44 & -40 \\ & 4 & -6 & -24 & 40 \\ \hline 2 & -3 & -12 & 20 & 0 \end{array}$$

The reduced equation determined by this synthetic division is

$$2x^3 - 3x^2 - 12x + 20 = 0.$$

Applying the rational root theorem to the reduced equation, we obtain the following list of possible rational roots:

$$\pm1, \ \pm2, \ \pm4, \ \pm5, \ \pm10, \ \pm20, \ \pm\frac{1}{2}, \ \pm\frac{5}{2}.$$

Since 2 could be a multiple root, we continue our search for rational roots by testing 2 in the reduced equation.

$$
\begin{array}{r|rrrr}
2 & 2 & -3 & -12 & 20 \\
  &   & 4 & 2   & -20 \\
\hline
  & 2 & 1 & -10 & 0
\end{array}
$$

Thus, 2 is a multiple root, having multiplicity at least 2. The new reduced equation

$$2x^2 + x - 10 = 0$$

can be factored to obtain the solutions $-\frac{5}{2}$ and 2. Hence the solution set of the original polynomial equation is

$$\left\{ -\frac{5}{2}, \ 2 \ (\text{multiplicity 3}) \right\}.$$

EXAMPLE 17

Find all rational roots (if any) of the equation $x^4 - \frac{13}{6}x^3 - \frac{31}{18}x^2 + \frac{4}{9}x + \frac{1}{9} = 0$. If possible, find all the roots of this equation.

Solution   We first multiply both sides of the given equation by 18 to obtain an equivalent equation that has integer coefficients:

$$18x^4 - 39x^3 - 31x^2 + 8x + 2 = 0. \qquad (8)$$

If c/d is a rational root of equation (8), then c is a factor of the constant term 2, and d is a factor of the leading coefficient 18. The possible values of c, d, and c/d are shown below.

$$
\begin{aligned}
c&: \quad \pm 1, \ \pm 2 \\
d&: \quad \pm 1, \ \pm 2, \ \pm 3, \ \pm 6, \ \pm 9, \ \pm 18 \\
\frac{c}{d}&: \quad \pm 1, \ \pm 2, \ \pm\frac{1}{2}, \ \pm\frac{1}{3}, \ \pm\frac{1}{6}, \ \pm\frac{1}{9}, \ \pm\frac{1}{18}, \ \pm\frac{2}{3}, \ \pm\frac{2}{9}.
\end{aligned}
$$

If we use synthetic division to test the values of c/d in their listed order, $\frac{1}{3}$ is the first root we find.

$$
\begin{array}{r|rrrrr}
\frac{1}{3} & 18 & -39 & -31 & 8 & 2 \\
  &   & 6 & -11 & -14 & -2 \\
\hline
  & 18 & -33 & -42 & -6 & 0
\end{array}
$$

The reduced equation determined by this synthetic division is

$$18x^3 - 33x^2 - 42x - 6 = 0.$$

Applying the rational root theorem to the equivalent equation

$$6x^3 - 11x^2 - 14x - 2 = 0, \qquad (9)$$

we see that any possible rational roots must come from the following list:

$$\pm 1, \ \pm 2, \ \pm\frac{1}{2}, \ \pm\frac{1}{3}, \ \pm\frac{1}{6}, \ \pm\frac{2}{3}.$$

Since the first six numbers in this list were eliminated above, and since $\frac{1}{3}$ could be a multiple root, our search for another rational root begins with $\frac{1}{3}$. Our test using synthetic division reveals that $\frac{1}{3}$ is not a root of equation (9), and the next rational root we encounter is $-\frac{1}{6}$.

$$
\begin{array}{r|rrrr}
-\frac{1}{6} & 6 & -11 & -14 & -2 \\
 & & -1 & 2 & 2 \\
\hline
 & 6 & -12 & -12 & 0
\end{array}
$$

From this synthetic division we obtain the reduced equation

$$6x^2 - 12x - 12 = 0,$$

or equivalently,

$$x^2 - 2x - 2 = 0.$$

Using the quadratic formula, we find that the solutions of the latter equation are $1 \pm \sqrt{3}$. Hence, the solution set of the original polynomial equation is

$$\left\{ -\frac{1}{6}, \frac{1}{3}, 1 - \sqrt{3}, 1 + \sqrt{3} \right\}.$$

EXERCISES 3.4

In Exercises 1–6, give the degree of each polynomial equation. In addition, determine all roots (including multiplicities, when appropriate) of each equation.

1. $(x - 1)(x + \frac{3}{4})^2(x - 5) = 0$

2. $(x + 2)^5(x - \frac{1}{2})^2(x + 3) = 0$

3. $(x - \sqrt{3})^3(x + 2i)(x - 2i) = 0$

4. $(x + 1 + i)(x + 1 - i)(x^2 - 5) = 0$

5. $(x - 2)^3(x^3 - 8)(x^2 - 2x + 1) = 0$

6. $(x - 2i)^2(x + 2i)^2(x^4 - 16) = 0$

In Exercises 7–12, show that the given root has the stated multiplicity.

7. 2 is a double root of $x^3 - 3x^2 + 4 = 0$.

8. 1 is a triple root of $x^5 - 5x^3 - x^2 + 8x = 4 - x^4$.

9. i is a double root of $x^6 + 4x^4 + 5x^2 = -2$.

10. $-\frac{1}{2}$ is a root of multiplicity 4 of $32x^5 + 16x^4 - 48x^3 - 56x^2 - 22x = 3$.

11. $\frac{1}{3}$ is a root of multiplicity 4 of $81x^6 - 108x^5 - 27x^4 + 96x^3 - 53x^2 + 12x - 1 = 0$.

12. $-2i$ is a root of multiplicity 4 of $x^6 + 12x^4 + 48x^2 + 64 = 0$.

In Exercises 13–16, find a polynomial $P(x)$ with leading coefficient 1 that has the given zeros and no others.

13. -1 (multiplicity 2), 3, and 5

14. $-\frac{1}{2}, \frac{1}{3}$ (multiplicity 3), and 6

15. 0, -3 (multiplicity 2), and $\frac{1}{2}$ (multiplicity 2)

16. $\frac{2}{3}, -\frac{5}{6}$, and -2 (multiplicity 3)

In Exercises 17–24, list all possible rational roots for each polynomial equation.

17. $2x^3 - 5x^2 + 2x - 5 = 0$

18. $x^3 + x^2 = 3$

19. $16t^2 - 5t^3 + 13t + 8 = 0$

20. $-a^5 + 2a^4 = 1 - a^2$

21. $\frac{1}{2}m^4 - \frac{3}{4}m^2 + 2m - 9 = 0$

22. $\frac{3}{4}x^2 - 1 = \frac{7}{8}x^4$

23. $r^6 - 2r^9 = r - 8r^4$

24. $12y^3 - 2y^2 + 8y - 16 = 0$

In Exercises 25–30, show that each polynomial has no rational zeros.

25. $2x^3 - 9x^2 - 2x + 5$

26. $5x^2 - 8 - x^4$

27. $y^3 + y^2 - 8$

28. $m^6 + 2m^4 - 11m^2 - 4$

29. $-x^3 + x - 6 + 3x^4$

30. $4t^3 + 9t^2 - 14t + 2$

In Exercises 31–38, find all rational solutions (if any) of each polynomial equation.

31. $x^4 - 4x^3 - 2x^2 + 21x - 18 = 0$

32. $8x^3 - 4x^2 - 2x + 1 = 0$

33. $2y^4 - 13y^3 + 30y^2 = 28y - 8$

34. $12t^3 - 16t^2 - 5t + 3 = 0$

35. $5x^3 + x^2 - 8x = x^5 + x^4 - 4$

36. $8x^3 + 1 = 4x^2 + 2x$

37. $2m^3 - \frac{16}{3}m^2 + \frac{1}{2}m + \frac{5}{6} = 0$

38. $4y^5 - \frac{1}{2}y^4 - \frac{1}{8}y^3 - 4y^2 + \frac{1}{2}y + \frac{1}{8} = 0$

In Exercises 39–46, find *all* solutions of each polynomial equation.

39. $2x^3 - 10x^2 + 12x - 4 = 0$

40. $x^4 + 3x^3 + 3x^2 + 3x + 2 = 0$

41. $3t^4 - 4t^3 = 8t^2 - 9t + 2$

42. $18y^3 - 21y^2 = 10y - 8$

43. $3x^3 + 8x^2 + 19x + 10 = 0$

44. $2x^5 - 3x^4 - 2x + 3 = 0$

45. $x^3 - \frac{11}{6}x^2 = \frac{2}{3}x - \frac{2}{3}$

46. $4x^4 + 3x^3 - \frac{15}{2}x^2 - 6x - 1 = 0$

In Exercises 47–50, express each polynomial in the form $k(x - r_1)(x - r_2) \cdots (x - r_n)$, where $r_1, r_2, \ldots, r_n$ are complex numbers and k is a nonzero constant.

47. $9x^3 - 7x + 2$

48. $x^4 - 7x^3 + 18x^2 - 20x + 8$

49. $3x^4 + 4x^3 + 2x^2 + 8x - 8$

50. $x^5 + 10x^2 - x - 10$

51. Consider the polynomial equation $x^2 - 2 = 0$.
   a. Verify that $-\sqrt{2}$ and $\sqrt{2}$ are the only solutions of this equation.
   b. Use the rational root theorem to prove that $\sqrt{2}$ is an irrational number.

52. Verify that $\sqrt{5}$ is an irrational number. [*Hint*: See Exercise 51.]

53. Find all integer values of k for which the equation $x^3 - k^2x^2 + 3kx - 1 = 0$ has a rational root.

54. If p is a prime, show that the equation $x^3 + px - p = 0$ has no rational root.

55. Find the x-intercepts (if any) of the graph of the function $f(x) = 6x^3 + 13x^2 + x - 2$.

56. Find the x-intercepts (if any) of the graph of the function $f(x) = x^4 + 2x^3 - 6x^2 - 16x - 8$.

In Exercises 57–60, find the point(s) of intersection (if any) of the graphs of the given functions.

57. $f(x) = 5x^3 - 6x^2 - x + 1$;
$g(x) = x^3 + 6x^2 - 10x + 3$

58. $f(x) = 5x^3 + 3x^2 - 2x - 10$;
$g(x) = 2x^4 + 4x^3 - 7x^2 + 2$

59. $f(x) = x^5 + 4x^3 - 3x - 5$;
$g(x) = 3x^3 + 2x^2 + 9x + 3$

60. $f(x) = x^5 + 4x^4 - 9x^3 - 21x + 13$;
$g(x) = 5x^5 - 12x^4 + 16x^3 - 25x^2 + 4$

3.5 FINDING THE ZEROS OF A POLYNOMIAL

In the previous section we discussed a method for finding all the rational zeros of a polynomial having real coefficients. Finding the irrational zeros, if any, of these polynomials can be a difficult task. In fact, we often have to settle for decimal approximations of such zeros in lieu of their exact values.

Suppose, for example, we wanted to find the real zeros of the polynomial

$$P(x) = x^3 - 4x^2 + 1;$$

that is, the real roots of the equation

$$x^3 - 4x^2 + 1 = 0. \tag{1}$$

By the rational root theorem, -1 and 1 are the only possible rational roots of equation (1). Since neither of these numbers is a solution, however, we must conclude that this equation has no rational roots.

Where do we go from here? While Theorem 3.2 guarantees that equation (1) has three roots (counting multiplicities), it does not tell us whether or not these roots are real. Since it is pointless to search for roots that do not exist, let us first determine whether or not this equation has any real roots.

As previously noted, the roots of equation (1) are the zeros of the polynomial function P defined by

$$P(x) = x^3 - 4x^2 + 1.$$

Since the real zeros of P (if any) occur at the x-intercepts of its graph, we can examine the graph to determine whether or not $P(x)$ has any real zeros. Noting that dom $(P) = \mathbb{R}$, we make a table of values and sketch the graph shown in Figure 3.1 (page 140).

FIGURE 3.1

x	$P(x)$
-2	-23
-1	-4
0	1
1	-2
2	-7
3	-8
4	1
5	26

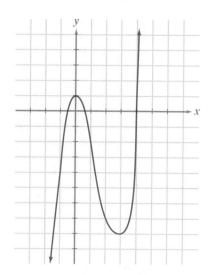

From the graph of $P(x) = x^3 - 4x^2 + 1$ in Figure 3.1, it appears that equation (1) has at least one real root in each of the intervals $(-1, 0)$, $(0, 1)$, and $(3, 4)$. The following theorem confirms our suspicions:

Intermediate Value Theorem for Polynomials

Let $P(x)$ be a polynomial and let a and b be real numbers such that $a < b$. If $P(a)$ and $P(b)$ have opposite signs, the equation $P(x) = 0$ has at least one real root in the interval (a, b).

The intermediate value theorem for polynomials enables us to locate each of the real roots of a polynomial equation within an interval on the x-axis. We can then approximate each root to any degree of accuracy by using a procedure called the **method of successive approximations**. We demonstrate this procedure by approximating one of the roots of equation (1).

EXAMPLE 18

Approximate the largest real root of the equation $x^3 - 4x^2 + 1 = 0$ to the nearest tenth.

Solution　From the previous discussion, we know that the largest real root of the equation

$$x^3 - 4x^2 + 1 = 0$$

lies in the interval $(3, 4)$. We begin the approximation process by locating the root between successive tenths in the interval $(3, 4)$. This is done by evaluating P at the points $3.1, 3.2, \ldots$ until we observe a sign change in the value of $P(x)$. Using a calculator, we find the following (approximate) values of $P(x)$:

x	3	3.1	3.2	3.3	3.4	3.5	3.6	3.7	3.8	3.9	4
$P(x)$	-8	-7.7	-7.2	-6.6	-5.9	-5.1	-4.2	-3.1	-1.9	-0.5	1

Since $P(3.9)$ and $P(4)$ have opposite signs, we know, by the intermediate value theorem for polynomials, that the root we seek lies in the interval $(3.9, 4)$. By evaluating $P(x)$ at 3.95, the midpoint of this interval, we can approximate this root to the nearest tenth. Since $P(3.95) \approx 0.2$ and $P(3.9) \approx -0.5$ have opposite signs, the root must lie in the interval $(3.9, 3.95)$. Hence the largest root of the equation, approximated to the nearest tenth, is 3.9.

If we wished to approximate this root with a greater degree of accuracy, say to the nearest hundredth, we would proceed as follows:

1. Evaluate $P(x)$ at the points 3.91, 3.92, . . . until a sign change occurs in the value of $P(x)$. This locates the root in an interval between successive hundredths. For convenience, call this interval (a, b).
2. Evaluate $P(x)$ at the midpoint m of the interval (a, b).
3. If $P(a)$ and $P(m)$ have opposite signs, $x = a$ is the desired approximation. If $P(a)$ and $P(m)$ agree in sign, choose $x = b$.

Bounds for Real Roots of Polynomial Equations

In Example 18, we used the graph of $P(x) = x^3 - 4x^2 + 1$ to determine that the largest real zero of this polynomial was located in the interval $(3, 4)$. As we shall see in the next section, it is not always practical to rely upon the graph of a polynomial to locate intervals containing its real zeros. We now investigate an alternate method for determining these intervals.

DEFINITION 3.3

Let $P(x)$ be a polynomial having real coefficients. If a and b are real numbers such that $a \leq r \leq b$, for all real roots r of the equation $P(x) = 0$, we say that a is a **lower bound** and b is an **upper bound** for the real roots of $P(x) = 0$.

Thus, if a is a lower bound and b is an upper bound for the real roots of a polynomial equation, then all the real roots of the equation must lie in the interval $[a, b]$. The following theorem tells us how we can use synthetic division to find upper and lower bounds for the real roots of a polynomial equation.

THEOREM 3.3

Let $P(x) = a_n x^n + a_{n-1} x^{n-1} + \cdots + a_1 x + a_0$, where $a_0, a_1, \ldots, a_n$ are real numbers and $a_n > 0$. Assume that b is a nonzero real number and that $P(x)$ is divided by $x - b$ using synthetic division.

(i) If $b > 0$ and all the numbers in the bottom row of the synthetic division are nonnegative, then b is an upper bound for the real roots of $P(x) = 0$.

(ii) If $b < 0$ and the numbers in the bottom row of the synthetic division alternate in sign (0 can be given either a plus sign or a minus sign), then b is a lower bound for the real roots of $P(x) = 0$.

Proof

(i) Assume that $P(x)$ is divided by $x - b$, using synthetic division, and that all the numbers in the bottom row of the division are nonnegative. By the division algorithm for polynomials, there exist a polynomial $Q(x)$ and a constant k such that

$$P(x) = Q(x)(x - b) + k. \qquad (2)$$

We know that k and the coefficients of $Q(x)$ are nonnegative numbers since they appear in the bottom row of the synthetic division.

To show that b is an upper bound for the real roots of $P(x) = 0$, we will show that no real number greater than b can be a root of $P(x) = 0$. Assume that r is a real number which is greater than b. Substituting $x = r$ into equation (2) gives us

$$P(r) = Q(r)(r - b) + k. \qquad (3)$$

We now make several observations:

1. Since b is positive and r is greater than b, both r and $r - b$ are positive.
2. $Q(r)$ is positive since the leading coefficient of $Q(x)$ is a_n, which is positive, and the remaining coefficients of $Q(x)$ are nonnegative.
3. The right side of equation (3) is positive since $Q(r)(r - b)$ is positive and k is nonnegative.

Since $P(r)$ is positive by our third observation, r cannot be a root of $P(x) = 0$. The proof of part (ii) is similar.

E X A M P L E   19

Find an upper bound and a lower bound for the real zeros of the polynomial $P(x) = -3x^3 - 2x + 7$.

Solution   We must find upper and lower bounds for the real roots of the polynomial equation

$$-3x^3 - 2x + 7 = 0.$$

Since Theorem 3.3 requires that $P(x)$ have a positive leading coefficient, we multiply both sides of this equation by -1, obtaining

$$3x^3 + 2x - 7 = 0. \qquad (4)$$

To search for an upper bound, we use synthetic division to divide $3x^3 + 2x - 7$ by $x - b$, where $b = 1, 2, \ldots$. When $b = 1$, we have

$$
\begin{array}{r|rrrr}
1 & 3 & 0 & 2 & -7 \\
  &   & 3 & 3 & 5 \\
\hline
  & 3 & 3 & 5 & -2
\end{array}
$$

Since a negative number appears in the bottom row, we continue with $b = 2$.

$$
\begin{array}{r|rrrr}
2 & 3 & 0 & 2 & -7 \\
  &   & 6 & 12 & 28 \\
\hline
  & 3 & 6 & 14 & 21
\end{array}
$$

Noting that no negative number appears in the bottom row, we conclude by Theorem 3.3 that 2 is an upper bound for the real roots of equation (4).

We now search for a lower bound using $b = -1, -2, \ldots$. For $b = -1$ we obtain

$$
\begin{array}{r|rrrr}
-1 & 3 & 0 & 2 & -7 \\
   &   & -3 & 3 & -5 \\
\hline
   & 3 & -3 & 5 & -12
\end{array}
$$

Since the numbers in the bottom row have alternating signs, -1 is a lower bound for the real roots of equation (4).

Thus -1 is a lower bound and 2 is an upper bound for the real zeros of $P(x) = -3x^3 - 2x + 7$, and we know that the real zeros of P (if any) must lie in the interval $[-1, 2]$. In general, if a is a lower bound and b is an upper bound for the real zeros of a polynomial P having real coefficients, then the real zeros (if any) of P must lie in the interval $[a, b]$.

Descartes' Rule of Signs

A theorem proved in 1637 by René Descartes (the French mathematician for whom the Cartesian plane is named) provides additional assistance in our search by telling us how many positive real zeros and how many negative real zeros a polynomial may have. Before we discuss this theorem, however, we must introduce some terminology.

When the terms of a polynomial having real coefficients are arranged in descending powers of the variable, we say that a **variation in sign** occurs when two successive terms have opposite signs. We ignore missing terms. For example, the polynomial

$$P(x) = 2x^5 - 3x^3 + 5x^2 - 4x - 1$$

has three variations in sign.

Descartes' Rule of Signs

Let $P(x)$ be a polynomial that has real coefficients.

(i) The number of positive real roots of $P(x) = 0$ either is equal to the number of variations in sign of $P(x)$ or is less than that number by an even integer.

(ii) The number of negative real roots of $P(x) = 0$ either is equal to the number of variations in sign of $P(-x)$ or is less than that number by an even integer.

The proof of Descartes' rule of signs is rather lengthy and will not be given here.

EXAMPLE 20

Use Descartes' rule of signs to discuss the nature of the roots of each of the following polynomial equations:

a. $x^5 - 3x^3 - 2x^2 + 1 = 0$ **b.** $2x^4 + 15x^3 + 60x - 32 = 0$

Solution

a. Let $P(x) = x^5 - 3x^3 - 2x^2 + 1$. Since $P(x)$ has two variations in sign, Descartes' rule of signs tells us that the equation

$$x^5 - 3x^3 - 2x^2 + 1 = 0 \qquad \qquad (5)$$

has either two or zero positive real roots.

Since $P(-x) = -x^5 + 3x^3 - 2x^2 + 1$ has three variations in sign, equation (5) above has either three or one negative real roots.

By Theorem 3.2, we know that equation (5) has exactly five roots (counting multiplicities). Noting that 0 is not a root (since $P(0) = 1$), we summarize the possibilities for the five roots in the table below.

Number of Positive Real Roots	Number of Negative Real Roots	Number of Nonreal Complex Roots
2	3	0
2	1	2
0	3	2
0	1	4

b. Since each of

$$P(x) = 2x^4 + 15x^3 + 60x - 32 \qquad \text{and}$$
$$P(-x) = 2x^4 - 15x^3 - 60x - 32$$

has only one variation in sign, we conclude by Descartes' rule of signs that the equation

$$2x^4 + 15x^3 + 60x - 32 = 0$$

has one positive real root and one negative real root. This means that the remaining two roots of this fourth-degree polynomial equation must be nonreal complex numbers.

_____

By following the procedure outlined in the solutions of Examples 16 and 17 of Section 3.4, we find that the zeros of the polynomial

$$P(x) = 2x^4 + 15x^3 + 60x - 32$$

are -3, $\frac{1}{2}$, $2i$, and $-2i$. We note that the two nonreal complex zeros are conjugates. As the following theorem states, the nonreal complex zeros of any nonconstant polynomial having real coefficients will always occur in conjugate pairs. The proof is left as an exercise (see Exercise 40).

//

THEOREM 3.4

Let $P(x)$ be a polynomial of degree $n > 0$ that has real coefficients. If $z = a + bi$ is a zero of $P(x)$, then the conjugate of z, $\bar{z} = a - bi$, is also a zero of $P(x)$.

EXAMPLE 21

Find the polynomial $P(x)$ of lowest degree that satisfies the following conditions:

a. $P(x)$ has real coefficients and a leading coefficient of 1;

b. $-1, 2i$, and $3 - \sqrt{2}i$ are zeros of $P(x)$.

Solution Since $P(x)$ has real coefficients, we know by Theorem 3.4 that the conjugates of $2i$ and $3 - \sqrt{2}i$, namely $-2i$ and $3 + \sqrt{2}i$, must also be zeros of $P(x)$. By the factor theorem,

$$P(x) = (x + 1)(x - 2i)(x + 2i)(x - (3 - \sqrt{2}i))(x - (3 + \sqrt{2}i)), \quad \text{or}$$
$$P(x) = x^5 - 5x^4 + 9x^3 - 9x^2 + 20x + 44.$$

EXAMPLE 22

Use Descartes' rule of signs to show that the polynomial equation $2x^5 + 3x^3 + 4x - 12 = 0$ has exactly one real root.

Solution Since $P(x) = 2x^5 + 3x^3 + 4x - 12$ has one variation in sign, the equation $P(x) = 0$ has exactly one positive real root. Since $P(-x) = -2x^5 - 3x^3 - 4x - 12$ has no variation in sign, there are no negative real roots. From the fact that $P(0) = -12$, we know that 0 is not a root. Thus the positive root is the only real root of the equation $2x^5 + 3x^3 + 4x - 12 = 0$.

Finding the Real Roots of a Polynomial Equation

We have now developed several methods for finding the real roots of a polynomial equation with real coefficients. We conclude this section with two examples that show how we can apply these methods systematically to find these roots.

EXAMPLE 23

Find the real zeros of the polynomial $P(x) = 8x^4 + 2x^2 - 24x - 9$. Approximate irrational zeros (if any) to the nearest tenth.

Solution Since $P(x) = 8x^4 + 2x^2 - 24x - 9$ has degree 4, $P(x)$ has four zeros (counting multiplicities). To determine the nature of these roots, we turn to Descartes' rule of signs. Since each of

$$P(x) = 8x^4 + 2x^2 - 24x - 9 \quad \text{and}$$
$$P(-x) = 8x^4 + 2x^2 + 24x - 9$$

has only one variation in sign, $P(x)$ has exactly two real zeros, one positive and one negative.

We can determine upper and lower bounds for these real zeros by applying Theorem 3.3. To find an upper bound, we divide $P(x)$ by $x - 1, x - 2, \ldots$, using synthetic division.

$$
\begin{array}{r|rrrrr}
1 & 8 & 0 & 2 & -24 & -9 \\
  &   & 8 & 8 & 10 & -14 \\
\hline
  & 8 & 8 & 10 & -14 & -23
\end{array}
\qquad (6)
$$

$$\begin{array}{r|rrrrr}
2 & 8 & 0 & 2 & -24 & -9 \\
  &   & 16 & 32 & 68 & 88 \\
\hline
  & 8 & 16 & 34 & 44 & 79
\end{array}$$
(7)

Since synthetic division by $x - 2$ produces only nonnegative numbers in the bottom row, we know that $x = 2$ is an upper bound for the real zeros.

To find a lower bound, we divide $P(x)$ by $x + 1$, $x + 2$, . . . , using synthetic division.

$$\begin{array}{r|rrrrr}
-1 & 8 & 0 & 2 & -24 & -9 \\
   &   & -8 & 8 & -10 & 34 \\
\hline
   & 8 & -8 & 10 & -34 & 25
\end{array}$$
(8)

The alternating signs in the bottom row tell us that -1 is a lower bound for the real zeros of $P(x)$.

For this particular polynomial, our synthetic divisions provide additional information about the real zeros. For example, we see from (6), (7), and (8) and the remainder theorem that $P(1) = -23$, $P(2) = 79$, and $P(-1) = 25$. Since $P(1)$ and $P(2)$ have opposite signs, the intermediate value theorem for polynomials guarantees that the positive real zero lies in the interval $(1, 2)$. Also, since $P(-1)$ and $P(0)$ have opposite signs, the negative real zero lies in the interval $(-1, 0)$.

If either of the real zeros is a rational number, the rational root theorem tells us that it must be one of the numbers listed below.

$$\pm 1, \ \pm 3, \ \pm 9, \ \pm \frac{1}{2}, \ \pm \frac{1}{4}, \ \pm \frac{1}{8}, \ \pm \frac{3}{2}, \ \pm \frac{3}{4}, \ \pm \frac{3}{8}, \ \pm \frac{9}{2}, \ \pm \frac{9}{4}, \ \pm \frac{9}{8}$$

Since the real zeros must lie in the intervals $(-1, 0)$ and $(1, 2)$, the list narrows to

$$-\frac{1}{2}, \ -\frac{1}{4}, \ -\frac{1}{8}, \ -\frac{3}{4}, \ -\frac{3}{8}, \frac{3}{2}, \frac{9}{8}.$$

Testing each of these zeros using synthetic division, we find that $\frac{3}{2}$ is the only rational zero.

We now know that the remaining real zero is an irrational number in the interval $(-1, 0)$. As in Example 18, we approximate this zero by evaluating $P(x) = 8x^4 + 2x^2 - 24x - 9$ at the points

$$-1, \ -0.9, \ -0.8, \ \cdots$$

until we observe a sign change in the value of $P(x)$.

x	-1	-0.9	-0.8	-0.7	-0.6	-0.5	-0.4	-0.3
$P(x)$	25	19.47	14.76	10.70	7.16	4	1.12	-1.55

From our table, we see that the zero lies in the interval $(-0.4, -0.3)$. Evaluating $P(x)$ at the midpoint of this interval, we find that $P(-0.35) = -0.2$. Since $P(-0.4)$ and $P(-0.35)$ have opposite signs, the zero must be closer to -0.4.

Thus the two real zeros of $P(x) = 8x^4 + 2x^2 - 24x - 9$ are -0.4 (rounded to the nearest tenth) and $\frac{3}{2}$.

EXAMPLE 24

Find all real roots (if any) of the polynomial equation $3x^3 - 8x^2 - 14x + 12 = 0$. Approximate irrational roots (if any) to the nearest tenth.

Solution Since $P(x) = 3x^3 - 8x^2 - 14x + 12$ has two variations in sign, the equation $P(x) = 0$ has either two or zero positive real roots. The single variation in sign in $P(-x) = -3x^3 - 8x^2 + 14x + 12$ means that $P(x) = 0$ has only one negative real root. Since 0 is not a root, the nature of the roots of $P(x) = 0$ is summarized in the table below.

Number of Positive Real Roots	Number of Negative Real Roots	Number of Nonreal Complex Roots
2	1	0
0	1	2

Next we apply Theorem 3.3 to find upper and lower bounds for the real roots. The synthetic division used to determine an upper bound is shown below.

$$
\begin{array}{r|rrrr}
1 & 3 & -8 & -14 & 12 \\
  &   & 3 & -5 & -19 \\
\hline
  & 3 & -5 & -19 & -7 \\
\end{array}
$$

$$
\begin{array}{r|rrrr}
2 & 3 & -8 & -14 & 12 \\
  &   & 6 & -4 & -36 \\
\hline
  & 3 & -2 & -18 & -24 \\
\end{array}
$$

$$
\begin{array}{r|rrrr}
3 & 3 & -8 & -14 & 12 \\
  &   & 9 & 3 & -33 \\
\hline
  & 3 & 1 & -11 & -21 \\
\end{array}
$$

$$
\begin{array}{r|rrrr}
4 & 3 & -8 & -14 & 12 \\
  &   & 12 & 16 & 8 \\
\hline
  & 3 & 4 & 2 & 20 \\
\end{array}
$$

Thus 4 is an upper bound, and we begin our search for a lower bound.

$$
\begin{array}{r|rrrr}
-1 & 3 & -8 & -14 & 12 \\
   &   & -3 & 11 & 3 \\
\hline
   & 3 & -11 & -3 & 15 \\
\end{array}
$$

$$
\begin{array}{r|rrrr}
-2 & 3 & -8 & -14 & 12 \\
   &   & -6 & 28 & -28 \\
\hline
   & 3 & -14 & 14 & -16 \\
\end{array}
$$

Since -2 is a lower bound, we know that the real roots must lie in the interval $(-2, 4)$.

As in the previous example, the synthetic division provides us with additional information. The values of $P(x)$ determined by the synthetic division, together with $P(0) = 12$, are shown below.

x	-2	-1	0	1	2	3	4
$P(x)$	-16	15	12	-7	-24	-19	20

The sign changes in $P(x)$ tell us that the negative real root lies in the interval $(-2, -1)$, while the two positive real roots lie in the intervals $(0, 1)$ and $(3, 4)$.

Next we apply the rational root theorem to determine whether or not the equation

$$3x^3 - 8x^2 - 14x + 12 = 0 \qquad (9)$$

has any rational roots. Since the constant term is 12 and the leading coefficient is 3, we have the following list of possible rational roots:

$$\pm 1, \ \pm 2, \ \pm 3, \ \pm 4, \ \pm 6, \ \pm 12, \ \pm\frac{1}{3}, \ \pm\frac{2}{3}, \ \pm\frac{4}{3}.$$

Knowing that the real roots must come from the intervals $(-2, -1)$, $(0, 1)$, and $(3, 4)$, narrows this list to

$$\frac{1}{3}, \frac{2}{3}, \ -\frac{4}{3}.$$

Using synthetic division, we find that $\frac{2}{3}$ is a root.

$$
\begin{array}{r|rrrr}
\frac{2}{3} & 3 & -8 & -14 & 12 \\
  &   & 2 & -4 & -12 \\
\hline
  & 3 & -6 & -18 & 0
\end{array}
$$

From the synthetic division, we obtain the reduced equation

$$3x^2 - 6x - 18 = 0.$$

Using the quadratic formula, we find that the solutions of the equivalent equation

$$x^2 - 2x - 6 = 0$$

are $x = 1 \pm \sqrt{7}$. Thus the three solutions of equation (9) above are $\frac{2}{3}$, $1 - \sqrt{7} \approx -1.6$, and $1 + \sqrt{7} \approx 3.6$.

EXERCISES 3.5

In Exercises 1–8, use Descartes' rule of signs to determine as much information as possible about the roots of each polynomial equation. Display the information in a table like the one in part (a) of Example 20 in this section.

1. $x^3 - 2x^2 - x - 4 = 0$
2. $4x^3 + 8x^2 - 11x - 15 = 0$
3. $x^5 - x^4 + x^3 - x^2 + x - 1 = 0$
4. $3x^4 - 4x^3 - x^2 + 8x - 2 = 0$

5. $2x^5 - x^3 + x^2 + 8 = 0$
6. $x^7 + 3x^3 + 12 = 0$
7. $\frac{1}{2}x^3 - 2\sqrt{2}x + 4 = 0$
8. $(-\sqrt{3}/2)x^8 + \frac{1}{2}x^4 - 4x - 2 = 0$

In Exercises 9–16, determine upper and lower bounds for the real zeros of each polynomial.

9. $P(x) = 2x^3 - 17x^2 + 12x - 2$

10. $P(x) = 5x^4 - 10x - 12$

11. $P(x) = 2x^4 - 3x^2 + 32$

12. $P(x) = 4x^3 - 10x^2 - 25x - 40$

13. $P(x) = 2x^5 - 5x^4 - 2x + 5$

14. $P(x) = x^4 - 6x^3 + 8x^2 + 2x - 1$

15. $P(x) = (\sqrt{2}/2)x^3 - 2x + 4$

16. $P(x) = (\sqrt{3}/2)x^5 - \frac{1}{2}x^2 + 3x + 16$

17. Without sketching the curve, show that the graph of $y = 2x^5 + 6x^3 + 3x - 1$ crosses the x-axis exactly once.

18. Without sketching the curve, show that the graph of $y = 3x^4 + 8x^2 + 1$ never crosses the x-axis.

In Exercises 19–26, find a polynomial of lowest degree with real coefficients that has the given numbers as zeros.

19. $-2, 2, i$

20. $0, 3 + i$

21. $-\frac{3}{2}, 4, -1 - i$

22. $\pm\sqrt{3}/2, i, -\sqrt{5}i$

23. $-\frac{1}{2}i, 1 + 4i, 0$

24. $\pm\sqrt{2}/2, 3 - 5i, \frac{5}{2}$

25. $(1 + \sqrt{3}i)/2, \sqrt{3}/2$

26. $(2 - \sqrt{10}i)/3, -1 + 3i$

In Exercises 27–32, find the real roots of each polynomial equation. Approximate irrational roots (if any) to the nearest tenth.

27. $2x^3 - 5x^2 + 1 = 0$

28. $2x^3 - 5x^2 + 6x - 3 = 0$

29. $x^4 - x^3 - 2 = 0$

30. $-3x^4 - x^3 + 2x^2 + 4 = 0$

31. $3x^5 + 2x + 9 = 0$

32. $2x^5 + x^3 - 8 = 0$

33. Approximate the irrational number $\sqrt[4]{7}$ to the nearest hundredth using the fact that $\sqrt[4]{7}$ is the positive real root of the equation $x^4 - 7 = 0$. Compare your result to the value given by your calculator for $\sqrt[4]{7}$.

34. Approximate $\sqrt[5]{12}$ to the nearest hundredth. [*Hint*: See Exercise 33.]

35. Show that i is a root of the polynomial equation $3x^4 - x^3 + 2ix^2 + x - 3 = 0$. Show, also, that $-i$, the conjugate of i, is *not* a solution of this equation. Is this a contradiction to Theorem 3.4? Explain.

36. Explain why every polynomial of odd degree that has real coefficients must have at least one real zero.

37. Use Descartes' rule of signs to show that the equation $x^3 - x^2 + 5x + 3 = 0$ has no positive real roots. [*Hint*: Multiply both sides of this equation by $x + 1$ and apply Descartes' rule to the resulting equation.]

38. Show that the equation $x^4 + x^3 + 2x^2 - 1 = 0$ has exactly one negative real root. [*Hint*: See Exercise 37.]

39. The cost C of producing compact discs is given by $C(x) = x^3 + 12x - 16$, where cost is measured in thousands of dollars and x (measured in hundreds) is the number of discs produced. If the revenue from the sale of these discs is expressed by $R(x) = 14x - 10$, in thousands of dollars, find the number of discs the company must manufacture and sell just to break even. Express your answer to the nearest ten discs.

40. Prove Theorem 3.4. [*Hint*: See Exercise 76 in Section 1.4.]

41. The displacement (in centimeters) at time t (in seconds) of a particle moving along a straight line is given by $s(t) = 2t^4 + 3t^3 + 6t^2 - t - 15$. At what time t ($t > 0$) is the displacement equal to 0? Express your answer to the nearest tenth of a second.

42. Using laws of physics, it can be shown that a spherical buoy of radius r centimeters and specific gravity s ($0 < s < 1$) will sink in water to a depth of x centimeters, as determined by the equation $x^3 - 2rx^2 - 4r^3s = 0$. How far will a styrofoam buoy of radius 2 centimeters and specific gravity 0.25 sink in water? Express your answer to the nearest tenth of a centimeter.

43. The following theorem is named for a student of Gauss: **Eisenstein's criteria**: Let $P(x) = a_nx^n + a_{n-1}x^{n-1} + \cdots + a_1x + a_0$ be a polynomial of degree $n > 0$ having integer coefficients. Assume that there exists a prime p satisfying:

p is a factor of $a_0, a_1, \ldots, a_{n-1}$;

p is not a factor of a_n; and

p^2 is not a factor of a_0.

Then $P(x)$ cannot be factored as the product of two polynomials having integer coefficients and positive degree. Show that each of the following polynomials satisfies the hypothesis of Eisenstein's criteria:

a. $P(x) = x^4 - 3x^3 + 6x^2 - 12$

b. $P(x) = 5x^3 + 12x^2 - 6x + 2$

44. Let $P(x) = c_1x^4 + c_2x^2 + c_3x - c_4$, where c_1, c_2, c_3, and c_4 are positive real numbers. Show that the equation $P(x) = 0$ has exactly two nonreal roots.

45. Let $A = (x, y)$ be the point in the first quadrant where the circle $x^2 + y^2 = 4$ intersects the graph of the equation $y = x^3$. Approximate the x-coordinate of A to the nearest tenth.

46. Let $A = (x, y)$ be the point in the first quadrant where the curves $y = x^2 + 2x + 3$ and $y = x^3$ intersect. Approximate the x-coordinate of A to the nearest tenth.

In calculus, we find that polynomial functions can often be used to approximate other functions. One such approximating polynomial is called a Taylor polynomial in honor of English mathematician Brook Taylor (1685–1731). In Exercises 47–49, three functions are approximated by Taylor polynomials. Evaluate each polynomial at $x = \frac{1}{2}$ and compare your result to the actual function value. Round each result to the nearest thousandth.

47. $\sqrt{1 + x} \approx 1 + \frac{1}{2}x - \frac{1}{8}x^2 + \frac{1}{16}x^3$

48. $1/(1 + x) \approx 1 - x + x^2 - x^3$

49. $(1 + x)^{3/2} \approx 1 + \frac{3}{2}x + \frac{3}{8}x^2 - \frac{1}{16}x^3$

50. Use your graphing calculator to approximate the indicated zero of each polynomial to the nearest tenth.

(a) $y = x^3 - 5x - 1$; smallest negative real zero

(b) $y = x^4 - 4x^3 - 2x^2 + 12x - 4$; largest positive real zero

3.6   GRAPHS OF POLYNOMIAL FUNCTIONS

In Chapter 2 we discussed the graphs of first-degree and second-degree polynomial functions (linear and quadratic functions). In this section we will study the graphs of polynomial functions of degree three or higher that have real coefficients.

Consider the polynomial function f defined as follows:

$$f(x) = a_n x^n + a_{n-1} x^{n-1} + \cdots + a_1 x + a_0,$$

where $a_0, a_1, \ldots, a_n$ are real numbers. Since $f(x)$ is a real number for each x in $\mathbb{R}$, it follows that dom $(f) = \mathbb{R}$. In principle, we can obtain the graph of f by making a table of values and plotting a sufficient number of points to reveal the important features of the graph. In practice, however, such point-by-point plotting can be tedious and can easily cause us to overlook significant features of the graph. In addition, this approach would not indicate any relationships between the graphs of different polynomial functions, nor help us understand why the graphs appear as they do. Instead, we will develop a more organized and more efficient approach to the graphing process. We first introduce some terminology that will simplify matters.

> We say that a real number t is **increasing without bound** and we write $t \to \infty$ when t is positive and getting larger and larger.
>
> On the other hand, when t is negative and $|t|$ is getting larger and larger, we say that t is **decreasing without bound** and we write $t \to -\infty$.
>
> Finally, we will write $|t| \to \infty$ to indicate that t is either increasing or decreasing without bound.

To see how this terminology can be used to describe the graph of a polynomial function, consider the graphs of f and g shown in Figure 3.2.

Examining the graph of f shown in Figure 3.2(a) we see that $f(x)$ decreases without bound as x decreases without bound, and we write

$$f(x) \to -\infty \quad \text{as} \quad x \to -\infty.$$

Since it is also true that $f(x)$ is decreasing without bound as x increases without bound, we can write

$$f(x) \to -\infty \quad \text{as} \quad x \to \infty.$$

In like manner, we see from the graph of g in Figure 3.2(b) that $g(x)$ increases without bound as x decreases without bound and decreases without bound as x increases without bound. Hence we can write

$$g(x) \to \infty \quad \text{as} \quad x \to -\infty \quad \text{and}$$
$$g(x) \to -\infty \quad \text{as} \quad x \to \infty.$$

The graphs of f and g in Figure 3.2 reveal several important features that the graphs of polynomials of degree three or higher have in common:

1. The graph is a smooth, unbroken curve that has no sharp corners.
2. The graph can have **turning points**, which are points at which the graph changes from rising to falling or vice versa. We note that the graph of f has turning points at x_2, x_4, and x_6, while the graph of g has no turning points.

FIGURE 3.2

a. $y = f(x)$

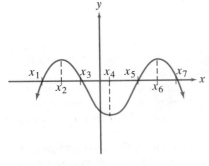

b. $y = g(x)$

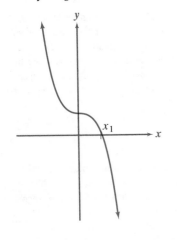

3. The graph lies entirely above the x-axis or entirely below the x-axis over the open intervals determined by adjacent real zeros of the function. For example, the graph of f lies entirely below the x-axis on $(-\infty, x_1)$, (x_3, x_5), and (x_7, ∞); and entirely above the x-axis on (x_1, x_3) and (x_5, x_7).

E X A M P L E 25

Determine which (if any) of the following graphs could represent the graph of a polynomial function:

a.

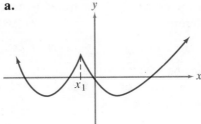

b.

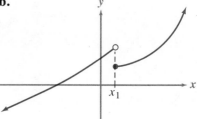

c.

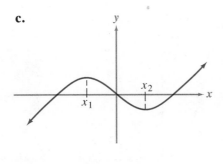

d.

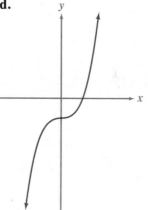

Solution   Since the graph in part (a) has a sharp corner at x_1, and the graph in part (b) has a break in it at x_1, neither can represent the graph of a polynomial function. The graphs shown in parts (c) and (d) can each represent the graph of a polynomial function. The graph in part (c) has turning points at x_1 and at x_2, while the graph in part (d) has no turning points.

The leading term of a polynomial determines the behavior of its graph as $|x| \to \infty$. This is due to the fact that, for large values of $|x|$, the leading term *dominates* the other terms of the polynomial. As an illustration, consider

$$f(x) = 4x^3 - x^2 - 8x + 12.$$

If $x \neq 0$, we can write

$$f(x) = x^3\left(4 - \frac{1}{x} - \frac{8}{x^2} + \frac{12}{x^3}\right).$$

As $|x| \rightarrow \infty$, the terms $1/x$, $8/x^2$, and $12/x^3$ are getting closer and closer to 0. This means that the quantity in parentheses, $4 - 1/x - 8/x^2 + 12/x^3$, is getting closer and closer to 4. Hence, for large values of $|x|$, the behavior of the graph of $f(x)$ is determined by its leading term, $4x^3$.

In general, the graph of

$$f(x) = a_n x^n + a_{n-1} x^{n-1} + \cdots + a_1 x + a_0, \qquad a_n \neq 0$$

will behave like the graph of

$$f(x) = a_n x^n,$$

for large values of $|x|$. We therefore direct our attention to graphs of functions of the form

$$f(x) = kx^n, \qquad k \neq 0,$$

beginning with the special case where $k = 1$.

The graphs of $y = x^2$ and $y = x^3$, which were discussed in Chapter 2, are shown in Figure 3.3.

FIGURE 3.3

a.   **b.**

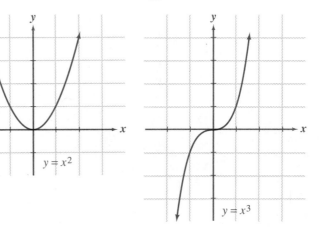

In Section 2.3 we learned that the graph of $y = x^2$ is a parabola. If n is an even integer greater than 2, the graph of $y = x^n$ (although not a parabola) is U-shaped and is symmetric with respect to the y-axis. As the value of n increases, the graphs of $y = x^n$ become *flatter* when x is between -1 and 1 and *steeper* when x is less than -1 or greater than 1. See Figure 3.4.

For odd integers $n > 3$, the graph of $y = x^n$ is symmetric with respect to the origin and closely resembles the graph of $y = x^3$. As was the case with even exponents, the graphs of $y = x^n$ become flatter when $|x| < 1$ and steeper when $|x| > 1$, as the exponent n gets larger and larger. See Figure 3.5.

If k is a positive real number, the graph of $f(x) = kx^n$ closely resembles the graph of $f(x) = x^n$, while the graph of $f(x) = -kx^n$ is simply the graph of $f(x) = kx^n$ reflected across the x-axis. Thus for any $k \neq 0$, the graph of $f(x) = kx^n$ has one of the forms shown in Figure 3.6.

FIGURE 3.4

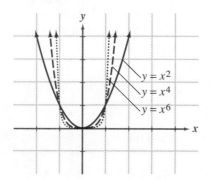

FIGURE 3.5

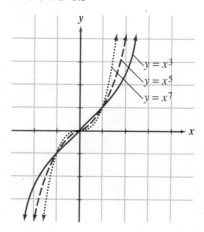

FIGURE 3.6

a. $f(x) = kx^n; k > 0$
   n even ($n \geq 2$)

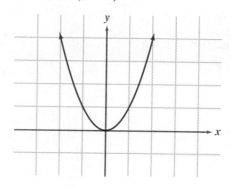

b. $f(x) = kx^n; k < 0$
   n even ($n \geq 2$)

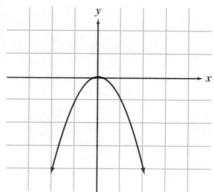

c. $f(x) = kx^n; k > 0$
   n odd ($n \geq 3$)

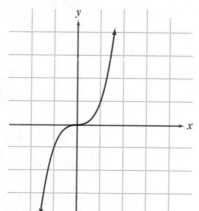

d. $f(x) = kx^n; k < 0$
   n odd ($n \geq 3$)

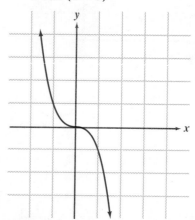

Since the graph of $f(x) = a_n x^n + a_{n-1} x^{n-1} + \cdots + a_1 x + a_0$ resembles the graph of $f(x) = a_n x^n$ as $|x| \to \infty$, the behavior of the graph of f for large values of $|x|$ can be described by one of the pictures shown in Figure 3.7.

FIGURE 3.7
Behavior of the graph of
$f(x) = a_n x^n + a_{n-1} x^{n-1} + \cdots + a_1 x + a_0$
for large values of $|x|$

a. n even and $n \geq 3$; $a_n > 0$

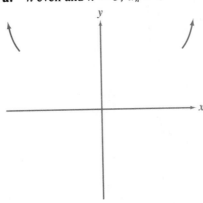

b. n even and $n \geq 2$; $a_n < 0$

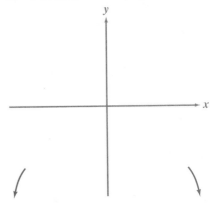

c. n odd and $n \geq 3$; $a_n > 0$

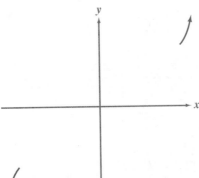

d. n odd and $n \geq 3$; $a_n < 0$

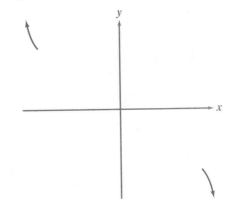

EXAMPLE 26

Which of these graphs could represent the graph of $y = -\frac{1}{2}x^5 + 3x^3 - 4x^2 + 13$?

a.

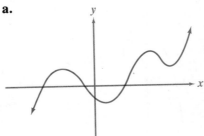

b.

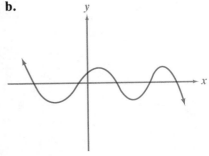

c.

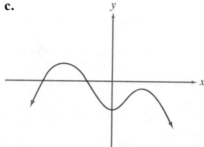

d.

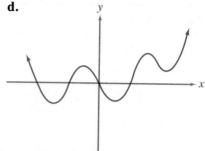

Solution   Since *f* has odd degree and a negative leading coefficient, the behavior of its graph for large values of |*x*| can be described by part (d) of Figure 3.7. Hence graph (b) is the only possibility.

EXAMPLE 27

Sketch the graph of $f(x) = x(x - 1)(x + 3)^2$.

Solution   Since $f(x)$ is in factored form, we can easily see that its real zeros are 0, 1, and -3 (multiplicity 2). As noted earlier in this section, the graph of *f* lies either entirely above or entirely below the *x*-axis between adjacent real zeros. Using the real zeros to divide the *x*-axis into intervals, we investigate the behavior of *f* on each of these intervals.

Interval	Test Point	Value of *f* at Test Point	Graph of *f* Lies
$(-\infty, -3)$	-4	20	above the *x*-axis
$(-3, 0)$	-1	8	above the *x*-axis
$(0, 1)$	$\frac{1}{2}$	$-\frac{49}{16}$	below the *x*-axis
$(1, \infty)$	2	50	above the *x*-axis

Since $f(x) = x(x - 1)(x + 3)^2 = x^4 + 5x^3 + 3x^2 - 9x$ has even degree and a positive leading coefficient, its graph will resemble Figure 3.7(a), for large values of |*x*|. This means that the values of $f(x)$ are increasing without bound as $x \to \infty$ and as $x \to -\infty$.

Using the information obtained thus far and plotting a few additional points, we obtain the graph of *f* shown in Figure 3.8. It is important to note that the turning points shown on our graph are merely educated guesses. The exact location of turning points cannot be determined without the aid of calculus.

FIGURE 3.8

x	-4	-3	-2	-1	0	$\frac{1}{2}$	1	2
f(x)	20	0	6	8	0	$-\frac{49}{16}$	0	50

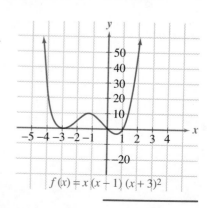

E X A M P L E · 28

Sketch the graph of $f(x) = 3x^3 - 8x^2 - 14x + 12$.

Solution   In Example 24 of Section 3.5, we found that $\frac{2}{3}$, $1 - \sqrt{7} \approx -1.6$, and $1 + \sqrt{7} \approx 3.6$ are the real zeros of f. In the accompanying table we investigate the behavior of f on the intervals of the x-axis determined by these zeros.

Interval	Test Point	Value of f at Test Point	Graph of f Lies
$(-\infty, 1 - \sqrt{7})$	-2	-16	below the x-axis
$(1 - \sqrt{7}, \frac{2}{3})$	0	12	above the x-axis
$(\frac{2}{3}, 1 + \sqrt{7})$	1	-7	below the x-axis
$(1 + \sqrt{7}, \infty)$	4	20	above the x-axis

Since f has odd degree and a positive leading coefficient, its graph resembles Figure 3.7(c) on page 154 for large values of $|x|$. Thus $f(x) \to -\infty$ as $x \to -\infty$, and $f(x) \to \infty$ as $x \to \infty$.

Plotting a few additional points, we obtain the graph of f shown in Figure 3.9.

FIGURE 3.9

x	-2	$1 - \sqrt{7}$	-1	0	1	2	3	$1 + \sqrt{7}$	4
$f(x)$	16	0	15	12	-7	-24	-21	0	20

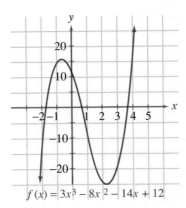

$f(x) = 3x^3 - 8x^2 - 14x + 12$

E X A M P L E 29

Sketch the graph of $f(x) = -3x^5 + 17x^4 - 24x^3 + 8x^2 - 21x - 9$.

Solution   Using the methods of Section 3.5, we find that the real zeros of f are $-\frac{1}{3}$ and 3 (a double root). The remaining two zeros are nonreal complex numbers. We use the real zeros to divide the x-axis into intervals and investigate the behavior of the graph on each interval.

Interval	Test Point	Value of f at Test Point	Graph of f Lies
$(-\infty, -\frac{1}{3})$	-1	64	above the x-axis
$(-\frac{1}{3}, 3)$	0	-9	below the x-axis
$(3, \infty)$	4	-221	below the x-axis

Since f has a negative leading coefficient and odd degree, its graph resembles Figure 3.7(d) on page 154 when $|x|$ is large. Thus the values of $f(x)$ are increasing without bound as $x \to -\infty$, and decreasing without bound as $x \to \infty$.

After determining some additional points, we sketch the graph of f as shown in Figure 3.10.

FIGURE 3.10

x	-2	-1	0	1	2	3	4
$f(x)$	625	64	-9	-32	-35	0	-221

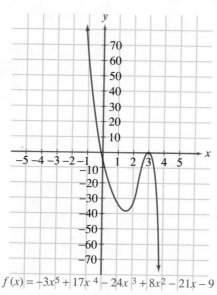

$$f(x) = -3x^5 + 17x^4 - 24x^3 + 8x^2 - 21x - 9$$

EXERCISES 3.6

In Exercises 1–8, determine which (if any) of the graphs shown represent the graph of a polynomial function.

1.

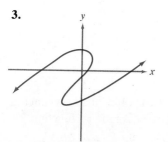

2.

5.

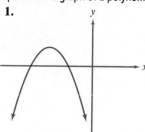

6.

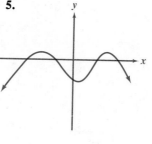

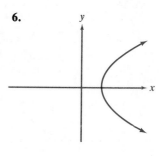

3.

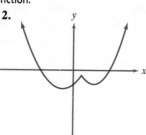

4.

7.

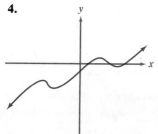

8.

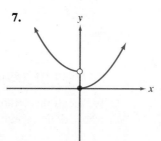

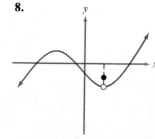

In Exercises 9–26, sketch the graph of each polynomial function.

9. $f(x) = -x^3$
10. $f(x) = \frac{1}{2}x^4$
11. $f(x) = 5x^2$
12. $f(x) = 12x^3$
13. $f(x) = -2x^5/3$
14. $f(x) = x^8$
15. $f(x) = 3x^6$
16. $f(x) = -\frac{1}{4}x^5$
17. $f(x) = 1 - x^4$
18. $f(x) = x^3 + 2$
19. $f(x) = \frac{1}{2}x^3 - 3$
20. $f(x) = (x - 4)^3$
21. $f(x) = (x + 1)^2 + 2$
22. $f(x) = 1 + (x - \frac{3}{2})^2$
23. $f(x) = 3 - (x + \frac{5}{2})^5$
24. $f(x) = 2 - (2 - x)^3$
25. $f(x) = \frac{4}{5}(x - 3)^3$
26. $f(x) = \frac{2}{3}(x - 4)^5$

In Exercises 27–32, match each polynomial function with the graph [(a) through (f)] that best represents it.

27. $f(x) = x^3 - x^2 + 1$
28. $f(x) = 2x^4 - x^2 - 2$
29. $f(x) = -2x^3 + 3x - 2$
30. $f(x) = x^5 + 4x^4 - 3x^3 - 18x^2$
31. $f(x) = x^6 - 7x^4 + 12x^2$
32. $f(x) = -x^6 + x^4 - x^2 + x$

(a)

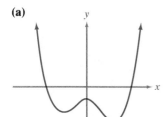

(b)

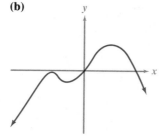

(c)

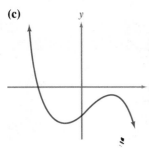

(d)

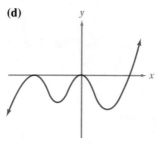

(e)

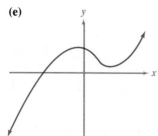

(f)

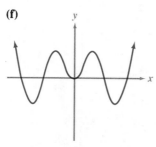

In Exercises 33–52, sketch the graph of each polynomial function.

33. $y = (x + 3)(x^2 - 1)$
34. $y = (x^2 - 9)(x + 1)$
35. $y = -2(x^2 - 1)(2x + 5)$
36. $y = \frac{1}{2}(2x - 3)(x + 2)$
37. $y = x(x^2 - 4)(x^2 + 2)$
38. $y = -x(x + 4)(x^2 - 4)$
39. $y = \frac{1}{4}(x^2 - 9)(x + 5)(x - 1)$
40. $y = \frac{1}{3}(x^2 + 2)(x^2 - 1)(2x - 5)$
41. $y = x^3 + 4x^2 - 4x - 16$
42. $y = 2x^3 - x^2 - 10x + 5$
43. $y = x^4 - 9x^2 + 8$
44. $y = x^5 + x^3 - 6x$
45. $y = -2x^3 - 8x^2 + 3x + 12$
46. $y = -x^4 - 3x^2 + 4$
47. $y = x^4 - 2x^3 - x^2 + 6x - 8$
48. $y = x^4 - x^3 - x - 6$
49. $y = -x^4 + 4x^3 - 3x^2 - x + 3$
50. $y = -3x^4 + 4x^3 + x^2 + 4x + 4$
51. $y = \frac{1}{2}x^5 - \frac{3}{2}x^3 - 2x$
52. $y = \frac{3}{2}x^5 - 6x^2 - 12x$

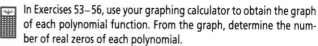

In Exercises 53–56, use your graphing calculator to obtain the graph of each polynomial function. From the graph, determine the number of real zeros of each polynomial.

53. $y = -x^4 + 3x^2 + 5$ **54.** $y = 1.2x^3 - 3.5x + 4$
55. $y = x^5 - 12x^2 + 16$ **56.** $y = -x^6 - x^3 + 8$

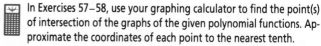

In Exercises 57–58, use your graphing calculator to find the point(s) of intersection of the graphs of the given polynomial functions. Approximate the coordinates of each point to the nearest tenth.

57. $y = 2x^3 - 3x + 1$ and $y = -x^3 + x^2 - 5$
58. $y = 3x^4 - 8x^2 - 4x - 2$ and $y = -2x^2 + 3x + 6$

In Exercises 59–60, use your graphing calculator to find the intervals on which each polynomial function is increasing and the intervals on which each is decreasing.

59. $y = \frac{1}{3}x^3 - x^2 - 3x + 4$
60. $y = -2x^3 + 3x^2 + 12x - 7$

3.7 GRAPHS OF RATIONAL FUNCTIONS

We recall that a rational expression is a quotient of two polynomials. A function whose defining equation is a rational expression is called a *rational function*.

> **DEFINITION 3.4   Definition of a Rational Function**
>
> A **rational function** is any function f whose rule has the form
>
> $$f(x) = \frac{P(x)}{Q(x)}$$
>
> where $P(x)$ and $Q(x)$ are polynomials and $Q(x) \neq 0$.

Unless otherwise specified, we will assume in this section that the rational function $f(x) = P(x)/Q(x)$ is in lowest terms and that $P(x)$ has positive degree. The domain of f is the set of all real numbers x such that $Q(x) \neq 0$ and the zeros of $f(x)$, if any, are the zeros of the polynomial $P(x)$.

Vertical and Horizontal Asymptotes

Unlike polynomial functions, whose graphs are smooth unbroken curves, the graphs of rational functions may have one or more breaks in them. Since the breaks (if any) in the graph of $f(x) = P(x)/Q(x)$ occur at the real zeros of $Q(x)$, we must pay close attention to the behavior of the graph near these zeros.

EXAMPLE 30

Sketch the graph of $f(x) = 1/(x - 3)$.

Solution   Noting that f is defined for all real numbers except $x = 3$, we first investigate the behavior of the graph for values of x that are close to 3. We begin by considering values of x that get closer and closer to 3, but remain larger than 3. We say that these values of x **approach 3 from the right**, and we write $x \to 3^+$. In Table 3.1, we see that the values of $f(x)$ increase without bound as x approaches 3 from the right. We denote this fact by writing

$$f(x) \to \infty, \quad \text{as} \quad x \to 3^+.$$

TABLE 3.1

x	4	3.1	3.01	3.001	3.0001	3.00001	3.000001
$f(x)$	1	10	100	1000	10,000	100,000	1,000,000

Now let us consider the behavior of $f(x)$ for values of x that get closer and closer to 3, but remain smaller than 3. We say that these values of x are **approaching 3 from the left**, and we write $x \to 3^-$. As we see from Table 3.2, the values of $f(x)$ are decreasing without bound as x approaches 3 from the left. We use the following notation to indicate this fact:

$$f(x) \to -\infty, \quad \text{as} \quad x \to 3^-.$$

TABLE 3.2

x	2	2.5	2.9	2.99	2.999	2.9999	2.99999	2.999999
$f(x)$	-1	-2	-10	-100	-1000	$-10,000$	$-100,000$	$-1,000,000$

From Tables 3.1 and 3.2, we obtain the portion of the graph of $f(x) = 1/(x - 3)$ shown in Figure 3.11 on page 160.

FIGURE 3.11
Behavior of $f(x) = 1/(x - 3)$, near $x = 3$

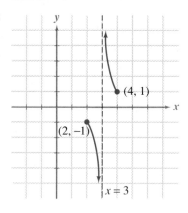

The line $x = 3$, shown dashed in Figure 3.11, is called a **vertical asymptote** of the graph. Although this vertical asymptote is not actually part of the graph, it is drawn to help us describe the behavior of the graph near $x = 3$. We note that the graph of f gets closer and closer to the asymptote, without actually touching it, as the values of x get closer and closer to 3 from either side.

Next we investigate the behavior of $f(x)$ for large values of $|x|$. We see from Table 3.3(a), that the values of $f(x)$ remain positive but get closer and closer to 0 as x increases without bound. We express this result symbolically as

$$f(x) \to 0, \quad \text{as} \quad x \to \infty.$$

On the other hand, we see from Table 3.3(b) that the values of $f(x)$ are negative and getting closer and closer to 0 as x decreases without bound. We indicate this relationship by writing

$$f(x) \to 0, \quad \text{as} \quad x \to -\infty.$$

Tables 3.3(a) and (b) indicate the behavior of the graph of $f(x) = 1/(x - 3)$ for large values of $|x|$. See Figure 3.12.

TABLE 3.3

a.

x	13	103	1003	10,003	100,003	1,000,003
$f(x)$	0.1	0.01	0.001	0.0001	0.00001	0.000001

b.

x	-13	-103	-1003	$-10,003$	$-100,003$	$-1,000,003$
$f(x)$	-0.1	-0.01	-0.001	-0.0001	-0.00001	-0.000001

FIGURE 3.12
Behavior of $f(x) = 1/(x - 3)$ for large values of $|x|$

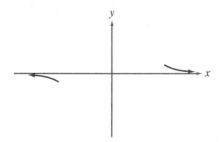

We see from Figure 3.12 that the graph of $f(x) = 1/(x - 3)$ is getting closer and closer to the x-axis ($y = 0$) as $|x| \to \infty$. We call the line $y = 0$ a **horizontal asymptote** of the graph of f. Since the numerator of $f(x) = 1/(x - 3)$ cannot equal 0, f has no real zeros. This means that the graph of f never crosses the x-axis. To obtain the graph shown in Figure 3.13, we use Figures 3.11 and 3.12 and plot the additional points shown in the accompanying table.

x	-10	-5	-1	0	1	5	10
$f(x)$	$-\frac{1}{13}$	$-\frac{1}{8}$	$-\frac{1}{4}$	$-\frac{1}{3}$	$-\frac{1}{2}$	$\frac{1}{2}$	$\frac{1}{7}$

It is worth noting that the graph of $f(x) = 1/(x - 3)$ can be obtained by translating the graph of $f(x) = 1/x$ three units to the right.

As we see in the following definition, two important ideas may be drawn from the solution of Example 30:

FIGURE 3.13

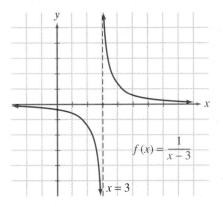

$f(x) = \dfrac{1}{x - 3}$

$x = 3$

DEFINITION 3.5   Vertical and Horizontal Asymptotes

Let f be a function and let a and b be real numbers.

(i)   The line $x = a$ is called a **vertical asymptote** of the graph of f provided

$$|f(x)| \to \infty, \quad \text{as} \quad x \to a^- \text{ or as } x \to a^+.$$

(ii)   The line $y = b$ is called a **horizontal asymptote** of the graph of f provided

$$f(x) \to b, \quad \text{as} \quad x \to -\infty \text{ or as } x \to \infty.$$

EXAMPLE 31

Write the equation of each asymptote shown below.

a.

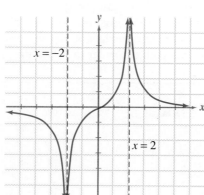

b.

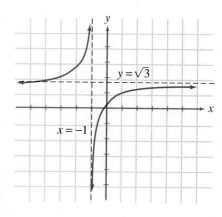

Solution

a. The lines $x = -2$ and $x = 2$ are vertical asymptotes, while $y = 0$ is a horizontal asymptote.

b. The line $x = -1$ is a vertical asymptote and $y = \sqrt{3}$ is a horizontal asymptote.

If $f(x) = P(x)/Q(x)$ is a rational function in lowest terms, and if a is any real zero of $Q(x)$, it can be shown that the line $x = a$ is a vertical asymptote of the graph of f. This means that we can locate the vertical asymptotes (if any) of the graph of f by finding the real roots of $Q(x) = 0$. The graph of f can have at most one horizontal asymptote, which can be found by using Definition 3.5.

EXAMPLE 32

Sketch the graph of $f(x) = (x^2 - 5x)/(x^2 - 4)$.

Solution   Since $x^2 - 4 = 0$ when $x = \pm 2$, the domain of f consists of all real numbers except -2 and 2. From the discussion preceding this example, we know that $x = -2$ and $x = 2$ are vertical asymptotes of the graph of f. Using Definition 3.5, we can determine whether or not the graph has a horizontal asymptote by determining the behavior of $f(x)$ for large values of $|x|$. As we shall see, we can simplify our work by dividing both numerator and denominator by the highest power of x appearing in the denominator. This gives us

$$f(x) = \frac{1 - \dfrac{5}{x}}{1 - \dfrac{4}{x^2}}.$$

Since the terms $5/x$ and $4/x^2$ each approach 0 for large values of $|x|$, we see that the values of $f(x)$ are approaching 1 as $|x| \to \infty$. This means that $y = 1$ is a horizontal asymptote of the graph.

The real zeros of f are the real zeros of the numerator $x^2 - 5x = x(x - 5)$, which are 0 and 5.

As we learned in Section 3.6, the graph of a polynomial lies entirely above or entirely below the x-axis between adjacent real zeros. In a similar manner, the graph of a rational function lies entirely above or entirely below the x-axis on the intervals of the x-axis determined by the real zeros of its numerator and denominator. For $f(x) = (x^2 - 5x)/(x^2 - 4)$, these intervals are $(-\infty, -2)$, $(-2, 0)$, $(0, 2)$, $(2, 5)$, and $(5, \infty)$.

From the information provided by Table 3.4, and our knowledge of the behavior of the graph near the vertical asymptotes, we need only plot a few additional points in order to sketch the graph of f. See Figure 3.14.

TABLE 3.4

Interval	Test Point	Value of f at Test Point	Graph of f Lies
$(-\infty, -2)$	-3	24	above the x-axis
$(-2, 0)$	-1	-2	below the x-axis
$(0, 2)$	1	$\frac{4}{3}$	above the x-axis
$(2, 5)$	3	$-\frac{6}{5}$	below the x-axis
$(5, \infty)$	6	$\frac{3}{16}$	above the x-axis

FIGURE 3.14

x	-6	-5	-4	-3	-1	0	1	3	4	5	6	7	8
$f(x)$	$\frac{33}{16}$	$\frac{50}{21}$	-3	24	-2	0	$\frac{4}{3}$	$-\frac{6}{5}$	$\frac{1}{3}$	0	$\frac{3}{16}$	$\frac{14}{45}$	$\frac{6}{15}$

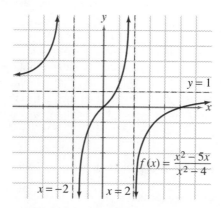

$$y = 1$$

$$f(x) = \frac{x^2 - 5x}{x^2 - 4}$$

$$x = -2 \qquad x = 2$$

We see from Figure 3.14 that the graph of $f(x) = (x^2 - 5x)/(x^2 - 4)$ crosses its horizontal asymptote $y = 1$. There is, however, no cause for alarm. Although the graph of a rational function can never cross a vertical asymptote, its graph may indeed cross a horizontal asymptote.

The method used to find the horizontal asymptote in Example 32 can be used to find the horizontal asymptote (if one exists) of any rational function. In fact, this method of finding horizontal asymptotes can be used to prove the following theorem (see Exercise 38):

THEOREM 3.5

Let

$$f(x) = \frac{a_n x^n + a_{n-1} x^{n-1} + \cdots + a_1 x + a_0}{b_m x^m + b_{m-1} x^{m-1} + \cdots + b_1 x + b_0}$$

be a rational function, where $a_0, a_1, \ldots, a_n$ and $b_0, b_1, \ldots, b_m$ are real numbers and $b_m \neq 0$.

(i)   If $n < m$, the x-axis is a horizontal asymptote of the graph of f.

(ii)   If $n = m$, the line $y = a_n/b_m$ is a horizontal asymptote of the graph of f.

(iii)   If $n > m$, the graph of f has no horizontal asymptote.

EXAMPLE 33

Determine the horizontal asymptote (if one exists) of the graph of each function.

a. $f(x) = \dfrac{1 - x^2}{(x - 2)^2}$

b. $f(x) = \dfrac{x^3 - 4x + 5}{3x^2 - 1}$

c. $f(x) = \dfrac{x^3 - 4x}{(x^2 - 1)^2}$

Solution

a.   Since $f(x) = (1 - x^2)/(x - 2)^2 = (-x^2 + 1)/(x^2 - 4x + 4)$, we see that the numerator and denominator of f have the same degree. Hence, by part (ii) of Theorem 3.5, $y = -1$ is a horizontal asymptote of the graph of f.

b.   Since the numerator of $f(x) = (x^3 - 4x + 5)/(3x^2 - 1)$ has degree 3, and the denominator has degree 2, we conclude by part (iii) of Theorem 3.5, that the graph of f has no horizontal asymptote.

c.   Since the numerator of $f(x) = (x^3 - 4x)/(x^2 - 1)^2$ $= (x^3 - 4x)/(x^4 - 2x^2 + 1)$ has lower degree than the denominator, part (i) of Theorem 3.5 tells us that the x-axis is a horizontal asymptote of the graph of f.

We are now in a position to summarize a general procedure for graphing a rational function $f(x) = P(x)/Q(x)$ where the degree of $P(x)$ is less than or equal to the degree of $Q(x)$.

Step 1.   Find the real zeros (if any) of $Q(x)$. It follows that
   a. $\text{dom}(f) = \{x \mid Q(x) \neq 0\}$
   b.   If $Q(a) = 0$, $x = a$ is a vertical asymptote of the graph of f.

Step 2.   Find the real zeros (if any) of f by solving $P(x) = 0$.

Step 3.   Use Theorem 3.5 to find the horizontal asymptote (if one exists).

Step 4.   If there is no horizontal asymptote, determine the behavior of the graph as $|x| \to \infty$. This may require plotting a few points.

Step 5.   Use the real zeros (if any) of the numerator and denominator of f to divide the x-axis into intervals. Make a table to determine the behavior of the graph over these intervals (see Example 32).

Step 6.   Plotting additional points if necessary, sketch a smooth curve that is consistent with the findings in steps 1–5 above. Remember that the only breaks in the graph occur at the vertical asymptotes and that it is possible for the graph to cross a horizontal asymptote.

EXAMPLE 34

Sketch the graph of $f(x) = \dfrac{1}{x^2 + 4x}$.

Solution   Since $1/(x^2 + 4x) = 1/x(x + 4)$, we see that $\text{dom}(f) = \{x \mid x \neq -4$ and $x \neq 0\}$ and that $x = -4$ and $x = 0$ are vertical asymptotes of the graph. Noting that the numerator of $f(x)$ is 1, we see that f has no real zeros. Since the degree of the numerator is less than the degree of the denominator, we know by part (i) of Theorem 3.5, that the x-axis is a horizontal asymptote of the graph.

Since the numerator of f has no real zeros, we investigate the behavior of the graph over the intervals determined by the zeros of the denominator.

Interval	Test Point	Value of f at Test Point	Graph of f Lies
$(-\infty, -4)$	-5	$\frac{1}{5}$	above the x-axis
$(-4, 0)$	-2	$-\frac{1}{6}$	below the x-axis
$(0, \infty)$	1	$\frac{1}{5}$	above the x-axis

Plotting some additional points, we obtain the graph of f shown in Figure 3.15.

FIGURE 3.15

x	-6	-5	-3	-2	-1	1	2
$f(x)$	$\frac{1}{12}$	$\frac{1}{5}$	$-\frac{1}{3}$	$-\frac{1}{4}$	$-\frac{1}{3}$	$\frac{1}{5}$	$\frac{1}{12}$

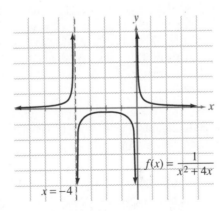

$$f(x) = \frac{1}{x^2 + 4x}$$

$x = -4$

EXAMPLE 35

Sketch the graph of $f(x) = \dfrac{1 - x^2}{(x - 2)^2}$.

Solution   Since 2 is the only solution of $(x - 2)^2 = 0$, $\text{dom}(f) = \{x \mid x \neq 2\}$ and the line $x = 2$ is the only vertical asymptote of the graph. As we saw in Example 33(a), the line $y = -1$ is a horizontal asymptote of the graph.

Since $1 - x^2 = (1 + x)(1 - x)$, the real zeros of f are -1 and 1. The real zeros of the numerator and denominator of f determine the following intervals: $(-\infty, -1)$, $(-1, 1)$, $(1, 2)$, and $(2, \infty)$. As in the previous example, we have:

Interval	Test Point	Value of f at Test Point	Graph of f Lies
$(-\infty, -1)$	-2	$-\frac{3}{16}$	below the x-axis
$(-1, 1)$	0	$\frac{1}{4}$	above the x-axis
$(1, 2)$	$\frac{3}{2}$	-5	below the x-axis
$(2, \infty)$	3	-8	below the x-axis

Using the information we have thus far, we need only plot a few additional points to obtain the graph of f. See Figure 3.16.

FIGURE 3.16

x	-4	-3	-2	-1	0	1	$\frac{3}{2}$	3	4	5
$f(x)$	$-\frac{15}{36}$	$-\frac{8}{25}$	$-\frac{3}{16}$	0	$\frac{1}{4}$	0	-5	-8	$-\frac{15}{4}$	$-\frac{24}{9}$

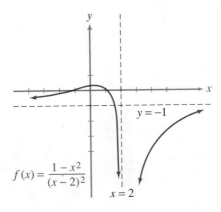

$$f(x) = \frac{1 - x^2}{(x - 2)^2}$$

$$y = -1$$

$$x = 2$$

Oblique Asymptotes

If the degree of the numerator of a rational function f exceeds the degree of its denominator by 1, the graph of f has an **oblique** or **slant asymptote**—an asymptote that is neither vertical nor horizontal. Consider, for example,

$$f(x) = \frac{x^2 - 1}{x}.$$

Dividing $x^2 - 1$ by x, we find that

$$f(x) = x - \frac{1}{x}.$$

Since $1/x \to 0$ for large values of $|x|$,

$$f(x) \to x, \quad \text{as} \quad |x| \to \infty.$$

This means that the line $y = x$ is an asymptote of the graph of f. Noting that $x = 0$ is a vertical asymptote, and that -1 and 1 are the real zeros of f, we sketch the graph of f shown in Figure 3.17.

FIGURE 3.17

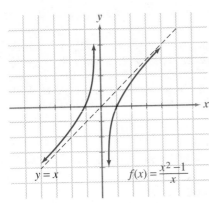

$$y = x \qquad f(x) = \frac{x^2 - 1}{x}$$

EXAMPLE 36

After finding the oblique asymptote, sketch the graph of $f(x) = \dfrac{x^2 + 2x}{x - 1}$.

Solution   Dividing $x^2 + 2x$ by $x - 1$ using synthetic division, we obtain

$$\begin{array}{r|rrr} 1 & 1 & 2 & 0 \\ & & 1 & 3 \\ \hline & 1 & 3 & 3 \end{array}$$

Thus $f(x) = x + 3 + 3/(x - 1)$. Since $3/(x - 1) \to 0$ as $|x| \to \infty$, the line $y = x + 3$ is an oblique asymptote of the graph of f.

Since $(x^2 + 2x)/(x - 1) = x(x + 2)/(x - 1)$, we see that $x = 1$ is a vertical asymptote and that the real zeros of f are -2 and 0. Plotting a few additional points, we obtain the graph of f shown in Figure 3.18.

FIGURE 3.18

x	-3	-2	-1	0	2	3	4
$f(x)$	$-\frac{3}{4}$	0	$\frac{1}{2}$	0	6	$\frac{15}{2}$	$\frac{20}{3}$

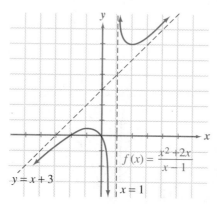

$$f(x) = \frac{x^2 + 2x}{x - 1}$$

$$y = x + 3$$

$$x = 1$$

EXERCISES 3.7

In Exercises 1–20, sketch the graph of each rational function.

1. $f(x) = -\dfrac{1}{x^3}$

2. $f(x) = \dfrac{1}{x^2 - 4}$

3. $f(x) = \dfrac{4}{x + 5}$

4. $f(x) = \dfrac{3}{x - 2}$

5. $f(x) = \dfrac{x - 3}{2x + 4}$

6. $f(x) = \dfrac{x}{2 - x}$

7. $f(x) = -\dfrac{2}{(x - 2)^2}$

8. $f(x) = -\dfrac{3}{(x - 4)^2}$

9. $f(x) = \dfrac{1}{(x + 3)(x - 2)}$

10. $f(x) = \dfrac{2}{x^2 - 9}$

11. $f(x) = \dfrac{8x}{(x + 1)^2}$

12. $f(x) = \dfrac{3x}{(x - 2)^2}$

13. $f(x) = \dfrac{x^2 - 1}{x(x + 4)}$

14. $f(x) = \dfrac{x + 1}{(x + 4)(x - 1)}$

15. $f(x) = \dfrac{8x^2}{(2x + 1)^2}$

16. $f(x) = \dfrac{16x^2}{(3 - 2x)^2}$

17. $f(x) = \dfrac{x^2}{x^2 - x - 2}$

18. $f(x) = \dfrac{x - 5}{x^2 - 8x + 12}$

19. $f(x) = \dfrac{x^4 - 1}{(x^2 + 1)^2}$

20. $f(x) = \dfrac{(x^2 - 1)^2}{x^2(x^2 + 4)}$

Consider the rational function $f(x) = (x^2 - 4)/(x - 2)$, whose domain contains all real numbers except 2. If $x \neq 2$, we can write $(x^2 - 4)/(x - 2) = (x + 2)(x - 2)/(x - 2) = x + 2$. Hence we can graph f as the line $y = x + 2$ with a "hole" where $x = 2$. In Exercises 21–26, follow a similar procedure to sketch the graph of each function.

21. $f(x) = \dfrac{x^2 - 1}{x + 1}$

22. $f(x) = \dfrac{x^2 - 9}{x - 3}$

23. $f(x) = \dfrac{x^2 - x - 6}{x - 3}$

24. $f(x) = \dfrac{3x^2 + 2x - 1}{3x - 1}$

25. $f(x) = \dfrac{x^3 - 8}{x - 2}$

26. $f(x) = \dfrac{x^2 - 16}{2x - 8}$

In Exercises 27–32, sketch the graph of each rational function after first determining the oblique asymptote.

27. $f(x) = \dfrac{x^2 - 4}{x + 1}$

28. $f(x) = \dfrac{4x^2 - 1}{2x}$

29. $f(x) = \dfrac{x^2 + 2x - 8}{x - 1}$

30. $f(x) = \dfrac{x^2 - x + 1}{x + 1}$

31. $f(x) = \dfrac{x^3 - 3x^2 - x}{x^2 - 1}$

32. $f(x) = \dfrac{x^3 - 1}{1 - x^2}$

33. The cost in dollars of producing x units of a certain product is given by $C(x) = x^2 + 2x + 15$, where $x \geq 0$. The **average cost** for producing x units is given by the function

$\overline{C}$, which is defined by $\overline{C}(x) = C(x)/x$, where $x > 0$. Find the average cost function $\overline{C}$ and sketch its graph.

34. A manufacturer finds that the cost of producing x items is given by $C(x) = 2x^2 - x + 2$, where $x \geq 0$. Determine the average cost function $\overline{C}$ (see Exercise 33) and sketch its graph.

35. A medical researcher finds that the function f, defined by $f(x) = (5x + 20)/x$, where $x > 0$, describes the number of milligrams of an experimental drug that remain in the bloodstream x hours after an intramuscular injection of the drug. Sketch the graph of f and interpret the meaning of the horizontal asymptote to the researcher.

36. In 1917, noted psychologist L. L. Thurstone proposed the function $f(x) = a(x + c)/[(x + c) + b]$, where $x \geq 0$, to describe the number of successful acts per unit of time a person could accomplish after x practice sessions. An electronics firm finds that the function $f(x) = 20(x + 1)/(x + 5)$, where $x \geq 0$, is Thurstone's equation for predicting the number of components a worker can assemble after x weeks of training. Sketch the graph of f and interpret the meaning of the horizontal asymptote to the electronics firm.

37. Is it possible for the graph of a rational function to have both a horizontal asymptote and an oblique asymptote? Explain.

38. Prove Theorem 3.5.

 In Exercises 39–40, use your graphing calculator to obtain the graph of each rational function.

39. $f(x) = \dfrac{x^2 - x - 6}{x^2 - 2x}$

40. $f(x) = \dfrac{-x^3 + x^2 + 4}{x^2}$

Photo Dwight Ellefsen/Freelance Photographer's Guild

CHAPTER 3 REVIEW EXERCISES

In Exercises 1–4, use long division to find the quotient and the remainder when the first polynomial is divided by the second.

1. $2x^2 - x + 5; 2x + 3$
2. $y^3 - 1; y + 3$
3. $x^3 + x^2 + x + 1; x^2 - x + 1$
4. $6x^3 + 10x^2 + x - 8; 2x^2 + 1$

In Exercises 5–12, use synthetic division to find the quotient and the remainder when the first polynomial is divided by the second.

5. $3x^4 + 2x^3 - 4x - 1; x + 3$
6. $5 - 3x + 2x^2 - x^3; x + 1$
7. $6x^3 - x^2 + 2x + 2; x - \frac{1}{3}$
8. $4x^4 + \frac{3}{4}x^2 + 1; x - \frac{1}{4}$
9. $x^3 + 2x - 4; x - i$
10. $x^3 - x^2 + 4x - 4; x + 2i$
11. $x^4 + ix^2 + 1; x - 1 + i$
12. $x^3 - 4x^2 + 5x; x - 2 - i$

In Exercises 13–16, use synthetic division to show that the second expression is a factor of the first.

13. $6x^3 - 23x^2 - 5x + 4; x - 4$
14. $2x^3 + x^2 + 8x + 4; x + \frac{1}{2}$
15. $x^4 - 3x^3 + 11x^2 - 27x + 18; x - 3i$
16. $x^3 - 5x^2 + 11x - 15; x - 1 + 2i$
17. Find the value of k for which $x - 2$ is a factor of $x^3 + kx + 4$.
18. Find a value of b that produces a remainder of 2 when $2x^4 - bx + 4$ is divided by $x - 1$.

In Exercises 19–22, use the remainder theorem to find the remainder when the first polynomial is divided by the second.

19. $2x^3 - 5x^2 + 3x - 4; x - 3$
20. $x^3 + 3x^2 - 3x + 1; x + \sqrt{3}$
21. $3x^4 - 2x^2 + 1; x + 2i$
22. $x^4 + 6x^2 + 8; x - 2i$

In Exercises 23–26, use the factor theorem to determine whether or not the second polynomial is a factor of the first.

23. $5x^3 + 17x^2 + 3x - 9; x + 3$
24. $x^3 + x^2 - 2x + 2; x - \sqrt{2}$
25. $8x^3 - 6x^2 + 5x - 3; x + \frac{1}{4}$
26. $2x^3 - x^2 + 2x - 1; x - i$

In Exercises 27–30, find a polynomial having integer coefficients that has the given degree and the given zeros.

27. degree 3; $\pm 1, 3$
28. degree 3; $-4, 0, \frac{1}{2}$
29. degree 4; $1 \pm \sqrt{2}, -2, 2$
30. degree 4; $3 - i, 3 + i, 0$ (a double root)

In Exercises 31–34, a polynomial equation is followed by one or more of its solutions. In each case, find all remaining solutions.

31. $6x^3 + 7x^2 - 1 = 0; -1$
32. $2x^3 - 4x^2 - 3x - 9 = 0; 3$
33. $x^3 - 2x^2 + 9x - 18 = 0; -3i$
34. $2x^4 - 13x^3 + 28x^2 - 11x = 0; 3 + \sqrt{2}i$
35. Show that $x + 1$ is a factor of $x^n - 1$, where n is an even positive integer.
36. Show that $x - c$ is a factor of $x^n - c^n$, where n is any positive integer.
37. Find a value of b such that $x + 1$ is a factor of $3x^3 + bx^2 - 2b + 5$.
38. Find the remainder when $2x^{51} - 13x^{34} + 2x^{17} - 8$ is divided by $x + 1$.

In Exercises 39–42, give the degree of each polynomial equation. In addition, determine all roots (including multiplicities when appropriate) of each equation.

39. $(x + 1)(2x - 1)^2(3x + 4) = 0$
40. $2x^3(x + 2)(x - 3)^4 = 0$
41. $5x(x + \sqrt{3}i)(x - \sqrt{3}i) = 0$
42. $(x - 2 + i)(x - 2 - i)(x + 1)^2 = 0$

In Exercises 43–46, find a polynomial $P(x)$ with leading coefficient 1 that has the given zeros and no others.

43. $-2, 1$ (multiplicity 2), 4
44. $-5, 4, \pm\sqrt{2}$
45. $-\frac{1}{3}, \frac{1}{2}$ (multiplicity 2)
46. $0, -\frac{2}{5}, 2$ (multiplicity 3)

In Exercises 47–50, list all possible rational roots for each polynomial equation.

47. $3x^4 + x^3 - 12x^2 + 5x - 3 = 0$
48. $8x^4 - 6x^3 + 9x^2 - 18 = 0$
49. $18x^4 - 2x^3 + 15x^2 - 12 = 0$
50. $\frac{3}{4}x^5 - 17x^2 + \frac{1}{2}x + 5 = 0$

In Exercises 51–56, find all rational solutions (if any) of each polynomial equation.

51. $6x^4 + 5x^3 - 10x^2 - 5x + 4 = 0$
52. $9x^4 - 3x^3 + 7x^2 - 3x - 2 = 0$
53. $2x^3 - 3x^2 + 5x - 1 = 0$
54. $4x^4 + 12x^3 + 19x^2 - 3x - 5 = 0$

55. $\frac{2}{3}x^3 + \frac{3}{2}x^2 - x - \frac{1}{2} = 0$
56. $\frac{1}{4}x^3 - \frac{6}{5}x^2 + \frac{3}{10}x + \frac{2}{5} = 0$
57. Show that $\sqrt[3]{2}$ is an irrational number. [*Hint:* Apply the rational root theorem to the equation $x^3 - 2 = 0$.]
58. Find the points of intersection (if any) of the graphs of the functions $f(x) = 2x^3 - x^2 - x + 3$ and $g(x) = 3x^3 - x^2 + 2x - 1$.
59. Find the x-intercepts (if any) of the graph of $f(x) = 3x^4 + 5x^2 - 8$.
60. The hypotenuse of a right triangle is two units longer than one of its legs. If the area of the triangle is $2\sqrt{3}$ square units, find the lengths of its sides.

In Exercises 61–64, use Descartes' rule of signs to determine as much information as possible about the nature of the roots of each polynomial. Display your results in a table.

61. $2x^4 - 5x^3 + 13x^2 - 1 = 0$
62. $4x^3 - 3x^2 + 2x - 7 = 0$
63. $2x^6 + 3x^4 - 2x^2 - 1 = 0$
64. $-3x^5 - 2x^4 + 5x^3 + 2x^2 - x - 4 = 0$

In Exercises 65–68, determine upper and lower bounds for the real roots of each polynomial.

65. $P(x) = -x^3 + 3x^2 + 8x + 3$
66. $P(x) = 3x^4 - 4x^3 - x^2 + 8x - 2$
67. $P(x) = 2x^5 - 3x^4 - 2x^2 + 4$
68. $P(x) = 2x^5 - x^4 - 2x^3 + 2x^2 - 4x - 1$

In Exercises 69–72, find the real roots of the given polynomial equation. Approximate irrational roots (if any) to the nearest tenth.

69. $5x^3 + 8x - 6 = 0$
70. $2x^4 - x^3 + 7x^2 - 4x - 4 = 0$
71. $-2x^4 + 2x^3 + 3x^2 + 4 = 0$
72. $x^5 + 2x^3 - 3x - 20 = 0$
73. Using the fact that $\sqrt[3]{5}$ is the positive real root of the equation $x^3 - 5 = 0$, approximate $\sqrt[3]{5}$ to the nearest hundredth.
74. A rectangular box is to be made from a piece of cardboard 6 inches wide and 14 inches long by cutting out squares of the same size from each corner and turning up the sides. If the volume of the resulting box is 40 cubic inches, what is the length of each side of the cut-out squares?
75. The displacement at time t of a particle moving along a straight line is given by $s(t) = t^3 - 2t^2 - 5t + 6$, where t is measured in seconds and displacement is measured in centimeters. After 1 second has elapsed, the displacement is 0 centimeters. At what other time is the displacement 0?
76. Evaluate the Taylor polynomial shown below at $x = \frac{1}{2}$, and compare your result with the actual value of $1/(x^2 + 1)$ at $x = \frac{1}{2}$. Round each result to the nearest thousandth.

$$\frac{1}{x^2 + 1} \approx 1 - x^2 + x^4 - x^6 + x^8$$

In Exercises 77–84, sketch the graph of each polynomial function.

77. $f(x) = -\frac{1}{2}x^3$

78. $f(x) = (x^2 - 1)(x + 3)^2$

79. $f(x) = 4x(x^2 - 1)$

80. $f(x) = -x^4 + 5x^2 - 4$

81. $f(x) = -\frac{5}{2}(x + 1)^2(x - 1)$

82. $f(x) = x^4 - 6x^3 + 11x^2 - 6x$

83. $f(x) = 2x^3 - 5x^2 - 4x + 3$

84. $f(x) = x^3 + 2x^2 - x - 2$

In Exercises 85–92, sketch the graph of each rational function.

85. $f(x) = \dfrac{x}{x - 2}$

86. $f(x) = \dfrac{3}{(x + 1)(x - 2)}$

87. $f(x) = \dfrac{2x}{1 - x^2}$

88. $f(x) = \dfrac{x - 2}{x + 2}$

89. $f(x) = \dfrac{3x - 1}{x^2 - 4}$

90. $f(x) = \dfrac{1}{x^2 - 9}$

91. $f(x) = \dfrac{3x^2 - 3x - 6}{x^2 + 8x + 16}$

92. $f(x) = \dfrac{x^2 - x - 2}{x^2 + 1}$

In Exercises 93–96, sketch the graph of each rational function after first determining the oblique asymptote.

93. $f(x) = \dfrac{x^2 - 4}{x}$

94. $f(x) = \dfrac{x^2 + 2}{x}$

95. $f(x) = \dfrac{x^2 - 8}{x + 1}$

96. $f(x) = \dfrac{x^2 - x - 2}{x - 1}$

97. The total cost of producing x units of a certain product is given by $C(x) = \frac{1}{2}x^2 + 2x + 5$. Determine the average cost function $\overline{C}$ for this product and sketch its graph.

98. A psychologist is conducting an experiment on memory retention. A group of subjects is asked to memorize a list of 25 special symbols each day for a period of 30 days. At the end of each day, a record is made of how many symbols a subject can recall. Based on the average number of symbols recalled each day, the psychologist finds that the function $f(t) = (4t + 12)/t$, where $t \geq 1$, gives a good approximation of the average number of symbols recalled after t days. Sketch the graph of f and explain the significance of the horizontal asymptote to the pscyhologist.

CHAPTER 4

Exponential and Logarithmic Functions

In Chapter 3, we studied both polynomial and rational functions. Such functions are part of a class of functions known as algebraic functions. An algebraic function has a defining equation that involves the operations of addition, subtraction, multiplication, division, or extraction of roots. In this chapter, we study two types of non-algebraic functions: exponential and logarithmic functions. After investigating properties of these functions and their graphs, we conclude by giving some of their many applications in such fields as medicine, engineering, and the life sciences.

4.1 EXPONENTIAL FUNCTIONS

An investor has \$10,000 to invest for five years. He has a choice of investing the money in an institution that pays 10% interest compounded quarterly or in an institution that pays $9\frac{3}{4}\%$ interest compounded continuously. In which institution should the money be invested to maximize the interest earned?

A newspaper article reports that bones found at an archaeological dig are approximately 5000 years old. What process was used to estimate the age of the bones?

Questions similar to the ones posed above cannot be answered using functions of the type we have studied thus far. A new class of functions, called **exponential functions**, is needed to solve these problems.

Recall the definition of b^r where r is a rational number: If m and n are integers such that $n > 0$ and $r = m/n$ is in lowest terms, then

$$b^r = b^{m/n} = (\sqrt[n]{b})^m,$$

provided $\sqrt[n]{b}$ exists. For example,

$$4^{5/2} = (\sqrt{4})^5 = 2^5 = 32 \qquad \text{and}$$

$$81^{-3/4} = (\sqrt[4]{81})^{-3} = 3^{-3} = \frac{1}{3^3} = \frac{1}{27}.$$

It is also possible, using concepts from calculus, to define b^k, where k is an irrational number. Such a definition preserves the usual properties of exponents and gives meaning to expressions such as $4^{\sqrt{3}}$ and $81^{-\pi}$. In the absence of such techniques, however, we can still give meaning to numbers such as $4^{\sqrt{3}}$, as we shall now see. Since the rational numbers 1.7, 1.73, 1.7320, 1.73205, 1.7320508, . . . approach the value of $\sqrt{3}$, does it not seem reasonable that the numbers $4^{1.7}$, $4^{1.73}$, $4^{1.7320}$, $4^{1.73205}$, $4^{1.7320508}$, . . . should approach $4^{\sqrt{3}}$? (See Table 4.1.) In fact, this is how $4^{\sqrt{3}}$ is often defined in advanced mathematics courses.

TABLE 4.1

x	1.7	1.73	1.7320	1.73205	1.7320508
4^x	10.556063	11.004335	11.034887	11.035652	11.035665

In general, if b is a positive number not equal to 1 and x is any real number, b^x is a unique real number. The properties stated in the following theorem are central to our study in this chapter:

THEOREM 4.1

Let b be a positive real number not equal to 1, and let x and y be real numbers. Then

(i) b^x is a unique real number;
(ii) $b^x = b^y$ if and only if $x = y$;
(iii) If $b > 1$ and $x < y$, then $b^x < b^y$;
(iv) If $0 < b < 1$ and $x < y$, then $b^x > b^y$.

In Theorem 4.1, observe that $b \neq 1$. (Remember that $1^x = 1$ for each real number x.) Illustrating part (iii) of the theorem, we have $2^3 < 2^\pi$ since $2 > 1$ and $3 < \pi$. Similarly, $4^{\sqrt{2}} < 4^2$. In addition, from part (iv) we see that $(\frac{1}{3})^3 > (\frac{1}{3})^\pi$ since $0 < \frac{1}{3} < 1$ and $3 < \pi$.

We can now define the exponential function with base b.

DEFINITION 4.1

Let b be a positive real number not equal to 1. The function f defined by $f(x) = b^x$ for each real number x is called the **exponential function with base b**.

Notice that the unknown in the defining equation of an exponential function occurs in the exponent, not in the base. From Theorem 4.1, we know that b^x is defined for all real numbers x. Hence, the domain of every exponential function is $\mathbb{R}$.

EXAMPLE 1

Sketch the graph of each function.

a. $f(x) = 2^x$ **b.** $g(x) = (\frac{1}{2})^x$

Solution

a. and b.   We know that dom $(f) = $ dom $(g) = \mathbb{R}$. The data in Table 4.2 will help us sketch the graphs of f and g.

TABLE 4.2

x	-3	-2	-1	0	1	2	3
$f(x) = 2^x$	$\frac{1}{8}$	$\frac{1}{4}$	$\frac{1}{2}$	1	2	4	8
$g(x) = (\frac{1}{2})^x$	8	4	2	1	$\frac{1}{2}$	$\frac{1}{4}$	$\frac{1}{8}$

FIGURE 4.1

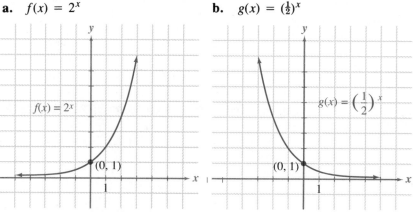

a. $f(x) = 2^x$ **b.** $g(x) = (\frac{1}{2})^x$

Observe from Table 4.2 that $g(x) = 1/f(x)$; that is, $(\frac{1}{2})^x = (2^{-1})^x = 2^{-x}$ $= \dfrac{1}{2^x}$. It follows that the graph of $g(x) = (\frac{1}{2})^x$ may be obtained by reflecting the graph of $f(x) = 2^x$ through the y-axis.

EXAMPLE 2

Sketch the graph of each function.

a. $f(x) = 3^x$ **b.** $g(x) = (\frac{1}{3})^x$

Solution

a. and b. As in Example 1, the graph of $g(x) = (\frac{1}{3})^x = 3^{-x}$ may be obtained by reflecting the graph of $f(x) = 3^x$ through the y-axis.

x	-3	-2	-1	0	1	2	3
3^x	$\frac{1}{27}$	$\frac{1}{9}$	$\frac{1}{3}$	1	3	9	27

FIGURE 4.2

a. $f(x) = 3^x$ **b.** $g(x) = (\frac{1}{3})^x$

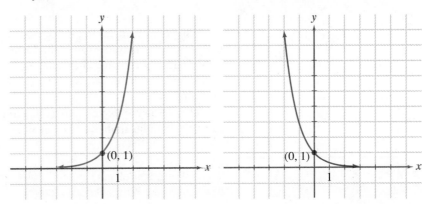

Figures 4.1 and 4.2 reveal key properties of exponential functions and their graphs. For example, both $f(x) = 2^x$ and $f(x) = 3^x$ are increasing functions, while both $g(x) = (\frac{1}{2})^x$ and $g(x) = (\frac{1}{3})^x$ are decreasing functions. Properties of exponential functions and their graphs are summarized in Table 4.3.

TABLE 4.3
Properties of $f(x) = b^x$

$f(x) = b^x$ $(b > 1)$ $f(x) = b^x$ $(0 < b < 1)$

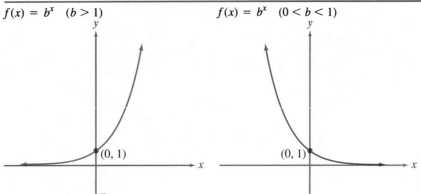

(i)   dom $(f) = \mathbb{R}$;
(ii) ran $(f) = (0, \infty)$;
(iii) f is one-to-one;
(iv) The y-intercept of the graph is $(0, 1)$;
(v)   There is no x-intercept;
(vi) The x-axis is a horizontal asymptote for the graph;
(vii) If $b > 1$, f is increasing, and if $0 < b < 1$, f is decreasing.

FIGURE 4.3

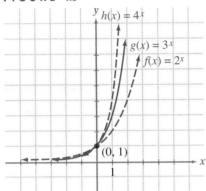

FIGURE 4.4

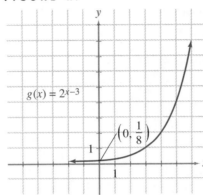

FIGURE 4.5

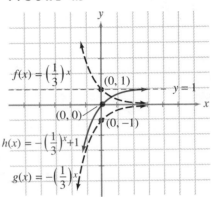

As we saw in Examples 1 and 2, $g(x) = (1/b)^x$ can be expressed as $g(x) = b^{-x}$. Therefore, we will focus our attention on exponential functions whose bases are greater than one. As we saw in Table 4.3, such functions are increasing. In Figure 4.3, we compare the graphs of three such exponential functions.

The following two examples illustrate how the techniques of graphing from Chapter 2 can be applied to exponential functions.

EXAMPLE 3

Sketch the graph of $g(x) = 2^{x-3}$. Give the domain and range of g, and give an equation of the horizontal asymptote.

Solution   Observe the relationship between $f(x) = 2^x$ and $g(x) = 2^{x-3}$. Since $g(x) = f(x - 3)$, the graph of g, shown in Figure 4.4, can be obtained by shifting the graph of f three units to the right.

We see from the graph that $\text{dom}(g) = \mathbb{R}$, $\text{ran}(g) = (0, \infty)$, and that the x-axis is a horizontal asymptote for the graph. We could have taken another approach to arrive at the graph of g. Noting that $2^{x-3} = 2^x \cdot 2^{-3} = \frac{1}{8} \cdot 2^x$, the graph of g could also be obtained from the graph of f by multiplying each y-coordinate by $\frac{1}{8}$.

EXAMPLE 4

Sketch the graph of $h(x) = -(\frac{1}{3})^x + 1$. Give the domain and range of h, and give an equation of the horizontal asymptote.

Solution   Again, using graphing techniques from Section 2.4, the graph of h can be obtained by first reflecting the graph of $f(x) = (\frac{1}{3})^x$ shown in Figure 4.2(b) through the x-axis, obtaining the graph of $g(x) = -(\frac{1}{3})^x$, and then shifting the graph up one unit. See Figure 4.5.

We see from the graph shown in Figure 4.5 that $\text{dom}(h) = \mathbb{R}$ and $\text{ran}(h) = (-\infty, 1)$. In addition, the line $y = 1$ is a horizontal asymptote for the graph.

EXAMPLE 5

Sketch the graph of the function defined by $f(x) = (1.06)^x$.

Solution   Since the base of this exponential function is $b = 1.06 > 1$, we know from Table 4.3 that f is an increasing function whose graph is similar to

FIGURE 4.6

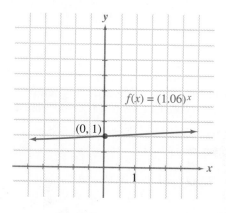

$f(x) = (1.06)^x$

$(0, 1)$

the graphs of $y = 2^x$ and $y = 3^x$. We construct a table of values with the help of a calculator that has a y^x key. The graph is sketched in Figure 4.6.

x	−3	−2	−1	0	1	2	3
$(1.06)^x$	0.8396	0.8900	0.9434	1	1.06	1.1236	1.1910

Applications

We shall now illustrate applications of exponential functions in the banking and investment industry. We shall use the next example as a vehicle for introducing some terminology. Suppose $2000 is invested for three years in a savings account that pays 8% interest **compounded annually**. This means that at the end of each year, the savings institution will contribute to the account 8%, or 0.08, of the amount in the account at that time. Interest accumulated in this fashion is called **compound interest**. The amount of money deposited initially is called the **principal P**. The annual interest rate is called the **nominal rate of interest** and, when expressed as a decimal, is denoted by r. Thus, in our example, $r = 0.08$. Assuming no withdrawals are made from the account, Table 4.4 reveals the amount A in the account after one, two, and three years.

TABLE 4.4

Time Period	Formula	Amount in Dollars
After 1 year	A = Principal + interest $= P + rP$ $= P(1 + r)$	$A = 2000(1 + 0.08)$ $= 2000\,(1.08)$ $= \$2160.00$
After 2 years	A = Principal + interest $= P(1 + r) + rP(1 + r)$ $= [P(1 + r)](1 + r)$ $= P(1 + r)^2$	$A = 2000(1.08)^2$ $= \$2332.80$
After 3 years	A = Principal + interest $= P(1 + r)^2 + rP(1 + r)^2$ $= [P(1 + r)^2](1 + r)$ $= P(1 + r)^3$	$A = 2000(1.08)^3$ $\approx \$2519.42$

In Table 4.4 the formulas for the amounts in the account after one, two, and three years lead to a formula for the amount in the account at the end of t years:

$$A = P(1 + r)^t. \qquad (1)$$

Formula (1) should be used when interest is compounded annually. It is common practice for savings institutions to compute interest more often. When interest is compounded more than one time each year, the following formula is applicable.

Compound Interest

If a principal of P dollars is invested at a nominal interest rate r—expressed as a decimal—compounded n times per year, then the amount A in the account after t years is given by

$$A = P\left(1 + \frac{r}{n}\right)^{nt} \qquad (2)$$

Observe that formula (2) reduces to formula (1) when $n = 1$, that is, when interest is compounded annually.

EXAMPLE 6

Assume that \$2000 is invested for five years in an account paying 8% annual interest (nominal rate). Assuming no withdrawals are made and the interest is invested in the account, find the value of this account, if interest is compounded

a. quarterly;   **b.** monthly;   **c.** daily (assume there are 365 days per year).

Solution   We use formula (2) with $P = \$2000$ and $r = 0.08$.

a. Here $n = 4$, and

$$A = 2000\left(1 + \frac{0.08}{4}\right)^{4(5)}$$
$$= 2000(1.02)^{20}$$
$$\approx 2000(1.4859)$$
$$= \$2971.80.$$

b. Since $n = 12$,

$$A = 2000\left(1 + \frac{0.08}{12}\right)^{12(5)}$$
$$\approx 2000(1.0066667)^{60}$$
$$\approx 2000(1.4898)$$
$$= \$2979.60.$$

c. Here $n = 365$, and

$$A = 2000\left(1 + \frac{0.08}{365}\right)^{365(5)}$$
$$\approx 2000(1.0002192)^{1825}$$
$$\approx 2000(1.4918)$$
$$= \$2983.60.$$

EXAMPLE 7

How much interest is earned by \$1000 invested for one year in an account paying 12% interest compounded monthly?

Solution   The amount in the account at the end of one year is

$$A = 1000\left(1 + \frac{0.12}{12}\right)^{12(1)}$$
$$= 1000(1.01)^{12}$$
$$\approx 1000(1.126825)$$
$$\approx \$1126.83.$$

Hence, the interest earned is

$$\$1126.83 - \$1000 = \$126.83.$$

In Example 7, notice that $126.83 is 12.683% of $1000. Thus, the stated, or nominal, rate of 12% does not reflect the rate at which interest is earned. The rate, when compounded annually, that yields the same accumulated amount as the nominal rate compounded n times per year is called the **effective rate of interest**. Thus, in Example 7, the effective rate is 12.683%. Unless otherwise stated, we will use only nominal interest rates.

The Number e

You may be aware that many savings institutions have saving plans where the interest is **compounded continuously**. Intuitively, compounding continuously means that interest is compounded at each instant. To gain insight into this process, suppose that one dollar is invested for one year at the rate of 100%. The amount in the account is given by

$$A = 1\left(1 + \frac{1}{n}\right)^n,$$

where n is the number of times interest is compounded annually. In Table 4.5 we see how the amount in the account changes as n becomes larger and larger.

TABLE 4.5

n	$\left(1 + \dfrac{1}{n}\right)^n$ (approximately)
1	2.00000
10	2.59374
100	2.70481
1,000	2.71692
10,000	2.71814
100,000	2.71827
1,000,000	2.71828

It can be proved that as n becomes larger and larger, the expression $\left(1 + \dfrac{1}{n}\right)^n$ gets closer and closer to a certain irrational number, denoted by e, which is approximately equal to 2.71828. The prolific Swiss mathematician, Leonhard Euler (1707–1783), is given credit for discovering the number e.

The Natural Exponential Function

The exponential function defined by $f(x) = e^x$, is called the **natural exponential function**. The natural exponential function, which is often used to help describe natural phenomena, is one of the most important functions in mathematics. Since $2 < e < 3$, it follows that $2^x < e^x < 3^x$ for all positive real numbers x, and $2^x > e^x > 3^x$ for all negative real numbers x as shown in Figure 4.7.

It is now possible for us to derive a formula for compounding interest continuously. Consider the expression $P\left(1 + \dfrac{r}{n}\right)^{nt}$ given by formula (2) on page 175. We will examine values of this expression as n becomes larger and larger. To simplify the calculations, let $m = n/r$. Hence, $n = mr$ and thus,

FIGURE 4.7

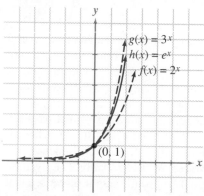

$$A = P \left(1 + \frac{r}{n}\right)^{nt} = P \left(1 + \frac{r}{mr}\right)^{mrt} = P \left[\left(1 + \frac{1}{m}\right)^{m}\right]^{rt}.$$

Since $m = n/r$ and r is fixed, as n becomes larger and larger, so does m. Thus, as n becomes larger and larger, $\left(1 + \frac{1}{m}\right)^{m}$ gets closer and closer to the number e. Hence, $\left[\left(1 + \frac{1}{m}\right)^{m}\right]^{rt}$ gets closer and closer to e^{rt}. Therefore, if interest is compounded continuously, the amount A in the account after t years is given by

$$A = Pe^{rt}, \tag{3}$$

where r is the annual interest rate expressed as a decimal. We can now solve one of the problems posed at the beginning of this section:

> An investor has $10,000 to invest for five years. He has a choice of investing the money in an institution that pays 10% interest compounded quarterly or in an institution that pays $9\frac{3}{4}$% interest compounded continuously. In which institution should the money be invested to maximize the interest earned?

Calculating the interest accrued on $10,000 invested for five years paying 10% interest compounded quarterly, we have

$$\begin{aligned} A &= 10,000 \left(1 + \frac{0.10}{4}\right)^{4(5)} \\ &= 10,000 \, (1.025)^{20} \\ &\approx 10,000 \, (1.6386) \\ &= \$16,386. \end{aligned}$$

Thus, the interest earned is $16,386 − $10,000 = $6,386.

To calculate the interest earned on $10,000 invested for five years at an interest rate of $9\frac{3}{4}$% compounded continuously, we have by equation (3)

$$\begin{aligned} A &= 10,000 e^{(0.0975)5} \\ &\approx 10,000 \, (1.6282) \\ &= \$16,282. \end{aligned}$$

Since the interest earned is $16,282 − $10,000 = $6,282, the money should be invested in the account paying 10% interest compounded quarterly.

Annuities

A sequence of equal deposits made into an account at regular intervals over a specified period of time is called an **annuity**. Money invested in the account earns interest at a fixed rate until the expiration of the annuity. The amount, or **future value**, of an annuity is the total amount of money, deposits and interest, in the account at the expiration date of the annuity.

Suppose you deposit $500 every six months into an account that pays 8% interest compounded semiannually. Over the next three years, if you make six deposits, one at the end of each interest payment period, how much money will be in the account after the last deposit is made? We use equation (2) given on page 175 to generate the data given in Table 4.6. From Table 4.6, we see that the total amount in the account after the sixth deposit is (far-right column),

TABLE 4.6

Years	1st Year		2nd Year		3rd Year	
Deposit Number	1	2	3	4	5	6

$500 $500 $500 $500 $500 $500

$$\begin{array}{l} \rightarrow \$500 \\ \rightarrow \$500(1.04) \\ \rightarrow \$500(1.04)^2 \\ \rightarrow \$500(1.04)^3 \\ \rightarrow \$500(1.04)^4 \\ \rightarrow \$500(1.04)^5 \end{array} \left.\begin{array}{l} \\ \\ \end{array}\right\} \text{Future Value}$$

$$A = 500 + 500(1.04) + 500(1.04)^2 + 500(1.04)^3 + 500(1.04)^4 + 500(1.04)^5. \quad \textbf{(4)}$$

We could evaluate each of the future values given in Table 4.6 using a calculator and then add the results to find A. However, we shall derive a formula for finding A that takes only a few steps even when a very large number of deposits have been made. We proceed by multiplying each side of equation (4) by 1.04 to obtain the equation

$$1.04\,A = 500(1.04) + 500(1.04)^2 + 500(1.04)^3 + 500(1.04)^4 + 500(1.04)^5 + 500(1.04)^6. \quad \textbf{(5)}$$

By subtracting corresponding sides of equation (4) from equation (5), we get

$$1.04A - A = 500(1.04)^6 - 500.$$

Therefore,

$$0.04A = 500[(1.04)^6 - 1],$$

or

$$A = \frac{500[(1.04)^6 - 1]}{0.04}.$$

Using a calculator, we find that

$$A \approx \$3316.49.$$

In general, if P is the periodic deposit, r is the interest rate *per period*—expressed as a decimal, and n is the number of conversion periods, then the future value of the annuity is given by

$$A = \frac{P[(1 + r)^n - 1]}{r}. \quad \textbf{(6)}$$

EXAMPLE 8

Every three months $100 is deposited into an account earning 12% interest compounded quarterly. What is the value of the annuity at the end of three years?

Solution   Using equation (6) with $P = \$100$, $r = 0.03$, and $n = 12$, we have

$$A = \frac{100[(1.03)^{12} - 1]}{0.03}$$

$$\approx \$1419.20.$$

Equations Involving Exponential Functions

Exponential functions are studied extensively in calculus. Just as in the case of polynomial functions, we are often interested in determining the zeros (if any) of an exponential function.

EXAMPLE 9

Find the zeros of the function defined by $f(x) = x^2 e^x - 4xe^x$.

Solution   To find the zeros of f, we must find the solutions of the equation

$$f(x) = 0, \quad \text{or} \quad x^2 e^x - 4xe^x = 0.$$

Factoring the last equation, we get

$$xe^x(x - 4) = 0.$$

Since $e^x > 0$ for every x, $f(x) = 0$ if and only if $x = 0$ or $x - 4 = 0$. Hence, the zeros of f are 0 and 4.

An equation such as the equation $3^{2x+1} = 1$, in which a variable occurs as an exponent, is called an **exponential equation**. By using properties of exponential functions, we can sometimes find the solution set of an exponential equation by using the techniques discussed in Chapter 1.

EXAMPLE 10

Find the solution set of the exponential equation

$$2^{x^2 - 3x} = 4^{1-x}.$$

Solution   Since $4^{1-x} = (2^2)^{1-x} = 2^{2-2x}$, the equation $2^{x^2-3x} = 4^{1-x}$ is equivalent to the equation $2^{x^2-3x} = 2^{2-2x}$. Furthermore, since the exponential function defined by $f(x) = 2^x$ is one-to-one,

$$2^{x^2-3x} = 2^{2-2x}$$

implies that

$$x^2 - 3x = 2 - 2x, \quad \text{or}$$

$$x^2 - x - 2 = 0.$$

Solving the last equation, we find that $x = -1$ or $x = 2$. Hence, the solution set of the given equation is $S = \{-1, 2\}$.

EXERCISES 4.1

In Exercises 1–26, sketch the graph of the function f. Give dom(f) ran(f) and if applicable, give the equation of any horizontal asymptote.

1. $f(x) = 10^x$
2. $f(x) = 6^x$
3. $f(x) = -3^x$
4. $f(x) = -5^x$
5. $f(x) = -2^{-x}$
6. $f(x) = -4^{-x}$
7. $f(x) = (\frac{1}{4})^x$
8. $f(x) = (\frac{2}{5})^x$
9. $f(x) = 3^{|x|}$
10. $f(x) = 3^{-|x|}$
11. $f(x) = 2^{x+4}$
12. $f(x) = (\frac{1}{2})^{x-2}$
13. $f(x) = 2 + 2^x$
14. $f(x) = 2^{-x} - 2$
15. $f(x) = e^{x+3}$
16. $f(x) = e^{(-x^2+2)}$
17. $f(x) = e^x + 3$
18. $f(x) = e^{(-x^2+2)} - 3$
19. $f(x) = x + 2^x$
20. $f(x) = 3x + 3^x$
21. $f(x) = e^{1/x}$
22. $f(x) = e^{-x/2}$
23. $f(x) = (1.04)^x$
24. $f(x) = (1.09)^x$
25. $f(x) = 4^{2-x^2}$
26. $f(x) = 3^{-(x^2-1)}$
27. Sketch the graphs of $f(x) = x^3$ and $g(x) = 3^x$ in the same coordinate system and discuss the differences in these graphs.

28. Sketch the graphs of the following functions in the same coordinate system:
    a. $f(x) = 4^{-x}$ **b.** $g(x) = 4^{-(x+2)}$
    c. $h(x) = 4^{-x} + 2$

29. Sketch the graphs of the following functions in the same coordinate system:
    a. $f(x) = 3^x$ **b.** $g(x) = 3^{x+1}$
    c. $h(x) = 3^{x+1} - 4$.

30. **a.** Sketch the graphs of $f(x) = e^x$ and $g(x) = e^{-x}$ in the same coordinate system.

    b. Sketch the graph of $h(x) = \dfrac{e^x + e^{-x}}{2}$. Show that h is even.

    c. Sketch the graph of $k(x) = \dfrac{e^x - e^{-x}}{2}$. Show that k is odd.

31. Use a calculator to approximate the values of 3^3, $3^{3.1}$, $3^{3.14}$, $3^{3.141}$, $3^{3.1415}$, $3^{3.14159}$. Then use the calculator to approximate the value of 3^π. What do you conclude from these approximations?

32. Use a calculator to approximate the values of 3^1, $3^{1.4}$, $3^{1.41}$, $3^{1.414}$, $3^{1.4142}$, $3^{1.41421}$. Then use the calculator to approximate the value of $3^{\sqrt{2}}$. What do you conclude from these approximations?

33. Sketch the graph of $f(x) = 10^x$ and its inverse in the same coordinate system. Give the domain and range of each.

34. Sketch the graph of $f(x) = 4^{x-1}$ and its inverse in the same coordinate system. Give the domain and range of each.

35. Sketch the graph of $f(x) = e^x$ and its inverse on the same coordinate system. Give the domain and range of each.

36. Let $f(x) = e^x$ and $g(x) = x - 4$.
    a. Give the defining equations of $f \circ g$ and $g \circ f$.
    b. Determine the domains of $f \circ g$ and $g \circ f$.
    c. Sketch the graphs of $f \circ g$ and $g \circ f$.

37. If $f(x) = e^x$, show that $\dfrac{f(x + h) - f(x)}{h}$
    $$= e^x \left(\frac{e^h - 1}{h} \right).$$

38. In calculus, it is shown that $F(x) = e^x$ can be approximated by $f(x) = 1 + x + \dfrac{x^2}{2} + \dfrac{x^3}{6}$ for values of x close to zero. Use a calculator to approximate the difference $F(x) - f(x)$ for the following values of x:
    a. $x = 1$ b. $x = 0.5$ c. $x = 0.1$
    d. $x = 0.01$ e. $x = 0.001$ f. $x = 0.0001$

In Exercises 39–41, assume that P dollars have been invested at a nominal annual interest rate r compounded as indicated for a term of t years. Calculate the amount in each account.

39. $P = \$2000$, $r = 7\%$, compounded semiannually, $t = 5$
40. $P = \$10,000$, $r = 9\frac{1}{4}\%$, compounded monthly, $t = 2$
41. $P = \$6000$, $r = 4\frac{1}{2}\%$, compounded continuously, $t = 3$
42. A person wishes to have $300,000 in an account for retirement 20 years from now. How much should be deposited

semiannually in an account paying 8% interest compounded semiannually?

43. Suppose Janice invests $5000 for one year in an account paying interest at an annual rate of 10% compounded daily, while Mary invests $5000 for one year in an account paying interest at a rate of 10% compounded continuously. Compare the interest earned by the two accounts.

44. In five years a couple would like to have $20,000 for a down payment on a house. What amount should be deposited each month into an account paying 9% interest compounded monthly?

45. Beginning in January, a person plans to deposit $200 at the end of each month into an account earning 9% interest compounded monthly. Each year taxes must be paid on the interest earned during that year. Find the interest earned during each of the first three years.

46. Mary can afford monthly deposits of only $150 into an account that pays 9.5% interest compounded monthly. Will Mary have enough money in her account after three years to buy a car costing $7500?

47. Find the zeros of each of the following functions:
    a. $f(x) = xe^x + 5x^2e^x$
    b. $g(x) = e^{2x-3} - e^{-x^2}$ [Hint: Solve $e^{2x-3} = e^{-x^2}$.]
    c. $h(x) = 4^x - 2^x - 56$ [Hint: See Example 10.]

48. Graph each of the following functions in the same coordinate system:
    a. $f(x) = 4^{1-\sqrt{x}}$ b. $g(x) = 4^{1-\sqrt{x-1}}$
    c. $h(x) = 4^{1-\sqrt{x-1}} + 3$

49. Graph each of the following functions in the same coordinate system:
    a. $f(x) = 2^{\sqrt{3}x}$ b. $g(x) = 2^{\sqrt{3}(x+2)}$
    c. $h(x) = 2^{\sqrt{3}(x+2)} - 1$

50. Sketch the graph of $y = 3^{x+2}$ and $y = e^{1-x}$ in the same coordinate. Approximate the point(s) of intersection. Give the approximate solution(s) of the equation $3^{x+2} = e^{1-x}$ to two decimal places.

51. Sketch the graph of $y = 2^x$ and $y = 2x + 5$ in the same coordinate system. Approximate the point(s) of intersection. Give the approximate solution(s) of the equation $2^x = 2x + 5$ to two decimal places.

4.2 LOGARITHMIC FUNCTIONS

In Section 4.1, we found that exponential functions are one-to-one, and thus each has an inverse function. Let us review our method for finding the equation for the inverse of a one-to-one function, which was outlined in Section 2.6. Consider, for example, the one-to-one function defined by $f(x) = 2x$. To obtain an equation for f^{-1}, we replace $f(x)$ by y and interchange x and y in the equation $y = 2x$ to obtain $x = 2y$. Solving the last equation for y yields $y = x/2$ so that $f^{-1}(x) = x/2$.

Can we find an equation for the inverse of an exponential function by using the method we just outlined? We begin with the equation

$$y = b^x \qquad \qquad (1)$$

where $b > 0$ and $b \neq 1$. Interchanging x and y in equation (1) yields

$$x = b^y. \qquad (2)$$

In the absence of algebraic techniques for solving equation (2) for y, we make the following definition:

DEFINITION 4.2   Definition of $\log_b x$

Let b be a positive number not equal to 1. If x is a positive real number, then

$$y = \log_b x \quad \text{if and only if} \quad b^y = x.$$

The symbol $\log_b x$ is read "the logarithm of x with base b" or "log base b of x."

The equation $y = \log_b x$ is said to be in **logarithmic form** while the equivalent equation $x = b^y$ is in **exponential form**. From Definition 4.2, we see that $\log_b x$ is the power to which the base b must be raised to yield x. For example,

$$\log_2 16 = 4, \quad \text{since } 2^4 = 16 \quad \text{and}$$

$$\log_5 \frac{1}{125} = -3, \quad \text{since } 5^{-3} = \frac{1}{125}.$$

In Table 4.7, we list several equations in exponential form and their equivalent logarithmic form.

TABLE 4.7

Exponential Form	Logarithmic Form
$27 = 3^3$	$\log_3 27 = 3$
$\frac{1}{32} = 2^{-5}$	$\log_2 \frac{1}{32} = -5$
$1 = e^0$	$\log_e 1 = 0$
$e = e^1$	$\log_e e = 1$

Using Definition 4.2, we can now solve equation (2) for y, to obtain

$$y = \log_b x.$$

Thus, $f(x) = \log_b x$ (where $x > 0$) is the inverse of $g(x) = b^x$ (where $x \in \mathbb{R}$). We call f the **logarithmic function with base b**. Since f is the inverse of g, $\text{dom}(f) = \text{ran}(g) = (0, \infty)$ and $\text{ran}(f) = \text{dom}(g) = \mathbb{R}$. The graph of f is shown in Figure 4.8. We can deduce some properties of logarithmic functions by studying the graphs shown in Figure 4.8. Many of these properties are summarized in Table 4.8.

TABLE 4.8
Properties of $f(x) = \log_b x$

(i) $\text{dom}(f) = (0, \infty)$;
(ii) $\text{ran}(f) = \mathbb{R}$;
(iii) The x-intercept of the graph is $(1, 0)$;
(iv)   The graph has no y-intercept;
(v)   The y-axis is a vertical asymptote for the graph;
(vi)   If $b > 1$, f is an increasing function, and
       if $0 < b < 1$, f is a decreasing function.

FIGURE 4.8
Graph of the logarithmic function
with base b

a. $f(x) = \log_b x$ $(b > 1)$

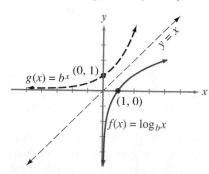

b. $f(x) = \log_b x$ $(0 < b < 1)$

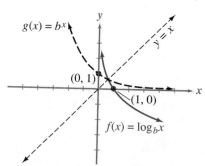

Natural and Common Logarithmic Functions

Perhaps the two most useful bases b for logarithmic functions are e and 10. The function f defined by $f(x) = \log_e x$ is called the **natural logarithmic function** and is usually denoted by $f(x) = \ln x$ (see Figure 4.9). In advanced mathematics, cumbersome formulas can sometimes be simplified using the natural logarithmic function or its inverse $g(x) = e^x$. The function defined by $f(x) = \log_{10} x$ is called the **common logarithmic function** and is denoted by $f(x) = \log x$. It is, of course, the inverse of $g(x) = 10^x$ (see Figure 4.10). The importance of base 10 stems from the fact that 10 is the base of our number system. Logarithms with base 10 were used extensively in arithmetic calculations before calculators were widely available. However, the common logarithmic function is still used in a wide range of applications.

FIGURE 4.9

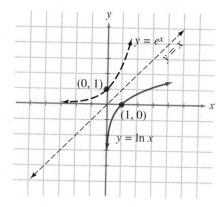

FIGURE 4.10

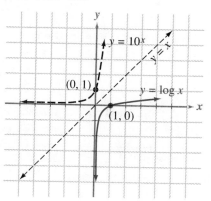

EXAMPLE 11

For each of the following functions,

    find the domain;

    sketch the graph; and

    give the equation of any vertical asymptote.

a. $f(x) = \ln(x + 2)$ **b.** $f(x) = \log |x|$ **c.** $f(x) = \log_2(-x)$

Solution

a. Since the natural logarithmic function is defined only for positive numbers, we require that $x + 2 > 0$. Hence $x > -2$, and dom$(f) = (-2, \infty)$.

FIGURE 4.11 $f(x) = \ln(x + 2)$

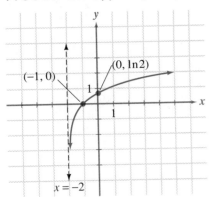

Notice that if $g(x) = \ln x$, then $f(x) = \ln(x + 2) = g(x + 2)$. Thus, the graph of f can be obtained by shifting the graph of g (see Figure 4.9) two units to the left as shown in Figure 4.11. From Figure 4.11, we see that the line $x = -2$ is a vertical asymptote for the graph.

b. To determine dom (f), we require that $|x| > 0$. Since the inequality $|x| > 0$ is equivalent to the compound inequality $x < 0$ or $x > 0$, dom $(f) = (-\infty, 0) \cup (0, \infty)$. Since $f(-x) = \log|-x| = \log|x| = f(x)$, f is an even function and its graph is symmetric with respect to the y-axis. Moreover, if $x > 0$, $\log|x| = \log x$. Thus, to sketch the graph of f, we sketch the graph of $g(x) = \log x$ and its reflection through the y-axis as shown in Figure 4.12. We see from Figure 4.12 that the line $x = 0$ (the y-axis) is a vertical asymptote for the graph.

FIGURE 4.12 $f(x) = \log|x|$

x	$\log x$
$\frac{1}{10}$	-1
1	0
10	1

FIGURE 4.13 $f(x) = \log_2(-x)$

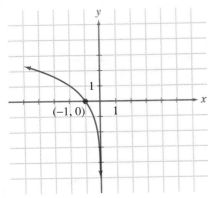

c. For $\log_2(-x)$ to be defined, $-x$ must be positive. Therefore, x must be negative, and hence, dom $(f) = (-\infty, 0)$. Since

$$\log_2(-x) = y \quad \text{if and only if} \quad 2^y = -x,$$

we can find points on the graph of f by substituting values for y into the equation $x = -2^y$. This leads to the following table:

y	-2	-1	0	1	2
$x = -2^y$	$-\frac{1}{4}$	$-\frac{1}{2}$	-1	-2	-4

Plotting these points, we obtain the graph of f as shown in Figure 4.13. Note that, as in part (b), the y-axis is a vertical asymptote for the graph.

Recall from Theorem 2.4 that if f and f^{-1} are inverse functions, then $(f \circ f^{-1})(x) = x$ for each $x \in$ dom (f^{-1}) and $(f^{-1} \circ f)(x) = x$ for each $x \in$ dom (f). For example, since $f(x) = \log_2 x$ is the inverse of $g(x) = 2^x$,

$$\log_2 2^x = x \quad \text{for } x \in \mathbb{R} \quad \text{and} \quad 2^{\log_2 x} = x \quad \text{for } x \in (0, \infty).$$

Also, since $f(x) = \log_e x = \ln x$ is the inverse of $g(x) = e^x$,

$$\log_e e^x = \ln e^x = x \quad \text{for } x \in \mathbb{R} \quad \text{and} \quad e^{\log_e x} = e^{\ln x} = x \quad \text{for } x \in (0, \infty).$$

In general,

$$\log_b b^x = x \quad \text{for } x \in \mathbb{R}, \tag{3}$$

and
$$b^{\log_b x} = x \quad \text{for } x \in (0, \infty). \tag{4}$$

Since $b^1 = b$, it follows from property (3) that

$$\log_b b = 1.$$

Similarly, since $b^0 = 1$,

$$\log_b 1 = 0.$$

Properties of logarithms are often useful when we solve equations involving exponential or logarithmic functions, as we see in the next examples.

E X A M P L E 12

Solve each of the following equations for x.

a. $10 = e^{4x}$ **b.** $15 = 10^{2-3x}$

Solution

a. Since an exponential function with base e appears in the equation $10 = e^{4x}$, we take the natural logarithm of each side and we apply property (3).

$$\begin{aligned}
10 &= e^{4x} \\
\ln 10 &= \ln e^{4x} \\
&= 4x. \qquad \text{Property (3)}
\end{aligned}$$

Hence,
$$x = \frac{\ln 10}{4}.$$

Using a calculator, we find that $x \approx 0.576$.

b. Since an exponential function with base 10 appears on the right side of equation $15 = 10^{2-3x}$, we take the common logarithm of each side to obtain the equation

$$\begin{aligned}
\log 15 &= \log 10^{2-3x} \\
&= 2 - 3x. \qquad \text{Property (3)}
\end{aligned}$$

Thus, $\log 15 = 2 - 3x$, and solving the latter equation for x yields

$$x = \frac{2 - \log 15}{3}.$$

Again, using a calculator, we find that $x \approx 0.275$.

E X A M P L E 13

Solve each of the following equations for x:

a. $\log_x 16 = -4.$ **b.** $\log_4 x^2 = 3.$ **c.** $\log(x^2 + 5x + 64) = 2.$

Solution

a. By Definition 4.2, the equation $\log_x 16 = -4$ is equivalent to the equation $x^{-4} = 16$. Thus, since $x > 0$,

$$x = 16^{-1/4} = \frac{1}{16^{1/4}} = \frac{1}{\sqrt[4]{16}} = \frac{1}{2}.$$

b. Since $\log_4 x^2 = 3$, it follows that $x^2 = 4^3 = 64$. Therefore $x = \pm 8$.

c. Since $\log(x^2 + 5x + 64) = 2$ if and only if $x^2 + 5x + 64 = 10^2$, it follows that

$$x^2 + 5x + 64 = 100,$$
$$x^2 + 5x - 36 = 0, \quad \text{or}$$
$$(x + 9)(x - 4) = 0.$$

Thus, $x = -9$ or $x = 4$.

Functions such as the exponential and logarithmic functions, which are not algebraic functions, are called **transcendental functions**. Such functions are widely used in applied mathematics. In Chapter 5, we will study the six basic trigonometric functions, which are also transcendental functions.

E X E R C I S E S 4.2

In Exercises 1–8, write each equation in the equivalent logarithmic form.

1. $6^3 = 216$

2. $8^2 = 64$

3. $(\frac{1}{3})^{-2} = 9$

4. $(\frac{1}{4})^{-5} = 1024$

5. $\sqrt{5} = 5^{1/2}$

6. $\sqrt[4]{10} = 10^{1/4}$

7. $e^{4t} = 10$

8. $e^{2-6t} = 14$

In Exercises 9–14, write each equation in the equivalent exponential form.

9. $\log_4 \frac{1}{4} = -1$

10. $\log_3 \frac{1}{81} = -4$

11. $\ln \sqrt[4]{e} = \frac{1}{4}$

12. $\ln \sqrt[5]{e} = \frac{1}{5}$

13. $\log_6 x = a$

14. $\log_b(2x + 3) = 5$

In Exercises 15–22, find the exact value of each expression using properties (3) and (4) on page 185.

15. $\log_2 128$

16. $\log_2 512$

17. $\log_6 1296$

18. $\log_6 \frac{1}{216}$

19. $\log_2 \dfrac{\sqrt[3]{32}}{8}$

20. $\log_4 \dfrac{\sqrt[3]{4}}{64}$

21. $5^{\log_5 25}$

22. $6^{\log_6 9^2}$

In Exercises 23–34,

a. Give the domain of each function f.

b. Sketch the graph of f.

c. Give the equation of any vertical asymptote of the graph of f.

23. $f(x) = \log_3(x - 2)$

24. $f(x) = \log_5(x + 3)$

25. $f(x) = 2 \ln |x|$

26. $f(x) = \ln |x - 1|$

27. $f(x) = 4 + \log(x - 2)$

28. $f(x) = 1 - \log(x + 5)$

29. $f(x) = \ln(-x) - 1$

30. $f(x) = 4[\ln(-2x) + 3]$

31. $f(x) = \sqrt{\log_3 x}$

32. $f(x) = \sqrt{\log_{1/2} x}$

33. $f(x) = \sqrt{\log x}$

34. $f(x) = \sqrt{\ln x}$

In Exercises 35–44, solve for x.

35. $\log_x 10{,}000 = 4$

36. $\log_x \frac{1}{256} = -8$

37. $\log_3 x = -2$

38. $\log_9 729 = x$

39. $\log_5 x^2 = 2$

40. $\log_2(x^2 + 2x) = 8$

41. $\log(x^2 - 5x + 14) = 1$

42. $\log_5(2x^2 - 3x + 26) = 2$

43. $\ln e^{x^2} = 6$

44. $e^{\ln |x|} = 2$

45. Find the domain of f where

a. $f(x) = \log_6(3x + 1)$

b. $f(x) = \log_7(8 - 3x)$

c. $f(x) = \ln^2 x$

d. $f(x) = \ln x^2$

46. Find the domain of f where

a. $f(x) = \log(6 + x - x^2)$

b. $f(x) = \ln \left| \dfrac{x + 2}{2x - 3} \right|$

c. $f(x) = \dfrac{6}{2 + \ln x}$

d. $f(x) = |\ln x - 1|$

In Exercises 47–54, find functions g and h such that $f = g \circ h$. Verify your answer by forming the composition.

47. $f(x) = \log_6(3x + 1)$

48. $f(x) = \ln(e^x + 1)$

49. $f(x) = \ln |x + 2|$

50. $f(x) = |\ln x - 1|$

51. $f(x) = \ln x^2$

52. $f(x) = \ln^2 x$

53. $f(x) = e^{x^2 + x - 1}$

54. $f(x) = e^{\log_2 (x + 1)}$

55. Chemists use a number, denoted by pH, to describe quantitatively the acidity or basicity of a substance. By definition,

$$\text{pH} = -\log[\text{H}^+]$$

where $[\text{H}^+]$ is the hydrogen ion concentration in moles per liter. Solutions with a pH of 7 are said to be neutral; a pH below 7 indicates an acid and a pH above 7 indicates a base.

a. Find the pH for stomach acid given that it has a hydrogen ion concentration of 1.2×10^{-3}.

b. Find $[\text{H}^+]$ for blood, given that its pH is 7.3.

c. Characterize carrots as an acid or a base given that for carrots, $[H^+] = 5.0 \times 10^{-9}$.

56. The loudness of a sound, as experienced by the human ear, is based upon intensity levels. The formula

$$S = 10 \log\left(\frac{I}{I_0}\right)$$

gives the loudness, in decibels, produced by a sound wave of intensity I watts per square meter at the eardrum. The constant I_0 is a special value of I agreed upon to be the weakest sound that can be detected by the ear under certain conditions. Find S if

a. I is 10 times as great as I_0.
b. I is 100 times as great as I_0.
c. I is 10,000 times as great as I_0. (This is the intensity level of the average voice.)

Using a graphing calculator, sketch the graph of each of the following functions in the same coordinate system:

57. $f(x) = \ln \sqrt{3}x$, $g(x) = \ln[\sqrt{3}(x + 0.2)]$, and $h(x) = \ln[\sqrt{3}(x + 0.2)] - 7$

58. $f(x) = \ln 3x$, $g(x) = \ln(3x^2 - 6\sqrt{2}x + 6)$, and $h(x) = \ln(3x^2 - 6\sqrt{2}x + 6) + e$

59. a. $f(x) = 6 - x$ and $g(x) = \ln(x^2 - 1)$.
    b. Approximate, to two decimal places, the solution of the equation $6 - x = \ln(x^2 - 1)$.

60. Using a graphing calculator, sketch the graph of each pair of functions in the same coordinate system. Determine a relationship between each given pair of functions.
    a. $f(x) = \ln x$; $g(x) = \ln(2x) - \ln 2$
    b. $f(x) = \ln x$; $g(x) = \ln(x/e) + 1$
    c. $f(x) = 2 \ln x$; $g(x) = \ln x^5 - \ln x^3$

4.3 PROPERTIES OF LOGARITHMS

In this section we translate properties of exponents into useful properties of logarithms by appealing to the relationship between exponents and logarithms. Following the proofs of some of these properties, we illustrate their uses in several examples.

North Wind Picture Archives

JOHN NAPIER
(1550–1617)

John Napier, Baron of Murchiston, was a Scottish laird who devoted much of his leisure time to mathematics, politics, and religion. In 1614, after twenty years of study, he published his discovery of logarithms. The word *logarithm* is derived from the Greek words *logos*, which means "ratio," and *arithmos*, which means "number." Napier's development of logarithms did not incorporate the

T H E O R E M 4.2   Properties of Logarithms

Let p, q, and b be positive real numbers with b not equal to 1. Then

(i) $\log_b pq = \log_b p + \log_b q$;
(The logarithm of the product of two positive numbers is the sum of the logarithms of the two numbers.)

(ii) $\log_b \dfrac{p}{q} = \log_b p - \log_b q$;
(The logarithm of the quotient of two positive numbers is the logarithm of the numerator minus the logarithm of the denominator.)

(iii) $\log_b p^r = r \log_b p$, where r is any real number.
(The logarithm of a positive number p raised to any real number r is the product of r and the logarithm of p.)

Proof

(i) First, let $M = \log_b p$ and $N = \log_b q$. Then by definition of logarithm,
$$b^M = p \quad \text{and} \quad b^N = q.$$
Hence, $\qquad p \cdot q = b^M \cdot b^N = b^{M+N}.$ **(1)**
Writing equation (1) in its equivalent logarithmic form yields
$$\log_b(p \cdot q) = M + N$$
$$= \log_b p + \log_b q,$$
which is our desired result.

idea of a base for logarithms. It was not until later that the English mathematician Henry Briggs (1561–1631) consulted with Napier and introduced the common logarithm. Euler introduced the concept of the natural logarithm more than one hundred years later. The work of Napier and Briggs led to the invention in 1620 of the slide rule, an instrument used until the 1960's to find products, quotients, powers, and roots using the laws of logarithms.

(ii)   The proof of property (ii) is similar to the proof of property (i) (see Exercise 61).

(iii) If we let $M = \log_b p$, then $p = b^M$.

Thus, $$p^r = (b^M)^r.$$

Taking logarithms of both sides of this equation and simplifying, we have

$$\begin{aligned}
\log_b p^r &= \log_b(b^M)^r \\
&= \log_b(b^{Mr}) \\
&= \log_b(b^{rM}) \\
&= rM \qquad \log_b b^k = k \\
&= r \log_b p.
\end{aligned}$$

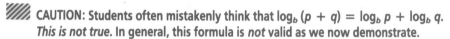 CAUTION: Students often mistakenly think that $\log_b (p + q) = \log_b p + \log_b q$. *This is not true.* In general, this formula is *not* valid as we now demonstrate.

$$\log_2(4 + 4) = \log_2 8 = 3, \quad \text{while} \quad \log_2 4 + \log_2 4 = 2 + 2 = 4.$$

In fact, there is no formula that expresses $\log_b(p + q)$ or $\log_b(p - q)$ in terms of $\log_b p$ and $\log_b q$ (see Exercise 57).

The following examples illustrate uses of the properties of logarithms.

EXAMPLE 14

If $\log_b 4 \approx 1.39$ and $\log_b 11 \approx 2.40$, find

a. $\log_b 44$ **b.** $\log_b \frac{4}{11}$ **c.** $\log_b \sqrt[3]{11}$ **d.** $\dfrac{\log_b 4}{\log_b 11}$

Solution

a. $\begin{aligned}[t]
\log_b 44 &= \log_b(4 \cdot 11) \\
&= \log_b 4 + \log_b 11 \qquad \text{Property (i)} \\
&\approx 1.39 + 2.40 = 3.79.
\end{aligned}$

b. $\begin{aligned}[t]
\log_b \tfrac{4}{11} &= \log_b 4 - \log_b 11 \qquad \text{Property (ii)} \\
&\approx 1.39 - 2.40 = -1.01.
\end{aligned}$

c. $\begin{aligned}[t]
\log_b \sqrt[3]{11} &= \log_b (11)^{1/3} \\
&= \tfrac{1}{3} \log_b 11 \qquad \text{Property (iii)} \\
&\approx \tfrac{1}{3}(2.40) \\
&= 0.80.
\end{aligned}$

 From properties (i) and (ii) of Theorem 4.2, we see that by using logarithms we can reduce the operations of multiplication and division to addition and subtraction. At the time Scottish mathematician John Napier (1550–1617) invented logarithms, the study of astronomy involved laborious mathematical computations. The use of logarithms so greatly simplified these calculations that French mathematician Pierre de Laplace is reported to have said that their invention doubled the life of the astronomer. Today, with the advances in computer technology, it is perhaps difficult to appreciate the impact that logarithms have had on mathematics.

d. Since there is no property of logarithms that allows us to simplify $\dfrac{\log_b 4}{\log_b 11}$, we must divide $\log_b 4 \approx 1.39$ by $\log_b 11 \approx 2.40$ to obtain the approximation 0.58. Compare this result with that in part (b).

EXAMPLE 15

If x, y, and z are positive real numbers, express $\log_b \dfrac{x^4 \sqrt[3]{y}}{z^2}$ in terms of logarithms of x, y, and z.

Solution

$$
\begin{aligned}
\log_b \frac{x^4 \sqrt[3]{y}}{z^2} &= \log_b (x^4 y^{1/3}) - \log_b z^2 \qquad &\text{Property (ii)} \\
&= \log_b x^4 + \log_b y^{1/3} - \log_b z^2 \qquad &\text{Property (i)} \\
&= 4 \log_b x + \frac{1}{3} \log_b y - 2 \log_b z. \qquad &\text{Property (iii)}
\end{aligned}
$$

EXAMPLE 16

Rewrite each expression as a single logarithm.

a. $\log_4 5 + \log_4 (x + 1)$ **b.** $3 \log x - \dfrac{1}{5}\log(x^2 - 4)$

Solution

a. $\log_4 5 + \log_4 (x + 1) = \log_4[5(x + 1)].$ Property (i)

b. $3 \log x - \dfrac{1}{5} \log(x^2 - 4) = \log x^3 - \log(x^2 - 4)^{1/5}$ Property (iii)

$$
\begin{aligned}
&= \log x^3 - \log \sqrt[5]{x^2 - 4} \\
&= \log \frac{x^3}{\sqrt[5]{x^2 - 4}}. \qquad \text{Property (ii)}
\end{aligned}
$$

Logarithmic and Exponential Equations

The following examples demonstrate the use of the properties of logarithms in solving equations.

EXAMPLE 17

Solve the equation

$$\log x + \log(x - 21) = 2.$$

Solution   First, we note that both x and $x - 21$ must be positive for $\log x$ and $\log(x - 21)$ to be defined. Hence, we require that $x > 21$. Since

$$\log x + \log (x - 21) = 2,$$

$$\log[x(x - 21)] = 2, \text{ by property (i).}$$

Thus, by the definition of logarithm, it follows that

$$x(x - 21) = 10^2,$$

or
$$x^2 - 21x = 100.$$ (2)

Therefore,
$$x^2 - 21x - 100 = 0.$$

Solving equation (2) by factoring, we find that its solutions are -4 and 25. Since x must be greater than 21, the solution for the given equation is 25.

EXAMPLE 18

Solve the equation

$$4 = 5e^{3-2x}.$$

Solution   Taking the natural logarithm of both sides of the equation $4 = 5e^{3-2x}$ and applying properties of logarithms, we have

$$4 = 5e^{3-2x}$$

$$\ln 4 = \ln(5e^{3-2x})$$

$$= \ln 5 + \ln e^{3-2x}   \qquad \textbf{Property (i)}$$

$$= \ln 5 + 3 - 2x.   \qquad \textbf{In } e^n = n$$

Therefore, $\ln 4 = \ln 5 + 3 - 2x.$ (3)

Solving equation (3) for x yields

$$2x = \ln 5 - \ln 4 + 3, \qquad \text{or} \qquad x = \frac{\ln 5 - \ln 4 + 3}{2}.$$

Using a calculator, we find that $x \approx 1.61$.

EXAMPLE 19

Solve the equation

$$\frac{3^x - 3^{-x}}{2} = 4.$$

Solution   We begin by multiplying both sides of the equation $\dfrac{3^x - 3^{-x}}{2} = 4$ by 2 to get

$$3^x - 3^{-x} = 8.$$

Multiplying both sides of the last equation by 3^x, we have

$$(3^x)^2 - 1 = 8(3^x), \qquad \text{or} \qquad (3^x)^2 - 8(3^x) - 1 = 0.$$

Letting $t = 3^x$, we obtain the quadratic equation

$$t^2 - 8t - 1 = 0   \qquad (4)$$

in the variable t. Using the quadratic formula to solve equation (4), we find that

$$t = 4 \pm \sqrt{17}.$$

Since $t = 3^x$ is always positive, it follows that

$$t = 4 + \sqrt{17}, \qquad \text{or} \qquad 3^x = 4 + \sqrt{17}.   \qquad (5)$$

Taking the natural logarithm of both sides of equation (5) and using property (iii) of logarithms, we have

$$\ln 3^x = \ln(4 + \sqrt{17})$$

$$x \ln 3 = \ln(4 + \sqrt{17}).$$

Hence,

$$x = \frac{\ln(4 + \sqrt{17})}{\ln 3}.$$

Using a calculator, we find that $x \approx 1.91$.

Change-of-Base Formula

Most calculators have common and natural logarithm keys, but they do not have a key for the general logarithm $\log_b x$ where $b \ne 10$ and $b \ne e$. The following example suggests a way of evaluating $f(x) = \log_b x$ in terms of $g(x) = \ln x$. (We could also use $\log x$ in place of $\ln x$.)

EXAMPLE 20

Express $\log_3 8$ in terms of natural logarithms.

Solution   Letting $t = \log_3 8$, we have $3^t = 8$, so that $\ln 3^t = \ln 8$. By property (iii) of logarithms, the last equation can be written as

$$t \ln 3 = \ln 8.$$

Since $\ln 3 \ne 0$ (Why?), we have

$$t = \frac{\ln 8}{\ln 3}, \qquad \text{or} \qquad \log_3 8 = \frac{\ln 8}{\ln 3}.$$

In general, the following **change-of-base** formula can be proved (see Exercise 59).

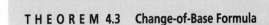

THEOREM 4.3   Change-of-Base Formula

Let a, b, and x be positive numbers, where $a \ne 1$ and $b \ne 1$. Then

$$\log_b x = \frac{\log_a x}{\log_a b}.$$

Illustrating Theorem 4.3, we have

$$\log x = \frac{\ln x}{\ln 10} \qquad \text{Changing base 10 to base } e$$

$$\ln x = \frac{\log x}{\log e} \qquad \text{Changing base } e \text{ to base 10}$$

$$\log_4 x = \frac{\log_3 x}{\log_3 4}. \qquad \text{Changing base 4 to base 3}$$

Theorem 4.3 tells us that for any positive number x, $\log_b x$ is equal to $\log_a x$

multiplied by the constant $\dfrac{1}{\log_a b}$. This fact is useful in the study of logarithmic functions in calculus.

EXERCISES 4.3

In Exercises 1–10, simplify each expression by using the definition of logarithm and properties of logarithms.

1. $\log_2 2$
2. $\log_4 \sqrt[4]{4} + 4\log_4 64$
3. $\log_8 72 - \log_8 9$
4. $\log_5 30 + \log_5 \frac{5}{6}$
5. $-\frac{1}{4} + \log \sqrt[4]{10}$
6. $\log_7 2 - \log_7 14$
7. $\ln e^4 - \ln 1$
8. $2^{\log_2 16 - \log_2 2^{16}}$
9. $\log_b b^b$
10. $\log_b b^{1/b}$
11. If $\log_b 5 \approx 0.90$ and $\log_b 12 \approx 1.39$, find
    a. $\log_b 60$ b. $\log_b \frac{5}{12}$ c. $\log_b \sqrt[5]{12}$
    d. $(\log_b 5)/(\log_b 12)$
12. Prove that $\log_b(xyz) = \log_b x + \log_b y + \log_b z$ where x, y, and z are positive real numbers.

In Exercises 13–20, rewrite each expression as a single logarithm in simplest form.

13. $4\ln 3x - \ln x$
14. $\log \dfrac{x^2}{y} + \log \dfrac{y^2}{2x}$
15. $\log_6 \frac{2}{3} - \log_6 4 + \log_6 32$
16. $\ln 2x - \ln x^2 - \ln \dfrac{4}{x}$
17. $\log_b \dfrac{x+y}{2z} - \log_b \dfrac{4}{x+y}$
18. $\log_2 \dfrac{x^2 + 2x - 8}{x^2 - 3x + 2} - \log_2 \dfrac{x^2 + 6x + 8}{x^2 - 1}$
19. $\log_5 x - \frac{1}{2}[\log_5(x-1) + \log_5(2x+3)]$
20. $\log_b 6 + 5[\log_b(2x) - 3\log_b(1-x)]$

In Exercises 21–28, write each expression in terms of logarithms of x, y, and z. Assume that x, y, and z are positive real numbers.

21. $\log_b \dfrac{x^2 y^2}{z^3}$
22. $\log_b \dfrac{xz^3}{y}$
23. $\log_2 \sqrt[5]{\dfrac{x^2}{yz}}$
24. $\log_4 x \sqrt[3]{\dfrac{y}{z}}$
25. $\ln \dfrac{2 - x}{(x+1)(2x-1)^2}$
26. $\ln \dfrac{1}{x^3 - x + 2}$
27. $\log_b xy\sqrt{z}$
28. $\log_b \sqrt[3]{xy^2} \sqrt[3]{z}$

In Exercises 29–40, solve each equation without using a calculator or a table. Approximate irrational solutions to two decimal places.

29. $2^x = 10$
30. $e^x = 15$
31. $4^{2-x} = 5^{x-3}$
32. $3^{-x^2} = \frac{1}{8}$
33. $\log(40x + 1) = 4 + \log(x - 2)$
34. $\ln x + \ln(x + 1) = \ln 6$
35. $\log_2(x^2 - 7) = 4$
36. $\log_3(2x^2 - 1) = 3$

37. $\log_x(15 - 10x) = 2$
38. $\log_x(x^2 + 3x - 3) = 3$
39. $\log_3 |4 - x^2| = 2$
40. $\log_5 |6 - 2x^2| = 3$

In Exercises 41–46, write each expression in terms of common logarithms.

41. $\log_5 6$
42. $\ln 4$
43. $\log_3 x$
44. $\log_4(x^2 - 1)$
45. $\ln(x - 1)$
46. $\log_b(5 - x)$

In Exercises 47–50, write each expression in terms of natural logarithms.

47. $\log 15$
48. $\log_2 x$
49. $\log_b(x^2 - 1)$
50. $\log_4(x^2 - 1)$

In Exercises 51–54, use natural logarithms to solve each equation for x in terms of y. See Example 19.

51. $y = \dfrac{e^x + e^{-x}}{2}$
52. $y = \dfrac{e^x - e^{-x}}{2}$
53. $y = \dfrac{e^x + e^{-x}}{e^x - e^{-x}}$
54. $y = \dfrac{e^x - e^{-x}}{e^x + e^{-x}}$

In Exercises 55–58, use your calculator to help verify each statement.

55. Show that $\dfrac{\log 29}{\log 42} \neq \log 29 - \log 42$.
56. Show that $\sqrt{\ln 6} \neq \ln \sqrt{6}$.
57. Show that $\ln(2^{10} + 5) \neq \ln 2^{10} + \ln 5$.
58. Show that $(\log 47)^3 \neq 3\log 47$.
59. Let a and b be positive numbers each different from 1. Prove that
    a. $\log_b x = \dfrac{\log_a x}{\log_a b}$ where $x > 0$, (Theorem 4.3)
    b. $\log_b a = \dfrac{1}{\log_a b}$.
60. Simplify $(\log_2 5)(\log_5 4)(\log_6 7)$ completely.
61. Prove property (ii) of Theorem 4.2.
62. If $f(x) = \ln(x + \sqrt{x^2 + 1})$, show that $f^{-1}(x) = \dfrac{e^x - e^{-x}}{2}$.
63. Using a graphing calculator, graph the functions defined by the given equations in the same coordinate system. [*Hint*: Use the change-of-base formula when necessary.]
    a. $f(x) = \ln x$ b. $g(x) = \log_{1.08} x$
    c. $h(x) = \log_7 x$
64. Using a graphing calculator, graph each pair of functions given in Exercise 60 in Section 4.2. Use properties of logarithms to explain the relationship between the graphs of each of the given pairs of functions.

4.4 APPLICATIONS

Now that we have introduced the general exponential and logarithmic functions, we consider a wide range of practical problems that can be described mathematically in terms of these functions. We begin by considering examples from the investment world.

EXAMPLE 21

A certain account pays interest at the rate of 8% compounded continuously. How long does it take for an investment to double in value at this rate?

Solution Assume P dollars are invested in the account. From equation (3) on page 178, the amount in the account after t years is

$$A(t) = Pe^{0.08t}.$$

To find the time it takes for the money to double, we must solve the equation

$$2P = Pe^{0.08t} \qquad (1)$$

for t.

Dividing both sides of equation (1) by P yields

$$e^{0.08t} = 2.$$

Taking the natural logarithm of both sides, we get

$$\ln e^{0.08t} = \ln 2, \qquad \text{or}$$
$$0.08t = \ln 2. \qquad \qquad \ln e^x = x$$

Hence $t = \dfrac{\ln 2}{0.08} \approx 8.664$ years, and the value of the account doubles in approximately 8.7 years.

EXAMPLE 22

How much money should be invested in a bank account that pays 10% interest, compounded continuously, to have a total of $50,000 in the account after 15 years?

Solution Using equation (3) on page 178, with P denoting the amount of money initially invested, we have

$$50,000 = Pe^{(0.10)(15)} = Pe^{1.5}.$$

Solving for P yields $\qquad P = \dfrac{50,000}{e^{1.5}} \approx \$11,156.51.$

Therefore, $11,156.51 invested in an account paying interest at the rate of 10% compounded continuously, yields $50,000 at the end of 15 years.

Observe that the equations formulated in the solutions of Examples 21 and 22 both have the form

$$y = y_0 e^{kt}, \quad \text{where } y_0 > 0. \qquad (2)$$

Equation (2) is said to represent **exponential growth** if $k > 0$ and **exponential decay** if $k < 0$. The value of the constant k determines how rapidly the growth or decay occurs. Thus, we see from Examples 21 and 22 that the amount of money in an account grows exponentially when interest is compounded continuously. As we shall now see, radioactive material provides a good example of exponential decay.

EXAMPLE 23

Polonium, a radioactive element discovered by the Polish-born scientist Marie Curie, is known to decay exponentially. If y_0 grams of polonium are present initially, the number of grams, y, present after t days is given by

$$y = y_0 e^{-0.005t}.$$

a. If $y_0 = 2$ grams, sketch a graph that shows the amount y, in grams, of polonium remaining after t days, for $0 \le t \le 365$.

b. How long will it take for one-half of the 2-gram sample to disintegrate?

FIGURE 4.14

t	$y = 2e^{-0.005t}$
0	2
100	1.2130
200	0.7358
300	0.4463
365	0.3224

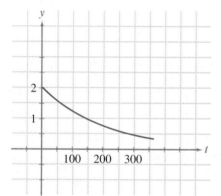

Solution

a. We use a calculator to find points on the graph of $y = 2e^{-0.005t}$ that is sketched in Figure 4.14.

b. To find how long it takes for one-half of the sample to disintegrate, we solve the equation $y_0/2 = y_0 e^{-0.005t}$ for t where $y_0 = 2$.

$$1 = 2e^{-0.005t}$$

$$\frac{1}{2} = e^{-0.005t}$$

$$\ln \frac{1}{2} = \ln e^{-0.005t} \qquad \textbf{taking the natural logarithm of both sides}$$

$$\ln \frac{1}{2} = -0.005t \qquad \textbf{ln } e^x = x$$

$$t = \frac{\ln \frac{1}{2}}{-0.005}$$

$$\approx 138.6 \text{ days.}$$

In part (b) of Example 23, we found that one-half of the 2-gram sample of polonium disintegrated in approximately 138.6 days. We say that the **half-life** of polonium is approximately 138.6 days. In general, the half-life of a radioactive substance is the time required for one-half of a given sample to disintegrate. The half-life is an intrinsic property of a substance and, therefore, does not depend on the sample size. For example, 10 grams of polonium will decay to 5 grams in approximately 138.6 days. (Verify this for yourself following the procedures outlined in part (b) of Example 23.)

EXAMPLE 24

A breeder reactor converts the relatively stable uranium 238 into the isotope plutonium 239. After ten years it is determined that 0.057% of the initial amount y_0 of the plutonium has disintegrated. Find the half-life of this isotope.

Solution   If 0.057% of the initial amount present has disintegrated after ten years, then 99.943% of the substance remains. Using equation (2) on page 193, we have

$$0.99943y_0 = y_0e^{10k}.$$

Solving the latter equation for k, we first divide both sides by y_0:

$$0.99943 = e^{10k}.$$

Thus,

$$\ln(0.99943) = 10k \quad \text{or}$$

$$k = \frac{\ln(0.99943)}{10}$$

$$\approx -0.00005702.$$

To find the half-life of this isotope, we solve the equation

$$\frac{y_0}{2} = y_0e^{-0.00005702t}$$

for t. Using the same procedure as above yields

$$\frac{1}{2} = e^{-0.00005702t}$$

$$\ln\frac{1}{2} = -0.00005702t$$

$$t = \frac{\ln\frac{1}{2}}{-0.00005702}$$

$$\approx 12{,}160.$$

Hence, the half-life of this isotope is approximately 12,160 years.

EXAMPLE 25

An archaeologist has found a fossil in which the ratio of carbon-14 to carbon-12 is one fourth the ratio found in the atmosphere. Approximately how old is the fossil?

Solution   Letting R_0 be the ratio of carbon-14 to carbon-12 found in the atmosphere, we use equation (2) on page 193 to write

$$\frac{1}{4}R_0 = R_0e^{kt}.$$

Carbon-12, which is stable, and carbon-14, which is radioactive, occur in a fixed ratio in the cells of all living plants and animals. When an organism dies, it ceases to absorb carbon-14. By comparing the proportionate amount of carbon-14 in a fossil with the constant ratio found in the atmosphere, it is possible to obtain a reasonable estimate of its age. This method, devised by Dr. Willard Libby in 1947, is based upon the knowledge that the half-life of carbon-14 is approximately 5600 years. (Dr. Libby was awarded the 1960 Nobel Peace Prize in chemistry for his work in carbon dating.) In 1988, scientists revealed that carbon-dating studies showed that the famous Shroud of Turin, believed by some to be the burial cloth of Jesus, is a medieval forgery.

Photo © Ross Rappaport 1991/FPG International

Solving this equation for t, we have

$$\frac{1}{4} = e^{kt}$$

$$\ln \frac{1}{4} = kt$$

$$t = \frac{\ln \frac{1}{4}}{k},$$

or
$$t = \frac{-\ln 4}{k}. \qquad (4)$$

Using the fact that the half-life of carbon-14 is approximately 5600 years (see box on page 195), we know that $\frac{1}{2}R_0 = R_0 e^{5600k}$. Thus,

$$\frac{1}{2} = e^{5600k},$$

and therefore,
$$\ln \frac{1}{2} = 5600k.$$

Hence,
$$k = \frac{\ln \frac{1}{2}}{5600} = \frac{-\ln 2}{5600}.$$

Substituting $\dfrac{-\ln 2}{5600}$ for k in equation (4) we have

$$t = \frac{-\ln 4}{\dfrac{-\ln 2}{5600}}$$

$$= \frac{2 \ln 2}{\dfrac{\ln 2}{5600}}$$

$$= 2(5600)$$

$$= 11,200.$$

Therefore, the fossil is approximately 11,200 years old.

The amount of a drug that remains in a person's bloodstream decreases exponentially with time. For example, if the initial concentration is C_0 milligrams per milliliter of blood, then the concentration t hours after the drug is administered is given by the equation

$$C(t) = C_0 e^{-kt},$$

where k is a constant that measures how rapidly the drug is absorbed.

Every drug has a minimum concentration below which it is not effective. When the concentration decreases to this level, another dose of the drug should be administered. Using methods we have developed in this chapter, we can determine how often the drug should be administered to maintain an effective concentration. (See Figure 4.15.)

FIGURE 4.15

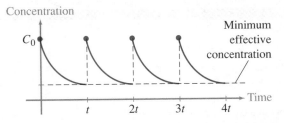

EXAMPLE 26

The rate of absorption of a certain drug is measured to be $k = 0.23$, and the minimum effective concentration is one milligram per milliliter of blood. If the original concentration is $C_0 = 4$, determine how often a dose of this drug should be administered to maintain an effective concentration.

Solution   The concentration of this drug t hours after a dose is administered is given by

$$C(t) = C_0 e^{-kt}$$

where $C_0 = 4$ and $k = 0.23$. Thus,

$$C(t) = 4e^{-0.23t}. \tag{5}$$

To find the time when this concentration reaches the minimum effective level of 1, we set $C(t) = 1$ in equation (5) and solve the resulting equation

$$1 = 4e^{-0.23t}$$

for t. Thus, we have

$$1 = 4e^{-0.23t}$$

$$\frac{1}{4} = e^{-0.23t}$$

$$\ln \frac{1}{4} = -0.23t$$

$$t = \frac{\ln \frac{1}{4}}{-0.23}$$

$$\approx 6.03.$$

Hence, the concentration will reach the minimum effective level in 6.03 hours. Therefore, a dose of this drug should be administered approximately every 6 hours. (See Figure 4.16.)

FIGURE 4.16

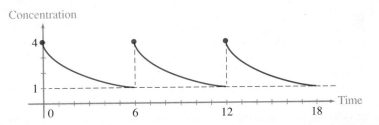

EXERCISES 4.4

1. If $1000 is invested at an interest rate of 7% compounded continuously, how much money is in the account after five years?

2. How much should you invest in a bank account that pays 8% interest, compounded continuously, to have a total of $2500 after ten years?

3. Tom invests $5000 in a savings account paying $7\frac{1}{2}\%$ interest compounded semiannually. Mary invests the same amount at the same rate, but compounded daily. How much larger is Mary's account after five years?

4. If $5000 is invested at an interest rate of 8% per year compounded quarterly, how long will it take the amount in the account to equal
   a. $7500?   b. $10,000?   c. $15,000?

5. The half-life of radium is 1600 years. If the initial amount is y_0 milligrams, then the quantity y remaining after t years is given by $y = y_0 e^{kt}$. Find k.

6. The number of bacteria in a certain culture increased from 200 to 600 between 9:00 A.M. and 11:00 A.M. If we assume the number of bacteria in the culture is growing exponentially, it can be shown that the number y of bacteria, t hours after 9:00 A.M., is given by $y = 200(3)^{t/2}$.
   a. Estimate the number of bacteria in the culture at 10:00 A.M.; 12:00 noon; 1:00 P.M.
   b. Sketch the graph of y for t in $[0, 4]$.

7. An archaeological expedition working in the Gobi desert came across some ancient objects. It was estimated that 20% of the original amount of carbon-14 was still present in the objects. Find the approximate age of the objects.

8. According to Newton's law of cooling, the rate at which an object cools is directly proportional to the difference in temperature between the object and the surrounding medium. You prepare an apple pie for your friends Bob, Ted, and Kimberly. It is removed from an oven at 400° and left to cool at a room temperature of 70°. In one-half hour the pie's temperature is 200°. It can be shown, using methods of calculus, that the temperature y of the pie after t hours of cooling, is given by $y = 330e^{-1.71t} + 70$. Eager to have a piece of the pie, Ted calculates the time when the temperature of the pie will be 100°. How long will he have to wait?

9. If an automobile is purchased for C dollars, then its trade-in value V at the end of t years is given by $V(t) = 0.75 \, C(0.83)^{t-1}$. If the original cost is $15,000, calculate the trade-in value to the nearest dollar after
   a. one year   b. three years   c. five years.

10. A department store requires its credit card customers to pay interest at the rate of 18% per year, compounded monthly, on the unpaid balance. If a person buys a stereo system for $1500 on credit and makes no payments for one year, how much does he owe at the end of the year?

11. If the concentration of a drug in a patient's bloodstream is C (milligrams per milliliter of blood), t hours later the concentration will be
$$C(t) = C_0 e^{-kt},$$
where k is a constant that measures how rapidly the drug is absorbed. If the original concentration is $C_0 = 5$ and $k = 0.12$, determine how often the drug should be administered to maintain an effective concentration of 1.3.

12. If the value of real estate increases at the rate of 5% per year, then after t years the value V of a house purchased for P dollars is given by $V(t) = P(1.05)^t$. If a house is purchased in 1988 for $150,000, what will it be worth in 1995?

13. After t days of advertising a new movie, the proportion P of moviegoers in a city who have seen the ad is
$$P(t) = 1 - e^{-0.05t}.$$
How long must the ad run to reach
   a. 75% of the moviegoers?
   b. 90% of the moviegoers?

14. When a flexible cord or chain is suspended from its ends, it hangs in a curve called a **catenary**, the equation of which has the form $y = \dfrac{a}{2}(e^{x/a} + e^{-x/a})$. If $a = 1$, the resulting catenary is the graph of the **hyperbolic cosine** function. Sketch the graph of the hyperbolic cosine function.

15. The population P of a small country in 1988 was 3,500,000. Its population after t years using ($t = 0$ for 1988) is given by the formula $P(t) = 3,500,000e^{0.02t}$. What is the projected population of this country in the year 2000?

16. A **normal probability density function** f is defined by
$$f(x) = \frac{1}{\sqrt{2\pi}} e^{-(x-\mu)^2/2\sigma^2}$$
where μ is a constant called the **mean**, and σ is a positive constant called the **standard deviation**. Sketch the graph of f for $\mu = 0$ and $\sigma = 1$.

17. Under certain conditions, the atmospheric pressure p (measured in inches of mercury) at altitude h (measured in feet) is given by $p(h) = 29e^{-0.000034h}$. What is the atmospheric pressure at an altitude of 20,000 feet?

18. The electric current I, in amperes, flowing in a series circuit having an inductance of L henrys, a resistance of R ohms, and a constant electromotive force of E volts, satisfies the equation
$$I = \frac{E}{R}(1 - e^{-Rt/L})$$
where t is the time in seconds after the current begins to

flow. Sketch the graph of I as a function of t for $E = 10$ volts, $R = 5$ ohms, and $L = 0.05$ henry.

19. The graph of a function defined by $Q(t) = \dfrac{B}{1 + Ae^{-Bkt}}$ is called a **logistic curve**. The Public Health Department indicates that t weeks after the outbreak of a certain disease, approximately $Q(t) = \dfrac{15}{1 + 14e^{-0.9t}}$ thousand people have caught the disease.

  a. How many people have the disease when it first breaks out?

  b. How many have caught the disease by the end of the second week?

  c. Sketch the graph of Q.

  d. If the trend continues, use the graph to estimate the total number of people who will contract the disease.

20. The magnitude M of an earthquake on the Richter scale (U.S. seismologist Charles Richter established the scale in 1935) is given by $M = 0.67 \cdot \log(0.37E) + 1.46$, where E is the energy of the earthquake in kilowatt-hours. Find the energy of an earthquake of magnitude 7.

21. The maximum permitted level of lead in fresh water, having a water hardness of h milligrams per liter, is $M(h) = e^p - 3.37$, where $p = 1.51(\ln h)$. Find $M(h)$ if

  a. $h = 8.3$ **b.** $h = 10$.

22. The percentage p of moisture that falls as snow rather than rain in the central Sierra Nevada mountains of California is approximated by

$$p(h) = 8.63 \ln (h - 680)$$

where h is the altitude given in feet ($h > 300$).

  a. Find p when h equals

    (1) 3000 feet   **(2)** 6000 feet   **(3)** 8000 feet

  b. Sketch the graph of p on the interval $[3{,}000, 12{,}000]$

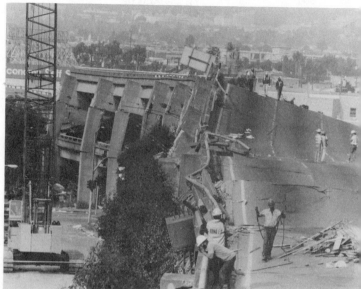

Photo AP/Wide World Photos

CHAPTER 4 REVIEW EXERCISES

In Exercises 1–10, sketch the graph of the function f. Give dom (f), ran (f), and if applicable, the equation of any horizontal or vertical asymptotes.

1. $f(x) = 7^x$

2. $f(x) = (\tfrac{1}{4})^x$

3. $f(x) = \log_4 x$

4. $f(x) = \log_3(x - 1)$

5. $f(x) = e^{x^2}$

6. $f(x) = x + e^{3x}$

7. $f(x) = \ln(x + 4)$

8. $f(x) = \ln(x^2 - 4) + 1$

9. $f(x) = (1.08)^x$

10. $f(x) = \dfrac{2}{1 + e^{2x}}$

In Exercises 11–18, find the solution set. Do not use a calculator.

11. $4^{2x+3} = 7^{x-2}$

12. $2^{5x+3} = 3^{2x+3}$

13. $\log(5x - 1) - \log(x - 3) = 2$

14. $\log \sqrt[4]{x + 1} = \tfrac{1}{2}$

15. $e^{\ln|x+2|} = 4$

16. $e^{1-4x} = e$

17. $\ln(x + 3) = \ln x + \ln 3$

18. $\ln(3x) = \ln 3 + \ln x$

In Exercises 19–24, simplify each expression completely.

19. $\log_2 \sqrt[3]{2}$

20. $\log_5 75 - \log_5 3$

21. $\dfrac{\ln 16}{\ln 4}$

22. $\log_2 2^7$

23. $3^{\log_3 4} - 6 \log_4 \sqrt[3]{4}$

24. $e^{\ln 2} - e^{\ln 3} + e^{\ln e}$

25. Express $\log x^3 \sqrt[4]{\dfrac{z}{y^3}}$ in terms of logarithms of x, y, and z.

26. Express $\ln \dfrac{x^2}{y^3} + 3 \ln y - 4 \ln xy$ as a single logarithm.

27. The Beer-Lambert Law asserts that the amount of light I that penetrates to a depth of x meters in seawater is given by

$$I = I_0 b^x \qquad (1)$$

where $0 < b < 1$ and I_0 is the amount of light at the surface.

  a. Use common logarithms to solve equation (1) for x in terms of I, I_0, and b.

  b. For $b = \tfrac{1}{3}$, find the depth x at which $I = 0.01 I_0$.

28. Find the hydrogen ion concentration $[H^+]$ in moles/liter of each substance (rounded to two decimal places). See Exercise 55 on page 186.

  a. vinegar: pH $= 3.1$

  b. rain: pH $= 5.6$

  c. beer: pH $= 4.3$.

29. At takeoff, a certain supersonic jet produces a sound wave of intensity 0.3 watt per square meter. Using $I_0 = 10^{-12}$ watt per square meter, find the loudness in decibels. (See Exercise 56, page 187.)

30. A manufacturing plant estimates that the value V, in dollars, of a machine is decreasing exponentially according to the equation

$$V(t) = 75,000e^{-0.12t}$$

where t is the number of years since the machine was put into service.
  a. Find the approximate value of the machine after ten years.
  b. Find the "half-life" of the machine.

31. In psychological tests it is often found that if a group of people memorize a list of nonsense words, the fraction P of these people who remember all the words x hours later is given by

$$P = 1 - c \cdot \ln(x - 1)$$

where c is a constant depending on the length of the list of words and other factors. A certain group was given memory lists and after 3 hours, only one-third of the group could remember all the words.
  a. Find the value of c for this experiment.
  b. Predict the approximate fraction of group members who will remember all the words after 4 hours.

32. The half-life of a radioactive substance is m days. If the initial sample weighs g grams, how long will it take until h grams are left?

33. You have just won first prize in a state lottery and you have your choice of the following:
  a. $500 will be placed in a savings account in your name, and the interest will be compounded continuously at an annual rate of 10% for ten years, or
  b. One penny will be placed in a fund in your name, and the amount in the fund will double every 6 months over the next ten years.
  Which plan do you choose? Why?

34. Which investment has the greater effective interest rate: 8.5% per year compounded quarterly or 8.4% per year compounded continuously?

35. How much should you invest now at an annual interest rate of 8% so that your balance ten years from now will be $25,000 if interest is compounded
  a. daily?   **b.** continuously?

36. The value of a certain asset depreciates in such a way that at the end of any 12-month period it is 80% of its value at the beginning of that period. If the initial value of the asset is V_0, find a formula for the value V in terms of time t.

37. The population of the world in 1975 was approximately 4 billion. Assuming that the world population grows exponentially at a rate of 10% per year, how many people will there be in the year 2000?

38. When learning a particular task, such as typing or skiing, a person progresses faster at the beginning and then levels off. This phenomenon is illustrated by a **learning curve** where the level of performance is considered as a function of time (see the accompanying figure). A learning curve can be closely approximated by an exponential equation of the form

$$y = a(1 - e^{-bx}),$$

where a and b are positive constants.

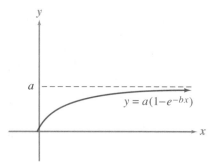

Suppose a particular person's performance level while learning to type is given by the equation

$$y = 75(1 - e^{-0.07x}),$$

where y gives the number of words per minute typed after x weeks of instruction. Approximately how many weeks did it take this person to learn to type 65 words per minute?

39. On a college campus of 1000 students, one student returned from vacation with a contagious flu virus. The spread of the virus through the student population is given by

$$p(t) = \frac{1000}{1 + 999e^{-0.6t}}$$

where $p(t)$ is the total number infected after t days. The college will cancel classes when at least 50% of the students are ill.
  a. How many students are infected after five days?
  b. After how many days will the college cancel classes?

40. The management at a factory has found that the maximum number of units a worker can produce in a day is 100. The learning curve for the number of units U produced per day after a new employee has worked t days is given by

$$U = 100(1 - e^{ct}).$$

After 50 days on the job, a particular worker produced 75 units.
  a. Find the learning curve for this worker.
  b. In how many days will this worker be producing 90 units per day?

CHAPTER 5

Trigonometric Functions

In the previous chapter, we learned that exponential and logarithmic functions are examples of transcendental functions. In this chapter, we continue our study of transcendental functions as we use the unit circle and the wrapping function to develop the trigonometric functions of real numbers. We present the graphs of the six basic trigonometric functions and discuss techniques for graphing functions that are variations of the basic six. The chapter concludes with a development of the inverse trigonometric functions and their graphs.

5.1 THE WRAPPING FUNCTION

Historically, the trigonometric functions were developed by the Greeks to solve geometric problems involving right triangles. The word *trigonometry* comes from the Greek words for *triangle measurement*. Although originally defined for acute angles in right triangles and not for arbitrary real numbers, the trigonometric functions have been extended so that they may be used in situations where angles are not involved. It is in this setting that such functions are studied in calculus, and thus it is appropriate that our development be of this more general type. We defer our study of the trigonometry of right triangles until Chapter 7.

In this section, we will introduce and discuss a function P called the **wrapping function**. In the following section, we will use the wrapping function to define the trigonometric functions of a real number t.

Recall from Section 2.1 that the graph of the equation $x^2 + y^2 = 1$ is the unit circle in the xy-plane with its center at the origin. In preparation for defining the wrapping function, we have graphed the unit circle and constructed a number line L that is tangent to the circle at the point $A = (1, 0)$ (see Figure 5.1). The unit scale used on the number line L is the same as the one used for the unit circle, and the number 0 on L coincides with the point A on the circle.

As we shall see, the wrapping function P *wraps* the number line L around the unit circle by associating with each real number on L a unique point on the unit circle. The portion of L that lies above the point $A = (1, 0)$, and thus contains the positive real numbers, is wrapped in a counterclockwise direction (see Figure 5.2). On the other hand, the portion of L lying below $A = (1, 0)$, and thus containing the negative real numbers, is wrapped in a clockwise direction (see Figure 5.3).

The Trigonometric Point $P(t)$ for a Real Number t

For a given real number t, we find the unique point $P(t) = (x, y)$ on the unit circle, which the wrapping function associates with t, as follows. We start at the *initial point* $A = (1, 0)$ on the unit circle. If $t > 0$, we measure an arc of length t on the unit circle in a *counterclockwise* direction (see Figure 5.4). If $t < 0$, we measure an arc of length $|t|$ on the unit circle in a *clockwise* direction (see

North Wind Picture Archives

HIPPARCHUS OF NICAEA
(ca. 180–ca. 125 B.C.)

Hipparchus of Nicaea, the famous Greek astronomer considered by many to be the founder of trigonometry, was the first to divide a circle into 360 parts or degrees. He also advocated the use of latitude and longitude to locate positions on the earth. For his work in astronomy, Hipparchus developed a type of spherical geometry and compiled the first trigonometric tables. His contributions to astronomy include more precise measurements of important astronomical constants, such as the length of the month and year, the size of the moon, and the precession of the equinoxes. His calculation of the length of the average lunar month was within one second of its presently accepted value.

FIGURE 5.1

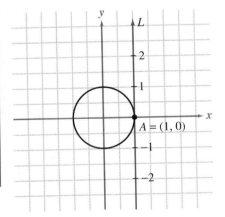

FIGURE 5.2

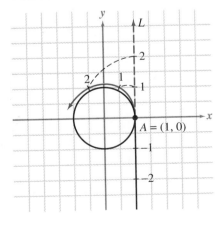

FIGURE 5.3

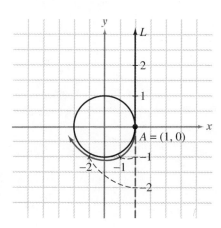

FIGURE 5.4
$P(t) = (x, y)$, where $t > 0$

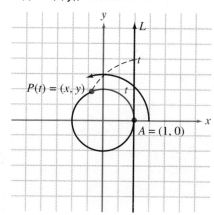

FIGURE 5.5
$P(t) = (x, y)$, where $t < 0$

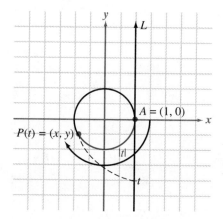

FIGURE 5.6

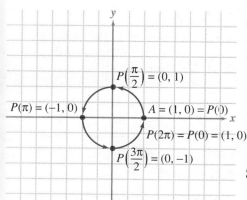

Figure 5.5). If $t = 0$, we stay at the initial point $A = (1, 0)$. Following this procedure, we arrive at a unique point on the unit circle called the **trigonometric point** $P(t)$ **associated with** t.

For some values of t, the exact coordinates of $P(t)$ are easy to find. For example, $P(0) = (1, 0)$, and since 2π is the circumference of the unit circle, $P(2\pi) = (1, 0)$. One-fourth, one-half, and three-fourths of the circumference are, respectively, $\pi/2$, π, and $3\pi/2$. It follows that

$$P\left(\frac{\pi}{2}\right) = (0, 1),$$

$$P(\pi) = (-1, 0), \quad \text{and}$$

$$P\left(\frac{3\pi}{2}\right) = (0, -1).$$

See Figure 5.6.

In our first example, we find the approximate location of $P(t)$ for some other values of t where $|t| \leq 2\pi$.

EXAMPLE 1

Give the approximate location of the point $P(t)$ on the unit circle and, if possible, find the exact coordinates of $P(t)$.

a. $P(2)$ **b.** $P(5)$ **c.** $P\left(-\frac{3\pi}{2}\right)$ **d.** $P(-2)$

Solution

a. To locate $P(2)$, we start at $A = (1, 0)$ and measure in a counterclockwise direction an arc of length $t = 2$. Since

$$\frac{\pi}{2} \approx 1.57 < 2 < 3.14 \approx \pi,$$

it follows that $P(2)$ is in quadrant II. See Figure 5.7 (page 204).

FIGURE 5.7

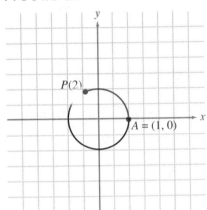

FIGURE 5.8

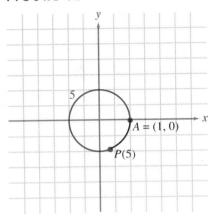

FIGURE 5.9

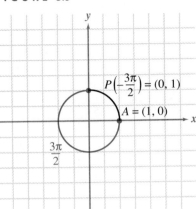

FIGURE 5.10

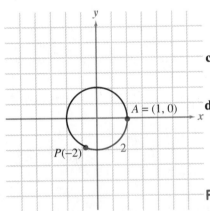

FIGURE 5.11

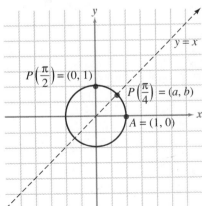

b. Since $5 > 0$, we proceed as in part (a). Noting that

$$\frac{3\pi}{2} \approx 4.71 < 5 < 6.28 \approx 2\pi,$$

we see that $P(5)$ is located in quadrant IV. See Figure 5.8.

c. To locate $P(-3\pi/2)$, we start at $A = (1, 0)$ and measure, in a clockwise direction, an arc of length $|-3\pi/2| = 3\pi/2$. Since $3\pi/2$ is three-fourths of the circumference, $P(-3\pi/2) = (0, 1)$. See Figure 5.9.

d. Since $-2 < 0$, we proceed as in part (c) above and measure, in a clockwise direction starting at $A = (1, 0)$, an arc of length $|-2| = 2$. Since $\pi/2 \approx 1.57 < 2 < 3.14 \approx \pi$, it follows that $P(-2)$ is in quadrant III. See Figure 5.10.

Finding the Trigonometric Point $P(t)$ for Special Values of t

The numbers $t = \pi/6$, $t = \pi/4$, and $t = \pi/3$ play a key role in the development of the trigonometric functions. It is therefore important that we obtain the coordinates of $P(t)$ for each of these numbers. We begin with $t = \pi/4$.

Let $P(\pi/4) = (a, b)$. Since $\pi/4$ is one-half of $\pi/2$, it follows that $P(\pi/4)$ is the midpoint of the arc of the unit circle in the first quadrant joining $A = (1, 0)$ and $P(\pi/2) = (0, 1)$. Hence, $P(\pi/4)$ is on the line $y = x$ and it follows that $a = b$. See Figure 5.11. Since (a, b) lies on the unit circle,

$$a^2 + b^2 = 1.$$

Substituting a for b, we have

$$2a^2 = 1, \qquad \text{or}$$

$$a = \pm\frac{1}{\sqrt{2}} = \pm\frac{\sqrt{2}}{2}.$$

Since $P(a, b)$ is in quadrant I, we know that a is positive. Thus

$$a = \frac{1}{\sqrt{2}} = \frac{\sqrt{2}}{2} = b, \qquad \text{and} \qquad P\left(\frac{\pi}{4}\right) = \left(\frac{\sqrt{2}}{2}, \frac{\sqrt{2}}{2}\right).$$

Before obtaining the coordinates of $P(\pi/3)$ and $P(\pi/6)$, we observe that, by the symmetry of the unit circle,

$$P\left(\frac{3\pi}{4}\right) = \left(-\frac{\sqrt{2}}{2}, \frac{\sqrt{2}}{2}\right),$$

$$P\left(\frac{5\pi}{4}\right) = \left(-\frac{\sqrt{2}}{2}, -\frac{\sqrt{2}}{2}\right), \quad \text{and} \quad P\left(\frac{7\pi}{4}\right) = \left(\frac{\sqrt{2}}{2}, -\frac{\sqrt{2}}{2}\right).$$

(See Figure 5.12.)

To find the coordinates of $P(\pi/3)$, we divide the upper semicircle from $A = (1, 0)$ to $P(\pi) = (-1, 0)$ into three equal parts, thereby obtaining the points $P(\pi/3)$ and $P(2\pi/3)$. Letting $P(\pi/3) = (a, b)$, from the symmetry of the unit circle, it follows that $P(2\pi/3) = (-a, b)$. See Figure 5.13.

To simplify our discussion, let $B = P(\pi/3)$ and $C = P(2\pi/3)$. Since each of the arcs $\overset{\frown}{AB}$ and $\overset{\frown}{BC}$ has length $\pi/3$, we know from plane geometry that the chords $\overline{AB}$ and $\overline{BC}$ must also have equal length. By the distance formula, the length of chord $\overline{AB}$ is

$$\begin{aligned}
d(A, B) &= \sqrt{(a - 1)^2 + (b - 0)^2} \\
&= \sqrt{a^2 - 2a + 1 + b^2} \\
&= \sqrt{a^2 + b^2 - 2a + 1}.
\end{aligned}$$

FIGURE 5.12

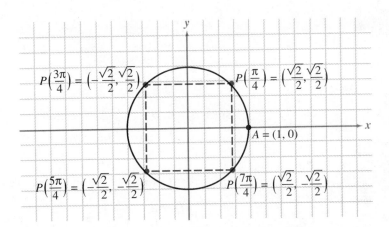

FIGURE 5.13

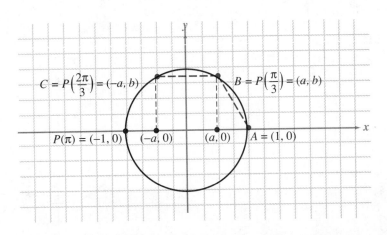

From Figure 5.13, we see that chord $\overline{BC}$ has length $2a$. Equating the lengths of the two chords, we have

$$\sqrt{a^2 + b^2 - 2a + 1} = 2a. \tag{1}$$

Squaring both sides of equation (1), we obtain

$$a^2 + b^2 - 2a + 1 = 4a^2. \tag{2}$$

Since $P(\pi/3) = (a, b)$ lies on the unit circle,

$$a^2 + b^2 = 1.$$

Substituting 1 for $a^2 + b^2$ in equation (2) yields

$$1 - 2a + 1 = 4a^2$$
$$4a^2 + 2a - 2 = 0$$
$$2a^2 + a - 1 = 0.$$

Factoring, we have

$$(2a - 1)(a + 1) = 0.$$

Hence, $a = \dfrac{1}{2}$ or $a = -1$.

Since $a > 0$, it follows that $a = \frac{1}{2}$ is the x-coordinate of $P(\pi/3)$. To find the y-coordinate b, we substitute $a = \frac{1}{2}$ into $a^2 + b^2 = 1$:

$$\left(\frac{1}{2}\right)^2 + b^2 = 1$$
$$b^2 = \frac{3}{4}$$
$$b = \pm\frac{\sqrt{3}}{2}.$$

Since $b > 0$, $b = \sqrt{3}/2$. Therefore,

$$P\left(\frac{\pi}{3}\right) = \left(\frac{1}{2}, \frac{\sqrt{3}}{2}\right); \quad \text{hence, } P\left(\frac{2\pi}{3}\right) = \left(-\frac{1}{2}, \frac{\sqrt{3}}{2}\right).$$

Using the symmetry of the unit circle, we see from Figure 5.14 that

$$P\left(\frac{4\pi}{3}\right) = \left(-\frac{1}{2}, -\frac{\sqrt{3}}{2}\right) \quad \text{and} \quad P\left(\frac{5\pi}{3}\right) = \left(\frac{1}{2}, -\frac{\sqrt{3}}{2}\right).$$

FIGURE 5.14

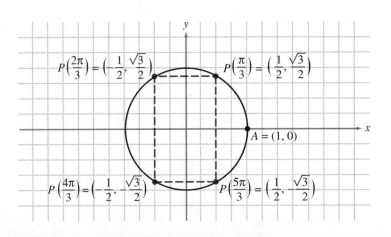

By an argument similar to that used to find the coordinates of $P(\pi/3)$, it can be shown that

$$P\left(\frac{\pi}{6}\right) = \left(\frac{\sqrt{3}}{2}, \frac{1}{2}\right).$$

(See Exercise 34.)

By the symmetry of the unit circle, we see from Figure 5.15 that

$$P\left(\frac{5\pi}{6}\right) = \left(-\frac{\sqrt{3}}{2}, \frac{1}{2}\right),$$

$$P\left(\frac{7\pi}{6}\right) = \left(-\frac{\sqrt{3}}{2}, -\frac{1}{2}\right), \quad \text{and} \quad P\left(\frac{11\pi}{6}\right) = \left(\frac{\sqrt{3}}{2}, -\frac{1}{2}\right).$$

For reference purposes, we summarize our results thus far in Figure 5.16.

FIGURE 5.15

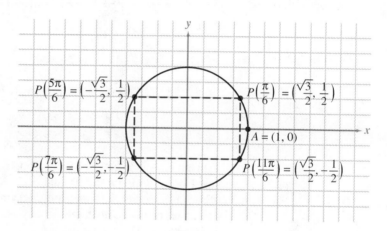

FIGURE 5.16

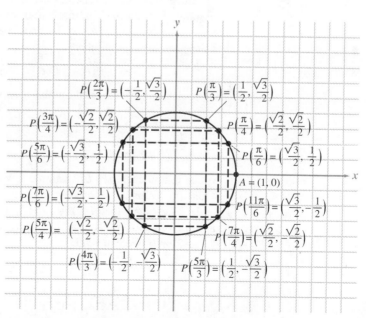

We should point out that each of the points labeled in Figure 5.16 is also the trigonometric point of a real number between -2π and 0. For example,

$$P\left(-\frac{\pi}{6}\right) = P\left(\frac{11\pi}{6}\right) \quad \text{and} \quad P\left(-\frac{7\pi}{4}\right) = P\left(\frac{\pi}{4}\right).$$

You should justify the labeling shown in Figure 5.17.

FIGURE 5.17

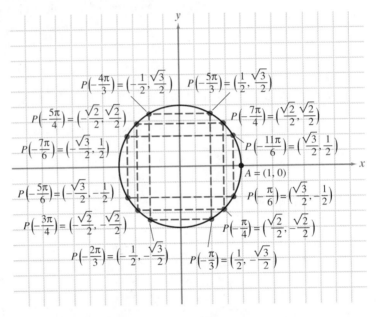

In practice it is not necessary, nor recommended, that you memorize the points labeled on Figures 5.16 and 5.17. You need only remember the points in quadrant I of Figure 5.16 and use the symmetry of the unit circle and the rules for signs in the quadrants to obtain any of the other points shown.

Since the circumference of the unit circle is 2π, it follows that if we start at any point $P(t)$ on the unit circle and travel a distance of 2π units (either clockwise or counterclockwise) along the circle, we will return to the same point $P(t)$. Hence, for any real number t,

$$P(t) = P(t - 2\pi) = P(t + 2\pi),$$

and in general, for any real number t and for any integer k,

$$P(t) = P[t + k(2\pi)]. \tag{3}$$

Finding $P(t)$ When $|t| > 2\pi$

Notice that in each part of Example 1 we were considering $P(t)$ where $|t| \le 2\pi$. When $|t| > 2\pi$, it is usually convenient to find a number t' in the interval $[-2\pi, 2\pi]$ such that $P(t) = P(t')$. If $t > 2\pi$, we can subtract an integral multiple of 2π from t so that the difference t' is between 0 and 2π. For example, if $t = 29\pi/6$, we can write

$$t = \frac{29\pi}{6} = \frac{5\pi}{6} + 2(2\pi),$$

so that

$$t' = \frac{29\pi}{6} - 2(2\pi) = \frac{5\pi}{6}.$$

Therefore, by equation (3),

$$P\left(\frac{29\pi}{6}\right) = P\left(\frac{5\pi}{6}\right) = \left(-\frac{\sqrt{3}}{2}, \frac{1}{2}\right).$$

(See Figure 5.16.)

In a similar manner, if $t < -2\pi$, we can express $P(t)$ as $P(t')$, where $-2\pi \le t' \le 0$. To illustrate, consider $t = -35\pi/4$.

Since

$$t = -\frac{35\pi}{4} = -\frac{3\pi}{4} - 4(2\pi),$$

it follows that

$$t' = -\frac{35\pi}{4} + 4(2\pi) = -\frac{3\pi}{4}.$$

Hence, by equation (3),

$$P\left(-\frac{35\pi}{4}\right) = P\left(-\frac{3\pi}{4}\right) = \left(-\frac{\sqrt{2}}{2}, -\frac{\sqrt{2}}{2}\right).$$

(See Figure 5.17.)

Our next example provides additional illustrations of this process.

EXAMPLE 2

Give the approximate location of $P(t)$ on the unit circle, and find the exact coordinates of $P(t)$ in each case.

a. $P\left(-\frac{13\pi}{2}\right)$ **b.** $P\left(\frac{19\pi}{6}\right)$

c. $P\left(-\frac{16\pi}{3}\right)$ **d.** $P\left(\frac{15\pi}{4}\right)$

Solution

a. Since $-13\pi/2 = (-\pi/2) - 3(2\pi)$, it follows that

$$P\left(-\frac{13\pi}{2}\right) = P\left(-\frac{\pi}{2}\right) = (0, -1).$$

(See Figure 5.18.)

b. Expressing $19\pi/6$ as $7\pi/6 + 2\pi$, we have

$$P\left(\frac{19\pi}{6}\right) = P\left(\frac{7\pi}{6}\right) = \left(-\frac{\sqrt{3}}{2}, -\frac{1}{2}\right).$$

(See Figure 5.19, page 210.)

c. Since $(-16\pi/3) = (-4\pi/3) - 2(2\pi)$, it follows that

$$P\left(\frac{-16\pi}{3}\right) = P\left(-\frac{4\pi}{3}\right) = \left(-\frac{1}{2}, \frac{\sqrt{3}}{2}\right).$$

(See Figure 5.20, page 210.)

FIGURE 5.18

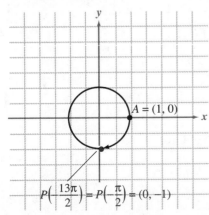

$$P\left(-\frac{13\pi}{2}\right) = P\left(-\frac{\pi}{2}\right) = (0, -1)$$

FIGURE 5.19

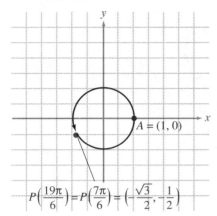

$$P\left(\frac{19\pi}{6}\right) = P\left(\frac{7\pi}{6}\right) = \left(-\frac{\sqrt{3}}{2}, -\frac{1}{2}\right)$$

FIGURE 5.20

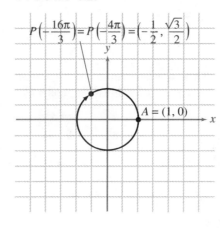

$$P\left(-\frac{16\pi}{3}\right) = P\left(-\frac{4\pi}{3}\right) = \left(-\frac{1}{2}, \frac{\sqrt{3}}{2}\right)$$

FIGURE 5.21

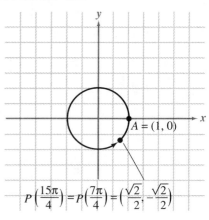

$$P\left(\frac{15\pi}{4}\right) = P\left(\frac{7\pi}{4}\right) = \left(\frac{\sqrt{2}}{2}, -\frac{\sqrt{2}}{2}\right)$$

d. Writing $15\pi/4$ as $7\pi/4 + 2\pi$, we have

$$P\left(\frac{15\pi}{4}\right) = P\left(\frac{7\pi}{4}\right) = \left(\frac{\sqrt{2}}{2}, -\frac{\sqrt{2}}{2}\right).$$

(See Figure 5.21.)

Notice that in each part of Example 2, t has the form $k\pi$, for some rational number k. When t cannot be expressed as $k\pi$ for some rational number k, we find t' by approximating π as 3.14 and proceeding as in Example 2.

EXAMPLE 3

Locate $P(t)$ on the unit circle in each case.

a. $P(36.5)$ **b.** $P(-23)$

Solution

a. Using a calculator to divide 36.5 by $2\pi \approx 6.28$, we find that the quotient is approximately 5.8. To find t', we subtract $5(2\pi) \approx 31.4$ from 36.5, to obtain 5.1. Therefore,

$$36.5 = 5(6.28) + 5.1 \approx 5.1 + 5(2\pi),$$

so that $P(36.5) \approx P(5.1).$

Since $\dfrac{3\pi}{2} \approx 4.71 < 5.1 < 6.28 \approx 2\pi,$

we see that $P(5.1)$ is in quadrant IV (see Figure 5.22).

b. Since $-23 = -3(6.28) - 4.16 \approx -4.16 - 3(2\pi),$

it follows that $P(-23) \approx P(-4.16).$

Since $\pi \approx 3.14 < 4.16 < 4.71 \approx \dfrac{3\pi}{2},$

$P(-4.16)$ is in quadrant II (see Figure 5.23).

FIGURE 5.22

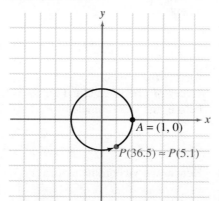

FIGURE 5.23

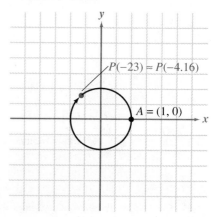

In Section 5.2 we define the trigonometric functions and investigate their domains. At that time, we will find the results of the following example to be very useful.

EXAMPLE 4

Find all real values of t for which $P(t)$ satisfies the stated conditions:

a. The x-coordinate of $P(t)$ is 0. **b.** The y-coordinate of $P(t)$ is 0.

Solution

a. As we see from Figure 5.24, there are exactly two values of t between 0 and 2π for which the x-coordinate of $P(t)$ is 0, namely $t = \pi/2$ and $t = 3\pi/2$.

Since $P(t) = P[t + n(2\pi)]$ for any integer n, the x-coordinate of $P(t)$ is 0 whenever

$$t = \frac{\pi}{2} + n(2\pi) = \frac{\pi}{2} + (2n)\pi \qquad (4)$$

or

$$t = \frac{3\pi}{2} + n(2\pi) = \frac{\pi}{2} + (2n + 1)\pi. \qquad (5)$$

FIGURE 5.24

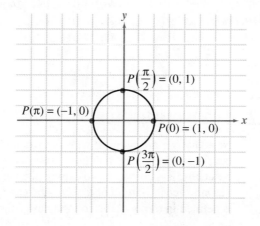

It follows from equations (4) and (5) that the x-coordinate of $P(t)$ is 0 whenever

$$t = \frac{\pi}{2} + k\pi, \quad \text{where } k \text{ is any integer.}$$

b. Again, from Figure 5.24, we see that $t = 0$ and $t = \pi$ are the only values of t between 0 and 2π for which the y-coordinate of $P(t)$ is 0. As before, the y-coordinate of $P(t)$ is 0 whenever either

$$t = 0 + n(2\pi) = (2n)\pi$$
or
$$t = \pi + n(2\pi) = (2n + 1)\pi.$$

Hence, the y-coordinate of $P(t)$ is 0 whenever $t = k\pi$, for any integer k.

EXERCISES 5.1

In Exercises 1–18, find the exact coordinates of $P(t)$ and display $P(t)$ on the unit circle.

1. $P(5\pi)$
2. $P(-10\pi)$
3. $P(-2\pi/3)$
4. $P(7\pi/4)$
5. $P(-7\pi/6)$
6. $P(9\pi)$
7. $P(-7\pi/4)$
8. $P(-5\pi/3)$
9. $P(23\pi/3)$
10. $P(-5\pi/6)$
11. $P(50\pi/6)$
12. $P(-2\pi/3)$
13. $P(-4\pi/3)$
14. $P(100\pi/4)$
15. $P(22\pi/8)$
16. $P(8\pi/3)$
17. $P(-25\pi/6)$
18. $P(-50\pi/6)$

In Exercises 19–33, find the approximate location of $P(t)$ on the unit circle. Use $\pi \approx 3.14$.

19. $P(5)$
20. $P(6)$
21. $P(1)$
22. $P(-30)$
23. $P(-3)$
24. $P(16)$
25. $P(100)$
26. $P(-60)$
27. $P(-25)$
28. $P(12)$
29. $P(-9\pi/10)$
30. $P(13\pi/7)$
31. $P(15)$
32. $P(-8)$
33. $P(-\pi/5)$

34. Show that $P(\pi/6) = (\sqrt{3}/2, \frac{1}{2})$.

In Exercises 35–40, find all values of $t \in [0, 2\pi]$ for which $P(t)$ satisfies the stated conditions.

35. The y-coordinate of $P(t)$ is $-\sqrt{3}/2$.
36. The x-coordinate of $P(t)$ is $\sqrt{2}/2$.
37. The x-coordinate of $P(t)$ is 1.
38. The y-coordinate of $P(t)$ is $-1/2$.
39. The y-coordinate of $P(t)$ is ± 1.
40. The x-coordinate of $P(t)$ is $-\sqrt{2}/2$.

In Exercises 41–48, $P(t)$ lies on the unit circle in the indicated quadrant. Find c.

41. $P(t) = (1/\sqrt{5}, c)$;   quadrant IV
42. $P(t) = (c, \sqrt{2}/3)$;   quadrant II
43. $P(t) = (-\frac{1}{3}, c)$;   quadrant II
44. $P(t) = (-\frac{2}{5}, c)$;   quadrant III
45. $P(t) = (4/c, 5/c)$;   quadrant I
46. $P(t) = (6/c, 3/c)$;   quadrant I
47. $P(t) = (c, 3c)$;   quadrant III
48. $P(t) = (2c, c)$;   quadrant III

5.2   TRIGONOMETRIC FUNCTIONS

Now that we are familiar with the wrapping function P, we are ready to define the six trigonometric functions: **sine, cosine, tangent, cotangent, secant**, and **cosecant**. It is customary to express the values of these functions at a real number t as sin t, cos t, tan t, cot t, sec t, and csc t, respectively. Each of these functions is defined in terms of the coordinates of the trigonometric point $P(t)$.

///

DEFINITION 5.1   Trigonometric Functions

Let t be a real number and let $P(t) = (x, y)$ be the trigonometric point associated with t by the wrapping function P. Then

$$\sin t = y \qquad\qquad \csc t = \frac{1}{y}, \quad y \neq 0$$

$$\cos t = x \qquad\qquad \sec t = \frac{1}{x}, \quad x \neq 0$$

$$\tan t = \frac{y}{x}, \quad x \neq 0 \qquad \cot t = \frac{x}{y}, \quad y \neq 0$$

How can we determine the domains of the six trigonometric functions? Since $\cos t$ and $\sin t$ are simply the x- and y-coordinates, respectively, of the trigonometric point $P(t)$, which is defined for each real number t, it follows that

$$\text{dom (sin)} = \text{dom (cos)} = \mathbb{R}.$$

From Definition 5.1, we see that the tangent and secant functions are only defined at those real numbers t for which the x-coordinate of $P(t)$ is nonzero. In Example 4 of Section 5.1, we found that the x-coordinate of $P(t)$ is 0 whenever $t = (\pi/2) + k\pi$, where k is an integer. Therefore,

$$\text{dom (tan)} = \text{dom (sec)} = \left\{ t \text{ in } \mathbb{R} \mid t \neq \frac{\pi}{2} + k\pi, \quad k \text{ any integer} \right\}.$$

Similarly, it follows from Definition 5.1 and Example 4 from Section 5.1 that

$$\text{dom (cot)} = \text{dom (csc)} = \{ t \text{ in } \mathbb{R} \mid t \neq k\pi, \quad k \text{ any integer} \}.$$

To summarize our results concerning the domains of the trigonometric functions:

///

Domains of the Trigonometric Functions

$$\text{dom (sin)} = \text{dom (cos)} = \mathbb{R}$$

$$\text{dom (tan)} = \text{dom (sec)} = \left\{ t \text{ in } \mathbb{R} \mid t \neq \frac{\pi}{2} + k\pi, \quad k \text{ any integer} \right\}$$

$$\text{dom (cot)} = \text{dom (csc)} = \{ t \text{ in } \mathbb{R} \mid t \neq k\pi, \quad k \text{ any integer} \}$$

To determine the ranges of the sine and cosine functions, we examine the graph of the unit circle shown in Figure 5.25, page 214. If (x, y) is any point on this circle, we note that

$$-1 \leq x \leq 1 \qquad \text{and} \qquad -1 \leq y \leq 1.$$

Since for any real number t, $P(t) = (x, y) = (\cos t, \sin t)$, it follows that

$$-1 \leq \cos t \leq 1 \qquad \text{and} \qquad -1 \leq \sin t \leq 1.$$

Hence, $$\text{ran (sin)} = \text{ran (cos)} = [-1, 1].$$

FIGURE 5.25

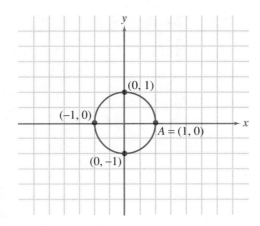

We discuss the ranges of the tangent, cotangent, secant, and cosecant functions in Section 5.3 when we sketch their graphs.

Evaluating Trigonometric Functions

In Section 5.1, we learned that $P(\pi/6) = (\sqrt{3}/2, 1/2)$. Therefore, it follows from Definition 5.1 that

$$\sin \frac{\pi}{6} = \frac{1}{2} \qquad\qquad \cos \frac{\pi}{6} = \frac{\sqrt{3}}{2}$$

$$\tan \frac{\pi}{6} = \frac{1/2}{\sqrt{3}/2} = \frac{1}{\sqrt{3}} = \frac{\sqrt{3}}{3} \qquad \cot \frac{\pi}{6} = \frac{\sqrt{3}/2}{1/2} = \sqrt{3}$$

$$\sec \frac{\pi}{6} = \frac{1}{\sqrt{3}/2} = \frac{2}{\sqrt{3}} = \frac{2\sqrt{3}}{3} \qquad \csc \frac{\pi}{6} = \frac{1}{1/2} = 2.$$

Proceeding in a similar manner, we obtain the other entries shown in Table 5.1 below.

If t and t' are any two real numbers such that

$$P(t) = P(t') = (x, y),$$

it follows from Definition 5.1 that each of the six trigonometric functions evaluated at t is the same as the corresponding function evaluated at t'. For example, if $t = 17\pi/4$, we can write $17\pi/4 = 2(2\pi) + (\pi/4)$. Hence,

$$P\left(\frac{17\pi}{4}\right) = P\left(\frac{\pi}{4}\right) = \left(\frac{\sqrt{2}}{2}, \frac{\sqrt{2}}{2}\right),$$

TABLE 5.1

t	$\sin t$	$\cos t$	$\tan t$	$\cot t$	$\sec t$	$\csc t$
$\dfrac{\pi}{6}$	$\dfrac{1}{2}$	$\dfrac{\sqrt{3}}{2}$	$\dfrac{\sqrt{3}}{3}$	$\sqrt{3}$	$\dfrac{2\sqrt{3}}{3}$	2
$\dfrac{\pi}{4}$	$\dfrac{\sqrt{2}}{2}$	$\dfrac{\sqrt{2}}{2}$	1	1	$\sqrt{2}$	$\sqrt{2}$
$\dfrac{\pi}{3}$	$\dfrac{\sqrt{3}}{2}$	$\dfrac{1}{2}$	$\sqrt{3}$	$\dfrac{\sqrt{3}}{3}$	2	$\dfrac{2\sqrt{3}}{3}$

and it follows that the trigonometric function values of $17\pi/4$ are the same as the trigonometric function values of $\pi/4$. From Table 5.1, we have

$$\sin \frac{17\pi}{4} = \sin \frac{\pi}{4} = \frac{\sqrt{2}}{2} \qquad \cos \frac{17\pi}{4} = \cos \frac{\pi}{4} = \frac{\sqrt{2}}{2}$$

$$\tan \frac{17\pi}{4} = \tan \frac{\pi}{4} = 1 \qquad \cot \frac{17\pi}{4} = \cot \frac{\pi}{4} = 1$$

$$\sec \frac{17\pi}{4} = \sec \frac{\pi}{4} = \sqrt{2} \qquad \csc \frac{17\pi}{4} = \csc \frac{\pi}{4} = \sqrt{2}.$$

EXAMPLE 5

Find the exact value, if possible, of each expression.

a. $\sin\left(-\dfrac{16\pi}{3}\right)$ **b.** $\sec \dfrac{19\pi}{4}$ **c.** $\tan\left(-\dfrac{17\pi}{2}\right)$

Solution

a. In Example 2 of Section 5.1, we found that

$$P\left(-\frac{16\pi}{3}\right) = \left(-\frac{1}{2}, \frac{\sqrt{3}}{2}\right).$$

Therefore, $\qquad \sin\left(-\dfrac{16\pi}{3}\right) = \dfrac{\sqrt{3}}{2}.$

b. Since $19\pi/4 = 2(2\pi) + 3\pi/4$, we have

$$P\left(\frac{19\pi}{4}\right) = P\left(\frac{3\pi}{4}\right) = \left(-\frac{\sqrt{2}}{2}, \frac{\sqrt{2}}{2}\right).$$

Hence, $\qquad \sec \dfrac{19\pi}{4} = \dfrac{1}{-\sqrt{2}/2} = -\sqrt{2}.$

c. Expressing $-17\pi/2$ as $(-\pi/2) - 4(2\pi)$, it follows that

$$P\left(-\frac{17\pi}{2}\right) = P\left(-\frac{\pi}{2}\right) = (0, -1).$$

Therefore, $\tan(-13\pi/2)$ is *undefined* (since division by 0 is undefined).

EXAMPLE 6

Show that the point $(-1/\sqrt{10}, 3/\sqrt{10})$ lies on the unit circle, and hence, is the trigonometric point $P(t)$ for some t in $\mathbb{R}$. Then find the exact value of $\cot t$ and $\sec t$.

Solution Since $x^2 + y^2 = (-1/\sqrt{10})^2 + (3/\sqrt{10})^2 = \frac{1}{10} + \frac{9}{10} = 1$, $(-1/\sqrt{10}, 3/\sqrt{10})$ lies on the unit circle. Hence, $(-1/\sqrt{10}, 3/\sqrt{10}) = P(t)$, for some t in $\mathbb{R}$, and we have

$$\cot t = \frac{-1/\sqrt{10}}{3/\sqrt{10}} = -\frac{1}{3} \quad \text{and} \quad \sec t = \frac{1}{-1/\sqrt{10}} = -\sqrt{10}.$$

When we evaluate any of the six trigonometric functions at a real number t that is not an integer multiple of $\pi/6$, $\pi/4$, $\pi/3$, or $\pi/2$, we typically use a calculator or a table to approximate the function value.

EXAMPLE 7

Use a calculator to approximate the indicated function values to four decimal places.

a. tan 23.7   **b.** $\cos(-25)$ **c.** sin 100

Solution

a. Placing the calculator in the radian mode, we find that

$$\tan 23.7 \approx -7.197.$$

b. Proceeding as in part (a), we have

$$\cos(-25) \approx 0.9912.$$

c. Some of the currently available calculators cannot handle values of t as large as 100. However, we can write

$$100 = 15(6.2832) + 5.752 \approx 15(2\pi) + 5.752,$$

and it follows that

$$\sin 100 \approx \sin 5.752 \approx -0.5066.$$

The Basic Trigonometric Identities

A number of significant relationships between the trigonometric functions follow immediately from their definitions. Equations involving trigonometric functions that are true for all permissible values of their variables are called **trigonometric identities**. The seven identities listed below are called the basic trigonometric identities because of their extensive use in our work with the trigonometric functions.

Basic Trigonometric Identities

For all real numbers t for which each of the following expressions is defined:

(i) $\sec t = \dfrac{1}{\cos t}$ 　　　　　(ii) $\csc t = \dfrac{1}{\sin t}$

(iii) $\tan t = \dfrac{\sin t}{\cos t}$ 　　　　　(iv) $\cot t = \dfrac{\cos t}{\sin t} = \dfrac{1}{\tan t}$

(v) $\cos^2 t + \sin^2 t = 1$ 　　(vi) $1 + \tan^2 t = \sec^2 t$

(vii) $\cot^2 t + 1 = \csc^2 t.$

We verify basic identities (iii), (v), and (vi), and leave the verification of those remaining as exercises (see Exercises 46–49). To begin, let $P(t) = (x, y) = (\cos t, \sin t)$. By Definition 5.1,

$$\tan t = \frac{y}{x} = \frac{\sin t}{\cos t}, \quad \text{for } x \neq 0.$$

Thus, basic identity (iii) holds. To verify basic identity (v), we note that since $P(t)$ is on the unit circle,

$$x^2 + y^2 = 1.$$

Thus, $\qquad (\cos t)^2 + (\sin t)^2 = 1, \qquad \text{or}$

$$\cos^2 t + \sin^2 t = 1.$$

To establish basic identity (vi), we multiply both sides of basic identity (v) by $1/\cos^2 t$, for $\cos t \neq 0$, to obtain

$$1 + \frac{\sin^2 t}{\cos^2 t} = \frac{1}{\cos^2 t}.$$

By basic identities (iii) and (i), it follows that

$$1 + \tan^2 t = \sec^2 t.$$

Even-Odd Properties of the Trigonometric Functions

One of the useful properties of the trigonometric functions is that each is either even or odd. To see this, let t be a real number. Without loss of generality, we can assume that $0 \leq t < 2\pi$. We will consider the case where $P(t)$ lies in quadrant II. From the unit circle shown in Figure 5.26, we note that $P(t)$ and $P(-t)$ are symmetric with respect to the x-axis. Thus if $P(t) = (x, y)$, then $P(-t) = (x, -y)$. Hence, by Definition 5.1,

$$\sin(-t) = -y = -\sin t,$$

$$\cos(-t) = x = \cos t, \qquad \text{and}$$

$$\tan(-t) = \frac{-y}{x} = -\frac{y}{x} = -\tan t.$$

It follows that sine and tangent are odd functions, while cosine is an even function. It can be shown that secant is an even function and that cosecant and cotangent are odd functions (see Exercises 43–45). Theorem 5.1 summarizes:

FIGURE 5.26

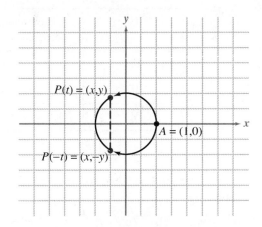

//

THEOREM 5.1

For all values of t in the domain of each function:

   (i) $\sin(-t) = -\sin t$ (ii) $\cos(-t) = \cos t$
  (iii) $\tan(-t) = -\tan t$ (iv) $\cot(-t) = -\cot t$
  (v) $\sec(-t) = \sec t$ (vi) $\csc(-t) = -\csc t$

(The cosine and secant functions are even; the other four are odd.)

Signs of the Values of the Trigonometric Functions

If t is a real number such that $\cos t = \sqrt{7}/3$, can we find a unique value for $\sin t$? We know from basic identity (v) that

$$\cos^2 t + \sin^2 t = 1.$$

Substituting $\sqrt{7}/3$ for $\cos t$ and solving for $\sin t$, we have

$$\left(\frac{\sqrt{7}}{3}\right)^2 + \sin^2 t = 1$$

$$\sin^2 t = 1 - \frac{7}{9} = \frac{2}{9}$$

$$\sin t = \pm\frac{\sqrt{2}}{3}.$$

Since $\sin t = \sqrt{2}/3$ or $\sin t = -\sqrt{2}/3$, a unique value for $\sin t$ cannot be determined without some additional information about t. Since $\cos t$ is positive, $P(t)$ lies in either quadrant I or quadrant IV. If $P(t)$ is in quadrant I, then $\sin t = \sqrt{2}/3$. On the other hand, if $P(t)$ is in quadrant IV, then $\sin t = -\sqrt{2}/3$.

As we have just seen, the quadrant that contains $P(t)$ determines the algebraic signs of $\sin t$ and $\cos t$ and hence of $\tan t$, $\cot t$, $\sec t$, and $\csc t$. We summarize the possibilities below.

**Signs of the Values of
the Trigonometric Functions**

Quadrant that Contains $P(t)$	Algebraic Signs of the Trigonometric Function Values of t
I	All six trigonometric function values of t are positive.
II	Only $\sin t$ and $\csc t$ are positive.
III	Only $\tan t$ and $\cot t$ are positive.
IV	Only $\cos t$ and $\sec t$ are positive.

EXAMPLE 8

If $\sin t = \sqrt{5}/3$ and $P(t)$ lies in quadrant II, find the exact value of each of the following:

  a. $\cos t$ **b.** $\cot(-t)$ **c.** $\sec(-t)$

Solution

a. Since $\cos^2 t + \sin^2 t = 1$ by basic identity (v) and since $\cos t < 0$ for $P(t)$ in quadrant II, we have

$$\cos t = -\sqrt{1 - \sin^2 t}$$

$$= -\sqrt{1 - \left(\frac{\sqrt{5}}{3}\right)^2}$$

$$= -\sqrt{1 - \frac{5}{9}} = -\sqrt{\frac{4}{9}} = -\frac{2}{3}.$$

b. Since cotangent is an odd function,

$$\cot(-t) = -\cot t$$

$$= -\frac{\cos t}{\sin t} \qquad \textbf{Basic identity (iv)}$$

$$= -\frac{-2/3}{\sqrt{5}/3} = \frac{2}{\sqrt{5}} = \frac{2\sqrt{5}}{5}.$$

c. Since secant is an even function,

$$\sec(-t) = \sec t$$

$$= \frac{1}{\cos t} \qquad \textbf{Basic identity (i)}$$

$$= -\frac{3}{2}.$$

EXAMPLE 9

If $\tan t = \frac{3}{4}$ and $P(t)$ is in quadrant III, find the exact value of each of the following:

a. $\cos t$ **b.** $\sin t$ **c.** $\csc(-t)$

Solution

a. Since $1 + \tan^2 t = \sec^2 t$ by basic identity (vi), and since $\sec t < 0$ when $P(t)$ is in quadrant III, we have

$$\sec t = -\sqrt{1 + \tan^2 t}$$

$$= -\sqrt{1 + \left(\frac{3}{4}\right)^2}$$

$$= -\sqrt{1 + \frac{9}{16}} = -\sqrt{\frac{25}{16}} = -\frac{5}{4}.$$

Hence, by basic identity (i),

$$\cos t = \frac{1}{\sec t} = -\frac{4}{5}.$$

b. Since $\cos t = -\frac{4}{5}$ and $\sin t < 0$ for $P(t)$ in quadrant III, we have by basic identity (v),

$$\sin t = -\sqrt{1 - \cos^2 t}$$

$$= -\sqrt{1 - \left(-\frac{4}{5}\right)^2}$$

$$= -\sqrt{1 - \frac{16}{25}} = -\sqrt{\frac{9}{25}} = -\frac{3}{5}.$$

c. Since cosecant is an odd function,

$$\csc(-t) = -\csc t$$

$$= -\frac{1}{\sin t} \qquad \text{Basic identity (ii)}$$

$$= \frac{5}{3}.$$

 CAUTION: In parts (a) and (b) of Example 9 we are given that $\tan t = \frac{3}{4}$. Since $\tan t = \sin t/\cos t$, it is tempting to conclude that $\sin t = 3$ and $\cos t = 4$. In general, it is incorrect to reason that if two fractions are equal, their numerators must be equal and their denominators must be equal. In this particular case, the conclusion that $\sin t = 3$ and $\cos t = 4$ is impossible since for all t in $\mathbb{R}$, $-1 \le \sin t \le 1$ and $-1 \le \cos t \le 1$.

EXERCISES 5.2

In Exercises 1–21, find the exact value of each expression, if possible.

1. $\tan(-7\pi/2)$
2. $\cot(-19\pi/3)$
3. $\sec(11\pi/4)$
4. $\csc(-23\pi/6)$
5. $\sin(-7\pi/3)$
6. $\cos(21\pi/4)$
7. $\cos(20\pi/3)$
8. $\sin(-13\pi/3)$
9. $\cot(-11\pi)$
10. $\sec(15\pi/2)$
11. $\tan(-15\pi/4)$
12. $\cot(25\pi/6)$
13. $\csc(-7\pi)$
14. $\sin(-50\pi/8)$
15. $\cos(-23\pi/4)$
16. $\sec(-29\pi/6)$
17. $\sec(-25\pi/3)$
18. $\sin(-300\pi)$
19. $\csc 250\pi$
20. $\tan(-58\pi/3)$
21. $\cos(51\pi/6)$

In Exercises 22–30, show that each point lies on the unit circle, and hence is the trigonometric point $P(t)$, for some t in $\mathbb{R}$. Then find the exact value of all six trigonometric functions at t.

22. $(-1/\sqrt{5}, -2/\sqrt{5})$
23. $(-\frac{3}{5}, -\frac{4}{5})$
24. $(-5/\sqrt{26}, -1/\sqrt{26})$
25. $(-4/\sqrt{41}, 5/\sqrt{41})$
26. $(-\frac{2}{3}, -\sqrt{5}/3)$
27. $(\frac{3}{4}, \sqrt{7}/4)$
28. $(\frac{5}{6}, \sqrt{11}/6)$
29. $(\sqrt{3}/5, -\sqrt{22}/5)$
30. $(-\sqrt{2}/3, \sqrt{7}/3)$

In Exercises 31–42, use a calculator in the radian mode to approximate the indicated function value to four decimal places.

31. $\sin 21.3$
32. $\cos(-35)$
33. $\sec(-10)$
34. $\tan(-100)$
35. $\cos 50$
36. $\cot 27$
37. $\csc(-\sqrt{5})$
38. $\cos[\tan(-3.6)]$
39. $\tan[\sin(-7)]$
40. $\sin(\sec 2)$
41. $\cos(\csc 5)$
42. $\tan[\cos(\sqrt{7})]$
43. Show that cotangent is an odd function.
44. Show that secant is an even function.
45. Show that cosecant is an odd function.
46. Verify that $\cot t = \cos t/\sin t = 1/\tan t$, for $\sin t \ne 0$ and $\cos t \ne 0$.
47. Verify that $\sec t = 1/\cos t$, for $\cos t \ne 0$.
48. Verify that $\csc t = 1/\sin t$, for $\sin t \ne 0$.
49. Verify that $\cot^2 t + 1 = \csc^2 t$.

In Exercises 50–61, use the given information to find the indicated function values at t in each case. Give exact answers.

50. If $\sin t = -\frac{2}{5}$ and $\tan t > 0$, find
    a. $\cos t$ **b.** $\cot t$.

51. If $\sec t = 10$ and $P(t)$ is in quadrant IV, find
a. $\sin t$ b. $\tan(-t)$.

52. If $\csc t = -3$ and $\cot t > 0$, find
a. $\cos t$ b. $\sin(-t)$.

53. If $\tan t = -5$ and $\cos t < 0$, find
a. $\sin t$ b. $\sec(-t)$.

54. If $\cot t = -7$ and $P(t)$ is in quadrant II, find
a. $\cos t$ b. $\csc(-t)$.

55. If $\cos t = -\frac{9}{10}$ and $P(t)$ is in quadrant III, find
a. $\csc(-t)$ b. $\tan t$.

56. If $\sin t = \frac{5}{7}$ and $\cos t < 0$, find
a. $\sec t$ b. $\cot(-t)$.

57. If $\sec t = 8$, and $\csc t < 0$, find
a. $\sin t$ b. $\tan(-t)$.

58. If $\csc t = -4$ and $P(t)$ is in quadrant III, find
a. $\sec t$ b. $\sin(-t)$.

59. If $\tan t = -\frac{5}{6}$ and $P(t)$ is in quadrant II, find
a. $\sin t$ b. $\cos(-t)$.

60. If $\cot t = -2$ and $\sin t < 0$, find
a. $\cos t$ b. $\sin(-t)$.

61. If $\cos t = 0.2$ and $\tan t < 0$, find
a. $\sin(-t)$ b. $\cos(-t)$.

62. If $P(t) = (3/c, 5/c)$ is a point on the unit circle in quadrant III, find the exact value of c and of all six trigonometric functions at t.

63. If $P(t) = (-1/c, \sqrt{2}/c)$ is a point on the unit circle in quadrant II, find the exact value of c and of all six trigonometric functions at t.

64. Show that each of the other five trigonometric function values at t can be expressed in terms of $\sin t$. [*Hint:* First, use basic identity (v) to show that $\cos t = \pm\sqrt{1 - \sin^2 t}$.]

65. Show that each of the other five trigonometric function values at t can be expressed in terms of $\cos t$. [*Hint:* First, use basic identity (v) to show that $\sin t = \pm\sqrt{1 - \cos^2 t}$.]

66. Show that if $u = 5 \sin t$ where $P(t)$ is in quadrant IV, then $\sqrt{25 - u^2} = 5 \cos t$.

67. Show that if $u = 6 \tan t$ where $P(t)$ is in quadrant I, then $\sqrt{36 + u^2} = 6 \sec t$.

68. Show that if $u = 3 \sin t$ where $P(t)$ is in quadrant I, then $\sqrt{9 - u^2} = 3 \cos t$.

69. Show that if $u = 4 \sec t$ where $P(t)$ is in quadrant II, then $\sqrt{u^2 - 16} = 4 \tan(-t)$.

70. Show that if $u = 10 \sec t$ where $P(t)$ is in quadrant IV, then $\sqrt{u^2 - 100} = 10 \tan(-t)$.

71. Show that if $u = 9 \sin t$ where $P(t)$ is in quadrant II, then $\sqrt{81 - u^2} = -9 \cos t$.

72. Use your calculator in the radian mode to complete the following table:

t	0.1	0.01	0.001	0.0001	0.00001	0.000001
$\sin t$						
$\dfrac{\sin t}{t}$						

Based on the table you just constructed, which real number is the expression $\sin t/t$ approaching as t *gets closer and closer to 0?*

73. Use your calculator in the radian mode to complete the following table:

t	0.1	0.01	0.001	0.0001	0.00001	0.000001
$\cos t$						
$\dfrac{1 - \cos t}{t}$						

Based on the table you just constructed, which real number is the expression $(1 - \cos t)/t$ approaching as t *gets closer and closer to 0?*

5.3 GRAPHS OF THE TRIGONOMETRIC FUNCTIONS

To further our understanding of the trigonometric functions, we now direct our attention to their graphs. The behavior of a wide variety of natural phenomena such as tides, sound waves, electricity, and molecular particle motion can be described using variations of the graphs of the trigonometric functions.

As we learned in Section 5.1, for any real number t and for any integer k,

$$P[t + k(2\pi)] = P(t). \tag{1}$$

It follows from equation (1) that the real numbers t, $t \pm 2\pi$, $t \pm 4\pi$, . . . all determine the same trigonometric point on the unit circle and thus, by Definition 5.1, have the same sine and cosine. That is, for any real number t and for any integer k,

$$\sin[t + k(2\pi)] = \sin t \qquad (2)$$

and $$\cos[t + k(2\pi)] = \cos t. \qquad (3)$$

Functions which have this kind of repetitive behavior are said to be periodic.

DEFINITION 5.2 A function f is **periodic** if and only if there exists a positive real number p such that

$$f(t + p) = f(t),$$

for all t in the domain of f. The smallest such number p, if it exists, is called the **period** of f.

THEOREM 5.2 The trigonometric functions are periodic. Sine, cosine, secant, and cosecant all have period 2π, while tangent and cotangent have period π.

The Sine Function

The portion of the graph of the sine function on the interval $[0, 2\pi]$, is called the **basic cycle of the sine function** or the **basic sine wave**. Plotting the points from the accompanying table of values, and connecting them with a smooth curve, we obtain the graph shown in Figure 5.27. Notice that we have marked off units on the t-axis in terms of π.

Since the sine function has domain $\mathbb{R}$ and period 2π, we obtain the complete graph of $y = \sin t$ by repeating the basic cycle every 2π units to the left and right as shown in Figure 5.28.

t	0	$\pi/6$	$\pi/3$	$\pi/2$	$2\pi/3$	$5\pi/6$	π	$7\pi/6$	$4\pi/3$	$3\pi/2$	$5\pi/3$	$11\pi/6$	2π
$\sin t$	0	$1/2$	$\sqrt{3}/2$	1	$\sqrt{3}/2$	$1/2$	0	$-1/2$	$-\sqrt{3}/2$	-1	$-\sqrt{3}/2$	$-1/2$	0

FIGURE 5.27
Basic cycle of the sine function

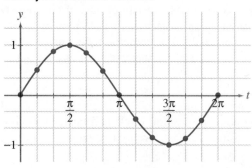

FIGURE 5.28
$y = \sin t$

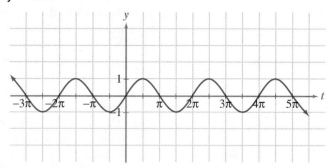

//

Properties of the Sine Function

   (i)   dom(sin) = $\mathbb{R}$;

  (ii) ran(sin) = $[-1, 1]$;

 (iii) The period of the sine function is 2π;

 (iv) The sine function is not one-to-one;

  (v) The sine function is odd (its graph is symmetric with respect to the origin).

The Cosine Function

Since the cosine function also has domain $\mathbb{R}$ and period 2π, we can sketch its graph in a similar manner. See Figures 5.29 and 5.30.

t	0	$\pi/6$	$\pi/3$	$\pi/2$	$2\pi/3$	$5\pi/6$	π	$7\pi/6$	$4\pi/3$	$3\pi/2$	$5\pi/3$	$11\pi/6$	2π
$\cos t$	1	$\sqrt{3}/2$	$1/2$	0	$-1/2$	$-\sqrt{3}/2$	-1	$-\sqrt{3}/2$	$-1/2$	0	$1/2$	$\sqrt{3}/2$	1

FIGURE 5.29
Basic cycle of the cosine function
(Basic cosine wave)

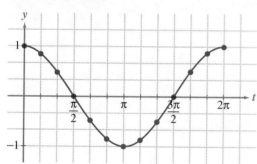

FIGURE 5.30
$y = \cos t$

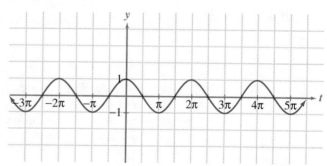

//

Properties of the Cosine Function

   (i)   dom(cos) = $\mathbb{R}$;

  (ii) ran(cos) = $[-1, 1]$;

 (iii) The period of the cosine function is 2π;

 (iv) The cosine function is not one-to-one;

  (v) The cosine function is even (its graph is symmetric with respect to the y-axis).

To display the graphs of the sine and cosine functions in the xy-plane, we use x, rather than t, on the horizontal axis. See Figure 5.31, page 224.

FIGURE 5.31

a. $y = \sin x$

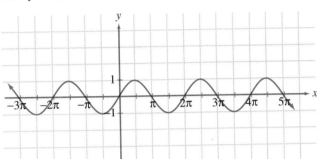

b. $y = \cos x$

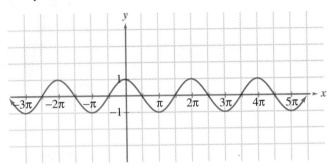

The Tangent Function

To graph $y = \tan x$, recall that

$$\text{dom (tan)} = \left\{ x \mid x \neq \frac{\pi}{2} + k\pi, \quad k \text{ any integer.} \right\}$$

Since the period of tangent is π, it seems natural to use the interval $[0, \pi]$ to obtain its basic cycle. However, this is impossible since tangent is undefined at $x = \pi/2$. Instead, it is customary to use the open interval $(-\pi/2, \pi/2)$. Several values of the tangent function for x in this interval are shown in Table 5.2.

TABLE 5.2

x	$-\pi/3$	$-\pi/4$	$-\pi/6$	0	$\pi/6$	$\pi/4$	$\pi/3$
$\tan x$	$-\sqrt{3}$	-1	$-\sqrt{3}/3$	0	$\sqrt{3}/3$	1	$\sqrt{3}$

Since the tangent function is undefined at $\pi/2$, we must investigate its behavior at some values of x in $(-\pi/2, \pi/2)$ that are close to $\pi/2$. We see from Figure 5.31 that if x is close to $\pi/2$ but remains less than $\pi/2$, then $\sin x$ is close to 1 and $\cos x$ is positive and close to 0. This means that

$$\tan x = \frac{\sin x}{\cos x}$$

is a positive number that gets larger and larger as x gets closer and closer to $\pi/2$. In Table 5.3 we have used a calculator to obtain the approximate values of $\tan x$ for some values of x near $\pi/2 \approx 1.5708$. Hence, as $x \rightarrow (\pi/2)^-$, $\tan x \rightarrow \infty$. This means that the line $x = \pi/2$ is a vertical asymptote of the graph of $y = \tan x$.

TABLE 5.3

x	1.5	1.52	1.54	1.57	1.5706	1.5707	1.57078	1.57079
$\tan x$	14.1	19.7	32.5	1,255.8	5,093.5	10,381.3	61,249	158,057.8

Similarly, it can be shown that as $x \rightarrow (-\pi/2)^+$, $\tan x \rightarrow -\infty$ so that the line $x = -\pi/2$ is also a vertical asymptote of the graph. Plotting the points from Table 5.2 and using the information about asymptotes, we obtain the basic cycle of the graph of the tangent function shown in Figure 5.32. To obtain the complete graph of the tangent function, we repeat the basic cycle every π units to the left and right. See Figure 5.33.

FIGURE 5.32
Basic cycle of $y = \tan x$

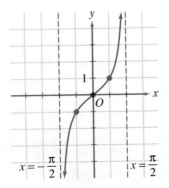

FIGURE 5.33 $y = \tan x$

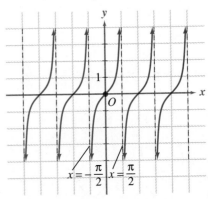

Properties of the Tangent Function

(i) $\text{dom(tan)} = \{x \mid x \neq (\pi/2) + n\pi, \quad n \text{ any integer}\}$;

(ii) $\text{ran(tan)} = \mathbb{R}$;

(iii) The period of the tangent function is π;

(iv) The tangent function is not one-to-one;

(v)   The tangent function is odd (its graph is symmetric with respect to the origin).

The Cotangent Function

In a similar manner, we can obtain the graph of the cotangent function that is shown in Figure 5.34 (see Exercise 1).

FIGURE 5.34 $y = \cot x$

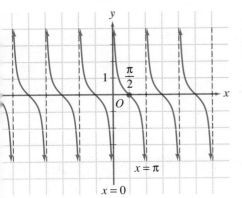

Properties of the Cotangent Function

(i) $\text{dom(cot)} = \{x \mid x \neq n\pi, \quad n \text{ any integer}\}$;

(ii) $\text{ran(cot)} = \mathbb{R}$;

(iii) The period of the cotangent function is π;

(iv) The cotangent function is not one-to-one;

(v)   The cotangent function is odd (its graph is symmetric with respect to the origin).

The Secant Function

To obtain the graph of the secant function, we use basic identity (i) on page 216 to write $\sec x = 1/\cos x$. Therefore, except for the values of x for which $\cos x = 0$ (that is, $x = (\pi/2) + n\pi$, n an integer), we can obtain $\sec x$ by taking the reciprocal of $\cos x$ (see the table on the next page).

x	0	$\pi/6$	$\pi/3$	$\pi/2$	$2\pi/3$	$5\pi/6$	π	$7\pi/6$	$4\pi/3$	$3\pi/2$	$5\pi/3$	$11\pi/6$	2π
$\cos x$	1	$\sqrt{3}/2$	$1/2$	0	$-1/2$	$-\sqrt{3}/2$	-1	$-\sqrt{3}/2$	$-1/2$	0	$1/2$	$\sqrt{3}/2$	1
$\sec x$	1	$2/\sqrt{3}$	2	undef.	-2	$-2/\sqrt{3}$	-1	$-2/\sqrt{3}$	-2	undef.	2	$2/\sqrt{3}$	1

FIGURE 5.35 $y = \sec x$

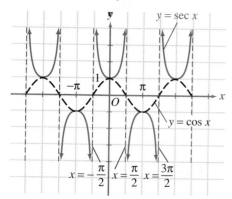

At the points where the graph of the cosine function crosses the x-axis, the graph of the secant function will have vertical asymptotes. In Figure 5.35, we have sketched the graph of $y = \sec x$, as well as the graph of $y = \cos x$ (shown dashed).

Properties of the Secant Function

(i) $\text{dom}(\sec) = \{x \mid x \neq (\pi/2) + n\pi, \quad n \text{ any integer}\}$;

(ii) $\text{ran}(\sec) = (-\infty, -1] \cup [1, \infty)$;

(iii) The period of the secant function is 2π;

(iv) The secant function is not one-to-one;

(v)   The secant function is even (its graph is symmetric with respect to the y-axis).

The Cosecant Function

Since $\csc x = 1/\sin x$, the graph of the cosecant function, shown in Figure 5.36, can be obtained from the graph of the sine function in a similar manner (see Exercise 3).

FIGURE 5.36 $y = \csc x$

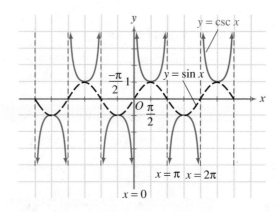

Properties of the Cosecant Function

(i) $\text{dom}(\csc) = \{x \mid x \neq n\pi, \quad n \text{ any integer}\}$;

(ii) $\text{ran}(\csc) = (-\infty, -1] \cup [1, \infty)$;

(iii) The cosecant function has period 2π;

(iv) The cosecant function is not one-to-one;

(v)   The cosecant function is odd (its graph is symmetric with respect to the origin).

Graphing Variations of Trigonometric Functions

We conclude this section with several examples involving functions whose graphs are variations of those presented earlier in this section.

EXAMPLE 10

Sketch the graph of the function $y = 3 \cos x$ and give its range.

Solution We can obtain the graph of $y = 3 \cos x$ from the graph of $y = \cos x$ by multiplying the y-coordinate of each point on the graph of the cosine function by 3. Thus, the factor 3 in $y = 3 \cos x$ affects the graph of $y = \cos x$ by "stretching" it vertically. The graph of $y = 3 \cos x$ appears in Figure 5.37 (along with the graph of $y = \cos x$, which is shown dashed). From the graph of $y = 3 \cos x$, we see that the range of this function is $[-3, 3]$.

FIGURE 5.37

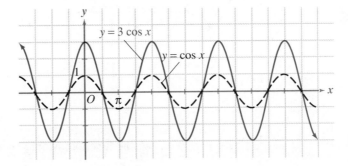

Amplitude

In Figure 5.37, we note that 3 is the largest value of y for any point on the graph of the function $y = 3 \cos x$. This maximum value of y is called the **amplitude** of this function. More generally, we have

Amplitude of Sine and Cosine

If a is any nonzero real number, then each of the functions $y = a \sin x$ and $y = a \cos x$ has amplitude $|a|$.

The geometric effect of the amplitude is to "stretch" (when $|a| > 1$) or "shrink" (when $|a| < 1$) vertically the corresponding graphs of $y = \sin x$ or $y = \cos x$.

EXAMPLE 11

Sketch the graph of the function $y = \sin 3x$. Give its amplitude and range.

Solution Observe that this function has amplitude 1. Since we obtain one complete cycle of $y = \sin x$ when

$$0 \le x \le 2\pi,$$

one complete cycle of $y = \sin 3x$ occurs when

$$0 \le 3x \le 2\pi \qquad \text{or} \qquad 0 \le x \le \frac{2\pi}{3}.$$

This means that the period for $y = \sin 3x$ is $2\pi/3$, which is one-third of the period for the sine function. The graph of $y = \sin 3x$ is sketched in Figure 5.38 (where the graph of the sine function is shown dashed). Observe from the graph of $y = \sin 3x$ that the range of this function is $[-1, 1]$.

FIGURE 5.38

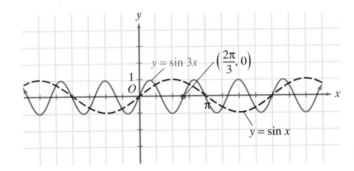

As we saw in Example 11, the period of $y = \sin 3x$ is $2\pi/3$, which is $\frac{1}{3}$ of the period of the sine function. More generally, we have the following theorem:

THEOREM 5.3 If $y = f(x)$ is a periodic function with period p, and b is a positive real number, then the function $y = f(bx)$ is periodic with period p/b.

To further illustrate Theorem 5.3, consider the functions $y = \cos 2x$, which has period $2\pi/2 = \pi$ (half the period of cosine) and $y = \cos(x/3)$ which has period $\dfrac{2\pi}{\frac{1}{3}} = 6\pi$ (three times the period of cosine). Thus we see that the effect of the positive number b from Theorem 5.3 is to "compress" (when $b > 1$) or to "stretch" (when $0 < b < 1$) the graph horizontally.

EXAMPLE 12

Sketch the graph of the function $g(x) = \sin[x + (\pi/2)]$. Compare the graph of g with the graph of the cosine function shown in Figure 5.31 (b).

Solution Since

$$g(x) = f\left(x + \frac{\pi}{2}\right),$$

where $f(x) = \sin x$, the graph of g is the graph of f translated horizontally a distance of $\pi/2$ units to the left. The graph of g and the graph of f (shown dashed) are sketched in Figure 5.39. Notice that the graph of g is the same as the graph of the cosine function that is shown in Figure 5.31 (b).

FIGURE 5.39

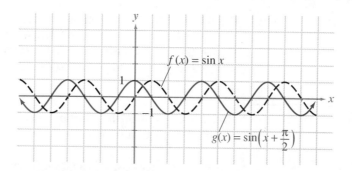

Phase Shift

In the solution of Example 12, the graph of $g(x) = \sin[x + (\pi/2)]$ was obtained by translating the graph of $f(x) = \sin x$ horizontally a distance of $\pi/2$ units to the left. The number $-\pi/2$ is called the phase shift. In general,

---

Phase Shift

If f is any one of the six basic trigonometric functions and h is any real number, then the function $g(x) = f(x + h)$ will have a **phase shift of** $-h$. Thus, the graph of g is the graph of f translated $|h|$ units to the right when $h < 0$ and to the left when $h > 0$.

We summarize our results for functions of the form $y = a \sin(bx + c)$ or $y = a \cos(bx + c)$ below.

Properties of Graphs of Functions of the Form
$y = a \sin(bx + c)$ or $y = a \cos(bx + c)$

Let a, b, and c be real numbers where $a \neq 0$ and $b > 0$. Then any function of the form

$$y = a \sin(bx + c) = a \sin\left[b\left(x + \frac{c}{b}\right)\right] \quad \text{or}$$

$$y = a \cos(bx + c) = a \cos\left[b\left(x + \frac{c}{b}\right)\right]$$

will have

$$\text{period} = \frac{2\pi}{b},$$

$$\text{amplitude} = |a|, \quad \text{and}$$

$$\text{phase shift} = -\frac{c}{b}$$

The graph is translated $|c/b|$ units $\begin{cases} \text{to the left if } c > 0 \\ \text{to the right if } c < 0. \end{cases}$

REMARK: When $b < 0$ for a function of the form $y = a \sin(bx + c)$ or $y = a \cos(bx + c)$, we use the fact that sine is odd and cosine is even to express the function in a form where $b > 0$. For example, $y = 2 \sin(1 - x)$ can be written as $y = -2 \sin(x - 1)$ and $y = 3 \cos(2 - 5x)$ can be written as $y = 3 \cos(5x - 2)$.

EXAMPLE 13

Sketch the graph of the function $f(x) = -2 \cos(3 - 4x)$. Give the period, amplitude, phase shift, and range.

Solution   Since cosine is an even function, we can write

$$\cos(3 - 4x) = \cos[-(3 - 4x)] = \cos(4x - 3).$$

Therefore, $f(x) = -2\cos(4x - 3),$

so that f has

$$\text{period} = \frac{2\pi}{4} = \frac{\pi}{2},$$

$$\text{amplitude} = |-2| = 2, \quad \text{and}$$

$$\text{phase shift} = \frac{3}{4}.$$

Since the basic cycle of the cosine function occurs when $0 \le x \le 2\pi$, it follows that one cycle of f occurs when

$$0 \le 4x - 3 \le 2\pi$$

$$3 \le 4x \le 3 + 2\pi$$

$$\frac{3}{4} \le x \le \frac{3}{4} + \frac{\pi}{2}.$$

Hence, one cosine wave of amplitude 2 (reflected across the x-axis since $a = -2 < 0$) occurs over the interval $[\frac{3}{4}, \frac{3}{4} + \pi/2]$. To complete the graph of f, we repeat this cycle to the right and left over intervals of length $\pi/2$. From the graph of f shown in Figure 5.40, we see that $\text{ran}(f) = [-2, 2]$.

FIGURE 5.40

$f(x) = -2 \cos(3 - 4x)$

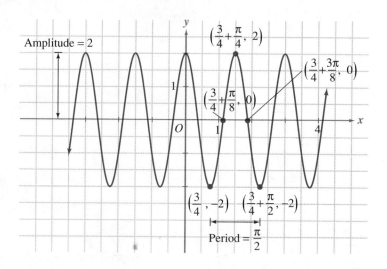

Proceeding as with sine and cosine, we obtain similar results for the tangent and cotangent functions except that amplitude is not defined for tangent or cotangent (see Exercise 4) and both have period π.

Properties of Graphs of Functions of the Form
$$y = a\tan(bx + c) \quad \text{or} \quad y = a\cot(bx + c)$$

Let a, b, and c be real numbers where $a \neq 0$ and $b > 0$. Then any function of the form

$$y = a\tan(bx + c) \quad \text{or} \quad y = a\cot(bx + c)$$

has

$$\text{period} = \frac{\pi}{b}$$

and

$$\text{phase shift} = -\frac{c}{b}$$

The graph is translated $|c/b|$ units $\begin{cases} \text{to the right if } c < 0 \\ \text{to the left if } c > 0. \end{cases}$

EXAMPLE 14

Sketch the graph of $f(x) = -3\tan(2x + 1)$. Give its period, phase shift, and range.

Solution   Since $f(x) = -3\tan(2x + 1)$, it follows that this function has

$$\text{period} = \frac{\pi}{b} = \frac{\pi}{2} \quad \text{and}$$

$$\text{phase shift} = -\frac{1}{2}.$$

Since the basic cycle of the tangent function occurs when $-\pi/2 < x < \pi/2$, it follows that one cycle of f occurs when

$$-\frac{\pi}{2} < 2x + 1 < \frac{\pi}{2}$$

$$-1 - \frac{\pi}{2} < 2x < -1 + \frac{\pi}{2}$$

$$-\frac{1}{2} - \frac{\pi}{4} < x < -\frac{1}{2} + \frac{\pi}{4}.$$

Thus, one tangent-type cycle (reflected across the x-axis since $a = -3 < 0$) occurs over the interval $(-\frac{1}{2} - \pi/4, -\frac{1}{2} + \pi/4)$. The vertical lines

$$x = -\frac{1}{2} - \frac{\pi}{4} \quad \text{and} \quad x = -\frac{1}{2} + \frac{\pi}{4}$$

are vertical asymptotes for the graph of f over this interval. To obtain the complete graph of f we repeat this cycle to the left and right over intervals of length $\pi/2$. From the graph (Figure 5.41, page 232) we see that $\text{ran}(f) = \mathbb{R}$.

FIGURE 5.41

$f(x) = -3 \tan(2x + 1)$

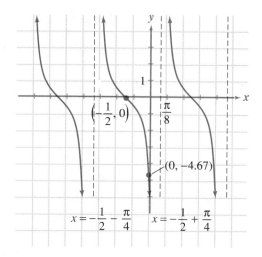

EXERCISES 5.3

1. Sketch the graph of the cotangent function by modifying the method followed in the text to obtain the graph of the tangent function.

2. Sketch the graph of the cotangent function by using the basic identity $\cot x = 1/\tan x$ and a modification of the method used in the text to obtain the graph of the secant function.

3. Sketch the graph of the cosecant function by modifying the method used in the text to obtain the graph of the secant function.

4. Explain why the tangent, cotangent, secant, and cosecant functions do not have an amplitude.

In Exercises 5–32, sketch at least one cycle of the graph of each function. Give the period, amplitude, phase shift, and range of each.

5. $y = 2 \sin x$
6. $y = -3 \cos x$
7. $y = \cos 3x$
8. $y = \sin(-2x)$
9. $y = -4 \sin(-x)$
10. $y = \pi \cos(2x + 1)$
11. $y = 2 \sin 3x$
12. $y = \cos 4x$
13. $y = (-\pi/2) \cos(-x)$
14. $y = 3 \cos[(x/2) - 1]$
15. $y = 2 \cos(\pi - x)$
16. $y = -\sin(-x - 3)$
17. $y = \frac{1}{3} \sin[(x/3) + 1]$
18. $y = \frac{2}{3} \cos[(3x/2) + 1]$
19. $y = -\frac{4}{3} \cos(4 - x)$
20. $y = 10 \sin[(2x/3) - 1]$
21. $y = -5 \sin(2x - \pi)$
22. $y = -\pi \cos[(x/\pi) + 1]$
23. $y = \frac{1}{2} \cos[(x/2) + 3]$
24. $y = \frac{1}{4} \sin[(3x/4) - \pi]$
25. $y = -\sin(\pi - 3x)$
26. $y = -\cos(\pi - 4x)$
27. $y = 2 \sin(2x + 2\pi)$
28. $y = 3 \cos(3x + 3)$
29. $y = -3 \cos(\pi x + 2\pi)$
30. $y = -\pi \cos(\pi x - 2\pi)$
31. $y = \pi \sin[1 - (x/2)]$
32. $y = \sqrt{3} \sin[2x - (2\pi/3)]$

In Exercises 33–50, sketch at least one cycle of the graph of each function. Give the period, phase shift, and range of each. Also, sketch any vertical asymptotes.

33. $y = -\cot x$
34. $y = -\tan x$
35. $y = \tan 2x$
36. $y = \cot 2x$
37. $y = -\tan(2x - 1)$
38. $y = -\cot(2x + 1)$
39. $y = \cot(1 - x)$
40. $y = \tan(2 - x)$
41. $y = -\tan(1 - 2x)$
42. $y = -\cot(1 - 3x)$
43. $y = \sec 2x$
44. $y = \csc 2x$
45. $y = \sec(x + \pi)$
46. $y = \csc(\pi - x)$
47. $y = -2 \csc[x - (\pi/2)]$
48. $y = -2 \sec[x + (\pi/2)]$
49. $y = \frac{1}{2} \tan[(x/2) + 1]$
50. $y = -\frac{1}{2} \cot[(x/2) - 1]$

51. Sketch the graph of $y = \cos[x + (\pi/2)]$. Compare this graph with the graph of $y = -\sin x$. What do you conclude?

52. Sketch the graphs of $y = \sec[(\pi/2) - x]$ and $y = \csc x$. How do these graphs compare?

53. Sketch the graphs of $y = \tan[x + (\pi/2)]$ and $y = -\cot x$. How do these graphs compare?

Functions of the form $y = a \sin(bt + c)$ or $y = a \cos(bt + c)$ where $a \neq 0$, $b > 0$, and t represents time are used to describe **simple harmonic motion**. Simple harmonic motion includes such natural phenomena as soundwaves, electricity, vibrating strings, and electrical particle motion. Using the formulas developed in this section we can find the period, amplitude, and phase shift of such functions. The **frequency** is defined to be $b/2\pi$, which is the reciprocal of the period. To illustrate, if time t is in seconds and the period is $\frac{1}{3}$, then three cycles are completed in one second. That is, the frequency counts the number of cycles per unit of time.

54. An object of mass 7 kilograms suspended from a spring with **spring constant** $k = 5$ oscillates in a simple harmonic motion described by the function

$$f(t) = 20 \cos\left(\sqrt{\frac{5}{7}}\, t\right)$$

where t is in seconds. Find
a. the period,   **b.** the amplitude,
c. the phase shift,   **d.** the frequency, and
e. the first time when $f(t) = 0$.

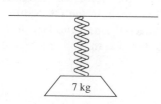

7 kg

Photo D. Gleiter 1991/FPG International

55. The path of a sound wave is given by the function

$$f(t) = 3 \sin[\pi(t_0 - 250t)],$$

where t_0 is a positive constant and t is in seconds. Find
a. the period,   **b.** the amplitude,
c. the phase shift,   **d.** the frequency, and
e. the first time when $f(t) = 0$.

56. A point on the end of a tuning fork moves in a simple harmonic motion which is given by the function

$$d = f(t) = 3 \sin\left(\frac{2\pi}{3}t - 1\right),$$

where t is in seconds and d is in centimeters. Find
a. the period,   **b.** the amplitude,

c. the phase shift,   **d.** the frequency, and
e. the second time when $f(t) = 0$.

57. A buoy oscillates in a simple harmonic motion which is described by the function

$$g(t) = 100 \cos(10\pi t - 2)$$

where t is in seconds. Find
a. the period,   **b.** the amplitude,
c. the phase shift,   **d.** the frequency, and
e. the second time when $g(t) = 0$.

58. The current in an electrical circuit is given by the function

$$I = f(t) = 60 \sin(120\pi t - \pi)$$

where I is measured in amperes and t is in seconds. Find
a. the period,   **b.** the amplitude,
c. the phase shift,   **d.** the frequency, and
e. the second time when $f(t) = 0$.

Photo © Fredrik D. Bodin

59. Use a graphing calculator to sketch the graph of each function.
a. $y = \sqrt{2}\sin(\sqrt{3}x - 1)$
b. $y = -\sqrt{3}\cos(\sqrt{5}x + 1)$

60. Use a graphing calculator to sketch the graph of each function.
a. $y = e^x \cos x$ **b.** $y = \sin(\ln x)$

61. Use a graphing calculator to sketch the graphs of the following functions over the interval $(-\pi/2, \pi/2)$ in the same coordinate system. How do the graphs compare?

$$y = \tan(-x) \quad \text{and} \quad y = -\tan x.$$

62. Use a graphing calculator to sketch the graphs of the following functions over the interval $[0, 2\pi]$ in the same coordinate system. How do the graphs compare?

$$y = \cos(-x) \quad \text{and} \quad y = \cos x.$$

5.4   INVERSE TRIGONOMETRIC FUNCTIONS

As we observed in Section 5.3, the sine function is not one-to-one and hence does not have an inverse function. It is possible, however, to restrict the domain of the sine function so that the resulting function is one-to-one and thus has an inverse. It is customary to restrict the domain to the interval $[-\pi/2, \pi/2]$ and to call the resulting function the **restricted sine function**. In Figure 5.42 the graph of the restricted sine function is shown as the solid portion of the sine wave.

FIGURE 5.42
Restricted sine function

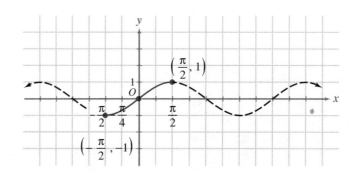

The Inverse Sine Function

We see from Figure 5.42 that the restricted sine function is one-to-one and that its range is the interval $[-1, 1]$, which is also the range of the sine function. Since it is one-to-one, the restricted sine function has an inverse called the inverse sine function. This function is denoted by

$$y = \sin^{-1} x$$

and is defined as follows:

DEFINITION 5.3   Inverse Sine Function

The **inverse sine function**, denoted by $\sin^{-1}$, is the inverse function of the restricted sine function. Thus,

$$y = \sin^{-1} x \quad \text{if and only if} \quad \sin y = x,$$
$$\text{for } -1 \le x \le 1 \quad \text{and} \quad -\frac{\pi}{2} \le y \le \frac{\pi}{2}.$$

It follows from Definition 5.3 that

$$\text{dom}(\sin^{-1}) = [-1, 1] \quad \text{and} \quad \text{ran}(\sin^{-1}) = \left[-\frac{\pi}{2}, \frac{\pi}{2}\right].$$

As in our discussion of inverse functions in Section 2.6, the graph of the inverse sine function can be obtained by reflecting the graph of the restricted sine function across the line $y = x$. See Figure 5.43.

We will also refer to the inverse sine function as the **arcsine function** and use the notation arcsin x in place of $\sin^{-1} x$.

FIGURE 5.43

a. Obtaining the graph of $y = \sin^{-1} x$ **b.** $y = \sin^{-1} x$

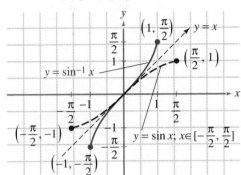

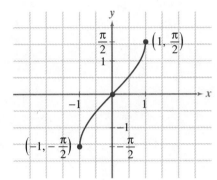

EXAMPLE 15

Find, if possible, each of the following:

a. $\arcsin(-\sqrt{3}/2)$ **b.** $\arcsin 5$ **c.** $\sin^{-1}(-\frac{9}{10})$

Solution

a. Let $y = \arcsin(-\sqrt{3}/2)$. Then

$$\sin y = -\frac{\sqrt{3}}{2} \quad \text{and} \quad -\frac{\pi}{2} \le y \le \frac{\pi}{2}.$$

Since $\sin(-\pi/3) = -\sqrt{3}/2$, it follows that $y = -\pi/3$. Therefore,

$$\arcsin\left(-\frac{\sqrt{3}}{2}\right) = -\frac{\pi}{3}.$$

b. Since 5 is *not* in the interval $[-1, 1]$, which is the domain of arcsin, arcsin 5 is undefined.

c. If $y = \sin^{-1}(-\frac{9}{10})$, then

$$\sin y = -\frac{9}{10} \quad \text{and} \quad -\frac{\pi}{2} \le y \le \frac{\pi}{2}.$$

Since we have not encountered a number whose sine is $-\frac{9}{10}$, we use a calculator (in the radian mode) to find that

$$\sin^{-1}\left(-\frac{9}{10}\right) \approx -1.12.$$

_____

We learned in Section 6 of Chapter 2 that if f has an inverse function f^{-1}, then

$$f[f^{-1}(x)] = x, \quad \text{for all } x \text{ in } \text{dom}(f^{-1}) \quad \text{and}$$
$$f^{-1}[f(x)] = x, \quad \text{for all } x \text{ in } \text{dom}(f).$$

Letting f be the restricted sine function and f^{-1} be the inverse sine function, we obtain the following identities:

$$\sin(\arcsin x) = \sin(\sin^{-1} x) = x, \quad \text{for all } x \text{ in } [-1, 1] \tag{1}$$

and $$\arcsin(\sin x) = \sin^{-1}(\sin x) = x, \quad \text{for all } x \text{ in } \left[-\frac{\pi}{2}, \frac{\pi}{2}\right]. \tag{2}$$

EXAMPLE 16

Find, if possible, each of the following:

a. arcsin[sin($-\pi/3$)]   b. $\sin^{-1}$[sin($7\pi/6$)]   c. $\sin^{-1}$[cos($3\pi/4$)]

Solution

a. Since $-\pi/3$ is in $[-\pi/2, \pi/2]$, identity (2) above implies that

$$\arcsin\left[\sin\left(-\frac{\pi}{3}\right)\right] = -\frac{\pi}{3}.$$

b. Since $7\pi/6$ is *not* in the domain of the restricted sine function, $\sin^{-1}[\sin(7\pi/6)]$ is *not* $7\pi/6$. However, since $\sin(7\pi/6) = -\frac{1}{2}$, it follows that

$$\sin^{-1}\left[\sin\left(\frac{7\pi}{6}\right)\right] = \sin^{-1}\left(-\frac{1}{2}\right).$$

Since $\sin(-\pi/6) = -\frac{1}{2}$, and $-\pi/2 \le -\pi/6 \le \pi/2$,

$$\sin^{-1}\left[\sin\left(\frac{7\pi}{6}\right)\right] = -\frac{\pi}{6}.$$

c. Since $\cos(3\pi/4) = -1/\sqrt{2}$, it follows that

$$\sin^{-1}\left[\cos\left(\frac{3\pi}{4}\right)\right] = \sin^{-1}\left(-\frac{1}{\sqrt{2}}\right).$$

Letting $y = \sin^{-1}(-1/\sqrt{2})$, we see that

$$\sin y = -\frac{1}{\sqrt{2}} \quad \text{and} \quad -\frac{\pi}{2} \le y \le \frac{\pi}{2}.$$

Hence, $y = -\pi/4$ and

$$\sin^{-1}\left[\cos\left(\frac{3\pi}{4}\right)\right] = -\frac{\pi}{4}.$$

We can define inverse functions for the remaining trigonometric functions by following a procedure similar to that used to define the inverse sine function. In each case, we determine a convenient subset of the domain so that a one-to-one function is obtained that has the same range as the original function.

The Inverse Cosine Function

If we restrict the domain of the cosine function to the interval $[0, \pi]$, we obtain a new function called the **restricted cosine function** whose graph is shown as the solid portion of the cosine wave in Figure 5.44. As can be seen from Figure 5.44, the restricted cosine function is one-to-one and its range is the interval $[-1, 1]$, which is also the range of the cosine function. The inverse of the restricted cosine function, called the inverse cosine function, is denoted by

$$y = \cos^{-1} x$$

and is defined as follows:

FIGURE 5.44
Restricted cosine function

FIGURE 5.44
Restricted cosine function

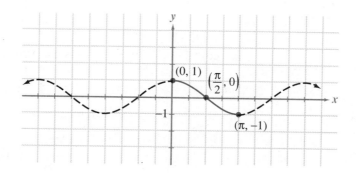

//

DEFINITION 5.4   Inverse Cosine Function

The **inverse cosine function** is the inverse function of the restricted cosine function. Thus

$$y = \cos^{-1} x \quad \text{if and only if} \quad \cos y = x,$$
$$\text{for } -1 \le x \le 1 \quad \text{and} \quad 0 \le y \le \pi.$$

It follows from Definition 5.4 that

$$\text{dom}(\cos^{-1}) = [-1, 1] \quad \text{and} \quad \text{ran}(\cos^{-1}) = [0, \pi].$$

As in the case of the inverse sine function, the graph of the inverse cosine function can be obtained by reflecting the graph of the restricted cosine function across the line $y = x$. See Figure 5.45.

We will also refer to the inverse cosine function as the **arccosine function** and use the notation arccos x in place of $\cos^{-1}x$. Since the restricted cosine function and the arccosine function are inverses of each other, the following identities hold:

$$\cos(\arccos x) = \cos(\cos^{-1} x) = x, \quad \text{for all } x \text{ in } [-1, 1] \qquad \textbf{(3)}$$

and
$$\arccos(\cos x) = \cos^{-1}(\cos x) = x, \quad \text{for all } x \text{ in } [0, \pi]. \qquad \textbf{(4)}$$

FIGURE 5.45

a. Obtaining the graph of $y = \cos^{-1} x$ **b.** $y = \cos^{-1} x$

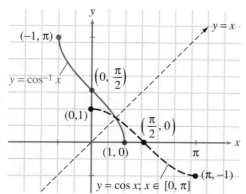

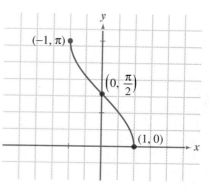

EXAMPLE 17

Find, if possible, each of the following:

a. $\cos^{-1}(-\sqrt{3}/2)$ **b.** $\arccos[\cos(5\pi/4)]$ **c.** $\sin[\arccos(-\sqrt{7}/3)]$

Solution

a. If $y = \cos^{-1}(-\sqrt{3}/2)$, then

$$\cos y = -\frac{\sqrt{3}}{2} \quad \text{and} \quad 0 \le y \le \pi.$$

Since $y = 5\pi/6$ is in $[0, \pi]$ and $\cos y = -\sqrt{3}/2$, it follows that $y = 5\pi/6$. Hence,

$$\cos^{-1}\left(-\frac{\sqrt{3}}{2}\right) = \frac{5\pi}{6}.$$

b. Since $5\pi/4$ is *not* in the domain of the restricted cosine function, $\arccos[\cos(5\pi/4)]$ is *not* $5\pi/4$. However, $\cos(5\pi/4) = \cos(3\pi/4) = -1/\sqrt{2}$, and $3\pi/4$ *is* in the domain of the restricted cosine function. Hence,

$$\arccos\left(\cos\frac{5\pi}{4}\right) = \arccos\left(\cos\frac{3\pi}{4}\right) = \frac{3\pi}{4}.$$

c. Let $y = \arccos(-\sqrt{7}/3)$. Then

$$\cos y = -\frac{\sqrt{7}}{3} \quad \text{and} \quad 0 \le y \le \pi.$$

Since y is in $[0, \pi]$, $\sin y \ge 0$, and it follows from basic identity (v) on page 216 that

$$\sin y = \sqrt{1 - \cos^2 y}$$
$$= \sqrt{1 - \left(-\frac{\sqrt{7}}{3}\right)^2}$$
$$= \sqrt{1 - \frac{7}{9}} = \sqrt{\frac{2}{9}} = \frac{\sqrt{2}}{3}.$$

Hence, $\sin\left[\arccos\left(-\frac{\sqrt{7}}{3}\right)\right] = \frac{\sqrt{2}}{3}.$

The Inverse Tangent Function

If we restrict the domain of the tangent function to the interval $(-\pi/2, \pi/2)$, we obtain the **restricted tangent function** whose graph is the solid portion of the graph of the tangent function shown in Figure 5.46.

We see from Figure 5.46 that the restricted tangent function is one-to-one and that its range is $\mathbb{R}$, which is also the range of the tangent function. The inverse of the restricted tangent function, called the inverse tangent function, is denoted by

$$y = \tan^{-1} x$$

and is defined as follows:

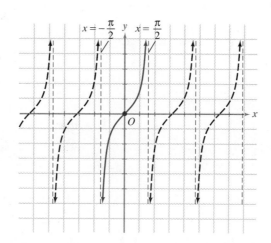

FIGURE 5.46
Restricted tangent function

DEFINITION 5.5   Inverse Tangent Function

The **inverse tangent function** is the inverse function of the restricted tangent function. Thus

$$y = \tan^{-1} x \quad \text{if and only if } \tan y = x,$$

where x is any real number and $-\pi/2 < y < \pi/2$.

The graph of the inverse tangent function is obtained in Figure 5.47 (a) by reflecting the graph of the restricted tangent function across the line $y = x$. As can be seen from Figure 5.47 (b),

$$\text{dom}(\tan^{-1}) = \mathbb{R} \quad \text{and} \quad \text{ran}(\tan^{-1}) = \left(-\frac{\pi}{2}, \frac{\pi}{2}\right).$$

We will also refer to the inverse tangent function as the **arctangent function** and use arctan x in place of $\tan^{-1} x$.

Since the restricted tangent function and the arctangent function are inverses of each other, we have the identities

$$\tan(\arctan x) = \tan(\tan^{-1} x) = x, \quad \text{for all } x \text{ in } \mathbb{R} \tag{5}$$

and $\quad \arctan(\tan x) = \tan^{-1}(\tan x) = x, \quad \text{for all } x \text{ in } \left(-\frac{\pi}{2}, \frac{\pi}{2}\right). \tag{6}$

FIGURE 5.47

a. Obtaining the graph of $y = \tan^{-1} x$ **b.** $y = \tan^{-1} x$

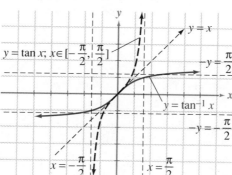

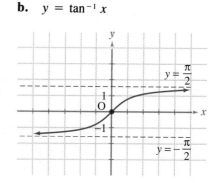

EXAMPLE 18

Find, if possible, each of the following:

a. $\tan^{-1}(-\sqrt{3}/3)$ b. $\arctan(\tan 7)$ c. $\arctan[\sin(3\pi/2)]$

Solution

a. If $y = \tan^{-1}(-\sqrt{3}/3)$, then

$$\tan y = -\sqrt{3}/3 \quad \text{where} \quad -\frac{\pi}{2} < y < \frac{\pi}{2}.$$

Since $y = -\pi/6$ is in $(-\pi/2, \pi/2)$ and $\tan y = -\sqrt{3}/3$, it follows that

$$\tan^{-1}\left(-\frac{\sqrt{3}}{3}\right) = -\frac{\pi}{6}.$$

b. Since 7 is *not* in the domain of the restricted tangent function, $\arctan(\tan 7)$ is *not* 7. However, using a calculator (in the radian mode) we find that

$$\tan 7 \approx 0.8714$$

and that

$$\arctan(\tan 7) \approx 0.7168.$$

c. Since $\sin(3\pi/2) = -1$, it follows that

$$\arctan\left(\sin \frac{3\pi}{2}\right) = \arctan(-1).$$

If $y = \arctan(-1)$, then

$$\tan y = -1 \quad \text{where} \quad -\frac{\pi}{2} < y < \frac{\pi}{2}.$$

Since $y = -\pi/4$ is in $(-\pi/2, \pi/2)$ and $\tan y = -1$,

$$\arctan(-1) = -\frac{\pi}{4}.$$

Hence, $\arctan\left(\sin \frac{3\pi}{2}\right) = -\frac{\pi}{4}.$

There is no general agreement on the domains of the restricted secant, cosecant, and cotangent functions. Some typical domain restrictions are found in Exercises 55–58.

Verifying Identities Involving Inverse Trigonometric Functions

We conclude this section with several examples involving manipulations of the inverse trigonometric functions.

EXAMPLE 19

Show that for all x in $[-1, 1]$, $\sin(\cos^{-1} x) = \sqrt{1 - x^2}$.

Solution Let x be in $[-1, 1]$ and let $y = \cos^{-1} x$. Then

$$\cos y = x \quad \text{and} \quad 0 \leq y \leq \pi.$$

Since $\sin y \geq 0$ for y in $[0, \pi]$, it follows from the basic identity $\sin^2 y + \cos^2 y = 1$, that

$$\begin{aligned} \sin y &= \sqrt{1 - \cos^2 y} \\ &= \sqrt{1 - x^2}. \end{aligned}$$

Hence, $\sin(\cos^{-1} x) = \sqrt{1 - x^2}$.

EXAMPLE 20

Show that $\sec^2(\tan^{-1} x) = 1 + x^2$, for all real numbers x.

Solution Let x be a real number and let $y = \tan^{-1} x$. Then

$$\tan y = x \quad \text{and} \quad -\frac{\pi}{2} < y < \frac{\pi}{2}.$$

By basic identity (vi) on page 216,

$$\sec^2 y = 1 + \tan^2 y.$$

Substituting x for $\tan y$ and $\tan^{-1} x$ for y, we obtain

$$\sec^2(\tan^{-1} x) = 1 + x^2.$$

EXAMPLE 21

Show that the inverse sine function is an odd function. That is, show that $\sin^{-1}(-x) = -\sin^{-1} x$, for all x in $[-1, 1]$.

Solution Let x be in $[-1, 1]$ and let $y = \sin^{-1}(-x)$. Then

$$\sin y = -x \quad \text{and} \quad -\frac{\pi}{2} \leq y \leq \frac{\pi}{2}.$$

Multiplying both sides of the equation above by -1 and each part of the inequality above by -1, we obtain

$$-\sin y = x \quad \text{and} \quad \frac{\pi}{2} \geq -y \geq -\frac{\pi}{2},$$

or $\quad -\sin y = x \quad \text{and} \quad -\frac{\pi}{2} \leq -y \leq \frac{\pi}{2}.$ (7)

Since sine is an odd function, $\sin(-y) = -\sin y$. Substituting $\sin(-y)$ for $-\sin y$ in the equation in statement (7) yields

$$\sin(-y) = x \quad \text{and} \quad -\frac{\pi}{2} \leq -y \leq \frac{\pi}{2}.$$

Hence, by definition of the inverse sine function,

$$-y = \sin^{-1} x, \quad \text{or}$$
$$y = -\sin^{-1} x.$$

Since $y = \sin^{-1}(-x)$, $\qquad \sin^{-1}(-x) = -\sin^{-1} x$,

and the inverse sine function is odd.

EXERCISES 5.4

In Exercises 1–27, find, if possible, each of the following. Do not use a calculator or tables.

1. $\sin^{-1}(-\sqrt{2}/2)$
2. $\cos^{-1}(0)$
3. $\cos^{-1}(\sqrt{3}/2)$
4. $\arctan(-\sqrt{3})$
5. $\sin^{-1}(-\frac{1}{2})$
6. $\arccos[\sin(4\pi/3)]$
7. $\sin[\arctan(-\sqrt{2}/2)]$
8. $\arcsin[\cos(3\pi/4)]$
9. $\sin^{-1}[\sin(11\pi/6)]$
10. $\cos^{-1}[\cos(5\pi/4)]$
11. $\arcsin[\sin(-\pi/6)]$
12. $\sin^{-1}[\sin(7\pi/4)]$
13. $\tan[\arctan(-25)]$
14. $\arctan[\tan(7\pi/6)]$
15. $\sin(\arccos \frac{3}{5})$
16. $\cos[\arcsin(-\frac{9}{10})]$
17. $\cos[\arctan(-2)]$
18. $\sin^{-1}[\sin(-1.4)]$
19. $\sin[\arccos(-1/\sqrt{3})]$
20. $\tan[\arccos(-\frac{1}{2})]$
21. $\sin(\sin^{-1} 4)$
22. $\cos(\cos^{-1} \pi)$
23. $\tan^{-1}(\tan 1.5)$
24. $\sin^{-1}(\sin \sqrt{2})$
25. $\tan(\tan^{-1} 100)$
26. $\cos[\arcsin(-\frac{7}{8})]$
27. $\sec(\arctan 50)$

In Exercises 28–36, use a calculator in the radian mode to approximate each function value to four decimal places.

28. $\sin^{-1}(0.35)$
29. $\arccos(-\sqrt{11}/20)$
30. $\arctan 13$
31. $\arcsin(-\frac{3}{4})$
32. $\cos^{-1}(-\pi/10)$
33. $\tan^{-1}(\sqrt{\pi})$
34. $\sin^{-1}(\sqrt{\pi} - 1)$
35. $\arccos(\frac{13}{14})$
36. $\arctan(-\sqrt{13})$

In Exercises 37–50, sketch the graph of each function.

37. $f(x) = -\sin^{-1} x$
38. $f(x) = 1 + \tan^{-1} x$
39. $f(x) = 2 + \cos^{-1} x$
40. $f(x) = 2 \sin^{-1} x + 1$
41. $f(x) = 1 - \tan^{-1} x$
42. $f(x) = \sin^{-1} x + 2$
43. $f(x) = -\cos^{-1} x$
44. $f(x) = 2 \tan^{-1} x$
45. $f(x) = 1 - \sin^{-1} x$
46. $f(x) = \tan^{-1}(x + 1)$
47. $f(x) = \cos^{-1}(\cos x)$
48. $f(x) = \sin(\sin^{-1} x)$
49. $f(x) = \tan(\arctan x)$
50. $f(x) = \cos(\cos^{-1} x)$

51. Show that the inverse tangent function is an odd function; that is, show that $\tan^{-1}(-x) = -\tan^{-1} x$, for all x in $\mathbb{R}$.

52. Show that for all x in $(-1, 1)$, $\tan(\sin^{-1} x) = \dfrac{x}{\sqrt{1 - x^2}}$.

53. Show that $\tan(\cos^{-1} x) = \dfrac{\sqrt{1 - x^2}}{x}$, for all x in $[-1, 0) \cup (0, 1]$.

54. Show that for all x in $\mathbb{R}$, $\cos(\tan^{-1} x) = 1/\sqrt{1 + x^2}$.

55. One way to define the inverse secant function is to restrict the domain of the secant function to $[0, \pi/2) \cup [\pi, 3\pi/2)$. Use this approach to define $\sec^{-1}$ and sketch the graph of $y = \sec^{-1} x$.

56. One way to define the inverse secant function is to restrict the domain of the secant function to $[0, \pi/2) \cup (\pi/2, \pi]$. Use this approach to define $\sec^{-1}$ and sketch the graph of $y = \sec^{-1} x$.

57. Obtain the inverse cotangent function by restricting the domain of the cotangent function to the interval $(0, \pi)$. Sketch the graph of $y = \cot^{-1} x$.

58. Obtain the inverse cosecant function by restricting the domain of the cosecant function to $[-\pi/2, 0) \cup (0, \pi/2]$. Sketch the graph of $y = \csc^{-1} x$.

59. If the domain of the cosine function is restricted to the interval $[-\pi/2, \pi/2]$ (as we restricted the sine function), will the restricted function have an inverse? Explain.

60. Explain why the graph of the inverse sine function is symmetric with respect to the origin.

61. Explain why the graph of the inverse tangent function is symmetric with respect to the origin.

In Exercises 62–68, use a variation of the process illustrated below to solve each equation for x in terms of y.

To solve the equation $y = 3 - 4 \cos^{-1}(2x - 6)$ for x in terms of y, we proceed as follows (you should supply a reason for each step shown):

$$y - 3 = -4 \cos^{-1}(2x - 6)$$

$$\frac{y - 3}{-4} = \cos^{-1}(2x - 6)$$

$$\frac{3 - y}{4} = \cos^{-1}(2x - 6)$$

$$\cos\left(\frac{3 - y}{4}\right) = 2x - 6$$

$$\cos\left(\frac{y - 3}{4}\right) = 2x - 6$$

$$2x = \cos\left(\frac{y - 3}{4}\right) + 6$$

$$x = \frac{1}{2} \cos\left(\frac{y - 3}{4}\right) + 3.$$

62. $y = 7 + 8 \sin^{-1}(5x + 2)$
63. $y = 4 - 5 \cos^{-1}(3x - 9)$

64. $y = \pi + 3 \tan^{-1}(4x + 5)$

65. $y = -3 - 4 \sin^{-1}(x - 9)$

66. $y = -6 - 7 \sin^{-1}(5 - 3x)$

67. $y = 3 + \tan^{-1}(2 - \pi x)$

68. $y = 2 - \cos^{-1}(\pi - 7x)$

In Exercises 69–74, find functions f and g such that $h = f \circ g$. Verify your answer by forming the composition.

69. $h(x) = \sin(\ln x)$

70. $h(x) = e^{\tan x}$

71. $h(x) = \cos(\tan^{-1} x)$

72. $h(x) = \ln(\cos x)$

73. $h(x) = e^{\sec x}$

74. $h(x) = \sin(e^x)$

75. Use a graphing calculator to sketch the graph of each function.

   a. $y = -2 \sin^{-1}(x - 1) + 1$

   b. $y = 2 \cos^{-1}(2x + 1) - 1$

76. Use a graphing calculator to sketch the graph of each function.

   a. $y = 1 - 2 \tan^{-1}(2x - 1)$

   b. $y = 3 \sin^{-1}(3x - 1) - 2$

CHAPTER 5 REVIEW EXERCISES

In Exercises 1–10, find the exact coordinates of $P(t)$ and display $P(t)$ on the unit circle.

1. $P(-9\pi)$

2. $P(-4\pi/3)$

3. $P(17\pi/3)$

4. $P(13\pi/2)$

5. $P(-25\pi/4)$

6. $P(19\pi/6)$

7. $P(-17\pi/6)$

8. $P(-19\pi/4)$

9. $P(25\pi/2)$

10. $P(28\pi)$

In Exercises 11–20, find the approximate location of $P(t)$ on the unit circle. Use $\pi \approx 3.14$.

11. $P(4)$

12. $P(-1)$

13. $P(-10)$

14. $P(25)$

15. $P(12)$

16. $P(-8)$

17. $P(-7\pi/8)$

18. $P(-3\pi/5)$

19. $P(-6.1)$

20. $P(13)$

In Exercises 21–25, find all values of t (if any) in $[0, 2\pi)$ for which $P(t)$ satisfies the stated conditions.

21. The x-coordinate of $P(t)$ is ± 1.

22. The y-coordinate of $P(t)$ is $\pm \sqrt{3}/2$.

23. The y-coordinate of $P(t)$ is -2.

24. The sum of the x- and y-coordinates of $P(t)$ is 0.

25. The x-coordinate of $P(t)$ is $\sqrt{3}/2$.

In Exercises 26–30, $P(t)$ lies on the unit circle in the indicated quadrant. Find c.

26. $P(t) = (-2/\sqrt{5}, c)$; quadrant III.

27. $P(t) = (c, \sqrt{3}/5)$; quadrant II.

28. $P(t) = (-\frac{2}{3}, c)$; quadrant II.

29. $P(t) = (15/c, 3/c)$; quadrant III.

30. $P(t) = (4c, c)$; quadrant III.

In Exercises 31–42, find the exact value of each expression, if possible.

31. $\tan(-5\pi/2)$

32. $\cot(5\pi/2)$

33. $\csc(-11\pi/3)$

34. $\sec(10\pi/3)$

35. $\sin(7\pi/4)$

36. $\cos(-7\pi/6)$

37. $\cos(23\pi/6)$

38. $\sin(11\pi/4)$

39. $\sec 13\pi$

40. $\csc 20\pi$

41. $\cot(-20\pi/6)$

42. $\tan(-30\pi/4)$

In Exercises 43–48, show that each point lies on the unit circle, and hence is $P(t)$, for some t in $\mathbb{R}$. Then find the exact value of all six trigonometric functions at t.

43. $(-\sqrt{6}/3, 1/\sqrt{3})$

44. $(2/\sqrt{5}, -1/\sqrt{5})$

45. $(\sqrt{22}/5, -\sqrt{3}/5)$

46. $(-\sqrt{11}/6, \frac{5}{6})$

47. $(-\sqrt{5}/3, -\frac{2}{3})$

48. $(\sqrt{7}/4, \frac{3}{4})$

In Exercises 49–58, use your calculator in the radian mode to approximate the indicated function value to four decimal places.

49. $\csc 15$

50. $\sin(-10)$

51. $\cos(-\sqrt{7})$

52. $\cos(\sqrt{13})$

53. $\tan 12$

54. $\cot(-16)$

55. $\sec(-21)$

56. $\csc 30$

57. $\sin(\cot 5)$

58. $\tan[\cos(-\sqrt{5})]$

In Exercises 59–64, use the given information to find the indicated function values at t in each case. Give exact answers.

59. If $\sin t = \sqrt{5}/3$ and $P(t)$ is in quadrant II, find $\tan(-t)$.

60. If $\tan t = -5$ and $P(t)$ is in quadrant IV, find $\sin t$.

61. If $\cos t = -\frac{1}{7}$, and $\tan t > 0$, find $\csc t$.

62. If $\cot t = 10$ and $\cos t < 0$, find $\sin(-t)$.

63. If $\sec t = 4$ and $P(t)$ is in quadrant IV, find $\cot(-t)$.

64. If $\csc t = -\sqrt{3}$ and $\cot t > 0$, find $\sec(-t)$.

65. If $P(t) = (\sqrt{5}/c, \sqrt{11}/c)$ lies on the unit circle in quadrant III, find the exact value of c and of all six trigonometric functions at t.

66. Show that if $u = 5 \sin t$ where $P(t)$ is in quadrant II, then $\sqrt{25 - u^2} = -5 \cos t$.

67. Show that if $u = 9 \sec t$ and $P(t)$ is in quadrant III, then $\sqrt{u^2 - 81} = 9 \tan t$.

68. Show that if $u = 4 \tan t$ and $P(t)$ is in quadrant I, then $\sqrt{16 + u^2} = 4 \sec(-t)$.

In Exercises 69–82, sketch at least one cycle of the graph of each function. Give the period, amplitude, phase shift, and range of each.

69. $y = 5 \cos x$

70. $y = -\pi \sin x$

71. $y = \sin 4x$

72. $y = \cos(-2x)$

73. $y = -3 \cos(x/3)$

74. $y = -2 \sin(-2x)$

75. $y = 2 \sin(3x/2)$

76. $y = 3 \cos(2x/3)$

77. $y = \frac{1}{2} \cos(x/2)$

78. $y = \frac{2}{3} \sin(2x/3)$

79. $y = -2 \cos(3x + 1)$

80. $y = -3 \cos(2x + 1)$

81. $y = 4 \sin(2 - 3x)$

82. $y = \frac{1}{3} \cos(4 - 2x)$

In Exercises 83–90, sketch at least one cycle of the graph of each function. Give the period, phase shift, and range of each. Also sketch any vertical asymptotes.

83. $y = \cot(-x)$

84. $y = \tan(-x)$

85. $y = 2 \tan(x/2)$

86. $y = 3 \cot(x/3)$

87. $y = \sec(2 - x)$

88. $y = 2 \csc(1 - 2x)$

89. $y = -2 \csc(2 - 4x)$

90. $y = -3 \sec(2x + 1)$

91. The current in an electrical circuit is given by the function

$$I = f(t) = 50 \cos(120\pi t - \pi),$$

where I is measured in amperes and t is in seconds. Find

a. the period,   **b.** the amplitude,

c. the phase shift,   **d.** the frequency, and

e. the first time when $f(t) = 0$.

92. The path of a sound wave is given by the function

$$f(t) = 2 \sin[\pi(t_0 - 360t)],$$

where t_0 is a positive constant and t is elapsed time in seconds. Find

a. the period,   **b.** the amplitude,

c. the phase shift,   **d.** the frequency, and

e. the second time when $f(t) = 0$.

In Exercises 93–110, find, if possible, each of the following. Do not use a calculator or tables.

93. $\sin^{-1}(-\sqrt{3}/2)$

94. $\cos^{-1}(-\sqrt{2}/2)$

95. $\arctan 1$

96. $\arccos(-\frac{1}{2})$

97. $\arccos(\sin 2\pi)$

98. $\arcsin(\cos 2\pi)$

99. $\sin^{-1}[\cos(3\pi/4)]$

100. $\cos^{-1}[\sin(5\pi/6)]$

101. $\tan^{-1}[\cot(5\pi/6)]$

102. $\sin^{-1}[\sin(5\pi/3)]$

103. $\tan[\arctan(-\pi)]$

104. $\sin[\cos^{-1}(-\frac{2}{3})]$

105. $\sin^{-1}[\sin(2\pi/3)]$

106. $\tan^{-1}[\tan(7\pi/6)]$

107. $\cos(\sin^{-1}\frac{3}{4})$

108. $\cos[\arcsin(-\frac{5}{6})]$

109. $\sec(\tan^{-1} 5)$

110. $\sin[\arcsin(-3)]$

In Exercises 111–116, use a calculator in the radian mode to approximate each function value to four decimal places.

111. $\arcsin(-0.56)$

112. $\cos^{-1}(-\sqrt{5}/3)$

113. $\tan^{-1} 25$

114. $\arcsin(-\pi/12)$

115. $\cos^{-1}(-\frac{15}{16})$

116. $\tan^{-1} \pi$

In Exercises 117–122, sketch the graph of each function.

117. $f(x) = -\cos^{-1} x$

118. $f(x) = 1 + \sin^{-1} x$

119. $f(x) = 1 + \tan^{-1} x$

120. $f(x) = 1 + \sin^{-1}(-x)$

121. $f(x) = 2 + \sin^{-1}(-x)$

122. $f(x) = 3 - \cos^{-1} x$

In Exercises 123–127, solve each equation for x in terms of y.

123. $y = 9 - 10 \sin^{-1}(11x + 12)$

124. $y = 2 + 5 \cos^{-1}(3x - 4)$

125. $y = -5 + 6 \tan^{-1}(\pi x + 2)$

126. $y = 1 - 8 \sin^{-1}(4x - 5)$

127. $y = 4 + 7 \cos^{-1}(8x - 2)$

CHAPTER 6

Trigonometric Identities and Equations

In the preceding chapter, we discussed the trigonometric functions of real numbers. In this chapter we define the trigonometric functions of angles and show that the functions obtained using these two approaches are essentially the same. We then review the fundamental trigonometric identities and verify a variety of additional trigonometric identities. The chapter concludes with a thorough treatment of trigonometric equations.

6.1   TRIGONOMETRIC FUNCTIONS OF ANGLES

In Chapter 5 we defined the trigonometric functions of a real number t by using the unit circle and the coordinates of the trigonometric point $P(t)$. In this section we present an alternate development in which the trigonometric functions are defined as functions of angles. It is in this context that these functions are often used in the applications of mathematics to surveying, navigation, astronomy, and other areas of study. As we shall see, the trigonometric functions defined for real numbers and the trigonometric functions defined for angles are essentially the same.

Recall from geometry that an **angle** is formed by rotating a half-line about its endpoint (called the **vertex** of the angle) from an initial position to a terminal position. The initial position of the half-line is called the **initial side** of the angle, and the terminal position of the half-line is called the **terminal side**. In Figure 6.1, angle BAC, denoted by $\angle BAC$, is formed by rotating the half-line AB about the vertex A from its initial position AB to its terminal position AC.

An angle formed by rotating a half-line counterclockwise is said to be **positive**; if the rotation is clockwise the angle is said to be **negative**. In Figure 6.2, $\angle BAC$ is positive, while $\angle FED$ is negative. An angle whose vertex is at the origin of a rectangular coordinate system and whose initial side coincides with the positive x-axis is said to be in **standard position**. An angle in standard position whose terminal side lies along a coordinate axis is called a **quadrantal angle**. If the terminal side does not lie along a coordinate axis, the angle is said to **lie in the quadrant** containing its terminal side. Two angles in standard position whose terminal sides coincide are called **coterminal angles**. For example, in Figure 6.3, angle α is a quadrantal angle, angle β lies in quadrant II, and angles γ and θ are coterminal angles that lie in quadrant III.

The Degree Measure of an Angle

One of the commonly used units for measuring angles is the degree. **One degree** (written $1°$) is the measure of a positive angle formed by $\frac{1}{360}$ of a complete rotation. Hence, the measure of a positive angle formed by one complete rotation is $360°$. A **right angle** is an angle whose measure is $90°$, while a **straight angle** has measure $180°$. An **acute angle** is an angle whose measure is greater than $0°$ but less than $90°$, while an **obtuse angle** is an angle whose measure is greater than $90°$ but less than $180°$. See Figure 6.4.

In general we will not distinguish between an angle and the measure of that angle. For example, we will use variables such as α, β, γ, and θ to represent either an angle or its measure. The context in which the variable is used will make it clear which applies. Thus, we will write $\theta = 75°$ and say that θ is a $75°$ angle when we mean that θ is an angle whose measure is $75°$.

A degree can be divided into smaller units using decimals. We can also divide a degree into smaller units called minutes and seconds. Each degree is divided into 60 equal parts called **minutes** ($'$), and each minute is divided into 60 equal parts called **seconds** ($''$). Thus $1° = 60'$ and $1' = 60''$. It follows that $1° = 3600''$.

In our work in trigonometry, we are often required to convert a degrees-minutes-seconds measurement of an angle into decimal degrees or vice versa. While many scientific calculators have special keys for making these conversions, our next example shows how such conversions can be made without using these special keys.

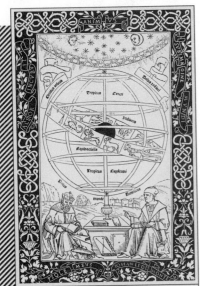

North Wind Picture Archives

CLAUDIUS PTOLEMY
(ca. A.D. 85–ca. 165)

Claudius Ptolemy, one of the preeminent astronomers of the second century A.D., did for astronomy what Euclid did for plane geometry. He collected, systematically organized, and published the discoveries of his predecessors in a thirteen-volume treatise. This work, the *Almagest*, was the most influential and significant trigonometric work of the time and was a standard of excellence for many centuries. Ptolemy elaborated the trigonometric tables used by Hipparchus and discovered theorems that are equivalent to the sum and difference formulas for sine and cosine. For this reason, these formulas are sometimes called Ptolemy's formulas. Like many mathematicians, he attempted to prove Euclid's parallel postulate from the other axioms and postulates stated in Euclid's *Elements*.

FIGURE 6.1 Angle *BAC*

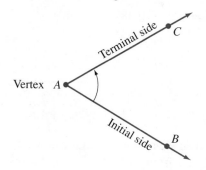

FIGURE 6.2
a. ∠*BAC* is positive **b.** ∠*FED* is negative

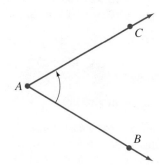
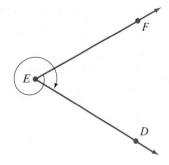

FIGURE 6.3
a. α is a quadrantal angle **b.** β lies in quadrant II **c.** γ and θ lie in quadrant III and are coterminal

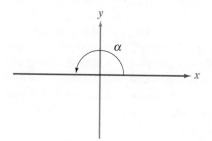
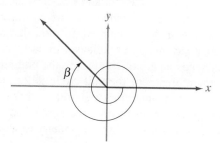
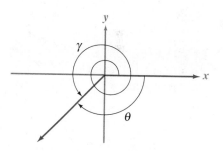

FIGURE 6.4
a. A right angle **b.** A straight angle **c.** An acute angle **d.** An obtuse angle

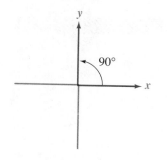
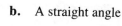
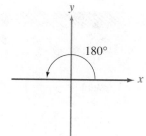
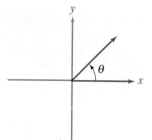
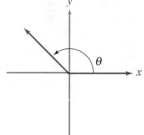

EXAMPLE 1

a. Express 82°18′25″ in decimal degrees to the nearest thousandth.
b. Express 123.523° in degrees, minutes, and seconds.

Solution

a. Since

$$18' = \left(\frac{18}{60}\right)^{\circ} \quad \text{and} \quad 25'' = \left(\frac{25}{3600}\right)^{\circ},$$

it follows that

$$82°18'25'' = \left(82 + \frac{18}{60} + \frac{25}{3600}\right)^{\circ} \approx 82.307°.$$

b. Since

$$0.523° = (0.523)(60') = 31.38' = 31' + 0.38', \quad \text{and}$$
$$0.38' = (0.38)(60'') = 22.8'' \approx 23'',$$

we see that

$$123.523° \approx 123°31'23''.$$

FIGURE 6.5

θ is an angle of measure 1 radian

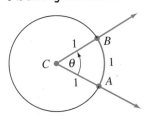

FIGURE 6.6

θ is an angle of 1 radian in standard position

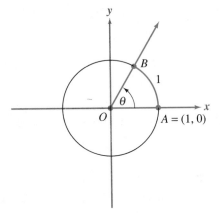

The Radian Measure of an Angle

There is another unit for measuring angles called the radian, which is commonly used in calculus and in scientific work. (Recall from geometry that a **central angle** is an angle whose vertex is at the center of a circle.) **One radian** is the measure of a positive angle which, when viewed as a central angle in a circle of radius 1, intercepts an arc of length 1. See Figure 6.5.

Thus, we see that an angle in standard position with measure 1 radian subtends an arc of length 1 on the unit circle. See Figure 6.6. More generally, a positive angle in standard position that subtends an arc of length t on the unit circle has measure t radians. Similarly, a negative angle in standard position that subtends an arc of length t on the unit circle has measure $-t$ radians. Since the circumference of the unit circle is 2π, it follows that an angle formed by one complete counterclockwise rotation has measure 2π radians. Since both $360°$ and 2π radians represent the measure of an angle formed by one complete counterclockwise rotation, it is customary to write

$$360° = 2\pi \text{ radians,}$$
$$\text{or} \quad 180° = \pi \text{ radians.} \qquad \textit{(1)}$$

Equation (1) leads us to the following useful conversion formulas:

Radian-Degree Conversion Formulas

$$1 \text{ radian} = \left(\frac{180}{\pi}\right)°$$

$$1° = \left(\frac{\pi}{180}\right) \text{ radians}$$

We illustrate these conversion formulas in our next example.

EXAMPLE 2

a. Find the radian measure of $\alpha = 420°$ and sketch α in standard position.
b. Find the degree measure of $\beta = -3\pi/10$ radians and sketch β in standard position.

Solution

a. Using the radian-degree conversion formula, we have

FIGURE 6.7

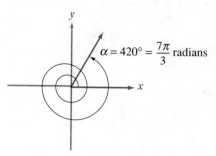

$$\alpha = 420° = 420\left(\frac{\pi}{180}\text{ radians}\right) = \frac{7\pi}{3}\text{ radians.}$$ (See Figure 6.7.)

b. Similarly,

$$\beta = -\frac{3\pi}{10}\text{ radians} = \left(-\frac{3\pi}{10}\right)\left(\frac{180}{\pi}\right)° = -54°.$$ (See Figure 6.8.)

FIGURE 6.8

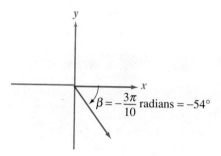

Table 6.1 displays both the degree and radian measure of several commonly used angles. The entries in the table may be confirmed by using the radian-degree conversion formulas.

TABLE 6.1

Degrees	0°	30°	45°	60°	90°	120°	135°	150°	180°
Radians	0	$\pi/6$	$\pi/4$	$\pi/3$	$\pi/2$	$2\pi/3$	$3\pi/4$	$5\pi/6$	π

It is customary to omit the word *radian* when the measure of an angle is expressed in radians. Therefore, when no units of measure are mentioned, we will assume that the angle is measured in radians.

Length of an Arc

As we shall now see, it is possible to define the radian measure of an angle using a circle other than the unit circle. In Figure 6.9, θ is a central angle of both the unit circle and a circle of radius $r > 1$. (A similar approach holds when $r < 1$.)

From Figure 6.9, we see that θ subtends an arc of length t on the unit circle and therefore has radian measure t. In addition, we see that θ subtends an arc of length s on the circle of radius r. Since the ratio of these arc lengths is the same as the ratio of the radii of the two circles, we know that

$$\frac{s}{t} = \frac{r}{1},\quad\text{or}\quad s = rt.$$

FIGURE 6.9

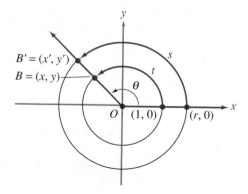

Since $t = \theta$,

$$s = r\theta.$$

We have thus proved the following theorem:

//

THEOREM 6.1

If a positive central angle θ, measured in radians, subtends an arc of length s on a circle of radius r, then

$$s = r\theta.$$

The equation $s = r\theta$ from Theorem 6.1 can also be written in the form

$$\theta = \frac{s}{r}.$$

Thus we have

//

COROLLARY 6.1

If a positive central angle θ (measured in radians) subtends an arc of length s on a circle of radius r, then the radian measure of θ is given by $\theta = s/r$.

We see from Corollary 6.1, that when $s = r$, $\theta = 1$ radian.

FIGURE 6.10

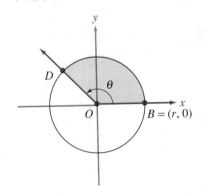

Area of a Circular Sector

When we know the radian measure of a central angle θ in a circle of radius r, we can use the equation $s = r\theta$ from Theorem 6.1 to find the length of the arc subtended by θ. We can also use the radian measure of θ to find the area of the circular sector determined by the angle θ. To see this, let θ be the radian measure of central angle BOD in the circle of radius r shown in Figure 6.10. We know that the ratio of the area A of the circular sector BOD to the area of the entire circle is equal to the ratio of the radian measure of angle BOD to the radian measure of the central angle subtended by the entire circle. Hence,

$$\frac{A}{\pi r^2} = \frac{\theta}{2\pi},$$

or

$$A = \frac{1}{2}r^2\theta. \tag{2}$$

Using Theorem 6.1, equation (2) can be written in an alternate form:

$$A = \frac{1}{2}r^2\theta = \frac{1}{2}r(r\theta) = \frac{1}{2}r(s).$$

Thus,

$$A = \frac{1}{2}rs. \tag{3}$$

//// CAUTION: When using Theorem 6.1 or either of formulas (2) and (3), the central angle θ must be measured in radians.

EXAMPLE 3

Find the length of the arc and the area of the circular sector subtended by a central angle of 74° in a circle of radius 36 centimeters.

Solution We must first convert $\theta = 74°$ to radian measure. Using the radian-degree conversion formula, we have

$$\theta = 74° = 74\left(\frac{\pi}{180}\right) = \frac{37\pi}{90} \text{ radians.}$$

Hence, by Theorem 6.1, the length of the arc subtended by θ is

$$s = r\theta = (36)\left(\frac{37\pi}{90}\right) = \frac{74\pi}{5} \approx 46.50 \text{ centimeters.}$$

By equation (3), the area of the circular sector is

FIGURE 6.11

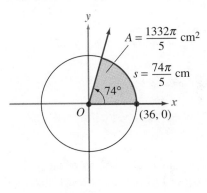

$$A = \frac{1}{2}rs$$
$$= \frac{1}{2}(36)\left(\frac{74\pi}{5}\right)$$
$$= \frac{1332\pi}{5} \text{ square centimeters}$$
$$\approx 836.92 \text{ square centimeters.}$$

(See Figure 6.11.)

EXAMPLE 4

The tip of the pendulum in a grandfather clock subtends an arc of $13\pi/15$ feet while subtending a central angle of 43.2°. How long is the pendulum?

Solution If we let r denote the length of the pendulum, we see that the tip of the pendulum subtends an arc of length $s = 13\pi/15$ on a circle of radius r. From the arc length formula $s = r\theta$, it follows that $r = s/\theta$. To find r, we must first express the measure of central angle θ in radians. We have

$$\theta = 43.2°$$
$$= 43.2\left(\frac{\pi}{180}\right)$$
$$= \frac{6\pi}{25} \text{ radians.}$$

Substituting for s and θ we obtain

$$r = \frac{s}{\theta}$$
$$= \frac{(13\pi/15)}{(6\pi/25)}$$
$$= \left(\frac{13\pi}{15}\right)\left(\frac{25}{6\pi}\right)$$
$$= \frac{65}{18}$$
$$\approx 3.61 \text{ feet.}$$

Thus, the pendulum is approximately 3.61 feet long.

FIGURE 6.12

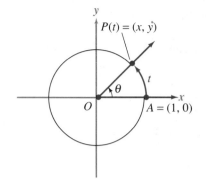

Trigonometric Functions of Angles

We are now in a position to define the trigonometric functions of an angle. Let θ (measured in radians) be a central angle of the unit circle which subtends an arc of length t. Let $P(t) = (x, y)$ be the trigonometric point associated with t by the wrapping function P. (See Figure 6.12.) Since $P(t) = (x, y) = (\cos t, \sin t)$, we define the trigonometric functions of the angle θ as follows.

D E F I N I T I O N 6.1   **Trigonometric Functions of an Angle θ with Radian Measure t**

If θ is an angle whose radian measure is t, then the value of each trigonometric function at angle θ is equal to the value of that trigonometric function at the real number t. That is,

$$\sin \theta = \sin t \qquad \csc \theta = \csc t$$
$$\cos \theta = \cos t \qquad \sec \theta = \sec t$$
$$\tan \theta = \tan t \qquad \cot \theta = \cot t.$$

We see from Definition 6.1 that the trigonometric functions of angles are essentially the same as the trigonometric functions of real numbers. Hence, all of the results obtained in Chapter 5 for the trigonometric functions of real numbers also hold for the trigonometric functions of angles. We can find the trigonometric functions of an angle θ measured in degrees by first converting the measure of θ to radians as illustrated in the next example.

EXAMPLE 5

Find the exact value of each of the following:

a. $\cos 45°$ **b.** $\tan 120°$ **c.** $\csc(-150°)$

Solution

a. Since $45° = 45(\pi/180) = \pi/4$ radians, it follows that $\cos 45° = \cos(\pi/4) = \sqrt{2}/2$.

b. Converting $120°$ to radian measure, we obtain $120° = 120(\pi/180) = 2\pi/3$. Hence, $\tan 120° = \tan(2\pi/3) = -\sqrt{3}$.

c. Since $-150° = -5\pi/6$, $\csc(-150°) = \csc(-5\pi/6) = -2$.

In Definition 6.1, the trigonometric functions of an angle θ are defined in terms of the coordinates of the point P where the terminal side of θ intersects the unit circle. It is possible to extend this definition so that the trigonometric functions of θ are expressed in terms of the coordinates of *any* point Q, other than the vertex, on the terminal side of θ. Let θ (measured in radians) be in standard position and $Q = (x, y)$ be any point, other than the origin O, on the terminal side of θ as shown in Figure 6.13. Let $r = \sqrt{x^2 + y^2}$ be the length of the line segment $\overline{OQ}$, and let $P = (\cos \theta, \sin \theta)$ be the point where the terminal side of θ intersects the unit circle.

FIGURE 6.13

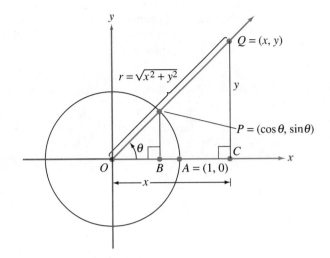

In Figure 6.13, we note that triangle OBP and triangle OCQ are similar triangles. Since corresponding sides of similar triangles are proportional, it follows that

$$\frac{d(B, P)}{d(O, P)} = \frac{y}{r} \quad \text{and} \quad \frac{d(O, B)}{d(O, P)} = \frac{x}{r}.$$

Since $d(B, P) = \sin \theta$, $d(O, P) = 1$, and $d(O, B) = \cos \theta$, it follows that

$$\frac{\sin \theta}{1} = \frac{y}{r} \quad \text{and} \quad \frac{\cos \theta}{1} = \frac{x}{r},$$

$$\text{or} \quad \sin \theta = \frac{y}{r} \quad \text{and} \quad \cos \theta = \frac{x}{r}.$$

This leads us to the following theorem:

THEOREM 6.2

Let θ be an angle in standard position. If $Q = (x, y)$ is any point, other than the origin O, on the terminal side of θ, and if $d(O, Q) = \sqrt{x^2 + y^2} = r$, then

$$\sin \theta = \frac{y}{r} \qquad\qquad \csc \theta = \frac{r}{y}, \ y \neq 0$$

$$\cos \theta = \frac{x}{r} \qquad\qquad \sec \theta = \frac{r}{x}, \ x \neq 0$$

$$\tan \theta = \frac{y}{x}, \quad x \neq 0 \qquad \cot \theta = \frac{x}{y}, \ y \neq 0.$$

EXAMPLE 6

Assume that angle θ is in standard position and that the terminal side of θ contains the point $Q = (-2, 3)$. Find the exact values of all six trigonometric functions of θ.

FIGURE 6.14

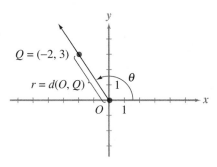

Solution   Figure 6.14 displays the smallest positive angle θ whose terminal side contains the point $Q = (-2, 3)$. Since $r = d(O, Q) = \sqrt{(-2)^2 + (3)^2} = \sqrt{13}$, it follows from Theorem 6.2 that

$$\sin \theta = \frac{y}{r} = \frac{3}{\sqrt{13}} = \frac{3\sqrt{13}}{13} \qquad \csc \theta = \frac{r}{y} = \frac{\sqrt{13}}{3}$$

$$\cos \theta = \frac{x}{r} = -\frac{2}{\sqrt{13}} = -\frac{2\sqrt{13}}{13} \qquad \sec \theta = \frac{r}{x} = -\frac{\sqrt{13}}{2}$$

$$\tan \theta = \frac{y}{x} = -\frac{3}{2} \qquad \cot \theta = \frac{x}{y} = -\frac{2}{3}.$$

FIGURE 6.15

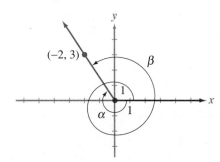

Consider the angle θ from Example 6. As Figure 6.15 shows, there are other angles in standard position whose terminal side contains the point $(-2, 3)$. Thus, we see that if $Q \neq (0, 0)$, there are many different angles in standard position whose terminal side contains the point Q. Since all of these angles are coterminal, it follows from Theorem 6.2 that these angles have the same trigonometric function values. In general, the values of the trigonometric functions are the same for

$$\theta \text{ and } \theta + k(2\pi), \text{ when } \theta \text{ is measured in radians,}$$
$$\text{or} \quad \theta \text{ and } \theta + k(360°), \text{ when } \theta \text{ is measured in degrees,}$$

for any integer k.

EXAMPLE 7

If $\cot \theta = \frac{3}{4}$ and θ lies in quadrant III, use Theorem 6.2 to find the exact values of the other five trigonometric functions of θ. Sketch θ where $0° \leq \theta < 360°$.

Solution   To apply Theorem 6.2, we must find a point $Q = (x, y)$ other than $(0, 0)$ on the terminal side of θ. Because θ lies in quadrant III, both the x- and y-coordinates of Q will be negative. Since $\cot \theta = x/y = \frac{3}{4}$, one possible choice is $x = -3$ and $y = -4$. This choice, of course, is not unique. (Why?) Hence, $Q = (-3, -4)$ is a point on the terminal side of θ, and

$$r = d(O, Q) = \sqrt{(-3)^2 + (-4)^2} = \sqrt{25} = 5.$$

(See Figure 6.16.)

FIGURE 6.16

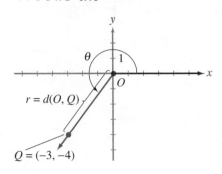

It follows that

$$\sin \theta = \frac{y}{r} = -\frac{4}{5} \qquad \csc \theta = \frac{r}{y} = -\frac{5}{4}$$

$$\cos \theta = \frac{x}{r} = -\frac{3}{5} \qquad \sec \theta = \frac{r}{x} = -\frac{5}{3}$$

$$\tan \theta = \frac{y}{x} = \frac{4}{3}.$$

EXERCISES 6.1

In Exercises 1–4, express the given degrees-minutes-seconds measurement in decimal degrees to the nearest thousandth.

1. $75°32'14''$

2. $102°51'19''$

3. $187°15'12''$

4. $301°23'51''$

In Exercises 5–8, express the given decimal degree measurement in degrees, minutes, and seconds.

5. $88.123°$

6. $99.568°$

7. $286.897°$

8. $199.543°$

In Exercises 9–18, convert the indicated degree measure to radians. Sketch each angle in standard position.

9. $95°10'13''$

10. $80°13'5''$

11. $(2\pi)°$

12. $-\pi°$

13. $-240°$

14. $300°$

15. $570°$

16. $-135°$

17. $-315°$

18. $-495°$

In Exercises 19–28, convert the indicated radian measure to decimal degrees to the nearest thousandth. Sketch each angle in standard position.

19. $-\dfrac{9\pi}{10}$

20. $-\dfrac{7\pi}{8}$

21. 5

22. -3

23. $-\dfrac{11\pi}{6}$

24. $-\dfrac{7\pi}{4}$

25. $-\dfrac{9\pi}{4}$

26. $\dfrac{13\pi}{12}$

27. $-\dfrac{11\pi}{24}$

28. $-\dfrac{13\pi}{15}$

In Exercises 29–34, find the length of the arc and the area of the circular sector subtended by the indicated central angle θ in a circle with the indicated radius r.

29. $r = 10$ centimeters; $\theta = 80°$

30. $r = 5$ inches; $\theta = 7$

31. $r = 8$ inches; $\theta = \pi°$

32. $r = 12$ meters; $\theta = 25°$

33. $r = 3$ feet; $\theta = 20$

34. $r = \pi$ feet; $\theta = (\pi/2)°$

In Exercises 35–42, find the *exact* value of each expression.

35. $\cos 210°$

36. $\sec(-45°)$

37. $\sin(-225°)$

38. $\cos 150°$

39. $\tan 300°$

40. $\cot(-240°)$

41. $\cot(-510°)$

42. $\tan(-585°)$

In Exercises 43–46, angle θ is in standard position with the given point Q on its terminal side. Use Theorem 6.2 to find the exact values of all six trigonometric functions of θ.

43. $Q = (-1, -4)$

44. $Q = (-3, -2)$

45. $Q = (-1, 5)$

46. $Q = (\sqrt{3}, -2)$

In Exercises 47–52, use the given information and Theorem 6.2 to find the *exact* values of the other five trigonometric functions of θ. Sketch θ where $0° \le \theta < 360°$.

47. $\cos \theta = \frac{5}{12}$; θ lies in quadrant I

48. $\cot \theta = \frac{5}{6}$; θ lies in quadrant III

49. $\sec \theta = 3$; θ lies in quadrant IV

50. $\csc \theta = -4$; θ lies in quadrant IV

51. $\sin \theta = -\frac{7}{9}$; θ lies in quadrant III

52. $\tan \theta = -\frac{9}{10}$; θ lies in quadrant II

53. How many revolutions will a bicycle wheel of diameter 24 inches make as the wheel travels 1 mile?

54. A truck wheel has a diameter of 48 inches and travels 4000π feet. How many revolutions does the wheel make?

55. Atlanta, Georgia and New Orleans, Louisiana are approximately 500 miles apart. Assuming that the radius of the earth is approximately 3960 miles, find the approximate radian measure of the central angle joining these two cities.

56. The minute hand of a certain clock is 10 inches long. How far does the tip of the minute hand travel in 75 minutes?

57. A tire is suspended by a chain from a tree. As the tire moves from its highest forward position to its highest backward position, the central angle subtended by the chain is $36°$ and the length of the subtended arc is 3π feet. How long is the chain?

Photo © Fredrik D. Bodin

58. You are served one-ninth of a circular pizza and the area of your slice is 16π in^2. Find the diameter of the pizza.

59. The hour hand of a clock is 5 inches long. How far does the tip of the hour hand travel in 2 hours and 20 minutes?

60. Let θ be the radian measure of a central angle in a circle of radius r and let A be the area of the circular sector subtended by θ. Show that $A = d^2\theta/8$, where d is the diameter of the circle.

61. Your friend is served a slice of a round pizza whose diameter is 12 inches. If the area of her slice is 18π square inches, find the measure of the central angle θ this slice determines.

62. Find both the radian and degree measure of the central angle θ that subtends an arc of 2 feet on a circle of diameter 120 inches.

63. If a central angle θ of measure $12°$ subtends an arc of length 21 centimeters on a circle of radius r, find r.

If a point moves along an arc of length s around a circle of radius r at a constant velocity v, then this **linear velocity v** is given by the equation

$$v = \frac{s}{T},$$

where T is the number of units of time required. For example, if s is measured in feet and T in seconds, then v is measured in feet per second (fps). Since $s = r\theta$, where θ is the radian measure of the subtended central angle, it follows that

$$v = \frac{s}{T} = \frac{r\theta}{T} = r\left(\frac{\theta}{T}\right).$$

The **angular velocity** ω (the Greek letter omega), which measures the rate of change of the central angle with respect to time is given by

$$\omega = \frac{\theta}{T}.$$

Hence the linear velocity v and the angular velocity ω are related by the equation

$$v = r\omega.$$

Suppose a wheel of diameter 4 feet is rotating at a rate of 1500 revolutions per minute (rpm). Then, since the number of revolutions per minute is 1500, and since each revolution corresponds to a central angle of radian measure 2π radians, the central angle generated in one minute has radian measure $\theta = (1500)(2\pi) = 3000\pi$. Hence the angular velocity of the wheel is

$$\omega = \frac{\theta}{T} = \frac{3000\pi}{1} = 3000\pi \text{ radians per minute.}$$

It follows that the linear velocity of the wheel is given by

$$v = r\omega = (2)(3000\pi) = 6000\pi \text{ feet per minute.}$$

In Exercises 64–67, for a wheel of the indicated radius r, which is rotating at the indicated rate, find both the linear velocity and the angular velocity of the wheel.

64. $r = 6$ feet; 2000 rpm   **65.** $r = 3$ inches; 1400 rpm

66. $r = 12$ inches; 1200 rpm   **67.** $r = 5$ feet; 100 rpm

68. A wheel 10 inches in diameter is rotating at a rate of 60 revolutions per minute. Find the velocity in feet per second of a point on the rim of the wheel.

69. The outside diameter of a bicycle wheel is 26 inches. Find the number of revolutions the wheel makes in 1 second if the bicycle is traveling at 20 miles per hour.

70. The radius of the front wheel of a scooter is 6 inches and the rear wheel has a radius of 8 inches. As the rear wheel turns through 24 radians, through how many radians does the front wheel turn?

71. A large pulley has a diameter of 5 feet and is used to lift building equipment. Through how many radians does the pulley turn when 12 feet of cable are pulled through it?

72. An automobile fan belt connects a pulley with a 10 centimeters diameter and a pulley with a 25 centimeters diameter. As the larger pulley turns through 12 radians, through how many radians does the smaller pulley turn?

73. Refer to Exercise 72. As the smaller pulley turns through 8 radians, through how many radians does the larger pulley turn?

6.2 FUNDAMENTAL IDENTITIES

One of the skills required in calculus and in more advanced mathematics courses is the ability to rewrite trigonometric expressions in alternate forms. This can often be done by using the seven basic trigonometric identities that were discussed in Section 5.2. We summarize these identities below where t represents either a real number or an angle.

Basic Trigonometric Identities

For all values of t for which each of the following expressions is defined:

(i) $\sec t = \dfrac{1}{\cos t}$ (ii) $\csc t = \dfrac{1}{\sin t}$

(iii) $\tan t = \dfrac{\sin t}{\cos t}$ (iv) $\cot t = \dfrac{\cos t}{\sin t} = \dfrac{1}{\tan t}$

(v) $\cos^2 t + \sin^2 t = 1$ (vi) $1 + \tan^2 t = \sec^2 t$

(vii) $\cot^2 t + 1 = \csc^2 t$

In this section, we illustrate how the basic identities can be used to simplify trigonometric expressions and to verify other trigonometric identities.

EXAMPLE 8

Use the basic identities to simplify each expression.

a. $\dfrac{1}{1 - \cos \theta} + \dfrac{1}{1 + \cos \theta}$

b. $\dfrac{\csc x}{\tan x + \cot x}$

Solution

a. We have

$$\dfrac{1}{1 - \cos \theta} + \dfrac{1}{1 + \cos \theta} = \dfrac{(1 + \cos \theta) + (1 - \cos \theta)}{(1 - \cos \theta)(1 + \cos \theta)} \quad \text{add fractions}$$

$$= \dfrac{2}{1 - \cos^2 \theta}$$

$$= \dfrac{2}{\sin^2 \theta} \quad \text{identity (v)}$$

$$= 2 \csc^2 \theta. \quad \text{identity (ii)}$$

b. We have

$$\dfrac{\csc x}{\tan x + \cot x} = \dfrac{\dfrac{1}{\sin x}}{\dfrac{\sin x}{\cos x} + \dfrac{\cos x}{\sin x}} \quad \text{identities (ii), (iii), and (iv)}$$

$$= \dfrac{\dfrac{1}{\sin x}}{\dfrac{\sin^2 x + \cos^2 x}{\cos x \sin x}} \quad \text{add fractions}$$

$$= \dfrac{\dfrac{1}{\sin x}}{\dfrac{1}{\cos x \sin x}} \quad \text{identity (v)}$$

$$= \dfrac{1}{\sin x} \cdot \dfrac{\cos x \sin x}{1}$$

$$= \cos x.$$

Although there is no well-defined set of rules to follow when verifying a trigonometric identity, the guidelines on page 258 should prove helpful.

Guidelines for Verifying Trigonometric Identities

(i) If one side of an identity is more complicated than the other, transform the more complicated side into the simpler side. To do this, use the basic identities and, when feasible, algebraic operations such as factoring, adding or multiplying fractions, combining like terms, reducing fractions, etc.

(ii) If both sides of an identity appear equally complicated, transform each side *separately* into the same expression.

(iii) It sometimes helps to begin by expressing the functions involved in terms of sines and cosines only.

E X A M P L E 9

Verify the identity $2 + \cot^2 t = \dfrac{\csc^4 t - 1}{\cot^2 t}$.

Solution   Here we start with the right side, which is more complicated.

$$\begin{aligned}
\frac{\csc^4 t - 1}{\cot^2 t} &= \frac{(\csc^2 t + 1)(\csc^2 t - 1)}{\cot^2 t} &&\text{factor the numerator}\\[2mm]
&= \frac{(\csc^2 t + 1)(\cot^2 t)}{\cot^2 t} &&\text{identity (vii)}\\[2mm]
&= \csc^2 t + 1\\[2mm]
&= (1 + \cot^2 t) + 1 &&\text{identity (vii)}\\[2mm]
&= 2 + \cot^2 t.
\end{aligned}$$

Notice that we verified the identity in Example 9 by factoring and using the basic identities. Sometimes the verification of an identity requires us to be a bit more resourceful, as the following examples illustrate.

E X A M P L E 10

Verify the identity $\sec t + \tan t = \dfrac{1}{\sec t - \tan t}$.

Solution   We begin on the right side and follow a procedure similar to that used to rationalize a denominator containing a difference of radicals.

$$\begin{aligned}
\frac{1}{\sec t - \tan t} &= \frac{1}{\sec t - \tan t} \cdot \frac{\sec t + \tan t}{\sec t + \tan t}\\[2mm]
&= \frac{\sec t + \tan t}{\sec^2 t - \tan^2 t}\\[2mm]
&= \frac{\sec t + \tan t}{(1 + \tan^2 t) - \tan^2 t} &&\text{identity (vi)}\\[2mm]
&= \frac{\sec t + \tan t}{1}\\[2mm]
&= \sec t + \tan t.
\end{aligned}$$

CAUTION: Notice that in Example 10, we verified the identity

$$\sec t + \tan t = \frac{1}{\sec t - \tan t}.$$

We are tempted to multiply both sides of this identity by $\sec t - \tan t$ to obtain

$$(\sec t + \tan t)(\sec t - \tan t) = 1$$
$$\sec^2 t - \tan^2 t = 1$$
$$(1 + \tan^2 t) - \tan^2 t = 1$$
$$1 = 1.$$

However, this approach is incorrect. When verifying an identity, it is incorrect to multiply both sides by the same expression or to add the same expression on both sides, since this approach is based on the assumption that the desired equality already holds.

EXAMPLE 11

Verify the identity $\dfrac{\tan \theta}{\sec \theta + 1} = \dfrac{1}{\cot \theta + \csc \theta}$.

Solution   Since both sides are equally complicated, we work with each side separately. Expressing the left side in terms of sines and cosines, we have

$$\frac{\tan \theta}{\sec \theta + 1} = \frac{\dfrac{\sin \theta}{\cos \theta}}{\dfrac{1}{\cos \theta} + 1} \qquad \text{identities (i) and (iii)}$$

$$= \frac{\dfrac{\sin \theta}{\cos \theta}}{\dfrac{1 + \cos \theta}{\cos \theta}}$$

$$= \frac{\sin \theta}{\cos \theta} \cdot \frac{\cos \theta}{1 + \cos \theta}$$

$$= \frac{\sin \theta}{1 + \cos \theta}.$$

Similarly, working with the right side, we have

$$\frac{1}{\cot \theta + \csc \theta} = \frac{1}{\dfrac{\cos \theta}{\sin \theta} + \dfrac{1}{\sin \theta}} \qquad \text{identities (iv) and (ii)}$$

$$= \frac{1}{\dfrac{\cos \theta + 1}{\sin \theta}}$$

$$= \frac{\sin \theta}{1 + \cos \theta}.$$

Since both sides of the original identity have been transformed into the same expression, the verification is complete.

EXAMPLE 12

Verify the identity $\dfrac{1 - \cos u}{\sin u} = -\dfrac{\sin(-u)}{1 + \cos u}$.

Solution   Working first with the right side and noting that sine is an odd function, we have

$$-\frac{\sin(-u)}{1 + \cos u} = \frac{-(-\sin u)}{1 + \cos u} = \frac{\sin u}{1 + \cos u}.$$

Working now with the left side and applying a technique similar to that used in Example 10, we have

$$\frac{1 - \cos u}{\sin u} = \frac{1 - \cos u}{\sin u} \cdot \frac{1 + \cos u}{1 + \cos u}$$

$$= \frac{1 - \cos^2 u}{\sin u(1 + \cos u)}$$

$$= \frac{\sin^2 u}{\sin u(1 + \cos u)} \qquad \text{identity (v)}$$

$$= \frac{\sin u}{1 + \cos u}.$$

Since both sides of the original identity have been transformed into the same expression, the verification is complete.

EXAMPLE 13

Verify the identity $\ln |\sec \theta + \tan \theta| = -\ln |\sec \theta + \tan(-\theta)|$.

Solution   Let us start with the right side:

$$-\ln |\sec \theta + \tan(-\theta)| = -\ln |\sec \theta - \tan \theta| \qquad \tan(-\theta) = -\tan \theta$$

$$= \ln |\sec \theta - \tan \theta|^{-1} \qquad p \ln x = \ln x^p$$

$$= \ln \frac{1}{|\sec \theta - \tan \theta|}$$

$$= \ln\left(\frac{1}{|\sec \theta - \tan \theta|} \cdot \frac{|\sec \theta + \tan \theta|}{|\sec \theta + \tan \theta|}\right)$$

$$= \ln \frac{|\sec \theta + \tan \theta|}{|\sec^2 \theta - \tan^2 \theta|}$$

$$= \ln \frac{|\sec \theta + \tan \theta|}{|1|} \qquad \text{identity (vi)}$$

$$= \ln |\sec \theta + \tan \theta|.$$

Trigonometric Substitution

In calculus, it it is sometimes convenient to change the form of algebraic expressions such as $\sqrt{a^2 - u^2}$, $\sqrt{a^2 + u^2}$, and $\sqrt{u^2 - a^2}$ using a technique known as **trigonometric substitution**, as the following example illustrates:

EXAMPLE 14

By making the substitution $u = a \tan \theta$, where $-\pi/2 < \theta < \pi/2$ and $a > 0$, write the expression $\sqrt{a^2 + u^2}$ in terms of trigonometric functions of θ not involving radicals.

Solution   Substituting $a \tan \theta$ for u and simplifying, we have

$$\sqrt{a^2 + u^2} = \sqrt{a^2 + (a^2 \tan^2 \theta)}$$
$$= \sqrt{a^2(1 + \tan^2 \theta)}$$
$$= \sqrt{a^2 \sec^2 \theta}.$$

Since $a > 0$, and $\sec \theta > 0$ when $-\pi/2 < \theta < \pi/2$,

$$\sqrt{a^2 \sec^2 \theta} = a \sec \theta.$$

Hence,

$$\sqrt{a^2 + u^2} = a \sec \theta.$$ ———

EXERCISES 6.2

In Exercises 1–12, use the basic identities to simplify each expression.

1. $(\tan t + \sec t)(\tan t - \sec t)$
2. $\sec t(\cot t - \cos t)$
3. $(\sin \theta - \cos \theta)^2 + (\sin \theta + \cos \theta)^2$
4. $\dfrac{1}{1 + 2 \cot^2 t + \cot^4 t}$
5. $\dfrac{1}{(\csc \theta + \cot \theta)(1 - \cos \theta)}$
6. $\dfrac{1}{1 + \csc t} + \dfrac{1}{1 - \csc t}$
7. $(\cot x + \tan x)\cos x$
8. $\cot^2 t(\sec^2 t - 1)$
9. $\dfrac{\tan^2 v}{\csc^2 v - 1}$
10. $\dfrac{\sec^2 t}{1 + \cot^2 t}$
11. $\dfrac{\csc^2 \theta - \cot^2 \theta}{\cos^2 \theta - 1}$
12. $\dfrac{\cos t(1 + \cot^2 t)}{\cot t}$

In Exercises 13–78, verify each identity.

13. $\cos t \csc t = \cot t$
14. $\cos^2 x - \sin^2 x = 1 - 2 \sin^2 x$
15. $\dfrac{\sin \theta + \cos \theta}{\cos \theta} = 1 + \tan \theta$
16. $\dfrac{\sin v}{1 - \sin^2 v} = \dfrac{\sec v}{\cot v}$
17. $\dfrac{\sin z}{\csc z} + \dfrac{\cos z}{\sec z} = 1$
18. $(\cos \theta - \sin \theta)^2 = 1 - 2 \sin \theta \cos \theta$
19. $\dfrac{1}{\tan x + \cot x} = \sin x \cos x$
20. $\csc x \sec x = \tan x + \cot x$
21. $\sec u - \cos u = \sin u \tan u$
22. $(1 - \sin^2 t)\sec t = \cos t$
23. $\dfrac{1 - \cos^2 v}{[1 + \sin(-v)](1 + \sin v)} = \tan^2 v$
24. $\dfrac{\csc \theta}{\cot \theta + \tan \theta} = \cos \theta$

25. $\dfrac{\sec x - \cos x}{\sec x + \cos x} = \dfrac{\sin^2 x}{1 + \cos^2 x}$
26. $\dfrac{1 + \tan^2 x}{\tan^2 x} = \csc^2 x$
27. $\dfrac{\tan \theta}{\sec \theta + 1} = \dfrac{1}{\cot \theta + \csc \theta}$
28. $\cos t - \dfrac{\cos t}{1 + \tan(-t)} = \dfrac{\sin t \cos t}{\sin t - \cos t}$
29. $\sec^4 \theta - \tan^4 \theta = 1 + 2 \tan^2 \theta$
30. $\dfrac{\tan^2 x}{\sec x + 1} = \dfrac{1 - \cos x}{\cos x}$
31. $\cot w + \sec(-w) = \dfrac{\cos(-w) + \tan w}{\sin w}$
32. $\dfrac{1}{1 - \cos \theta} + \dfrac{1}{1 + \cos \theta} = 2 \csc^2 \theta$
33. $\sec t + \csc t = (\tan t + \cot t)(\sin t + \cos t)$
34. $\dfrac{1 + \sec x}{\sec x} = \dfrac{\sin^2 x}{1 - \cos x}$
35. $\dfrac{1 + \sin v}{\cos v} = \dfrac{\cos v}{\sin(-v) + 1}$
36. $(\tan^2 \theta + 1)(1 + \cos^2 \theta) = \tan^2 \theta + 2$
37. $\dfrac{\cos \theta \cot \theta}{1 + \sin(-\theta)} - 1 = \csc \theta$
38. $(\sec x - \tan x)^2 = \dfrac{1 - \sin x}{1 + \sin x}$
39. $\dfrac{\sin \theta}{1 + \cos \theta} + \dfrac{1 + \cos \theta}{\sin \theta} = 2 \csc \theta$
40. $\dfrac{1 + \sin t}{1 - \sin t} - \dfrac{1 - \sin t}{1 + \sin t} = 4 \tan t \sec t$
41. $\dfrac{\tan t}{\sec t + 1} = \dfrac{1}{\cot t + \csc t}$

42. $(\cot u - \csc u)^2 = \dfrac{1 - \cos u}{1 + \cos u}$

43. $\cos^4 \theta + \sin^2 \theta = \cos^2 \theta + \sin^4 \theta$

44. $\dfrac{\sin^3 x - \cos^3 x}{\sin x - \cos x} = 1 + \sin x \cos x$

45. $\tan^4 m - \sec^4 m = 1 - 2 \sec^2 m$

46. $\dfrac{\sec y}{\sin y} + \cot(-y) = \tan y$

47. $\dfrac{\csc^4 x - 1}{\cot^2 x} = 2 + \cot^2 x$

48. $\dfrac{\cos^3 t + \sin^3 t}{\cos t + \sin t} = 1 - \sin t \cos t$

49. $\dfrac{1 + \sin v}{\cos v} = \dfrac{\cos v}{1 - \sin v}$

50. $(\tan x + \cot x)(\sin x \cos x) = 1$

51. $\sin^4 x + 2 \sin^2 x \cos^2 x + \cos^4 x = 1$

52. $\dfrac{1 - \sin \theta}{1 + \sin \theta} = (\sec \theta - \tan \theta)^2$

53. $\dfrac{1}{1 + \cos t} = \csc^2 t - \csc t \cot t$

54. $\dfrac{\tan v - \sin v}{\sin^3 v} = \dfrac{\sec v}{1 + \cos v}$

55. $\dfrac{\cot u - 1}{\cot u + 1} = \dfrac{1 - \tan u}{1 + \tan u}$

56. $\sec^4 x - \sec^2 x = \tan^4 x + \tan^2 x$

57. $\dfrac{1 + \sec(-t)}{\sin(-t) + \tan(-t)} = -\csc t$

58. $\dfrac{\cos(-x)}{1 + \sin(-x)} = \sec x + \tan x$

59. $\sqrt{\dfrac{1 - \sin \theta}{1 + \sin \theta}} = \dfrac{|1 - \sin \theta|}{|\cos \theta|}$

60. $\sqrt{\dfrac{1 - \cos t}{1 + \cos t}} = \dfrac{|1 - \cos t|}{|\sin t|}$

61. $(1 - \tan^2 y)^2 = \sec^4 y - 4 \tan^2 y$

62. $(\cot u - \csc u)^2 = \dfrac{1 - \cos u}{1 + \cos u}$

63. $\dfrac{\cos w}{1 - \tan w} + \dfrac{\sin w}{1 - \cot w} = \cos w + \sin w$

64. $\dfrac{\cos \theta - \sin \theta + 1}{\cos \theta + \sin \theta - 1} = \dfrac{\cos \theta + 1}{\sin \theta}$

65. $\dfrac{1 + \sin x + \cos x}{1 + \cos x - \sin x} = \sec x + \tan x$

66. $\ln |\sec \theta| = -\ln |\cos \theta|$

67. $\ln |\tan \theta| = \ln |\sin \theta| - \ln |\cos \theta|$

68. $\ln |\sec^2 \theta - \tan^2 \theta| = 0$ **69.** $\ln |\sin x| + \ln |\csc x| = 0$

70. $\ln |\tan x| + \ln(1 + \cot^2 x) = \ln |\tan x + \cot x|$

71. $-\ln |1 + \cos \theta| = \ln |1 - \cos \theta| - 2 \ln |\sin \theta|$

72. $10^{\log |\cos t|} = |\cos t|$ **73.** $\ln |\cot x| = -\ln |\tan x|$

74. $\log |\csc x + \cot x| + \log |\csc x - \cot x| = 0$

75. $\dfrac{\sin x + \cos y}{\cos x + \sin y} = \dfrac{\cos x - \sin y}{\cos y - \sin x}$

76. $\dfrac{\tan x + \tan y}{\cot x + \cot y} = \dfrac{\tan x \tan y - 1}{1 - \cot x \cot y}$

77. $(\sin t + \cos t)^2 - (\sin u + \cos u)^2$
$= 2(\sin t \cos t - \sin u \cos u)$

78. $(\sin t - \sin u)^2 + (\cos t - \cos u)^2$
$= 2(1 - \sin t \sin u - \cos t \cos u)$

In Exercises 79–88, show that each equation is *not* an identity by finding a value of x for which the equation is false.

79. $\sin x = \sqrt{1 - \cos^2 x}$ **80.** $\sec x = \sqrt{1 + \tan^2 x}$

81. $\cot x = \tan x$ **82.** $\sin x = \cos x$

83. $\tan(\cot x) = 1$ **84.** $\sec(\cos x) = 1$

85. $(\sin x + \cos x)^2 = 1$ **86.** $\dfrac{\cos x}{1 + \sin x} = \dfrac{1 + \sin x}{\cos x}$

87. $\ln \left| \dfrac{1}{\cos x} \right| = \dfrac{1}{\ln |\cos x|}$

88. $\ln(\sin^2 x) + \ln(\cos^2 x) = 0$

In Exercises 89–96, use the indicated trigonometric substitution to simplify each expression. Refer to Example 14.

89. $\sqrt{a^2 - u^2}$; $u = a \sin \theta$; $-\dfrac{\pi}{2} \le \theta \le \dfrac{\pi}{2}$ and $a > 0$

90. $\sqrt{4 + u^2}$; $u = 2 \tan \theta$; $-\dfrac{\pi}{2} < \theta < \dfrac{\pi}{2}$

91. $\dfrac{1}{\sqrt{u^2 + a^2}}$; $u = a \tan \theta$; $-\dfrac{\pi}{2} < \theta < \dfrac{\pi}{2}$ and $a > 0$

92. $\sqrt{u^2 - a^2}$; $u = a \sec \theta$; $0 < \theta < \dfrac{\pi}{2}$ and $a > 0$

93. $\dfrac{2}{\sqrt{x^2 - 9}}$; $x = 3 \sec \theta$ $0 < \theta < \dfrac{\pi}{2}$

94. $x^2 \sqrt{x^2 - 25}$; $x = 5 \sec \theta$; $0 < \theta < \dfrac{\pi}{2}$

95. $\dfrac{1}{\sqrt{9 + x^2}}$; $x = 3 \tan \theta$; $-\dfrac{\pi}{2} < \theta < \dfrac{\pi}{2}$

96. $\dfrac{2}{\sqrt{5 - x^2}}$; $x = \sqrt{5} \sin \theta$; $-\dfrac{\pi}{2} < \theta < \dfrac{\pi}{2}$

97. Use a graphing calculator to sketch the graphs of the following functions over the interval $(-\pi/2, \pi/2)$ in the same coordinate system. How do the graphs compare?

$$y = \dfrac{\sin^2 2x}{\cos^2 2x} \quad \text{and} \quad y = \tan^2 2x.$$

98. Use a graphing calculator to sketch the graphs of the following functions over the interval $(0, \pi)$ in the same coordinate system. How do the graphs compare?

$$y = \dfrac{\cos^2 2x}{\sin^2 2x} \quad \text{and} \quad y = \dfrac{1}{\tan^2 2x}.$$

6.3 SUM, DIFFERENCE, AND COFUNCTION IDENTITIES

So far, we have been able to find the exact values of the trigonometric functions of real numbers that are multiples of

$$\frac{\pi}{6}, \quad \frac{\pi}{4}, \quad \frac{\pi}{3}, \quad \text{or} \quad \frac{\pi}{2}$$

(or angles that are multiples of 30°, 45°, 60°, or 90°). In this section, we will develop formulas that enable us to evaluate expressions such as

$$\cos \frac{7\pi}{12}, \quad \sin \frac{3\pi}{8}, \quad \text{and} \quad \tan \frac{5\pi}{12}.$$

For example, since

$$\frac{7\pi}{12} = \frac{\pi}{4} + \frac{\pi}{3},$$

we know that

$$\cos \frac{7\pi}{12} = \cos\left(\frac{\pi}{4} + \frac{\pi}{3}\right).$$

Although it might seem that

$$\cos\left(\frac{\pi}{4} + \frac{\pi}{3}\right)$$

should be the same as

$$\cos \frac{\pi}{4} + \cos \frac{\pi}{3},$$

this is not the case:

$$\cos \frac{\pi}{4} + \cos \frac{\pi}{3} = \frac{\sqrt{2}}{2} + \frac{1}{2} = \frac{\sqrt{2} + 1}{2} > 0,$$

while $\cos(7\pi/12) < 0$, since $7\pi/12$ lies in quadrant II. However, as we shall see, there are formulas for evaluating the trigonometric functions of the sum and difference of two real numbers that can be used to evaluate expressions such as $\cos(\pi/4 + \pi/3)$.

Sum and Difference Formulas for Cosine

We begin with the sum formula for cosine.

Sum Formula for Cosine

$$\cos(u + v) = \cos u \cos v - \sin u \sin v$$

To verify the sum formula for cosine, we assume, for convenience, that $0 < u < \pi/2$ and that $\pi/2 < v < \pi$. A similar argument holds for any values

FIGURE 6.17

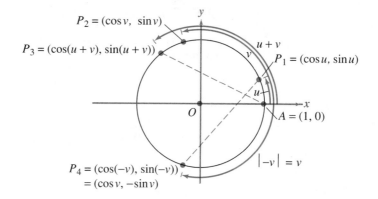

of u and v and the formula holds for all real numbers and angles in either radian or degree measure. On the unit circle, shown in Figure 6.17, we have located the trigonometric points $P_1 = P(u) = (\cos u, \sin u)$, $P_2 = P(v) = (\cos v, \sin v)$, $P_3 = P(u + v) = (\cos(u + v), \sin(u + v))$, and $P_4 = P(-v) = (\cos(-v), \sin(-v))$. Notice that the coordinates of P_4 can be written as $P_4 = (\cos v, -\sin v)$, since $\cos(-v) = \cos v$ and $\sin(-v) = -\sin v$.

Since the length of the arc $\overset{\frown}{AP_3}$ is $u + v$, which is the same as the length of the arc $\overset{\frown}{P_1P_4}$, and since equal arcs subtend equal chords, it follows that

$$d(A, P_3) = d(P_1, P_4).$$

By the distance formula,

$$d(A, P_3) = \sqrt{[\cos(u + v) - 1]^2 + [\sin(u + v) - 0]^2}.$$

Squaring both sides of the last equation, we get

$$
\begin{aligned}
[d(A, P_3)]^2 &= [\cos(u + v) - 1]^2 + [\sin(u + v) - 0]^2 \\
&= [\cos^2(u + v) - 2\cos(u + v) + 1] + \sin^2(u + v) \\
&= [\cos^2(u + v) + \sin^2(u + v)] - 2\cos(u + v) + 1 \\
&= 1 - 2\cos(u + v) + 1 \\
&= 2 - 2\cos(u + v).
\end{aligned}
$$

Similarly, since

$$d(P_1, P_4) = \sqrt{(\cos v - \cos u)^2 + (-\sin v - \sin u)^2},$$

$$
\begin{aligned}
[d(P_1, P_4)]^2 &= (\cos v - \cos u)^2 + (-\sin v - \sin u)^2 \\
&= (\cos^2 v - 2\cos v \cos u + \cos^2 u) \\
&\quad + (\sin^2 v + 2\sin v \sin u + \sin^2 u) \\
&= (\cos^2 v + \sin^2 v) + (\cos^2 u + \sin^2 u) \\
&\quad - 2\cos v \cos u + 2\sin v \sin u \\
&= 1 + 1 - 2\cos v \cos u + 2\sin v \sin u \\
&= 2 - 2\cos v \cos u + 2\sin v \sin u.
\end{aligned}
$$

Since $[d(A, P_3)]^2 = [d(P_1, P_4)]^2$, it follows that

$$2 - 2\cos(u + v) = 2 - 2\cos v \cos u + 2\sin v \sin u.$$

Thus, $-2\cos(u + v) = -2(\cos u \cos v - \sin u \sin v)$,   or

$$\cos(u + v) = \cos u \cos v - \sin u \sin v.$$

Using the sum formula for cosine, we can now find $\cos(7\pi/12)$:

$$\cos\frac{7\pi}{12} = \cos\left(\frac{\pi}{4} + \frac{\pi}{3}\right)$$

$$= \cos\frac{\pi}{4}\cos\frac{\pi}{3} - \sin\frac{\pi}{4}\sin\frac{\pi}{3}$$

$$= \left(\frac{\sqrt{2}}{2}\right)\left(\frac{1}{2}\right) - \left(\frac{\sqrt{2}}{2}\right)\left(\frac{\sqrt{3}}{2}\right)$$

$$= \frac{\sqrt{2} - \sqrt{6}}{4}.$$

Notice that $\cos(7\pi/12) < 0$, as we observed earlier.

We can use the sum formula for cosine to derive a formula for $\cos(u - v)$. Since $u - v = u + (-v)$ and since $\cos(-v) = \cos v$ and $\sin(-v) = -\sin v$, it follows from the sum formula for cosine that

$$\cos(u - v) = \cos[(u + (-v)]$$

$$= \cos u \cos(-v) - \sin u \sin(-v)$$

$$= \cos u \cos v - \sin u(-\sin v)$$

$$= \cos u \cos v + \sin u \sin v.$$

We have thus established the following identity:

Difference Formula for Cosine

$$\cos(u - v) = \cos u \cos v + \sin u \sin v.$$

We use the sum and difference formulas for cosine in our next example.

EXAMPLE 15

a.   Use the sum formula for cosine to simplify

$$\cos 20° \cos 50° - \sin 20° \sin 50°.$$

b.   Given that $5\pi/12 = (2\pi/3) - (\pi/4)$, use the difference formula for cosine to evaluate $\cos(5\pi/12)$.

Solution

a.   From the sum formula for cosine, we obtain

$$\cos 20° \cos 50° - \sin 20° \sin 50° = \cos(20° + 50°)$$

$$= \cos 70°.$$

b. Since $\dfrac{5\pi}{12} = \dfrac{2\pi}{3} - \dfrac{\pi}{4}$, it follows from the difference formula for cosine that

$$\cos \frac{5\pi}{12} = \cos\left(\frac{2\pi}{3} - \frac{\pi}{4}\right)$$

$$= \cos \frac{2\pi}{3} \cos \frac{\pi}{4} + \sin \frac{2\pi}{3} \sin \frac{\pi}{4}$$

$$= \left(-\frac{1}{2}\right)\left(\frac{\sqrt{2}}{2}\right) + \left(\frac{\sqrt{3}}{2}\right)\left(\frac{\sqrt{2}}{2}\right)$$

$$= \frac{-\sqrt{2} + \sqrt{6}}{4}.$$

Cofunction Identities

It is customary to refer to the sine and cosine functions as **cofunctions**. Similarly, tangent and cotangent are cofunctions as are secant and cosecant. Using the difference formula for cosine, we now derive some important identities involving cofunctions. Letting $u = \pi/2$ in the difference formula for cosine, we obtain

$$\cos\left(\frac{\pi}{2} - v\right) = \cos \frac{\pi}{2} \cos v + \sin \frac{\pi}{2} \sin v$$

$$= (0)(\cos v) + (1)(\sin v)$$

$$= \sin v.$$

Hence, $\cos\left(\dfrac{\pi}{2} - v\right) = \sin v.$ *(1)*

Substituting $(\pi/2) - u$ for v in equation (1), we obtain

$$\cos\left[\frac{\pi}{2} - \left(\frac{\pi}{2} - u\right)\right] = \sin\left(\frac{\pi}{2} - u\right),$$

or $\cos u = \sin\left(\dfrac{\pi}{2} - u\right).$ *(2)*

Since $\tan t = \dfrac{\sin t}{\cos t}$, it follows from equations (1) and (2) that

$$\tan\left(\frac{\pi}{2} - v\right) = \frac{\sin[(\pi/2) - v]}{\cos[(\pi/2) - v]}$$

$$= \frac{\cos v}{\sin v}$$

$$= \cot v.$$

In summary, we have

Cofunction Identities

(i) $\cos\left(\dfrac{\pi}{2} - v\right) = \sin v$ (ii) $\sin\left(\dfrac{\pi}{2} - v\right) = \cos v$

(iii) $\tan\left(\dfrac{\pi}{2} - v\right) = \cot v$

Sum and Difference Formulas for Sine

Using cofunction identities (i) and (ii), we now derive a formula for $\sin(u + v)$. By cofunction identity (i), we can write

$$\sin(u + v) = \cos\left[\frac{\pi}{2} - (u + v)\right]$$

$$= \cos\left[\left(\frac{\pi}{2} - u\right) - v\right].$$

Now, using the difference formula for cosine and cofunction identities (i) and (ii), we have

$$\cos\left[\left(\frac{\pi}{2} - u\right) - v\right] = \cos\left(\frac{\pi}{2} - u\right)\cos v + \sin\left(\frac{\pi}{2} - u\right)\sin v$$

$$= \sin u \cos v + \cos u \sin v.$$

Thus, we have

Sum Formula for Sine

$$\sin(u + v) = \sin u \cos v + \cos u \sin v.$$

Since $u - v = u + (-v)$, it follows that

$$
\begin{aligned}
\sin(u - v) &= \sin[u + (-v)] \\
&= \sin u \cos(-v) + \cos u \sin(-v) \qquad \text{sum formula for sine} \\
&= \sin u \cos v + \cos u \,(-\sin v) \qquad \cos(-v) = \cos v \text{ and} \\
&= \sin u \cos v - \cos u \sin v. \qquad\qquad \sin(-v) = -\sin v
\end{aligned}
$$

Thus we have established the following identity:

Difference Formula for Sine

$$\sin(u - v) = \sin u \cos v - \cos u \sin v.$$

EXAMPLE 16

Use the appropriate sum or difference formula to simplify each expression.

a. $\sin(180° - \theta)$

b. $\sin\dfrac{3\pi}{10}\cos\dfrac{\pi}{5} + \cos\dfrac{3\pi}{10}\sin\dfrac{\pi}{5}$

Solution

a. By the difference formula for sine,

$$\sin(180° - \theta) = \sin 180° \cos \theta - \cos 180° \sin \theta$$
$$= (0)(\cos \theta) - (-1)(\sin \theta)$$
$$= \sin \theta.$$

b. From the sum formula for sine, we see that

$$\sin \frac{3\pi}{10} \cos \frac{\pi}{5} + \cos \frac{3\pi}{10} \sin \frac{\pi}{5} = \sin\left(\frac{3\pi}{10} + \frac{\pi}{5}\right)$$
$$= \sin \frac{5\pi}{10}$$
$$= \sin \frac{\pi}{2} = 1.$$

EXAMPLE 17

If $\sin u = \sqrt{2}/3$ where u is in quadrant II, and $\cos v = \sqrt{15}/4$ where v is in quadrant I, find the exact value of $\sin(u + v)$ and $\cos(u + v)$. Also give the quadrant containing $u + v$.

Solution In order to use the sum formula for sine, we must first find the values of $\cos u$ and $\sin v$. Since $\sin^2 u + \cos^2 u = 1$ and since $\cos u < 0$ when u is in quadrant II, it follows that

$$\cos u = -\sqrt{1 - \sin^2 u}$$
$$= -\sqrt{1 - \left(\frac{\sqrt{2}}{3}\right)^2}$$
$$= -\sqrt{1 - \frac{2}{9}} = -\sqrt{\frac{7}{9}} = -\frac{\sqrt{7}}{3}.$$

Similarly, since $\sin v > 0$ when v is in quadrant I,

$$\sin v = \sqrt{1 - \cos^2 v}$$
$$= \sqrt{1 - \left(\frac{\sqrt{15}}{4}\right)^2}$$
$$= \sqrt{1 - \frac{15}{16}} = \sqrt{\frac{1}{16}} = \frac{1}{4}.$$

Hence,

$$\sin(u + v) = \sin u \cos v + \cos u \sin v$$
$$= \left(\frac{\sqrt{2}}{3}\right)\left(\frac{\sqrt{15}}{4}\right) + \left(-\frac{\sqrt{7}}{3}\right)\left(\frac{1}{4}\right)$$
$$= \frac{\sqrt{30}}{12} - \frac{\sqrt{7}}{12} = \frac{\sqrt{30} - \sqrt{7}}{12}.$$

Similarly,

$$\cos(u + v) = \cos u \cos v - \sin u \sin v$$
$$= \left(-\frac{\sqrt{7}}{3}\right)\left(\frac{\sqrt{15}}{4}\right) - \left(\frac{\sqrt{2}}{3}\right)\left(\frac{1}{4}\right)$$

$$= -\frac{\sqrt{105}}{12} - \frac{\sqrt{2}}{12} = -\frac{\sqrt{105} + \sqrt{2}}{12}.$$

Finally, since $\sin(u + v) > 0$ and $\cos(u + v) < 0$, we see that $u + v$ lies in quadrant II.

Sum and Difference Formulas for Tangent

Using the identity $\tan t = \dfrac{\sin t}{\cos t}$ and the sum formulas for sine and cosine, we can derive the sum formula for tangent as follows:

$$\tan(u + v) = \frac{\sin(u + v)}{\cos(u + v)}$$

$$= \frac{\sin u \cos v + \cos u \sin v}{\cos u \cos v - \sin u \sin v}$$

$$= \frac{\dfrac{\sin u \cos v}{\cos u \cos v} + \dfrac{\cos u \sin v}{\cos u \cos v}}{\dfrac{\cos u \cos v}{\cos u \cos v} - \dfrac{\sin u \sin v}{\cos u \cos v}} \qquad \begin{array}{l}\text{dividing numerator and}\\ \text{denominator by } \cos u \cos v\\ \text{where } \cos u \cos v \neq 0\end{array}$$

$$= \frac{\dfrac{\sin u}{\cos u} + \dfrac{\sin v}{\cos v}}{1 - \left(\dfrac{\sin u}{\cos u}\right)\left(\dfrac{\sin v}{\cos v}\right)}$$

$$= \frac{\tan u + \tan v}{1 - \tan u \tan v}.$$

Thus, we have

Sum Formula for Tangent

$$\tan(u + v) = \frac{\tan u + \tan v}{1 - \tan u \tan v}.$$

We leave the derivation of the difference formula for tangent as an exercise (see Exercise 68).

Difference Formula for Tangent

$$\tan(u - v) = \frac{\tan u - \tan v}{1 + \tan u \tan v}.$$

We summarize the sum and difference formulas for sine, cosine, and tangent in the box on page 270.

//

Sum and Difference Formulas

$$\cos(u + v) = \cos u \cos v - \sin u \sin v$$
$$\cos(u - v) = \cos u \cos v + \sin u \sin v$$
$$\sin(u + v) = \sin u \cos v + \cos u \sin v$$
$$\sin(u - v) = \sin u \cos v - \cos u \sin v$$
$$\tan(u + v) = \frac{\tan u + \tan v}{1 - \tan u \tan v}$$
$$\tan(u - v) = \frac{\tan u - \tan v}{1 + \tan u \tan v}$$

E X A M P L E   18

If $\sin u = \sqrt{2}/3$ where u is in quadrant II, and $\cos v = \sqrt{15}/4$ where v is in quadrant I, find the exact value of $\tan(u - v)$ and approximate your result to two decimal places.

Solution   In order to use the difference formula for tangent, we must first find the values of $\tan u$ and $\tan v$. As we found in the solution of Example 17,

$$\cos u = -\frac{\sqrt{7}}{3} \quad \text{and} \quad \sin v = \frac{1}{4}.$$

Therefore, $\tan u = \dfrac{\sin u}{\cos u}$

$$= \frac{\dfrac{\sqrt{2}}{3}}{-\dfrac{\sqrt{7}}{3}} = -\frac{\sqrt{14}}{7},$$

and $\tan v = \dfrac{\dfrac{1}{4}}{\dfrac{\sqrt{15}}{4}} = \dfrac{\sqrt{15}}{15}.$

Now, by the difference formula for tangent,

$$\tan(u - v) = \frac{\tan u - \tan v}{1 + \tan u \tan v}$$

$$= \frac{-\sqrt{14}/7 - \sqrt{15}/15}{1 + (-\sqrt{14}/7)(\sqrt{15}/15)}$$

$$= \frac{\dfrac{-15\sqrt{14} - 7\sqrt{15}}{105}}{\dfrac{105 - \sqrt{210}}{105}}$$

$$= \frac{15\sqrt{14} + 7\sqrt{15}}{\sqrt{210} - 105} \approx -0.92.$$

EXAMPLE 19

Verify the identity $\tan[(\pi/4) + x] = \dfrac{1 + \tan x}{1 - \tan x}$.

Solution   Using the sum formula for tangent, we have

$$\tan[(\pi/4) + x] = \frac{\tan \dfrac{\pi}{4} + \tan x}{1 - \tan \dfrac{\pi}{4} \tan x}$$

$$= \frac{1 + \tan x}{1 - (1)\tan x}$$

$$= \frac{1 + \tan x}{1 - \tan x}.$$

EXAMPLE 20

Find the exact value of $\cos[\sin^{-1}(-\tfrac{1}{3}) - \tan^{-1}(-1)]$ and approximate the value to two decimal places.

Solution   Notice that the given expression

$$\cos\left[\sin^{-1}\left(-\frac{1}{3}\right) - \tan^{-1}(-1) \right]$$

has the form $\cos(u - v)$, where $u = \sin^{-1}(-\tfrac{1}{3})$ and $v = \tan^{-1}(-1)$. Since $u = \sin^{-1}(-\tfrac{1}{3})$, it follows from the definition of the inverse sine function that

$$\sin u = -\frac{1}{3} \quad \text{and} \quad -\frac{\pi}{2} \le u \le \frac{\pi}{2}.$$

Since $\sin^2 u + \cos^2 u = 1$ and $\cos u \ge 0$ when $-\pi/2 \le u \le \pi/2$, we see that

$$\cos u = \sqrt{1 - \sin^2 u}$$

$$= \sqrt{1 - \left(-\frac{1}{3}\right)^2}$$

$$= \sqrt{1 - \frac{1}{9}}$$

$$= \sqrt{\frac{8}{9}} = \frac{2\sqrt{2}}{3}.$$

Since $v = \tan^{-1}(-1)$, it follows that

$$\tan v = -1 \quad \text{and} \quad -\frac{\pi}{2} < v < \frac{\pi}{2}.$$

Therefore,

$$v = -\frac{\pi}{4},$$

from which it follows that

$$\cos v = \frac{\sqrt{2}}{2} \quad \text{and} \quad \sin v = -\frac{\sqrt{2}}{2}.$$

Finally, by the difference formula for cosine,

$$\cos\left[\sin^{-1}\left(-\frac{1}{3}\right) - \tan^{-1}(-1)\right] = \cos(u - v)$$

$$= \cos u \cos v + \sin u \sin v$$

$$= \left(\frac{2\sqrt{2}}{3}\right)\left(\frac{\sqrt{2}}{2}\right) + \left(-\frac{1}{3}\right)\left(-\frac{\sqrt{2}}{2}\right)$$

$$= \frac{2}{3} + \frac{\sqrt{2}}{6}$$

$$= \frac{4 + \sqrt{2}}{6} \approx 0.90.$$

EXERCISES 6.3

In Exercises 1–12, use the appropriate sum or difference formula to find the exact value of each expression.

1. $\sin(7\pi/12)$

2. $\cos(-15°)$

3. $\tan 15°$

4. $\sin 75°$

5. $\sec(5\pi/12)$

6. $\tan(7\pi/12)$

7. $\cos(11\pi/12)$

8. $\sin 195°$

9. $\sin(-\pi/12)$

10. $\tan(11\pi/12)$.

11. $\cos 195°$

12. $\cos 345°$

In Exercises 13–26, use the appropriate sum or difference formula to find the exact value of each expression.

13. $\cos(2\pi/3)\cos(\pi/6) + \sin(2\pi/3)\sin(\pi/6)$

14. $\sin 40° \cos 10° - \cos 40° \sin 10°$

15. $\sin 21° \cos 24° + \cos 21° \sin 24°$

16. $\cos 220° \cos 80° - \sin 220° \sin 80°$

17. $\dfrac{\tan 160° - \tan 25°}{1 + \tan 160° \tan 25°}$

18. $\sin 53° \cos 82° + \cos 53° \sin 82°$

19. $\sin 173° \cos 53° - \cos 173° \sin 53°$

20. $\dfrac{\tan 115° + \tan 110°}{1 - \tan 115° \tan 110°}$

21. $\cos(3\pi/10)\cos(\pi/5) - \sin(3\pi/10)\sin(\pi/5)$

22. $\cos 280° \cos 70° + \sin 280° \sin 70°$

23. $\dfrac{\tan 200° + \tan 100°}{1 - \tan 200° \tan 100°}$

24. $\dfrac{\tan 260° - \tan 50°}{1 + \tan 260° \tan 50°}$

25. $\sin[(\pi/2) - v]\tan v - \sin v$

26. $\tan u \tan[(\pi/2) - u] - 1$

In Exercises 27–32, without using a calculator, use the given information to find each of the following:

a. $\sin(u + v)$

b. $\cos(u + v)$

c. $\sin(u - v)$

d. $\cos(u - v)$

e. $\tan(u + v)$

f. $\tan(u - v)$

g. The quadrant containing $u + v$

h. The quadrant containing $u - v$

27. $\sin u = \frac{1}{4}$ with u in quadrant II; $\cos v = -\sqrt{2}/3$ with v in quadrant III

28. $\cos u = -\frac{1}{3}$ with u in quadrant III; $\sin v = \sqrt{3}/5$ with v in quadrant II

29. $\tan u = -\frac{1}{2}$ with u in quadrant II; $\cos v = \frac{1}{6}$ with v in quadrant IV

30. $\sec u = -\frac{3}{2}$ with u in quadrant II; $\sin v = -\frac{1}{5}$ with v in quadrant III

31. $\csc u = -3$ with u in quadrant IV; $\cot v = -\frac{2}{3}$ with v in quadrant II

32. $\cot u = 5$ with u in quadrant I; $\sec v = -5$ with v in quadrant II

In Exercises 33–44, use a sum or difference identity to simplify each expression.

33. $\cos(270° - \theta)$

34. $\sec(180° - v)$

35. $\sin(180° + t)$

36. $\sin(90° - v)$

37. $\sin(u - \pi)$

38. $\csc(u + 270°)$

39. $\tan(\pi - v)$

40. $\cos(3\pi + v)$

41. $\cos(180° + w)$

42. $\tan(5\pi + v)$

43. $\tan[u + (\pi/6)]$

44. $\cot[u - (\pi/3)]$

In Exercises 45–50, find the exact value of each expression. Also approximate each answer to two decimal places.

45. $\sin[\sin^{-1}(-\frac{1}{3}) - \tan^{-1}\frac{1}{4}]$

46. $\sin(\sin^{-1}\frac{2}{3} + \cos^{-1}\frac{3}{5})$

47. $\cos[\sin^{-1}(-\frac{2}{3}) + \cos^{-1}\frac{3}{4}]$

48. $\cos[\sin^{-1}(\sqrt{3}/4) - \cos^{-1}(-\sqrt{3}/5)]$

49. $\tan(\sin^{-1}\frac{1}{5} + \cos^{-1}\frac{4}{5})$

50. $\tan[\cos^{-1}\frac{1}{5} - \tan^{-1}(-\frac{2}{3})]$

In Exercises 51–67, verify each identity.

51. $\sin[(2\pi/3) + v] = (\frac{1}{2})(\sqrt{3}\cos v - \sin v)$

52. $\sin[(5\pi/6) - v] = (\frac{1}{2})(\cos v + \sqrt{3}\sin v)$

53. $\cos[(\pi/6) - v] = \sin[(\pi/3) + v]$

54. $\cos[(3\pi/4) + v] = (-\sqrt{2}/2)(\cos v + \sin v)$

55. $\tan[u - (3\pi/2)] = -\cot u$

56. $\sin 2v = 2 \sin v \cos v$

57. $\cos 2v = \cos^2 v - \sin^2 v$

58. $\cos 2v = 1 - 2 \sin^2 v$

59. $\cos 2v = 2 \cos^2 v - 1$

60. $\tan 2v = \dfrac{2 \tan v}{2 - \sec^2 v}$

61. $\dfrac{\sin(x + h) - \sin x}{h} = \sin x\left(\dfrac{\cos h - 1}{h}\right)$
$+ \cos x\left(\dfrac{\sin h}{h}\right)$

62. $\dfrac{\cos(x + h) - \cos x}{h} = \cos x\left(\dfrac{\cos h - 1}{h}\right)$
$- \sin x\left(\dfrac{\sin h}{h}\right)$

63. $\dfrac{\sin(u + v)}{\sin(u - v)} = \dfrac{\tan u + \tan v}{\tan u + \tan(-v)}$

64. $\dfrac{\cos(u + v)}{\cos(u - v)} = \dfrac{\cot u \cot v - 1}{\cot u \cot v + 1}$

65. $\dfrac{\sin(u - v)}{\sin(u + v)} = \dfrac{1 - \cot u \tan v}{1 + \cot u \tan v}$

66. $\dfrac{\cos(u + v)}{\sin u \cos v} = \cot u - \tan v$

67. $\cos(u + v)\cos(u - v) = \cos^2 u - \sin^2 v$

68. Verify the difference formula for tangent.

69. Use a graphing calculator to sketch the graphs of the following functions over the interval $[0, 2\pi)$ in the same coordinate system. How do the graphs compare?

$$y = \sin(\pi - x) \qquad \text{and} \qquad y = \sin x$$

70. Use a graphing calculator to sketch the graphs of the following functions over the interval $[0, 2\pi)$ in the same coordinate system. How do the graphs compare?

$$y = \cos(\pi + x) \qquad \text{and} \qquad y = -\cos x$$

71. Use a graphing calculator to sketch the graphs of the following functions over the interval $[0, 2\pi)$ in the same coordinate system. How do the graphs compare?

$$y = \cos\left(\frac{\pi}{6} - x\right) \qquad \text{and} \qquad y = \sin\left(\frac{\pi}{3} + x\right)$$

72. Use a graphing calculator to sketch the graphs of the following functions over the interval $[0, 2\pi)$ in the same coordinate system. How do the graphs compare?

$$y = \sin\left(\frac{\pi}{2} + x\right) \qquad \text{and} \qquad y = \cos x$$

6.4 DOUBLE-ANGLE AND HALF-ANGLE IDENTITIES

In this section, we develop identities that are consequences of the sum formulas for sine, cosine, and tangent that we discussed in Section 6.3. For example, if we let $u = v$ in the sum formula for sine,

$$\sin(u + v) = \sin u \cos v + \cos u \sin v,$$

we obtain

$$\sin(v + v) = \sin v \cos v + \cos v \sin v,$$

or
$$\sin 2v = 2 \sin v \cos v. \qquad (1)$$

This formula, which expresses $\sin 2v$ in terms of $\sin v$ and $\cos v$, is called the **double-angle formula** for the sine function.

Similarly, we can obtain the double-angle formula for cosine by letting $u = v$ in the sum formula for cosine,

$$\cos(u + v) = \cos u \cos v - \sin u \sin v,$$

to obtain

$$\cos(v + v) = \cos v \cos v - \sin v \sin v,$$

or
$$\cos 2v = \cos^2 v - \sin^2 v. \qquad (2)$$

Using the identity $\cos^2 v + \sin^2 v = 1$, the double-angle formula for cosine can be written in two alternate forms. Substituting $1 - \sin^2 v$ for $\cos^2 v$ in equation (2), we obtain

$$\cos 2v = (1 - \sin^2 v) - \sin^2 v, \quad \text{or}$$
$$\cos 2v = 1 - 2 \sin^2 v.$$

Similarly, replacing $\sin^2 v$ with $1 - \cos^2 v$ yields

$$\cos 2v = \cos^2 v - (1 - \cos^2 v), \quad \text{or}$$
$$\cos 2v = 2 \cos^2 v - 1.$$

Finally, letting $u = v$ in the sum formula for tangent,

$$\tan(u + v) = \frac{\tan u + \tan v}{1 - \tan u \tan v},$$

we obtain the double-angle formula for tangent:

$$\tan(v + v) = \frac{\tan v + \tan v}{1 - \tan v \tan v}, \quad \text{or}$$
$$\tan 2v = \frac{2 \tan v}{1 - \tan^2 v}.$$

The double-angle formulas are summarized below.

Double-Angle Formulas

$$\sin 2v = 2 \sin v \cos v;$$
$$\cos 2v = \cos^2 v - \sin^2 v = 1 - 2 \sin^2 v = 2 \cos^2 v - 1;$$
$$\tan 2v = \frac{2 \tan v}{1 - \tan^2 v}.$$

EXAMPLE 21

Using the double-angle formulas, write each expression as a single trigonometric function value.

a. $2 \sin 40° \cos 40°$ **b.** $1 - 2 \sin^2 7$ **c.** $\dfrac{2 \tan 10°}{1 - \tan^2 10°}.$

Solution

a. Using the double-angle formula for sine, we have

$$2 \sin 40° \cos 40° = \sin(2 \cdot 40°) = \sin 80°.$$

b. By the double-angle formula for cosine,

$$1 - 2 \sin^2 7 = \cos(2 \cdot 7) = \cos 14.$$

c. From the double-angle formula for tangent, we see that

$$\frac{2 \tan 10°}{1 - \tan^2 10°} = \tan(2 \cdot 10°) = \tan 20°.$$

E X A M P L E   22

If $\sin v = \sqrt{5}/3$ and v is in quadrant II, find the exact values of $\sin 2v$, $\cos 2v$, and $\tan 2v$. Also, determine the quadrant in which $2v$ lies.

Solution   First, we find the value of $\cos v$. Since $\sin^2 v + \cos^2 v = 1$, and since $\cos v < 0$ when v is in quadrant II, it follows that

$$
\begin{aligned}
\cos v &= -\sqrt{1 - \sin^2 v} \\
&= -\sqrt{1 - \left(\frac{\sqrt{5}}{3}\right)^2} \\
&= -\sqrt{1 - \frac{5}{9}} \\
&= -\sqrt{\frac{4}{9}} \\
&= -\frac{2}{3}.
\end{aligned}
$$

Now, using the double-angle formula for sine, we have

$$
\begin{aligned}
\sin 2v &= 2 \sin v \cos v \\
&= 2\left(\frac{\sqrt{5}}{3}\right)\left(-\frac{2}{3}\right) \\
&= -\frac{4\sqrt{5}}{9}.
\end{aligned}
$$

Similarly, by the cosine double-angle formula,

$$
\begin{aligned}
\cos 2v &= \cos^2 v - \sin^2 v \\
&= \left(-\frac{2}{3}\right)^2 - \left(\frac{\sqrt{5}}{3}\right)^2 \\
&= \frac{4}{9} - \frac{5}{9} \\
&= -\frac{1}{9}.
\end{aligned}
$$

Therefore,

$$
\begin{aligned}
\tan 2v &= \frac{\sin 2v}{\cos 2v} \\
&= \frac{-\dfrac{4\sqrt{5}}{9}}{-\dfrac{1}{9}} \\
&= 4\sqrt{5}.
\end{aligned}
$$

Finally, since both $\sin 2v$ and $\cos 2v$ are negative, we know that $2v$ lies in quadrant III.

CAUTION: Do not make the mistake of assuming that sin 2v is the same as 2 sin v. In the solution of Example 22 above, we found that

$$\sin 2v = -\frac{4\sqrt{5}}{9},$$

while

$$2 \sin v = 2\left(\frac{\sqrt{5}}{3}\right) = \frac{2\sqrt{5}}{3}.$$

Similarly, cos 2v ≠ 2 cos v and tan 2v ≠ 2 tan v.

E X A M P L E   23

Verify the identity $\dfrac{1}{\cot v - \tan v} = \dfrac{\tan 2v}{2}$.

Solution   We work with each side separately. Expressing the left side in terms of sines and cosines, we have

$$\frac{1}{\cot v - \tan v} = \frac{1}{\dfrac{\cos v}{\sin v} - \dfrac{\sin v}{\cos v}}$$

$$= \frac{1}{\dfrac{\cos^2 v - \sin^2 v}{\sin v \cos v}}$$

$$= \frac{\sin v \cos v}{\cos^2 v - \sin^2 v}$$

$$= \frac{\sin v \cos v}{\cos 2v}. \qquad \text{double-angle formula for cosine}$$

Working with the right side, we have

$$\frac{\tan 2v}{2} = \frac{1}{2}\left(\frac{\sin 2v}{\cos 2v}\right)$$

$$= \frac{1}{2}\left(\frac{2 \sin v \cos v}{\cos 2v}\right) \qquad \text{double-angle formula for sine}$$

$$= \frac{\sin v \cos v}{\cos 2v}.$$

Since both sides of the original identity have been transformed into the same expression, the verification is complete.

E X A M P L E   24

Verify the identity $2 \cot 2w = \cot w - \tan w$.

Solution   Starting with left side, we have

$$2 \cot 2w = 2\left(\frac{1}{\tan 2w}\right)$$

$$= 2\left(\frac{1 - \tan^2 w}{2 \tan w}\right) \qquad \text{double-angle formula for tangent}$$

$$= \frac{1 - \tan^2 w}{\tan w}$$

$$= \frac{1}{\tan w} - \frac{\tan^2 w}{\tan w}$$

$$= \cot w - \tan w.$$

EXAMPLE 25

Verify the identity $\sin 3v = 3 \sin v - 4 \sin^3 v$.

Solution Starting on the left side, we have

$$
\begin{aligned}
\sin 3v &= \sin(2v + v) \\
&= \sin 2v \cos v + \cos 2v \sin v && \text{sum formula for sine} \\
&= (2 \sin v \cos v) \cos v && \text{double-angle formulas} \\
&\quad + (\cos^2 v - \sin^2 v) \sin v && \text{for sine and cosine} \\
&= 2 \sin v \cos^2 v \\
&\quad + \cos^2 v \sin v - \sin^3 v \\
&= 3 \sin v \cos^2 v - \sin^3 v \\
&= 3 \sin v (1 - \sin^2 v) - \sin^3 v \\
&= (3 \sin v - 3 \sin^3 v) - \sin^3 v \\
&= 3 \sin v - 4 \sin^3 v.
\end{aligned}
$$

We will now use the double-angle formula for cosine to obtain three additional identities. Solving the equation

$$\cos 2v = 1 - 2 \sin^2 v$$

for $\sin^2 v$, we obtain

$$2 \sin^2 v = 1 - \cos 2v,$$

or $\qquad\qquad \sin^2 v = \dfrac{1 - \cos 2v}{2}.$ $\qquad$ **(3)**

Similarly, solving the equation

$$\cos 2v = 2 \cos^2 v - 1$$

for $\cos^2 v$, we have

$$2 \cos^2 v = 1 + \cos 2v,$$

or $\qquad\qquad \cos^2 v = \dfrac{1 + \cos 2v}{2}.$ $\qquad$ **(4)**

Finally, using identities (3) and (4) and the fact that $\tan^2 v = \dfrac{\sin^2 v}{\cos^2 v}$, we obtain the following identity (see Exercise 53):

$$\tan^2 v = \frac{1 - \cos 2v}{1 + \cos 2v}. \qquad \textbf{(5)}$$

We use identities (3) and (4) in our next example.

EXAMPLE 26

Verify the identity $\sin^4 v = (\frac{1}{8})(3 - 4\cos 2v + \cos 4v)$.

Solution   Starting on the left side, we have

$$\sin^4 v = (\sin^2 v)^2$$

$$= \left(\frac{1 - \cos 2v}{2}\right)^2 \qquad \text{identity (3)}$$

$$= \frac{1 - 2\cos 2v + \cos^2 2v}{4}$$

$$= \frac{1}{4}(1 - 2\cos 2v + \cos^2 2v)$$

$$= \frac{1}{4}\left(1 - 2\cos 2v + \frac{1 + \cos 4v}{2}\right) \qquad \text{identity (4)}$$

$$= \frac{1}{4}\left(\frac{2 - 4\cos 2v + 1 + \cos 4v}{2}\right)$$

$$= \frac{1}{8}(3 - 4\cos 2v + \cos 4v).$$

The Half-Angle Formulas

Identity (3) can also be used to verify the half-angle formula for the sine function:

$$\sin\frac{v}{2} = \pm\sqrt{\frac{1 - \cos v}{2}}.$$

To see this, we first replace v with $v/2$ in identity (3), to obtain

$$\sin^2\frac{v}{2} = \frac{1 - \cos v}{2}.$$

Taking the square root of each side of the last equation, we get

$$\sqrt{\sin^2\frac{v}{2}} = \sqrt{\frac{1 - \cos v}{2}}, \qquad \text{or} \qquad \left|\sin\frac{v}{2}\right| = \sqrt{\frac{1 - \cos v}{2}}.$$

It follows that

$$\sin\frac{v}{2} = \pm\sqrt{\frac{1 - \cos v}{2}}, \tag{6}$$

where the sign is determined by the quadrant in which $v/2$ lies. Thus, if $v/2$ lies in quadrants I or II, $\sin(v/2) > 0$ and

$$\sin\frac{v}{2} = \sqrt{\frac{1 - \cos v}{2}}.$$

On the other hand, if $v/2$ lies in quadrants III or IV, $\sin(v/2) < 0$ and

$$\sin\frac{v}{2} = -\sqrt{\frac{1 - \cos v}{2}}.$$

In a similar way (see Exercises 54–55), we can use identities (4) and (5) to derive the following half-angle formulas for cosine and tangent, respectively:

$$\cos \frac{v}{2} = \pm \sqrt{\frac{1 + \cos v}{2}}; \qquad (7)$$

$$\tan \frac{v}{2} = \pm \sqrt{\frac{1 - \cos v}{1 + \cos v}}. \qquad (8)$$

We summarize the half-angle formulas below:

Half-Angle Formulas

$$\sin \frac{v}{2} = \pm \sqrt{\frac{1 - \cos v}{2}}$$

$$\cos \frac{v}{2} = \pm \sqrt{\frac{1 + \cos v}{2}}$$

$$\tan \frac{v}{2} = \pm \sqrt{\frac{1 - \cos v}{1 + \cos v}}$$

The sign on the right side of each equation is determined by the quadrant in which $v/2$ lies.

EXAMPLE 27

Use the half-angle formula for cosine to find the exact value of cos 105°.

Solution   Since $105° = 210°/2$ lies in quadrant II, it follows that

$$\cos 105° = \cos \frac{210°}{2} = -\sqrt{\frac{1 + \cos 210°}{2}}$$

$$= -\sqrt{\frac{1 + (-\sqrt{3}/2)}{2}}$$

$$= -\frac{\sqrt{2 - \sqrt{3}}}{2}.$$

EXAMPLE 28

Use the fact that $\frac{105°}{2} = 52.5°$, to find the exact value of sin 52.5°.

Solution   In Example 27, we found the exact value of cos 105° to be

$$\cos 105° = -\frac{\sqrt{2 - \sqrt{3}}}{2}.$$

Since $52.5° = \frac{105°}{2}$ lies in quadrant I, the half-angle formula for sine yields

$$\sin 52.5° = \sqrt{\frac{1 - \cos 105°}{2}}$$

$$= \sqrt{\frac{1 - (-\sqrt{2} - \sqrt{3}/2)}{2}}$$

$$= \sqrt{\frac{2 + \sqrt{2} - \sqrt{3}}{4}}$$

$$= \frac{\sqrt{2 + \sqrt{2} - \sqrt{3}}}{2}.$$

There are two alternate forms for tan $(v/2)$ that do not involve a $\pm$ sign. To help us obtain these alternate forms, we verify the following identity:

$$\tan v = \frac{\sin 2v}{1 + \cos 2v}. \tag{9}$$

Starting on the right side of equation (9) and applying the double-angle formulas for sine and cosine, we have

$$\frac{\sin 2v}{1 + \cos 2v} = \frac{2 \sin v \cos v}{1 + (2 \cos^2 v - 1)}$$

$$= \frac{2 \sin v \cos v}{2 \cos^2 v}$$

$$= \frac{\sin v}{\cos v} = \tan v.$$

Thus, we have verified identity (9) above.

Now, replacing v by $v/2$ in identity (9), we obtain

$$\tan \frac{v}{2} = \frac{\sin v}{1 + \cos v}. \tag{10}$$

If we multiply the numerator and the denominator of the expression on the right side of equation (10) by $1 - \cos v$, we get

$$\frac{\sin v}{1 + \cos v} \cdot \frac{1 - \cos v}{1 - \cos v} = \frac{\sin v (1 - \cos v)}{1 - \cos^2 v}$$

$$= \frac{\sin v(1 - \cos v)}{\sin^2 v}$$

$$= \frac{1 - \cos v}{\sin v}.$$

Therefore,
$$\tan \frac{v}{2} = \frac{1 - \cos v}{\sin v}. \tag{11}$$

We summarize our results below.

///

Alternate Half-Angle Formulas for Tangent

$$\tan \frac{v}{2} = \frac{\sin v}{1 + \cos v}$$

$$\tan \frac{v}{2} = \frac{1 - \cos v}{\sin v}$$

EXAMPLE 29

Find the exact value of tan $(\pi/12)$.

Solution   Since $\pi/12 = \frac{1}{2} \cdot (\pi/6)$, we have by formula (10) above,

$$\tan \frac{\pi}{12} = \frac{\sin \dfrac{\pi}{6}}{1 + \cos \dfrac{\pi}{6}}$$

$$= \frac{\dfrac{1}{2}}{1 + \dfrac{\sqrt{3}}{2}}$$

$$= \frac{1}{2 + \sqrt{3}} \, .$$

EXAMPLE 30

If $\sin v = -\frac{1}{9}$, where $3\pi/2 < v < 2\pi$, find the exact value of cos $(v/2)$ and approximate your answer to two decimal places.

Solution   To apply the half-angle formula for cosine, we must first find the value of cos v. Since $3\pi/2 < v < 2\pi$, we know that cos $v > 0$. Therefore, from the identity $\sin^2 v + \cos^2 v = 1$, it follows that

$$\cos v = \sqrt{1 - \sin^2 v}$$

$$= \sqrt{1 - \left(-\frac{1}{9}\right)^2}$$

$$= \sqrt{1 - \frac{1}{81}}$$

$$= \sqrt{\frac{80}{81}} = \frac{4\sqrt{5}}{9} \, .$$

From the inequality $3\pi/2 < v < 2\pi$, it follows that $3\pi/4 < v/2 < \pi$. Hence, $v/2$ lies in quadrant II and cos $(v/2) < 0$. Therefore, by the half-angle formula for cosine, we have

$$\cos \frac{v}{2} = -\sqrt{\frac{1 + \cos v}{2}}$$

$$= -\sqrt{\frac{1 + \dfrac{4\sqrt{5}}{9}}{2}}$$

$$= -\sqrt{\frac{9 + 4\sqrt{5}}{18}}$$

$$= -\frac{\sqrt{9 + 4\sqrt{5}}}{3\sqrt{2}} \approx -1.00.$$

EXERCISES 6.4

In Exercises 1–10, use the appropriate double-angle or half-angle formula to write each expression as a single trigonometric function value.

1. $2 \sin 12 \cos 12$

2. $\cos^2 10 - \sin^2 10$

3. $2 \cos^2 50° - 1$

4. $\dfrac{2 \tan 80°}{1 - \tan^2 80°}$

5. $\pm\sqrt{\dfrac{1 + \cos 10t}{2}}$

6. $\pm\sqrt{\dfrac{1 - \cos 6x}{2}}$

7. $\pm\sqrt{\dfrac{1 - \cos 50x}{1 + \cos 50x}}$

8. $1 - 2 \sin^2 6$

9. $\dfrac{1 - \cos 100°}{\sin 100°}$

10. $\dfrac{\sin 200°}{1 + \cos 200°}$

In Exercises 11–20, use the given information to find the exact value of $\sin 2v$, $\cos 2v$, and $\tan 2v$. Also, determine the quadrant in which $2v$ lies.

11. $\sin v = -\sqrt{3}/5$ and v is in quadrant IV

12. $\cos v = \frac{2}{3}$ and v is in quadrant I

13. $\tan v = -\frac{1}{2}$ and v is in quadrant II

14. $\sin v = -\frac{3}{4}$ and v is in quadrant III

15. $\sec v = -5$ and v is in quadrant III

16. $\csc v = 10$ and v is in quadrant II

17. $\tan v = \frac{4}{3}$ and $\sin v < 0$

18. $\cot v = -\frac{3}{5}$ and $\cos v > 0$

19. $\csc v = -6$ and $\tan v > 0$

20. $\sec v = 9$ and $\tan v < 0$

In Exercises 21–36, find the exact value of each expression.

21. $\cos (5\pi/8)$

22. $\sin (\pi/8)$

23. $\cos (5\pi/16)$
[*Hint*: Use the result from Exercise 21.]

24. $\tan(-22.5°)$

25. $\tan (7\pi/12)$

26. $\cos (7\pi/12)$

27. $\sin 165°$

28. $\cos(-75°)$

29. $\tan(-67.5°)$

30. $\sin (5\pi/8)$

31. $\cos (7\pi/8)$

32. $\tan (11\pi/12)$

33. $\sin(-22.5°)$

34. $\cos 112.5°$

35. $\tan 15°$

36. $\sin 157.5°$

In Exercises 37–44, use the given information to find the exact value of $\sin(v/2)$, $\cos(v/2)$, and $\tan(v/2)$.

37. $\sin v = -\frac{1}{6}$ and $3\pi/2 < v < 2\pi$

38. $\cos v = \frac{5}{6}$ and $0 < v < \pi/2$

39. $\cos v = -\frac{3}{8}$ and $\pi/2 < v < \pi$

40. $\sin v = \frac{5}{7}$ and $\pi/2 < v < \pi$

41. $\tan v = 3$ and $\pi < v < 3\pi/2$

42. $\tan v = 4$ and $\pi < v < 3\pi/2$

43. $\sec v = 7$ and $0 < v < \pi/2$

44. $\csc v = -3$ and $3\pi/2 < v < 2\pi$

In Exercises 45–52, use the appropriate double-angle formula to find the exact value of each expression.

45. $\sin[2 \cos^{-1}(0.2)]$ [*Hint*: Let $t = \cos^{-1}(0.2)$. Then use the definition of $\cos^{-1}$ and the identity $\cos^2 t + \sin^2 t = 1$.]

46. $\tan[2 \cos^{-1}(-\frac{2}{7})]$

47. $\cos[2 \sin^{-1}(-1)]$

48. $\cos(2 \sin^{-1}\frac{1}{4})$

49. $\tan[2 \tan^{-1}(-\frac{2}{3})]$

50. $\sin[2 \sin^{-1}(-\frac{3}{5})]$

51. $\sin[2 \sin^{-1}(-\frac{6}{7})]$

52. $\cos[2 \sin^{-1}(-\frac{5}{6})]$

In Exercises 53–72, verify each identity.

53. $\tan^2 v = \dfrac{1 - \cos 2v}{1 + \cos 2v}$

54. $\cos \dfrac{v}{2} = \pm\sqrt{\dfrac{1 + \cos v}{2}}$

55. $\tan \dfrac{v}{2} = \pm\sqrt{\dfrac{1 - \cos v}{1 + \cos v}}$

56. $\dfrac{1 + \tan x}{1 - \tan x} = \dfrac{\cos 2x}{1 - \sin 2x}$

57. $\cos^4 x = \frac{1}{8}(3 + 4 \cos 2x + \cos 4x)$

58. $\dfrac{\sin 2v}{2 \tan v} = \cos^2 v$

59. $\cos 3x = 4 \cos^3 x - 3 \cos x$

60. $\sec 2x = \dfrac{1 + \tan^2 x}{1 - \tan^2 x}$

61. $[\sin (v/2) - \cos (v/2)]^2 - 1 = \sin(-v)$

62. $\cot 2t = \dfrac{\cot^2 t - 1}{2 \cot t}$

63. $\dfrac{\sin^3 v + \cos^3 v}{\sin v + \cos v} = \dfrac{2 - \sin 2v}{2}$

64. $\tan (v/2) = \csc v - \cot v$

65. $\tan v = \dfrac{\sin 2v}{1 + \cos 2v}$

66. $\dfrac{2 \sin 2v}{\sin 4v} = \sec 2v$

67. $\cos^4 (v/2) - \sin^4 (v/2) = \cos v$

68. $\sin 2v = \dfrac{2 \tan v}{\sec^2 v}$

69. $\frac{1}{2} \cos v \cot (v/2) \sec^2 (v/2) = \cot v$

70. $\dfrac{\cos 3v}{\cos v} - \dfrac{\sin 3v}{\sin v} = -2$

71. $2 \sin x \cos x + 2 \sin x \cos x \tan^2 x = \sec^2 x \sin 2x$

72. $\sin^6 x = \frac{1}{16}(5 - 7 \cos 2x + 3 \cos 4x - \cos 4x \cos 2x)$

6.5 PRODUCT-TO-SUM AND SUM-TO-PRODUCT IDENTITIES

In certain applications, it is necessary to express a product of trigonometric functions as a sum or difference of these functions. In other applications, we are required to express a sum or difference of trigonometric functions as a product of these functions. The identities that are developed in this section are useful in such applications.

Product-to-Sum Identities

Recall from Section 6.3 the sum and difference formulas for cosine:

$$\cos(u + v) = \cos u \cos v - \sin u \sin v \qquad (1)$$

and
$$\cos(u - v) = \cos u \cos v + \sin u \sin v. \qquad (2)$$

If we add the left and right sides of equation (1) to the corresponding sides of equation (2), we obtain

$$\cos(u + v) + \cos(u - v) = 2 \cos u \cos v, \qquad \text{or}$$

$$\cos u \cos v = \frac{1}{2}[\cos(u + v) + \cos(u - v)].$$

On the other hand, subtracting the left and right sides of equation (1) from the corresponding sides of equation (2) yields

$$\cos(u - v) - \cos(u + v) = 2 \sin u \sin v, \qquad \text{or}$$

$$\sin u \sin v = \frac{1}{2}[\cos(u - v) - \cos(u + v)].$$

Similarly, using the sum and difference formulas for sine, we can obtain two additional **product-to-sum identities** that appear as identities (iii) and (iv) in the summary below (see Exercise 45).

Product-to-Sum Identities

(i) $\cos u \cos v = \frac{1}{2}[\cos(u + v) + \cos(u - v)]$

(ii) $\sin u \sin v = \frac{1}{2}[\cos(u - v) - \cos(u + v)]$

(iii) $\sin u \cos v = \frac{1}{2}[\sin(u + v) + \sin(u - v)]$

(iv) $\cos u \sin v = \frac{1}{2}[\sin(u + v) - \sin(u - v)]$

EXAMPLE 31

Use the appropriate product-to-sum identity to express each product as a sum or difference.

a. $\cos 3t \cos 5t$ **b.** $\cos 2\theta \sin 7\theta$.

Solution

a. Using product-to-sum identity (i) with $u = 3t$ and $v = 5t$, we have

$$\cos 3t \cos 5t = \frac{1}{2}[\cos(3t + 5t) + \cos(3t - 5t)]$$

$$= \frac{1}{2}[\cos 8t + \cos(-2t)]$$

$$= \frac{1}{2}(\cos 8t + \cos 2t). \qquad\qquad \cos(-s) = \cos s$$

b.   Applying product-to-sum identity (iv), we obtain

$$\cos 2\theta \sin 7\theta = \frac{1}{2}[\sin(2\theta + 7\theta) - \sin(2\theta - 7\theta)]$$

$$= \frac{1}{2}[\sin 9\theta - \sin(-5\theta)]$$

$$= \frac{1}{2}(\sin 9\theta + \sin 5\theta). \qquad\qquad \sin(-s) = -\sin s$$

Sum-to-Product Identities

The product-to-sum identities can be used to obtain the following **sum-to-product identities** that express a sum or difference of two functions as a product.

Sum-to-Product Identities

$$\text{(i)} \quad \cos u + \cos v = 2 \cos\left(\frac{u + v}{2}\right) \cos\left(\frac{u - v}{2}\right)$$

$$\text{(ii)} \quad \cos u - \cos v = -2 \sin\left(\frac{u + v}{2}\right) \sin\left(\frac{u - v}{2}\right)$$

$$\text{(iii)} \quad \sin u + \sin v = 2 \sin\left(\frac{u + v}{2}\right) \cos\left(\frac{u - v}{2}\right)$$

$$\text{(iv)} \quad \sin u - \sin v = 2 \cos\left(\frac{u + v}{2}\right) \sin\left(\frac{u - v}{2}\right)$$

We verify sum-to-product identity (i) and leave the remaining verifications as exercises (see Exercise 48). By product-to-sum identity (i), we can write

$$\cos x \cos y = \frac{1}{2}[\cos(x + y) + \cos(x - y)]. \qquad\qquad (3)$$

If we let

$$u = x + y \qquad \text{and} \qquad v = x - y,$$

then

$$u + v = 2x \qquad \text{and} \qquad u - v = 2y.$$

Hence,

$$x = \frac{u + v}{2} \qquad \text{and} \qquad y = \frac{u - v}{2}. \qquad\qquad (4)$$

Substituting equations (4) into equation (3) yields

$$\cos\left(\frac{u + v}{2}\right) \cos\left(\frac{u - v}{2}\right) = \frac{1}{2}(\cos u + \cos v), \qquad \text{or}$$

$$\cos u + \cos v = 2 \cos \left(\frac{u + v}{2} \right) \cos \left(\frac{u - v}{2} \right).$$

EXAMPLE 32

Use the appropriate sum-to-product identity to write each expression as a product.

a. $\sin 5x + \sin 9x$ **b.** $\cos 34° - \cos 10°$

Solution

a. Letting $u = 5x$ and $v = 9x$ in sum-to-product identity (iii), we have

$$\sin 5x + \sin 9x = 2 \sin \left(\frac{5x + 9x}{2} \right) \cos \left(\frac{5x - 9x}{2} \right)$$

$$= 2 \sin 7x \cos(-2x)$$

$$= 2 \sin 7x \cos 2x. \qquad\qquad \cos(-t) = \cos t$$

b. By sum-to-product identity (ii),

$$\cos 34° - \cos 10° = -2 \sin \left(\frac{34° + 10°}{2} \right) \sin \left(\frac{34° - 10°}{2} \right)$$

$$= -2 \sin 22° \sin 12°.$$

EXAMPLE 33

Use an appropriate sum-to-product or product-to-sum identity to find the exact value of each expression.

a. $\sin 165° - \sin 75°$ **b.** $\left(\cos \dfrac{13\pi}{12} \right) \cdot \left(\cos \dfrac{11\pi}{12} \right)$

Solution

a. Using sum-to-product identity (iv), we have

$$\sin 165° - \sin 75° = 2 \cos \left(\frac{165° + 75°}{2} \right) \sin \left(\frac{165° - 75°}{2} \right)$$

$$= 2 \cos 120° \sin 45°$$

$$= 2 \left(-\frac{1}{2} \right) \left(\frac{\sqrt{2}}{2} \right)$$

$$= -\frac{\sqrt{2}}{2}.$$

b. By product-to-sum identity (i),

$$\left(\cos \frac{13\pi}{12} \right) \cdot \left(\cos \frac{11\pi}{12} \right) = \frac{1}{2} \left[\cos \left(\frac{13\pi}{12} + \frac{11\pi}{12} \right) + \cos \left(\frac{13\pi}{12} - \frac{11\pi}{12} \right) \right]$$

$$= \frac{1}{2} \left(\cos 2\pi + \cos \frac{\pi}{6} \right)$$

$$= \frac{1}{2} \left(1 + \frac{\sqrt{3}}{2} \right) = \frac{2 + \sqrt{3}}{4}.$$

EXAMPLE 34

Verify the identity $\dfrac{\sin \theta + \sin 3\theta}{\cos \theta - \cos 3\theta} = \cot \theta$.

Solution   Starting on the left side and applying sum-to-product identities (iii) and (ii), we have

$$\frac{\sin \theta + \sin 3\theta}{\cos \theta - \cos 3\theta} = \frac{2 \sin \left(\dfrac{\theta + 3\theta}{2}\right) \cos \left(\dfrac{\theta - 3\theta}{2}\right)}{-2 \sin \left(\dfrac{\theta + 3\theta}{2}\right) \sin \left(\dfrac{\theta - 3\theta}{2}\right)}$$

$$= \frac{2 \sin 2\theta \cos(-\theta)}{-2 \sin 2\theta \sin(-\theta)}$$

$$= \frac{\cos \theta}{-(-\sin \theta)} \qquad \begin{array}{l} \cos(-\theta) = \cos \theta \text{ and} \\ \sin(-\theta) = -\sin \theta \end{array}$$

$$= \frac{\cos \theta}{\sin \theta} = \cot \theta.$$

EXAMPLE 35

Verify the identity $\dfrac{\cos x - \cos 3x}{\sin 3x - \sin x} = \dfrac{\cos x - \cos 5x}{\sin x + \sin 5x}$.

Solution   We work with each side separately. Starting on the left side and applying sum-to-product identities (ii) and (iv), we have

$$\frac{\cos x - \cos 3x}{\sin 3x - \sin x} = \frac{-2 \sin \left(\dfrac{x + 3x}{2}\right) \sin \left(\dfrac{x - 3x}{2}\right)}{2 \cos \left(\dfrac{3x + x}{2}\right) \sin \left(\dfrac{3x - x}{2}\right)}$$

$$= \frac{-2 \sin 2x \sin(-x)}{2 \cos 2x \sin x}$$

$$= \frac{-\sin 2x(-\sin x)}{\cos 2x \sin x} \qquad \sin(-x) = -\sin x$$

$$= \frac{\sin 2x}{\cos 2x} = \tan 2x.$$

Working now with the right side and applying sum-to-product identities (ii) and (iii), we have

$$\frac{\cos x - \cos 5x}{\sin x + \sin 5x} = \frac{-2 \sin \left(\dfrac{x + 5x}{2}\right) \sin \left(\dfrac{x - 5x}{2}\right)}{2 \sin \left(\dfrac{x + 5x}{2}\right) \cos \left(\dfrac{x - 5x}{2}\right)}$$

$$= \frac{-2 \sin 3x \sin(-2x)}{2 \sin 3x \cos(-2x)}$$

$$= \frac{-(-\sin 2x)}{\cos 2x} \qquad \begin{array}{l} \sin(-t) = -\sin t \text{ and} \\ \cos(-t) = \cos t. \end{array}$$

$$= \frac{\sin 2x}{\cos 2x} = \tan 2x.$$

Since both sides of the original identity have been transformed into the same expression, the verification is complete. ──────

EXERCISES 6.5

In Exercises 1–10, express each product as a sum or difference.

1. $\cos 2x \cos 3x$
2. $\sin 2v \sin 8v$
3. $\sin 20° \sin 50°$
4. $\cos 15° \cos 10°$
5. $\sin 3\theta \cos 5\theta$
6. $\cos 6\theta \sin 10\theta$
7. $\cos 10t \sin 8t$
8. $\sin 12x \cos 6x$
9. $10 \cos 20t \cos 30t$
10. $4 \sin 5t \sin 9t$

In Exercises 11–20, write each expression as a product.

11. $\sin 3\theta + \sin 6\theta$
12. $\sin 5\theta - \sin 10\theta$
13. $\cos 12° - \cos 62°$
14. $\cos 20° + \cos 38°$
15. $\cos 7x + \cos 8x$
16. $\cos 2t - \cos 5t$
17. $\sin 12v - \sin 20v$
18. $\sin 3v + \sin 4v$
19. $\sin 2y + \sin(-3y)$
20. $\cos 6w + \cos(-7w)$

In Exercises 21–34, use the appropriate sum-to-product or product-to-sum identity to find the exact value of each expression.

21. $\cos 195° - \cos 105°$
22. $\cos 195° + \cos 105°$
23. $\sin 285° - \sin 195°$
24. $\sin 285° + \sin 195°$
25. $\sin 165° \cos 105°$
26. $\sin 165° \sin 105°$
27. $\cos 15° \sin 75°$
28. $\sin 15° \cos 75°$
29. $\cos (5\pi/12) - \cos (\pi/12)$
30. $\cos 165° - \cos 75°$
31. $\cos (\pi/3) \sin (\pi/6)$
32. $\cos (\pi/3) \cos (\pi/6)$
33. $\cos (7\pi/8) \cos (5\pi/8)$
34. $\sin (7\pi/8) \sin (5\pi/8)$

In Exercises 35–52, verify each identity.

35. $\cos t - \cos 3t = 4 \sin^2 t \cos t$
36. $\dfrac{\sin t + \sin v}{\cos t - \cos v} = \cot \left(\dfrac{v - t}{2} \right)$
37. $\dfrac{\sin v - \sin u}{\cos v + \cos u} = \tan \left(\dfrac{v - u}{2} \right)$

38. $\dfrac{\cos 3x - \cos 5x}{\sin 3x + \sin 5x} = \tan x$
39. $\dfrac{\cos v - \cos u}{\cos v + \cos u} = \tan \left(\dfrac{u + v}{2} \right) \tan \left(\dfrac{u - v}{2} \right)$
40. $\dfrac{\cos u + \cos v}{\sin u + \sin v} = \cot \left(\dfrac{v + u}{2} \right)$
41. $\dfrac{\cos 50° - \cos 10°}{\sin 10° - \sin 50°} = \dfrac{\sqrt{3}}{3}$
42. $\dfrac{\cos 6x - \cos 4x}{\sin 6x + \sin 4x} = -\tan x$
43. $\dfrac{\sin 6x - \sin 10x}{\cos 10x - \cos 6x} = \dfrac{\cos 6x + \cos 10x}{\sin 6x + \sin 10x}$
44. $\dfrac{\cos u + \cos 3u}{\sin u + \sin 3u} = \cot 2u$
45. Verify product-to-sum identities (iii) and (iv).
46. $\dfrac{\sin 2\theta + \sin 4\theta}{\cos 2\theta + \cos 4\theta} = \tan 3\theta$
47. $\dfrac{2 \sin 2u}{\sin 3u + \sin u} = \sec u$
48. Verify sum-to-product identities (ii), (iii), and (iv).
49. $\dfrac{\cos 6t + \cos 10t}{\sin 6t + \sin 10t} = \cot 8t$
50. $\dfrac{2 \cos 2v}{\cos 3v + \cos v} = \sec v$
51. $\sin 70° - \sin 110° = 0$
52. $\dfrac{\sin x + \sin 3x}{\cos 3x - \cos x} = -\cot x$
53. Express $\cos(ax + b) \sin(ax + b)$ as a sum.

6.6 TRIGONOMETRIC EQUATIONS

In Sections 6.2 through 6.5, we studied trigonometric identities that are trigonometric equations true for all permissible values of the variables they involve. In this section, we will study **conditional trigonometric equations** that are equations that may be true for some values of the variables they involve, but false for other values. The methods we will use to solve trigonometric equations are similar to those used to solve algebraic equations. Since no particular approach will work in all cases, our examples will present a variety of techniques.

The equation

$$\cos x = \frac{1}{2}$$

FIGURE 6.18

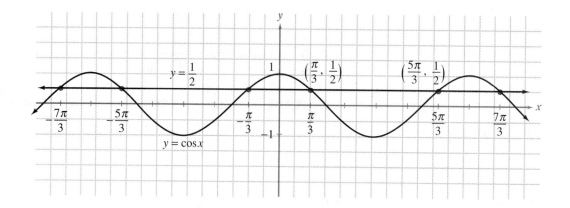

is a conditional equation, since it is true when $x = \pi/3$ ($\cos (\pi/3) = \frac{1}{2}$) but false when $x = \pi$ ($\cos \pi = -1 \neq \frac{1}{2}$). As illustrated in Figure 6.18, there are infinitely many different values of x for which $\cos x = \frac{1}{2}$. We will find all of the solutions of this equation in our next example.

EXAMPLE 36

Find the solution set of the equation $\cos x = \frac{1}{2}$.

Solution   We first find the solutions of the equation

$$\cos x = \frac{1}{2} \qquad (1)$$

in the interval $[0, 2\pi)$. Since $\cos x$ is positive for x in quadrants I or IV, there are two solutions of equation (1) in $[0, 2\pi)$:

$$\frac{\pi}{3} \quad \text{and} \quad \frac{5\pi}{3}. \qquad \text{(See Figure 6.18.)}$$

Since the cosine function has period 2π, it follows that the solution set of equation (1) is

$$S = \left\{ x \mid x = \frac{\pi}{3} + 2k\pi \text{ or } x = \frac{5\pi}{3} + 2k\pi, \quad \text{where } k \text{ is any integer} \right\}.$$

Reference Angles

To help us solve conditional trigonometric equations, it is convenient to introduce the concept of reference angles.

DEFINITION 6.2

If θ is a nonquadrantal angle in standard position, then the **reference angle for** θ is the positive acute angle θ' formed by the terminal side of θ and the x-axis.

FIGURE 6.19

a. $0 < \theta < \dfrac{\pi}{2}; \theta' = \theta$

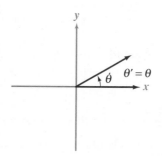

b. $\dfrac{\pi}{2} < \theta < \pi; \theta' = \pi - \theta$

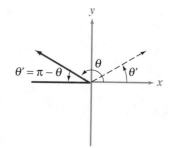

c. $\pi < \theta < \dfrac{3\pi}{2}; \theta' = \theta - \pi$

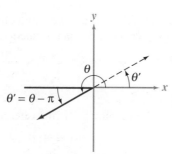

d. $\dfrac{3\pi}{2} < \theta < 2\pi; \theta' = 2\pi - \theta$

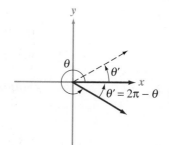

Figure 6.19 displays the reference angles θ' for positive angles θ lying in quadrants I, II, III, and IV.

Since the reference angle θ' for any angle θ is acute, all six trigonometric functions of θ' are positive. The following theorem describes the relationship between the trigonometric functions of θ and those of its reference angle θ'.

THEOREM 6.3   Reference Angle Theorem

If θ is a nonquadrantal angle in standard position with reference angle θ', then

(i) $\sin \theta' = |\sin \theta|$ (iv) $\csc \theta' = |\csc \theta|$
(ii) $\cos \theta' = |\cos \theta|$ (v) $\sec \theta' = |\sec \theta|$
(iii) $\tan \theta' = |\tan \theta|$ (vi) $\cot \theta' = |\cot \theta|.$

FIGURE 6.20

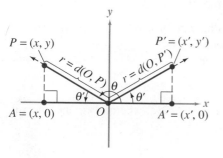

We outline the proof of Theorem 6.3 for the case when $\pi/2 < \theta < \pi$. The other cases can be proved in a similar manner (see Exercise 69). In Figure 6.20, θ is in standard position and its reference angle θ' is also shown in standard position. Choose any point $P = (x, y)$, other than the origin O, on the terminal side of θ and let r be the distance from O to P. Now choose the point $P' = (x', y')$ on the terminal side of θ' so that the distance from O to P' is also r. Then, as we see from Figure 6.20, triangles OAP and $OA'P'$ are congruent triangles. It follows that $x = -x'$ and that $y = y'$. Hence, by Theorem 6.2 (see page 253),

$$\sin \theta = \frac{y}{r} = \frac{y'}{r} = \sin \theta',$$

$$\cos \theta = \frac{x}{r} = \frac{-x'}{r} = -\cos \theta', \quad \text{and} \quad \tan \theta = \frac{y}{x} = \frac{y'}{-x'} = -\tan \theta'.$$

Thus, $\sin \theta' = |\sin \theta|$,

$$\cos \theta' = |\cos \theta|, \quad \text{and} \quad \tan \theta' = |\tan \theta|.$$

Similarly,

$$\csc \theta' = |\csc \theta|, \; \sec \theta' = |\sec \theta|, \; \text{and} \; \cot \theta' = |\cot \theta|.$$

The next example illustrates how to use the reference angle theorem.

EXAMPLE 37

Using Theorem 6.3, express the value of the given function of the given angle θ in terms of the value of the same function of its reference angle θ'.

a. $\cos 520°$ **b.** $\tan(-2.79)$

Solution

FIGURE 6.21

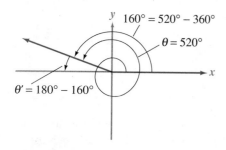

a. As we see in Figure 6.21, the reference angle for $\theta = 520°$ is

$$\theta' = 180° - 160° = 20°.$$

From the reference angle theorem, it follows that

$$\cos 20° = |\cos 520°|.$$

Since $\theta = 520°$ lies in quadrant II, we know that $\cos 520° < 0$. Therefore,

$$\cos 20° = -\cos 520°, \quad \text{or} \quad \cos 520° = -\cos 20°.$$

FIGURE 6.22

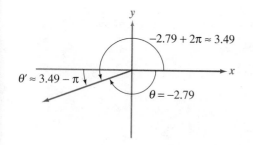

b. From Figure 6.22, we see that the reference angle for $\theta = -2.79$ is

$$\theta' \approx 3.49 - \pi \approx 0.35.$$

Hence, $\tan 0.35 \approx |\tan(-2.79)|$.

Because $\theta = -2.79$ lies in quadrant III, we know that $\tan(-2.79) > 0$. Thus,

$$\tan 0.35 \approx \tan(-2.79), \quad \text{or} \quad \tan(-2.79) \approx \tan 0.35.$$

Solving Trigonometric Equations Using Reference Angles

The remaining examples in this section illustrate how reference angles can be used when we solve trigonometric equations.

EXAMPLE 38

Find the solution set for the equation $\cos t = -\sqrt{2}/2$, where t is in the interval $[0, 2\pi)$.

Solution

Since cos t is negative for t in quadrants II or III, we know that there are two values of t—one in quadrant II and one in quadrant III—that satisfy the equation cos $t = -\sqrt{2}/2$. In either case, if t' is the reference angle for t, then the reference angle theorem tells us that

$$\cos t' = |\cos t| = \frac{\sqrt{2}}{2}.$$

Since t' is an acute angle, it follows that $t' = \pi/4$. If t lies in quadrant II, then

$$t' = \pi - t,$$

and, as we see from Figure 6.23,

$$t = \pi - t'$$
$$= \pi - \frac{\pi}{4} = \frac{3\pi}{4}.$$

On the other hand, if t lies in quadrant III, then

$$t' = t - \pi,$$

which implies (see Figure 6.24) that

$$t = \pi + t'$$
$$= \pi + \frac{\pi}{4} = \frac{5\pi}{4}.$$

Therefore, the solution set is

$$\left\{ \frac{3\pi}{4}, \frac{5\pi}{4} \right\}.$$

FIGURE 6.23

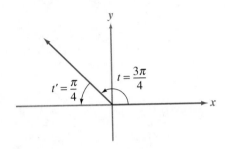

FIGURE 6.24

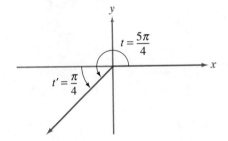

EXAMPLE 39

Find the solution set for the equation tan $2t = -\sqrt{3}$, where t is in the interval $[0, 2\pi)$.

Solution For convenience, we begin by letting $u = 2t$ and solving the equation

$$\tan u = -\sqrt{3}. \tag{2}$$

Since $0 \le t < 2\pi$, it follows that

$$0 \le 2t < 4\pi, \quad \text{or} \quad 0 \le u < 4\pi.$$

Since the period of the tangent function is π, we first find the solutions of equation (2) that lie in the interval $[0, \pi)$. We can then obtain the remaining solutions between 0 and 4π by adding multiples of π to these solutions.

To find the solutions of equation (2) that lie in the interval $[0, \pi)$, we first note that, since tan $u < 0$, u lies in quadrant II. Since tan $u = -\sqrt{3}$, and tan $(\pi/3) = \sqrt{3}$, we see that the reference angle for u is $u' = \pi/3$. Since u lies

in quadrant II it follows that

$$u' = \pi - u.$$

Hence,
$$u = \pi - u'$$
$$= \pi - \frac{\pi}{3} = \frac{2\pi}{3}. \qquad \text{(See Figure 6.25.)}$$

FIGURE 6.25

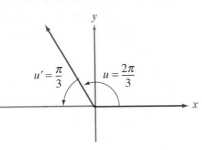

By adding π, 2π, and 3π to the solution $2\pi/3$, we obtain the remaining solutions of equation (2) in the interval $[0, 4\pi)$:

$$\frac{2\pi}{3} + \pi = \frac{5\pi}{3},$$

$$\frac{2\pi}{3} + 2\pi = \frac{8\pi}{3}, \qquad \text{and}$$

$$\frac{2\pi}{3} + 3\pi = \frac{11\pi}{3}.$$

Hence, the solutions of equation (2) in the interval $[0, 4\pi)$ are

$$\frac{2\pi}{3}, \qquad \frac{5\pi}{3}, \qquad \frac{8\pi}{3}, \qquad \text{and} \qquad \frac{11\pi}{3}.$$

Since $u = 2t$, we divide these values of u by 2 to obtain the following values of t:

$$\frac{\pi}{3}, \qquad \frac{5\pi}{6}, \qquad \frac{4\pi}{3}, \qquad \text{and} \qquad \frac{11\pi}{6}.$$

Therefore, the solution set for the original equation is

$$\left\{ \frac{\pi}{3}, \frac{5\pi}{6}, \frac{4\pi}{3}, \frac{11\pi}{6} \right\}.$$

E X A M P L E 40

Find the solution set for the equation $\sin 2\theta = \sqrt{3} \cos \theta$, where $0° \leq \theta < 360°$.

Solution   Notice that this equation involves both θ and 2θ. Using the identity $\sin 2\theta = 2 \sin \theta \cos \theta$, we can obtain an equivalent equation involving only θ:

$$\sin 2\theta = \sqrt{3} \cos \theta$$
$$2 \sin \theta \cos \theta = \sqrt{3} \cos \theta.$$

Subtracting $\sqrt{3} \cos \theta$ from both sides of the last equation and factoring we obtain

$$2 \sin \theta \cos \theta - \sqrt{3} \cos \theta = 0$$
$$\cos \theta (2 \sin \theta - \sqrt{3}) = 0$$
$$\cos \theta = 0 \quad \text{or} \quad 2 \sin \theta - \sqrt{3} = 0.$$

If $\cos \theta = 0$, then

$$\theta = 90° \quad \text{or} \quad \theta = 270°.$$

On the other hand, if $2 \sin \theta - \sqrt{3} = 0$, then

$$2 \sin \theta = \sqrt{3},$$

or
$$\sin \theta = \frac{\sqrt{3}}{2}. \tag{3}$$

Since $\sin 60° = \sqrt{3}/2$ and since $\sin \theta$ is positive for θ in quadrants I or II, it follows that the values of θ, where $0° \leq \theta < 360°$, that satisfy equation (3) are

$$60° \qquad \text{and} \qquad 120°.$$

Hence, the solution set for the original equation is

$$\{60°, 90°, 120°, 270°\}.$$

 CAUTION: In the solution of Example 40, we encountered the equation

$$2 \sin \theta \cos \theta = \sqrt{3} \cos \theta.$$

We may be tempted to divide both sides of this equation by $\cos \theta$ to obtain

$$2 \sin \theta = \sqrt{3}.$$

However, this approach should *not* be followed, since it assumes that $\cos \theta$ is not equal to zero, and thus two solutions, 90° and 270°, are lost.

In all of the previous examples, the solutions we obtained were either quadrantal angles or angles having reference angles $\pi/6$, $\pi/4$, or $\pi/3$. When a trigonometric equation has solutions that are not of this form, we can solve it by using a calculator as the next example illustrates.

EXAMPLE 41

Find the solutions of the equation $-6 \sin^2 t - \cos t + 4 = 0$, where t is in the interval $[0, 2\pi)$. Round off any approximate solutions to one decimal place.

Solution Substituting $1 - \cos^2 t$ for $\sin^2 t$ in the equation

$$-6 \sin^2 t - \cos t + 4 = 0,$$

we obtain

$$-6(1 - \cos^2 t) - \cos t + 4 = 0$$
$$-6 + 6 \cos^2 t - \cos t + 4 = 0$$
$$6 \cos^2 t - \cos t - 2 = 0.$$

The last equation is a quadratic equation in $\cos t$ that can be solved by factoring:

$$(2 \cos t + 1)(3 \cos t - 2) = 0$$
$$2 \cos t + 1 = 0 \qquad \text{or} \qquad 3 \cos t - 2 = 0. \qquad \textbf{(4)}$$

If $2 \cos t + 1 = 0$, then

$$\cos t = -\frac{1}{2}.$$

FIGURE 6.26

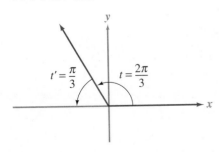

Since $\cos t$ is negative, we know that t lies in either quadrant II or quadrant III. In either case, if t' is the reference angle for t, then

$$\cos t' = |\cos t| = \frac{1}{2}.$$

Thus, the reference angle is $t' = \pi/3$. If t lies in quadrant II, then

$$t = \pi - t'$$
$$= \pi - \frac{\pi}{3} = \frac{2\pi}{3}. \qquad \text{(See Figure 6.26.)}$$

FIGURE 6.27

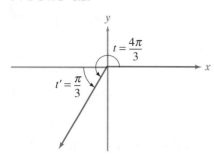

If t lies in quadrant III, then

$$t = \pi + t'$$
$$= \pi + \frac{\pi}{3}$$
$$= \frac{4\pi}{3}. \quad \text{(See Figure 6.27.)}$$

We next consider the equation (from statement (4))

$$3 \cos t - 2 = 0$$

which is equivalent to

$$\cos t = \frac{2}{3}.$$

Since $\cos t$ is positive, we know that t lies in either quadrant I or quadrant IV. In either case, the reference angle t' for t satisfies the equation

$$\cos t' = |\cos t| = \frac{2}{3}.$$

Hence, $\qquad\qquad t' = \cos^{-1} \frac{2}{3} \approx 0.84.$

If t lies in quadrant I, then

$$t = t' \approx 0.84.$$

On the other hand, if t lies in quadrant IV,

$$t = 2\pi - t'$$
$$\approx 6.28 - 0.84$$
$$= 5.44 \quad \text{(See Figure 6.28.)}$$

FIGURE 6.28

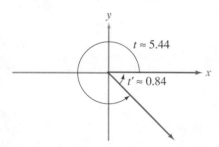

It follows that the solutions of the original equation that are in the interval $[0, 2\pi)$ are

$$\frac{2\pi}{3}, \qquad \frac{4\pi}{3}, \qquad 0.8, \qquad \text{and} \qquad 5.4,$$

with the approximate solutions rounded to one decimal place.

E X A M P L E   42

Find the solutions of the equation $\cos 2\theta = \sin \theta - 1$, where $0° \le \theta < 360°$. Round off any approximate solutions to one decimal place.

Solution   As in Example 40, the given equation

$$\cos 2\theta = \sin \theta - 1$$

involves both θ and 2θ. Using the identity $\cos 2\theta = 1 - 2 \sin^2 \theta$, we can obtain an equivalent equation involving only θ. Thus,

$$\cos 2\theta = \sin \theta - 1$$
$$1 - 2 \sin^2 \theta = \sin \theta - 1$$
$$2 \sin^2 \theta + \sin \theta - 2 = 0.$$

This last equation is a quadratic equation in $\sin \theta$. Using the quadratic formula, we find that

$$\sin \theta = \frac{-1 \pm \sqrt{17}}{4}.$$

If $\sin \theta = (-1 + \sqrt{17})/4 \approx 0.7808$, then θ lies in either quadrant I or quadrant II and, in either case, has

$$\theta' = \sin^{-1}(0.7808) \approx 51.3°$$

as its reference angle. If θ lies in quadrant I,

$$\theta = \theta' \approx 51.3°.$$

On the other hand, if θ lies in quadrant II,

$$\begin{aligned}
\theta &= 180° - \theta' \\
&\approx 180° - 51.3° \\
&= 128.7°. \qquad \text{(See Figure 6.29.)}
\end{aligned}$$

Since $\sin \theta = (-1 - \sqrt{17})/4 \approx -1.281$ is impossible (Why?), it follows that the solutions (rounded to one decimal place) of the original equation, that satisfy $0° \le \theta < 360°$, are

$$51.3° \qquad \text{and} \qquad 128.7°. \qquad\rule{2cm}{0.4pt}$$

When solving a trigonometric equation, it is sometimes helpful to square both sides, as the following example illustrates.

FIGURE 6.29

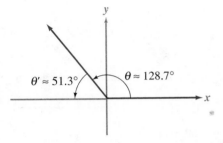

$\theta' \approx 51.3°$ $\theta \approx 128.7°$

EXAMPLE 43

Find the solutions of the equation $\cot x - \sqrt{3} = \csc x$, where x is in the interval $[0, 2\pi)$. Round off any approximate solutions to one decimal place.

Solution   Squaring both sides of the equation

$$\cot x - \sqrt{3} = \csc x, \qquad\qquad (5)$$

we have

$$(\cot x - \sqrt{3})^2 = \csc^2 x$$
$$\cot^2 x - 2\sqrt{3} \cot x + 3 = \csc^2 x.$$

Substituting $1 + \cot^2 x$ for $\csc^2 x$ in the last equation, we obtain

$$\begin{aligned}
\cot^2 x - 2\sqrt{3} \cot x + 3 &= 1 + \cot^2 x \\
-2\sqrt{3} \cot x &= -2 \\
\cot x &= \frac{1}{\sqrt{3}}.
\end{aligned}$$

Since $\cot x$ is positive, we know that x lies in either quadrant I or quadrant III, and since $\cot (\pi/3) = 1/\sqrt{3}$, we see that $x' = \pi/3$ is the reference angle for x in either case. Therefore,

$$x = \frac{\pi}{3} \qquad \text{or} \qquad x = \frac{4\pi}{3}.$$

However, since we squared both sides of the original equation, we must check both of these potential solutions in the original equation. Substituting $\frac{\pi}{3}$ for x on the left side of equation (5), we have

$$\cot x - \sqrt{3} = \cot \frac{\pi}{3} - \sqrt{3} = \frac{1}{\sqrt{3}} - \sqrt{3} = \frac{1 - 3}{\sqrt{3}} = -\frac{2}{\sqrt{3}},$$

while substitution on the right side of equation (5) yields

$$\csc x = \csc \frac{\pi}{3} = \frac{2}{\sqrt{3}}.$$

Since $-\frac{2}{\sqrt{3}} \neq \frac{2}{\sqrt{3}}$, $\frac{\pi}{3}$ is not a solution. Similarly, substituting $\frac{4\pi}{3}$ for x on both sides of equation (5), we have

$$\cot \frac{4\pi}{3} - \sqrt{3} = \frac{1}{\sqrt{3}} - \sqrt{3} = -\frac{2}{\sqrt{3}} \quad \text{and} \quad \csc \frac{4\pi}{3} = -\frac{2}{\sqrt{3}}.$$

Therefore, $\frac{4\pi}{3}$ is the only solution of the original equation in the interval $[0, 2\pi)$.

The last example in this section uses the following result:

Let a and b be real numbers, not both zero, and let $r = \sqrt{a^2 + b^2}$. Let u be the angle in standard position with the point $Q = (a, b)$ on its terminal side, where $0 \le u < 2\pi$. Then, for any angle v,

$$a \sin v + b \cos v = r \sin(u + v). \tag{6}$$

The verification of the result, which involves a variation of the steps shown in the solution of the example, is left as an exercise (see Exercise 70).

EXAMPLE 44

Find the solutions of the equation $-2 \sin x + 2 \cos x = 1$, where x is in the interval $[0, 2\pi)$. Round off any approximate solutions to one decimal place.

Solution We can use the result (6) as follows. Comparing the left side of the given equation

$$-2 \sin x + 2 \cos x = 1 \tag{7}$$

with the left side of the equation

$$a \sin v + b \cos v = r \sin(u + v) \tag{8}$$

from result (6), we see that

$$a = -2, \quad b = 2, \quad \text{and} \quad v = x.$$

In Figure 6.30, angle u is in standard position with the point

$$Q = (a, b) = (-2, 2)$$

on its terminal side. From Figure 6.30 we see that u lies in quadrant II and that

$$r = \sqrt{a^2 + b^2} = \sqrt{(-2)^2 + (2)^2} = \sqrt{8} = 2\sqrt{2}$$

is the distance from O to Q.

FIGURE 6.30

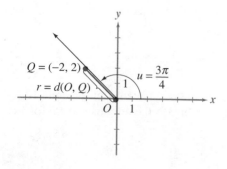

Therefore, using Theorem 6.2 (see page 253), we can write

$$\cos u = \frac{a}{r} = \frac{-2}{2\sqrt{2}} = -\frac{1}{\sqrt{2}} \qquad \text{and}$$

$$\sin u = \frac{b}{r} = \frac{2}{2\sqrt{2}} = \frac{1}{\sqrt{2}}.$$

Since $\cos u = -1/\sqrt{2}$ and $\sin u = 1/\sqrt{2}$, it follows that u has $\pi/4$ as its reference angle. Because $0 \le u < 2\pi$ and u lies in quadrant II, we know that $u = 3\pi/4$. Using equation (8), we can now rewrite the left side of equation (7):

$$-2 \sin x + 2 \cos x = r \sin(u + v)$$
$$= 2\sqrt{2} \sin\left(\frac{3\pi}{4} + x\right).$$

It thus follows from equation (7) that

$$2\sqrt{2} \sin\left(\frac{3\pi}{4} + x\right) = 1,$$

or $$\sin\left(\frac{3\pi}{4} + x\right) = \frac{1}{2\sqrt{2}}. \qquad (9)$$

We can solve equation (9) by letting $\theta = (3\pi/4) + x$ and solving the equation

$$\sin \theta = \frac{1}{2\sqrt{2}}. \qquad (10)$$

Since $\sin \theta$ is positive, θ lies in either quadrant I or quadrant II and, in either case, has

$$\theta' = \sin^{-1}\left(\frac{1}{2\sqrt{2}}\right) \approx 0.36$$

as its reference angle. If θ lies in quadrant I,

$$\theta = \theta' \approx 0.36.$$

On the other hand, if θ lies in quadrant II,

$$\theta = \pi - \theta'$$
$$\approx 3.14 - 0.36$$
$$= 2.78.$$

Thus the solutions of equation (10) are

$$\theta \approx 0.36 \qquad \text{or} \qquad \theta \approx 2.78.$$

Hence, $$\frac{3\pi}{4} + x \approx 0.36 \qquad \text{or} \qquad \frac{3\pi}{4} + x \approx 2.78$$

$$x \approx -2.00 \qquad \text{or} \qquad x \approx 0.42.$$

Notice that 0.42 is in the interval $[0, 2\pi)$, but -2.00 is not in this interval. Since there are two solutions of equation (9) in the interval $[0, 2\pi)$, we must find the value of x in $[0, 2\pi)$ that corresponds to the solution -2.00 found above. Using the methods we learned in Section 5.3 (see page 229), we know that the function

$$f(x) = \sin\left(\frac{3\pi}{4} + x\right)$$

from equation (9) has period 2π. Therefore, the value of x in $[0, 2\pi)$ that corresponds to the solution -2.00 is approximately

$$-2.00 + 2\pi \approx 4.28.$$

It follows that the solutions of the original equation in the interval $[0, 2\pi)$, rounded off to one decimal place, are

$$4.3 \quad \text{and} \quad 0.4. \quad \underline{\hspace{3cm}}$$

EXERCISES 6.6

In Exercises 1–6, use Theorem 6.3 to express the value of the given function of the given angle θ in terms of the value of the same function of its reference angle θ'. (Refer to Example 37.)

1. $\sin 200°$
2. $\cos 253°$
3. $\cos(-256°)$
4. $\sin(-460°)$
5. $\tan 11.2$
6. $\tan 1.68$

In Exercises 7–20, find the solution set S of each equation. Express S in a form similar to that used in Example 36.

7. $\sin x = \frac{1}{2}$
8. $\sin x = -\sqrt{3}/2$
9. $\tan x = 1$
10. $\cot x = -1$
11. $\sin x = -\sqrt{2}/2$
12. $\csc x = -\sqrt{2}$
13. $\cos x = -\sqrt{3}/2$
14. $\sec x = 2/\sqrt{3}$
15. $\tan x = -1/\sqrt{3}$
16. $\cos x = -\frac{1}{2}$
17. $\sin x = 1$
18. $\tan x = \sqrt{3}$
19. $\sec x = -\sqrt{2}$
20. $\sin x = -1$

In Exercises 21–30, find the solution set for each equation where t is in the interval $[0, 2\pi)$.

21. $\cos t = -\frac{1}{2}$
22. $\sin t = \sqrt{2}/2$
23. $\tan 2t = -1$
24. $\tan 2t = -1/\sqrt{3}$
25. $\sin t = \sqrt{3}/2$
26. $\cos 3t = -1$
27. $\cos 2t = 0$
28. $\sin 2t = 0$
29. $\sec t = -2/\sqrt{3}$
30. $\csc t = -2$

In Exercises 31–40, find the solution set for each equation where $0° \le \theta < 360°$.

31. $\sin \theta = -\sqrt{3}/2$
32. $\cos \theta = -\frac{1}{2}$
33. $\cos \theta = \sqrt{2}/2$
34. $\sin \theta = -\sqrt{2}/2$
35. $\tan 2\theta = \sqrt{3}$
36. $\cot 2\theta = -\sqrt{3}$
37. $\csc 2\theta = -2$
38. $\cot^2 \theta = \cot \theta$
39. $\sin 2\theta = \sqrt{2} \cos \theta$
40. $\sin 2\theta = -\sqrt{3} \cos \theta$

In Exercises 41–68, find the solutions of each equation that are in the interval $[0, 2\pi)$. Round off any approximate solutions to one decimal place.

41. $\sin 3t = -1$
42. $\cos 3t = 0$
43. $\sin 2t \sin t = -\cos t$
44. $\tan^2 x = \tan x$
45. $\sin 2t = \sqrt{2} \sin t$
46. $\cot^2 t = 1$
47. $\sin 2x = \cos x$
48. $\tan^2 t = -\tan t$
49. $2 \sin^2 t + \sin t = 0$
50. $2 \cos^2 x - 3 \cos x = 0$
51. $2 \cos^2 t - 9 \cos t - 5 = 0$
52. $\cot^4 x - 1 = 0$
53. $2 \sin^2 x - 5 \sin x - 3 = 0$
54. $6 \sin^2 x - \sin x - 1 = 0$
55. $6 \cos^2 t - \sin t - 4 = 0$
56. $10 \cos^2 x + 3 \cos x - 1 = 0$
57. $\sin 2t \cos t = \sin t$
58. $6 \sin^2 x + \sin x - 1 = 0$
59. $\cos 2x = 4 \sin x$
60. $15 \tan^2 x + 2 \tan x - 1 = 0$
61. $3 \tan^2 t - 5 \tan t + 1 = 0$
62. $10 \cos^2 x - 3 \cos x - 1 = 0$
63. $\tan t + \sqrt{3} = \sec t$
64. $1 + \sin x = \cos x$
65. $\sin x + \sqrt{3} \cos x = 1$
66. $\sqrt{3} \cos x - \sin x = 1$
67. $\sin x - \cos x = 1$
68. $\sec x - \tan x = -\sqrt{2}$

69. Complete the proof of Theorem 6.3 by considering the cases when $0 < \theta < \pi/2$, $\pi < \theta < 3\pi/2$, and $3\pi/2 < \theta < 2\pi$.

70. Verify the result (6), which is stated on page 296 just before Example 44.

71. The current in an electrical circuit is given by the function

$$I = f(t) = 60 \cos(120\pi t - \pi),$$

where I is measured in amperes and t is in seconds. Find the first time when $I = 20$ amperes. Round off the solution to two decimal places.

72. A buoy oscillates in a simple harmonic motion, which is described by the function

$$g(t) = 200 \sin(30\pi t - 3),$$

where t is in seconds. Find the second time when $g(t) = 50$. Round off the solution to two decimal places.

Photo © Fredrik D. Bodin

73. The displacement d of an ocean wave is represented by the equation

$$d = 2 \sin\left[\frac{2\pi}{3}(6 - 4t)\right],$$

where d is in meters and t is in seconds.
a. Find the first time when $d = 2$.
b. Find the second time when $d = 1$.
Round off the solutions to two decimal places.

74. A point on the end of a tuning fork moves in a simple harmonic motion, which is given by the function

$$d = f(t) = 5 \cos\left(\frac{2\pi}{3}t - 3\right),$$

where t is in seconds and d is in centimeters.
a. Find the second time when $d = 2$.
b. Find the first time when $d = 3$.
Round off the solutions to two decimal places.

75. An object of mass 10 kilograms suspended from a spring with spring constant $k = 3$ oscillates in a harmonic motion described by the function

$$f(t) = 8 \sin\left(\sqrt{\frac{3}{10}}\,t\right)$$

where t is in seconds.
a. Find the first time when $f(t) = 6$.
b. Find the second time when $f(t) = 3$.
Round off the solutions to two decimal places.

According to **Snell's law** from physics, if a ray of light striking a surface between two mediums forms an **angle of incidence** θ_1 with the perpendicular to the surface (see the accompanying figure), the ray will be bent, or **refracted**, so that it forms an **angle of refraction** θ_2 with the perpendicular to the surface.

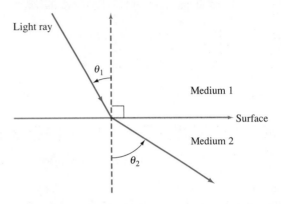

Snell's law states that the ratio of the sine of the angle of incidence θ_1 to the sine of the angle of refraction θ_2 is a constant N (called the **index of refraction**) for any two mediums. Thus,

$$\frac{\sin \theta_1}{\sin \theta_2} = N.$$

In Exercises 76–80, use Snell's law to solve each. Round off all answers to the nearest tenth of a degree.

76. A ray of light travels from air into water. Find the angle of refraction of the ray if the angle of incidence is 25° and the index of refraction is 1.33.

77. A ray of light travels from air into ethyl alcohol. Find the angle of incidence of the ray if the angle of refraction is 17.6° and the index of refraction is 1.36.

78. A ray of light travels from air into flint glass. Find the angle of refraction of the ray if the angle of incidence is 10.2° and the index of refraction is 1.65.

79. A ray of light travels from air into a diamond. Find the angle of incidence of the ray if the angle of refraction is 23.7° and the index of refraction is 2.42.

80. A ray of light travels from air into carbon disulfide. Find the angle of refraction of the ray if the angle of incidence is 50.3° and the index of refraction is 1.63.

81. Graph $y = 3 \cos x$ and $y = 2$ over the interval $[0, 2\pi)$ in the same coordinate system. Estimate the x-coordinates of the points of intersection of the two graphs (to the nearest tenth). Compare your answers to two of those obtained in Example 41.

82. Graph $y = 2 \cos x$ and $y = -1$ over the interval $[0, 2\pi)$ in the same coordinate system. Estimate the x-coordinates of the points of intersection of the two graphs (to the nearest tenth). Compare your answers to two of those obtained in Example 41.

83. Graph $y = 4 \sin x$ and $y = 3$ over the interval $[0, 2\pi)$ in the same coordinate system. Estimate the *x*-coordinates of the points of intersection of the two graphs (to the nearest tenth). Use the results to approximate the solutions in $[0, 2\pi)$ to the equation $\sin x = \frac{3}{4}$.

84. Graph $y = 2 \cos x$ and $y = -0.3$ over the interval $[0, 2\pi)$ in the same coordinate system. Estimate the *x*-coordinates of the points of intersection of the two graphs (to the nearest tenth). Use the results to approximate the solutions in $[0, 2\pi)$ to the equation $\cos x = -0.15$.

85. Graph $y = \cos x$ and $y = e^{-x/5}$ over the interval $[0, 2\pi)$ in the same coordinate system. Estimate the *x*-coordinates of the points of intersection of the two graphs (to the nearest tenth). Use the results to approximate the solutions in $[0, 2\pi)$ to the equation $\cos x = e^{-x/5}$.

86. Graph $y = \sin x$ and $y = \ln x$ over the interval $(0, 2\pi)$ in the same coordinate system. Estimate the *x*-coordinates of the points of intersection of the two graphs (to the nearest tenth). Use the results to approximate the solutions in $(0, 2\pi)$ to the equation $\sin x = \ln x$.

CHAPTER 6 REVIEW EXERCISES

1. Express the given degrees-minutes-seconds measurement in decimal degrees to the nearest thousandth: $202°43'29''$.

2. Express the given decimal degree measurement in degrees, minutes, and seconds: $112.453°$.

In Exercises 3–6, convert the indicated degree measure to radians. Sketch each angle in standard position.

3. $100°$ **4.** $210°$

5. $-225°$ **6.** $-120°$

In Exercises 7–10, convert the indicated radian measure to decimal degrees to the nearest thousandth. Sketch each angle in standard position.

7. $-\dfrac{7\pi}{10}$ **8.** $-\dfrac{3\pi}{8}$

9. $\dfrac{13\pi}{24}$ **10.** $\dfrac{13\pi}{15}$

In Exercises 11–12, find the length of the arc and the area of the circular sector subtended by the indicated central angle θ in a circle with the indicated radius r.

11. $r = 12$ cm; $\theta = 35°$

12. $r = 8''$; $\theta = 9$

In Exercises 13–16, find the *exact* value of each expression.

13. $\sin(-210°)$ **14.** $\csc(-60°)$

15. $\cos 315°$ **16.** $\tan 240°$

In Exercises 17–20, use the given information and Theorem 6.2 to find the *exact* values of the other five trigonometric functions of θ. Sketch θ where $0° \le \theta < 360°$.

17. $\tan \theta = \frac{3}{2}$; θ lies in quadrant III

18. $\cos \theta = -\frac{3}{4}$; θ lies in quadrant II

19. $\sin \theta = -\frac{4}{9}$; θ lies in quadrant IV

20. $\sec \theta = -5$; θ lies in quadrant III

21. How many revolutions will a bicycle wheel of diameter $36''$ make as the wheel travels 1 mile?

22. Chicago, Illinois and Boston, Massachusetts are approximately 1000 miles apart. Assuming that the radius of the earth is approximately 3960 miles, find the approximate radian measure of the central angle joining these two cities.

23. The minute hand of a certain clock is 12 inches long. How far does the tip of the minute hand travel in 45 minutes?

24. A tire is suspended by a chain from a tree. As the tire moves from its highest forward position to its highest backward position, the central angle subtended by the chain is $7°$ and the length of the subtended arc is $\pi/2$ feet. How long is the chain?

25. If a central angle θ of measure $18°$ subtends an arc of length 30 centimeters on a circle of radius r, find r.

In Exercises 26–27, for a wheel of the indicated radius r, which is rotating at the indicated rate, find both the linear velocity and the angular velocity of the wheel.

26. $r = 5$ feet; 1000 rpm   **27.** $r = 6$ inches; 1500 rpm

28. A wheel 18 inches in diameter is rotating at a rate of 50 revolutions per minute. Find the velocity in feet per second of a point on the rim of the wheel.

29. The outside diameter of a bicycle wheel is 26 inches. Find the number of revolutions the wheel makes in 1 second if the bicycle is traveling at 5 miles per hour.

30. An automobile fan belt connects a pulley with a 12 centimeter diameter and a pulley with an 18 centimeter diameter. As the larger pulley turns through 10 radians, through how many radians does the smaller pulley turn?

In Exercises 31–34, use the basic identities to simplify each expression.

31. $(\cot t + \csc t)(\cot t - \csc t)$

32. $\csc t(\tan t - \sin t)$

33. $\dfrac{1}{1 + 2 \tan^2 t + \tan^4 t}$

34. $\dfrac{1}{1 + \sec t} + \dfrac{1}{1 - \sec t}$

In Exercises 35–36, show that each equation is *not* an identity by finding a value of x for which the equation is false.

35. $\cos x = \sqrt{1 - \sin^2 x}$ **36.** $\sin x + \cos x = 1$

In Exercises 37–38, use the indicated trigonometric substitution to simplify each expression.

37. $\sqrt{9 + u^2}$; $\quad u = 3 \tan \theta$; $\quad -\dfrac{\pi}{2} < \theta < \dfrac{\pi}{2}$

38. $\dfrac{2}{\sqrt{25 - x^2}}$; $\quad x = 5 \sin \theta$; $\quad -\dfrac{\pi}{2} < \theta < \dfrac{\pi}{2}$

In Exercises 39–48, use the appropriate sum or difference formula to find the exact value of each expression.

39. $\cos (7\pi/12)$

40. $\sin(-15°)$

41. $\sin 195°$

42. $\tan (5\pi/12)$

43. $\sin (2\pi/3) \cos (\pi/6) - \cos (2\pi/3) \sin (\pi/6)$

44. $\cos 40° \cos 10° + \sin 40° \sin 10°$

45. $\cos 201° \cos 24° - \sin 201° \sin 24°$

46. $\dfrac{\tan 200° - \tan 80°}{1 + \tan 200° \tan 80°}$

47. $\sin 32° \cos 58° + \cos 32° \sin 58°$

48. $\dfrac{\tan 110° + \tan 100°}{1 - \tan 110° \tan 100°}$

In Exercises 49–50, without using a calculator, use the given information to find each of the following:

a. $\sin(u + v)$ **b.** $\cos(u + v)$

c. $\sin(u - v)$ **d.** $\cos(u - v)$

e. $\tan(u + v)$ **f.** $\tan(u - v)$

g. The quadrant containing $u + v$

h. The quadrant containing $u - v$

49. $\cos u = -\frac{1}{4}$ with u in quadrant III; $\quad \sin v = \sqrt{2}/3$ with v in quadrant II

50. $\tan u = \frac{1}{2}$ with u in quadrant III; $\quad \cos v = \frac{1}{5}$ with v in quadrant IV

In Exercises 51–54, use a sum or difference identity to simplify each expression.

51. $\sin(270° + \theta)$

52. $\csc(180° - v)$

53. $\cos[u - (\pi/6)]$

54. $\tan[u + (\pi/3)]$

In Exercises 55–56, find the exact value of each expression. Also approximate each answer to two decimal places.

55. $\cos[\sin^{-1}(\frac{1}{2}) - \tan^{-1}(-\frac{1}{4})]$

56. $\sin[\sin^{-1}(-\frac{2}{3}) + \cos^{-1}(-\frac{1}{2})]$

In Exercises 57–60, use the appropriate double-angle or half-angle formula to write each expression as a single trigonometric function value.

57. $2 \sin 25 \cos 25$

58. $2 \cos^2 83 - 1$

59. $1 - 2 \sin^2 23°$

60. $\dfrac{\sin 100°}{1 + \cos 100°}$

In Exercises 61–62, use the given information to find the exact value of sin 2v, cos 2v, and tan 2v. Also, determine the quadrant in which 2v lies.

61. $\sin v = -\sqrt{2}/5$ and v is in quadrant IV

62. $\tan v = \frac{1}{2}$ and v is in quadrant III

In Exercises 63–66, find the exact value of each expression.

63. $\sin (5\pi/8)$

64. $\cos (\pi/8)$

65. $\tan 67.5°$

66. $\cos(-22.5°)$

In Exercises 67–68, use the given information to find the exact value of sin(v/2), cos(v/2), and tan(v/2).

67. $\cos v = \frac{1}{6}$ and $3\pi/2 < v < 2\pi$

68. $\tan v = 4$ and $\pi < v < 3\pi/2$

In Exercises 69–70, find the exact value of each expression.

69. $\cos[2 \cos^{-1}(-\frac{1}{2})]$

70. $\sin[2 \sin^{-1}(\frac{3}{4})]$

In Exercises 71–74, express each product as a sum or difference.

71. $\sin 3x \sin 2x$

72. $\cos 8v \cos 2v$

73. $\cos 50° \sin 20°$

74. $\sin 10° \cos 15°$

In Exercises 75–78, write each expression as a product.

75. $\cos 6\theta + \cos 3\theta$

76. $\sin 10\theta + \sin 5\theta$

77. $\sin 62° - \sin 12°$

78. $\cos 38° - \cos 20°$

In Exercises 79–82, use the appropriate sum-to-product or product-to-sum identity to find the exact value of each expression.

79. $\sin 105° - \sin 195°$

80. $\sin 195° + \sin 285°$

81. $\cos (7\pi/8) \cos (\pi/8)$

82. $\sin (3\pi/8) \cos (\pi/8)$

In Exercises 83–86, find the solution set S of each equation. Express S in a form similar to that used in Example 36.

83. $\sin x = -\frac{1}{2}$

84. $\cos x = -\sqrt{3}/2$

85. $\tan x = -1$

86. $\cot x = -\sqrt{3}$

In Exercises 87–90, find the solution set for each equation where t is in the interval $[0, 2\pi)$.

87. $\cos 2t = -1$

88. $\cos 3t = -\frac{1}{2}$

89. $\csc t = -\sqrt{2}$

90. $\sin 2t = -\sqrt{3}/2$

In Exercises 91–94, find the solution set for each equation where $0° \le \theta < 360°$.

91. $\sin 3\theta = -\sqrt{2}/2$

92. $\cos \theta = -1/\sqrt{2}$

93. $\sin 2\theta = -\sqrt{2} \sin \theta$

94. $\sin 2\theta = \sqrt{3} \cos \theta$

In Exercises 95–104, find the solutions of each equation that are in the interval $[0, 2\pi)$. Round off any approximate solutions to one decimal place.

95. $\sin 3t = -1$

96. $\cos 3t = -\sqrt{3}/2$

97. $2 \sin^2 x + \sin x = 0$

98. $2 \cos^2 x - 5 \cos x - 3 = 0$

99. $3 \cot^2 t - 1 = 0$

100. $1 + \sin x = \cos x$

101. $2 \tan^2 x - 3 \tan x - 15 = 0$

102. $15 \cos^2 t - 2 \cos t - 1 = 0$

103. $2 \sin^2 t - 9 \sin t - 5 = 0$

104. $\cot x - \sqrt{3} = -\csc x$

105. If the current in an electrical circuit is given by the function

$$I = f(t) = 50 \sin(100\pi t - 3\pi),$$

where I is measured in amperes and t is in seconds, find the first time when $I = 30$ amperes. Round off the solution to two decimal places.

106. A buoy oscillates in a simple harmonic motion, which is described by the function

$$g(t) = 100 \cos \left(\tfrac{2}{3} \pi t - 1\right),$$

where t is in seconds. Find the second time when $g(t) = 25$. Round off the solution to two decimal places.

107. A point on the end of a tuning fork moves in a simple harmonic motion, which is given by the function

$$d = f(t) = 6 \sin \left(\frac{3\pi}{4} t - 2\right),$$

where t is in seconds and d is in centimeters. Find the first time when $d = 4$. Round off the solution to two decimal places.

In Exercises 108–109, use Snell's law to solve each. Round off all answers to the nearest tenth of a degree.

108. A ray of light travels from air into sodium chloride. Find the angle of refraction of the ray if the angle of incidence is $34°$ and the index of refraction is 1.53.

109. A ray of light travels from air into a quartz crystal. Find the angle of incidence of the ray if the angle of refraction is $22.7°$ and the index of refraction is 1.45.

In Exercises 110–132, verify each identity.

110. $\dfrac{\cos t}{1 - \cos^2 t} = \dfrac{\csc t}{\tan t}$

111. $\dfrac{1 + \tan t}{\csc t + \sec t} = \sin t$

112. $\dfrac{1}{\sin x \cos x} = \tan x + \cot x$

113. $(1 - \cos^2 t) \csc t = \sin t$

114. $\dfrac{[1 + \sin(-v)](1 + \sin v)}{1 - \cos^2 v} = \cot^2 v$

115. $\dfrac{\tan^2 x}{1 + \tan^2 x} = \sin^2 x$

116. $\dfrac{\sec t + 1}{\tan t} = \cot t + \csc t$

117. $\dfrac{\sec x + 1}{\tan^2 x} = \dfrac{\cos x}{1 - \cos x}$

118. $\dfrac{\sec x}{1 + \sec x} = \dfrac{1 - \cos x}{\sin^2 x}$

119. $\dfrac{\cos v}{1 + \sin v} = \dfrac{\sin(-v) + 1}{\cos v}$

120. $\dfrac{1}{(\sec x - \tan x)^2} = \dfrac{1 + \sin x}{1 - \sin x}$

121. $\dfrac{\cos^3 t + \sin^3 t}{\cos t + \sin t} = 1 - \tfrac{1}{2} \sin 2t$

122. $\dfrac{\sin^3 v}{\tan v - \sin v} = \dfrac{1 + \cos v}{\sec v}$

123. $\dfrac{1 + \sin(-x)}{\cos(-x)} = \dfrac{1}{\sec x + \tan x}$

124. $\cot t \csc t = \dfrac{1}{(1 + \sec t)(1 - \cos t)}$

125. $\dfrac{\cos \theta + \sin \theta - 1}{\cos \theta - \sin \theta + 1} = \dfrac{\sin \theta}{1 + \cos \theta}$

126. $\dfrac{\cot x + \cot y}{\tan x + \tan y} = \dfrac{1 - \cot x \cot y}{\tan x \tan y - 1}$

127. $\sin(2\pi - t) = \cos \left(\dfrac{\pi}{2} + t\right)$

128. $\dfrac{\sin(v - u)}{\sin u \sin v} = \cot u + \cot(-v)$

129. $\dfrac{\cos(u + v) + \cos(u - v)}{\sin(u + v) + \sin(u - v)} = \cot u$

130. $\dfrac{\sin(u + v)}{\sin(u - v)} = \dfrac{1 + \cot u \tan v}{1 - \cot u \tan v}$

131. $\dfrac{\sin 3x + \sin 5x}{\cos 3x - \cos 5x} = \cot x$

132. $-\dfrac{\cos 10x - \cos 6x}{\sin 10x - \sin 6x} = \dfrac{\sin 6x + \sin 10x}{\cos 6x + \cos 10x}$

CHAPTER **7**

Applications of
Trigonometry

In this chapter, we explore a variety of applications using the tools of trigonometry developed in the previous two chapters. We first consider problems whose solutions require us to find the sides and angles of a right triangle. We then solve problems involving triangles that do not contain a right angle by developing the Law of Sines and the Law of Cosines. After using trigonometry to develop methods for determining the roots and powers of a complex number, we examine vectors in the plane from both a geometric and algebraic standpoint. In the final section of the chapter, we define the polar coordinate system and discuss the graphs of polar equations.

North Wind Picture Archives

PYTHAGORAS
(ca. 582–ca. 500 B.C.)

Born on the Greek island of Samos, Pythagoras eventually settled in Croton, which is located in what is now southern Italy. There, he established a secret society (the Pythagoreans) dedicated to the study of mathematics and religious philosophy. Pythagoras directed his disciples to focus on four subjects he considered to be fundamental for an educated person: number theory, music, astronomy, and geometry. Due to the secrecy of the Pythagoreans, it is difficult to attribute credit to any individual for a particular discovery. However, the society attributed the proof of the Pythagorean theorem to their leader, who is said to have sacrificed one hundred oxen to the gods as a token of gratitude. In addition to the famous theorem, Pythagoras is credited with discovering irrational numbers, developing the theory of proportionals, and introducing the word *proof* into mathematical reasoning.

7.1   RIGHT TRIANGLE TRIGONOMETRY

A cruise ship leaves Miami at 11:00 A.M. and sails due east for $2\frac{1}{4}$ hours. The ship then changes course and sails due south until 7:00 P.M., when it radios the coast guard that its engine has failed. If the ship has maintained a constant cruising speed of 40 kilometers/hour throughout its voyage, how long will it take a rescue helicopter, flying from Miami at 130 kilometers/hour, to reach the disabled vessel?

This problem is typical of a large class of physical problems whose solutions involve applications of trigonometry in finding the measures of sides and angles of a triangle. In this section we investigate problems whose solutions involve right triangles, and in Section 7.2 we will consider problems involving triangles that do not contain a right angle.

A triangle whose vertices are the points A, B, and C will be called *triangle ABC* and will be labeled as shown in Figure 7.1. We will refer to α, β, γ, a, b, and c as the *parts* of triangle ABC. When triangle ABC is a right triangle, it is customary to place the right angle at vertex C. Thus $\gamma = 90°$ and right triangle ABC is labeled as shown in Figure 7.2.

When solving problems that involve right triangles, it is helpful for us to view the definitions of the trigonometric functions in a different light. We begin by placing right triangle ABC in a rectangular coordinate system, so that angle α is in standard position and vertex C lies on the positive x-axis. Notice, in Figure 7.3, that the point $B = (b, a)$ lies on the terminal side of angle α, and that $d(A, B) = c$. If we denote a and b (the lengths of the sides opposite and adjacent to angle α) by opp and adj, and c (the length of the hypotenuse) by hyp, we see from Theorem 6.2 that

FIGURE 7.1   Standard labeling for triangle *ABC*

Vertex	Angle	Length of Opposite Side
A	α	a
B	β	b
C	γ	c

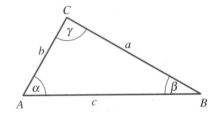

FIGURE 7.2
Standard labeling for right triangle *ABC*

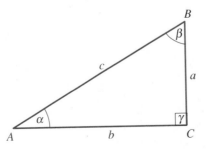

FIGURE 7.3

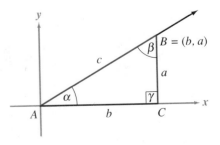

$$\sin \alpha = \frac{a}{c} = \frac{\text{opp}}{\text{hyp}}; \qquad \csc \alpha = \frac{c}{a} = \frac{\text{hyp}}{\text{opp}};$$

$$\cos \alpha = \frac{b}{c} = \frac{\text{adj}}{\text{hyp}}; \qquad \sec \alpha = \frac{c}{b} = \frac{\text{hyp}}{\text{adj}};$$

$$\tan \alpha = \frac{a}{b} = \frac{\text{opp}}{\text{adj}}; \qquad \cot \alpha = \frac{b}{a} = \frac{\text{adj}}{\text{opp}}.$$

Thus the trigonometric functions of an acute angle of a right triangle can be defined in terms of ratios of lengths of sides of the triangle. We generalize this result for an acute angle θ of an arbitrary right triangle as follows:

Trigonometric Functions of an Acute Angle θ of a Right Triangle

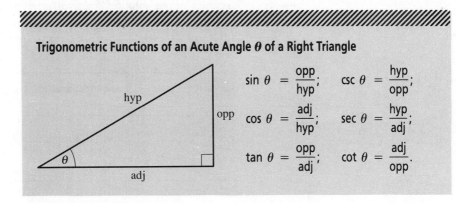

$$\sin \theta = \frac{\text{opp}}{\text{hyp}}; \qquad \csc \theta = \frac{\text{hyp}}{\text{opp}};$$

$$\cos \theta = \frac{\text{adj}}{\text{hyp}}; \qquad \sec \theta = \frac{\text{hyp}}{\text{adj}};$$

$$\tan \theta = \frac{\text{opp}}{\text{adj}}; \qquad \cot \theta = \frac{\text{adj}}{\text{opp}}.$$

Solving Right Triangles

If the measures of two sides, or the measures of one side and an acute angle of a right triangle are known, we can always find the measures of the remaining sides and angles of the triangle. We illustrate with several examples.

EXAMPLE 1

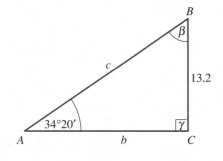

In right triangle ABC, $\alpha = 34°20'$, and $a = 13.2$. Find the remaining parts of the triangle.

Solution We begin by sketching right triangle ABC and labeling its parts. Since $\alpha + \beta = 90°$, it follows that

$$\beta = 90° - \alpha = 90° - 34°20' = 55°40'.$$

We can find b by using the fact that $\tan \beta = \text{opp}/\text{adj} = b/a$ to write

$$b = a \tan \beta = 13.2 \, (\tan 55°40') \approx 19.3.$$

We note that there are several other ways to find b. For example, we could use any one of the following relationships:

$$\cot \beta = \frac{\text{adj}}{\text{opp}} = \frac{a}{b}, \qquad \tan \alpha = \frac{\text{opp}}{\text{adj}} = \frac{a}{b}, \qquad \text{or} \qquad \cot \alpha = \frac{\text{adj}}{\text{opp}} = \frac{b}{a}.$$

To find c, we can use $\sin \alpha = \text{opp}/\text{hyp} = a/c$ to obtain

$$c = \frac{a}{\sin \alpha} = \frac{13.2}{\sin 34°20'} \approx 23.4.$$

We can also find c by using the Pythagorean theorem:

$$c = \sqrt{a^2 + b^2} = \sqrt{(13.2)^2 + (19.3)^2} \approx 23.4.$$

In the solution of Example 1, we were required to find the values of tan 55°40′ and sin 34°20′. These values can be found by using a calculator or by using the table of trigonometric function values in Appendix B. Since the table gives function values rounded to four significant digits, while calculators usually provide eight or ten significant digits, our choice of table or calculator can have a marked influence on the accuracy of our final answer. Moreover, the accuracy of our solution is also affected by the number of significant digits in the measures of any given side lengths or angles. As a matter of policy, we will use Table 7.1 as a guide when we perform our calculations.

TABLE 7.1

Number of Significant Digits in Sides	Round Angle Measurement to Nearest
2	1°
3	0.1° or 10′
4	0.01° or 1′
5	0.001° or 10″

EXAMPLE 2

In right triangle ABC, $b = 13,560$ and $c = 22,420$. Find the remaining parts of the triangle.

Solution   First we sketch and label triangle ABC. Since cos α = adj/hyp = 13,560/22,420 ≈ 0.6048, it follows that

$$\alpha = \cos^{-1} 0.6048 \approx 52.79°.$$

Subtracting angle α from 90°, we find that

$$\beta \approx 90° - 52.79° = 37.21°.$$

To find a, we note that

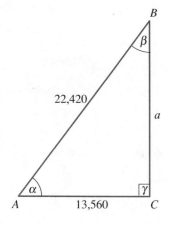

$$\sin \alpha = \frac{\text{opp}}{\text{hyp}} = \frac{a}{c}.$$

Thus,

$$\sin 52.79° = \frac{a}{22,420}, \qquad \text{or}$$

$$a = 22,420 \sin 52.79° \approx 17,860.$$

Applications

We now answer the question posed at the beginning of this section:

A cruise ship leaves Miami at 11:00 A.M. and sails due east for $2\frac{1}{4}$ hours. The ship then changes course and sails due south until 7:00 P.M., when it radios the Coast Guard station at Miami that its engine has failed. If the ship has maintained a constant cruising

FIGURE 7.4

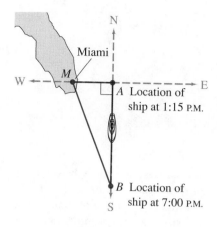

A Location of ship at 1:15 P.M.

B Location of ship at 7:00 P.M.

speed of 40 kilometers/hour throughout its voyage, how long will it take a rescue helicopter flying from Miami at 130 kilometers/hour, to reach the disabled vessel?

In Figure 7.4, *M* denotes Miami, while *A* and *B* denote the location of the ship at 1:15 P.M. (time the ship changes course) and at 7:00 P.M., respectively. To solve this problem we must find the distance between *M* and *B*, which is the length of the hypotenuse of right triangle *MAB*. We begin by finding the lengths of *MA* and *MB*, the legs of this triangle. Since the ship has maintained a constant cruising speed of 40 kilometers/hour, we can use the relationship distance = rate · time, to obtain

length of *MA* = (40 kilometers/hour)(2.25 hour) = 90 kilometers   and
length of *AB* = (40 kilometers/hour)(5.75 hour) = 230 kilometers.

Thus, by the Pythagorean theorem,

$$\text{length of } MB = \sqrt{(90)^2 + (230)^2} \approx 250 \text{ kilometers.}$$

Since the rescue helicopter flies at 130 kilometers/hour, its trip from Miami will take $\frac{250}{130} \approx 1.9$ hours, or approximately 1 hour and 54 minutes.

EXAMPLE 3

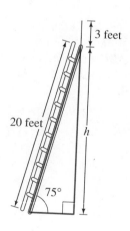

A safety label on a 20-foot extension ladder warns that the ladder is unsafe if the angle it makes with the ground exceeds 75°. If a painter can reach 3 feet above the top of the ladder when it leans against a wall, what is the maximum height this ladder will allow the painter to reach?

Solution   Assuming the ladder is placed on level ground, we can sketch a right triangle that displays the given information. Since sin 75° = opp/hyp = *h*/20, it follows that

$$h = 20(\sin 75°) \approx 19 \text{ feet.}$$

Thus the maximum height the painter can reach is approximately 19 feet + 3 feet = 22 feet.

In road and bridge construction, distances and angles must be measured with considerable precision. To make such measurements, engineers and surveyors often use an instrument called a *transit*. By making use of a telescope, a level, and a device called a vernier [named for French mathematician and inventor Pierre Vernier (1580–1637)], a transit can measure vertical and horizontal angles to the nearest 10″.

A transit is especially useful in determining distances that are difficult to measure directly, such as the distances

Photo © Fredrik D. Bodin

FIGURE 7.5 Angle of elevation and angle of depression

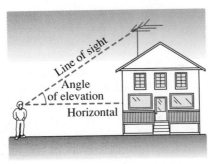
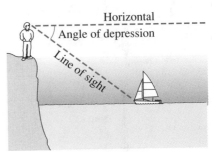

across canyons, rivers, and lakes. In such instances, an angle between the line-of-sight and the horizontal is often used to measure the desired distance. Such an angle is called either the **angle of elevation** or the **angle of depression**, depending on whether the line-of-sight is above or below the horizontal (see Figure 7.5).

EXAMPLE 4

From a helicopter hovering at 5500 feet above a lake, a civil engineer finds that the angles of depression to points A and B on opposite sides of a lake below are 67.8° and 42.1°, respectively (see Figure 7.6). Approximate the distance across the lake.

Solution   Let C denote the position of the helicopter, and let D denote the point on the ground directly under the helicopter. We must find y, the distance across the lake. From Figure 7.6, we obtain right triangles ADC and BDC shown in Figure 7.7. In right triangle ADC, we see from Figure 7.6, that $\angle ACD$ has measure $90° - 67.8° = 22.2°$. Using Figure 7.7, we see that

FIGURE 7.6

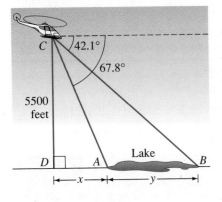

$$\tan(\angle ACD) = \tan 22.2° = \frac{\text{opp}}{\text{adj}} = \frac{x}{5500},$$

which means that

$$x = 5500(\tan 22.2°) \approx 2240.$$

We next consider right triangle BDC. Again, using Figure 7.6, we see that $\angle BCD = 90° - 42.1° = 47.9°$. Since $\tan(\angle BCD) = \text{opp}/\text{adj}$, we see from Figure 7.7 that

$$\tan 47.9° = \frac{x + y}{5500}.$$

FIGURE 7.7

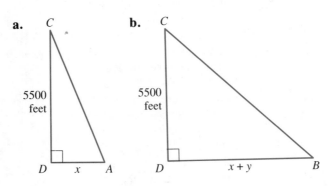

a.   **b.**

Substituting 2240 for x in the latter equation, we obtain

$$2240 + y = 5500(\tan 47.9°),$$
so that
$$y = 5500(\tan 47.9°) - 2240 \approx 3850.$$

Thus the distance across the lake is approximately 3850 feet.

In surveying and navigation, directions are often expressed in terms of *bearings*. The **bearing** from a point A to a point B is the measure of the acute angle formed by the half-line from A through B and the north-south line passing through A. Figure 7.8 illustrates four such measurements. We read the bearing N42°E as "42 degrees east of north," and the bearing S12°W as "12 degrees west of south." The remaining bearings are read in a similar manner.

FIGURE 7.8

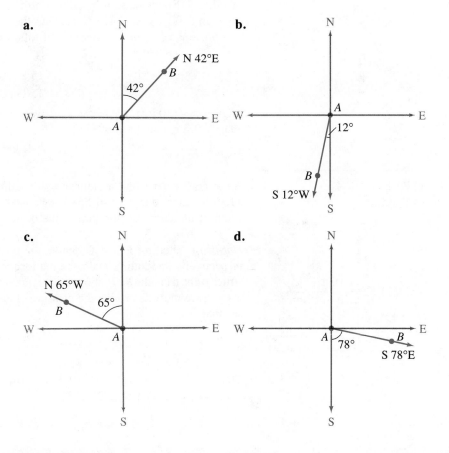

EXAMPLE 5

A ship sailing due north at 30 knots (1 knot = 1 nautical mile per hour $\approx$ 1.15 miles per hour) measures the bearing to a lighthouse as N31°50′W at 4:00 P.M. If the ship is located due east of the lighthouse at 4:30 P.M., approximate the distance in nautical miles between the ship and the lighthouse at that time.

Solution   Letting A and B denote the position of the ship at 4:00 P.M. and 4:30 P.M., respectively, and letting L denote the location of the lighthouse, we construct right triangle ABL as shown in Figure 7.9. We must find x, the length

FIGURE 7.9

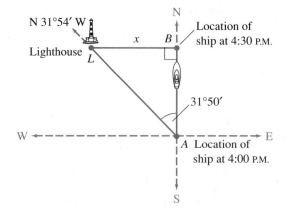

of side LB. Since the ship is sailing north at 30 knots, it travels 15 nautical miles between 4:00 P.M. and 4:30 P.M. Thus the length of side AB is 15 nautical miles. Since $\tan(\angle LAB) = $ opp/adj, it follows that

$$\tan 31°50' = \frac{x}{15}, \qquad \text{or}$$

$$x = 15(\tan 31°50') \approx 9.3$$

Hence, at 4:30 P.M. the distance between the ship and the lighthouse is approximately 9.3 nautical miles.

EXAMPLE 6
FIGURE 7.10

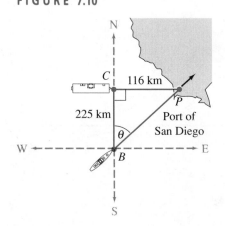

An aircraft carrier and a battleship are sailing into San Diego. If the carrier is 116 kilometers due west of San Diego, and the battleship is 225 kilometers due south of the carrier, approximate the bearing from the battleship to San Diego.

Solution   Letting B and C denote the positions of the battleship and carrier, respectively, and letting P denote the location of the port (San Diego), we construct right triangle BCP shown in Figure 7.10. To determine the bearing from B to P, we must determine the measure of angle θ. From right triangle BCP, we see that

$$\tan \theta = \frac{116}{225} \approx 0.5155.$$

Thus $\theta = \tan^{-1} 0.5155 \approx 27.3°$, and the bearing from the battleship to San Diego is approximately N27.3°E.

EXERCISES 7.1

In Exercises 1–16, find the remaining parts of right triangle ABC.

1. $c = 32,\ \alpha = 71.4°$
2. $b = 55.4,\ \beta = 43°20'$
3. $c = 824,\ b = 115$
4. $\beta = 19°30',\ c = 78.5$
5. $a = 2.35,\ b = 4.18$
6. $\alpha = 60.5°,\ b = 10.2$
7. $a = 3250,\ \beta = 26°50'$
8. $\alpha = 8°10',\ c = 80.4$
9. $\beta = 50°20',\ a = 3.14$
10. $a = 0.248,\ c = 0.565$
11. $c = 19,500,\ \beta = 45.2°$
12. $a = 17.1,\ b = 36.9$
13. $b = 2.6,\ \alpha = 28°$
14. $b = 1250,\ c = 2100$
15. $\beta = 10.4°,\ b = 72.6$
16. $a = 78,400,$
    $b = 128,000$

In Exercises 17–24, find the exact values of the remaining parts of right triangle ABC.

17. $\alpha = 60°, b = 12$

18. $\beta = 45°, a = 50$

19. $\beta = 30°, b = 144$

20. $\alpha = 45°, c = 18$

21. $a = 5\sqrt{3}, c = 10$

22. $\beta = 60°, c = 108$

23. $\alpha = 45°, a = 17$

24. $c = 7\sqrt{2}, a = 7$

25. A guy wire stretched from the top of a radio station's transmission tower is anchored to the ground at a point 225 feet (measured along level ground) from the base of the tower. If the guy wire makes an angle of 78.5° with the ground, approximate the height of the tower.

26. If each edge of a cube measures 12 centimeters, approximate the length of a diagonal of the cube.

27. A rectangular swimming pool is 3 feet deep at the shallow end and 12 feet deep at the deep end. If the pool is 30 feet long, approximate the angle θ between the bottom of the pool and the horizontal. See the accompanying figure.

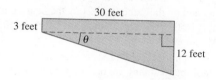

28. A retractible extension ladder on the back of a fire truck can be extended to a maximum length of 125 feet. If the base of the ladder is 6 feet above the ground and 20 feet from the front of a building, approximate the maximum height the ladder will reach up the front of the building.

29. A piece of wire 188 centimeters long is bent into the shape of an isosceles triangle. If the largest angle of the triangle measures 74.5°, approximate the length of all three sides.

30. A belt is wound tightly around two pulleys whose centers are 36.5 inches apart. If the radii of the two pulleys are 8.0 inches and 12.0 inches, find the distance along the belt between the points P_1 and P_2, which are located at the top of each pulley. See the accompanying figure.

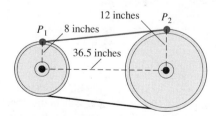

31. A pendulum on a grandfather clock is 84.0 centimeters in length. When the pendulum is vertical, its bob is 25.4 centimeters above the floor, and when the pendulum is swinging, its bob sweeps an arc of 36.2° from side to side. Approximate the maximum distance between the pendulum bob and the floor.

32. If a pentagon (5-sided regular polygon) is inscribed in a circle of radius 16 inches, approximate the area of the pentagon.

33. A manufacturer of paper cups wants each of its conical cups to have a top diameter of 4.25 inches. Approximate the angle θ (see figure) that will give the cup a capacity of 81 cubic inches.

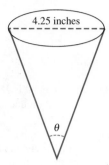

34. An astronomer observes the planet Jupiter through his telescope. If Jupiter, which is approximately 390 million miles from earth, subtends an angle of 47″ to the astronomer's viewfinder, approximate its diameter.

35. Show that the isosceles triangle shown has area $A = \frac{1}{4}b^2 \cot(\frac{1}{2}\theta)$.

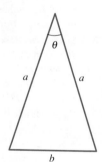

36. In physics, it is shown that the velocity of a ball rolling down an incline plane is given by $v = gt \sin \theta$, where $g \approx 32$ feet/second² is the gravitational constant, t is time and θ is the angle of inclination of the plane. If a rubber ball is started at the top of the plane, which is 4.5 feet high, it attains a velocity of 21.2 feet/second when it reaches the bottom of the plane 3.8 seconds later. Find the approximate length of the plane's inclined side.

37. A circle is tangent to all three sides of an isosceles triangle as shown in the accompanying figure. Approximate the radius of the circle.

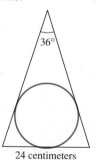

38. A steel rod 20 feet long is being carried through a hallway that is 6 feet wide. Show that the rod, if carried horizontally, cannot turn the corner shown in the figure on page 312.

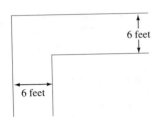

39. If a tree that is 48.0 feet tall casts a shadow 30.0 feet long on level ground, approximate the angle of elevation of the sun.

40. From the top of a building 200 feet high, the angle of elevation to the top of a taller building across the street is 47°40′. If the two buildings are 135 feet apart, approximate the height of the taller building.

41. From atop a fire tower, a forest ranger finds that the angle of depression to a distant fire in the valley below is 8°10′. If the fire tower is 2470 meters above sea level, and the valley below has an elevation of 515 meters, approximate the distance from the tower to the fire.

42. From a car traveling along a level highway, the angle of elevation of a hot air balloon rising vertically above the highway is 33°30′. In the time it takes the car to move 3600 feet closer to the balloon, the angle of elevation changes to 42°10′ and the balloon rises 288 feet vertically. Approximate the height of the balloon at this time.

43. The angle of elevation to the north pole, as measured from a communications satellite in orbit 23,000 miles directly above the equator, is 8.4° (see accompanying figure). Approximate the radius of the earth.

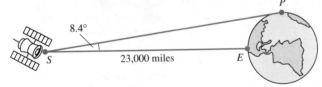

44. A weather balloon released by a meteorologist is being blown 15.0 feet/second horizontally as it rises vertically at the rate of 20.0 feet/second. Approximate the balloon's angle of elevation, as measured by the meteorologist, 10 minutes after the balloon's release.

45. From a sheer cliff 2400 feet above a lake, the angles of depression to two boats on the lake below are 55°20′ and 42°30′. Approximate the distance between the two boats.

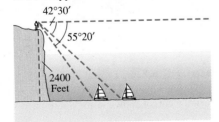

46. A large helium-filled balloon is anchored to the ground at a point P directly beneath the balloon. From the balloon's shadow, which is located at a point Q, 264 feet from P on level ground, the angle of elevation of the balloon is 72°40′. Approximate the height of the balloon above the ground.

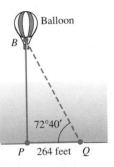

47. A jet flying at 600 miles per hour climbs steadily from an altitude of 35,000 feet to an altitude of 40,000 feet in 30 seconds. Assuming the jet faces a constant headwind of 50 miles per hour, approximate its angle of ascent. [*Hint*: 600 miles per hour ≈ 880 feet/second.]

48. From a point P that is due south of the Washington Monument and on the same level as its base, the angle of elevation to the top is 37.5°. From a point Q that is 888 feet due east of point P, and on the same level as P, the bearing to the monument is N50.8°W. Approximate the height of the Washington Monument.

49. Highway engineers must tunnel through a mountain that lies in the path of the road they are building. From a point P atop the 6130-foot mountain, they find that the angles of depression to points A and B on opposite sides of the mountain's base are 68°20′ and 59°50′, respectively. If the tunnel they plan to construct is level and runs from point A to point B, approximate its length. See the accompanying figure.

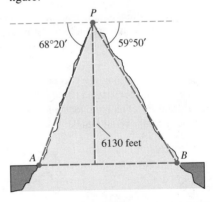

50. A ship leaves port at 10:00 A.M. and sails in the direction S35.5°W at the rate of 30 kilometers/hour. Another ship leaves the same port at 10:30 A.M. and sails in the direction S54.5°E at the rate of 20 kilometers/hour. Find the distance between the ships at 1:30 P.M. and the bearing from the first ship to the second ship at that time.

51. The bearing from a ship to its home port is N51°20′W. If the bearing from the ship to a lighthouse located 14.2 miles due east of the port is N10°40′E, approximate the distance from the ship to its port.

52. A sailboat travels 16.0 miles due south from Honolulu and then 24.2 miles due east before it runs aground on a reef. If a rescue ship is sent from Honolulu, what bearing should it take?

53. While playing in the woods, some young boys discover an old treasure map. According to the map, one must walk 120 paces (yards) due east of a large oak tree and then 224 paces due south in order to locate the buried treasure. When the boys attempt to follow these directions, they find that a private estate, surrounded by a brick wall, has been built 100 yards due east of the oak tree. To help the boys overcome this obstacle, find the distance and bearing from the tree to the treasure.

54. An airplane leaves a private airfield and flies 73.7 miles on a bearing of S27°10′E. The plane then changes direction and flies 122.4 miles due west to the airport in Lubbock, Texas. Approximate the distance between the private airfield and Lubbock.

55. To measure the width of a river, a surveyor first locates a point B directly across the river from his position at point A (see figure to the right). After walking 100 meters to a third point C, which is located on a line that is perpendicular to the line through A and B, he measures the bearing from C to B as N15°50′W. Approximate the width of the river.

56. In air navigation, an angle measured from due north in a clockwise direction is called an **azimuth**. If a plane flies due east of an airport for 235 miles and then flies due south for 340 miles, approximate its distance and azimuth from the airport.

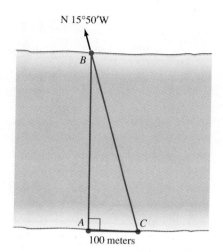

100 meters

57. A commercial jet, flying at a speed of 540 miles per hour, flies on an azimuth of 127°40′ from Atlanta for 45 minutes. The jet then changes course and flies on an azimuth of 217°40′ for 30 minutes more. Find the approximate distance and the azimuth of the jet from Atlanta. (See Exercise 56.)

58. A small plane and a helicopter leave O'Hare Field in Chicago at the same time. If the plane flies 650 miles due east and the helicopter flies 188 miles due north, approximate the azimuth from the helicopter to the plane. (See Exercise 56.)

7.2 LAW OF SINES AND LAW OF COSINES

FIGURE 7.11

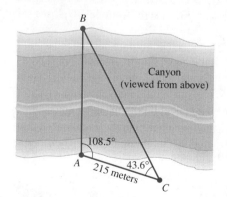

To measure the distance across a canyon, a surveyor first locates a point B on the opposite side of the canyon from his location at point A (see Figure 7.11). He then locates a point C on his side of the canyon, 215 meters from point A. Using a transit, the surveyor finds that the measures of angles BAC and BCA are 108.5° and 43.6°, respectively. Using the surveyor's measurements, can we approximate the distance across the canyon?

We see from Figure 7.11, that the distance across the canyon is the length of side AB of triangle ABC. We note, however, that triangle ABC is not a right triangle. Thus we cannot use the methods of Section 7.1 to find the length of this side. A triangle such as triangle ABC, which does not contain a right angle, is called an **oblique triangle**. To solve problems involving oblique triangles, we introduce in this section two important results called the law of sines and the law of cosines.

The Law of Sines

If, for a given oblique triangle, we know the measures of two angles and a side, or two sides and an angle opposite one of them, we can find the measures of the remaining sides and angles by using the following theorem:

Law of Sines

In any triangle, the sines of the angles are proportional to the lengths of the opposite sides. In particular, for triangle ABC, we have the following proportions:

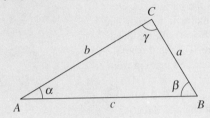

$$\frac{a}{\sin \alpha} = \frac{b}{\sin \beta} = \frac{c}{\sin \gamma}.$$

Proof Since an oblique triangle can contain at most one obtuse angle, there are two cases to consider:

(i) α, β and γ are all acute angles.   **(ii)** One angle (α) is obtuse.

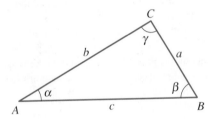
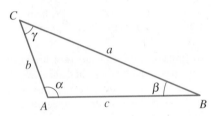

FIGURE 7.12

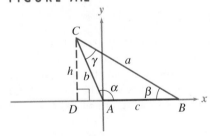

We will prove the theorem for case (ii), where angle α is obtuse. The proof for case (i) is similar (see Exercise 66). We begin by placing triangle ABC in a rectangular coordinate system with angle α in standard position and vertex B on the positive x-axis (see Figure 7.12). Next, we construct segment CD perpendicular to the x-axis and let h denote its length. From right triangle BDC, we see that

$$\sin \beta = \frac{h}{a},$$

or

$$h = a \sin \beta. \tag{1}$$

From right triangle ADC in Figure 7.12, we see that the measure of angle DAC is $180° - \alpha$. Since $\sin (180° - \alpha) = \sin \alpha$, it follows that

$$\sin (\angle DAC) = \frac{h}{b} = \sin \alpha,$$

so that

$$h = b \sin \alpha. \tag{2}$$

From equations (1) and (2), we have

$$a \sin \beta = b \sin \alpha,$$

or

$$\frac{a}{\sin \alpha} = \frac{b}{\sin \beta}. \tag{3}$$

If we place triangle ABC so that angle γ is in standard position and vertex

B lies on the positive x-axis, a similar argument will show that

$$\frac{b}{\sin \beta} = \frac{c}{\sin \gamma}. \tag{4}$$

By combining equations (3) and (4), we obtain the desired result

$$\frac{a}{\sin \alpha} = \frac{b}{\sin \beta} = \frac{c}{\sin \gamma}.$$

In the following example, we use the law of sines to find the remaining parts of a triangle in which we know two angles and a side.

EXAMPLE 7

In triangle ABC, $\alpha = 27°20'$, $\beta = 84°30'$, and $c = 62.5$. Find the remaining parts.

Solution Since $\alpha = 27°20'$ and $\beta = 84°30'$, it follows that

$$\gamma = 180° - (27°20' + 84°30') = 68°10'.$$

We use the law of sines to find a and b as follows:

$$\frac{a}{\sin \alpha} = \frac{c}{\sin \gamma}$$

$$\frac{a}{\sin 27°20'} = \frac{62.5}{\sin 68°10'}$$

$$a = \frac{62.5(\sin 27°20')}{\sin 68°10'} \approx 30.9,$$

and

$$\frac{b}{\sin \beta} = \frac{c}{\sin \gamma}$$

$$\frac{b}{\sin 84°30'} = \frac{62.5}{\sin 68°10'}$$

$$b = \frac{62.5(\sin 84°30')}{\sin 68°10'} \approx 67.0.$$

Using the law of sines, we can solve the problem posed at the beginning of this section:

To measure the distance across a canyon, a surveyor first locates a point B on the opposite side of the canyon from his location at point A (see Figure 7.11, page 313). He then locates a point C on his side of the canyon, 215 meters from point A. Using a transit, the surveyor finds that the measures of angles BAC and BCA are 108.5° and 43.6°, respectively. Using the surveyor's measurements, approximate the distance across the canyon.

We begin by sketching triangle ABC from Figure 7.11, and letting x denote the distance across the canyon (see the figure to the left). Noting that $\beta = 180° - (108.5° + 43.6°) = 27.9°$, we use the law of sines to find the distance across the canyon as follows:

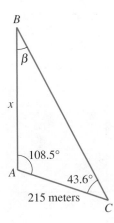

$$\frac{215}{\sin 27.9°} = \frac{x}{\sin 43.6°},$$

or

$$x = \frac{215(\sin 43.6°)}{\sin 27.9°} \approx 317 \text{ meters.}$$

Thus, the canyon is approximately 317 meters wide.

The Ambiguous Case of the Law of Sines

Whenever two angles and a side of a triangle are given, as in the previous two examples, a unique triangle is always determined. However, when two sides of a triangle and the angle opposite one of them are given, a unique triangle is not always determined. For example, in triangle ABC, if a, b, and angle α are given, we have the various possibilities shown in Figure 7.13. In this figure, we have placed angle α in standard position with vertex C on its terminal side so that the segment AC has length b. In each case, the length of the perpendicular line segment from C to the x-axis is denoted by h.

From Figure 7.13, we see that knowing two sides of a triangle and the angle opposite one of them leads to one of the following three possibilities:

No triangle is determined [cases (a) and (e)].

One triangle is determined [cases (b), (d) and (f)].

Two triangles are determined [case (c)].

Because of the numerous possibilities, the case in which two sides and the angle opposite one of them are given is called the **ambiguous case** of the law of sines. Fortunately, we do not need to remember the facts associated with each possible outcome for the ambiguous case. In a specific problem, the case that occurs will become evident as we solve the problem.

FIGURE 7.13

a. If $a < h$, no triangle is determined.

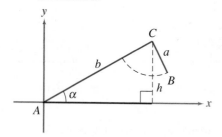

b. If $a = h$, one (right) triangle is determined.

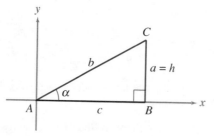

c. If $h < a < b$, the two triangles AB_1C and AB_2C are determined.

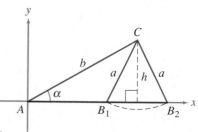

d. If $a > b$, one triangle is determined.

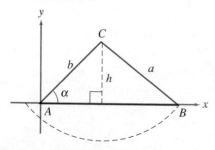

e. If $a \le b$, no triangle is determined.

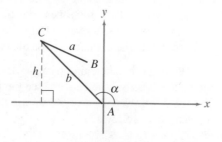

f. If $a > b$, one triangle is determined.

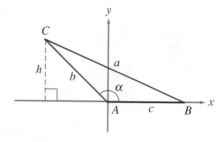

EXAMPLE 8

In triangle ABC, $b = 12.2$, $c = 21.7$, and $\beta = 47.4°$. Find the remaining parts.

Solution By the law of sines,

$$\frac{b}{\sin \beta} = \frac{c}{\sin \gamma}.$$

Substituting the given values, we obtain

$$\frac{12.2}{\sin 47.4°} = \frac{21.7}{\sin \gamma}, \quad \text{or}$$

$$\sin \gamma = \frac{21.7(\sin 47.4°)}{12.2} \approx 1.31.$$

Since $\sin \gamma > 1$ is impossible (Why?), there is no triangle having the given parts.

EXAMPLE 9

In triangle ABC, $a = 108.8$, $b = 183.7$, and $\alpha = 28.8°$. Find the remaining parts.

Solution By the law of sines,

$$\frac{a}{\sin \alpha} = \frac{b}{\sin \beta}, \quad \text{or} \quad \frac{108.8}{\sin 28.8°} = \frac{183.7}{\sin \beta}.$$

Hence, $\sin \beta = \frac{(183.7)(\sin 28.8°)}{108.8} \approx 0.8134.$

There are two values of β between $0°$ and $180°$ for which $\sin \beta = 0.8134$, namely

$$\beta_1 = \sin^{-1}(0.8134) \approx 54.4° \quad \text{and} \quad \beta_2 = 180° - \beta_1 \approx 125.6°.$$

Since both $\alpha + \beta_1 \approx 83.2°$ and $\alpha + \beta_2 \approx 154.4°$ are less than $180°$, there are two triangles, AB_1C and AB_2C, that satisfy the given conditions (see Figure 7.14). We must find the remaining parts of triangles AB_1C and AB_2C (shown in the figures below). We find the remaining angles in each triangle as follows:

In triangle AB_1C,

$$\gamma_1 \approx 180° - (28.8° + 54.4°) = 96.8°;$$

and in triangle AB_2C,

$$\gamma_2 \approx 180° - (28.8° + 125.6°) = 25.6°.$$

FIGURE 7.14

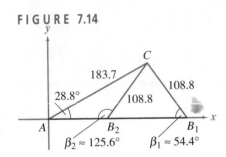

(i) Triangle AB_1C

(ii) Triangle AB_2C

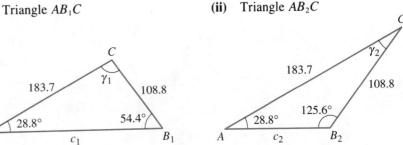

To find the remaining side of each triangle, we use the law of sines. In triangle AB_1C,

$$\frac{a}{\sin \alpha} = \frac{c_1}{\sin \gamma_1}.$$

Hence, $$c_1 = \frac{a \sin \gamma_1}{\sin \alpha} \approx \frac{108.8(\sin 96.8°)}{\sin 28.8°} \approx 224.3.$$

Similarly, in triangle AB_2C, we use

$$\frac{a}{\sin \alpha} = \frac{c_2}{\sin \gamma_2}$$

to obtain

$$c_2 \approx 97.6.$$

Hence, the remaining parts of triangle AB_1C are

$$\beta_1 \approx 54.4°, \; \gamma_1 \approx 96.8°, \text{ and } c_1 \approx 224.3,$$

and the remaining parts of triangle AB_2C are

$$\beta_2 \approx 125.6°, \; \gamma_2 \approx 25.6°, \text{ and } c_2 \approx 97.6.$$

EXAMPLE 10

In triangle ABC, $a = 12.3$, $c = 15.4$, and $\gamma = 58.4°$. Find the remaining parts.

Solution   Using the law of sines, we find $\sin \alpha$ as follows:

$$\frac{12.3}{\sin \alpha} = \frac{15.4}{\sin 58.4°}, \quad \text{or}$$

$$\sin \alpha = \frac{(12.3)(\sin 58.4°)}{15.4} \approx 0.6803.$$

Even though there are two values of α between $0°$ and $180°$ for which $\sin \alpha = 0.6803$, namely

$$\alpha_1 = \sin^{-1}(0.6803) \approx 42.9° \quad \text{and} \quad \alpha_2 = 180° - \alpha_1 \approx 137.1°,$$

there is only one triangle that satisfies the given conditions. To see this, we need only observe that

$$\gamma + \alpha_1 \approx 58.4° + 42.9° = 101.3° \leq 180°,$$

while $$\gamma + \alpha_2 \approx 58.4° + 137.1° = 195.5° > 180°.$$

Since $\alpha \approx 42.9°$ is the only possibility,

$$\beta \approx 180° - (58.4° + 42.9°) = 78.7°.$$

By the law of sines,

$$\frac{a}{\sin \alpha} = \frac{b}{\sin \beta},$$

or $$b = \frac{a \sin \beta}{\sin \alpha} \approx \frac{12.3(\sin 78.7°)}{\sin 42.9°} \approx 17.7.$$

Hence, the remaining parts of triangle ABC are

$$\alpha \approx 42.9°, \quad \beta \approx 78.7°, \quad \text{and} \quad b \approx 17.7.$$

The Law of Cosines

When we know two sides and the included angle or three sides of a triangle, the law of sines cannot be applied directly to find the remaining parts. In these cases, we use the following theorem:

Law of Cosines

In any triangle, the square of the length of any side is equal to the sum of the squares of the lengths of the other two sides minus twice the product of the lengths of the other two sides and the cosine of their included angle. In particular, for triangle ABC, we have

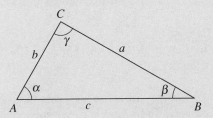

$$a^2 = b^2 + c^2 - 2bc \cos \alpha;$$
$$b^2 = a^2 + c^2 - 2ac \cos \beta;$$
$$c^2 = a^2 + b^2 - 2ab \cos \gamma.$$

Proof   We will verify that $a^2 = b^2 + c^2 - 2bc \cos \alpha$. The verifications of the remaining two equations are analogous. As in the proof of the law of sines, there are two cases to consider:

(i) α, β, and γ are acute angles.   **(ii)**   One angle (α) is obtuse.

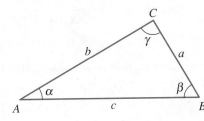   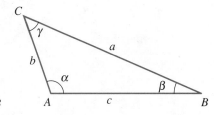

We will prove the theorem for case (ii) and leave the proof of case (i) as an exercise (see Exercise 66). We begin by placing triangle ABC in a rectangular coordinate system, so that angle α is in standard position and vertex B lies along the positive x-axis (see Figure 7.15). Next, we construct segment CD perpendicular to the x-axis and denote its length by h. Letting k denote the x-coordinate of the point D, we see that vertex C is the point (k, h). Applying the Pythagorean theorem to right triangle BDC, we have

$$a^2 = [d(D, B)]^2 + h^2.$$

Substituting $c - k$ for $d(D, B)$, we obtain

$$a^2 = (c - k)^2 + h^2,$$
or
$$a^2 = c^2 - 2ck + k^2 + h^2. \tag{5}$$

Similarly, applying the Pythagorean theorem to right triangle ADC, we get

$$b^2 = k^2 + h^2.$$

FIGURE 7.15

Substituting b^2 for $k^2 + h^2$ in equation (5), we have

$$a^2 = c^2 - 2ck + b^2,$$

or
$$a^2 = b^2 + c^2 - 2ck. \tag{6}$$

Since the point $C = (k, h)$ lies on the terminal side of angle α, and since $d(A, C) = b$, it follows from Theorem 6.2 that

$$\cos \alpha = \frac{k}{b}, \quad \text{or} \quad k = b \cos \alpha.$$

Replacing k by $b \cos \alpha$ in equation (6), we obtain the desired result

$$a^2 = b^2 + c^2 - 2bc \cos \alpha.$$

E X A M P L E   11

In triangle ABC, $a = 185$, $b = 104$, and $c = 128$. Find all three angles of this triangle.

Solution   Since we are given three sides of triangle ABC, we apply the law of cosines. To find angle α, we solve the equation

$$a^2 = b^2 + c^2 - 2bc \cos \alpha$$

for $\cos \alpha$ to obtain

$$\cos \alpha = \frac{b^2 + c^2 - a^2}{2bc}.$$

Thus, $\cos \alpha = \dfrac{(104)^2 + (128)^2 - (185)^2}{2(104)(128)} \approx -0.2639$ and

$$\alpha \approx \cos^{-1}(-0.2639) \approx 105.3°.$$

To find angle β, we turn again to the law of cosines. From the equation

$$b^2 = a^2 + c^2 - 2ac \cos \beta,$$

we find that

$$\cos \beta = \frac{a^2 + c^2 - b^2}{2ac}.$$

Hence, $\cos \beta = \dfrac{(185)^2 + (128)^2 - (104)^2}{2(185)(128)} \approx 0.8402,$

so that $\beta \approx \cos^{-1} 0.8402 \approx 32.8°.$

Finally, we have

$$\gamma = 180° - (\alpha + \beta)$$
$$\approx 180° - 138.1° = 41.9°.$$

Thus the three angles of triangle ABC are

$$\alpha \approx 105.3°, \qquad \beta \approx 32.8°, \qquad \text{and} \qquad \gamma \approx 41.9°.$$

E X A M P L E 12

In triangle ABC, $b = 24.7$, $c = 42.0$, and $\alpha = 48°40'$. Find the remaining parts.

Solution   Since we are given two sides of triangle ABC and the included angle, we apply the law of cosines. To find a, we use the equation

$$a^2 = b^2 + c^2 - 2bc \cos \alpha$$

to obtain

$$a = \sqrt{(24.7)^2 + (42.0)^2 - 2(24.7)(42.0)(\cos 48°40')} \approx 31.7.$$

To find β, we solve the equation

$$b^2 = a^2 + c^2 - 2ac \cos \beta$$

for $\cos \beta$ to get

$$\cos \beta = \frac{a^2 + c^2 - b^2}{2ac} \approx \frac{(31.7)^2 + (42.0)^2 - (24.7)^2}{2(31.7)(42.0)} \approx 0.8107.$$

Thus, $\beta \approx \cos^{-1}(0.8107) \approx 35°50'$,

and the remaining parts of triangle ABC are

$$a \approx 31.7, \qquad \beta \approx 35°50', \qquad \text{and} \qquad \gamma = 180° - (\alpha + \beta) \approx 95°30'.$$

E X A M P L E 13

A ship leaves port A and sails due east for 450 miles. The ship then changes direction and follows a course of N44.2°E for 325 miles to port B. Find the distance between the two ports.

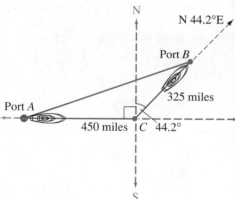

Solution   Letting C denote the point at which the ship changes course, we obtain triangle ABC shown in the accompanying figure. From the figure, we see that the angle at vertex C has measure $90° + 44.2° = 134.2°$. Since we know two sides and the included angle of triangle ABC, we use the law of cosines to obtain c.

$$c^2 = a^2 + b^2 - 2ab \cos \gamma$$
$$= (325)^2 + (450)^2 - 2(325)(450)(\cos 134.2°) \approx 512,000,$$

so that $c \approx \sqrt{512,000} \approx 716.$

Hence, the two ports are approximately 716 miles apart.

The Area of a Triangle

When the base b and altitude h of a triangle are known, we can use the familiar formula

$$\mathcal{A} = \frac{1}{2} bh$$

to calculate its area. We now present two formulas that can be used to find the area of a triangle when the height h is not given.

FIGURE 7.16

a. α, β and γ are acute angles.

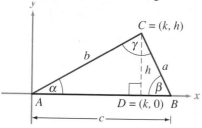

b.   Angle α is obtuse.

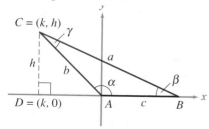

Since, as noted earlier, a given oblique triangle can contain at most one obtuse angle, we consider the two cases shown in Figure 7.16. In each case, we have placed angle α in standard position and constructed segment CD perpendicular to the x-axis. We see from Figure 7.16 that, in both cases, the area $\mathcal{A}$ of triangle ABC is given by

$$\mathcal{A} = \frac{1}{2} (\text{base})(\text{height}) = \frac{1}{2} ch. \qquad (7)$$

Since angle α is in standard position, and $d(A, C) = b$, it follows from Theorem 6.2, that

$$\sin \alpha = \frac{h}{b}, \qquad \text{or} \qquad h = b \sin \alpha.$$

Replacing h by $b \sin \alpha$ in equation (7), we obtain

$$\mathcal{A} = \frac{1}{2} bc \sin \alpha.$$

Using similar arguments with each of β and γ in standard position, we can obtain the formulas

$$\mathcal{A} = \frac{1}{2} ac \sin \beta \qquad \text{and} \qquad \mathcal{A} = \frac{1}{2} ab \sin \gamma.$$

We have thus proved the following theorem, which can be used to find the area of a triangle when we know two sides and the included angle:

THEOREM 7.1   Area of a Triangle

The area of a triangle is equal to one-half the product of the lengths of any two sides and the sine of the included angle. Thus the area $\mathcal{A}$ of triangle ABC is given by each of the following formulas:

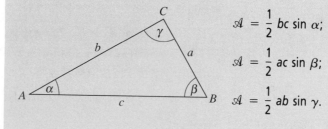

$$\mathcal{A} = \frac{1}{2} bc \sin \alpha;$$

$$\mathcal{A} = \frac{1}{2} ac \sin \beta;$$

$$\mathcal{A} = \frac{1}{2} ab \sin \gamma.$$

EXAMPLE 14

Find the area of a regular pentagon that is inscribed in a circle of radius 18.4 centimeters.

Solution   To find the area of this pentagon, we divide it into 5 congruent triangles (see the accompanying figure) and apply Theorem 7.1 to each. Since the area $\mathcal{A}$ of each of these triangles is given by

$$\mathcal{A} = \frac{1}{2}(18.4)(18.4)(\sin 72°) \approx 161 \text{ square centimeters,}$$

the area of the pentagon is approximately $5(161) = 805$ square centimeters.

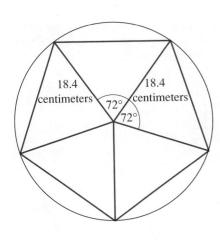

We now state (without proof) a formula that can be used to find the area of a triangle when we know the lengths of all three sides. This formula is known as Heron's formula in honor of Greek mathematician Heron of Alexandria.

Heron's Formula for the Area of a Triangle

The area $\mathcal{A}$ of triangle ABC is given by the formula

$$\mathcal{A} = \sqrt{s(s - a)(s - b)(s - c)},$$

where $s = \dfrac{a + b + c}{2}$.

EXAMPLE 15

A farmer's field is in the shape of a triangle with sides of length 680 feet, 840 feet, and 1020 feet. Given that one acre contains 43,560 square feet, find the approximate number of acres in this field.

Solution Since we know all three sides of this triangular field, we can use Heron's formula to find its area. Since

$$s = \frac{a + b + c}{2} = \frac{680 + 840 + 1020}{2} = 1270,$$

it follows that

$$\mathcal{A} = \sqrt{s(s - a)(s - b)(s - c)} = \sqrt{1270(590)(430)(250)}$$
$$\approx 283,800 \text{ square feet.}$$

Thus the field contains $\dfrac{283,800 \text{ square feet}}{43,560 \text{ square feet}} \approx 6.52$ acres.

EXERCISES 7.2

In Exercises 1–24, find the remaining parts of all triangles, if any, that satisfy the given conditions.

1. $\beta = 19.8°, \gamma = 102.7°, c = 37.5$
2. $\alpha = 26.3°, \beta = 94.5°, c = 9.18$
3. $\gamma = 37°10', b = 1250, c = 875$
4. $\alpha = 86.4°, \gamma = 17.5°, a = 42.6$
5. $\alpha = 18°40', a = 235, b = 702$
6. $b = 91.2, c = 87.1, \alpha = 43.6°$
7. $a = 25.2, b = 81.4, c = 68.7$
8. $\beta = 74.4°, b = 5.15, c = 4.63$
9. $\alpha = 5°25', \beta = 78°40', b = 9.75$
10. $\gamma = 27.2°, a = 52.3, c = 36.7$

11. $\beta = 31°50'$, $a = 12.0$, $b = 7.5$
12. $a = 152$, $b = 236$, $c = 357$
13. $a = 12.14$, $b = 16.35$, $\beta = 33°42'$
14. $\beta = 11.7°$, $a = 32.4$, $c = 38.8$
15. $\gamma = 120°30'$, $a = 56.4$, $b = 102.9$
16. $a = 15.2$, $b = 29.8$, $c = 13.4$
17. $\beta = 115.1°$, $a = 2.36$, $c = 1.84$
18. $a = 78.9$, $c = 125.0$, $\beta = 128°10'$
19. $\alpha = 65°$, $a = 24$, $b = 30$
20. $\gamma = 17.2°$, $a = 2250$, $c = 1372$
21. $\beta = 57°50'$, $a = 4.68$, $b = 5.03$
22. $a = 124.7$, $b = 17.3$, $c = 95.7$
23. $a = 30.2$, $c = 42.8$, $\beta = 132.3°$
24. $a = 0.2086$, $b = 0.3757$, $c = 0.4142$

In Exercises 25–36, find the area of triangle ABC.

25. $\alpha = 102.4°$, $b = 46.8$, $c = 82.5$
26. $\gamma = 41°20'$, $a = 7.18$, $b = 10.40$
27. $a = 33$, $b = 46$, $c = 65$
28. $a = 3.48$, $b = 5.19$, $c = 7.04$
29. $b = 2110$, $c = 3540$, $\alpha = 82.2°$
30. $\beta = 75°40'$, $a = 625$, $c = 948$
31. $a = 1.48$, $b = 0.75$, $c = 1.28$
32. $\gamma = 112.9°$, $a = 32.5$, $b = 44.6$
33. $\alpha = 13°35'$, $b = 256.7$, $c = 244.1$
34. $\beta = 150°$, $a = 14.8$, $c = 31.5$
35. $a = 1280$, $b = 1645$, $c = 1090$
36. $a = 156,400$, $b = 314,000$, $c = 175,500$
37. A TV antenna is supported by two guy wires on opposite sides but in the same vertical plane. One wire is 53.5 meters long and makes an angle of $45°40'$ with the ground. If the other wire is 47.8 meters long, find the angle it makes with the ground.
38. From a point A at street level, an observer finds that the angle of elevation to the top of a building is $68.6°$. If she walks 22.5 feet directly toward the building to a point B, the angle of elevation changes to $74.8°$. Find the height of the building.
39. A telephone pole is tilted $10.6°$ from the vertical. From a point on level ground 135 feet from the base of the pole, the angle of elevation to the top is $18.1°$ (see the accompanying figure). Find the height of the pole.

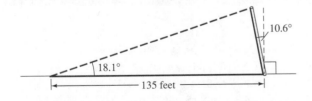

40. A flagpole is located at the top of a hill that makes an angle of $20.2°$ with the horizontal (see figure). From a

point 125 feet directly down the hill from the base of the flagpole, the angle of elevation to the top of the pole is $58.4°$. Find the height of the flagpole.

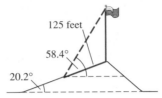

41. A ship leaves port and sails 135.0 kilometers on a bearing of S48°20'W. It then changes course and sails 88.4 kilometers on a bearing of N12°10'W. Find the distance and bearing from the ship to its port.
42. A forest fire is spotted simultaneously by two rangers in two fire towers that are 16.5 miles apart. If the bearing from tower A to tower B is N65.3°W, and if the bearings of the fire from A and B are N27.5°W and N15.8°E, respectively, find the distance from each tower to the fire.
43. A straight road makes an angle of $12.5°$ with the horizontal. At 4:00 P.M., a tree beside the road casts a shadow 89.5 feet long parallel to the road (see figure). If the tree is 48.0 feet high, find the angle of elevation of the sun at 4:00 P.M.

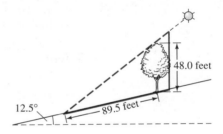

44. A vertical pole 37.5 meters high stands on a hillside that makes an angle of $24°40'$ with the horizontal (see figure). Find the length of a support wire that runs from the top of the pole to a point that is 24.2 meters directly down the hillside from the base of the pole.

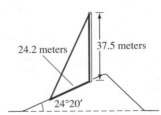

45. After being hit by a truck, a telephone pole 52.5 feet tall is tilted slightly from the vertical. When the angle of elevation of the sun is $68.4°$, the pole casts a shadow 32.4 feet long in the direction it is tilted (see figure on next page). Find the angle between the pole and the vertical.

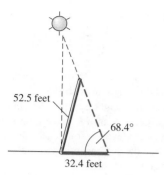

52.5 feet

68.4°

32.4 feet

46. A cruise ship leaves port at 2:00 P.M. and sails on a bearing of N35.1°E at the rate of 32 miles per hour. Another ship leaves the same port at 2:45 P.M. and sails on a bearing of N22.5°W at 24 miles per hour. Find the distance between the two ships at 4:00 P.M.

47. Two people are standing on level ground 100 feet apart and in line with the base of a tower. The angle of elevation to the top of the tower from one person is 38°40′, and from the other person 30°20′. Find the height of the tower.

48. The angles of elevation of a hot-air balloon from points A and B on level ground are 25.2° and 48.9°, respectively. The points A and B are 7.65 miles apart and the balloon is between the two points in the same vertical plane. Find the height of the balloon above the ground.

49. A triangular-shaped garden plot has sides measuring 112 meters, 130 meters, and 164 meters. Find the area of the garden plot.

50. A field in the shape of a triangle has sides of length 625 yards, 830 yards, and 1220 yards. Given that one acre contains 4840 square yards, determine the number of acres in this field.

51. Quadrilateral $ABCD$ has angles measuring 72.8°, 84.3°, and 103.4° at vertices A, B, and C, respectively (see figure). If sides AB and BC measure 108.5 inches and 95.1 inches, respectively, find the area of the quadrilateral. [*Hint*: Divide the quadrilateral into two triangles.]

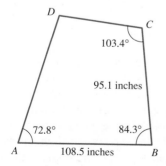

D

C

103.4°

95.1 inches

72.8°

84.3°

A

108.5 inches

B

52. The longer and shorter sides of a parallelogram measure 28.8 centimeters and 17.5 centimeters, respectively. If the angle at one vertex measures 48°40′, find the area of the parallelogram.

53. Ship B is 2.75 kilometers due east of ship A. A lighthouse 7.25 kilometers from ship A is on a bearing of N18°10′E from ship A. Find the distance and the bearing from ship B to the lighthouse.

54. An airplane flew 465 miles from airport A to airport B on a bearing of N65.4°E. It then flew on a bearing of S48.7°E to airport C. If the bearing from airport A to airport C is S62.8°E, find the distance between airports A and C.

55. To measure the distance across a swamp, a surveyor walks 1250 meters from point A to point B, makes a turn of 55.2°, and walks 1060 meters to point C (see the accompanying figure). Using the surveyor's measurements, find the distance across the swamp.

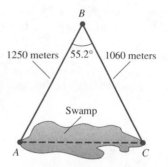

B

1250 meters 55.2° 1060 meters

Swamp

A

C

56. To measure the distance across a lake, a surveyor determines the measurements shown in the figure. Use the surveyor's measurements to find the distance across the lake.

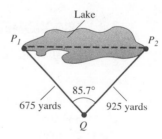

Lake

P_1

P_2

85.7°

675 yards

925 yards

Q

57. A ship's navigator sights a lighthouse at a bearing of N36°20′E. After the ship travels 8.25 miles on a bearing of N27°50′W, the bearing from the ship to the lighthouse has changed to S81°30′E. Find the distance between the ship and the lighthouse at the second sighting.

58. A tower is located at the top of a hill. From a point A located 138.5 feet down the hillside from the base of the tower, the angle to the top of the tower is 32.45°. From point B, which is located on a line between A and the base of the tower, the angle to the top of the tower is 51.30°. If the distance between points A and B is 47.2 feet, find the height of the tower (see figure on next page).

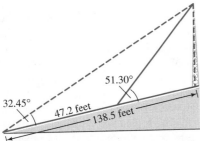

59. The alternate vertices of a regular hexagon (6-sided polygon) are connected by straight line segments to form a triangle. If each side of the hexagon measures 18.4 centimeters, find the length of each side of the triangle and the area of the triangle.

60. A baseball diamond is actually a 90-foot square whose vertices are home plate and the three bases. If the distance from the pitcher's mound to home plate is 60.5 feet, find the distance from the pitcher's mound to each of the three bases.

61. As shown in the figure, triangle ABC is inscribed in a rectangular box that is 20 inches long, 12 inches wide and 8 inches high. Find the measure of angle α and the area of triangle ABC.

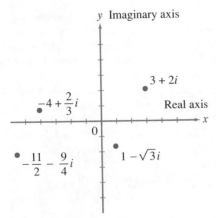

62. Use the law of cosines to show that for any triangle ABC,
$$\frac{a^2 + b^2 + c^2}{2abc} = \frac{\cos \alpha}{a} + \frac{\cos \beta}{b} + \frac{\cos \gamma}{c}.$$

63. The area of an equilateral triangle is 36 square inches. Use Heron's formula to find the length of each side.

64. In surveying, an inaccessible height h can be determined using a line segment perpendicular to h called a *base line*. If α and β are the angles of elevation from each end of a base line of length m (see the accompanying figure), show that the height h is given by the formula
$$h = m \left[\frac{\sin \alpha \sin \beta}{\sin(\beta - \alpha)} \right].$$

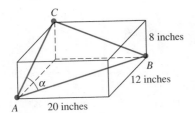

65. If a circle of radius r is inscribed in triangle ABC, show that
$$r = \sqrt{\frac{(s - a)(s - b)(s - c)}{s}},$$
where $s = \dfrac{a + b + c}{2}$.

66. **a.** Verify the law of sines for the case where all three angles of triangle ABC are acute.
b. Verify the law of cosines for the case where all three angles of triangle ABC are acute.

7.3 TRIGONOMETRIC FORM OF A COMPLEX NUMBER

Recall from Section 1.4 that the set $\mathscr{C}$ of complex numbers is defined by
$$\mathscr{C} = \{a + bi \mid a \text{ and } b \text{ are real numbers and } i^2 = -1\}.$$

In this section we explore some important relationships between trigonometry and the complex number system. We begin by reviewing some properties of complex numbers.

The Geometric Representation of a Complex Number

If $z = a + bi$ is a complex number, the real numbers a and b are called the *real part* and *imaginary part*, respectively, of z and we say that z is in rectangular form. By referring to the x- and y-axes as the real and imaginary axes, respectively, we can use the natural association between the complex number $a + bi$ and the ordered pair (a, b) to describe the xy-plane as the complex plane.

Since each complex number $z = a + bi$ can be represented geometrically as the point $P = (a, b)$ in the complex plane, we call the point P the **geometric representation** of z. Recall that the modulus (or absolute value) of $z = a + bi$ is given by
$$|z| = \sqrt{a^2 + b^2}.$$

The Complex Plane

As we noted in Section 1.4, the modulus of z represents the distance between z and the origin O of the complex plane.

The Trigonometric Form of a Complex Number

FIGURE 7.17

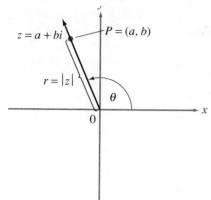

As we shall see, the geometric representation of complex numbers as points in the plane provides a means for representing complex numbers in terms of trigonometric functions. Let $z = a + bi$ be any nonzero complex number as illustrated in Figure 7.17. Let $P = (a, b)$ denote the geometric representation of z, and let θ be any angle in standard position whose terminal side contains P.

Letting $r = |z| = \sqrt{a^2 + b^2}$, we know by Theorem 6.2 that

$$\cos \theta = \frac{a}{r} \quad \text{and} \quad \sin \theta = \frac{b}{r}, \quad \text{or}$$

$$a = r \cos \theta \quad \text{and} \quad b = r \sin \theta.$$

Substituting these values for a and b in $z = a + bi$, we obtain

$$z = r \cos \theta + (r \sin \theta)i, \quad \text{or}$$

$$z = r(\cos \theta + i \sin \theta).$$

The latter equation, which can be written in the abbreviated form

$$z = r \text{ cis } \theta,$$

expresses the complex number z in trigonometric (or polar) form.

Trigonometric Form of a Complex Number

Let $z = a + bi$ be any nonzero complex number. If $r = |z| = \sqrt{a^2 + b^2}$, and if θ is any angle in standard position whose terminal side contains the point $P = (a, b)$, the **trigonometric form** of z is given by

$$z = r(\cos \theta + i \sin \theta) = r \text{ cis } \theta. \tag{1}$$

The number r is called the **modulus** of z and the angle θ is called an **argument** of z.

We note that the trigonometric form of a complex number is not unique since the angle θ in equation (1) can be replaced by any angle that is coterminal with θ. However, unless otherwise noted, we will choose the smallest positive angle θ as the argument whenever we write a complex number in trigonometric form. We also note that two complex numbers are equal if and only if their moduli are equal and their arguments are coterminal angles.

EXAMPLE 16

Express each of the following complex numbers in trignometric form:

a. $\sqrt{3} + i$ **b.** $5 - 5i$ **c.** $-\frac{7}{2}i$

Solution

a. To express $\sqrt{3} + i$ in the form $r(\cos \theta + i \sin \theta)$, we must determine the values of both r and θ, where, as we noted above, $0 \le \theta < 2\pi$. We begin by plotting the point $P = (\sqrt{3}, 1)$, which is the geometric representation of $\sqrt{3} + i$ (see Figure 7.18). We then choose the smallest positive angle θ, in standard position, whose terminal side contains P. Thus $r = \sqrt{(\sqrt{3})^2 + (1)^2} = \sqrt{4} = 2$, and we see from Figure 7.18 that

$$\cos \theta = \frac{a}{r} = \frac{\sqrt{3}}{2} \quad \text{and} \quad \sin \theta = \frac{b}{r} = \frac{1}{2}.$$

From the values of $\sin \theta$ and $\cos \theta$, we know that $\theta = \pi/6$. Hence, in trigonometric form,

$$\sqrt{3} + i = 2 \left(\cos \frac{\pi}{6} + i \sin \frac{\pi}{6} \right) = 2 \operatorname{cis} \frac{\pi}{6}.$$

b. As in part (a), we plot $P = (5, -5)$ and choose the smallest possible angle θ, $0 \le \theta < 2\pi$, whose terminal side contains P. Hence, $r = \sqrt{(5)^2 + (-5)^2} = \sqrt{50} = 5\sqrt{2}$, and we see from Figure 7.19 that

$$\cos \theta = \frac{5}{5\sqrt{2}} = \frac{1}{\sqrt{2}} \quad \text{and} \quad \sin \theta = -\frac{5}{5\sqrt{2}} = -\frac{1}{\sqrt{2}}.$$

Since θ lies in quadrant IV, we know that $\theta = 7\pi/4$. Thus, in trigonometric form,

$$5 - 5i = 5\sqrt{2} \left(\cos \frac{7\pi}{4} + i \sin \frac{7\pi}{4} \right) = 5\sqrt{2} \operatorname{cis} \frac{7\pi}{4}.$$

c. The geometric representation of the complex number $-\frac{7}{2}i$ is the point $P = (0, -\frac{7}{2})$ that lies on the negative y-axis (see Figure 7.20). Thus $r = \frac{7}{2}$, and $\theta = 3\pi/2$ is the smallest positive angle whose terminal side contains $P = (0, -\frac{7}{2})$. It follows that

$$-\frac{7}{2}i = \frac{7}{2} \left(\cos \frac{3\pi}{2} + i \sin \frac{3\pi}{2} \right) = \frac{7}{2} \operatorname{cis} \frac{3\pi}{2}.$$

FIGURE 7.18

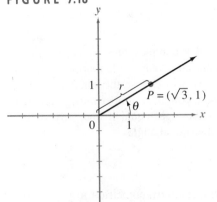

FIGURE 7.19

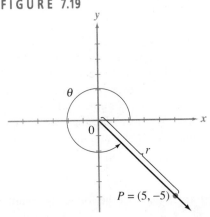

FIGURE 7.20

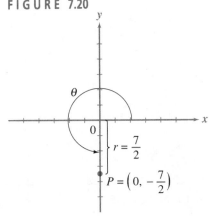

EXAMPLE 17

Express each of the following numbers in the form $a + bi$, where a and b are real numbers:

a. $\dfrac{5}{3} \operatorname{cis} \dfrac{3\pi}{4}$ **b.** $2\sqrt{3} \operatorname{cis} \dfrac{7\pi}{6}$

Solution

a. Since $\cos(3\pi/4) = -\sqrt{2}/2$, and $\sin(3\pi/4) = \sqrt{2}/2$, it follows that

$$\frac{5}{3} \operatorname{cis} \frac{3\pi}{4} = \frac{5}{3}\left(-\frac{\sqrt{2}}{2} + \frac{\sqrt{2}}{2}\,i\right) = -\frac{5\sqrt{2}}{6} + \frac{5\sqrt{2}}{6}\,i.$$

b. Since $\cos(7\pi/6) = -\sqrt{3}/2$, and $\sin(7\pi/6) = -\frac{1}{2}$,

$$2\sqrt{3} \operatorname{cis} \frac{7\pi}{6} = 2\sqrt{3}\left(-\frac{\sqrt{3}}{2} - \frac{1}{2}\,i\right) = -3 - \sqrt{3}i.$$

In the preceding example, we determined trigonometric forms for complex numbers whose arguments were multiples of the familiar angles $\pi/4$ and $\pi/6$. When this is not the case, we can use the inverse tangent function to determine arguments, as our next example illustrates.

EXAMPLE 18

Express each of the following complex numbers in trigonometric form:

a. $-4 - 2i$ **b.** $-4 + \sqrt{5}i$

Solution

FIGURE 7.21

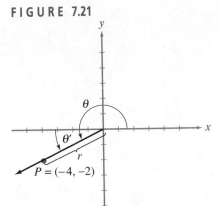

a. As we see in Figure 7.21, both $P = (-4, -2)$ and the smallest positive angle θ, whose terminal side contains P, lie in quadrant III. Since $\tan\theta = \dfrac{-2}{-4} = \dfrac{1}{2}$, and θ lies in quadrant III, we use the reference angle θ' to obtain

$$\tan\theta' = |\tan\theta| = \frac{1}{2}.$$

Thus, $$\theta' = \tan^{-1}\frac{1}{2} \approx 26.6°$$

and $$\theta = 180° + \theta' \approx 206.6°.$$

Noting that

$$r = \sqrt{(-4)^2 + (-2)^2}$$
$$= \sqrt{20} = 2\sqrt{5} \approx 4.5,$$

we see that, in trigonometric form,

$$-4 - 2i = 2\sqrt{5} \operatorname{cis}\left(180° + \tan^{-1}\frac{1}{2}\right)$$

$$\approx 4.5 \operatorname{cis} 206.6°.$$

b. The point $P = (-4, \sqrt{5})$ and the smallest positive angle θ whose ter-

FIGURE 7.22

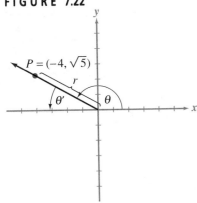

minal side contains P are shown in Figure 7.22. We see from Figure 7.22 that

$$r = \sqrt{(-4)^2 + (\sqrt{5})^2} = \sqrt{21} \approx 4.6,$$

and that

$$\tan \theta = \frac{\sqrt{5}}{-4} = -\frac{\sqrt{5}}{4}.$$

Using the reference angle θ', we see that

$$\tan \theta' = |\tan \theta| = \frac{\sqrt{5}}{4},$$

so that

$$\theta' = \tan^{-1}\left(\frac{\sqrt{5}}{4}\right) \approx 29.2°.$$

Hence,

$$\theta = 180° - \theta' \approx 150.8°,$$

and, in trigonometric form,

$$-4 + \sqrt{5}i = \sqrt{21} \operatorname{cis}\left(180° - \tan^{-1}\frac{\sqrt{5}}{4}\right) \approx 4.6 \operatorname{cis} 150.8°.$$

Products and Quotients of Complex Numbers in Trigonometric Form

The following theorem tells us how to multiply and divide complex numbers in trigonometric form:

THEOREM 7.2

Let $z_1 = r_1 \operatorname{cis} \theta_1$ and let $z_2 = r_2 \operatorname{cis} \theta_2$ be complex numbers. Then

(i) $z_1 z_2 = r_1 r_2 \operatorname{cis}(\theta_1 + \theta_2)$;

(ii) $\dfrac{z_1}{z_2} = \dfrac{r_1}{r_2} \operatorname{cis}(\theta_1 - \theta_2)$, if $z_2 \neq 0$.

Proof

(i) We find the product of z_1 and z_2 as follows:

$$\begin{aligned}
z_1 z_2 &= (r_1 \operatorname{cis} \theta_1)(r_2 \operatorname{cis} \theta_2) \\
&= [r_1(\cos \theta_1 + i \sin \theta_1)][r_2(\cos \theta_2 + i \sin \theta_2)] \\
&= r_1 r_2 [(\cos \theta_1 \cos \theta_2 - \sin \theta_1 \sin \theta_2) \\
&\quad + i(\sin \theta_1 \cos \theta_2 + \cos \theta_1 \sin \theta_2)].
\end{aligned}$$

Using the sum formulas for sine and cosine, the latter equation becomes

$$z_1 z_2 = r_1 r_2 [\cos(\theta_1 + \theta_2) + i \sin(\theta_1 + \theta_2)], \quad \text{or}$$
$$z_1 z_2 = r_1 r_2 \operatorname{cis}(\theta_1 + \theta_2).$$

(ii) The verification of this equation is similar and is left as an exercise (see Exercise 70).

We see from Theorem 7.2 that the product of two complex numbers in trigonometric form is found by multiplying their moduli and adding their arguments. Similarly, to find the quotient of two complex numbers in trigonometric form, we find the quotient of their moduli and the difference in their arguments. We conclude this section with an example that illustrates this theorem.

EXAMPLE 19

If $z_1 = 5 \text{ cis}(3\pi/4)$, and $z_2 = 2 \text{ cis}(\pi/6)$, find $z_1 z_2$, and z_1/z_2. Leave each result in trigonometric form.

Solution   By equation (i) of Theorem 7.2,

$$z_1 z_2 = (5)(2) \text{ cis}\left(\frac{3\pi}{4} + \frac{\pi}{6}\right) = 10 \text{ cis } \frac{11\pi}{12}.$$

Similarly, by equation (ii) of Theorem 7.2,

$$\frac{z_1}{z_2} = \frac{5}{2} \text{ cis}\left(\frac{3\pi}{4} - \frac{\pi}{6}\right) = \frac{5}{2} \text{ cis } \frac{7\pi}{12}.$$

EXERCISES 7.3

In Exercises 1–24, express the given complex number in trigonometric form.

1. $1 + \sqrt{3}i$
2. $2 - 2i$
3. $-4 + 4i$
4. $\sqrt{3} - i$
5. $-5i$
6. $-1 - i$
7. $(\sqrt{3}/2) - \frac{1}{2}i$
8. 9
9. 0
10. $i/2$
11. $-3 + \sqrt{3}i$
12. $\sqrt{2} + 2i$
13. $3 - 2i$
14. $-1 + i$
15. $-\sqrt{3} + i$
16. $-1 + \sqrt{2}i$
17. $-2 - 2\sqrt{2}i$
18. $2\sqrt{3} - 2i$
19. $-5 + 12i$
20. $3 + 5i$
21. $-\sqrt{3}/2$
22. $-9i$
23. $2i(\sqrt{3} - i)$
24. $-3i(\sqrt{2} + i)$

In Exercises 25–44, express each number in the form $a + bi$, where a and b are real numbers.

25. $2 \text{ cis}(3\pi/4)$
26. $3 \text{ cis}(\pi/3)$
27. $\sqrt{5} \text{ cis}(\pi/2)$
28. $\sqrt{2} \text{ cis}(3\pi/2)$
29. $3\sqrt{2} \text{ cis}(2\pi/3)$
30. $8 \text{ cis}(5\pi/4)$
31. $6 \text{ cis}(5\pi/3)$
32. $4 \text{ cis}(5\pi/6)$
33. $12 \text{ cis}(7\pi/6)$
34. $\sqrt{17} \text{ cis } 0$
35. $(5/2) \text{ cis } \pi$
36. $\frac{3}{4} \text{ cis}(7\pi/4)$
37. $15 \text{ cis}(4\pi/3)$
38. $2\sqrt{3} \text{ cis}(5\pi/3)$
39. $(\sqrt{3}/2) \text{ cis } 330°$
40. $24 \text{ cis } 225°$
41. $4 \text{ cis}(5\pi/12)$
42. $5 \text{ cis}(7\pi/12)$
43. $8 \text{ cis } 315°$
44. $10 \text{ cis } 105°$

In Exercises 45–52, use Theorem 7.2 to perform the indicated operations. Leave each result in trigonometric form.

45. $[2 \text{ cis}(\pi/3)][4 \text{ cis}(2\pi/3)]$
46. $[6 \text{ cis}(3\pi/2)][2 \text{ cis}(\pi/2)]$
47. $[5 \text{ cis}(5\pi/6)][\frac{1}{2} \text{ cis}(\pi/3)]$
48. $[12 \text{ cis}(2\pi/3)][(\sqrt{3}/2) \text{ cis}(\pi/6)]$
49. $\dfrac{18 \text{ cis}(3\pi/4)}{3 \text{ cis}(\pi/2)}$
50. $\dfrac{2 \text{ cis}(5\pi/6)}{12 \text{ cis}(2\pi/3)}$
51. $\dfrac{3\sqrt{3} \text{ cis } 420°}{6 \text{ cis } 210°}$
52. $\dfrac{\sqrt{6} \text{ cis } 225°}{\sqrt{2} \text{ cis } 75°}$

In Exercises 53–60, use Theorem 7.2 to perform the indicated operations. Express the results in the form $a + bi$, where a and b are real numbers.

53. $[3\sqrt{2} \text{ cis}(2\pi/3)][5 \text{ cis}(\pi/6)]$
54. $[\frac{3}{4} \text{ cis}(\pi/3)][8 \text{ cis}(5\pi/6)]$
55. $[16 \text{ cis}(5\pi/4)][\frac{3}{8} \text{ cis}(\pi/2)]$
56. $[(2\sqrt{2} \text{ cis}(\pi/6)][3\sqrt{2} \text{ cis}(5\pi/3)]$
57. $\dfrac{\sqrt{3}/2 \text{ cis}(7\pi/6)}{\frac{2}{3} \text{ cis}(2\pi/3)}$
58. $\dfrac{12 \text{ cis}(5\pi/3)}{\frac{5}{4} \text{ cis}(5\pi/6)}$
59. $\dfrac{25 \text{ cis } 315°}{2 \text{ cis } 75°}$
60. $\dfrac{7\sqrt{2} \text{ cis } 220°}{3 \text{ cis } 85°}$
61. If $z = r(\cos \theta + i \sin \theta)$, show that the conjugate of z is $\bar{z} = r(\cos \theta - i \sin \theta)$.

62. If $z = r(\cos \theta + i \sin \theta)$, show that $1/z = (1/r)(\cos \theta - i \sin \theta)$. [*Hint*: Express 1 in trigonometric form and use Theorem 7.2.]

63. Show that the negative of $r \operatorname{cis} \theta$ is $r \operatorname{cis}(\theta + \pi)$.

64. If $z = r \operatorname{cis} \theta$, find a formula for z^2 and z^3. Do your results suggest a general formula for z^n?

65. In the complex plane, sketch the graph of all complex numbers $z = a + bi$ for which $|z| = 1$.

66. In the complex plane, sketch the graph of all complex numbers $z = r \operatorname{cis} \theta$, where $\theta = \pi/3$.

67. The **complex exponential function** is defined by
$$e^z = e^x(\cos y + i \sin y),$$
where x and y are real numbers and $z = x + yi$.

a. Evaluate e^0, $e^{\pi/2i}$, and $e^{\pi i}$.

b. Show that $e^{z + 2\pi i} = e^z$.   **c.** Show that $|e^z| = e^x$.

68. The **complex logarithmic function** is defined by
$$\log z = \ln r + i\theta,$$
where $z = r \operatorname{cis} \theta$, and $0 \le \theta < 2\pi$.

a. Evaluate $\log i$, $\log(-1)$ and $\log(-i)$.

b. Show that $e^{\log z} = z$. (See Exercise 67.)

69. Extend part (i) of Theorem 7.2 to the case of three complex numbers.

70. Prove part (ii) of Theorem 7.2.

7.4 DE MOIVRE'S THEOREM AND THE nTH ROOTS OF A COMPLEX NUMBER

As we noted in the previous section, the product of two complex numbers in trigonometric form can be found by multiplying their moduli and adding their arguments. In this section, we see how this procedure can simplify the process of finding powers and roots of complex numbers.

If, for example, we apply part (i) of Theorem 7.2 to the product $z \cdot z$, where $z = r \operatorname{cis} \theta$, we find that

$$z^2 = z \cdot z = (r \operatorname{cis} \theta)(r \operatorname{cis} \theta) = r^2(\operatorname{cis} 2\theta).$$

Applying the same theorem to the product $z \cdot z^2$, we obtain

$$z^3 = z \cdot z^2 = (r \operatorname{cis} \theta)(r^2 \operatorname{cis} 2\theta) = r^3 \operatorname{cis} 3\theta.$$

Repeated applications of this theorem lead us to a useful formula for finding the nth power of a complex number, where n is a positive integer. The result is the following theorem, which is credited to French mathematician Abraham DeMoivre (1667–1754).

DeMoivre's Theorem

If $z = r \operatorname{cis} \theta$, then

$$z^n = (r \operatorname{cis} \theta)^n = r^n \operatorname{cis} n\theta$$

for every positive integer n.

Although the discussion preceding DeMoivre's theorem appears to lead directly to the given formula, a complete proof requires the principle of mathematical induction. Since mathematical induction is not discussed until Section 10.4, we will defer the proof until that time.

EXAMPLE 20

Use DeMoivre's theorem to evaluate each of the following. Express each solution in the form $a + bi$, where a and b are real numbers.

a. $(2 \operatorname{cis} 48°)^5$ **b.** $\left(-\dfrac{3\sqrt{3}}{2} + \dfrac{3}{2}i \right)^3$ **c.** $(1 - \sqrt{3}i)^4$

Solution

a. Applying DeMoivre's theorem and simplifying, we obtain

$$\begin{aligned}(2 \text{ cis } 48°)^5 &= 2^5 \text{ cis}(5 \cdot 48°) \\ &= 32 \text{ cis } 240° \\ &= 32\left(-\frac{1}{2} - \frac{\sqrt{3}}{2}i\right) \\ &= -16 - 16\sqrt{3}i.\end{aligned}$$

b. Expressing $-3\sqrt{3}/2 + \frac{3}{2}i$ in trigonometric form, we find that

$$-\frac{3\sqrt{3}}{2} + \frac{3}{2}i = 3 \text{ cis } \frac{5\pi}{6}.$$

Hence, by DeMoivre's Theorem, we have

$$\begin{aligned}\left(-\frac{3\sqrt{3}}{2} + \frac{3}{2}i\right)^3 &= 3^3 \text{ cis}\left(3 \cdot \frac{5\pi}{6}\right) \\ &= 27 \text{ cis } \frac{5\pi}{2} \\ &= 27(0 + i) \\ &= 27i.\end{aligned}$$

c. As in part (b), we begin by expressing $1 - \sqrt{3}i$ in trigonometric form.

$$1 - \sqrt{3}i = 2 \text{ cis } \frac{5\pi}{3}.$$

Applying DeMoivre's theorem, we find that

$$\begin{aligned}(1 - \sqrt{3}i)^4 &= 2^4 \text{ cis}\left(4 \cdot \frac{5\pi}{3}\right) \\ &= 16 \text{ cis } \frac{20\pi}{3} \\ &= 16\left(-\frac{1}{2} + \frac{\sqrt{3}}{2}i\right) \\ &= -8 + 8\sqrt{3}i.\end{aligned}$$

EXAMPLE 21

Use DeMoivre's theorem to approximate $(-4 + \sqrt{5}i)^5$. Express the answer in trigonometric form.

Solution   In the solution of part (b) of Example 18 in the previous section, we found that

$$-4 + \sqrt{5}i = \sqrt{21} \text{ cis}\left(180° - \tan^{-1}\frac{\sqrt{5}}{4}\right) \approx 4.6 \text{ cis } 150.8°.$$

Hence,

$$\begin{aligned}(-4 + \sqrt{5}i)^5 &\approx (4.6)^5 \text{ cis}(5 \cdot 150.8°) \\ &\approx 2060 \text{ cis } 754° \\ &= 2060 \text{ cis } 34°\end{aligned}$$

The *n*th Roots of a Complex Number

As we shall now see, DeMoivre's theorem can also be used to find roots of complex numbers.

DEFINITION 7.1

Let z be a complex number and let n be a positive integer. If u is a complex number such that $u^n = z$, then we say that u is an **nth root** of z.

In part (b) of Example 20, we found that

$$\left(-\frac{3\sqrt{3}}{2} + \frac{3}{2}i\right)^3 = 27i.$$

This means that $-3\sqrt{3}/2 + \frac{3}{2}i$ is a cube root of the complex number $27i$. At this point, it seems natural for us to ask the question: Is $-(3\sqrt{3}/2) + \frac{3}{2}i$ the only cube root of $27i$? This question, in turn, leads to a more general question, namely: How many nth roots does a given complex number have, and how can they be found? The answer to each of these questions is provided by DeMoivre's theorem. We begin with the first.

If the complex number u is a cube root of $27i$, we know from Definition 7.1 that

$$u^3 = 27i.$$

If $u = r \operatorname{cis} \theta$, in trigonometric form, the latter equation becomes

$$(r \operatorname{cis} \theta)^3 = 27i. \tag{1}$$

Applying DeMoivre's theorem to the left side of equation (1), and expressing $27i$ in trigonometric form, we obtain

$$r^3 \operatorname{cis} 3\theta = 27 \operatorname{cis} \frac{\pi}{2}. \tag{2}$$

Since two complex numbers that are equal have equal moduli, we see from equation (2) that

$$r^3 = 27, \quad \text{or} \quad r = 3.$$

Since the arguments of two complex numbers that are equal must be coterminal angles, we know also from equation (2) that

$$3\theta = \frac{\pi}{2} + k(2\pi), \quad \text{where } k \text{ is an integer.}$$

Dividing both sides of this equation by 3, we find that an argument θ of any cube root of $27i$ must satisfy the equation

$$\theta = \frac{\pi}{6} + k\left(\frac{2\pi}{3}\right), \quad \text{where } k \text{ is an integer.} \tag{3}$$

We can summarize our findings to this point by saying that a complex number is a cube root of $27i$ if and only if it has a modulus of 3 and an argument θ, that is given by equation (3). Thus any cube root of $27i$ must have the form

$$u_k = 3 \operatorname{cis}\left[\frac{\pi}{6} + \left(k \cdot \frac{2\pi}{3}\right)\right], \quad \text{where } k \text{ is an integer.} \qquad (4)$$

Although it appears from equation (4) that $27i$ has infinitely many cube roots, further investigation will reveal that this is not the case. If we substitute 0 for k in equation (4), we obtain the cube root

$$u_0 = 3 \operatorname{cis}\frac{\pi}{6} = 3\left(\frac{\sqrt{3}}{2} + \frac{1}{2}i\right) = \frac{3\sqrt{3}}{2} + \frac{3}{2}i.$$

(Note that the subscript on u indicates the value of k that is used.) If, in like manner, we replace k in equation (4) by 1 and 2, respectively, we obtain two additional cube roots of $27i$ as shown below.

$$u_1 = 3 \operatorname{cis}\left(\frac{\pi}{6} + \frac{2\pi}{3}\right) = 3 \operatorname{cis}\frac{5\pi}{6} = -\frac{3\sqrt{3}}{2} + \frac{3}{2}i \quad \text{and}$$

$$u_2 = 3 \operatorname{cis}\left(\frac{\pi}{6} + \frac{4\pi}{3}\right) = 3 \operatorname{cis}\frac{3\pi}{2} = -3i.$$

(Note that u_1 is the cube root of $27i$ from part (b) of Example 20.) If we continue this process and replace k by 3 in equation (4), we get

$$u_3 = 3 \operatorname{cis}\left(\frac{\pi}{6} + 2\pi\right).$$

Since both the sine and cosine functions have period 2π, this equation becomes

$$u_3 = 3 \operatorname{cis}\frac{\pi}{6}.$$

Thus we see that the cube root u_3 is the same as the cube root u_0. Similarly, when k is replaced (successively) by 4 and 5, the roots obtained are precisely the roots u_1 and u_2 we obtained earlier. In fact, when k is replaced by any integer other than 0, 1, and 2, we find that no additional cube roots of $27i$ are obtained. We thus conclude that the complex number $27i$ has exactly three cube roots, namely u_0, u_1, and u_2, as described above.

We observe that the three cube roots of $27i$ are the roots of the polynomial equation

$$x^3 - 27i = 0.$$

Our results are thus in keeping with Theorem 3.2 (on page 135), which states that every polynomial equation of degree n has exactly n roots.

The cube roots of the complex number $27i$ have an interesting geometric representation. Since each root has a modulus of 3, each can be represented in the complex plane as a point on the circle with center at the origin and radius 3. The three roots, moreover, are equally spaced on the circle since the arguments of successive roots differ by $2\pi/3$ radians. See Figure 7.23.

The following theorem, which answers the second of the questions posed earlier in this section, generalizes our results for the cube roots of $27i$ to the *n*th roots of any complex number z, where n is a positive integer. The proof is left as an exercise (see Exercise 70).

FIGURE 7.23

Geometric representation of the three cube roots of $27i$

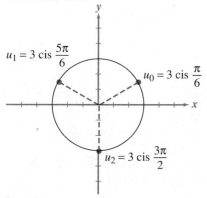

///

THEOREM 7.3   nth Roots of a Complex Number

Let $z = r \operatorname{cis} \theta$ be a nonzero complex number. For each positive integer n, z has exactly n distinct nth roots, which are given by the formula

$$u_k = r^{1/n} \operatorname{cis}\left[\frac{\theta + k(2\pi)}{n}\right],$$

where $k = 0, 1, 2, \ldots, n - 1$.

The nth roots of a complex number $z = r \operatorname{cis} \theta$ can be represented geometrically as points on a circle of radius $r^{1/n}$ centered at the origin of the complex plane. The roots are equally spaced around the circle since the arguments of successive nth roots differ by $2\pi/n$ radians.

EXAMPLE 22

Find the sixth roots of $-2 + 2i$ and represent them geometrically.

Solution   Expressing $-2 + 2i$ in trigonometric form, we find that

$$-2 + 2i = 2^{3/2} \operatorname{cis} \frac{3\pi}{4}.$$

By Theorem 7.3, the sixth roots of $-2 + 2i$ are given by

$$u_k = (2^{3/2})^{1/6} \operatorname{cis}\left[\frac{(3\pi/4) + k(2\pi)}{6}\right]$$

$$= 2^{1/4} \operatorname{cis}\left[\frac{\pi}{8} + k\left(\frac{\pi}{3}\right)\right],$$

where $k = 0, 1, 2, 3, 4,$ and 5. Thus the six sixth roots of $-2 + 2i$ are

$$u_0 = 2^{1/4} \operatorname{cis} \frac{\pi}{8}$$

$$u_1 = 2^{1/4} \operatorname{cis}\left(\frac{\pi}{8} + \frac{\pi}{3}\right) = 2^{1/4} \operatorname{cis} \frac{11\pi}{24}$$

$$u_2 = 2^{1/4} \operatorname{cis}\left(\frac{\pi}{8} + \frac{2\pi}{3}\right) = 2^{1/4} \operatorname{cis} \frac{19\pi}{24}$$

$$u_3 = 2^{1/4} \operatorname{cis}\left(\frac{\pi}{8} + \pi\right) = 2^{1/4} \operatorname{cis} \frac{9\pi}{8}$$

$$u_4 = 2^{1/4} \operatorname{cis}\left(\frac{\pi}{8} + \frac{4\pi}{3}\right) = 2^{1/4} \operatorname{cis} \frac{35\pi}{24}$$

$$u_5 = 2^{1/4} \operatorname{cis}\left(\frac{\pi}{8} + \frac{5\pi}{3}\right) = 2^{1/4} \operatorname{cis} \frac{43\pi}{24}$$

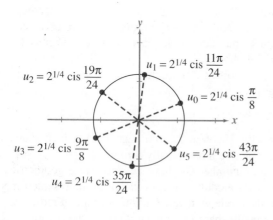

From the discussion following Theorem 7.3, we know that these six roots are equally spaced on a circle centered at the origin with radius $\sqrt[4]{2} \approx 1.19$. After locating u_0, whose argument is $\pi/8$ radians, we can readily locate the remaining five roots using angular increments of $\pi/3$ radians.

EXAMPLE 23

Find the fourth roots of 16 cis 148° and represent them geometrically.

Solution Replacing 2π by 360° in Theorem 7.3, the fourth roots of 16 cis 148° are given by

$$u_k = (16)^{1/4} \text{ cis}\left[\frac{148° + k(360°)}{4}\right] = 2 \text{ cis}[37° + k(90°)],$$

where $k = 0, 1, 2,$ and 3.

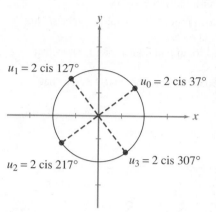

$u_1 = 2$ cis 127°

$u_0 = 2$ cis 37°

$u_2 = 2$ cis 217°

$u_3 = 2$ cis 307°

Thus $u_0 = 2 \text{ cis } 37°,$ $u_1 = 2 \text{ cis } 127°,$

$$u_2 = 2 \text{ cis } 217°,   \text{and}   u_3 = 2 \text{ cis } 307°.$$

Starting with $u_0 = 2 \text{ cis } 37°$, the four roots are spaced in increments of 90° around a circle of radius 2 centered at the origin.

The *n* distinct *n*th roots of 1, which are called the **nth roots of unity**, are of special importance in higher mathematics. These roots can be represented geometrically as points that are equally spaced along the unit circle in the complex plane.

EXAMPLE 24

Find the fifth roots of unity and represent them geometrically.

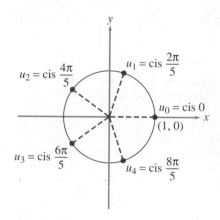

$u_2 = \text{cis } \dfrac{4\pi}{5}$

$u_1 = \text{cis } \dfrac{2\pi}{5}$

$u_0 = \text{cis } 0$

(1, 0)

$u_3 = \text{cis } \dfrac{6\pi}{5}$

$u_4 = \text{cis } \dfrac{8\pi}{5}$

Solution Since $1 = \text{cis } 0$ in trigonometric form, it follows from Theorem 7.3 that the fifth roots of unity are given by

$$u_k = \text{cis}\left[\frac{0 + k(2\pi)}{5}\right] = \text{cis } k\left(\frac{2\pi}{5}\right),$$

where $k = 0, 1, 2, 3,$ and 4. Substituting 0, 1, 2, 3, and 4 for k, we find that

$$u_0 = \text{cis } 0 = 1, \ u_1 = \text{cis } \frac{2\pi}{5}, \ u_2 = \text{cis } \frac{4\pi}{5}, \ u_3 = \frac{6\pi}{5}, \text{ and } u_4 = \frac{8\pi}{5}.$$

After plotting $u_0 = (1, 0)$, the remaining four roots are equally spaced on the unit circle at angular increments of $2\pi/5$ radians $= 72°$.

EXERCISES 7.4

In Exercises 1–24, use DeMoivre's theorem to find the indicated powers of the given complex numbers. Express your solutions in the form $a + bi$.

1. $[2 \text{ cis } (7\pi/12)]^4$

2. $[3 \text{ cis } (5\pi/12)]^6$

3. $(3 \text{ cis } 120°)^3$

4. $(2 \text{ cis } 135°)^8$

5. $[\text{cis } (\pi/8)]^{10}$

6. $[\text{cis } (3\pi/8)]^{12}$

7. $(\sqrt{2} \text{ cis } 30°)^9$

8. $(2\sqrt{3} \text{ cis } 45°)^5$

9. $[\sqrt[4]{3} \text{ cis } (5\pi/12)]^{16}$

10. $[\sqrt[5]{2} \text{ cis } (2\pi/5)]^{10}$

11. $[(\sqrt{3}/2) \text{ cis } 105°]^6$

12. $[(\sqrt{2}/2) \text{ cis } 22.5°]^8$

13. $(1 + \sqrt{3}i)^4$

14. $(\sqrt{3} - i)^3$

15. $[\frac{1}{2} - (\sqrt{3}/2)i]^7$

16. $(5 + 5i)^4$

17. $(-1 + i)^5$

18. $(1 - i)^8$

19. $[(-\sqrt{2}/2) - (\sqrt{2}/2)i]^6$

20. $[(-\sqrt{2}/2) + (\sqrt{2}/2)i]^{11}$

21. $(-3 + 3i)^4$

22. $(\sqrt{3} - \sqrt{3}i)^7$

23. $(2\sqrt{3} - 2i)^{10}$

24. $(-5 + 5\sqrt{3}i)^{12}$

In Exercises 25–36, use DeMoivre's theorem to find the indicated powers of the given complex numbers. Express your solutions in trigonometric form.

25. $(\sqrt{5} \text{ cis } 20°)^7$

26. $(3\sqrt{2} \text{ cis } 12°)^6$

27. $\left[\frac{7}{4} \text{ cis } (5\pi/7)\right]^{12}$

28. $\left[\frac{1}{2} \text{ cis } (3\pi/5)\right]^{7}$

29. $(-2\sqrt{2} - 2i)^{6}$

30. $(-3 + 4i)^{10}$

31. $(\frac{1}{2} - \frac{1}{3}i)^{4}$

32. $(\frac{1}{4} - \frac{3}{4}i)^{3}$

33. $(3 - 2i)^{4}$

34. $(-1 + \sqrt{2}i)^{6}$

35. $(-5 + 12i)^{3}$

36. $(1 + 3i)^{5}$

In Exercises 37–54, find the indicated roots and represent them geometrically.

37. fifth roots of $\sqrt{3} - i$

38. cube roots of $-1 + \sqrt{3}i$

39. cube roots of $-\sqrt{2} - \sqrt{2}i$

40. fourth roots of $8 - 8i$

41. fourth roots of $-2 + 2\sqrt{3}i$

42. fifth roots of $-4\sqrt{3} - 4i$

43. square roots of $16i$

44. square roots of $-9i$

45. sixth roots of unity

46. fourth roots of unity

47. fourth roots of -16

48. fifth roots of 32

49. cube roots of $8 \text{ cis}(2\pi/3)$

50. cube roots of $27 \text{ cis}(5\pi/4)$

51. fourth roots of $81 \text{ cis } 225°$

52. fifth roots of $64 \text{ cis } 120°$

53. fifth roots of $32 \text{ cis } 230°$

54. cube roots of $125 \text{ cis } 87°$

In Exercises 55–60, find all solutions of the given equation. Express your answers in trigonometric form.

55. $x^3 + 27 = 0$

56. $x^3 - 1 = 0$

57. $x^5 - i = 0$

58. $x^4 + 2i = 0$

59. $x^4 - 1 + i = 0$

60. $x^5 - \sqrt{3} + i = 0$

61. DeMoivre's theorem can be extended to apply to all integers. In other words,

$$(r \text{ cis } \theta)^n = r^n \text{ cis } n\theta,$$

where n is any integer. Use the extended version of DeMoivre's theorem to evaluate $(1 - \sqrt{3}i)^{-4}$. Express your answer in the form $a + bi$.

62. Using the extended version of DeMoivre's theorem (see Exercise 61) to evaluate the following expression. Give the

solution in the form $a + bi$.

$$\frac{(1 - i)^{-2}}{(-\sqrt{3} + i)^4}$$

63. If $-1 + i$ is a fourth root of the complex number z, find the remaining fourth roots of z.

64. If $\sqrt{3} - i$ is a sixth root of the complex number z, find the remaining sixth roots of z.

65. If u_k is any cube root of unity other than 1, show that $1 + u_k + u_k^2 = 0$.

66. If u_k is any nth root of unity other than 1, show that $1 + u_k + u_k^2 + \cdots + u_k^{n-1} = 0$.

67. If

$$\cos \theta = 1 - \frac{\theta^2}{2!} + \frac{\theta^4}{4!} - \frac{\theta^6}{6!} + \cdots$$
$$+ \frac{(-1)^n \theta^{2n}}{(2n)!} + \cdots,$$

$$\sin \theta = \theta - \frac{\theta^3}{3!} + \frac{\theta^5}{5!} - \frac{\theta^7}{7!} + \cdots$$
$$+ \frac{(-1)^n \theta^{2n+1}}{(2n+1)!} + \cdots,$$

and

$$e^{i\theta} = 1 + (i\theta) + \frac{(i\theta)^2}{2!} + \frac{(i\theta)^3}{3!} + \frac{(i\theta)^4}{4!} + \cdots$$
$$+ \frac{(i\theta)^n}{n!} + \cdots,$$

show that

$$e^{i\theta} = \cos \theta + i \sin \theta.$$

The latter equation is known as **Euler's formula**.

68. Using Euler's formula from Exercise 67, verify that $e^{i\pi} = -1$.

69. Evaluate $(\cos \theta + i \sin \theta)^2$ by using DeMoivre's theorem and by multiplying the two binomial factors. By equating the imaginary parts of the two results, show that $\sin 2\theta = 2 \sin \theta \cos \theta$.

70. Prove Theorem 7.3.

7.5 VECTORS IN THE PLANE

Quantities such as area, time, and distance, which can be completely characterized by a single real number, or **scalar**, are called **scalar quantities**. On the other hand, many physical quantities—such as force, velocity, and acceleration—involve both magnitude (size) and direction, and hence cannot be completely characterized by a single real number. We call such two-dimensional quantities **vector quantities**, and we represent them by mathematical entities called **vectors**.

Geometric Representation of a Vector

Geometrically, a vector can be viewed as a directed line segment represented by an arrow in the plane. The length and direction of the arrow indicate the magnitude and direction, respectively, of the vector quantity it represents. For example, to represent a wind velocity of 20 miles per hour from the southwest, we can use an arrow 20 units long pointing to the northeast as shown in Figure 7.24.

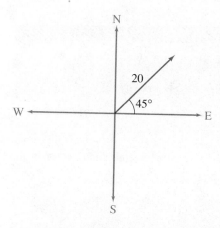

FIGURE 7.24 Vector representing a wind velocity of 20 mph from the southwest

If A and B are points in the plane, the vector represented by the directed line segment from A to B is denoted by $\overrightarrow{AB}$ (see Figure 7.25). The point A is called the **initial point** of $\overrightarrow{AB}$, while B is called the **terminal point**. Vectors whose endpoints are not specified will be represented by lowercase boldface letters such as **u**, **v**, and **w**. (In handwritten work, the symbols $\vec{u}$, $\vec{v}$, and $\vec{w}$ can be used to represent the vectors **u**, **v**, and **w**.)

The **magnitude** of a vector $\mathbf{v} = \overrightarrow{AB}$ is the length of the line segment AB and is denoted by $\|\mathbf{v}\|$ or $\|\overrightarrow{AB}\|$. Vectors having the same magnitude and the same direction are said to be **equal**. For example, all of the vectors shown in Figure 7.26 are equal.

FIGURE 7.26 Equal vectors

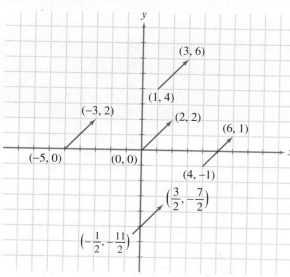

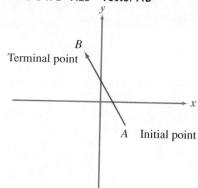

FIGURE 7.25 Vector $\overrightarrow{AB}$

Algebraic Representation of a Vector

From the definition of equal vectors, we see that a vector is determined by its magnitude and direction only, and not by its location in the plane. Thus we can freely translate (move) any vector in the plane from one location to another, provided neither the magnitude nor the direction of the vector is changed. This means that any vector in the plane is equal to a vector of the same length that has its initial point at the origin. A vector having its initial point at the origin is called a **position vector**, or **radius vector**. For the remainder of this section, unless otherwise specified, we will deal exclusively with position vectors. Since a position vector is uniquely determined by its terminal point, we can give the following algebraic representation of a vector.

DEFINITION 7.2   Algebraic Representation of a Vector

Let **v** be a position vector whose terminal point is (v_1, v_2). The **algebraic representation** of **v** is given by

$$\mathbf{v} = \langle v_1, v_2 \rangle,$$

where v_1 and v_2 are called the **components** of **v**.

From the algebraic representation of a vector, we see that the magnitude of $\mathbf{v} = \langle v_1, v_2 \rangle$ is $\sqrt{v_1^2 + v_2^2}$, the distance from the origin to the point $P = (v_1, v_2)$. Thus we have the following theorem:

THEOREM 7.4   Magnitude of a Vector

The **magnitude** (length) of the vector $\mathbf{v} = \langle v_1, v_2 \rangle$ is given by

$$\|\mathbf{v}\| = \sqrt{v_1^2 + v_2^2}.$$

EXAMPLE 25

Find the magnitude of each of the following vectors. Sketch each vector in the xy-plane.

a. $\mathbf{u} = \langle 2, -5 \rangle$ b. $\mathbf{v} = \langle -\frac{5}{2}, 4 \rangle$

Solution

a.   By Theorem 7.4, the magnitude of $\mathbf{u}$ is given by
$$\|\mathbf{u}\| = \sqrt{(2)^2 + (-5)^2}$$
$$= \sqrt{29} \approx 5.4.$$

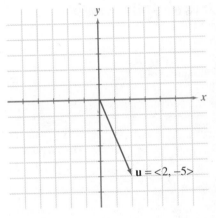

b.   Again by Theorem 7.4,
$$\|\mathbf{v}\| = \sqrt{(-\tfrac{5}{2})^2 + (4)^2}$$
$$= \sqrt{89}/2 \approx 4.7.$$

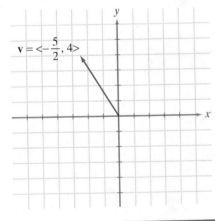

Operations on Vectors

The process of multiplying a vector by a real number (*scalar*) is known as **scalar multiplication**. When a vector is expressed in algebraic form, scalar multiplication is readily performed using the components of the vector.

DEFINITION 7.3 **Scalar Multiplication**

If $\mathbf{v} = \langle v_1, v_2 \rangle$ is a vector and k is a real number, then

$$k\mathbf{v} = k\langle v_1, v_2 \rangle = \langle kv_1, kv_2 \rangle.$$

The vector $k\mathbf{v}$ is called a **scalar multiple** of the vector $\mathbf{v}$.

EXAMPLE 26

Let $\mathbf{v} = \langle 2, 3 \rangle$.

a. Find $\|\mathbf{v}\|$.   **b.** Find $3\mathbf{v}$ and $\|3\mathbf{v}\|$.   **c.** Find $-\frac{1}{2}\mathbf{v}$ and $\|-\frac{1}{2}\mathbf{v}\|$.

d. Sketch $\mathbf{v}$, $3\mathbf{v}$, and $-\frac{1}{2}\mathbf{v}$ in the xy-plane.

Solution

a. By Theorem 7.4, $\|\mathbf{v}\| = \sqrt{(2)^2 + (3)^2} = \sqrt{13} \approx 3.6$.

b. By Definition 7.3, $3\mathbf{v} = 3\langle 2, 3 \rangle = \langle 6, 9 \rangle$, and by Theorem 7.4,

$$\|3\mathbf{v}\| = \sqrt{(6)^2 + (9)^2} = 3\sqrt{13} \approx 10.8.$$

c. As in part (b), $-\frac{1}{2}\mathbf{v} = -\frac{1}{2}\langle 2, 3 \rangle = \langle -1, -\frac{3}{2} \rangle$ and

$$\left\| -\frac{1}{2}\mathbf{v} \right\| = \sqrt{(-1)^2 + \left(-\frac{3}{2} \right)^2} = \frac{\sqrt{13}}{2} \approx 1.8.$$

d. The vectors $\mathbf{v}$, $3\mathbf{v}$, and $-\frac{1}{2}\mathbf{v}$ are shown in Figure 7.27.

FIGURE 7.27

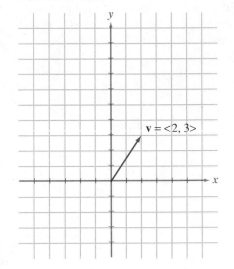

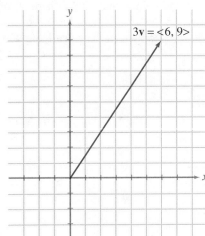

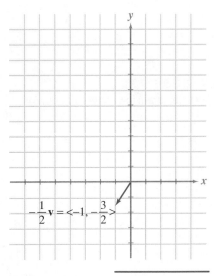

In the solution of Example 26, we found that

$$\|3\mathbf{v}\| = 3\|\mathbf{v}\| \qquad \text{and} \qquad \left\| -\frac{1}{2}\mathbf{v} \right\| = \frac{1}{2}\|\mathbf{v}\| = \left| -\frac{1}{2} \right| \|\mathbf{v}\|.$$

Also, we see from Figure 7.27 that the vectors $\mathbf{v}$ and $3\mathbf{v}$ have the same direction, while $\mathbf{v}$ and $-\frac{1}{2}\mathbf{v}$ have opposite directions. More generally, we have the result shown in the box on the next page.

///

Geometric Implications of Scalar Multiplication

Let **v** be a vector and let k be a real number.

(i) The magnitude of the vector $k\mathbf{v}$ is equal to $|k|$ times the magnitude of **v**; that is
$$\|k\mathbf{v}\| = |k|\,\|\mathbf{v}\|.$$

(ii) The vectors $k\mathbf{v}$ and **v** have the same direction when k is positive and opposite directions when k is negative.

As was the case for scalar multiplication, the addition of vectors is readily carried out when vectors are expressed in algebraic form.

///

DEFINITION 7.4   Vector Addition

Let $\mathbf{u} = \langle u_1, u_2 \rangle$ and $\mathbf{v} = \langle v_1, v_2 \rangle$ be vectors. The **sum** of **u** and **v** is the vector
$$\mathbf{u} + \mathbf{v} = \langle u_1, u_2 \rangle + \langle v_1, v_2 \rangle = \langle u_1 + v_1, u_2 + v_2 \rangle.$$

According to Definition 7.4, we add vectors by adding their corresponding components. When the vectors **u** and **v** have different (but not opposite) directions, it can be shown that their sum $\mathbf{u} + \mathbf{v}$ is the vector emanating from the origin and coinciding with a diagonal of the parallelogram having **u** and **v** as its adjacent sides (see Figure 7.28). For this reason, the geometric version of vector addition is known as the **parallelogram law**.

FIGURE 7.28
Parallelogram law for vector addition

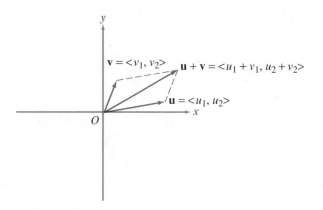

EXAMPLE 27

If $\mathbf{u} = \langle -\frac{5}{2}, -4 \rangle$ and $\mathbf{v} = \langle -3, 7 \rangle$, find $\mathbf{u} + \mathbf{v}$. Use the vectors **u** and **v** to illustrate the parallelogram law.

Solution   By Definition 7.4, we have
$$\mathbf{u} + \mathbf{v} = \left\langle -\frac{5}{2}, -4 \right\rangle + \langle -3, 7 \rangle = \left\langle -\frac{5}{2} - 3, -4 + 7 \right\rangle = \left\langle -\frac{11}{2}, 3 \right\rangle.$$

FIGURE 7.29

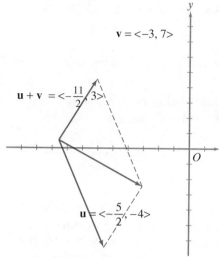

$v = <-3, 7>$

$u + v = <-\dfrac{11}{2}, 3>$

$u = <-\dfrac{5}{2}, -4>$

In Figure 7.29, we illustrate the parallelogram law using the vectors **u** and **v**.

Vectors, under the operations of addition and scalar multiplication, share many of the properties of real numbers under ordinary addition and multiplication. For example, the vector $\langle 0, 0 \rangle$ plays the role of the additive identity for vector addition since, for any vector $\mathbf{v} = \langle v_1, v_2 \rangle$,

$$\langle v_1, v_2 \rangle + \langle 0, 0 \rangle = \langle v_1, v_2 \rangle.$$

The vector $\langle 0, 0 \rangle$ is called the **zero vector** and is denoted by **0**. In a similar manner, the vector $-\mathbf{v} = -1\langle v_1, v_2 \rangle = \langle -v_1, -v_2 \rangle$ serves as the additive inverse of **v** since

$$\mathbf{v} + (-\mathbf{v}) = \langle v_1, v_2 \rangle + \langle -v_1, -v_2 \rangle = \langle 0, 0 \rangle = \mathbf{0}.$$

We summarize properties of vector addition and scalar multiplication in the following box:

Properties of Vector Addition and Scalar Multiplication

Let **u**, **v**, and **w** be vectors and let a and b be real numbers. Then

(i) $\mathbf{u} + \mathbf{v} = \mathbf{v} + \mathbf{u}$ (vi) $\mathbf{v} + \mathbf{0} = \mathbf{v}$

(ii) $(\mathbf{u} + \mathbf{v}) + \mathbf{w} = \mathbf{u} + (\mathbf{v} + \mathbf{w})$ (vii) $1(\mathbf{v}) = \mathbf{v}$

(iii) $(ab)\mathbf{v} = a(b\mathbf{v})$ (viii) $\mathbf{v} + (-\mathbf{v}) = \mathbf{0}$

(iv) $a(\mathbf{u} + \mathbf{v}) = a\mathbf{u} + a\mathbf{v}$ (ix) $0(\mathbf{v}) = \mathbf{0}$

(v) $(a + b)\mathbf{v} = a\mathbf{v} + b\mathbf{v}$ (x) $a(\mathbf{0}) = \mathbf{0}$

Property (i), which states that vector addition is commutative, can be verified as follows: Let $\mathbf{u} = \langle u_1, u_2 \rangle$ and $\mathbf{v} = \langle v_1, v_2 \rangle$ be vectors. Then

$\mathbf{u} + \mathbf{v} = \langle u_1 + v_1, u_2 + v_2 \rangle$ Definition 7.4

$\qquad = \langle v_1 + u_1, v_2 + u_2 \rangle$ commutative law for addition of real numbers

$\qquad = \mathbf{v} + \mathbf{u}.$ Definition 7.4

The remaining properties can be verified in a similar manner (see Exercise 70).

As in the case of real numbers, we can define vector subtraction in terms of vector addition. Specifically, if **u** and **v** are vectors,

$$\mathbf{u} - \mathbf{v} = \mathbf{u} + (-\mathbf{v}).$$

In terms of components, vector subtraction is defined as follows:

Vector Subtraction

Let $\mathbf{u} = \langle u_1, u_2 \rangle$ and $\mathbf{v} = \langle v_1, v_2 \rangle$ be vectors. Then

$$\mathbf{u} - \mathbf{v} = \langle u_1 - v_1, u_2 - v_2 \rangle.$$

EXAMPLE 28

If $\mathbf{u} = \langle 8, -\frac{2}{3} \rangle$ and $\mathbf{v} = \langle -5, 11 \rangle$, find each of the following vectors:

a. $6\mathbf{u} + \mathbf{v}$ b. $\mathbf{u} - 2\mathbf{v}$ c. $-\frac{1}{2}\mathbf{u} + 3\mathbf{v}$

Solution   Using the definitions of scalar multiplication, vector addition, and vector subtraction, we have

a. $6\mathbf{u} + \mathbf{v} = 6\langle 8, -\frac{2}{3} \rangle + \langle -5, 11 \rangle = \langle 48, -4 \rangle + \langle -5, 11 \rangle = \langle 43, 7 \rangle$
b. $\mathbf{u} - 2\mathbf{v} = \langle 8, -\frac{2}{3} \rangle - 2\langle -5, 11 \rangle = \langle 8, -\frac{2}{3} \rangle - \langle -10, 22 \rangle = \langle 18, -\frac{68}{3} \rangle$
c. $-\frac{1}{2}\mathbf{u} + 3\mathbf{v} = -\frac{1}{2}\langle 8, -\frac{2}{3} \rangle + 3\langle -5, 11 \rangle = \langle -4, \frac{1}{3} \rangle + \langle -15, 33 \rangle = \langle -19, \frac{100}{3} \rangle$

The Direction Angle for a Vector

To this point, little has been said about the direction of a vector. However, in problems involving physical quantities, such as force and velocity, both the magnitude and the direction of a vector must be determined. To enable us to describe the direction of a nonzero vector, we introduce the concept of direction angle. (It is customary not to assign a direction to the zero vector.)

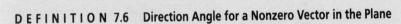

DEFINITION 7.6   Direction Angle for a Nonzero Vector in the Plane

Let $\mathbf{v} = \langle v_1, v_2 \rangle$ be a nonzero vector. The smallest positive angle θ, measured from the positive x-axis to the vector $\mathbf{v}$, is called the **direction angle** for $\mathbf{v}$.

The following theorem, which relates the components of a nonzero vector to its magnitude and direction, will enable us to find the direction angle for a given nonzero vector:

THEOREM 7.5

Let $\mathbf{v} = \langle v_1, v_2 \rangle$ be any nonzero vector. If θ is the direction angle for $\mathbf{v}$, then

$$v_1 = \|\mathbf{v}\| \cos \theta \quad \text{and} \quad v_2 = \|\mathbf{v}\| \sin \theta.$$

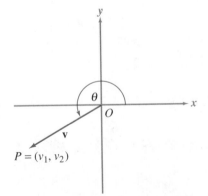

Proof   Let $\mathbf{v} = \langle v_1, v_2 \rangle$ be a nonzero vector and let θ be its direction angle. We will prove the theorem for the case where the terminal point of $\mathbf{v}$, $P = (v_1, v_2)$, lies in the third quadrant. A similar approach holds in general. Since $P = (v_1, v_2)$ is a point on the terminal side of angle θ (in standard position), and since $d(O, P) = \|\mathbf{v}\|$, it follows from Theorem 6.2, that

$$\cos \theta = \frac{v_1}{\|\mathbf{v}\|} \quad \text{and} \quad \sin \theta = \frac{v_2}{\|\mathbf{v}\|}.$$

(Since $\mathbf{v} \neq \mathbf{0}$, it follows that $\|\mathbf{v}\| > 0$. See Exercise 68.) Multiplying each of these equations by $\|\mathbf{v}\|$, we obtain the desired results

$$v_1 = \|\mathbf{v}\| \cos \theta, \quad \text{and} \quad v_2 = \|\mathbf{v}\| \sin \theta.$$

Let $\mathbf{v} = \langle v_1, v_2 \rangle$ be a vector such that $v_1 \neq 0$. If θ is the direction angle for $\mathbf{v}$, then $\cos \theta \neq 0$, and we see from Theorem 7.5 that

$$\tan \theta = \frac{\sin \theta}{\cos \theta} = \frac{\|\mathbf{v}\| \sin \theta}{\|\mathbf{v}\| \cos \theta} = \frac{v_2}{v_1}.$$

As the following example illustrates, we can use the equation

$$\tan \theta = \frac{v_2}{v_1}. \qquad (1)$$

and the inverse tangent function to determine the direction angle for a given nonzero vector:

E X A M P L E 29

Find the direction angle for each of the following vectors:

a. $\mathbf{u} = \langle -3, 3 \rangle$ **b.** $\mathbf{v} = \langle -2, -5 \rangle$

Solution

a. We begin by sketching the vector $\mathbf{u}$ in the plane. By equation (1), the direction angle θ for $\mathbf{u} = \langle -3, 3 \rangle$ satisfies

$$\tan \theta = \frac{3}{-3} = -1.$$

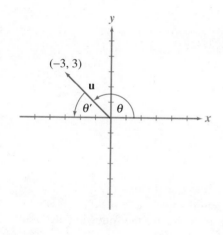

Since θ lies in quadrant II, we use the reference angle θ' to obtain

$$\tan \theta' = |\tan \theta| = 1.$$

Hence,

$$\theta' = \tan^{-1} 1 = \frac{\pi}{4},$$

and it follows that

$$\theta = \pi - \theta' = \frac{3\pi}{4}.$$

Thus, the direction angle for the vector $\mathbf{u} = \langle -3, 3 \rangle$ is $3\pi/4$ or $135°$.

b. As in part (a), we begin by sketching the vector $\mathbf{v}$. If θ is the direction angle for $\mathbf{v} = \langle -2, -5 \rangle$, we know from equation (1) that

$$\tan \theta = \frac{-5}{-2} = \frac{5}{2}.$$

Noting that θ lies in quadrant III, we again use the reference angle θ' and find that

$$\tan \theta' = |\tan \theta| = \frac{5}{2}.$$

Therefore,

$$\theta' = \tan^{-1} \frac{5}{2} \approx 68.2°,$$

and the direction angle for $\mathbf{v}$ is

$$\theta = 180° + \theta' \approx 248.2°.$$

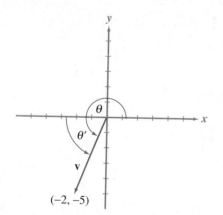

Applications

The next three examples illustrate the use of vectors, in conjunction with trigonometry to solve force and velocity problems.

EXAMPLE 30

Find the magnitude and direction of the **resultant** (vector sum) of the following two forces:

Force A: 450 kilograms in the direction S72.3°E

Force B: 325 kilograms in the direction N31.5°E

Solution   Vectors **a** and **b**, which represent the forces A and B, respectively, are shown in Figure 7.30. We must find the magnitude and direction of the vector sum **a** + **b**. To do this, we first use Theorem 7.5 to express the vectors **a** and **b** in algebraic form.

FIGURE 7.30

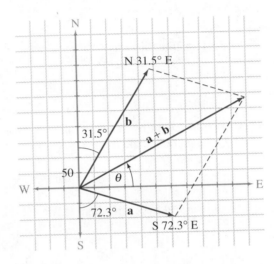

From Figure 7.31(a), we see that the direction angle for **a** is $270° + 72.3° = 342.3°$, and from Figure 7.31(b), the direction angle for **b** is $90° - 31.5° = 58.5°$. Thus, by Theorem 7.5, the vectors **a** and **b** have the following algebraic representations:

$$\mathbf{a} = \langle 450 \cos 342.3°, \ 450 \sin 342.3° \rangle \approx \langle 428.7, \ -136.8 \rangle;$$
$$\mathbf{b} = \langle 325 \cos 58.5°, \ 325 \sin 58.5° \rangle \approx \langle 169.8, \ 277.1 \rangle.$$

Adding the components of **a** and **b**, we find that

$$\mathbf{a} + \mathbf{b} \approx \langle 598.5, \ 140.3 \rangle.$$

FIGURE 7.31

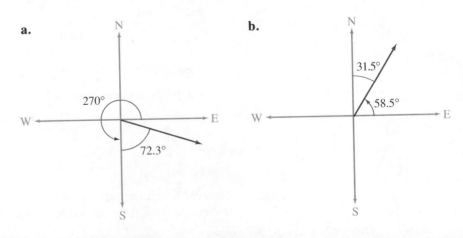

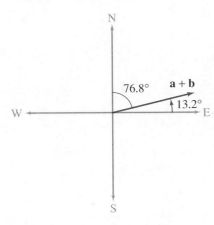

Hence, the magnitude of **a** + **b** is given by

$$\|\mathbf{a} + \mathbf{b}\| \approx \sqrt{(598.5)^2 + (140.3)^2} \approx 615.$$

The direction angle θ for **a** + **b** can be found using equation (1) on page 345.

Since

$$\tan \theta \approx \frac{140.3}{598.5} \approx 0.2344,$$

and since θ lies in quadrant I (see Figure 7.30), it follows that

$$\theta \approx \tan^{-1} 0.2344 \approx 13.2°.$$

To obtain the direction of **a** + **b**, we must subtract the direction angle θ from 90° (see Figure 7.32): $90° - 13.2° = 76.8°$. Thus the resultant of forces A and B is a force of approximately 615 kilograms in the (approximate) direction N76.8°E.

EXAMPLE 31

A car weighing 3250 pounds is parked on an incline of 24.1°. Find the amount of force that must be exerted to keep the car from rolling down the incline.

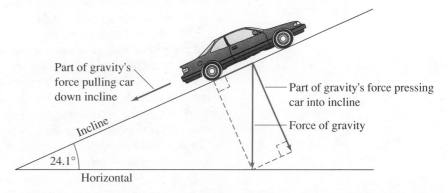

Solution The force of gravity, which pulls directly downward on the car, can be expressed as the sum of two forces acting on the car, as shown in Figure 7.33. One of these forces is parallel to the incline and tends to pull the car down the incline. The other force is perpendicular to the incline and tends to press the car into the incline. To simplify our work, we place the car (actually its center of

FIGURE 7.33

FIGURE 7.34

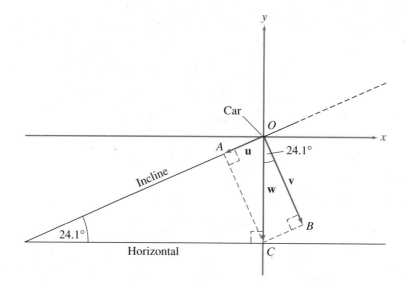

gravity) at the origin of a rectangular coordinate system as shown in Figure 7.34. Gravity, which is denoted by the vector **w**, pulls directly downward on the car with a force of 3250 pounds (the car's weight). The force that tends to pull the car down the incline is represented by the vector **u**, and the remaining force, which presses the car into the incline, is represented by the vector **v**.

To keep the car from rolling down the incline, a force having the same magnitude as the vector **u** must be exerted in the direction which is opposite that of **u**. To find $\|\mathbf{u}\|$, the magnitude of **u**, we first note from Figure 7.34 that the angle between the vectors **w** and **v** has the same measure (24.1°) as the inclination of the ramp. (Why?) The right triangle OBC from Figure 7.34 is reproduced in Figure 7.35.

FIGURE 7.35

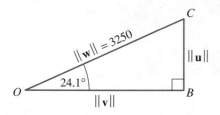

Since
$$\sin 24.1° = \frac{\|\mathbf{u}\|}{3250},$$
it follows that
$$\|\mathbf{u}\| = 3250\,(\sin 24.1°) \approx 1327.$$

Thus a force of approximately 1327 pounds must be exerted up the incline to keep the car from rolling down the incline. —————

A vector whose magnitude and direction represent the speed and direction of an object in motion is called a **velocity vector**. Such vectors play a central role in aviation. For example, when a pilot plots a course, the velocities of both the airplane and the wind must be considered. The **air speed** of an airplane is the speed at which it would fly in still air, while the **ground speed** is the plane's speed that takes into account the effect of the wind.

FIGURE 7.36

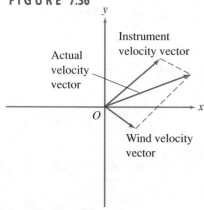

When an airplane is in the air, its instruments measure the speed and direction of the plane, with no allowance for the effect of the wind. For this reason, the velocity vector whose magnitude and direction are the plane's air speed and compass bearing, is called the **instrument velocity vector** or **apparent velocity vector**. The plane's **actual velocity vector**, which accounts for the effect of the wind, is the vector whose magnitude is the plane's ground speed and whose direction is the direction the plane is actually traveling relative to the ground.

To determine an airplane's course, a pilot must calculate the air speed and compass bearing of an instrument velocity vector which, when added to the wind velocity vector, will result in the desired actual velocity vector (see Figure 7.36).

EXAMPLE 32

A pilot wishes to fly her small plane from Kansas City, Missouri, to Bismarck, North Dakota. According to her map, the bearing from Kansas City to Bismarck is N38.8°W, and the distance between the two cities is 740 miles. If a wind of 65 miles per hour is blowing from the direction N57.4°E, determine the compass bearing and air speed she must maintain to make the trip in 4 hours.

FIGURE 7.37

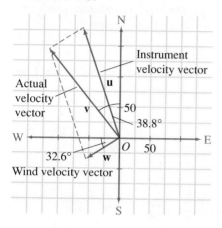

Solution We begin by letting **u** represent the (unknown) instrument velocity vector, **v** the actual velocity vector, and **w** the wind velocity vector. Since the pilot wishes to make the trip in 4 hours, the magnitude of the actual velocity vector **v** is $\frac{740}{4} = 185$ miles per hour. Also, since the actual velocity vector is the sum of the instrument and wind velocity vectors, we can represent the vector **u** graphically by using the parallelogram law. The vectors **u**, **v**, and **w** are shown in Figure 7.37. Since **v** = **u** + **w**, it follows that **u** = **v** − **w**. Hence, we can find the instrument velocity vector **u** by expressing the vectors **v** and **w** in algebraic form and subtracting **w** from **v**.

We see from Figure 7.38 that the direction angle of the actual velocity vector **v** is $90° + 38.8° = 128.8°$. Since the magnitude of **v** is 185, we know by Theorem 7.5 that the components of $\mathbf{v} = \langle v_1, v_2 \rangle$ are

$$v_1 = \|\mathbf{v}\| \cos 128.8° \approx (185)(-0.6627) \approx -116 \quad \text{and}$$

$$v_2 = \|\mathbf{v}\| \sin 128.8° \approx (185)(0.7793) \approx 144.$$

Hence, $\mathbf{v} \approx \langle -116, 144 \rangle$.

Since the wind is blowing from the direction N57.4°E, the wind velocity vector **w** makes an angle of 32.6° with the negative x-axis (see Figure 7.37). Thus the direction angle for **w** is $180° + 32.6° = 212.6°$. By Theorem 7.5, the components of the vector $\mathbf{w} = \langle w_1, w_2 \rangle$ are

$$w_1 = \|\mathbf{w}\| \cos 212.6° \approx (65)(-0.8424) \approx -55 \quad \text{and}$$

$$w_2 = \|\mathbf{w}\| \sin 212.6° \approx (65)(-0.5388) \approx -35.$$

Hence, $\mathbf{w} \approx \langle -55, -35 \rangle$, and the instrument velocity vector **u** is given by

$$\mathbf{u} = \mathbf{v} - \mathbf{w} \approx \langle -116, 144 \rangle - \langle -55, -35 \rangle = \langle -61, 179 \rangle.$$

FIGURE 7.38

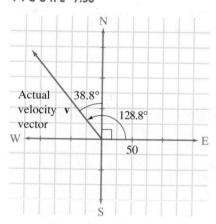

The vector **u** is graphed in Figure 7.39.

To find the plane's required compass bearing, we must first determine the direction angle θ for the vector $\mathbf{u} \approx \langle -61, 179 \rangle$. From equation (1), on page 345, we know that

$$\tan \theta = \frac{u_2}{u_1} \approx \frac{179}{-61} \approx -2.93.$$

Noting that θ lies in quadrant II, we use the reference angle θ' and find that

$$\tan \theta' = |\tan \theta| \approx 2.93.$$

Thus,
$$\theta' \approx \tan^{-1} 2.93 \approx 71°,$$

which means that

$$\theta = 180° - \theta' \approx 109°.$$

FIGURE 7.39

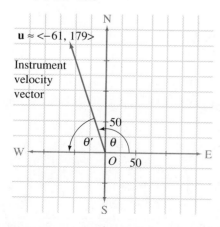

Since the plane's compass bearing is the angle between the vector **u** and the north-south line (y-axis), we subtract 90° from θ, obtaining $109° - 90° = 19°$. Thus the pilot should maintain a compass bearing of approximately N19°W.

Finally, by Theorem 7.4, the magnitude of the vector **u** is

$$\|\mathbf{u}\| = \sqrt{u_1^2 + u_2^2} \approx \sqrt{(-61)^2 + (179)^2} \approx 189,$$

which means that the pilot must maintain an air speed of approximately 189 miles per hour.

A vector whose magnitude is 1 is called a **unit vector**. For example, $\mathbf{u} = \langle \frac{4}{5}, \frac{3}{5} \rangle$, $\mathbf{v} = \langle 0, -1 \rangle$, and $\mathbf{w} = \langle -\sqrt{3}/2, \frac{1}{2} \rangle$ are all unit vectors since it is easily verified that $\|\mathbf{u}\| = \|\mathbf{v}\| = \|\mathbf{w}\| = 1$. Two special unit vectors

$$\mathbf{i} = \langle 1, 0 \rangle \qquad \text{and} \qquad \mathbf{j} = \langle 0, 1 \rangle$$

provide us with an alternate way to represent vectors in the plane (see Figure 7.40(a)). Using vector addition and scalar multiplication, any vector $\mathbf{v} = \langle v_1, v_2 \rangle$ can be expressed in terms of the unit vectors **i** and **j**. To see this, we write

$$\begin{aligned} \mathbf{v} = \langle v_1, v_2 \rangle &= \langle v_1, 0 \rangle + \langle 0, v_2 \rangle \\ &= v_1 \langle 1, 0 \rangle + v_2 \langle 0, 1 \rangle \\ &= v_1 \mathbf{i} + v_2 \mathbf{j}. \end{aligned}$$

Thus any vector $\mathbf{v} = \langle v_1, v_2 \rangle$ can be written in the form

$$\mathbf{v} = v_1 \mathbf{i} + v_2 \mathbf{j},$$

where $\mathbf{i} = \langle 1, 0 \rangle$ and $\mathbf{j} = \langle 0, 1 \rangle$. The expression $v_1 \mathbf{i} + v_2 \mathbf{j}$ is called a **linear combination** of the unit vectors **i** and **j**. For example, if $\mathbf{v} = \langle 2, 3 \rangle$ then $\mathbf{v} = 2\mathbf{i} + 3\mathbf{j}$, as illustrated in Figure 7.40(b).

The operations of vector addition, vector subtraction, and scalar multiplication can be expressed in terms of the unit vectors **i** and **j** as follows:

FIGURE 7.40

a.   Unit vectors **i** and **j**

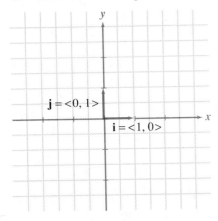

b. $\langle 2, 3 \rangle = 2\mathbf{i} + 3\mathbf{j}$

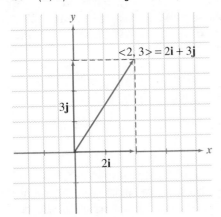

THEOREM 7.6   Vector Operations in Terms of Unit Vectors

If $\mathbf{u} = u_1 \mathbf{i} + u_2 \mathbf{j}$ and $\mathbf{v} = v_1 \mathbf{i} + v_2 \mathbf{j}$ are vectors and k is any real number, then

(i) $\mathbf{u} + \mathbf{v} = (u_1 \mathbf{i} + u_2 \mathbf{j}) + (v_1 \mathbf{i} + v_2 \mathbf{j}) = (u_1 + v_1)\mathbf{i} + (u_2 + v_2)\mathbf{j}$;

(ii) $\mathbf{u} - \mathbf{v} = (u_1 \mathbf{i} + u_2 \mathbf{j}) - (v_1 \mathbf{i} + v_2 \mathbf{j}) = (u_1 - v_1)\mathbf{i} + (u_2 - v_2)\mathbf{j}$;

(iii) $k\mathbf{v} = k(v_1 \mathbf{i} + v_2 \mathbf{j}) = kv_1 \mathbf{i} + kv_2 \mathbf{j}$.

EXAMPLE 33

Let $\mathbf{u} = 4\mathbf{i} - 7\mathbf{j}$ and $\mathbf{v} = -\frac{2}{5}\mathbf{i} + \frac{3}{2}\mathbf{j}$. Express each of the following vectors as a linear combination of the unit vectors **i** and **j**:

a. $3\mathbf{u} - \mathbf{v}$ **b.** $\frac{1}{2}\mathbf{u} + 5\mathbf{v}$

Solution   We apply Theorem 7.6 in each case.

a.
$$\begin{aligned} 3\mathbf{u} - \mathbf{v} &= 3(4\mathbf{i} - 7\mathbf{j}) - \left(-\frac{2}{5}\mathbf{i} + \frac{3}{2}\mathbf{j} \right) \\ &= (12\mathbf{i} - 21\mathbf{j}) - \left(-\frac{2}{5}\mathbf{i} + \frac{3}{2}\mathbf{j} \right) \end{aligned}$$

$$= \left(12 + \frac{2}{5} \right) \mathbf{i} + \left(-21 - \frac{3}{2} \right) \mathbf{j}$$

$$= \frac{62}{5} \mathbf{i} - \frac{45}{2} \mathbf{j}.$$

b.
$$\frac{1}{2} \mathbf{u} + 5 \mathbf{v} = \frac{1}{2} (4 \mathbf{i} - 7 \mathbf{j}) + 5 \left(-\frac{2}{5} \mathbf{i} + \frac{3}{2} \mathbf{j} \right)$$

$$= \left(2 \mathbf{i} - \frac{7}{2} \mathbf{j} \right) + \left(-2 \mathbf{i} + \frac{15}{2} \mathbf{j} \right)$$

$$= (2 - 2) \mathbf{i} + \left(-\frac{7}{2} + \frac{15}{2} \right) \mathbf{j}$$

$$= 4 \mathbf{j}.$$

EXERCISES 7.5

In Exercises 1–8, represent each vector as a position vector in the plane. Find the magnitude of each vector.

1. $\mathbf{v} = \langle 2, -3 \rangle$
2. $\mathbf{v} = \langle -1, 4 \rangle$
3. $\mathbf{v} = \langle -\frac{3}{2}, -\frac{7}{4} \rangle$
4. $\mathbf{v} = \langle -\frac{5}{2}, -\frac{3}{4} \rangle$
5. $\mathbf{v} = \langle 0, -6 \rangle$
6. $\mathbf{v} = \langle -5, 0 \rangle$
7. $\mathbf{v} = \langle 2\sqrt{3}, 4 \rangle$
8. $\mathbf{v} = \langle -\sqrt{3}/2, \frac{1}{2} \rangle$

In Exercises 9–12, determine the vectors $-2\mathbf{v}$ and $\frac{1}{3}\mathbf{v}$. Sketch $\mathbf{v}$, $-2\mathbf{v}$, and $\frac{1}{3}\mathbf{v}$ in the same coordinate system.

9. $\mathbf{v} = \langle -3, 6 \rangle$
10. $\mathbf{v} = \langle 6, -2 \rangle$
11. $\mathbf{v} = \langle -2, -\frac{15}{4} \rangle$
12. $\mathbf{v} = \langle -\frac{9}{4}, -12 \rangle$

In Exercises 13–18, find the sum of the given vectors. In each case, use the vectors $\mathbf{u}$ and $\mathbf{v}$ to illustrate the parallelogram law for vector addition.

13. $\mathbf{u} = \langle 1, -4 \rangle$, $\mathbf{v} = \langle -2, -3 \rangle$
14. $\mathbf{u} = \langle -3, 1 \rangle$, $\mathbf{v} = \langle 2, -5 \rangle$
15. $\mathbf{u} = \langle -5, \frac{7}{2} \rangle$, $\mathbf{v} = \langle 6, -8 \rangle$
16. $\mathbf{u} = \langle -\frac{5}{2}, 4 \rangle$, $\mathbf{v} = \langle -6, -2 \rangle$
17. $\mathbf{u} = \langle \frac{12}{5}, 0 \rangle$, $\mathbf{v} = \langle -9, 3 \rangle$
18. $\mathbf{u} = \langle 8, -\frac{3}{4} \rangle$, $\mathbf{v} = \langle 0, -5 \rangle$

In Exercises 19–30, find the indicated vectors, where $\mathbf{u} = \langle 3, -4 \rangle$ and $\mathbf{v} = \langle -1, 6 \rangle$.

19. $2\mathbf{u} - \mathbf{v}$
20. $-\mathbf{u} + 3\mathbf{v}$
21. $-\frac{1}{4}\mathbf{u} + \mathbf{v}$
22. $3\mathbf{u} - \frac{5}{2}\mathbf{v}$
23. $(\sqrt{2}/2)\mathbf{u} - (\sqrt{2}/2)\mathbf{v}$
24. $(\sqrt{3}/2)\mathbf{u} + \frac{1}{2}\mathbf{v}$
25. $-50\mathbf{u} + 80\mathbf{v}$
26. $12\mathbf{u} - 32\mathbf{v}$
27. $-3\mathbf{v} - \frac{1}{2}\mathbf{u}$
28. $-\frac{7}{2}\mathbf{v} + 2\mathbf{u}$
29. $\|\mathbf{u}\| (\mathbf{u} - \mathbf{v})$
30. $\|\mathbf{v}\| (\mathbf{u} + \mathbf{v})$

In Exercises 31–38, find the direction angle θ for each vector $\mathbf{v}$.

31. $\mathbf{v} = \langle 4, -4 \rangle$
32. $\mathbf{v} = \langle -3, -3 \rangle$
33. $\mathbf{v} = \langle -\sqrt{3}, -1 \rangle$
34. $\mathbf{v} = \langle -\frac{1}{2}, \sqrt{3}/2 \rangle$

35. $\mathbf{v} = \langle -6, 8 \rangle$
36. $\mathbf{v} = \langle 3, 2 \rangle$
37. $\mathbf{v} = \langle 2, -5 \rangle$
38. $\mathbf{v} = \langle -\frac{1}{2}, \frac{3}{4} \rangle$

In each of Exercises 39–44, the magnitude and direction angle of a vector $\mathbf{v}$ are given. In each case, find the algebraic representation of $\mathbf{v}$.

39. $\|\mathbf{v}\| = 3$, $\theta = \pi/3$
40. $\|\mathbf{v}\| = 4$, $\theta = \pi/4$
41. $\|\mathbf{v}\| = 24$, $\theta = 225°$
42. $\|\mathbf{v}\| = 15$, $\theta = 300°$
43. $\|\mathbf{v}\| = 9.52$, $\theta = 137°$
44. $\|\mathbf{v}\| = 37.4$, $\theta = 258°$

In Exercises 45–48, give the magnitude and direction of the resultant of each pair of vectors.

45. Force A: 175 pounds in the direction N15.6°E
Force B: 240 pounds in the direction N78.3°W

46. Force A: 44.3 kilograms in the direction S32.4°W
Force B: 28.6 kilograms in the direction N51.5°W

47. Velocity A: 815 kilometers per hour in the direction N41°20′W
Velocity B: 520 kilometers per hour in the direction S29°30′W

48. Velocity A: 450 miles per hour in the direction S64°45′W
Velocity B: 675 miles per hour in the direction S40°15′E

In Exercises 49–54, $\mathbf{u} = -\frac{1}{2}\mathbf{i} + 3\mathbf{j}$ and $\mathbf{v} = 5\mathbf{i} - \frac{3}{4}\mathbf{j}$. In each case, express the given vector as a linear combination of the unit vectors $\mathbf{i}$ and $\mathbf{j}$.

49. $3\mathbf{u} + 4\mathbf{v}$
50. $2\mathbf{u} - 5\mathbf{v}$
51. $-5\mathbf{u} + \frac{1}{2}\mathbf{v}$
52. $-\frac{4}{3}\mathbf{u} - \frac{1}{3}\mathbf{v}$
53. $-\sqrt{3}\mathbf{u} - \frac{5}{8}\mathbf{v}$
54. $(\sqrt{2}/2)\mathbf{v} + \mathbf{u}$

55. A commercial airliner is flying on a compass bearing of S48.6°E with an air speed of 565 miles per hour. If the wind is blowing in the direction N12.8°E at 48 miles per hour, find the magnitude and direction of the airliner's actual velocity vector.

56. At 4:15 P.M. an airport control tower receives a message from the pilot of a small plane that has crashed. Before blacking out, the injured pilot is able to report that his plane had been flying on a bearing of S65.5°E, at a speed of 215 miles per hour since its departure from the airport at 3:15 P.M. The control tower checks with the national weather service and finds that the wind along the plane's flight path had been 45 miles per hour from the direction S24.8°W. Find the distance and bearing from the airport to the downed aircraft.

57. A pilot wishes to fly his plane from Salt Lake City to Los Angeles. According to his map, the bearing from Salt Lake City to Los Angeles is S71.2°W, and the two cities are 714 miles apart. If the wind is blowing in the direction N24.3°W at 55 miles per hour and the pilot plans to make the trip in three hours, find the compass bearing and air speed he must maintain.

58. An airplane is flying on a compass bearing of N25.8°W with an air speed of 710 kilometers per hour. If the plane's actual velocity vector has a magnitude of 715 kilometers per hour in the direction N17.6°W, determine the speed and direction of the wind.

59. A person in a small boat, which can travel 12 miles per hour in still water, heads directly across a river. If the river has a current of 4.5 miles per hour that is parallel to the banks, determine the angle α between the boat's actual path and its intended path (see figure).

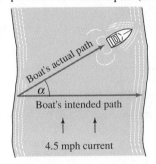

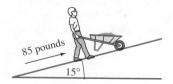

60. In order to travel directly north across a river, a boat must aim for a point on the opposite side that has a bearing of N11.6°E. If the boat can maintain a speed of 25 miles per hour in still water, determine the speed of the east-to-west flowing current.

61. Find the force required to keep a 4850-pound truck from rolling down an incline of 12.5°.

62. A force of 115 pounds is required to pull a garden tractor weighing 670 pounds up a ramp. Find the angle the ramp makes with the horizontal.

63. A man pushing a wheelbarrow load of dirt up a 15° incline pauses to rest (see figure). If he must exert a force of 85

pounds to keep the wheelbarrow from rolling down the incline, find the combined weight of the wheelbarrow and dirt.

64. A force of 825 pounds is required to pull a boat and trailer up a ramp that makes an angle of 17.5° with the horizontal (see figure). Find the combined weight of the boat and trailer.

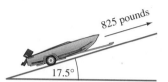

65. The resultant of forces $F_1, F_2, \ldots, F_n$ acting on a point P is the vector sum $F = F_1 + F_2 + \cdots + F_n$. If $F = \mathbf{0}$, we say that the system of forces is **in equilibrium**. If F_1 is a force of 150 pounds in the direction S42.9°W, and F_2 is a force of 230 pounds in the direction S26.7°E, find a force F_3 that will produce a system of three force vectors in equilibrium.

66. A 450-kilogram weight is supported by three cables as shown in the figure. Determine the amount of **tension** (magnitude of the stretching force) in each cable. [*Hint:* Treat the cables **u**, **v**, and **w** as a system of three position vectors in equilibrium and refer to Exercise 65.]

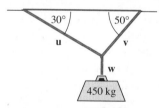

67. If $\mathbf{u} = \langle u_1, u_2 \rangle$ and $\mathbf{v} = \langle v_1, v_2 \rangle$ are vectors, the **dot product (scalar product)** of **u** and **v** is defined by

$$\mathbf{u} \cdot \mathbf{v} = u_1 v_1 + u_2 v_2.$$

Find the dot product $\mathbf{u} \cdot \mathbf{v}$ for the given vectors **u** and **v**.
a. $\mathbf{u} = \langle -4, 1 \rangle$, $\mathbf{v} = \langle 2, -3 \rangle$
b. $\mathbf{u} = \langle 0, -8 \rangle$, $\mathbf{v} = \langle -4, \frac{1}{2} \rangle$
c. $\mathbf{u} = \langle -\frac{5}{2}, -\frac{4}{3} \rangle$, $\mathbf{v} = \langle 6, 0 \rangle$

68. If **v** is a nonzero vector, show that $\|\mathbf{v}\| > 0$.

69. Let **u** and **v** be nonzero vectors. If α is the angle between **u** and **v**, it can be shown that

$$\cos \alpha = \frac{\mathbf{u} \cdot \mathbf{v}}{\|\mathbf{u}\| \, \|\mathbf{v}\|},$$

where $\mathbf{u} \cdot \mathbf{v}$ is the dot product of **u** and **v** defined in Exercise 67. Find the angle α between the vectors **u** and **v** given in parts (a), (b), and (c) of Exercise 67.

70. Verify properties (ii)–(x) of scalar multiplication and vector addition (see page 343).

71. If **v** is a vector in the plane with initial point $A = (x_1, y_1)$ and terminal point $B = (x_2, y_2)$, it can be shown that the vector $\mathbf{v} = \overrightarrow{AB}$ is equal to the position vector $\langle x_2 - x_1, y_2 - y_1 \rangle$. Find a position vector that is equal to the vector $\overrightarrow{AB}$ whose initial point A and terminal point B are given below.

a. $A = (-3, 4)$, $B = (5, -8)$
b. $A = (\frac{5}{12}, -2)$, $B = (7, -\frac{3}{4})$

72. a. If $\mathbf{v} = \langle v_1, v_2 \rangle$ is a nonzero vector, show that $\mathbf{u} = (1/\|\mathbf{v}\|)\mathbf{v}$ is a unit vector. The vector $\mathbf{u}$ is called the **unit vector in the direction of v**.

b. Find the unit vector in the direction of each vector v given in Exercises 1–8.

7.6 POLAR COORDINATES

In Chapter 1, we introduced a rectangular coordinate system for representing points in the plane by ordered pairs of real numbers. In this section we will discuss another coordinate system, called a **polar coordinate system**, which provides us with another useful way of representing points in the plane.

To define a polar coordinate system, we begin with a fixed ray, called the **polar axis**, which emanates from a fixed point O called the **pole**. For convenience, we will use the origin $(0, 0)$ of a rectangular coordinate system as the pole, and the positive x-axis as the polar axis. (As we shall see, superimposing the polar coordinate system over a rectangular coordinate system in this manner enables us to use the features of both systems simultaneously.) Using this convention, any point P in the plane can be represented by an ordered pair of the form (r, θ) where $r = d(O, P)$, the distance between P and the pole, and θ is any angle (measured in degrees or radians) having the polar axis as its initial side and the ray $\overrightarrow{OP}$ as its terminal side (see Figure 7.41). We call r and θ **polar coordinates** of the point P, and say that (r, θ) is a **polar representation** of P.

While each point in the plane can be uniquely represented by an ordered pair (x, y) of real numbers, points in the plane do not have unique representations of the form (r, θ). In fact, since all angles of the form $\theta + 2n\pi$ (where n is an integer) are coterminal with θ, it is easy to see that each point in the plane has infinitely many representations of the form (r, θ). For example, in Figure 7.42 we give two alternate polar representations of the point $P = (3, \pi/3)$, namely

FIGURE 7.41
Polar representation of a point P

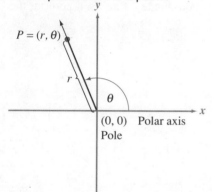

FIGURE 7.42 Three different polar representations of a point P

a.

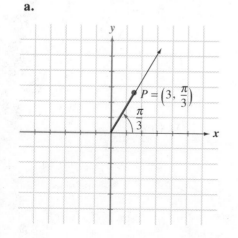

b.

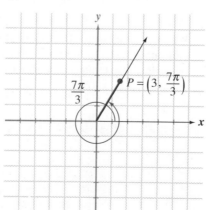

c.

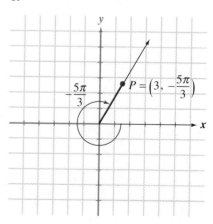

FIGURE 7.43

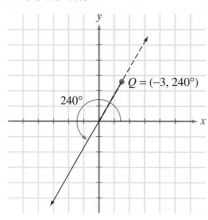

(3, $7\pi/3$) and (3, $-5\pi/3$). The pole, moreover, can be represented by an ordered pair of the form (0, θ), where θ is *any* angle.

We will also find it convenient to consider polar coordinates of a point P in which r is negative. If $r < 0$, we locate the point (r, θ) by measuring a distance of $|r|$ units from the pole along the ray formed by extending the terminal side of angle θ through the pole. To illustrate, the point $Q = (-3, 240°)$ is plotted in Figure 7.43. Note that $Q = (-3, 240°)$ gives yet another polar representation for the point $P = (3, \pi/3)$.

EXAMPLE 34

The following points are expressed in polar coordinates. Plot each point in the plane.

a. (2, $\pi/4$)   **b.** ($2\sqrt{3}$, 570°)
c. (5, $-2\pi/3$)   **d.** (-3, 135°)

Solution

a. The point (2, $\pi/4$) is located 2 units from the pole along the terminal side of the angle $\theta = \pi/4$.

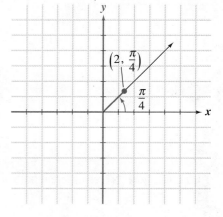

b. We locate the point ($2\sqrt{3}$, 570°) by measuring $2\sqrt{3} \approx 3.5$ units from the pole along the terminal side of the angle $\theta = 570°$.

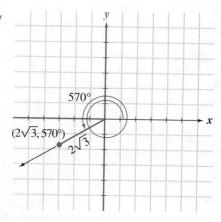

c. The point $(5, -2\pi/3)$ is found 5 units from the pole measured along the terminal side of $\theta = -2\pi/3$.

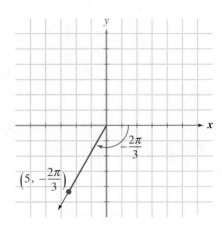

d. As noted earlier, when r is negative, we locate the point (r, θ) a distance of $|r|$ units from the pole, along the ray formed by extending the terminal side of angle θ through the pole. Thus the point $(-3, 135°)$ is found $|-3| = 3$ units from the pole, measured along the ray formed by extending the terminal side of $\theta = 135°$ through the pole.

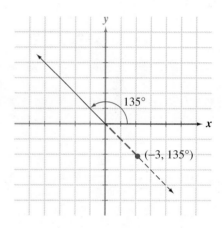

Relationships Between Polar and Rectangular Coordinates

FIGURE 7.44

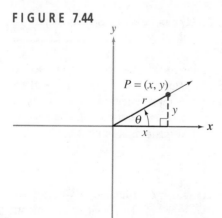

Sometimes in our work it is convenient to convert polar coordinates into rectangular coordinates, or vice versa. To help us make these conversions, we now observe several important relationships between the polar and rectangular coordinates of a given point P. In Figure 7.44, we have plotted the point $P = (x, y) \neq (0, 0)$ and have sketched an angle θ, in standard position, whose terminal side contains P. If $r = d(O, P)$, it follows that (r, θ) is a polar representation of the point P. Hence, from Theorem 6.2 (on page 253), we know that

$$\sin \theta = \frac{y}{r}, \tag{1}$$

$$\cos \theta = \frac{x}{r}, \quad \text{and} \tag{2}$$

$$\tan \theta = \frac{y}{x}, \quad x \neq 0. \tag{3}$$

In addition, since $r = \sqrt{x^2 + y^2}$, we have

$$x^2 + y^2 = r^2. \tag{4}$$

Although the point P in Figure 7.44 is located in quadrant I, equations (1)–(4) are valid for any point $P \neq (0, 0)$. Moreover, it can also be shown that these equations are valid when r is negative. Thus we have

///

Relationships Between Polar and Rectangular Coordinates

Let $P = (x, y) \neq (0, 0)$ be a point in the plane. If (r, θ) is a polar representation of P, then

(i) $x = r \cos \theta$ (ii) $y = r \sin \theta$;

(iii) $\tan \theta = \dfrac{y}{x} (x \neq 0)$;   (iv) $r^2 = x^2 + y^2$.

EXAMPLE 35

The following points are expressed in polar coordinates. Express each point in rectangular coordinates:

a. $(3, 5\pi/6)$ **b.** $(\sqrt{5}, 295°)$ **c.** $(-4, 5\pi/4)$

Solution   In each case, we use the equations $x = r \cos \theta$ and $y = r \sin \theta$.

a.   Since $x = r \cos \theta = 3\left(\cos \dfrac{5\pi}{6} \right) = 3\left(-\dfrac{\sqrt{3}}{2} \right) = -\dfrac{3\sqrt{3}}{2}$,

and $y = r \sin \theta = 3\left(\sin \dfrac{5\pi}{6} \right) = 3(\tfrac{1}{2}) = \tfrac{3}{2}$, the point $P = \left(3, \dfrac{5\pi}{6} \right)$

can be expressed in rectangular coordinates as $\left(-\dfrac{3\sqrt{3}}{2}, \dfrac{3}{2} \right)$.

b. $x = r \cos \theta = \sqrt{5} (\cos 295°) \approx 0.9$, and $y = \sqrt{5} (\sin 295°) \approx -2.0$. Thus, the point $(\sqrt{5}, 295°)$ can be expressed in rectangular coordinates as $(\sqrt{5} \cos 295°, \sqrt{5} \sin 295°) \approx (0.9, -2.0)$.

c.   Since $\cos \dfrac{5\pi}{4} = \sin \dfrac{5\pi}{4} = -\dfrac{\sqrt{2}}{2}$, it follows that $x = (-4)\left(-\dfrac{\sqrt{2}}{2} \right)$

$= 2\sqrt{2} = y$. Hence the point $(-4, 5\pi/4)$ can be expressed in rectangular coordinates as $(2\sqrt{2}, 2\sqrt{2})$.

The equations relating polar and rectangular coordinates can also be used to convert rectangular coordinates to polar coordinates. Since, as we noted earlier, each point in the plane has infinitely many polar representations, the results of such conversions will not be unique. We can, however, obtain unique results by requiring that $r \geq 0$ and by choosing the smallest positive argument θ.

EXAMPLE 36

The following points are expressed in rectangular coordinates. Express each point in polar coordinates, where $r \geq 0$ and θ is the smallest positive argument.

a. $P = (-4, 4)$ **b.** $P = (-2, -2\sqrt{3})$ **c.** $P = (2, -5)$

Solution

a.   We begin by plotting the point $P = (-4, 4)$ and sketching the ray $\overrightarrow{OP}$, as shown in Figure 7.45. Let θ be the smallest positive angle, in standard

FIGURE 7.45

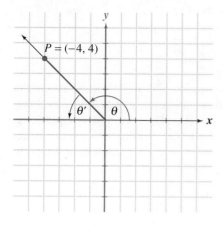

position, whose terminal side is the ray $\overrightarrow{OP}$. From Figure 7.45, we see that

$$\tan \theta = \frac{y}{x} = \frac{4}{-4} = -1.$$

If θ' is the reference angle for θ, we know that

$$\tan \theta' = |\tan \theta| = 1.$$

Thus,

$$\theta' = \tan^{-1} 1 = \frac{\pi}{4},$$

so that

$$\theta = \pi - \theta' = \frac{3\pi}{4}.$$

From the equation $r^2 = x^2 + y^2$, we have

$$r^2 = (-4)^2 + (4)^2 = 32.$$

Since $r \geq 0$, we see that

$$r = \sqrt{32} = 4\sqrt{2}.$$

Thus, the point $P = (-4, 4)$ can be expressed in polar coordinates as $(4\sqrt{2}, 3\pi/4)$.

FIGURE 7.46

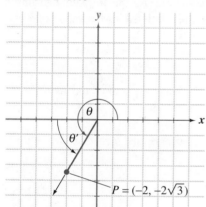

b. As in part (a), we plot the point $P = (-2, -2\sqrt{3})$ and sketch the smallest positive angle θ whose terminal side contains P. We see from Figure 7.46, that

$$\tan \theta = \frac{-2\sqrt{3}}{-2} = \sqrt{3}.$$

Thus, if θ' is the reference angle for θ,

$$\tan \theta' = |\tan \theta| = \sqrt{3}.$$

It follows that

$$\theta' = \tan^{-1}\sqrt{3} = \frac{\pi}{3},$$

so that

$$\theta = \pi + \theta' = \frac{4\pi}{3}.$$

Since $r^2 = (-2)^2 + (-2\sqrt{3})^2 = 16$, and $r \geq 0$, it follows that $r = 4$. Hence the point $P = (-2, -2\sqrt{3})$ can be expressed in polar coordinates as $(4, 4\pi/3)$.

FIGURE 7.47

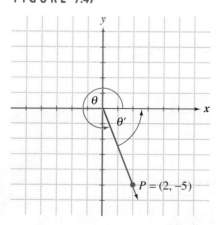

c. As we see in Figure 7.47, both $P = (2, -5)$ and the smallest positive angle θ whose terminal side contains P lie in quadrant IV. Thus

$$\tan \theta = -\frac{5}{2} = -2.5,$$

and if θ' is the reference angle for θ,

$$\tan \theta' = |\tan \theta| = 2.5.$$

It follows that

$$\theta' = \tan^{-1} 2.5 \approx 68.2°,$$

so that

$$\theta = 360° - \theta' \approx 291.8°.$$

Since $r \geq 0$, $r = \sqrt{(2)^2 + (-5)^2} = \sqrt{29} \approx 5.4$, and the point $P = (2, -5)$ can be expressed in polar coordinates as $(\sqrt{29}, 360° - \tan^{-1} 2.5) \approx (5.4, 291.8°)$.

Polar Equations

Equations such as $r = 3 \cos \theta + 1$, or $r^2 \sin 2\theta - 4 = 0$, which involve the variables r and θ, are called **polar equations**. In contrast, equations involving the variables x and y are called **Cartesian equations** or **rectangular equations**. A **solution** of a polar equation is an ordered pair (r_0, θ_0) that satisfies that equation. For example, the ordered pair $(r, \theta) = (\frac{5}{2}, \pi/3)$ is a solution of the polar equation $r = 3 \cos \theta + 1$, since

$$3 \cos \frac{\pi}{3} + 1 = 3 \left(\frac{1}{2} \right) + 1 = \frac{5}{2}.$$

The **graph** of a polar equation is the set of all points P in the plane that have at least one polar representation $P = (r, \theta)$ which satisfies the given equation. In general, we obtain the graph of a polar equation by plotting a sufficient number of points that satisfy the equation to get a reasonable idea about the shape of the curve. Sometimes, however, it is easier to determine the graph by first converting the polar equation to a Cartesian equation. As the following example illustrates, these conversions are done using the relationships between polar and rectangular coordinates given earlier in this section.

EXAMPLE 37

Convert the following polar equations to Cartesian equations and identify the curve each represents.

a. $r = \dfrac{3}{2 \sin \theta - \cos \theta}$ **b.** $r = 1$ **c.** $r^2 \cos^2 \theta = 6 - 3r \sin \theta$

Solution

a. Multiplying both sides of the equation

$$r = \frac{3}{2 \sin \theta - \cos \theta}$$

by $2 \sin \theta - \cos \theta$, we get

$$2r \sin \theta - r \cos \theta = 3.$$

Replacing $r \cos \theta$ by x and $r \sin \theta$ by y, the latter equation becomes

$$2y - x = 3,$$

which we recognize as the equation of a line.

b. Since $r \geq 0$, we can square both sides of the equation $r = 1$ to obtain

$$r^2 = 1, \quad \text{or}$$

$$x^2 + y^2 = 1.$$

We recognize the latter equation as that of the unit circle.

c. Replacing $r \cos \theta$ by x and $r \sin \theta$ by y, the equation

$$r^2 \cos^2 \theta = 6 - 3r \sin \theta$$

becomes $x^2 = 6 - 3y.$

Solving this equation for y, we obtain the equation

$$y = -\frac{1}{3}x^2 + 2,$$

whose graph is a parabola.   ———————

We can also use the formulas relating polar and rectangular coordinates to convert Cartesian equations to polar equations.

EXAMPLE 38

Express each of the following Cartesian equations as a polar equation:

a. $(x^2 + y^2)^{3/2} = 6xy$ **b.** $x^4 + y^4 = 9x^2 - 2x^2y^2 - 9y^2$

Solution

a. Replacing $x^2 + y^2$ by r^2, x by $r \cos \theta$ and y by $r \sin \theta$, the equation

$$(x^2 + y^2)^{3/2} = 6xy$$

becomes $(r^2)^{3/2} = 6(r \cos \theta)(r \sin \theta),$

or $r^3 = 6r^2 \sin \theta \cos \theta.$ **(5)**

If $r \neq 0$, we can divide both sides of equation (5) by r^2 to obtain

$$r = 6 \sin \theta \cos \theta, \quad \text{or}$$
$$r = 3(2 \sin \theta \cos \theta).$$

Using the identity $\sin 2\theta = 2 \sin \theta \cos \theta$, the latter equation becomes

$$r = 3 \sin 2\theta.$$

We should point out that our assumption that $r \neq 0$ in equation (5) is allowable since the only point (r, θ) for which $r = 0$ is the pole, and the pole satisfies both $r = 3 \sin 2\theta$ and equation (5).

b. Rearranging terms in the given equation and factoring, we have

$$x^4 + y^4 = 9x^2 - 2x^2y^2 - 9y^2$$
$$x^4 + 2x^2y^2 + y^4 = 9x^2 - 9y^2$$
$$(x^2 + y^2)^2 = 9x^2 - 9y^2.$$ **(6)**

Substituting r^2 for $x^2 + y^2$, $r \cos \theta$ for x and $r \sin \theta$ for y in equation (6), we obtain

$$(r^2)^2 = 9(r \cos \theta)^2 - 9(r \sin \theta)^2,$$

or $r^4 = 9(r^2\cos^2\theta - r^2\sin^2\theta).$ **(7)**

If $r \neq 0$, we can divide both sides of equation (7) by r^2 to get

$$r^2 = 9(\cos^2 \theta - \sin^2 \theta).$$

Using the identity $\cos 2\theta = \cos^2 \theta - \sin^2 \theta$, the latter equation becomes

$$r^2 = 9 \cos 2\theta.$$

Once again, our assumption that $r \neq 0$ is allowable since the representation $(0, \pi/4)$ of the pole satisfies both the latter equation and equation (7).

Tests for Symmetry

We now return to the problem of graphing polar equations in general. As in the case of graphing equations in x and y, we can often save time and effort by determining whether the graph of a polar equation possesses some type of symmetry. Three commonly used symmetry tests for the graphs of polar equations are given in Table 7.2, and illustrated in Figure 7.48.

TABLE 7.2
Symmetry Tests for the Graphs of Polar Equations

Graph is symmetric with respect to	If an equivalent equation results when
the x-axis	θ is replaced by $-\theta$
the y-axis	θ is replaced by $\pi - \theta$
the origin	r is replaced by $-r$

FIGURE 7.48

a. Symmetry with respect to the x-axis

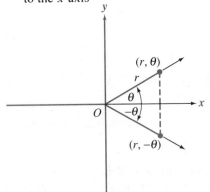

b. Symmetry with respect to the y-axis

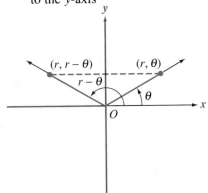

c. Symmetry with respect to the origin

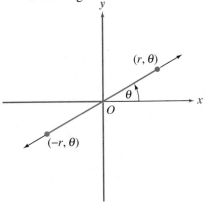

 CAUTION: The graph of a polar equation may possess one (or more) of these three types of symmetry, even though the equation fails the symmetry test(s) given in Table 7.2. Because of the multiple representations of points using polar coordinates, a symmetry may be present and yet not be detected by one of these three tests.

Cardioids and Limacons

EXAMPLE 39

Sketch the graph of the polar equation $r = 2 - 2 \sin \theta$.

Solution   We begin by applying the symmetry tests from Table 7.2.
    Replacing θ by $-\theta$ in the equation

$$r = 2 - 2 \sin \theta, \tag{8}$$

we have

$$r = 2 - 2 \sin(-\theta),$$

or

$$r = 2 + 2 \sin \theta. \tag{9}$$

Since equations (8) and (9) are not equivalent, equation (8) fails our test for symmetry with respect to the x-axis.

Replacing θ by $\pi - \theta$ in equation (8), we get

$$r = 2 - 2\sin(\pi - \theta)$$
$$= 2 - 2(\sin\pi\cos\theta - \cos\pi\sin\theta) \qquad \text{difference formula for sine}$$
$$= 2 - 2[(0)(\cos\theta) - (-1)(\sin\theta)]$$
$$= 2 - 2\sin\theta.$$

Thus the graph of the equation $r = 2 - 2\sin\theta$ is symmetric with respect to the y-axis.

Finally, replacing r by $-r$ in equation (8) yields

$$-r = 2 - 2\sin\theta, \qquad \text{or}$$
$$r = -2 + 2\sin\theta.$$

Since the latter equation is not equivalent to equation (8), equation (8) fails our test for symmetry with respect to the origin.

Next we use the equation $r = 2 - 2\sin\theta$ to determine the polar coordinates of some points on the graph. Because the graph is symmetric with respect to the y-axis, we need only determine points (r, θ), for values of θ in the intervals $[0, \pi/2]$ and $[3\pi/2, 2\pi]$.

θ	0	$\dfrac{\pi}{6}$	$\dfrac{\pi}{4}$	$\dfrac{\pi}{3}$	$\dfrac{\pi}{2}$	$\dfrac{3\pi}{2}$	$\dfrac{5\pi}{3}$	$\dfrac{7\pi}{4}$	$\dfrac{11\pi}{6}$	2π
$r = 2 - 2\sin\theta$	2	1	$2 - \sqrt{2} \approx 0.6$	$2 - \sqrt{3} \approx 0.3$	0	4	$2 + \sqrt{3} \approx 3.7$	$2 + \sqrt{2} \approx 3.4$	3	2

Plotting the points from this table, we obtain the portion of the graph shown in Figure 7.49(a). Using the y-axis symmetry, we reflect this portion of the graph across the y-axis to obtain the complete graph of the equation $r = 2 - 2\sin\theta$, which is shown in Figure 7.49(b). The resulting heart-shaped graph is called a *cardioid*, from the Greek word *kardia*, which means heart.

FIGURE 7.49

a.

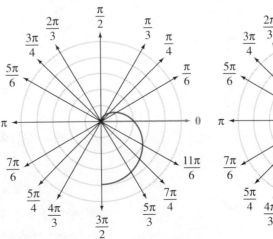

b. $r = 2 - 2\sin\theta$ (a cardioid)

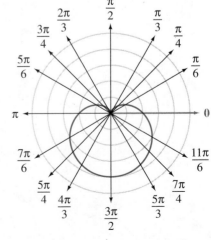

E X A M P L E 40

Sketch the graph of the polar equation $r = 2 + 3 \cos \theta$.

Solution   Since $\cos(-\theta) = \cos \theta$, an equivalent equation results when θ is replaced by $-\theta$ in the equation

$$r = 2 + 3 \cos \theta.$$

Thus the graph of this equation is symmetric with respect to the x-axis. You should verify that this equation fails our tests for symmetry with respect to the y-axis and the origin. Since the graph is symmetric with respect to the x-axis, we need only find points (r, θ) for values of θ in the interval $[0, \pi]$.

θ	0	$\dfrac{\pi}{6}$	$\dfrac{\pi}{4}$	$\dfrac{\pi}{3}$	$\dfrac{\pi}{2}$	$\dfrac{2\pi}{3}$	$\dfrac{3\pi}{4}$	$\dfrac{5\pi}{6}$	π
$r = 2 + 3 \cos \theta$	5	$2 + \dfrac{3\sqrt{3}}{2} \approx 4.6$	$2 + \dfrac{3\sqrt{2}}{2} \approx 4.1$	3.5	2	0.5	$2 - \dfrac{3\sqrt{2}}{2} \approx -0.1$	$2 - \dfrac{3\sqrt{3}}{2} \approx -0.6$	-1

Plotting the points from the table, we obtain the portion of the graph shown in Figure 7.50(a). Reflecting this portion across the x-axis, using x-axis symmetry, we obtain the complete graph shown in Figure 7.50(b). A graph such as this is called a *limacon*, or in this case, a *limacon with loop*. The word *limacon* comes from the Latin word *limacem*, which means snail.

FIGURE 7.50

a.

b. $r = 2 + 3 \cos \theta$ (limacon with loop)

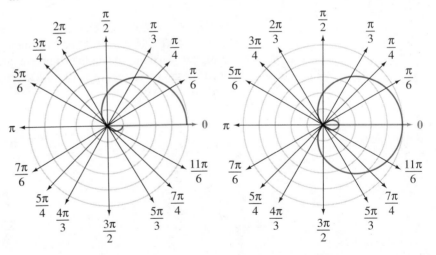

Cardioids

If a is a nonzero real number, the graph of a polar equation of the form

$$r = a(1 \pm \cos \theta) \quad \text{or} \quad r = a(1 \pm \sin \theta)$$

is a heart-shaped curve called a **cardioid**.

//

Limacons

If a and b are nonzero real numbers and $a \neq b$, the graph of a polar equation of the form

$$r = a + b \cos \theta \quad \text{or} \quad r = a + b \sin \theta$$

is called a **limacon**. The graph of a limacon has the heartlike shape of a cardioid, but may contain an additional loop.

n-Leaf Roses

EXAMPLE 41

Sketch the graph of the polar equation $r = 3 \sin 2\theta$.

Solution It can be verified that the equation $r = 3 \sin 2\theta$ fails all three of our tests for symmetry. We thus make a table for values of θ between 0 and 2π.

θ	0	$\dfrac{\pi}{6}$	$\dfrac{\pi}{4}$	$\dfrac{\pi}{3}$	$\dfrac{\pi}{2}$	$\dfrac{2\pi}{3}$	$\dfrac{3\pi}{4}$	$\dfrac{5\pi}{6}$	π
$r = 3 \sin 2\theta$	0	$\dfrac{3\sqrt{3}}{2} \approx 2.6$	3	$\dfrac{3\sqrt{3}}{2} \approx 2.6$	0	$-\dfrac{3\sqrt{3}}{2} \approx -2.6$	-3	$-\dfrac{3\sqrt{3}}{2} \approx -2.6$	0

θ	$\dfrac{7\pi}{6}$	$\dfrac{5\pi}{4}$	$\dfrac{4\pi}{3}$	$\dfrac{3\pi}{2}$	$\dfrac{5\pi}{3}$	$\dfrac{7\pi}{4}$	$\dfrac{11\pi}{6}$	2π
$r = 3 \sin 2\theta$	$\dfrac{3\sqrt{3}}{2} \approx 2.6$	3	$\dfrac{3\sqrt{3}}{2} \approx 2.6$	0	$-\dfrac{3\sqrt{3}}{2} \approx -2.6$	-3	$-\dfrac{3\sqrt{3}}{2} \approx -2.6$	0

Plotting the points from the table, we obtain the graph shown in Figure 7.51. (We note that this is the graph of the polar equation that is the solution of part (a) of Example 38.) A graph such as this is called a *4-leaf rose*. We note that the graph is symmetric with respect to the *x*-axis, *y*-axis, and origin, even though the equation $r = 3 \sin 2\theta$ fails all three of our symmetry tests.

FIGURE 7.51

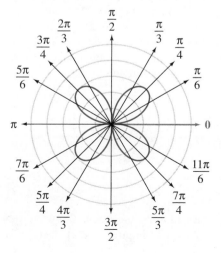

$r = 3 \sin 2\theta$ (a 4-leaf rose)

//

n-Leaf Roses

If a is a nonzero real number and if n is a positive integer greater than 1, the graph of a polar equation of the form

$$r = a \sin n\theta \quad \text{or} \quad r = a \cos n\theta$$

is called an **n-leaf rose**. If n is odd, the graph has n leaves and if n is even, the graph has 2n leaves. In either case, each leaf has length $|a|$.

Lemniscates

E X A M P L E   42

Sketch the graph of the polar equation $r^2 = 9 \cos 2\theta$.

Solution   The equation $r^2 = 9 \cos 2\theta$ satisfies all three of our tests for symmetry:

Replacing θ by $-\theta$ in the equation $r^2 = 9 \cos 2\theta$ results in an equivalent equation since

$$\cos 2(-\theta) = \cos(-2\theta) = \cos 2\theta.$$

Hence the graph is symmetric with respect to the x-axis.

When $\pi - \theta$ is substituted for θ in the equation $r^2 = 9 \cos 2\theta$, an equivalent equation results since

$$
\begin{aligned}
\cos 2(\pi - \theta) &= \cos(2\pi - 2\theta) \\
&= \cos 2\pi \cos 2\theta + \sin 2\pi \sin 2\theta \quad \text{difference formula for cosine} \\
&= (1)(\cos 2\theta) + (0)(\sin 2\theta) \\
&= \cos 2\theta.
\end{aligned}
$$

Thus the graph is also symmetric with respect to the y-axis.

Since the graph is symmetric with respect to both the x-axis and y-axis, it is also symmetric with respect to the origin.

Because of the symmetries, we need only consider values of θ in the interval $[0, \pi/2]$. However, the fact that r^2 is nonnegative means that $\cos 2\theta$ must also be nonnegative. Since $\cos 2\theta$ is negative over the interval $(\pi/4, \pi/2]$, we can further limit our values of θ to the interval $[0, \pi/4]$. Using the equation $r^2 = 9 \cos 2\theta$ in the equivalent form $r = \pm 3\sqrt{\cos 2\theta}$, we obtain the values shown in the following table:

θ	0	$\dfrac{\pi}{12}$	$\dfrac{\pi}{6}$	$\dfrac{\pi}{4}$
$r = \pm 3\sqrt{\cos 2\theta}$	± 3	$\pm 3\sqrt{\dfrac{\sqrt{3}}{2}} \approx \pm 2.8$	$\pm\dfrac{3}{\sqrt{2}} \approx \pm 2.1$	0

By plotting the points from the table, we obtain the portion of the graph shown in Figure 7.52(a). Using symmetry, we obtain the complete graph shown in part (b) of Figure 7.52. (We note that this is the graph of the polar equation obtained in part (b) of Example 38.) The resulting figure-eight-shaped graph is called a *lemniscate* from the Greek word *lemniskos*, which means ribbon.

FIGURE 7.52

a.   **b.** $r^2 = 9 \cos 2\theta$ (a lemniscate)

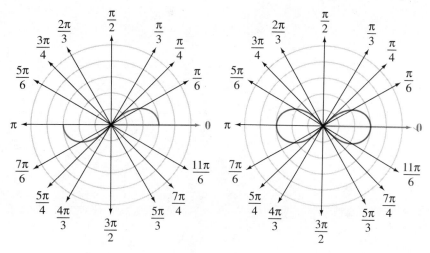

Lemniscates

If a is a nonzero real number, the graph of a polar equation of the form

$$r^2 = a^2 \cos 2\theta \quad \text{or} \quad r^2 = a^2 \sin 2\theta$$

is a figure-eight-shaped curve known as a **lemniscate**.

Spirals of Archimedes

EXAMPLE 43

Sketch the graph of the polar equation $r = \theta/\pi$, for values of θ in the interval $[0, 5\pi]$.

Solution   It can readily be shown that the equation $r = \theta/\pi$ fails all three of our symmetry tests. We therefore begin by making a table of values.

θ	0	$\dfrac{\pi}{4}$	$\dfrac{\pi}{2}$	$\dfrac{3\pi}{4}$	π	$\dfrac{5\pi}{4}$	$\dfrac{3\pi}{2}$	$\dfrac{7\pi}{4}$	2π	$\dfrac{9\pi}{4}$	$\dfrac{5\pi}{2}$	3π	$\dfrac{7\pi}{2}$	4π	$\dfrac{9\pi}{2}$	5π
$r = \dfrac{\theta}{\pi}$	0	$\dfrac{1}{4}$	$\dfrac{1}{2}$	$\dfrac{3}{4}$	1	$\dfrac{5}{4}$	$\dfrac{3}{2}$	$\dfrac{7}{4}$	2	$\dfrac{9}{4}$	$\dfrac{5}{2}$	3	$\dfrac{7}{2}$	4	$\dfrac{9}{2}$	5

Plotting the points from our table, we obtain the graph shown in Figure 7.53. The graphs of polar equations of the form $r = a\theta$, where $\theta \geq 0$, are called spirals of Archimedes, in honor of Archimedes who studied such curves in great detail.

FIGURE 7.53

$$r = \frac{\theta}{\pi} \text{ (a spiral of Archimedes)}$$

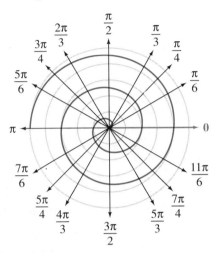

//

EXERCISES 7.6

The points in Exercises 1–12 are expressed in polar coordinates. Plot each point in the plane.

1. $(3, \pi/3)$
2. $(2, \pi/6)$
3. $(\sqrt{3}/2, 11\pi/6)$
4. $(3\sqrt{2}, 5\pi/4)$
5. $(0.9, -240°)$
6. $(1.4, -330°)$
7. $(\frac{5}{3}, \pi/12)$
8. $(\frac{5}{2}, 5\pi/12)$
9. $(-1.5, 135°)$
10. $(-2.7, 240°)$
11. $(-4, -7\pi/2)$
12. $(-3.5, -11\pi/3)$

The points in Exercises 13–22 are expressed in polar coordinates. Express each point in rectangular coordinates.

13. $(2, \pi/4)$
14. $(5, \pi/3)$
15. $(\sqrt{5}, 5\pi/3)$
16. $(\sqrt{3}/2, 7\pi/6)$
17. $(4, 75°)$
18. $(2, 15°)$
19. $(-3, 630°)$
20. $(-4, 135°)$
21. $(-\frac{8}{5}, -5\pi/4)$
22. $(-\frac{13}{4}, -7\pi/2)$

The points in Exercises 23–32 are expressed in rectangular coordinates. Express each point in polar coordinates, where $r \geq 0$ and θ is the smallest positive argument.

23. $(2\sqrt{3}, 2)$
24. $(3, -3)$
25. $(-4, 0)$
26. $(0, 5)$
27. $(3, -4)$
28. $(-2, 3)$
29. $(-1, -\sqrt{2}/2)$
30. $(-3\sqrt{2}, -2)$
31. $(2.7, -3.6)$
32. $(-\pi/3, \pi/4)$

In Exercises 33–38, plot each point in the plane. In each case, give four additional polar representations of the point (r, θ), two with r positive and two with r negative.

33. $(5, \pi/4)$
34. $(2, \pi/3)$
35. $(-3, 7\pi/6)$
36. $(-4, \pi/4)$
37. $(-2.5, -2\pi/3)$
38. $(-5.2, -5\pi/6)$

39. Determine which, if any, of the following points satisfy the polar equation $r^2 - 4\cos\theta = 0$:
    a. $(-\sqrt{2}, 5\pi/3)$ b. $(2, 3\pi/2)$ c. $(\sqrt{2}, -7\pi/3)$

40. Determine which, if any, of the following points satisfy the polar equation $r^2 = 9\sin 2\theta$:
    a. $(-3/\sqrt{2}, \pi/6)$ b. $(3, 5\pi/4)$
    c. $(3/\sqrt{2}, -7\pi/12)$

In Exercises 41–50, convert each polar equation to a Cartesian equation.

41. $r = 3$
42. $r = 5$
43. $\theta = 2\pi/3$
44. $\theta = 7\pi/6$
45. $r = \dfrac{5}{2\cos\theta + \sin\theta}$
46. $r = \dfrac{3}{\cos\theta - 5\sin\theta}$
47. $r\sin\theta = 4r^2\cos^2\theta - 1$
48. $r^2\cos^2\theta = 5 + 2r\sin\theta$
49. $r^2 - 9 = 8r\cos\theta$
50. $r^2 = 3 - 2r\sin\theta$

In Exercises 51–58, convert each Cartesian equation to a polar equation.

51. $2xy = 1$
52. $x^3 = y^2$
53. $x^2 + 4y^2 = 4$
54. $9x^2 - 4y^2 = 36$

55. $(x^2 + y^2)^{3/2} = 8xy$

56. $(x^2 + y^2)^{3/2} - 18xy = 0$

57. $x^4 + y^4 = 2xy - 2x^2y^2$

58. $x^4 + y^4 - 8xy + 2x^2y^2 = 0$

In Exercises 59–82, sketch the graph of each polar equation. Identify each graph as a cardioid, limacon, lemniscate, n-leaf rose, or spiral of Archimedes.

59. $r = 3 + 2\cos\theta$ **60.** $r = 1 - 4\sin\theta$

61. $r = 2\sin\theta - 3$ **62.** $r = 5\cos\theta + 2$

63. $r^2 = \cos 2\theta$ **64.** $r^2 = \sin 2\theta$

65. $r^2 = 9\sin 2\theta$ **66.** $r^2 = 4\cos 2\theta$

67. $r = 4 + 4\sin\theta$ **68.** $r = 2\sin\theta - 2$

69. $r = 2 - 3\cos\theta$ **70.** $r = 3 - 2\cos\theta$

71. $r = 4\sin 2\theta$ **72.** $r = 9\cos 2\theta$

73. $r = 2 - 2\cos\theta$ **74.** $r = 5 - 5\sin\theta$

75. $r^2 = 4\sin 4\theta$ **76.** $r^2 = 4\cos 4\theta$

77. $r^2 = 8\cos 2\theta$ **78.** $r^2 = 16\sin 2\theta$

79. $r = 2\theta, \quad \theta \geq 0$ **80.** $r = 3\theta, \quad \theta \geq 0$

81. $r = 2\theta/\pi, \quad \theta \geq 0$ **82.** $r = 4\theta/\pi, \quad \theta \geq 0$

In Exercises 83–90, sketch the graph of each polar equation.

83. $r = 3 + \sec\theta$ (a conchoid)

84. $r = 2 - \sec\theta$ (a conchoid)

85. $r = 1/\theta; \quad \theta > 0$ (a reciprocal spiral)

86. $r = -3/\theta; \quad \theta > 0$ (a reciprocal spiral)

87. $r = \sin\theta \tan\theta$ (a cissoid)

88. $r = 2\sin\theta \tan\theta$ (a cissoid)

89. $r = e^\theta; \quad \theta \geq 0$ (a logarithmic spiral)

90. $r = e^{\theta/2}; \quad \theta \geq 0$ (a logarithmic spiral)

91. It can be shown that the graph of a polar equation is symmetric with respect to the y-axis if the replacement of r by $-r$ and θ by $-\theta$ produces an equivalent equation. Use this test of symmetry to show that the graph of the equation $r = 3\sin 2\theta$ is symmetric with respect to the y-axis.

92. Use the law of cosines to verify that the distance between the points $P_1 = (r_1, \theta_1)$ and $P_2 = (r_2, \theta_2)$ in the plane is given by the formula below. [*Hint*: The accompanying figure illustrates one possible case.]

$$d(P_1, P_2) = \sqrt{r_1^2 + r_2^2 - 2r_1r_2 \cos(\theta_2 - \theta_1)}$$

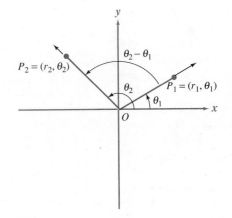

93. Use the formula from Exercise 92 to find the distance between $P_1 = (2, 5\pi/6)$ and $P_2 = (-5, 4\pi/3)$ in the plane. Also, express each point in rectangular coordinates and find $d(P_1, P_2)$ using the usual distance formula.

94. The graphs of the polar equations $r = 1 - \sin\theta$ and $r = 1 + \cos\theta$ intersect in exactly three points. Two of the points of intersection can be found by equating the r values and solving the resulting equation, $1 - \sin\theta = 1 + \cos\theta$, for θ. The remaining point will be obvious when the two cardioids are graphed. Find all three points of intersection, and show that the point determined from the graphs does indeed satisfy both equations.

 In Exercises 95–98, use your graphing calculator to obtain the graphs of the given curves.

95. $r = 1 - \sqrt{2}\cos\theta$ (a limacon)

96. $r = \sqrt{5}\sin 2\theta$ (a 4-leaf rose)

97. $r^2 = 3\cos 2\theta$ (a lemniscate)

98. $r = e\theta$ (a spiral of Archimedes)

 In Exercises 99–100, use your graphing calculator to find the polar coordinates of the points of intersection of the given curves.

99. $r = 1 + \sin\theta$ and $r = 1 + \cos 2\theta$, where $0° \leq \theta < 360°$.

100. $r = 1 + \cos\theta$ and $r = 2\theta$,   where $0° \leq \theta < 360°$.

CHAPTER 7 REVIEW EXERCISES

In Exercises 1–8, find the remaining parts of right triangle *ABC*.

1. $a = 15, c = 17$ **2.** $b = 5, c = 13$

3. $b = 78.5, \alpha = 53°50'$ **4.** $c = 5.14, \beta = 24°20'$

5. $a = 1120, b = 1450$ **6.** $a = \frac{3}{4}, b = \frac{5}{8}$

7. $b = 275, \beta = 17.1°$ **8.** $b = 1050, \alpha = 73.9°$

In Exercises 9–22, find the remaining parts of all possible triangles, if any, that satisfy the given conditions.

9. $\alpha = 23°20', \gamma = 14°30', b = 105$

10. $\beta = 58°42', \gamma = 87°19', b = 428$

11. $b = 12.2, c = 10.4, \beta = 85.1°$

12. $\alpha = 26.8°$, $a = 2.25$, $c = 1.84$
13. $a = 73.5$, $b = 46.8$, $c = 59.7$
14. $a = 1140$, $b = 1075$, $c = 835$
15. $\alpha = 50°10'$, $b = 17.3$, $c = 14.7$
16. $\beta = 88.5°$, $a = 33.2$, $c = 48.4$
17. $a = 52.4$, $b = 68.2$, $\alpha = 75.6°$
18. $b = 817$, $c = 670$, $\gamma = 52.3°$
19. $\gamma = 42°40'$, $a = 14.3$, $b = 17.8$
20. $\beta = 115°50'$, $a = 2.75$, $c = 1.91$
21. $a = 1016$, $c = 1589$, $\alpha = 15.6°$
22. $b = 18.4$, $c = 10.1$, $\alpha = 41.7°$

23. An airplane is flying at 600 miles per hour. One minute after the plane flies directly over an airport, its angle of elevation from the airport is 36.2°. Find the altitude of the airplane to the nearest 100 feet.

24. If a flagpole 48.5 feet tall casts a shadow 62.4 feet long, determine the angle of elevation of the sun.

25. Find the area of a regular decagon (10-sided polygon) inscribed in a circle of radius 12.2 centimeters.

26. A 24-foot ladder leans against a building. If the top of the ladder reaches 18 feet up the side of the building, find the angle between the ladder and the ground.

27. A ramp for handicap access is being built beside the steps to a building. If the total vertical rise of the steps is 4.5 feet, and the ramp is to be inclined at an angle of 15°, find the length of the ramp.

28. A ship leaves port and sails 780 kilometers on a bearing of N17.6°W. The ship then changes course and sails 540 kilometers on a bearing of N34.5°E. Find the distance from the ship to the port.

29. A ship is sailing due north at the rate of 24 miles per hour. At 5:00 P.M., a lighthouse is sighted at a bearing of N16.5°W. At 7:00 P.M., the bearing to the same lighthouse is S71.2°W. Find the distance from the ship to the lighthouse at 7:00 P.M.

30. A submarine on the surface of the ocean begins to dive with an angle of depression of 6.5°. If the submarine is traveling at 9.0 miles per hour, how long will it take to reach a depth of 1320 feet?

31. From a cliff overlooking a lake, the angles of depression to two boats on the lake are 42°10' and 48°40' (see figure). If the top of the cliff is 75 feet above the surface of the lake, find the distance between the two boats.

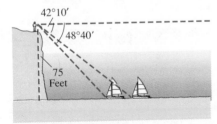

32. Two sides and the included angle of a parallelogram have measures of 24.6 centimeters, 32.9 centimeters, and 44°20', respectively. Find the lengths of the diagonals of the parallelogram and the area of the parallelogram.

33. A tract of land in the shape of a triangle has sides measuring 1480 meters, 1760 meters, and 2050 meters. Find its area.

34. To measure the height of a cliff, a surveyor first measures the angle of elevation to the top of the cliff from a point A which is located on ground that is level with the base of the cliff (see figure). She then walks 250 feet directly toward the cliff to a point B, where she again measures the angle of elevation to the top. If the successive measurements are 34°20' and 42°50', determine the height of the cliff.

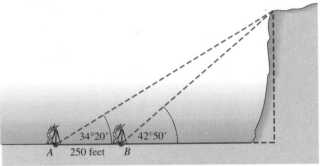

In Exercises 35–42, express the given complex number in trigonometric form.

35. $2 - 2i$
36. $-5 + 5i$
37. $-\sqrt{3} - i$
38. $-2\sqrt{3} + i$
39. $-\frac{5}{2}i$
40. 7
41. $2 - i$
42. $-4 - 7i$

In Exercises 43–50, express each number in the form $a + bi$, where a and b are real numbers.

43. $4 \operatorname{cis}(2\pi/3)$
44. $3 \operatorname{cis}(3\pi/4)$
45. $16 \operatorname{cis}(5\pi/4)$
46. $12 \operatorname{cis}(7\pi/6)$
47. $(\sqrt{3}/2) \operatorname{cis} 330°$
48. $7 \operatorname{cis} 240°$
49. $5 \operatorname{cis}(\pi/12)$
50. $(13/2) \operatorname{cis}(5\pi/12)$

In Exercises 51–54, perform the indicated operations. Leave each answer in trigonometric form.

51. $[3 \operatorname{cis}(\pi/5)][15 \operatorname{cis}(\pi/3)]$
52. $[4 \operatorname{cis}(3\pi/4)][12 \operatorname{cis}(\pi/2)]$
53. $\dfrac{3\sqrt{2} \operatorname{cis}(5\pi/3)}{6 \operatorname{cis}(\pi/3)}$
54. $\dfrac{2\sqrt{3} \operatorname{cis} 225°}{8 \operatorname{cis} 105°}$

In Exercises 55–58, perform the indicated operations. Leave each answer in the form $a + bi$.

55. $[12 \operatorname{cis}(7\pi/6)][(\sqrt{3}/2) \operatorname{cis}(2\pi/3)]$
56. $[5 \operatorname{cis}(2\pi/3)][2 \operatorname{cis}(5\pi/6)]$
57. $\dfrac{14 \operatorname{cis} 225°}{2 \operatorname{cis} 75°}$
58. $\dfrac{3\sqrt{2} \operatorname{cis} 315°}{9 \operatorname{cis} 135°}$

In Exercises 59–68, use DeMoivre's theorem to find the indicated powers of the given complex numbers. Express each answer in the form $a + bi$.

59. $[\sqrt{5} \operatorname{cis}(7\pi/12)]^6$
60. $[\sqrt{7} \operatorname{cis}(5\pi/12)]^8$

61. $[(\sqrt{2}/2)\text{ cis }20°]^{12}$ **62.** $[(\sqrt{3}/2)\text{ cis }15°]^{14}$

63. $[(\sqrt{2}/2) - (\sqrt{2}/2)i]^8$ **64.** $(4 - 4\sqrt{3}i)^6$

65. $[(\sqrt{3}/2) - \frac{1}{2}i]^5$ **66.** $(-3 + 3i)^4$

67. $(-8 - 8i)^6$ **68.** $[(5\sqrt{3}/2) + \frac{5}{2}i]^3$

In Exercises 69–74, use DeMoivre's theorem to find the indicated powers of the given complex numbers. Express each answer in trigonometric form.

69. $(\frac{2}{3}\text{ cis }32°)^8$ **70.** $[(\sqrt{2}/2)\text{ cis }24°]^6$

71. $(-\frac{1}{2} + \frac{3}{4}i)^6$ **72.** $(2\sqrt{3} - i)^3$

73. $(5 - 12i)^4$ **74.** $(-3 + 4i)^5$

In Exercises 75–80, find the indicated roots and represent them geometrically.

75. fifth roots of $2\sqrt{3} - 2i$

76. fourth roots of $-\sqrt{2}/2 + (\sqrt{2}/2)i$

77. cube roots of unity   **78.** eighth roots of unity

79. fourth roots of $81\text{ cis }72°$ **80.** sixth roots of $64\text{ cis }108°$

In Exercises 81–84, represent each vector as a position vector in the plane. Find the magnitude of each vector.

81. $\mathbf{v} = \langle -1, 4 \rangle$ **82.** $\mathbf{v} = \langle -3, -5 \rangle$

83. $\mathbf{v} = \langle \frac{7}{2}, -\frac{13}{4} \rangle$ **84.** $\mathbf{v} = \langle \sqrt{3}/2, -\frac{1}{2} \rangle$

In Exercises 85–86, find the sum of the given vectors. In each case, use the vectors $\mathbf{u}$ and $\mathbf{v}$ to illustrate the parallelogram law for vector addition.

85. $\mathbf{u} = \langle -2, 3 \rangle, \mathbf{v} = \langle 4, 5 \rangle$

86. $\mathbf{u} = \langle -\frac{3}{2}, -8 \rangle, \mathbf{v} = \langle \frac{5}{2}, 6 \rangle$

In Exercises 87–90, find the indicated vectors, where $\mathbf{u} = \langle -3, 5 \rangle$ and $\mathbf{v} = \langle 2, -4 \rangle$.

87. $-3\mathbf{u} + \frac{1}{2}\mathbf{v}$ **88.** $\frac{5}{3}\mathbf{u} - 2\mathbf{v}$

89. $-12\mathbf{u} - 4\mathbf{v}$ **90.** $(\sqrt{3}/2)\mathbf{u} - \frac{1}{2}\mathbf{v}$

In Exercises 91–94, find the direction angle for each vector $\mathbf{v}$.

91. $\mathbf{v} = \langle -5, -5 \rangle$ **92.** $\mathbf{v} = \langle \sqrt{3}/2, -\frac{1}{2} \rangle$

93. $\mathbf{v} = \langle -4, 3 \rangle$ **94.** $\mathbf{v} = \langle -2, 5 \rangle$

In Exercises 95–96, the magnitude and direction angle of a vector $\mathbf{v}$ are given. Find the algebraic representation of $\mathbf{v}$ in each case.

95. $\|\mathbf{v}\| = 5, \theta = 135°$ **96.** $\|\mathbf{v}\| = 12, \theta = 210°$

In Exercises 97–98, $\mathbf{u} = \frac{3}{4}\mathbf{i} - \mathbf{j}$ and $\mathbf{v} = -2\mathbf{i} + \frac{5}{2}\mathbf{j}$. In each case, express the given vector as a linear combination of the unit vectors $\mathbf{i}$ and $\mathbf{j}$.

97. $2\mathbf{u} - 4\mathbf{v}$ **98.** $-\frac{3}{2}\mathbf{u} + \frac{13}{4}\mathbf{v}$

In Exercises 99–100, give the magnitude and direction of the resultant of each pair of vectors.

99. Force A: 225 pounds in the direction S24.2°E
Force B: 280 pounds in the direction S41.9°W

100. Velocity A: 425 kilometers per hour in the direction N63.2°E
Velocity B: 310 kilometers per hour in the direction N17.8°E

101. Find the force required to keep a 2640-pound car from rolling down a 13.2° incline.

102. A boat that travels 15 miles per hour in still water attempts to go directly across a river in which the current is 6.5 miles per hour. Determine the angle between the boat's actual path and its intended path.

103. An airliner is flying on a compass bearing of S24.5°W at an airspeed of 515 miles per hour. If the wind is blowing from the direction N38.2°W at 62.5 miles per hour, determine the magnitude and direction of the plane's actual velocity vector.

104. A 620-pound weight is supported by two cables as shown in the figure. Find the amount of tension (stretching force) in each cable.

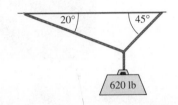

The points in Exercises 105–108 are expressed in polar coordinates. Express each point in rectangular coordinates.

105. $\left(3, \frac{\pi}{3}\right)$ **106.** $(3.25, 570°)$

107. $(-\frac{12}{5}, 120°)$ **108.** $\left(-4, -\frac{3\pi}{4}\right)$

The points in Exercises 109–112 are expressed in rectangular coordinates. Express each point in polar coordinates, where $r \geq 0$ and θ is the smallest positive argument.

109. $(-2, -2\sqrt{3})$ **110.** $(-5, 5)$

111. $(-3, 4)$ **112.** $(0, -\frac{7}{2})$

In Exercises 113–116, convert each polar equation to a Cartesian equation.

113. $r = 5$ **114.** $\theta = 3\pi/4$

115. $r^2 \sin 2\theta = 2$ **116.** $r = 12 \cos \theta$

In Exercises 117–120, convert each Cartesian equation to a polar equation.

117. $3xy = 1$ **118.** $4x^2 - 9y^2 = 36$

119. $x^2 + 9y^2 = 9$ **120.** $(x^2 + y^2)^{3/2} = 32xy$

In Exercises 121–132, sketch the graph of each polar equation. Identify each graph as a cardioid, limacon, lemniscate, n-leaf rose, or spiral of Archimedes.

121. $r = 3 + 3 \sin \theta$ **122.** $r = 3 + 5 \sin \theta$

123. $r = 16 \cos 2\theta$ **124.** $r = 9 \sin 2\theta$

125. $r = 5 - 2 \cos \theta$ **126.** $r = 4 - 4 \cos \theta$

127. $r = 1 - 3 \sin \theta$ **128.** $r = 5 - \cos \theta$

129. $r^2 = 16 \cos 2\theta$ **130.** $r^2 = 4 \sin 2\theta$

131. $r = \frac{1}{2}\theta, \quad \theta \geq 0$ **132.** $r = 2\theta, \quad \theta \geq 0$

CHAPTER **8**

Conic Sections

Our earlier study of functions included quadratic functions and their graphs, which are called parabolas. Parabolas, as well as circles, ellipses, and hyperbolas, belong to a class of curves known as the conic sections. In this chapter, we use the methods of analytic geometry to study the graphs of conic sections and consider a variety of applications of conic sections to everyday life. The chapter concludes with a discussion of parametric equations, which provide an alternative approach to defining and graphing equations in the xy-plane.

8.1 CONICS: AN INTRODUCTION

In Chapter 2, we encountered equations whose graphs are circles and parabolas. Such curves are called conic sections. To see why, we begin by defining a right circular cone.

FIGURE 8.1
Right Circular Cone

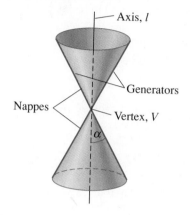

DEFINITION 8.1   Right Circular Cone

Let *l* be a line, *V* be a point on *l* and α be a fixed acute angle. The surface formed by all straight lines passing through *V* and making an angle α with *l* is called a **right circular cone** (see Figure 8.1). The line *l* is called the **axis** of the cone, and the point *V* is called its **vertex**. Each line passing through *V* and making an angle α with *l* is called a **generator** of the cone. The two parts of the cone, which are separated by the vertex, are called the **nappes** of the cone.

Henceforth, when we use the word *cone*, we mean a right circular cone.

The intersection of a plane and a cone is called a **conic section.** In Figure 8.2, we illustrate the four basic conic sections: the circle, the ellipse, the parabola, and the hyperbola. In each of these four cases, notice that the intersecting plane does not pass through the vertex of the cone.

FIGURE 8.2

a. Circle   **b.** Ellipse   **c.** Parabola   **d.** Hyperbola

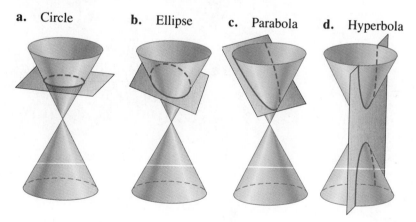

The Greek mathematician Apollonius of Perga (ca. 262–ca. 190 B.C.), who was known as "The Great Geometer," wrote an eight-book treatise called *The Conic Sections*. Apollonius is credited as the first mathematician to develop the conic sections using a two-napped right circular cone. In addition, it was Apollonius who gave the names *ellipse, parabola* and *hyperbola* to these three conic sections.

Apollonius had available only methods from Euclidian geometry in his study of conic sections. We shall, however, use both algebraic and geometric methods to study the conics. Such methods arise from a field of mathematics known as analytic geometry, which was founded by René Descartes.

We summarize these results below.

Conic Sections

If a cone is intersected by a plane that does not contain the vertex, then

(i)   a **circle** is formed if the plane is perpendicular to the axis of the cone;
(ii) an **ellipse** is formed if the plane is not perpendicular to the axis but passes entirely through one nappe of the cone without intersecting the other nappe;
(iii) a **parabola** is formed if the plane is parallel to a generator of the cone;
(iv) a **hyperbola** is formed if the plane is parallel to the axis of the cone.

The intersection of a cone and a plane that contains the vertex of the cone is a point, a line, or two intersecting lines (see Figure 8.3). These curves are called **degenerate conic sections**.

FIGURE 8.3

a. Point **b.** Line **c.** Two intersecting lines

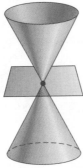  

8.2   PARABOLAS

In Section 2.3, we studied functions whose graphs are parabolas. However, we did not give a formal definition of a parabola at that time. We begin this section with a geometric definition of a parabola.

DEFINITION 8.2

A **parabola** is the set of all points in a plane that are equidistant from a fixed line *l* and a fixed point *F* not on *l*. The line *l* is called the **directrix** of the parabola, while the point *F* is called the **focus**.

FIGURE 8.4
a.

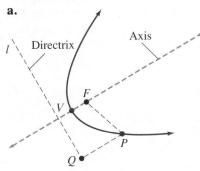

b.

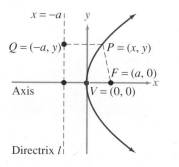

The line passing through the focus and perpendicular to the directrix is called the **axis** of the parabola. As we see in Figure 8.4(a), a parabola is symmetric with respect to its axis. The point of intersection of a parabola and its axis is called the **vertex**. V denotes the vertex in Figure 8.4(a). It follows from Definition 8.2 that the vertex is located halfway between the focus and the directrix.

In this section, we will study the case where the directrix of a parabola is parallel to either the x-axis or the y-axis. In Section 8.5, we consider the case where the directrix is not parallel to either coordinate axis.

Let us now derive an equation for a parabola. For convenience, we use a coordinate system positioned so that the axis of the parabola lies along one of the coordinate axes and the vertex of the parabola is located at the origin $(0, 0)$. First, we consider the case where the focus F is located on the positive x-axis; that is, $F = (a, 0)$ for some positive number a, as shown in Figure 8.4(b). Thus, the distance from the vertex $V = (0, 0)$ to the focus $F = (a, 0)$ is equal to a. Hence, the distance from the vertex to the directrix l must also be equal to a. Since the directrix is perpendicular to the axis of the parabola (the x-axis in this case), the directrix is the vertical line whose equation is $x = -a$. Recall from Definition 8.2 that a point P lies on the parabola with focus F and directrix l if and only if

$$d(P, F) = d(P, l). \qquad (1)$$

From Figure 8.4(b), we see that $d(P, l)$ is the distance between the points $P = (x, y)$ and $Q = (-a, y)$. Hence,

$$d(P, l) = x - (-a) = x + a.$$

Since $d(P, F) = \sqrt{(x - a)^2 + (y - 0)^2}$, we see from equation (1) that

$$\sqrt{(x - a)^2 + (y - 0)^2} = x + a.$$

Squaring both sides of this equation and simplifying, we get

$$(x - a)^2 + y^2 = (x + a)^2$$
$$x^2 - 2ax + a^2 + y^2 = x^2 + 2ax + a^2$$
$$y^2 = 4ax. \qquad (2)$$

If the focus F is placed on the negative x-axis at the point $(-a, 0)$, where $a > 0$, and the vertex V is located at the origin $(0, 0)$, then the equation of the directrix l is $x = a$. It can be shown that the equation of the parabola is

$$y^2 = -4ax. \qquad (3)$$

Hence, from equations (2) and (3), we see that the equation of a parabola whose vertex is located at the origin and whose axis coincides with the x-axis has the **standard form**

$$y^2 = 4px,$$

where $p = \pm a$. The parabola opens to the right if $p > 0$ and opens to the left if $p < 0$. In a similar manner, it can be shown that the equation of a parabola whose vertex lies at the origin $(0, 0)$ and whose axis coincides with the y-axis has the **standard form**

$$x^2 = 4py,$$

where p is a nonzero real number (see Exercise 31). The parabola opens upward if $p > 0$ and opens downward if $p < 0$. From Figure 8.5, we see that, in each case, the distance from the vertex V to the focus F is $|p|$, and the distance from the vertex to the directrix is $|p|$.

FIGURE 8.5

 $y^2 = 4px$

a. $p > 0$

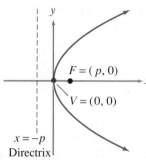

b. $p < 0$

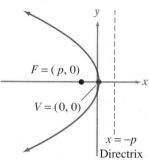

$x^2 = 4py$

c. $p > 0$

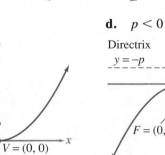

d. $p < 0$

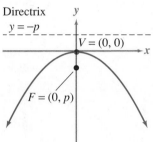

We summarize properties of parabolas in Table 8.1.

TABLE 8.1
Standard Forms for Equations of Parabolas with Vertex (0, 0) and Axis One of Coordinate Axes

Equation	Vertex	Axis	Focus	Directrix	Parabola Opens
$y^2 = 4px$	$(0, 0)$	x-axis	$(p, 0)$	$x = -p$	To the right, if $p > 0$ To the left, if $p < 0$
$x^2 = 4py$	$(0, 0)$	y-axis	$(0, p)$	$y = -p$	Upward, if $p > 0$ Downward, if $p < 0$

EXAMPLE 1

Find the equation, in standard form, of the parabola that has its vertex at the origin, opens to the left, and passes through the point $P = (-2, -4)$. Sketch the graph of this parabola.

Solution Since the parabola opens to the left and has its vertex at the origin, we see from Table 8.1 that its equation has the form

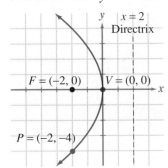

FIGURE 8.6 $y^2 = -8x$

$$y^2 = 4px. \tag{4}$$

Since the point $P = (-2, -4)$ is on the parabola, $(x, y) = (-2, -4)$ is a solution of equation (4). Thus,

$$(-4)^2 = 4p(-2)$$
$$16 = -8p$$
$$p = -2.$$

Hence, the equation of the parabola is

$$y^2 = -8x.$$

The graph of this parabola is shown in Figure 8.6.

EXAMPLE 2

Find the focus and directrix of the parabola with equation $x^2 = -2y$. Sketch the graph of this parabola.

Solution   The equation of the parabola has the form $x^2 = 4py$. Thus,

$$4p = -2, \quad \text{or} \quad p = -\frac{1}{2}.$$

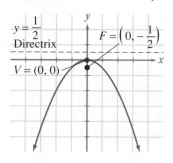

FIGURE 8.7 $x^2 = -2y$

From Table 8.1 we see that the focus is the point $F = (0, -\frac{1}{2})$ and the directrix is the horizontal line $y = \frac{1}{2}$. The graph of this parabola is sketched in Figure 8.7.

Let us now consider a parabola whose vertex is not located at the origin. For example, suppose the parabola

$$x^2 = -2y, \tag{5}$$

which we considered in Example 2, is shifted 3 units to the right and 2 units upward. The result is the parabola whose vertex is the point $V = (3, 2)$ and whose axis is the vertical line $x = 3$, as shown in Figure 8.8. The equation for this new parabola can be obtained from equation (5) by replacing x by $x - 3$ and y by $y - 2$. Hence, the equation of the parabola with vertex $V = (3, 2)$ and axis $x = 3$ is

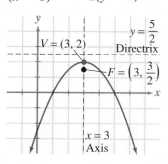

FIGURE 8.8
$(x - 3)^2 = -2(y - 2)$

$$(x - 3)^2 = -2(y - 2).$$

More generally, the equation, **in standard form**, of a parabola whose vertex is the point (h, k) and whose axis is the vertical line $x = h$ is

$$(x - h)^2 = 4p(y - k). \tag{6}$$

In addition, the equation, **in standard form**, of a parabola whose vertex is the point (h, k) and whose axis is the horizontal line $y = k$ is

$$(y - k)^2 = 4p(x - h). \tag{7}$$

Graphs of parabolas whose equations have the form of equation (6) or equation (7) are displayed in Figure 8.9 (page 376). Properties of such parabolas are summarized in Table 8.2.

FIGURE 8.9

a. $p > 0$ **b.** $p < 0$ **c.** $p > 0$ **d.** $p < 0$

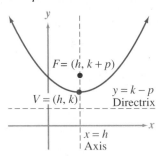

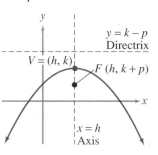

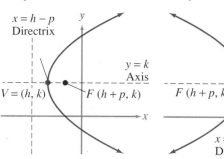

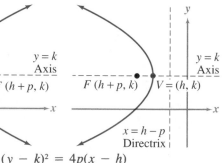

$$(x - h)^2 = 4p(y - k) \qquad\qquad (y - k)^2 = 4p(x - h)$$

TABLE 8.2
Standard Forms of Equations for Parabolas with Vertex (h,k) and Axis Parallel to a Coordinate Axis

Equation	Vertex	Axis	Focus	Directrix	Parabola Opens
$(x - h)^2 = 4p(y - k)$	(h, k)	$x = h$	$(h, k + p)$	$y = k - p$	Upward, if $p > 0$ Downward, if $p < 0$
$(y - k)^2 = 4p(x - h)$	(h, k)	$y = k$	$(h + p, k)$	$x = h - p$	To the right, if $p > 0$ To the left, if $p < 0$

EXAMPLE 3

FIGURE 8.10
$(x + 2)^2 = 16(y - 1)$

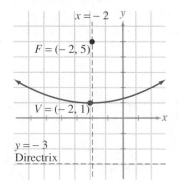

Find the standard form of the equation of the parabola with vertex $V = (-2, 1)$ and directrix $y = -3$. Sketch the graph of this parabola.

Solution   Since the vertex is the point $(-2, 1)$ and the directrix is the horizontal line $y = -3$, we see from Table 8.2 that the standard form is

$$(x + 2)^2 = 4p(y - 1).$$

Furthermore, since the vertex $(-2, 1) = (h, k)$, it follows that $k = 1$. Also, since the directrix is the line $y = -3$ and since $y = k - p$,

$$-3 = k - p$$
$$= 1 - p.$$

Thus, $p = 4.$

Hence, the standard form of the equation of this parabola is

$$(x + 2)^2 = 4(4)(y - 1) \qquad \text{or} \qquad (x + 2)^2 = 16(y - 1).$$

The graph is sketched in Figure 8.10.

By expanding the left side of the equation $(x - h)^2 = 4p(y - k)$ and solving the resulting equation for y, we obtain an equation of the form

$$y = ax^2 + bx + c,$$

where $a \neq 0$. You may recall that in Section 2.3 we showed that if $a \neq 0$, the graph of the equation $y = ax^2 + bx + c$ is a parabola whose axis is a vertical line. In a similar fashion, it can be shown that the equation $(y - k)^2 = 4p(x - h)$ can be written in the form

$$x = ay^2 + by + c, \tag{8}$$

where $a \neq 0$. Conversely, by completing the square in y, equation (8) can be expressed in the form

$$(y - k)^2 = 4p(x - h).$$

Hence, the graph of equation (8) is also a parabola.

EXAMPLE 4

Find the vertex, focus, axis, and directrix of the parabola whose equation is $y^2 + 6y + x + 13 = 0$. Graph this parabola.

Solution   We begin by writing the equation

$$y^2 + 6y + x + 13 = 0 \tag{9}$$

in the standard form

$$(y - k)^2 = 4p(x - h).$$

Completing the square in y in equation (9), we get

$$y + 6y + 9 = -x - 13 + 9$$
$$(y + 3)^2 = -x - 4,$$

or $$(y + 3)^2 = -(x + 4). \tag{10}$$

From equation (10) we see that $h = -4$, $k = -3$, and $4p = -1$, or $p = -\frac{1}{4}$. Hence, the vertex of this parabola is $V = (-4, -3)$. Using Table 8.2, we see that the focus is $F = (-4 - \frac{1}{4}, -3) = (-\frac{17}{4}, -3)$, the axis is the horizontal line $y = -3$, and the directrix is the vertical line $x = -4 + \frac{1}{4} = -\frac{15}{4}$.

Since $p < 0$, the parabola opens to the left. If we let $y = 0$ in equation (9), we see that $x = -13$. Thus, the point $(-13, 0)$ is on the graph of the parabola. Moreover, since the parabola is symmetric with respect to its axis $y = -3$, we know that the point $(-13, -6)$ is also on the graph as shown in Figure 8.11.

FIGURE 8.11
$y^2 + 6y + x + 13 = 0$

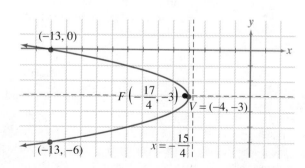

Reflective Property of Parabolas

Parabolas have a reflective property that is used in optics. Lines that are tangent to a parabola play a central role in this reflective property. A line is said to be **tangent** to a parabola at a point P on the parabola if the line intersects the

FIGURE 8.12

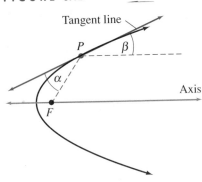

Tangent line

parabola only at P and is not parallel to the axis of the parabola (see Figure 8.12). The reflective property of a parabola is a consequence of the following property of tangent lines to parabolas:

Reflective Property of a Parabola

The tangent line to a parabola at a point P makes equal angles with the line through P parallel to the axis of the parabola and the line through P and the focus of the parabola.

The headlamps of an automobile, which are parabolic mirrors formed by rotating a parabola around its axis, illustrate the reflective property of parabolas (see Figure 8.13). The light from a bulb placed at the focus reflects from the surface in beams that are parallel to the axis of the parabola. In a similar manner, rays of light coming into a parabolic surface parallel to its axis are reflected through the focus (see Figure 8.14).

FIGURE 8.13

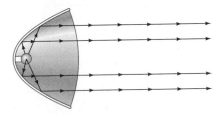

FIGURE 8.14

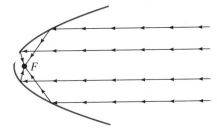

EXAMPLE 5

An automobile headlamp is designed so that a cross section through its axis is a parabola and the light source is placed at the focus. If the headlamp is 8 inches across and 6 inches deep, find the location of the light source.

The ancient Greeks were aware of the reflecting properties of a parabola. In the second century B.C., the Greek philosopher Diocles suggested in a book, titled *On Burning Mirrors*, that if victims were to be sacrificed in public, parabolic mirrors could be used to provide a visible burning spot on their bodies. Moreover, it was reported by the Greek historian Plutarch that Archimedes of Syracuse

(287–212 B.C.) helped save Syracuse from the Romans in 214 B.C. by designing and using "burning mirrors." As the story goes, these burning mirrors were parabolic in form and were used to concentrate the rays of the sun to burn Roman ships. It is interesting to note that the word *focus* comes from a Latin word meaning "fireplace."

FIGURE 8.15

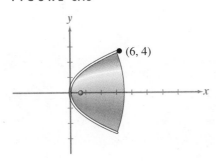

Solution In Figure 8.15, we have placed the parabolic cross section of the headlamp in a coordinate system so that the vertex is at the origin, the axis of the parabola is the x-axis, and the parabola opens to the right. Therefore, the equation of the parabola has the form

$$y^2 = 4px.$$

Since the parabola passes through the point $(6, 4)$, we know that

$$4^2 = 4p(6) \quad \text{or} \quad p = \frac{2}{3}.$$

Thus, the focus is the point $(\frac{2}{3}, 0)$ and the light source is located on the axis $\frac{2}{3}$ of an inch to the right of the vertex. _____

We used the concept of tangent lines to help explain the reflective property of parabolas. In calculus, tangent lines to graphs are defined in general. However, the equations of tangent lines to parabolas can be found without using techniques of calculus, as the following example illustrates.

EXAMPLE 6

FIGURE 8.16

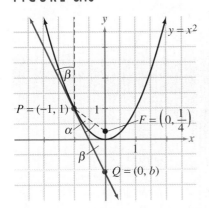

Find the equation of the tangent line to the parabola $y = x^2$ at the point $P = (-1, 1)$.

Solution This parabola has an equation of the form $x^2 = 4py$, where $4p = 1$. Thus, $p = \frac{1}{4}$ and the focus is $F = (0, \frac{1}{4})$. In Figure 8.16, we see that $\triangle PFQ$ is an isosceles triangle since $\angle FPQ = \alpha = \beta = \angle FQP$. Hence,

$$d(F, P) = d(F, Q).$$

Now

$$d(F, P) = \sqrt{(-1 - 0)^2 + \left(1 - \frac{1}{4}\right)^2} = \frac{5}{4}.$$

Also,

$$d(F, Q) = \frac{1}{4} - b.$$

Thus,

$$\frac{5}{4} = \frac{1}{4} - b \quad \text{or} \quad b = -1.$$

The reflective properties of parabolas are used in the design of mirrors for telescopes, radio antennae, and satellite and microwave dish receivers. Civil engineers have incorporated twin parabolic cables in the design of some bridges such as the Golden Gate Bridge in San Francisco and the George Washington Bridge in New York City.

The Italian scientist Galileo Galilei (1564–1643) discovered that a projectile shot into the air takes a parabolic path if gravity is the only force acting upon it. For example, the path of a football thrown in a nonvertical direction is parabolic. Water sprayed nonvertically into the air from a garden hose follows a parabolic path.

Therefore, the slope of the tangent line is

$$m = \frac{1 - (-1)}{-1 - 0}$$

$$= -2,$$

and the equation of the tangent line, in slope-intercept form, is

$$y = -2x - 1.$$

E X E R C I S E S 8.2

In Exercises 1–20, find the vertex, focus, axis, and directrix of the given parabola and sketch its graph.

1. $x^2 = 16y$
2. $x^2 = -2y$
3. $y^2 = 6x$
4. $y^2 = 10x$
5. $x^2 + 10y = 0$
6. $2y^2 - x = 10$
7. $2y^2 - 5x = 0$
8. $3x^2 + 5y = 0$
9. $(x - 1)^2 = -8(y - 2)$
10. $(y - 3)^2 = -8(x + 5)$
11. $y^2 - 2y - 8x + 25 = 0$
12. $x^2 - 2x + 12y + 13 = 0$
13. $4x^2 - 10x + y - 1 = 0$
14. $y^2 - 4x - 4y + 8 = 0$
15. $y - 9 = x^2 + 4$
16. $y + 5 = x^2 + 4x$
17. $x = y^2 - 6y - 3$
18. $x = y^2 - 2y - 2$
19. $6y = 4x^2 + 4x + 7$
20. $2x = y^2 + 2y + 5$

In Exercises 21–30, find the equation, in standard form, of the parabola satisfying the given conditions.

21. Vertex: $(0, 0)$
Focus: $(0, -\frac{1}{4})$
22. Vertex: $(0, 0)$
Focus: $(0, \sqrt{2})$
23. Vertex: $(-2, 3)$
Focus: $(-3, 3)$
24. Vertex: $(3, -1)$
Focus: $(5, -1)$
25. Focus: $(0, 0)$
Directrix: $x = -2$
26. Focus: $(0, 0)$
Directrix: $y = 6$
27. Vertex: $(-4, -2)$
Directrix: $y = -5$
28. Vertex: $(-1, 6)$
Directrix: $x = 2$
29. Vertex: $(0, 0)$
Axis is the x-axis
Contains $P = (-2, 4)$
30. Vertex: $(0, 0)$
Axis is the y-axis
Contains $P = (-1, -6)$

31. Derive the equation of the parabola whose vertex is the point $(0, 0)$, whose axis is the y-axis, and whose focus is
 a. the point $(0, a)$, where $a > 0$.
 b. the point $(0, -a)$, where $a > 0$.

32. Derive the equation of the parabola with vertex $V = (h, k)$ and focus $F = (h + p, k)$, where $p < 0$.

33. A reflecting telescope has a parabolic mirror that is 12 feet across at the top and 4 feet deep at the center. Where should the eyepiece be located?

34. A spotlight is designed so that a cross section through its axis is a parabola. The light source, which is placed at the focus, is 1 foot from the vertex. If the spotlight is 2 feet deep, what is the diameter of the spotlight at the opening?

35. Water gushing from the end of a horizontal pipe, located 10 feet from the ground, follows a parabolic path with vertex at the end of the pipe. At a point one foot below the end of the pipe, the horizontal distance from the water to a vertical line through the end of the pipe is 2 feet (see the accompanying figure). Where does the water strike the ground?

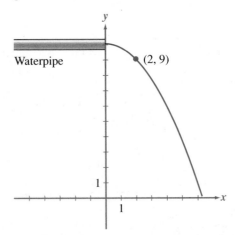

Waterpipe

36. The vertical position in feet of a projectile is given by the equation $y = -16t^2 + 64t$, and the horizontal position is given by the equation $x = 10t$ for $t \geq 0$ (t in seconds).
 a. Show that the path of the projectile is parabolic.
 b. Determine the position of the projectile when
  (i) $t = 1$ (ii) $t = 2$ (iii) $t = 3$
 c. Determine the maximum height attained by the projectile.
 d. At what time t does the projectile hit the ground?

37. Find an equation of the tangent line to the graph of the parabola at the given point. Sketch the parabola and the tangent line in each case.
 a. $x^2 = y + 1$; $(1, 0)$ **b.** $y^2 = x - 1$; $(2, 1)$
 c. $4y = 3 + 2x - x^2$; $(5, -3)$
 d. $y^2 = -4(x + 1)$; $(-2, 2)$

38. Suppose a golf ball is driven so that it travels a distance of 600 feet as measured along flat ground and reaches an altitude of 200 feet. Find an equation for the path of the golf ball assuming that the path is parabolic.

39. The cable of a suspension bridge, which is illustrated in the accompanying figure, is in the form of a parabola having a vertical axis. The bridge is 300 feet long. The longest of the vertical supporting cables is 100 feet and the shortest is 50 feet. Find the length of the supporting cable that is 50 feet from the middle.

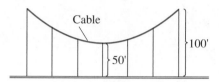

Cable

50' 100'

40. Sound waves are effectively transmitted using parabola reflections. Suppose two parabolic reflecting surfaces face one another as shown below. This double reflector device, known as a "whisper gallery," has the property that sound emitted at one focus will be reflected off the parabolas and concentrated at the other focus. (One such gallery is found at the Exploratorium in San Francisco, at which the reflectors have diameters of 8 feet and are placed 50 feet apart. A person talking softly at one focus can be clearly heard by a person at the other focus.) In the accompanying figure, the paths of two typical sound waves are shown. Use Definition 8.2 to show that all the sound waves will travel the same distance.

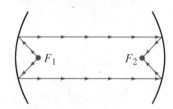

F_1 F_2

41. Sketch the region determined by the graphs of each of the following sets of equations.
  a. $y = x - 1; y^2 = 2x + 6$
  b. $y = x^2; y = 2x - x^2$
  c. $y = x^2 + 3; y = x; x = -1; x = 1$
  d. $x^2 + 2x + y = 0; x + y + 2 = 0$

42. The line segment passing through the focus of a parabola, perpendicular to its axis, and having endpoints on the parabola is called the **focal chord** or **latus rectum**. The **focal width** is the length of the focal chord. (See accompanying figure.)
  a. Find the focal width of the parabola $y^2 = 8x$.
  b. Show that the focal width of the parabolas $x^2 = 4py$ and $y^2 = 4px$ is $4|p|$.

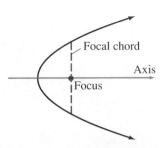

Focal chord

Axis

Focus

  c. For the parabolas $x^2 = 4py$ and $y^2 = 4px$, show that the endpoints of the focal chord are $(\pm 2p, p)$ and $(p, \pm 2p)$, respectively.

43. Prove that the vertex is the point on a parabola closest to the focus.

44. Let V and F denote the vertex and the focus, respectively, of the parabola $x^2 = 4py$. Suppose P and Q are points on the parabola such that the line segment $\overline{PV}$ is perpendicular to the line segment $\overline{QV}$. If R is the point where the line segment $\overline{PQ}$ meets the axis of the parabola, show that $d(R, V) = 4|p|$.

45. Show that the graph of the equation

$$Ax^2 + Dx + Ey + F = 0, \quad \text{where } A \neq 0,$$

  a. is a parabola if $E \neq 0$;
  b. is a vertical line if $E = 0$ and $D^2 - 4AF = 0$;
  c. is two vertical lines if $E = 0$ and $D^2 - 4AF > 0$;
  d. contains no points if $E = 0$ and $D^2 - 4AF < 0$.

46. Show that the graph of the equation

$$Cy^2 + Dx + Ey + F = 0, \quad \text{where } C \neq 0,$$

  a. is a parabola if $D \neq 0$;
  b. is a horizontal line if $D = 0$ and $E^2 - 4CF = 0$;
  c. is two horizontal lines if $D = 0$ and $E^2 - 4CF > 0$.
  d. contains no points if $D = 0$ and $E^2 - 4CF < 0$.

47. **a.** Use your graphing calculator to sketch the graph of each of the following equations using the same coordinate system.
  (i) $20x - 9y + 10 = 0$
  (ii) $x^2 + 4x - 4y = 0$
  b. Approximate, to two decimal places, the coordinates of the points of intersection.

48. **a.** Use your graphing calculator to sketch the graph of each of the following equations using the same coordinate system.
  (i) $x^2 + 2x + 2y - 1 = 0$
  (ii) $x^2 + 2x - 2y + 2 = 0$
  b. Approximate, to two decimal places, the coordinates of the points of intersection.

8.3 ELLIPSES

In Section 8.1, we stated that an ellipse is formed when a cone is intersected by a plane that is not perpendicular to the axis and passes entirely through one nappe of the cone without intersecting the other nappe. In particular, if β is the angle

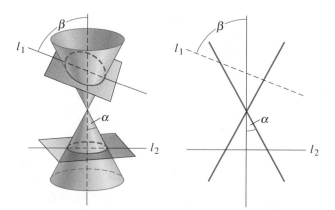

at which the plane intersects the axis of the cone and α is the angle between the axis and a generator of the cone, as shown in Figure 8.17, then an ellipse is formed when $\alpha < \beta < 90°$. We note that if $\beta = 90°$, a circle is formed. An ellipse is defined as follows:

F I G U R E   8.18

a.

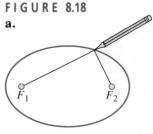

b.

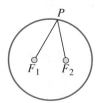

DEFINITION 8.3

An **ellipse** is the set of points in a plane, the sum of whose distances from two fixed points F_1 and F_2 is a positive constant. Each of the fixed points is called a **focus** (plural: **foci**). The midpoint of the line segment joining the foci is called the **center** of the ellipse.

By using a piece of string and two tacks, we can visualize the definition of an ellipse in the following way. Mark the foci F_1 and F_2 on a board. Now take a piece of string whose length is greater than the distance from F_1 to F_2 and tack the ends of the string to the foci. Trace a curve with a pencil held tightly against the string. The curve traced is an ellipse since for each point P on the curve, $d(P, F_1) + d(P, F_2)$ is a positive constant; namely, the length of the string. When the distance between the foci is nearly equal to the length of the string, the ellipse is relatively flat as shown in Figure 8.18(a). (In this case, the measure of angle β shown in Figure 8.17 is close to 0.) On the other hand, if the distance between the foci is small compared to the length of the string, the ellipse resembles a circle as shown in Figure 8.18(b).

From Figure 8.19, we see that an ellipse is oval-shaped and has two axes of symmetry: the **focal axis**, which is the line containing the foci, and the line that

Apollonius wrote his treatise on conics during the golden age of Greek mathematics, which was the period from about 300 to 200 B.C. Approximately 1800 years later, Johannes Kepler (1571–1630) brought conics back into prominence with his discovery that the planets revolve about the sun in elliptical orbits. Kepler's discovery discredited the earlier hypothesis of astronomer Nicholas Copernicus (1473–1543), who had proposed that the orbits of the planets were circular. It is known today that some comets also have elliptical orbits.

FIGURE 8.19

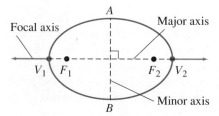

FIGURE 8.20

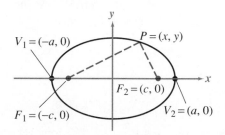

passes through the center and that is perpendicular to the focal axis. The **vertices** (plural of vertex) are the points where the focal axis intersects the ellipse. The vertices for the ellipse shown in Figure 8.19 are the points V_1 and V_2. The line segment joining the vertices is called the **major axis**, and the line segment that is perpendicular to the major axis at the center and has endpoints on the ellipse is called the **minor axis**.

We now derive the standard form for the equation of an ellipse with foci $F_1 = (-c, 0)$ and $F_2 = (c, 0)$, where $c > 0$. The center of this ellipse is located at the origin, as shown in Figure 8.20. For convenience, we let $2a$ denote the sum of the distances from any point $P = (x, y)$ on the ellipse to the foci. If a point $P = (x, y)$ is on the ellipse, then by Definition 8.3, the distance from P to one focus plus the distance from P to the second focus equals $2a$; that is,

$$d(P, F_1) + d(P, F_2) = 2a.$$

Thus, a point $P = (x, y)$ is on the ellipse with foci $F_1 = (-c, 0)$ and $F_2 = (c, 0)$ if and only if

$$d(P, F_1) + d(P, F_2) = 2a;$$

that is, if and only if

$$\sqrt{(x + c)^2 + (y - 0)^2} + \sqrt{(x - c)^2 + (y - 0)^2} = 2a. \qquad (1)$$

Rewriting equation (1), we have

$$\sqrt{(x - c)^2 + y^2} = 2a - \sqrt{(x + c)^2 + y^2}. \qquad (2)$$

Next, we square both sides of equation (2) and simplify as follows:

$$(\sqrt{(x - c)^2 + y^2})^2 = (2a - \sqrt{(x + c)^2 + y^2})^2$$
$$(x - c)^2 + y^2 = 4a^2 - 4a\sqrt{(x + c)^2 + y^2}$$
$$+ (x + c)^2 + y^2$$
$$x^2 - 2cx + c^2 + y^2 = 4a^2 - 4a\sqrt{(x + c)^2 + y^2}$$
$$+ x^2 + 2cx + c^2 + y^2$$
$$-4cx - 4a^2 = -4a\sqrt{(x + c)^2 + y^2}. \qquad (3)$$

Dividing each side of equation (3) by -4 and then squaring each side of the resulting equation, we get

$$(cx + a^2)^2 = (a\sqrt{(x + c)^2 + y^2})^2,$$

or

$$c^2x^2 + 2a^2cx + a^4 = a^2x^2 + 2a^2cx + a^2c^2 + a^2y^2. \qquad (4)$$

Equation (4) can be written

$$-a^2x^2 + c^2x^2 - a^2y^2 = -a^4 + a^2c^2.$$

Factoring, we get

$$-(a^2 - c^2)x^2 - a^2y^2 = -a^2(a^2 - c^2). \qquad (5)$$

Dividing each side of equation (5) by $-a^2(a^2 - c^2)$ yields

$$\frac{x^2}{a^2} + \frac{y^2}{a^2 - c^2} = 1. \qquad (6)$$

From Figure 8.20, we see that $d(F_1, F_2) = 2c < d(P, F_1) + d(P, F_2) = 2a$. Thus, $2c < 2a$ and therefore, $c < a$.

Since $a > c > 0$, it follows that $a^2 - c^2 > 0$. If we let

$$b = \sqrt{a^2 - c^2}, \quad \text{then } b^2 = a^2 - c^2.$$

Substituting b^2 for $a^2 - c^2$ in equation (6), we have

$$\frac{x^2}{a^2} + \frac{y^2}{b^2} = 1. \tag{7}$$

Since $c > 0$ and $b^2 = a^2 - c^2$, it follows that $a^2 > b^2$. Since a and b are both positive, we know that $a > b$. We have thus shown that the coordinates of every point $P = (x, y)$ on the ellipse satisfy equation (7). Conversely, it can be shown that if the coordinates of a point $P = (x, y)$ satisfy equation (7), then the point is on the ellipse. Equation (7) is called the **standard form** for the equation of an ellipse with foci $F_1 = (-c, 0)$ and $F_2 = (c, 0)$. Using the tests for symmetry, we see that the graph of an ellipse defined by equation (7) is symmetric with respect to the x-axis, the y-axis, and the origin.

The focal axis for this ellipse is the x-axis. Hence, the vertices are the x-intercepts of the ellipse. To find the x-intercepts, we set $y = 0$ in equation (7) to obtain

$$\frac{x^2}{a^2} = 1$$
$$x^2 = a^2$$
$$x = \pm a.$$

Therefore, the vertices are the points $V_1 = (-a, 0)$ and $V_2 = (a, 0)$. Moreover, the length of the major axis is

$$d(V_1, V_2) = a - (-a) = 2a.$$

The y-intercepts of the ellipse are found by setting $x = 0$ in equation (7).

$$\frac{y^2}{b^2} = 1$$
$$y^2 = b^2$$
$$y = \pm b.$$

Hence, the y-intercepts are the points $(0, -b)$ and $(0, b)$. Since, as we can see in Figure 8.21, the y-intercepts are the endpoints of the minor axis, it follows that the length of the minor axis is $2b$. Notice that, since $a > b$, the major axis is longer than the minor axis.

We summarize our results in Table 8.3.

FIGURE 8.21

$\dfrac{x^2}{a^2} + \dfrac{y^2}{b^2} = 1,$ $a > b$

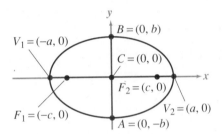

TABLE 8.3

Properties of the Ellipse with Equation

$\dfrac{x^2}{a^2} + \dfrac{y^2}{b^2} = 1,$ $a > b$

Foci:	$(\pm c, 0)$ where $c^2 = a^2 - b^2.$
Center:	$(0, 0)$
Vertices:	$(\pm a, 0)$
Endpoints of the minor axis:	$(0, \pm b)$
Focal axis:	the x-axis
Length of major axis:	$2a$
Length of minor axis:	$2b$

FIGURE 8.22

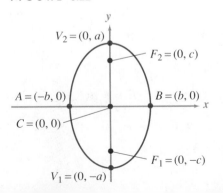

If the foci of an ellipse are located on the y-axis at the points $(0, -c)$ and $(0, c)$, as shown in Figure 8.22, then using an argument similar to the one given above (see Exercise 33), we obtain the **standard form**:

$$\frac{x^2}{b^2} + \frac{y^2}{a^2} = 1, \qquad a > b. \tag{8}$$

Some properties of an ellipse whose equation in standard form has the form of equation (8) are summarized in Table 8.4.

TABLE 8.4
Properties of the Ellipse with Equation

$$\frac{x^2}{b^2} + \frac{y^2}{a^2} = 1, \quad a > b$$

Foci:	$(0, \pm c)$ where $c^2 = a^2 - b^2$.
Center:	$(0, 0)$
Vertices:	$(0, \pm a)$
Endpoints of the minor axis:	$(\pm b, 0)$
Focal axis:	the y-axis
Length of major axis:	$2a$
Length of minor axis:	$2b$

We now have two standard forms for the equation of an ellipse depending upon the location of the foci:

1. Foci on the x-axis at $(\pm c, 0)$; $\dfrac{x^2}{a^2} + \dfrac{y^2}{b^2} = 1.$

2. Foci on the y-axis at $(0, \pm c)$; $\dfrac{x^2}{b^2} + \dfrac{y^2}{a^2} = 1.$

The relationship $a > b$ is the key to identifying which of the two standard forms applies to a given ellipse. Consider, for example, the equation $(x^2/9) + (y^2/16) = 1$. Since $16 > 9$, we know that $a^2 = 16$, and, therefore, the foci lie on the y-axis. On the other hand, for the equation $(x^2/6) + (y^2/5) = 1$, $a^2 = 6$, and the foci are located on the x-axis.

EXAMPLE 7

Find the vertices, foci, and the endpoints of the minor axis of the ellipse given by the equation $4x^2 + 3y^2 = 12$. Sketch the graph of this ellipse.

Solution   First, we divide each side of the equation

$$4x^2 + 3y^2 = 12$$

by 12 to express the equation in standard form:

$$\frac{x^2}{3} + \frac{y^2}{4} = 1.$$

FIGURE 8.23 $4x^2 + 3y^2 = 12$

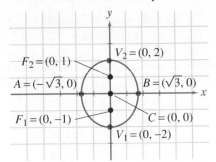

Since $4 > 3$, the denominator of the y^2-term is a^2 and the foci of the ellipse lie on the y-axis. Since $a^2 = 4$, $a = 2$; therefore, the vertices are $V_1 = (0, -2)$ and $V_2 = (0, 2)$. Moreover, since

$$c^2 = a^2 - b^2 = 4 - 3 = 1,$$

$c = 1$, and the foci are $F_1 = (0, -1)$ and $F_2 = (0, 1)$. Since $b^2 = 3$, $b = \sqrt{3}$, and the endpoints of the minor axis are $(-\sqrt{3}, 0)$ and $(\sqrt{3}, 0)$. The graph of this ellipse is sketched in Figure 8.23.

EXAMPLE 8

Find the equation, in standard form, of the ellipse whose foci are $F_1 = (-\sqrt{2}, 0)$ and $F_2 = (\sqrt{2}, 0)$ and whose major axis has length 6. Sketch the graph.

Solution   Since the foci are located on the x-axis, we see from Table 8.3 that the equation has the form

$$\frac{x^2}{a^2} + \frac{y^2}{b^2} = 1.$$

FIGURE 8.24 $\dfrac{x^2}{9} + \dfrac{y^2}{7} = 1$

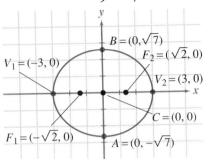

FIGURE 8.25
$\dfrac{(x-3)^2}{9} + \dfrac{(y+2)^2}{7} = 1$

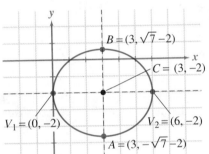

Moreover, since the length of the major axis is $2a = 6$, it follows that $a = 3$. To find b^2, we use the equation

$$b^2 = a^2 - c^2.$$

Since $c = \sqrt{2}$, it follows that

$$b^2 = 9 - 2 = 7.$$

Thus, the equation of the ellipse is

$$\frac{x^2}{9} + \frac{y^2}{7} = 1.$$

This ellipse is graphed in Figure 8.24.

If the ellipse $(x^2/9) + (y^2/7) = 1$ from Example 8 is shifted three units to the right and then 2 units downward, the center of the resulting ellipse is the point $(3, -2)$ (see Figure 8.25). The equation of this new ellipse can be found by replacing x by $x - 3$ and y by $y + 2$ in the equation $(x^2/9) + (y^2/7) = 1$ to get

$$\frac{(x-3)^2}{9} + \frac{(y+2)^2}{7} = 1.$$

More generally, it can be shown that the equation of an ellipse with center at (h, k) and major axis parallel to the x-axis has the standard form

$$\frac{(x-h)^2}{a^2} + \frac{(y-k)^2}{b^2} = 1, \quad \text{where } a > b. \tag{9}$$

(See Exercise 34.) Similarly, the equation of an ellipse with major axis parallel to the y-axis and having its center at the point (h, k) has the standard form

$$\frac{(x-h)^2}{b^2} + \frac{(y-k)^2}{a^2} = 1, \quad \text{where } a > b. \tag{10}$$

Properties of an ellipse having center (h, k) and major axis parallel to a coordinate axis are illustrated in Figure 8.26 and are summarized in Table 8.5.

As we see in Figure 8.26(a), a is the distance from the center (h, k) to the vertices, b is the distance from the center to the endpoints of the minor axis, and c is the distance from the center to the foci. We use these facts in Example 9.

FIGURE 8.26

a. $\dfrac{(x-h)^2}{a^2} + \dfrac{(y-k)^2}{b^2} = 1,$ $\quad a > b$

b. $\dfrac{(x-h)^2}{b^2} + \dfrac{(y-k)^2}{a^2} = 1,$ $\quad a > b$

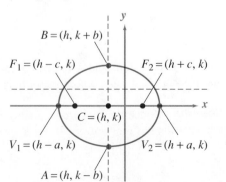

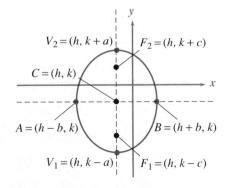

TABLE 8.5
Ellipses with Center at (h, k) and Major Axis Parallel to a Coordinate Axis

Equation	$\dfrac{(x-h)^2}{a^2} + \dfrac{(y-k)^2}{b^2} = 1,$ $a > b$	$\dfrac{(x-h)^2}{b^2} + \dfrac{(y-k)^2}{a^2} = 1,$ $a > b$
Center	(h, k)	(h, k)
Major Axis	Parallel to x-axis	Parallel to y-axis
Foci	$(h \pm c, k)$, where $c^2 = a^2 - b^2$	$(h, k, \pm c)$, where $c^2 = a^2 - b^2$
Vertices	$(h \pm a, k)$	$(h, k \pm a)$
Endpoints of Minor Axis	$(h, k \pm b)$	$(h \pm b, k)$

EXAMPLE 9

An ellipse has center at $(-3, 2)$, a vertical major axis of length 10 and a minor axis of length 5. Find an equation, in standard form, for this ellipse and sketch its graph.

Solution   Since the length of the major axis is $2a = 10$, we know that $a = 5$. Moreover, since the length of the minor axis is $2b = 5$, it follows that $b = \frac{5}{2}$. Since the major axis is vertical, or parallel to the y-axis, we see from Table 8.5 that the equation has the form

$$\frac{(x-h)^2}{b^2} + \frac{(y-k)^2}{a^2} = 1.$$

Thus, the equation for this ellipse, in standard form, is

$$\frac{(x+3)^2}{25/4} + \frac{(y-2)^2}{25} = 1.$$

Given that the major axis is vertical and that the center of the ellipse is the point $(-3, 2)$, we know that the major axis lies along the vertical line $x = -3$. Since the distance between the center $(-3, 2)$ and each vertex is $a = 5$, the vertices are located at the points $V_1 = (-3, 2 - 5) = (-3, -3)$ and $V_2 = (-3, 2 + 5) = (-3, 7)$. Similarly, since the minor axis is horizontal, the endpoints of the minor axis are the points $(-3 - \frac{5}{2}, 2) = (-\frac{11}{2}, 2)$ and $(-3 + \frac{5}{2}, 2) = (-\frac{1}{2}, 2)$. This ellipse is sketched in Figure 8.27.

FIGURE 8.27
$$\frac{(x+3)^2}{\frac{25}{4}} + \frac{(y-2)^2}{25} = 1$$

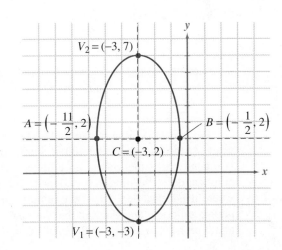

EXAMPLE 10

Find the center, vertices, and foci of the ellipse whose equation is

$$4x^2 + 9y^2 - 8x + 36y + 4 = 0.$$

Sketch the graph of this ellipse.

Solution　To write this equation in standard form, we must complete the square in both x and y. We proceed as follows:

$$4x^2 + 9y^2 - 8x + 36y + 4 = 0$$
$$4(x^2 - 2x \quad) + 9(y^2 + 4y \quad) = -4$$
$$4(x^2 - 2x + 1) + 9(y^2 + 4y + 4) = -4 + 4(1) + 9(4)$$
$$4(x - 1)^2 + 9(y + 2)^2 = 36.$$

FIGURE 8.28

$4x^2 + 9y^2 - 8x + 36y + 4 = 0$

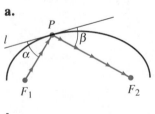

Dividing both sides of the last equation by 36 yields

$$\frac{(x - 1)^2}{9} + \frac{(y + 2)^2}{4} = 1.$$

From the last equation, we see that the center of the ellipse is located at the point $(1, -2)$, and that the larger denominator 9 occurs under the expression $(x - 1)^2$. Thus, the major axis is parallel to the x-axis. Since $a^2 = 9$, $a = 3$, and the vertices are the points $V_1 = (1 - 3, -2) = (-2, -2)$ and $V_2 = (1 + 3, -2) = (4, -2)$. For this ellipse, $b^2 = 4$, and since $c^2 = a^2 - b^2$, $c^2 = 9 - 4 = 5$. Thus $c = \sqrt{5}$, and the foci are the points $F_1 = (1 - \sqrt{5}, -2)$ and $F_2 = (1 + \sqrt{5}, -2)$. The graph is shown in Figure 8.28.

At the beginning of this section we noted that the planets follow elliptical paths about the sun. These planetary paths have the sun located at one focus, and with the exception of Mercury and Pluto, the planets travel in orbits that are nearly circular.

Examples of ellipses are common occurrences in our everyday lives. For example, if one tilts a circular pitcher partially filled with water, the surface of the liquid forms an ellipse. Also, it can easily be verified that a spherical object held in front of a light casts an elliptical shadow.

FIGURE 8.29

a.

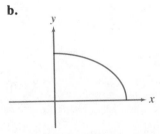

b.

Reflective Property of an Ellipse

An ellipse has a reflective property similar to that of a parabola. To help us define this reflective property, we first define a **tangent to an ellipse** as a line that intersects the ellipse in exactly one point. Let P be a point on an ellipse with foci F_1 and F_2, and let l denote the tangent line to the ellipse at the point P (see Figure 8.29(a)). If α is the angle between $\overline{F_1P}$ and l, and if β is the angle between $\overline{F_2P}$ and l, then $\alpha = \beta$.

Architects have used this reflective property of the ellipse for hundreds of years. If the portion of the ellipse shown in the first quadrant in Figure 8.29(b) is rotated about the y-axis, an elliptical dome is formed. In a room having an elliptical domed ceiling, a sound made at one focus is reflected to the other focus where it is heard very clearly. Examples of such rooms, which are called whispering galleries, are found in the National Statuary Hall at the U.S. Capitol (see photo on next page) and at St. Paul's Cathedral in London.

Photo © Craig Aurness. All Rights Reserved. Woodfin Camp & Assoc.

Physicians use the reflective property of the ellipse in a procedure that breaks up kidney stones without surgery. In this procedure, shock waves are emitted from an electrode at one focus of a semi-ellipsoid (one-half of a three-dimensional ellipse). The shock waves reflect off the sides of the semi-ellipsoid and are focused at the ellipsoid's second focus at which the kidney stone is located. The shock waves pulverize the stones into particles small enough to pass through the urinary tract.

EXERCISES 8.3

In Exercises 1–22, find the center, foci, vertices, and the endpoints of the minor axis of each ellipse. Sketch the graph of each ellipse.

1. $\dfrac{x^2}{36} + \dfrac{y^2}{25} = 1$

2. $\dfrac{x^2}{81} + \dfrac{y^2}{225} = 1$

3. $\dfrac{x^2}{25} + \dfrac{y^2}{36} = 1$

4. $\dfrac{x^2}{225} + \dfrac{y^2}{81} = 1$

5. $\dfrac{x^2}{16} + \dfrac{y^2}{7} = 1$

6. $\dfrac{x^2}{3} + \dfrac{y^2}{15} = 1$

7. $\dfrac{(x + 2)^2}{9} + \dfrac{(y - 3)^2}{4} = 1$

8. $\dfrac{(x - 5)^2}{10} + \dfrac{(y + 1)^2}{12} = 1$

9. $x^2 + \dfrac{(y + 5)^2}{2} = 1$

10. $\dfrac{(x + \pi)^2}{3} + y^2 = 1$

11. $\dfrac{4x^2}{3} + \dfrac{9y^2}{5} = 1$

12. $\dfrac{2x^2}{3} + \dfrac{y^2}{2} = 8$

13. $4x^2 + 3y^2 = 12$

14. $4x^2 + 5y^2 = 20$

15. $9x^2 + 64y^2 = 144$

16. $12x^2 + 25y^2 = 300$

17. $(x + 2)^2 + 5(y - 1)^2 = 25$

18. $9(x + 1)^2 + (y - 2)^2 = 18$

19. $5x^2 + 9y^2 - 20x + 36y + 11 = 0$

20. $7x^2 + 16y^2 - 28x - 32y - 68 = 0$

21. $x^2 + 2y^2 - 4x - 8y - 6 = 0$

22. $64x^2 + 48y^2 - 384y - 2304 = 0$

In Exercises 23–32, write the equation, in standard form, of the ellipse that satisfies the given conditions.

23. Foci $(0, \pm 3)$, vertices $(0, \pm 5)$

24. Foci $(\pm 4, 0)$, vertices $(\pm 5, 0)$

25. Foci $(\pm 1, 2)$, major axis of length 6

26. Focus $(0, 2)$, center $(2, 2)$, vertex $(5, 2)$

27. Center $(3, 2)$, vertex $(-5, 2)$, minor axis of length 6.

28. Vertices $(\pm 5, 0)$ and passes through the point $(-3, 2)$

29. Vertices $(4, \pm 5)$ and containing the origin

30. Center $(-1, -1)$, vertical major axis of length 10, and minor axis of length 3

31. Focus $(-1, 3)$, center $(-1, 1)$, and minor axis of length 4

32. Vertex $(4, -3)$, center $(4, -5)$, and minor axis of length 1

33. Using the definition of an ellipse, derive equation (8) given on page 384.

34. Using the definition of an ellipse, derive equation (9) given on page 386.

35. An arch in the shape of the upper half of an ellipse is used to support a bridge that is to span a river 60 feet wide. (See the figure on the next page.) The center of the arch is 20 feet above the middle of the river. Write an equation of

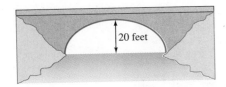

20 feet

the ellipse. How high is this elliptical arch 15 feet from its center?

36. The point in a lunar orbit nearest the surface of the moon is called the **perilune** and the point farthest from the surface is called the **apolune.** If a spacecraft is placed into an elliptical orbit with a perilune of 100 kilometers and an apolune of 310 kilometers, find an equation of its elliptical orbit. Assume that the radius of the moon is approximately 1700 kilometers and that the center of the moon is located at the center of the elliptical orbit.

37. The Ellipse is a park in Washington, D.C. that is located between the White House and the Washington Monument. The park is surrounded by an elliptical walkway with a major axis of length 1500 feet and a minor axis of length 1280 feet. Determine the distance between a vertex and the closer focus.

38. Consider the equation

$$Ax^2 + Cy^2 + Dx + Ey + F = 0, \qquad \textbf{(11)}$$

where A and C are positive and $A \neq C$. If

$$r = \frac{D^2}{4A} + \frac{E^2}{4C} - F,$$

show that the graph of equation (11)
a. is an ellipse when $r > 0$.
b. consists of a single point when $r = 0$.
c. contains no points when $r < 0$.
[*Hint*: Complete the square in both x and y.]

39. A line segment that passes through a focus of an ellipse, which is perpendicular to the major axis of the ellipse, and which has its endpoints on the ellipse is called a **focal chord**. Show that an ellipse whose equation has the form $(x^2/a^2) + (y^2/b^2) = 1$ has a focal chord of length $2b^2/a$.

40. If $P = (x_1, y_1)$ is a point on the ellipse with equation

$$\frac{x^2}{a^2} + \frac{y^2}{b^2} = 1,$$

then it can be shown that an equation of the tangent line to

the ellipse at P is

$$\frac{x_1 x}{a^2} + \frac{y_1 y}{b^2} = 1.$$

Consider the ellipse with equation

$$x^2 + 3y^2 = 12.$$

a. Verify that the point $P = (3, 1)$ lies on the ellipse.
b. Write an equation, in slope-intercept form, of the tangent line to the ellipse at the point $P = (3, 1)$.

41. Consider the equation

$$Ax^2 + Cy^2 + F = 0,$$

where $AC > 0$ (that is, A and C are of the same sign). Show that the graph of this equation is
a. an ellipse with center at $(0, 0)$ if $A \neq C$.
b. a circle with center at $(0, 0)$ if $A = C$.

42. Consider the equation

$$Ax^2 + Cy^2 + Dx + Ey + F = 0,$$

where $AC > 0$.
a. If $\dfrac{D^2}{4A} + \dfrac{E^2}{4C} - F$ has the same sign as A, show that the graph is an ellipse.
b. If $\dfrac{D^2}{4A} + \dfrac{E^2}{4C} - F = 0$, show that the graph is a point.
c. If $\dfrac{D^2}{4A} + \dfrac{E^2}{4C} - F$ and A have opposite signs, show that the graph contains no points.

43. **a.** Using your graphing calculator, sketch the graph of each of the following equations in the same coordinate system:

$$x^2 + 7y^2 = 26 \quad \text{and} \quad x^2 - x - y - 3 = 0.$$

b. Approximate, to two decimal places, the coordinates of the points of intersection.

44. **a.** Using your graphing calculator, sketch the graph of each of the following equations in the same coordinate system:

$$3(x + 2)^2 + y^2 = 4 \quad \text{and} \quad (x + 1)^2 + 4y^2 = 16.$$

b. Approximate, to two decimal places, the coordinates of the points of intersection.

8.4 HYPERBOLAS

A hyperbola is a conic that is formed when a plane intersects both nappes of a cone but does not contain the vertex of the cone. It follows that a hyperbola is formed when the plane is parallel to the axis of the cone or when the angle β at which the plane intersects the axis of the cone is less than the angle α between the axis and a generator of the cone (see Figure 8.30).

The definition of a hyperbola is similar to that of an ellipse. The distinction is that, for an ellipse, the *sum* of the distances between two fixed points and a

FIGURE 8.30

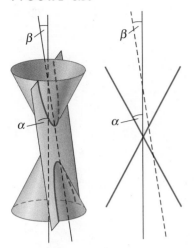

FIGURE 8.31

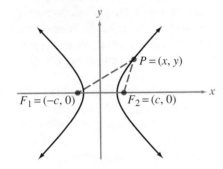

point on the ellipse is fixed; for a hyperbola, the *difference* of these two distances is fixed.

DEFINITION 8.4

A **hyperbola** is the set of points in the plane, the difference of whose distances from two fixed points F_1 and F_2 is a positive constant. Each of the fixed points, F_1 and F_2, is called a **focus**. The midpoint of the line segment joining the foci is called the **center** of the hyperbola.

To derive an equation for a hyperbola, we initially assume that the foci are located at the points $F_1 = (-c, 0)$ and $F_2 = (c, 0)$, where $c > 0$, and that $2a$ denotes the positive constant difference of the distances from any point $P = (x, y)$ on the hyperbola to the foci F_1 and F_2. From Figure 8.31, we see that a point P is on the hyperbola if and only if one of the following is true:

$$d(P, F_1) - d(P, F_2) = 2a, \tag{1}$$
or
$$d(P, F_2) - d(P, F_1) = 2a. \tag{2}$$

To obtain a hyperbola, the constant a must be less than the constant c (see Exercise 42). Equations (1) and (2) can be expressed as the single equation

$$d(P, F_2) - d(P, F_1) = \pm 2a.$$

Using the distance formula, we can write this equation in the form

$$\sqrt{(x - c)^2 + (y - 0)^2} - \sqrt{(x + c)^2 + (y - 0)^2} = \pm 2a,$$
or
$$\sqrt{(x - c)^2 + y^2} - \sqrt{(x + c)^2 + y^2} = \pm 2a. \tag{3}$$

Simplifying equation (3) by employing the procedure used to derive the equation of an ellipse in the previous section, we obtain the equation

$$(c^2 - a^2)x^2 - a^2y^2 = a^2(c^2 - a^2). \tag{4}$$

Since $c > a > 0$, it follows that $c^2 - a^2 > 0$. If we let $b = \sqrt{c^2 - a^2}$, then $b^2 = c^2 - a^2$, and equation (4) can be written in the form

$$b^2x^2 - a^2y^2 = a^2b^2.$$

Dividing each side of this equation by a^2b^2, we obtain the equation

$$\frac{x^2}{a^2} - \frac{y^2}{b^2} = 1. \tag{5}$$

We have now shown that the coordinates of every point $P = (x, y)$ on the hyperbola satisfy equation (5). Conversely, it can be shown that if the coordinates of a point $P = (x, y)$ satisfy equation (5), then the point lies on the hyperbola. Equation (5) is called the **standard form** for the equation of a hyperbola with foci $F_1 = (-c, 0)$ and $F_2 = (c, 0)$. From Figure 8.32 on page 392, we see that a hyperbola consists of two distinct curves, called **branches**. Using the tests for symmetry, we see that the graph of a hyperbola defined by equation (5) is symmetric with respect to the x-axis, the y-axis, and the origin.

The line containing the foci of a hyperbola is called the **focal axis**. From Figure 8.32, we see that the x-axis, which is the focal axis, intersects the hyperbola at two points V_1 and V_2, which are called the **vertices**. In this case, the

FIGURE 8.32 $\dfrac{x^2}{a^2} - \dfrac{y^2}{b^2} = 1$

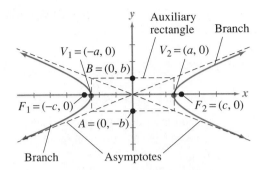

vertices are the x-intercepts of the graph. To find the x-coordinates of the vertices, we set $y = 0$ in equation (5) and solve for x.

$$\frac{x^2}{a^2} = 1$$

$$x^2 = a^2, \qquad \text{or} \qquad x = \pm a.$$

Hence, the vertices are the points $V_1 = (-a, 0)$ and $V_2 = (a, 0)$. There are no y-intercepts of the graph since the equation $-(y^2/b^2) = 1$ has no solutions that are real numbers. The line segment joining the vertices is called the **transverse axis** of the hyperbola. The line segment passing through the center of the hyperbola and having the points $A = (0, -b)$ and $B = (0, b)$ as its endpoints is called the **conjugate axis**.

As an aid in sketching the graph of the hyperbola defined by equation (5), we consider the **auxiliary rectangle** of width $2a$ and height $2b$ centered at the origin. The two lines that contain the diagonals of this rectangle are called the **asymptotes** of the hyperbola. These asymptotes serve as a guide for graphing the hyperbola. It can be shown that the equations of the asymptotes are

$$y = -\frac{b}{a} x \qquad \text{and} \qquad y = \frac{b}{a} x.$$

(See Exercise 41.)

E X A M P L E 11

Find the vertices, the foci, the lengths of the transverse and conjugate axes, and the equations of the asymptotes of the hyperbola whose equation is $25x^2 - 4y^2 = 100$. Sketch the graph of this hyperbola using the asymptotes as guides.

Solution   First, we convert the equation $25x^2 - 4y^2 = 100$ to standard form by dividing each side by 100. This yields

$$\frac{x^2}{4} - \frac{y^2}{25} = 1.$$

Comparing this equation with equation (5) on page 391, we see that $a^2 = 4$ and $b^2 = 25$. Hence, $a = 2$ and $b = 5$. Since $b^2 = c^2 - a^2$, it follows that

$$c^2 = a^2 + b^2$$
$$= 4 + 25$$
$$= 29,$$

and therefore, $c = \sqrt{29}.$

FIGURE 8.33 $\dfrac{x^2}{4} - \dfrac{y^2}{25} = 1$

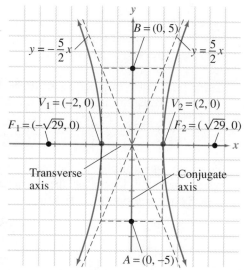

Hence, the vertices are $V_1 = (-2, 0)$ and $V_2 = (2, 0)$, and the foci are $F_1 = (-\sqrt{29}, 0)$ and $F_2 = (\sqrt{29}, 0)$. The length of the transverse axis is $2a = 4$, and the length of the conjugate axis is $2b = 10$. Lastly, the equations of the asymptotes are $y = -(b/a)x = -\frac{5}{2}x$ and $y = (b/a)x = \frac{5}{2}x$. The graph of this hyperbola is shown in Figure 8.33.

If the foci of a hyperbola are the points $F_1 = (0, -c)$ and $F_2 = (0, c)$, which are located on the y-axis, then the **standard form** of the equation is

$$\frac{y^2}{a^2} - \frac{x^2}{b^2} = 1, \quad \text{where } c^2 = a^2 + b^2. \tag{6}$$

In this case, the vertices of the hyperbola are the points $V_1 = (0, -a)$ and $V_2 = (0, a)$. In addition, the endpoints of the conjugate axis of the hyperbola defined by equation (6) are $A = (-b, 0)$ and $B = (b, 0)$, and the equations for the asymptotes are

$$y = -\frac{a}{b}x \quad \text{and} \quad y = \frac{a}{b}x.$$

Note the difference between these equations and the equations

$$y = -\frac{b}{a}x \quad \text{and} \quad y = \frac{b}{a}x,$$

which are the equations of the asymptotes of a hyperbola with defining equation (5). In either case, the asymptotes can be sketched using the auxiliary rectangle and it is not necessary to learn the equations of the asymptotes. We summarize several properties of hyperbolas in Table 8.6.

TABLE 8.6
Hyperbolas with Center at (0, 0) and Transverse Axis Lying Along One of the Coordinate Axes

Equation	$\dfrac{x^2}{a^2} - \dfrac{y^2}{b^2} = 1$	$\dfrac{y^2}{a^2} - \dfrac{x^2}{b^2} = 1$
Transverse Axis	lies on the x-axis	lies on the y-axis
Foci	$(\pm c, 0)$, where $c^2 = a^2 + b^2$	$(0, \pm c)$, where $c^2 = a^2 + b^2$
Vertices	$(\pm a, 0)$	$(0, \pm a)$

EXAMPLE 12

Give the equation, in standard form, of the hyperbola with foci $F_1 = (0, -4)$ and $F_2 = (0, 4)$ and conjugate axis of length 6. Sketch the graph of this hyperbola using the asymptotes as guides.

Solution   Since the foci $F_1 = (0, -4)$ and $F_2 = (0, 4)$ are located on the y-axis, the equation, in standard form, of this hyperbola is

$$\frac{y^2}{a^2} - \frac{x^2}{b^2} = 1.$$

From the y-coordinates of the foci, we know that $c = 4$, or $c^2 = 16$. Moreover, since the length of the conjugate axis is $2b = 6$, we know that $b = 3$ and therefore, $b^2 = 9$. Since $c^2 = a^2 + b^2$, it follows that

$$a^2 = c^2 - b^2$$
$$= 16 - 9$$
$$= 7.$$

FIGURE 8.34

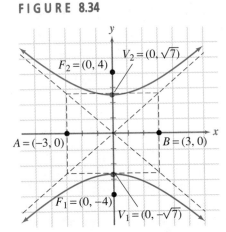

Hence, the equation of the hyperbola is

$$\frac{y^2}{7} - \frac{x^2}{9} = 1.$$

The graph of this hyperbola is shown in Figure 8.34.

Just as in the case of an ellipse, we are often interested in hyperbolas whose centers are located at points other than the origin. The equation of a hyperbola with center at (h, k) and transverse axis parallel to the x-axis has the form

$$\frac{(x - h)^2}{a^2} - \frac{(y - k)^2}{b^2} = 1. \tag{7}$$

(See Exercise 47.)
Similarly, the equation of a hyperbola with center at (h, k) and transverse axis parallel to the y-axis has the form

$$\frac{(y - k)^2}{a^2} - \frac{(x - h)^2}{b^2} = 1. \tag{8}$$

(See Exercise 48.)
In Table 8.7, we list some important properties of such hyperbolas. The two equations for hyperbolas shown in Table 8.7 are said to be in **standard form**. In

TABLE 8.7
Hyperbolas with Center at (h, k) and Transverse Axis Parallel to a Coordinate Axis

Equation	$\dfrac{(x - h)^2}{a^2} - \dfrac{(y - k)^2}{b^2} = 1$	$\dfrac{(y - k)^2}{a^2} - \dfrac{(x - h)^2}{b^2} = 1$
Center	(h, k)	(h, k)
Transverse Axis	Parallel to x-axis	Parallel to y-axis
Foci	$(h \pm c, k)$	$(h, k \pm c)$
	where $c^2 = a^2 + b^2$	where $c^2 = a^2 + b^2$
Vertices	$(h \pm a, k)$	$(h, k \pm a)$

FIGURE 8.35

a. $\dfrac{(x - h)^2}{a^2} - \dfrac{(y - k)^2}{b^2} = 1$ **b.** $\dfrac{(y - k)^2}{a^2} - \dfrac{(x - h)^2}{b^2} = 1$

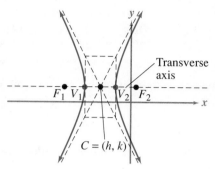
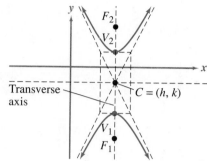

Figure 8.35, we illustrate the properties of a hyperbola with center at the point (h, k) and transverse axis parallel to a coordinate axis.

EXAMPLE 13

Find the equation, in standard form, of the hyperbola with center at $C = (-3, -2)$, focus at $F_1 = (-3, -5)$, and vertex at $V_1 = (-3, -4)$. Sketch the graph of this hyperbola using the asymptotes as a guide.

Solution   Since the center is at $C = (h, k) = (-3, -2)$, it follows that $h = -3$ and $k = -2$. Moreover, since -3 is the x-coordinate of the center, the focus F_1, and the vertex V_1, it follows that the transverse axis of this hyperbola is the vertical line $x = -3$. Since the distance from the center to a vertex is a, we see that

$$a = -2 - (-4) = 2.$$

(See Figure 8.36.)

Similarly, since the distance from the center to a focus is c, we see that

$$c = -2 - (-5) = 3.$$

Therefore, $c^2 = 9$ and $a^2 = 4$, and since $b^2 = c^2 - a^2$, $b^2 = 9 - 4 = 5$. Since the transverse axis is parallel to the y-axis, from Table 8.7 we see that the equation, in standard form, of this hyperbola is

$$\frac{(y + 2)^2}{4} - \frac{(x + 3)^2}{5} = 1.$$

The graph of this hyperbola is sketched in Figure 8.36.

FIGURE 8.36

$$\frac{(y + 2)^2}{4} - \frac{(x + 3)^2}{5} = 1$$

EXAMPLE 14

Find the center, the foci, the vertices, and the lengths of the transverse and conjugate axes of the hyperbola whose equation is

$$3x^2 - y^2 - 30x + 14y + 23 = 0.$$

Sketch the graph of this hyperbola using the asymptotes as guides.

Solution   To write the equation of this hyperbola in standard form, we must complete the squares in both x and y:

$$3x^2 - y^2 - 30x + 14y + 23 = 0$$
$$3(x^2 - 10x \qquad) - (y^2 - 14y \qquad) = -23$$
$$3(x^2 - 10x + 25) - (y^2 - 14y + 49) = -23 + 3(25) - 49$$
$$3(x - 5)^2 - (y - 7)^2 = 3.$$

Dividing both sides of this equation by 3 yields

$$(x - 5)^2 - \frac{(y - 7)^2}{3} = 1.$$

Since the latter equation has the form

$$\frac{(x - h)^2}{a^2} - \frac{(y - k)^2}{b^2} = 1,$$

we see from Table 8.7 that the center of the hyperbola is located at the point $(5, 7)$ and that the transverse axis is parallel to the x-axis. It follows that $a^2 = 1$ or $a = 1$. Hence, the vertices are the points $V_1 = (5 - 1, 7) = (4, 7)$ and $V_2 = (5 + 1, 7) = (6, 7)$. For this hyperbola, $b^2 = 3$, and since $c^2 = a^2 + b^2$, $c^2 = 1 + 3 = 4$. Hence, $c = 2$, and the foci are the points $F_1 = (5 - 2, 7) = (3, 7)$ and $F_2 = (5 + 2, 7) = (7, 7)$. The length of the transverse axis is $2a = 2(1) = 2$, and the length of the conjugate axis is $2b = 2\sqrt{3}$. The graph of this hyperbola is sketched in Figure 8.37.

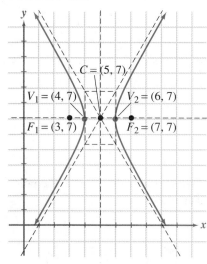

FIGURE 8.37

We previously noted that some comets, such as Halley's comet, have elliptical orbits. Other comets have hyperbolic orbits with the sun at a focus. Such comets pass near the sun only once.

Reflective Property of a Hyperbola

Like parabolas and ellipses, hyperbolas have a reflective property that has interesting applications. The reflective property of a hyperbola is a consequence of the following property of a hyperbola: If P is a point on a hyperbola, the tangent line at P makes equal angles with the two lines joining P and the two foci (see Figure 8.38(a)). As a result of this reflective property, light directed toward one focus of a hyperbolic mirror is reflected toward the other focus (see Figure 8.38(b)). The design of modern telescopes often incorporates the reflective properties of parabolas, ellipses, and hyperbolas.

FIGURE 8.38

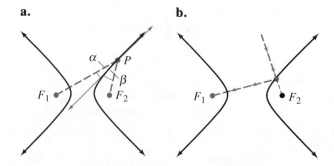

a.   **b.**

FIGURE 8.39

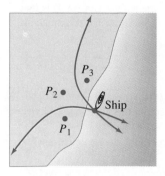

The LORAN System

The LORAN (an acronym for Long Range Navigation) system of navigation is based on properties of a hyperbola. To illustrate, suppose radio transmitters, located at the points P_1, P_2, and P_3 as shown in Figure 8.39, transmit signals simultaneously. If a ship with a receiver records the difference in the arrival times of the signals from the transmitters located at P_1 and P_2, the difference of the distances from the ship to the points P_1 and P_2 can be determined. It can be shown that the ship lies on one branch of a hyperbola with foci P_1 and P_2. Repeating this process using the transmitters located at P_2 and P_3, we see that the ship is also located on a branch of a hyperbola with foci P_2 and P_3. Thus, the position of the ship is found by determining the intersection of these two branches of the two hyperbolas (see Exercise 43).

EXERCISES 8.4

In Exercises 1–20, find the center, the vertices, the foci, and the lengths of the transverse and conjugate axes. Sketch the graph of the hyperbola using the asymptotes as guides.

1. $\dfrac{x^2}{16} - \dfrac{y^2}{9} = 1$ **2.** $\dfrac{y^2}{16} - \dfrac{x^2}{9} = 1$

3. $\dfrac{y^2}{25} - \dfrac{x^2}{144} = 1$ **4.** $x^2 - \dfrac{y^2}{49} = 1$

5. $9x^2 - 4y^2 = 25$ **6.** $y^2 - 25x^2 = 16$

7. $(x + 3)^2 - 6y^2 = 6$ **8.** $2x^2 - 5(y + 1)^2 = 25$

9. $(y - 3)^2 - \dfrac{(x + 2)^2}{4} = 1$

10. $\dfrac{(x - 1)^2}{2} - \dfrac{(y + 2)^2}{4} = 1$

11. $(x - 4)^2 - (y - 2)^2 = 1$

12. $(y + 1)^2 - (x + 1)^2 = 4$

13. $4(x + 2)^2 - 25(y - 1)^2 = 100$

14. $16(y - 1)^2 - 9(x - 5)^2 = 144$

15. $9x^2 - 16y^2 + 36x + 32y + 164 = 0$

16. $9x^2 - 4y^2 - 72x + 8y + 176 = 0$

17. $2y^2 - 3x^2 - 4y + 12x + 8 = 0$

18. $y^2 - 4x^2 + 4y - 8x - 9 = 0$

19. $12x^2 - 3y^2 + 24y - 84 = 0$

20. $4x^2 - 16x - 9y^2 - 54y - 101 = 0$

In Exercises 21–30, find the equation, in standard form, of the hyperbola satisfying the given conditions.

21. Vertices: $(0, \pm 4)$
    Foci: $(0, \pm 6)$

22. Vertices: $(\pm 1, 0)$
    Foci: $(\pm 3, 0)$

23. Vertices: $(2, 1)$, $(2, 5)$
    Foci: $(2, -1)$, $(2, 7)$

24. Vertices: $(-4, -1)$, $(0, -1)$
    Foci: $(-5, -1)$, $(1, -1)$

25. Foci: $(1, 2)$, $(1, 8)$
    conjugate axis of length 4

26. Foci: $(-3, -4)$, $(5, -4)$
    conjugate axis of length 7

27. Vertices: $(0, \pm\sqrt{5})$
    passing through the
    point $P = (4, 3)$

28. Vertices: $(\pm 2, 0)$
    passing through the
    point $P = (\sqrt{20}, 8)$

29. Center $(-2, 4)$
    transverse axis of
    length 2 and parallel
    to the x-axis; con-
    jugate axis of length 6

30. Center $(-1, -3)$
    transverse axis of
    length 10 and parallel
    to the y-axis; con-
    jugate axis of length 5

31. For what value of c does the hyperbola with defining equation

$$\frac{(y + 2)^2}{9} - \frac{(x - 1)^2}{4} = c$$

pass through the origin?

32. Describe the graph of the equation

$$\frac{x^2}{c} + \frac{y^2}{c - 9} = 1$$

in each of the following cases:
    a. $c > 9$ **b.** $0 < c < 9$ **c.** $c < 0$

In Exercises 33–40, classify the graph of each equation as a circle, a parabola, an ellipse, or a hyperbola.

33. $x^2 + 4x - 16y - 28 = 0$

34. $x^2 + y^2 + 2x - 4y - 15 = 0$

35. $9x^2 + 4y^2 - 54x + 16y + 29 = 0$

36. $x^2 + 9y^2 + 6x - 18y + 9 = 0$

37. $10x^2 - 8y^2 - 60x - 32y - 22 = 0$

38. $9x^2 + 8y^2 - 54x + 9 = 0$

39. $x^2 + y^2 + 8x - 2y + 6 = 0$

40. $y^2 - 4x - 4y + 8 = 0$

41. Show that $y = -\dfrac{b}{a}x$ and $y = \dfrac{b}{a}x$ are the equations of the asymptotes of the hyperbola whose defining equation is

$$\frac{x^2}{a^2} - \frac{y^2}{b^2} = 1.$$

42. Recall that a point $P = (x, y)$ lies on a hyperbola with foci $F_1 = (-c, 0)$ and $F_2 = (c, 0)$, as shown in the accompanying figure, if and only if there exists a positive constant a such that

$$d(P, F_1) - d(P, F_2) = 2a \quad \text{or}$$
$$d(P, F_2) - d(P, F_1) = 2a.$$

Show that $a < c$.

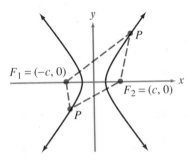

43. Two radio towers, R_1 and R_2, are positioned along the coast of Florida, 300 miles apart. A signal is sent simultaneously from each tower to a ship near the coastline. The signal from tower R_1 is received by the ship 250 microseconds after the signal from tower R_2. If the radio signal travels 0.2 miles per microsecond, determine the location of the ship.

44. The line segment that passes through a focus of a hyperbola, that is perpendicular to the transverse axis, and that has its endpoints on the hyperbola is called a **focal chord**. Show that if an equation of the hyperbola is

$$\frac{x^2}{a^2} - \frac{y^2}{b^2} = 1,$$

then the length of a focal chord of the hyperbola is $2b^2/a$.

45. Find an equation for the set of points such that the distance from each point in the set to the point $P_0 = (0, 4)$ is twice the distance to the line $y = -4$. Show that the equation represents a hyperbola.

46. Consider the equation

$$Ax^2 + Cy^2 + F = 0,$$

where $F \neq 0$, and $AC < 0$ (that is, A and C are of opposite sign). Show that the graph of this equation is a hyperbola with center $(0, 0)$.

47. Using the definition of a hyperbola, derive equation (7) given on page 394.

48. Using the definition of a hyperbola, derive equation (8) given on page 394.

49. Consider the equation

$$Ax^2 + Cy^2 + Dx + Ey + F = 0,$$

where $AC < 0$.

a. If $\dfrac{D^2}{4A} + \dfrac{E^2}{4C} - F \neq 0$, show that the graph of the given equation is a hyperbola.

b. If $\dfrac{D^2}{4A} + \dfrac{E^2}{4C} - F = 0$, show that the graph of the given equation is two intersecting lines.

50. **a.** Using your graphing calculator, sketch the graph of each of the following equations in the same coordinate system:

$$y = x^2 - 8x + 8 \quad \text{and}$$
$$x^2 - y^2 - 8x + 4y + 3 = 0.$$

b. Approximate, to two decimal places, the coordinates of the points of intersection.

8.5   ROTATION OF AXES

In the previous three sections we studied conics whose axes (or axis) were parallel to the coordinate axes. The defining equations for such conics can be written in the general form

$$Ax^2 + Cy^2 + Dx + Ey + F = 0. \tag{1}$$

Excluding degenerate cases, the graph of equation (1) is

a parabola, if $AC = 0$ and A and C are not both zero;

an ellipse, if $AC > 0$ and $A \neq C$;

a circle, if $AC > 0$ and $A = C$;

a hyperbola, if $AC < 0$.

(See Exercises 45 and 46 in Section 8.2, Exercises 41 and 42 in Section 8.3, and Exercises 46 and 49 in Section 8.4.)

In this section we investigate the equations of conics whose axes are not parallel to the x-axis or the y-axis. The defining equations of such conics can be written in the form

FIGURE 8.40

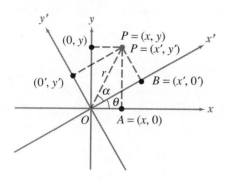

$$Ax^2 + Bxy + Cy^2 + Dx + Ey + F = 0. \qquad (2)$$

It is now clear why the letter B is not used in equation (1). It is reserved for the coefficient of the xy-term in equation (2). It is important to note that equation (1) has the form of equation (2), where $B = 0$.

Using a process known as **rotation of axes**, it is possible to find an equation that is equivalent to equation (2) but does not contain an xy-term. To do so, we introduce a new pair of coordinate axes, the x'- and y'-axes, by rotating the x- and y-axes about the origin, which remains fixed. As we see in Figure 8.40, each point P in the plane has two pairs of coordinates: one pair (x, y) relative to the x- and y-axes and one pair (x', y') relative to the x'- and y'-axes.

Equations of Rotation

Let us now determine how these two pairs of coordinates of P are related. From right triangle OAP, shown in Figure 8.40, we see that

$$\cos(\alpha + \theta) = \frac{x}{r} \quad \text{or}$$

$$x = r\cos(\alpha + \theta)$$
$$= r(\cos\alpha\cos\theta - \sin\alpha\sin\theta)$$
$$= (r\cos\alpha)\cos\theta - (r\sin\alpha)\sin\theta.$$

Thus
$$x = (r\cos\alpha)\cos\theta - (r\sin\alpha)\sin\theta. \qquad (3)$$

Using a similar argument, it can be shown that

$$y = (r\cos\alpha)\sin\theta + (r\sin\alpha)\cos\theta. \qquad (4)$$

From right triangle OBP, we see that

$$x' = r\cos\alpha \quad \text{and} \quad y' = r\sin\alpha.$$

Substituting x' for $r\cos\alpha$ and y' for $r\sin\alpha$ in equations (3) and (4) we have

$$x = x'\cos\theta - y'\sin\theta \quad \text{and}$$
$$y = x'\sin\theta + y'\cos\theta.$$

Summarizing our work to this point, we have

Equations of Rotation

If the x- and y-axes are rotated about the origin through an acute angle θ, then the coordinates (x, y) and (x', y') of a point P in the xy- and $x'y'$-planes are related as follows:

$$x = x'\cos\theta - y'\sin\theta \qquad (5)$$
$$y = x'\sin\theta + y'\cos\theta. \qquad (6)$$

These equations are called the **equations of rotation**.

The function f defined by

$$f(x) = \frac{1}{x} \qquad (7)$$

plays an important role in calculus. Replacing $f(x)$ by y in equation (7), we have $y = 1/x$, or $xy - 1 = 0$. Notice that the last equation has the same form as equation (2) where

$$A = C = D = E = 0, \qquad B = 1, \qquad \text{and} \qquad F = -1.$$

In the following example, we use the equations of rotation to verify that the graph of the function defined by equation (7), or equivalently, the equation

$$xy - 1 = 0,$$

is a hyperbola.

EXAMPLE 15

Express the equation $xy - 1 = 0$ in terms of new $x'y'$-coordinates by rotating the x- and y-axes through an angle of 45°. Sketch the graph of this equation.

Solution   Letting $\theta = 45°$ in equations (5) and (6) we have

$$x = x' \cos 45° - y' \sin 45° \qquad \text{and}$$
$$y = x' \sin 45° + y' \cos 45°.$$

Substituting $\sqrt{2}/2$ for $\sin 45°$ and for $\cos 45°$, we get

$$x = x'\left(\frac{\sqrt{2}}{2}\right) - y'\left(\frac{\sqrt{2}}{2}\right) = \frac{\sqrt{2}}{2}(x' - y') \qquad \text{and}$$

$$y = x'\left(\frac{\sqrt{2}}{2}\right) + y'\left(\frac{\sqrt{2}}{2}\right) = \frac{\sqrt{2}}{2}(x' + y').$$

FIGURE 8.41 $xy - 1 = 0$

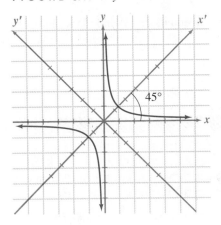

Substituting these expressions for x and y in the equation $xy - 1 = 0$ and simplifying, we have

$$\left[\frac{\sqrt{2}}{2}(x' - y')\right]\left[\frac{\sqrt{2}}{2}(x' + y')\right] - 1 = 0$$

$$\frac{1}{2}[(x')^2 - (y')^2] - 1 = 0$$

$$\frac{(x')^2}{2} - \frac{(y')^2}{2} = 1. \qquad (8)$$

We recognize equation (8) as the equation of a hyperbola with vertices $V_1 = (-\sqrt{2}, 0)$ and $V_2 = (\sqrt{2}, 0)$ located on the x'-axis. The graph of this hyperbola is sketched in Figure 8.41. Note that asymptotes of the hyperbola are the x-and y-axes.

Angle of Rotation

In Example 15, we see that by rotating the x- and y-axes through an angle of 45°, the equation $xy - 1 = 0$ is transformed into an equation in x' and y' in which no $x'y'$-term appears. Therefore, the choice of the angle $\theta = 45°$ was an appropriate one.

In general, how do we know what choice of θ will eliminate the xy-term in equation (2) on page 399? We begin by substituting the expression

$$x' \cos \theta - y' \sin \theta$$

for x and the expression

$$x' \sin \theta + y' \cos \theta$$

for y in equation (2) and then simplifying to obtain an equation of the form

$$a(x')^2 + b(x')(y') + c(y')^2 + d(x') + e(y') + f = 0, \qquad (9)$$

where a, b, c, d, e, and f are constants that depend upon the angle θ. To eliminate the $x'y'$-term in equation (9), its coefficient b must be zero. It can be shown that

$$b = 2(C - A)(\sin \theta \cos \theta) + B(\cos^2 \theta - \sin^2 \theta).$$

(See Exercise 21.)

Thus, we require that

$$2(C - A)(\sin \theta \cos \theta) + B(\cos^2 \theta - \sin^2 \theta) = 0.$$

Using the double-angle formulas for sine and cosine, this equation can be written as

$$(C - A) \sin 2\theta + B \cos 2\theta = 0, \qquad \text{or}$$

$$B \cos 2\theta = (A - C) \sin 2\theta.$$

Hence,

$$\frac{\cos 2\theta}{\sin 2\theta} = \frac{A - C}{B}, \qquad \text{or}$$

$$\cot 2\theta = \frac{A - C}{B}, \qquad B \neq 0.$$

We summarize our results below.

Angle of Rotation

To transform the equation

$$Ax^2 + Bxy + Cy^2 + Dx + Ey + F = 0, \qquad \text{where } B \neq 0,$$

into an equation in x' and y' with no $x'y'$-term, rotate the axis through an angle θ, where $0° < \theta < 90°$ and

$$\cot 2\theta = \frac{A - C}{B}. \qquad (10)$$

Since equation (10) has infinitely many solutions, any rotation through an angle θ that satisfies equation (10) will eliminate the xy-term. However, by selecting the positive acute angle θ satisfying equation (10), there are only two possibilities:

If $\cot 2\theta \geq 0$, then $0° < 2\theta \leq 90°$, and hence $0° < \theta \leq 45°$.

If $\cot 2\theta < 0$, then $90° < 2\theta < 180°$, and hence $45° < \theta < 90°$.

EXAMPLE 16

Make a rotation of axes to eliminate the xy-term in the equation

$$3x^2 + 2\sqrt{3}xy + y^2 - 2x + 2\sqrt{3}y = 0. \qquad (11)$$

Identify the conic defined by this equation and sketch its graph.

Solution   Equation (11) has the form of equation (2) with $A = 3$, $B = 2\sqrt{3}$, and $C = 1$. Hence,

$$\cot 2\theta = \frac{A - C}{B}$$

$$= \frac{3 - 1}{2\sqrt{3}} = \frac{1}{\sqrt{3}} = \frac{\sqrt{3}}{3}.$$

Therefore,

$$2\theta = 60°,$$

and thus,

$$\theta = 30°.$$

Substituting 30° for θ in equations (5) and (6) on page 399 yields

$$x = x'\left(\frac{\sqrt{3}}{2}\right) - y'\left(\frac{1}{2}\right) = \frac{\sqrt{3}x' - y'}{2} \qquad \text{and}$$

$$y = x'\left(\frac{1}{2}\right) + y'\left(\frac{\sqrt{3}}{2}\right) = \frac{x' + \sqrt{3}y'}{2}.$$

Substituting these values for x and y in the given equation, $3x^2 + 2\sqrt{3}xy + y^2 - 2x + 2\sqrt{3}y = 0$, we have

$$3\left[\frac{\sqrt{3}x' - y'}{2}\right]^2 + 2\sqrt{3}\left[\frac{\sqrt{3}x' - y'}{2}\right]\left[\frac{x' + \sqrt{3}y'}{2}\right]$$

$$+ \left[\frac{x' + \sqrt{3}y'}{2}\right]^2 - 2\left[\frac{\sqrt{3}x' - y'}{2}\right] + 2\sqrt{3}\left[\frac{x' + \sqrt{3}y'}{2}\right] = 0.$$

Multiplying both sides of the latter equation by 4 and expanding the terms, we get

$$9(x')^2 - 6\sqrt{3}x'y' + 3(y')^2 + 6(x')^2 + 4\sqrt{3}x'y' - 6(y')^2 + (x')^2$$

$$+ 2\sqrt{3}x'y' + 3(y')^2 - 4\sqrt{3}x' + 4y' + 4\sqrt{3}x' + 12y' = 0.$$

Simplifying the latter equation yields

$$16(x')^2 + 16y' = 0$$

$$y' = -(x')^2. \qquad (12)$$

or

We recognize equation (12) as the equation of a parabola whose vertex and focus relative to the $x'y'$-coordinate system are $V = (0, 0)$ and $F = (0, -\frac{1}{4})$, respectively. The directrix of this parabola, whose graph is shown in Figure 8.42, is parallel to the x'-axis.

FIGURE 8.42

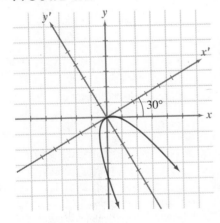

In Example 16, we found the angle of rotation to be 30°. However, many second-degree equations do not yield solutions such as 30°, 45°, and 60° to the equation

$$\cot 2\theta = \frac{A - C}{B}.$$

In such cases, as our next example illustrates, we can determine the appropriate equations of rotation by applying half-angle identities.

EXAMPLE 17

Make a rotation of axes to eliminate the xy-term in the equation

$$5x^2 - 3xy + y^2 + 65x - 25y + 203 = 0. \qquad (13)$$

Identify the conic defined by this equation and sketch its graph.

Solution Equation (13) has the form of equation (2) with $A = 5$, $B = -3$, and $C = 1$. Hence,

$$\cot 2\theta = \frac{A - C}{B}$$

$$= \frac{5 - 1}{-3}$$

$$= -\frac{4}{3}.$$

To determine the equations of rotation (5) and (6), we must find $\sin \theta$ and $\cos \theta$. Since we are assuming θ is a positive acute angle, we know that $\sin \theta > 0$ and $\cos \theta > 0$. Thus, by the half-angle identities for sine and cosine, we know that

$$\sin \theta = \sqrt{\frac{1 - \cos 2\theta}{2}} \quad \text{and} \quad \cos \theta = \sqrt{\frac{1 + \cos 2\theta}{2}}.$$

Therefore, to find $\sin \theta$ and $\cos \theta$, we must determine the value of $\cos 2\theta$. Since $\cot 2\theta = -\frac{4}{3} < 0$, we chose $90° < 2\theta < 180°$. Using basic identities, it can be shown that $\cos 2\theta = -\frac{4}{5}$.

Thus,
$$\sin \theta = \sqrt{\frac{1 - \cos 2\theta}{2}} = \sqrt{\frac{1 - (-\frac{4}{5})}{2}}$$

$$= \sqrt{\frac{9}{10}} = \frac{3}{\sqrt{10}} = \frac{3\sqrt{10}}{10}$$

and
$$\cos \theta = \sqrt{\frac{1 + \cos 2\theta}{2}} = \sqrt{\frac{1 + (-\frac{4}{5})}{2}}$$

$$= \sqrt{\frac{1}{10}} = \frac{1}{\sqrt{10}} = \frac{\sqrt{10}}{10}.$$

Substituting these values for $\sin \theta$ and $\cos \theta$ into equations (5) and (6), we find the equations of rotations are

$$x = x'\left(\frac{\sqrt{10}}{10}\right) - y'\left(\frac{3\sqrt{10}}{10}\right) = \frac{\sqrt{10}}{10}(x' - 3y') \quad \text{and}$$

$$y = x'\left(\frac{3\sqrt{10}}{10}\right) + y'\left(\frac{\sqrt{10}}{10}\right) = \frac{\sqrt{10}}{10}(3x' + y').$$

Substituting these values of x and y into equation (13), we have

$$5x^2 - 3xy + y^2 + 65x - 25y + 203 = 0$$

$$5\left[\frac{\sqrt{10}}{10}(x' - 3y')\right]^2 - 3\left[\frac{\sqrt{10}}{10}(x' - 3y') \cdot \frac{\sqrt{10}}{10}(3x' + y')\right]$$

$$+ \left[\frac{\sqrt{10}}{10}(3x' + y')\right]^2 + 65\left[\frac{\sqrt{10}}{10}(x' - 3y')\right]$$

$$- 25\left[\frac{\sqrt{10}}{10}(3x' + y')\right] + 203 = 0.$$

Multiplying both sides of the latter equation by 10 and expanding the terms,

$$5(x')^2 - 30x'y' + 45(y')^2 - 9(x')^2 + 24x'y' + 9(y')^2 + 9(x')^2 + 6x'y'$$
$$+ (y')^2 + 65\sqrt{10}x' - 195\sqrt{10}y' - 75\sqrt{10}x' - 25\sqrt{10}y' + 2030 = 0.$$

Simplifying the latter equation, we have

$$5(x')^2 + 55(y')^2 - 10\sqrt{10}x' - 220\sqrt{10}y' + 2030 = 0.$$

Completing the squares in x' and y' yields

$$5[(x')^2 - 2\sqrt{10}x' \quad \quad] + 55[(y')^2 - 4\sqrt{10}y' \quad \quad] = -2030$$
$$5[(x')^2 - 2\sqrt{10}x' + 10] + 55[(y')^2 - 4\sqrt{10}y' + 40] = -2030 + 50 + 2200$$
$$5(x' - \sqrt{10})^2 + 55(y' - 2\sqrt{10})^2 = 220$$

FIGURE 8.43

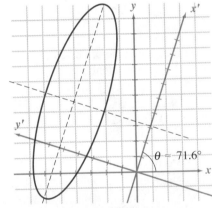

or
$$\frac{(x' - \sqrt{10})^2}{44} + \frac{(y' - 2\sqrt{10})^2}{4} = 1. \tag{14}$$

We recognize equation (14) as that of an ellipse with major axis parallel to the x'-axis. The center and vertices in the $x'y'$-coordinate system are

$$C = (\sqrt{10}, 2\sqrt{10}), \quad V_1 = (\sqrt{10} - 2\sqrt{11}, 2\sqrt{10}) \quad \text{and}$$
$$V_2 = (\sqrt{10} + 2\sqrt{11}, 2\sqrt{10}).$$

Since $\sin \theta = 3\sqrt{10}/10 > 0$, we know that

$$\theta = \arcsin \frac{3\sqrt{10}}{10}.$$

Using a calculator, we find that $\theta \approx 71.6°$. The graph is shown in Figure 8.43.

EXERCISES 8.5

In Exercises 1–20, make a rotation of axes to eliminate the xy-term in the given equation. Identify the conic (or degenerate conic) defined by each equation and sketch its graph.

1. $x^2 - xy + y^2 = 2$ **2.** $xy + 1 = 0$

3. $x^2 + 2xy + y^2 + x - y = 0$

4. $8x^2 - 4xy + 5y^2 = 36$ **5.** $2x^2 + 3xy - 2y^2 = 25$

6. $8x^2 + 12xy + 13y^2 = 885$ **7.** $x^2 - xy + y^2 = 6$

8. $73x^2 - 72xy + 52y^2 - 30x - 40y - 75 = 0$

9. $11x^2 + 4\sqrt{3}xy + 7y^2 - 1 = 0$

10. $34x^2 - 24xy + 41y^2 - 40x - 30y - 25 = 0$

11. $x^2 + 2\sqrt{3}xy + 3y^2 + 2\sqrt{3}x - 2y = 0$

12. $36x^2 + 96xy + 64y^2 + 20x - 15y + 25 = 0$

13. $2\sqrt{3}xy - 2y^2 - \sqrt{3}x - y = 0$

14. $3x^2 + 10xy + 3y^2 - 2x - 14y - 5 = 0$

15. $3x^2 - 6xy + 3y^2 + 2x = 7$

16. $4x^2 - 12xy + 9y^2 - 52x + 26y + 81 = 0$

17. $x^2 - 2\sqrt{3}xy + 3y^2 - 16\sqrt{3}x - 16y = 0$

18. $31x^2 + 10\sqrt{3}xy + 21y^2 - (124 - 40\sqrt{3})x$
      $+ (168 - 20\sqrt{3})y + 316 - 80\sqrt{3} = 0$

19. $5x^2 - 6xy - 3y^2 - 22x + 6y = 24$

20. $161x^2 + 480xy - 161y^2 - 510x - 272y = 0$

21. Consider the equation

$$Ax^2 + Bxy + Cy^2 + Dx + Ey + F = 0.$$

Substituting $x' \cos \theta - y' \sin \theta$ and $x' \sin \theta + y' \cos \theta$ for x and y, respectively, in the above equation results in an equation of the form

$$a(x')^2 + bx'y' + c(y')^2 + dx' + ey' + f = 0.$$

a. Show that $b = 2(C - A)(\sin \theta \cos \theta) + B(\cos^2\theta - \sin^2\theta)$

b. Show that $B^2 - 4AC = b^2 - 4ac$. Thus, we say that the expression $B^2 - 4AC$, which is called the **discriminant**, is **invariant under rotation**; that is, its value does not change under a rotation of axes.

c. Show that F and $A + C$ are invariant under rotation.

22. Show that, except for degenerate cases, the graph of the equation

$$Ax^2 + Bxy + Cy^2 + Dx + Ey + F = 0$$

is

a. a parabola, if $B^2 - 4AC = 0$.
b. an ellipse (or a circle), if $B^2 - 4AC < 0$.
c. a hyperbola, if $B^2 - 4AC > 0$.
d. a hyperbola, if $B^2 - 4AC > 0$.

In Exercises 23–32, using the results of Exercise 22, identify the graph of each equation without applying a rotation of axes.

23. $2x^2 - xy + 3y^2 - 3x + 4y = 6$
24. $2x^2 - xy + 3y^2 + 2x + 6y = 10$
25. $7x^2 - 5xy - 6y^2 + 78x + 52y + 26 = 0$
26. $4x^2 + 4xy + y^2 - 3y = 6$
27. $27x^2 + 72xy + 48y^2 - 200x - 225y + 275 = 0$
28. $2x^2 + 3xy + 6y^2 + 17x + 2 = 0$

29. $x^2 + 4xy + 4y^2 - 2x + y - 9 = 0$
30. $12x^2 + 12xy + 3y^2 - 9x + 435y + 72 = 0$
31. $-xy + y^2 - 3x + 2y = 0$
32. $4x^2 + 4xy + y^2 + x - 2y = 9$
33. Using the quadratic formula, it is possible to solve the equation

$$2x^2 - xy + 3y^2 - 3x + 4y = 6$$

for y. We get

$$2x^2 - xy + 3y^2 - 3x + 4y = 6$$
$$3y^2 + (4 - x)y + (2x^2 - 3x - 6) = 0$$
$$y = \frac{x - 4 \pm \sqrt{(4 - x)^2 - 12(2x^2 - 3x - 6)}}{6}.$$

Simplifying, we have

$$y = \frac{x - 4 + \sqrt{88 + 28x - 23x^2}}{6} \qquad (15)$$

or

$$y = \frac{x - 4 - \sqrt{88 + 28x - 23x^2}}{6}. \qquad (16)$$

a. Use your graphing calculator to graph equations (15) and (16) in the same coordinate system.

b. Compare your result with your answer to Exercise 23.

In Exercises 34–39, use your graphing calculator to graph the equation taken from the indicated exercise. Compare your result with the answer you obtained for the indicated exercise (see Exercise 33).

34. Exercise 25   **35.** Exercise 26
36. Exercise 29   **37.** Exercise 28
38. Exercise 31   **39.** Exercise 30

8.6 PARAMETRIC EQUATIONS

To this point, we have studied a variety of equations involving the two variables x and y. In particular, our study has included equations whose graphs in the xy-plane are lines, circles, parabolas, ellipses, or hyperbolas. In this section we will present an alternate—and sometimes more convenient—approach for defining and graphing such curves. In this alternate approach, a single equation in x and y that defines a curve is replaced by a pair of equations in which the variables x and y are each defined as a function of a third variable, t.

DEFINITION 8.5

Let I be an interval and let f and g be functions defined for all t in I. The graph of any relation of the form

$$C = \{(x, y) \mid x = f(t) \text{ and } y = g(t), \text{ where } t \in I\}$$

is called a **plane curve**. The variable t is called a **parameter**, and the equations $x = f(t)$ and $y = g(t)$ are called **parametric equations**. Any curve that is defined in this manner is said to be **defined parametrically**.

To illustrate this method of defining a curve, we recall from our study of trigonometry that, for each point (x, y) on the unit circle $x^2 + y^2 = 1$, there is a real number t in the interval $[0, 2\pi)$ such that $\cos t = x$ and $\sin t = y$. This means that the unit circle $x^2 + y^2 = 1$ can be defined parametrically as

$$\{(x, y) \mid x = \cos t \text{ and } y = \sin t, \text{ where } t \in [0, 2\pi)\}$$

To further illustrate this idea, consider the curve defined parametrically by the following relation:

$$W = \{(x, y) \mid x = t^2 + 2t - 3 \text{ and } y = t + 1, \text{ where } t \in \mathbb{R}\}$$

To obtain some points on the graph of W, we have determined the values of x and y for the integral values of t between -3 and 3.

t	$x = t^2 + 2t - 3$	$y = t + 1$	(x, y)
-3	0	-2	$(0, -2)$
-2	-3	-1	$(-3, -1)$
-1	-4	0	$(-4, 0)$
0	-3	1	$(-3, 1)$
1	0	2	$(0, 2)$
2	5	3	$(5, 3)$
3	12	4	$(12, 4)$

If we plot the points (x, y) from the table and connect them with a smooth curve, we obtain the graph shown in Figure 8.44.

Eliminating the Parameter

The curve shown in Figure 8.44 appears to be a parabola. To see that this is indeed the case, we obtain an equation in x and y for this curve as follows: Solving the equation $y = t + 1$ for t, we get

$$t = y - 1.$$

Substituting $y - 1$ for t in the equation $x = t^2 + 2t - 3$, we find that

$$x = (y - 1)^2 + 2(y - 1) - 3$$
$$= y^2 - 2y + 1 + 2y - 2 - 3$$
$$= y^2 - 4.$$

FIGURE 8.44

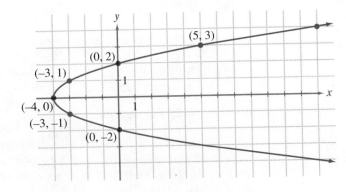

If we write the equation $x = y^2 - 4$ in the form

$$y^2 = x + 4,$$

we have the equation, in standard form, of a parabola that opens to the right and has $(-4, 0)$ as its vertex. In general, the process of obtaining an equation in x and y for a curve that is defined parametrically is called **eliminating the parameter**.

EXAMPLE 18

Sketch the graph of the curve C, which is defined parametrically as follows:

$$x = t^3 - t + 1 \quad \text{and} \quad y = t^2 + t, \quad \text{where } 0 \le t \le 3.$$

Solution   We begin by making a table of values.

FIGURE 8.45

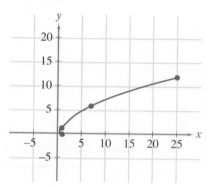

t	$x = t^3 - t + 1$	$y = t^2 + t$	(x, y)
0	1	0	$(1, 0)$
1	1	2	$(1, 2)$
2	7	6	$(7, 6)$
3	25	12	$(25, 12)$

Plotting the points (x, y) from the table and connecting them with a smooth curve, we obtain the graph shown in Figure 8.45. We note that it would be difficult to eliminate the parameter t in order to obtain an equation in x and y for this curve.

EXAMPLE 19

A curve C is defined parametrically as follows:

$$x = 3 \sin t \quad \text{and} \quad y = 4 \cos t, \quad \text{where } t \in \mathbb{R}.$$

Find an equation in x and y for C by using the identity $\sin^2 t + \cos^2 t = 1$ to eliminate the parameter, and sketch the graph of this curve.

Solution   Solving the equations

$$x = 3 \sin t \quad \text{and} \quad y = 4 \cos t$$

for $\sin t$ and $\cos t$, respectively, we find that

FIGURE 8.46

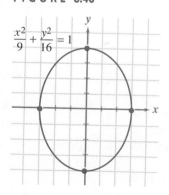

$$\sin t = \frac{x}{3} \quad \text{and} \quad \cos t = \frac{y}{4}.$$

From the identity $\sin^2 t + \cos^2 t = 1$, we obtain

$$\left(\frac{x}{3}\right)^2 + \left(\frac{y}{4}\right)^2 = 1 \quad \text{or} \quad \frac{x^2}{9} + \frac{y^2}{16} = 1.$$

We have obtained the equation, in standard form, of an ellipse with center at the origin, vertices $(0, \pm 4)$ and x-intercepts $(\pm 3, 0)$. The graph of this ellipse is shown in Figure 8.46.

As we see in our next example, when we eliminate the parameter to obtain an equation in x and y for a given curve, we must take into consideration any restrictions that are placed on x and y by the given parametric equations.

EXAMPLE 20

A curve C is defined parametrically as follows:

$$x = 2^t \quad \text{and} \quad y = 2^{t+2} - 3, \quad \text{where } t \in \mathbb{R}.$$

Obtain an equation in x and y for C by eliminating the parameter, and sketch the graph of C.

Solution   We begin by rewriting the defining equation for y as follows:

$$y = 2^{t+2} - 3$$
$$= 2^t \cdot 2^2 - 3$$
$$= 4 \cdot 2^t - 3.$$

Since $x = 2^t$, the equation $y = 4 \cdot 2^t - 3$ can be written as

$$y = 4x - 3.$$

It is important to note, however, that the graph of this linear equation is not the graph of C. Since

$$x = 2^t > 0 \quad \text{and} \quad y = 2^{t+2} - 3 > -3,$$

the graph of C is only part of the line $y = 4x - 3$, as shown in Figure 8.47.

FIGURE 8.47

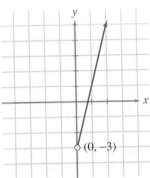

$(0, -3)$

Applications

FIGURE 8.48

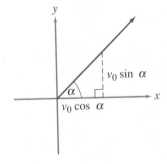

$v_0 \sin \alpha$

α

$v_0 \cos \alpha$

Many physical applications make use of parametric equations in which the parameter t represents time. Such equations are particularly useful for describing the path followed by an object in motion. For example, using Newton's laws of motion (with air friction neglected), it can be shown that the path of a projectile can be described by the following parametric equations:

$$x = (v_0 \cos \alpha)t \quad \text{and} \quad y = (v_0 \sin \alpha)t - 16t^2, \quad \text{where } t \geq 0.$$

In these equations, v_0 is the **initial velocity** of the projectile (velocity when $t = 0$), and α is an angle between $0°$ and $180°$ that indicates the initial direction of the projectile (see Figure 8.48). The parameter t, which represents time, is measured in seconds, and x and y are measured in feet.

EXAMPLE 21

Assume that a projectile is fired with an initial velocity of 800 feet per second at an angle of $60°$ with the horizontal.

a. Find parametric equations that describe the path of this projectile.
b. Find the length of time the projectile is in flight.
c. Find the range of the projectile (value of x at the point the projectile lands).
d. Find an equation in x and y for the path of the projectile.

Solution

a. Substituting 800 for v_0, and 60° for α in the equations

$$x = (v_0 \cos \alpha)t \quad \text{and} \quad y = (v_0 \sin \alpha)t - 16t^2,$$

we get

$$x = 800 \cdot \cos 60° \cdot t = 800 \cdot \frac{1}{2} \cdot t = 400\,t, \quad \text{and}$$

$$y = 800 \cdot \sin 60° \cdot t - 16t^2 = 800 \cdot \frac{\sqrt{3}}{2} \cdot t - 16t^2 = 400\sqrt{3}t - 16t^2.$$

Thus, parametric equations that describe the path of this projectile are

$$x = 400\,t \quad \text{and} \quad y = 400\sqrt{3}t - 16t^2, \quad \text{where } t \geq 0.$$

b. To find the length of time the projectile is in flight, we must find the value(s) of t for which $y = 0$. Therefore, we must solve the equation

$$400\sqrt{3}t - 16t^2 = 0,$$

for t. Factoring, we find that

$$16t(25\sqrt{3} - t) = 0$$
$$16t = 0 \quad \text{or} \quad 25\sqrt{3} - t = 0$$
$$t = 0 \quad \text{or} \quad t = 25\sqrt{3}.$$

Since $t = 0$ when the projectile is fired, the projectile is in flight for $25\sqrt{3} \approx 43.3$ seconds.

c. From part (b) we know that the projectile is airborne for $25\sqrt{3}$ seconds. Thus, the range of the projectile is

$$x = 400t = 400(25\sqrt{3}) = 10,000\sqrt{3} \approx 17,300 \text{ feet.}$$

d. Since $x = 400t$, it follows that $t = x/400$. Substituting $x/400$ for t in the equation

$$y = 400\sqrt{3}t - 16t^2,$$

we get

$$y = 400\sqrt{3}\left(\frac{x}{400}\right) - 16\left(\frac{x}{400}\right)^2 \quad \text{or}$$

$$y = \sqrt{3}x - \frac{x^2}{10,000}.$$

Thus, the path of this projectile is given by the equation

$$y = -\frac{x^2}{10,000} + \sqrt{3}x,$$

which we recognize as the equation of a parabola. _____

FIGURE 8.49
Curve traced by point *P* is called a cycloid

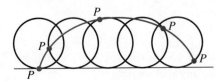

In our final example, we will consider the curve traced by a fixed point *P* on a circle that rolls along a line. Parametric equations provide an excellent means of describing such curves, which are called **cycloids**. See Figure 8.49.

EXAMPLE 22

If a circle of radius r rolls along the x-axis of a rectangular coordinate system, find parametric equations for the curve traced by a fixed point P that lies on the circle.

FIGURE 8.50

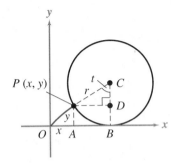

Solution   For convenience we will assume that the point P is first located at the origin and that the circle is rolling to the right along the x-axis. In Figure 8.50, we picture the circle after it has rolled along the x-axis for a distance that is equal to the length of the arc from P to point B. From this assumption, it follows that the distance $d(O, B)$ is equal to the length of arc $\widehat{PB}$.

If we let t denote the radian measure of $\angle PCB$, we know from Corollary 6.1 in Section 6.2, that

$$t = \frac{\text{length of arc } \widehat{PB}}{r}, \qquad \text{or}$$

$$\text{length of arc } \widehat{PB} = rt.$$

Since, as noted earlier, the length of arc $\widehat{PB}$ is equal to the distance between the points O and B, it follows that

$$d(O, B) = rt. \tag{1}$$

From right triangle PDC in Figure 8.50, we observe that

$$\sin t = \frac{d(P, D)}{r} \qquad \text{and} \qquad \cos t = \frac{d(D, C)}{r}, \quad \text{where } 0 \le t < \frac{\pi}{2}.$$

Therefore, it follows that

$$d(P, D) = r \sin t, \quad \text{where } 0 \le t < \frac{\pi}{2} \tag{2}$$

and

$$d(D, C) = r \cos t, \quad \text{where } 0 \le t < \frac{\pi}{2}. \tag{3}$$

We can now determine parametric equations for the coordinates of the point $P = (x, y)$. We proceed as follows:

$$\begin{aligned}
x &= d(O, A) \\
&= d(O, B) - d(A, B) \\
&= d(O, B) - d(P, D) \\
&= rt - r \sin t, \qquad \textbf{By equations (1) and (2)}
\end{aligned}$$

and

$$\begin{aligned}
y &= d(A, P) \\
&= d(B, D) \\
&= d(B, C) - d(D, C) \\
&= r - d(D, C) \\
&= r - r \cos t. \qquad \textbf{By equation (3)}
\end{aligned}$$

Although we derived the parametric equations

$$x = rt - r \sin t \qquad \text{and} \qquad y = r - r \cos t$$

under the restriction that $0 \le t < \pi/2$, it can be shown that these equations generate the entire cycloid if we allow t to be any real number.

FIGURE 8.51
Inverted cycloid is curve of
quickest descent

A cycloid has several interesting physical properties. For example, it can be shown that the path followed by an object sliding without friction from a point A to a point B lower than A, but not on the same vertical line as A, will arrive at B in the shortest amount of time by traveling along an inverted cycloid (see Figure 8.51). For this reason, a cycloid is sometimes referred to as the **curve of quickest descent**. For additional applications involving cycloids, see Exercises 45 and 46.

EXERCISES 8.6

In each of Exercises 1–20, a curve is defined parametrically. Sketch the graph of each curve by plotting points.

1. $x = 2t - 1$ and $y = 1 - t$; $t \in \mathbb{R}$
2. $x = 2 - 3t$ and $y = t + 1$; $t \in \mathbb{R}$
3. $x = t^2 + 1$ and $y = 3t^2 - 5$; $t \in \mathbb{R}$
4. $x = t^2 - 2t + 1$ and $y = 1 - t^2$; $t \in \mathbb{R}$
5. $x = \sqrt{1 - t^2}$ and $y = t^3$; $-1 \le t \le 1$
6. $x = t^2 + 1$ and $y = \sqrt{t - 1}$; $t \ge 1$
7. $x = 1/t$ and $y = t^2 - 1$; $t \ne 0$
8. $x = t^2$ and $y = 1/(1 - t)$; $t \ne 1$
9. $x = e^t$ and $y = t + 1$; $t \in \mathbb{R}$
10. $x = \ln t$ and $y = 1/t$; $t > 0$
11. $x = \cos t$ and $y = \sin t$; $0 \le t \le \pi$
12. $x = \cos t$ and $y = \sin t$; $\pi \le t \le 2\pi$
13. $x = \cos t - 1$ and $y = \sin t + 2$; $0 \le t < 2\pi$
14. $x = \cos t + 1$ and $y = \sin t - 2$; $0 \le t < 2\pi$
15. $x = \sec t$ and $y = \tan t$; $-\pi/2 < t < \pi/2$
16. $x = \csc t$ and $y = \cot t$; $0 < t < \pi$
17. $x = t - \sin t$ and $y = 1 - \cos t$; $0 \le t < 2\pi$
18. $x = 2t - 2\sin t$ and $y = 2 - 2\cos t$; $0 \le t < 2\pi$
19. $x = 2t - \sin t$ and $y = 2 - \cos t$; $0 \le t < 2\pi$
20. $x = t - 2\sin t$ and $y = 1 - 2\cos t$; $0 \le t < 2\pi$

In each of Exercises 21–40, a curve is defined parametrically. Find an equation in x and y for each curve by eliminating the parameter and sketch the graph of each curve.

21. $x = 2t + 1$ and $y = 4 - 3t$; $t \in \mathbb{R}$
22. $x = 3 - 2t$ and $y = \frac{1}{2}t + 1$; $t \in \mathbb{R}$
23. $x = (t + 1)/3$ and $y = t^2 + 5t - 1$; $t \in \mathbb{R}$
24. $x = (t - 1)/3$ and $y = t^2 - 2t + 3$; $t \in \mathbb{R}$
25. $x = t^3$ and $y = t^2 - 1$; $t \in \mathbb{R}$
26. $x = t^2$ and $y = t^3 + 1$; $t \in \mathbb{R}$
27. $x = 2\cos t$ and $y = 2\sin t$; $0 \le t < 2\pi$
28. $x = 3\cos t$ and $y = 3\sin t$; $0 \le t \le \pi$
29. $x = 3^t$ and $y = 3^{t+1}$; $t \in \mathbb{R}$
30. $x = 2^{t-1}$ and $y = 2^t - 2$; $t \in \mathbb{R}$
31. $x = e^t$ and $y = e^{-t} + 1$; $t \in \mathbb{R}$
32. $x = e^{-t}$ and $y = e^t - 2$; $t \in \mathbb{R}$
33. $x = 3\cos t$ and $y = \sin t$; $0 \le t < 2\pi$

34. $x = 2\cos t$ and $y = 3\sin t$; $0 \le t < 2\pi$
35. $x = 3\cos t$ and $y = 2\sin t - 1$; $0 \le t < 2\pi$
36. $x = 2\cos t - 3$ and $y = 4\sin t + 1$; $0 \le t < 2\pi$
37. $x = t$ and $y = \sqrt{t^2 - 1}$; $|t| \ge 1$
38. $x = t - 1$ and $y = \sqrt{t}$; $t \ge 0$
39. $x = 2/(t^2 + 1)$ and $y = 2t/(t^2 + 1)$; $t \in \mathbb{R}$
40. $x = 4t/(t^2 + 4)$ and $y = 8/(t^2 + 4)$; $t \in \mathbb{R}$

41. A cannonball is fired from a cannon with an initial velocity of 1250 feet per second. If the barrel of the cannon makes an angle of 28.6° with the horizontal, find the length of time the cannonball is in the air, and the distance from the cannon to the point the cannonball lands. Assume that air friction is neglected.

42. A bullet is fired from a rifle with a muzzle velocity of 680 feet per second. If the barrel of the rifle makes an angle of 67.5° with the horizontal when the rifle is fired, find the amount of time the bullet is in the air, and the horizontal distance it travels. Assume that air friction is neglected.

43. Let $P = (x, y)$ be any point other than the origin on the parabola $y^2 = 4px$. Find parametric equations for this parabola using the slope of the line through O and P as the parameter t.

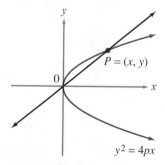

44. Let $P_1 = (x_1, y_1)$ and $P_2 = (x_2, y_2)$ be two points in the xy-plane. Show that the equations

$$x = x_1 + (x_2 - x_1)t \quad \text{and}$$
$$y = y_1 + (y_2 - y_1)t, \quad \text{where } t \in \mathbb{R},$$

are parametric equations for the line containing P_1 and P_2.

45. Assume that a circle of radius r rolls along the x-axis of a

rectangular coordinate system. Let C be the center of the circle and let l be a line containing C. If P is a fixed point on the line l such that $d(C, P) = a < r$, show that the curve traced by P can be described by the parametric equations

$$x = rt - a \sin t \quad \text{and}$$
$$y = r - a \cos t, \quad \text{where } t \in \mathbb{R}.$$

The resulting curve is called a **curtate cycloid**.

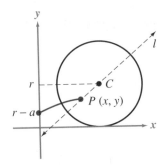

46. Assume that a circle of radius r rolls along the x-axis of a rectangular coordinate system. Let C be the center of the circle and let l be a line containing C. If P is a fixed point

on the line l such that $d(C, P) = a > r$, show that the curve traced by P can be described by the parametric equations

$$x = rt - a \sin t \quad \text{and}$$
$$y = r - a \cos t, \quad \text{where } t \in \mathbb{R}.$$

The resulting curve is called a **prolate cycloid**.

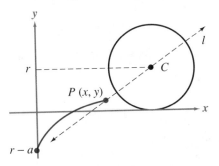

In Exercises 47–48, eliminate the parameter to obtain an equation in x and y for each curve. In each case, graph the resulting curve using your graphing calculator.

47. $x = \sin^3 t$ and $y = \cos^3 t$; $\quad 0 \le t < 2\pi$
48. $x = \tan t - 1$ and $y = \sec t + 2$; $\quad -\pi/2 < t < \pi/2$

CHAPTER 8 REVIEW EXERCISES

In Exercises 1–20, identify the conic defined by the given equation. If it is a parabola, give its vertex, focus, axis, and directrix; if it is an ellipse, give its center, vertices, foci, and the endpoints of its minor axis; if it is a hyperbola, give its center, vertices, foci, and the endpoints of its conjugate axis. Sketch the graph of each equation.

1. $x^2 = 2y$
2. $y^2 = -3x$
3. $x^2 - \dfrac{y^2}{2} = 1$
4. $x^2 + \dfrac{y^2}{2} = 1$
5. $\dfrac{x^2}{10} + \dfrac{y^2}{6} = 1$
6. $y^2 - \dfrac{x^2}{4} = 1$
7. $4x^2 + 4y^2 = 9$
8. $2x^2 - 3y^2 = 5$
9. $9x^2 - y^2 = 1$
10. $2x^2 + y - 4x = 0$
11. $\dfrac{(x + 2)^2}{9} - \dfrac{(y - 5)^2}{16} = 1$
12. $(x - 2)^2 + \dfrac{(y + 2)^2}{4} = 1$
13. $(x + 2)^2 = 4y - 8$
14. $2y = 4x^2 - 1$
15. $x^2 + 4y^2 - 6x + 5 = 0$
16. $25(x + 3)^2 + 16(y - 2)^2 = 400$
17. $49x^2 - 9y^2 + 98x - 36x - 428 = 0$
18. $9x^2 + 4y^2 - 36x + 8y + 4 = 0$
19. $3y^2 - x^2 + 4x - 12y = 19$
20. $16x^2 + 25y^2 - 32x + 100y = 284$

In Exercises 21–28, obtain an equation of the conic described.

21. Parabola; focus $(-3, -1)$ and directrix the line $x = -5$
22. Ellipse; vertices $(-4, 3)$ and $(-2, 3)$, and length of minor axis is 2.
23. Hyperbola; center $(0, 0)$, focus $(0, -4)$, and vertex $(0, 2)$
24. Parabola; vertex $(2, 3)$, and focus $(2, -1)$
25. Ellipse; center $(-1, -4)$, major axis of length 10 and parallel to the x-axis, and minor axis of length 5
26. Hyperbola; vertices $(1, 4)$ and $(1, -3)$, and asymptote the line $3x - 2y - 1 = 0$
27. Parabola; vertex $(-1, -1)$, symmetric with respect to the line $x = -1$, and passing through the point $(-3, 6)$
28. Ellipse; center $(0, 0)$, focus $(0, C)$, length of the major axis is 3 times the length of the minor axis and passes through the point $(-3, 3)$

In Exercises 29–38, make a rotation of axes to eliminate the xy term in the given equation. Identify the conic (or degenerate conic) defined by each equation and sketch its graph.

29. $5x^2 - 2xy + 5y^2 = 6$
30. $5x^2 - 2xy - 5y^2 = 6$
31. $6x^2 + 5xy - 6y^2 - 7 = 0$
32. $6x^2 + 5xy - 6y^2 + 7 = 0$
33. $x^2 - 4xy + 4y^2 - 4 = 0$

34. $9x^2 + 24xy + 16y^2 + 90x - 130y = 0$

35. $xy + x - 2y = 3$

36. $xy - 2x + y - 10 = 0$

37. $16x^2 + 24xy + 9y^2 - 30x + 40y = 0$

38. $17x^2 - 12xy + 8y^2 - 68x + 24y - 12 = 0$

In each of Exercises 39–44, a curve is defined parametrically. Sketch the graph of each curve by plotting points.

39. $x = t^2 - 1$ and $y = \sqrt{t - 1}$; $t \geq 1$

40. $x = 3t^2 + 1$ and $y = 2 - \sqrt{t}$; $t \geq 0$

41. $x = 2t$ and $y = t^2 - t - 1$; $t \in \mathbb{R}$

42. $x = e^{t-1}$ and $y = 3t^2$; $t \in \mathbb{R}$

43. $x = \frac{1}{2} \sec t$ and $y = 2 \tan t$; $-\pi/2 < t < \pi/2$

44. $x = 1 - \csc t$ and $y = \cot t$; $0 < t < \pi$

In each of Exercises 45–50, a curve is defined parametrically. Find an equation in x and y for each curve by eliminating the parameter and sketch the graph of each curve.

45. $x = \sqrt{t}$ and $y = 3t - 2$; $t \geq 0$

46. $x = 2t + 1$ and $y = \sqrt{t + 1}$; $t \geq -1$

47. $x = t - 1$ and $y = 2/t$; $t \neq 0$

48. $x = 1/t - 1$ and $y = t^2 - 4$; $t \neq 0$

49. $x = 4 - 4 \sin t$ and $y = 4 - 4 \cos t$; $0 \leq t < 2\pi$

50. $x = 1 - t^2$ and $y = 3t - 1$; $t \in \mathbb{R}$

51. A projectile is fired with an initial velocity of 500 feet per second at an angle of 52.5° with the horizontal. Find the total amount of time the projectile is in the air and the range of the projectile.

52. Show that an ellipse with center at $C = (h, k)$, having a major axis of length $2a$ and a minor axis of length $2b$ can be described by the parametric equations $x = a \cos t + h$ and $y = b \sin t + k$, where $0 \leq t < 2\pi$.

53. Use the definition of a parabola and the distance formula to find an equation of the parabola with vertex $(2, -3)$ and directrix $y = -1$.

54. Find an equation of the set of points in a plane each of whose distance from the point $(0, 4)$ is $\frac{2}{3}$ its distance from the line $y = 9$.

55. Find an equation of the ellipse whose foci are the vertices of the hyperbola $y^2 - 5x^2 = 25$ and whose vertices are the foci of this hyperbola. Sketch the graph of each.

56. Find an equation of the hyperbola whose foci are the vertices of the ellipse $9x^2 + 16y = 144$ and whose vertices are the foci of this ellipse. Sketch the graph of each.

CHAPTER 9

Systems of Equations and Inequalities

In the first part of this chapter, we discuss methods for solving systems of linear equations. We use matrices to develop techniques that are easily adapted for computer use in solving large-scale systems of linear equations and give a complete discussion of the algebra of matrices. We define the inverse of a matrix and show how it can be used to find the solution set of a system of linear equations. After introducing determinants and discussing Cramer's rule, we study methods for solving systems of nonlinear equations. We end this chapter with an introduction to a relatively new field of mathematics called *linear programming*.

9.1 SOLVING SYSTEMS OF LINEAR EQUATIONS BY SUBSTITUTION AND ELIMINATION

In Chapter 2, we learned that the graph of an equation of the form

$$ax + by = c,$$

where a and b are not both 0, is a line. For this reason, such an equation is called a **linear equation in the variables x and y**. A solution of this equation is any ordered pair (x, y) of numbers satisfying the equation. For example, $(-2, 3)$ is a solution of the equation

$$x + 2y = 4 \tag{1}$$

since $-2 + 2(3) = -2 + 6 = 4$. On the other hand, $(1, -4)$ is not a solution since

$$1 + 2(-4) = 1 - 8 = -7 \neq 4.$$

Some other solutions of equation (1) are $(0, 2)$, $(1, \frac{3}{2})$, $(2, 1)$, $(4, 0)$, and $(-1, \frac{5}{2})$. In fact, the coordinates of each point on the graph of the linear equation $x + 2y = 4$ satisfy the equation.

We can also verify, for example, that $(-2, -7)$, $(0, -3)$, $(1, -1)$, $(2, 1)$, and $(3, 3)$ all satisfy the equation

$$2x - y = 3. \tag{2}$$

Since the ordered pair $(2, 1)$ satisfies both equation (1) and equation (2), we say that $(2, 1)$ is a solution of the system of two linear equations

$$\begin{cases} x + 2y = 4 \\ 2x - y = 3. \end{cases} \tag{3}$$

The brace is used to emphasize that the two equations are to be treated simultaneously. In general, an ordered pair (x_0, y_0) is said to be a **solution of the system of two linear equations**

$$\begin{cases} ax + by = c \\ dx + ey = f \end{cases}$$

provided (x_0, y_0) is a solution of *each* equation in this system.

As shown above, $(2, 1)$ is a solution of system (3), which is formed by using equations (1) and (2). This means that the point $(2, 1)$ lies on both lines. It is natural to ask if there are other solutions of this system. From Figure 9.1, we see that there are no other solutions since $(2, 1)$ is the only point common to both lines. It follows that the solution set of system (3) is $\{(2, 1)\}$.

FIGURE 9.1

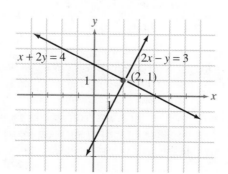

EXAMPLE 1

FIGURE 9.2

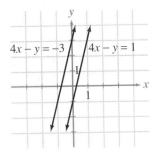

Find the solution set of the system

$$\begin{cases} 4x - y = -3 \\ 4x - y = 1. \end{cases}$$

Solution   We previously noted that any solution of a system of linear equations is a point on the graph of each equation in the system. These equations can be written $y = 4x + 3$ and $y = 4x - 1$, and the graph of each is a line with slope 4. In addition, note that $(0, 3)$ is the y-intercept of the first line, while $(0, -1)$ is the y-intercept of the second. Hence, the graphs of these two equations are the parallel lines shown in Figure 9.2. Therefore, these lines do not intersect and, thus, the solution set for this system is the empty set $\varnothing$.

EXAMPLE 2

FIGURE 9.3

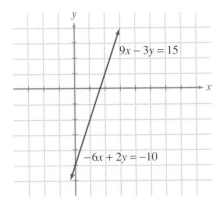

Solve the following system:

$$\begin{cases} 9x - 3y = 15 \\ -6x + 2y = -10. \end{cases} \quad (4)$$

Solution   If we multiply each side of the first equation in system (4) by 2, or multiply each side of the second equation by -3, we obtain the equation

$$18x - 6y = 30.$$

Therefore, the two equations in system (4) are equations for the same line, and hence, every point on this line is a solution of the system (see Figure 9.3). If we let $x = t$, and solve either equation for y, we obtain the equation $y = 3t - 5$. Thus, for any real number t, the ordered pair $(t, 3t - 5)$ is a solution of system (4). Hence, the solution set can be expressed as $\{(t, 3t - 5) \mid t \in \mathbb{R}\}$.

The variable t introduced in Example 2 is called a **parameter**. A particular solution of system (4) can be found by replacing t with a real number. For example if $t = -2$, we get the solution $(-2, -11)$.

The three possibilities just illustrated are the only possibilities for the solutions of a system of two linear equations in two variables. We summarize these possibilities below and give additional illustrations in Figure 9.4.

Solutions for a System of Two Linear Equations in Two Variables

For a system of two linear equations in two variables, exactly one of the following must hold:

(i)   The two equations represent lines that intersect in exactly one point. Thus, there is exactly one solution of the system. Such a system is said to be **consistent and independent**.

(ii) The two equations represent the same line. Since each point on the line is a solution of the system, there are infinitely many solutions of the system. In this case, the system is said to be **consistent and dependent**.

(iii) The two equations represent two parallel lines. Since the two lines do not intersect, the system has no solution. Such a system is said to be **inconsistent**.

FIGURE 9.4

a. Consistent and independent system

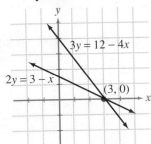

b. Consistent and dependent system

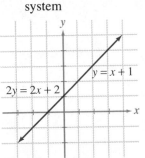

c. Inconsistent system

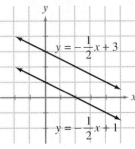

Although we can sometimes find the exact solutions of a system of two linear equations by graphing the two lines, in general, graphical methods give us only rough approximations of the solutions. Precise solutions of a system can, however, be found algebraically. We now develop two algebraic methods for solving systems of two linear equations in two variables. These methods are the **substitution method** and the **elimination method**.

The Substitution Method

We illustrate the substitution method by finding the solution set of the system

$$\begin{cases} 2x - y = 5 \\ 3x + 2y = 4. \end{cases} \tag{5}$$

We begin by choosing one of the two equations and using it to express one of the variables in terms of the other. To avoid fractions, we select the first equation and solve it for y in terms of x getting

$$y = 2x - 5. \tag{6}$$

Substituting the value of y given in equation (6) into the second equation in system (5), we get

$$3x + 2(2x - 5) = 4$$
$$3x + 4x - 10 = 4$$
$$7x = 14$$
$$x = 2.$$

FIGURE 9.5

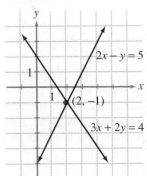

Substituting 2 for x in equation (6), we find that

$$y = 2(2) - 5$$
$$= -1.$$

Thus, we have found that $x = 2$ and $y = -1$. It is easily verified that the ordered pair $(2, -1)$ satisfies both equations in system (5). Hence, the solution set of system (5) is $\{(2, -1)\}$, and this system is consistent and independent. See Figure 9.5.

EXAMPLE 3

Use the substitution method to find the solution set of the system

$$\begin{cases} -\dfrac{2}{3}x + 2y = 10 \\ 2x - 6y = -3. \end{cases}$$

Solution Solving the second equation for x, we have

$$2x = 6y - 3, \quad \text{or} \quad x = 3y - \dfrac{3}{2}.$$

Substituting $3y - \frac{3}{2}$ for x in the first equation of the given system, we find that

$$-\dfrac{2}{3}\left(3y - \dfrac{3}{2}\right) + 2y = 10$$
$$-2y + 1 + 2y = 10$$
$$1 = 10 \qquad \text{False}$$

Since 1 is not equal to 10, this system is inconsistent and its solution set is the empty set.

The Elimination Method

A shortcoming of the substitution method is that it is not systematic and thus cannot easily be programmed on a computer. The elimination method, on the other hand, is systematic and will lead directly to matrix methods for finding the solution of a system of linear equations. These matrix methods, which we study in Section 9.3, are easily performed by a computer.

The elimination method involves the replacing of a system of equations by a simpler equivalent system until a system is obtained whose solution is obvious. **Equivalent systems**, as you might expect, are systems whose solution sets are identical. In Theorem 9.1, we list operations that produce equivalent systems of linear equations.

THEOREM 9.1

For a given system of linear equations, an equivalent system of linear equations is obtained when

 (i) two equations are interchanged;
 (ii) an equation is multiplied by a nonzero constant; or
 (iii) a constant multiple of one equation is added to another equation.

When using the elimination method, we apply Theorem 9.1 to replace a given system by an equivalent system in which the coefficients of one variable in two equations are additive inverses of each other. Then addition of the two equations will eliminate that variable in one of the equations. We illustrate the elimination method with several examples.

EXAMPLE 4

Use the elimination method to find the solution set of the system

$$\begin{cases} 9x - 3y = 24 \\ 11x + 2y = 1. \end{cases} \tag{7}$$

Solution If we multiply the first equation by 2 and the second equation by 3, we obtain an equivalent system; namely,

$$\begin{cases} 18x - 6y = 48 \\ 33x + 6y = 3. \end{cases} \tag{8}$$

Observe that the coefficients of y in system (8) are additive inverses of one another. We now add the first equation in system (8) to the second, obtaining the equivalent system

$$\begin{cases} 18x - 6y = 48 \\ 51x \quad\quad = 51. \end{cases}$$

Solving the latter equation for x, we find that $x = 1$. Substituting 1 for x in the equation $18x - 6y = 48$, we have

$$18(1) - 6y = 48$$
$$-6y = 30$$
$$y = -5.$$

Thus, the solution set for this system is $\{(1, -5)\}$, which implies that the system is consistent and independent.

EXAMPLE 5

Solve by elimination

$$\begin{cases} -3x + 4y = -8 \\ \dfrac{3}{2}x - 2y = 4. \end{cases}$$

Solution To eliminate x, we multiply both sides of the second equation by 2 and add the result to the first equation to obtain the equivalent system

$$\begin{cases} 0 = 0 \\ \dfrac{3}{2}x - 2y = 4. \end{cases} \tag{9}$$

The equation, $0 = 0$, in system (9) imposes no restrictions on x and y. Thus, the given system is consistent and dependent having a solution set consisting of all points (x, y) on the line $\frac{3}{2}x - 2y = 4$. Solving for y, we find that

$$y = \frac{3}{4}x - 2.$$

Letting $x = t$, the solution set is $\{(t, \frac{3}{4}t - 2) \mid t \in \mathbb{R}\}$.

Systems of two linear equations in two variables often arise in the solution of real-world problems, as the following example illustrates:

EXAMPLE 6

A woodworking shop manufactures two models of folding chairs. The standard model requires $12 in materials and takes a worker 40 minutes to construct. The deluxe model requires $15 in materials and takes a worker one hour to construct. How many chairs of each type can a worker produce in an 8-hour work day if the total cost of materials used is $126?

Solution   Let x represent the number of standard model chairs constructed and y represent the number of deluxe model chairs constructed by a worker in an eight-hour workday. Since each standard model chair requires $12 in materials and each deluxe model requires $15 in materials, and since the total cost of materials is $126, we are led to the equation

$$12x + 15y = 126. \qquad \text{Cost equation}$$

We know that an eight-hour workday has $8 \cdot 60 = 480$ minutes. Since each standard chair requires 40 minutes of construction time and each deluxe model chair requires 60 minutes, we have

$$40x + 60y = 480. \qquad \text{Time equation}$$

Pairing the cost equation and the time equation gives us the system

$$\begin{cases} 12x + 15y = 126 \\ 40x + 60y = 480. \end{cases}$$

You should verify that the results are $x = 3$ and $y = 6$. This means that, in an eight-hour workday, a worker can construct 3 standard model and 6 deluxe model folding chairs using exactly $126 in materials. ──────────

Equations involving more than two variables are frequently encountered. An equation of the form

$$ax + by + cz = d,$$

where a, b, c, and d are constants and a, b, and c are not all 0, is called a **linear equation in three variables**. Linear equations with four or more variables are similarly defined.

We now consider systems of three linear equations in three variables. Generalizing the case of two linear equations in two variables, we define the solution set of a system of the form

$$\begin{cases} a_1x + b_1y + c_1z = d_1 \\ a_2x + b_2y + c_2z = d_2 \\ a_3x + b_3y + c_3z = d_3 \end{cases}$$

to consist of all **ordered triples** (x_0, y_0, z_0) satisfying *each* equation in the system. While both the substitution method and the elimination method can be used to find the solution set, we will consider only the elimination method. (Recall, as we mentioned earlier, that this method is easily programmed for computer use.) We illustrate the elimination method in the following example:

EXAMPLE 7

Use the elimination method to find the solution set of the system

$$\begin{cases} x - 2y - z = 0 \\ 2x - 3y - 4z = 8 \\ 3x + y + 2z = -1. \end{cases}$$

Solution   By multiplying the first equation by -2 and adding the result to the second equation, we eliminate the variable x from the second equation.

$$\begin{cases} x - 2y - z = 0 \\ y - 2z = 8 \\ 3x + y + 2z = -1 \end{cases}$$

We now eliminate the variable x from the third equation by multiplying the first equation by -3 and adding the result to the third equation.

$$\begin{cases} x - 2y - z = 0 \\ y - 2z = 8 \\ 7y + 5z = -1. \end{cases}$$

If we multiply the second equation by -7 and add the result to the third equation, we eliminate the variable y from the third equation.

$$\begin{cases} x - 2y - z = 0 \\ y - 2z = 8 \\ 19z = -57. \end{cases} \qquad (10)$$

Multiplying the third equation by $\frac{1}{19}$ yields the equivalent system

$$\begin{cases} x - 2y - z = 0 \\ y - 2z = 8 \\ z = -3. \end{cases}$$

From the third equation, we readily see that $z = -3$. Substituting -3 for z in the second equation, we find that

$$\begin{aligned} y - 2(-3) &= 8 \\ y + 6 &= 8 \\ y &= 2. \end{aligned}$$

Finally, substituting 2 for y and -3 for z in the first equation, we have

$$\begin{aligned} x - 2(2) - (-3) &= 0 \\ x - 4 + 3 &= 0 \\ x &= 1. \end{aligned}$$

Hence, the solution set for the given system is $\{(1, 2, -3)\}$.

System (10), found in the solution of Example 7, is said to be in **triangular form**. With that system in triangular form we readily found the value of z from the third equation. Then we found the values of y and x by substituting known values of the variables into the other two equations. This process is called **back substitution**.

You may be aware that a linear equation in three variables represents a plane in three-dimensional space. Thus, the solution set of a system of linear equations in three variables consists of the points of intersection of the planes represented by the equations. It can be shown that *any* such system of linear equations must have exactly one solution, infinitely many solutions, or no solution. Such systems are described as consistent and independent, consistent and dependent, and inconsistent, just as they are in the case of a system of two linear equations in two variables. In Example 7, the three equations represent three planes intersecting at the point $(1, 2, -3)$.

EXAMPLE 8

Using the elimination method, find the solution set of the system

$$\begin{cases} 3u - v - 7w = 3 \\ 2u + v - 16w = 1 \\ 4u - 3v + 2w = 5. \end{cases}$$

Solution   After investigating the coefficients of the variables, we choose to eliminate the variable v from the second and third equations. To accomplish this, we add the first equation to the second and add -3 times the first equation to the third equation to get

$$\begin{cases} 3u - v - 7w = 3 \\ 5u \quad\quad - 23w = 4 \\ -5u \quad\quad + 23w = -4. \end{cases}$$

Adding the second equation to the third gives us the equivalent system

$$\begin{cases} 3u - v - 7w = 3 \\ 5u \quad\quad - 23w = 4 \\ \quad\quad\quad 0 = 0. \end{cases}$$

Since the third equation, $0 = 0$, is true for all values of u, v, and w, the solution of this system consists of all ordered triples (u, v, w) satisfying the first two equations

$$\begin{cases} 3u - v - 7w = 3 \\ 5u \quad\quad - 23w = 4. \end{cases} \qquad (11)$$

Solving the second equation in system (11) for u, we obtain

$$u = \frac{23}{5}w + \frac{4}{5}.$$

Replacing u by $\frac{23}{5}w + \frac{4}{5}$ in the first equation of system (11), and solving for v, we find that

$$3\left(\frac{23}{5}w + \frac{4}{5}\right) - v - 7w = 3$$

$$\frac{69}{5}w + \frac{12}{5} - v - 7w = 3$$

$$v = \frac{34}{5}w - \frac{3}{5}.$$

We now have both u and v expressed in terms of w. Since w can be chosen arbitrarily, we can express the solution set as

$$\left\{ \left(\frac{23}{5}w + \frac{4}{5}, \frac{34}{5}w - \frac{3}{5}, w\right) \,\middle|\, w \in \mathbb{R} \right\}.$$

In addition, there are other ways in which the solution set can be written. For example, if we solve the second equation in system (11) for w in terms of u, we find that $w = \frac{5}{23}u - \frac{4}{23}$. In turn, $v = \frac{34}{23}u - \frac{41}{23}$. Thus, the solution set may also be expressed as

$$\left\{ \left(u, \frac{34}{23}u - \frac{41}{23}, \frac{5}{23}u - \frac{4}{23}\right) \,\middle|\, u \in \mathbb{R} \right\}.$$

Similarly, if v is chosen arbitrarily, the solution set can be written as

$$\left\{\left(\frac{23}{34}v + \frac{41}{34}, \ v, \ \frac{5}{34}v + \frac{69}{782}\right)\ \middle|\ v \in \mathbb{R}\right\}.$$

It can be shown that the solution set for this system is a line.

EXAMPLE 9

Find the solution set of the following system using the method of elimination.

$$\begin{cases} r + 2s + 3t = 5 \\ 2r - 2s - 4t = 7 \\ 2r + 4s + 6t = 11. \end{cases}$$

Solution   To eliminate the variable r from the second and third equations, we multiply the first equation by -2 and add the result to each of the other two equations. This gives us

$$\begin{cases} r + 2s + 3t = 5 \\ \qquad - 6s - 10t = -3 \\ \qquad\qquad\qquad 0 = 1. \quad \textbf{False} \end{cases}$$

Since the third equation is impossible, we have an inconsistent system whose solution set is the empty set $\varnothing$.

EXERCISES 9.1

In Exercises 1–12, use the substitution method to find the solution set of each system.

1. $\begin{cases} x + 2y = -1 \\ 2x - 3y = 12 \end{cases}$

2. $\begin{cases} x + 2y = 1 \\ 5x - 4y = -23 \end{cases}$

3. $\begin{cases} 2x + 6y = -3 \\ x - 3y = 2 \end{cases}$

4. $\begin{cases} 3u - v = 5 \\ 2u + 5v = 9 \end{cases}$

5. $\begin{cases} 5r + 2s = 3 \\ 2r + 5s = 3 \end{cases}$

6. $\begin{cases} 3x + 5y = 0 \\ x - 4y = 0 \end{cases}$

7. $\begin{cases} 3x + 4y - 7 = 0 \\ 6x + 8y - 14 = 0 \end{cases}$

8. $\begin{cases} 2a - b - 3 = 0 \\ 4a - b - 4 = 0 \end{cases}$

9. $\begin{cases} \frac{1}{5}m - \frac{1}{3}n = \frac{1}{5} \\ m + n = 0 \end{cases}$

10. $\begin{cases} 8r + 16s = 20 \\ 16r + 50s = 55 \end{cases}$

11. $\begin{cases} 3x + 7y = 0 \\ -6x + y + 5 = 0 \end{cases}$

12. $\begin{cases} 1.8x + 1.2y = 4 \\ 9x + 6y = 3 \end{cases}$

In Exercises 13–44, use the elimination method to find the solution set of each system.

13. $\begin{cases} 3x + 4y = 3 \\ x - 2y = -4 \end{cases}$

14. $\begin{cases} x - 2y = 7 \\ 3x - 2y = 9 \end{cases}$

15. $\begin{cases} \frac{2}{3}x - \frac{1}{2}y = 3 \\ \frac{1}{3}x - \frac{1}{4}y = \frac{3}{2} \end{cases}$

16. $\begin{cases} \frac{2}{5}u - \frac{1}{2}v = \frac{13}{2} \\ \frac{3}{4}u - \frac{1}{5}v = \frac{17}{2} \end{cases}$

17. $\begin{cases} 2x - 3y = 7 \\ 4x - 6y = 5 \end{cases}$

18. $\begin{cases} 4s + 3t = 5 \\ 2s - 8t = -5 \end{cases}$

19. $\begin{cases} 0.12m + 0.05n = 0.70 \\ 0.11m - 0.03n = 0.25 \end{cases}$

20. $\begin{cases} 0.01x + 0.03y = -0.09 \\ 0.02x - 0.8y = 2.3 \end{cases}$

21. $\begin{cases} 14x - 10y = 2 \\ 7x - 5y = -3 \end{cases}$

22. $\begin{cases} 4a + 3b = 11 \\ -3a + 2b = -4 \end{cases}$

23. $\begin{cases} 2x - 4y = k - 1 \\ 3x + 2y = 3k \end{cases}$
(k is constant)

24. $\begin{cases} \frac{2}{3}x - \frac{1}{2}y = 5 \\ 4x + y = 54 \end{cases}$

25. $\begin{cases} x + 3y + z = 6 \\ 3x + y - z = -2 \\ 2x + 2y - z = 1 \end{cases}$

26. $\begin{cases} 3x + y + 2z = 1 \\ 4x - y + 4z = 4 \\ x + y - 2z = -5 \end{cases}$

27. $\begin{cases} 2a + 8b + 6c = 20 \\ 4a + 2b - 2c = -2 \\ 6a - 4b + 10c = 24 \end{cases}$

28. $\begin{cases} 2x - 3y + 3z = -15 \\ 3x + 2y - 5z = 19 \\ 5x - 4y - 2z = -2 \end{cases}$

29. $\begin{cases} 5x - 3y + 2z = 3 \\ 2x + 4y - z = 7 \\ x - 11y + 4z = 3 \end{cases}$

30. $\begin{cases} x + y - z = 7 \\ 4x + y - 5z = 4 \\ 6x + y + 3z = 18 \end{cases}$

31. $\begin{cases} 4u + v - w = 6 \\ 3u - v + 2w = 7 \\ 5u + 3v - 4w = 2 \end{cases}$

32. $\begin{cases} x - 2y + 3z = 4 \\ 2x + y - 4z = 3 \\ -3x + 4y - z = -2 \end{cases}$

33. $\begin{cases} 2y + 5z = 6 \\ x - 2z = 4 \\ 2x + 4y = -2 \end{cases}$

34. $\begin{cases} p + q - 3r = -1 \\ q - r = 0 \\ -p + 2q = 1 \end{cases}$

35. $\begin{cases} x + 4y - z = 10 \\ 3x + 2y + z = 4 \\ 2x - 3y + 2z = -7 \end{cases}$

36. $\begin{cases} x - 2y + 3z = 9 \\ -x + 3y = -4 \\ 2x - 5y + 5z = 17 \end{cases}$

37. $\begin{cases} 2r + 4s + 3t = 6 \\ r - 3s + 2t = -7 \\ -r + 2s - t = 5 \end{cases}$

38. $\begin{cases} x + 10y - 11z = 0 \\ 2x - y + 5z = 5 \\ x + 3y - 2z = 1 \end{cases}$

39. $\begin{cases} x + y - z = 0 \\ 2x - 4y + 3z = 0 \\ x - 7y + 6z = 0 \end{cases}$

40. $\begin{cases} 2x + 6z = -9 \\ 3x - 2y + 11z = -16 \\ 3x - y + 7z = -11 \end{cases}$

41. $\begin{cases} \dfrac{1}{x} + \dfrac{1}{y} = 1 \\ \dfrac{2}{x} + \dfrac{4}{y} = 0 \end{cases}$

42. $\begin{cases} \dfrac{1}{u} - \dfrac{1}{v} = 3 \\ \dfrac{3}{u} - \dfrac{1}{v} = -1 \end{cases}$

[*Hint:* Let $u = 1/x$ and $v = 1/y$.]

43. $\begin{cases} \dfrac{1}{x} + \dfrac{1}{y} - \dfrac{1}{z} = -3 \\ \dfrac{1}{x} - \dfrac{1}{y} - \dfrac{1}{z} = 1 \\ \dfrac{1}{x} - \dfrac{2}{y} + \dfrac{2}{z} = 12 \end{cases}$

44. $\begin{cases} \dfrac{1}{u} + \dfrac{1}{v} - \dfrac{1}{z} = 1 \\ \dfrac{2}{u} - \dfrac{1}{v} - \dfrac{2}{z} = -8 \\ -\dfrac{1}{u} - \dfrac{1}{v} + \dfrac{1}{z} = -5 \end{cases}$

45. The length of a rectangle with perimeter 38 meters is 6 meters greater than the width. Find the dimensions of the rectangle.

46. The general form of a system of two linear equations in two variables can be written as

$$\begin{cases} ax + by = e \\ cx + dy = f. \end{cases}$$

Use the elimination method to show that if $ad - bc \neq 0$, this system has the unique solution (x_0, y_0), where

$$x_0 = \frac{de - bf}{ad - bc} \quad \text{and} \quad y_0 = \frac{af - ce}{ad - bc}.$$

47. Two solutions of the equation $ax + by = 1$ are $(2, -1)$ and $(-3, 4)$. Find a and b.

48. An automobile radiator contains 15 liters of antifreeze and water. This mixture is 25% antifreeze. How much of this mixture should be drained and replaced with pure antifreeze so that the new mixture is 50% antifreeze?

49. Refer to Example 6 on page 420. Suppose the standard model chair requires $10 in materials and takes a worker 30 minutes to construct, and the deluxe model requires $12 in materials and takes 50 minutes to construct. How many chairs of each type can a worker produce in an 8-hour day if the total cost of materials used is $118?

Photo © Fredrik D. Bodin

50. A total of $100,000 is invested in three different securities for which the annual dividends are computed at 8%, 10%, and 14%. If the total income from these securities is $10,000, and the income from the 10% security is $800 more than from the 8% security, determine the amount invested in each security.

51. A dietitian prepares a meal of chicken, rice, and broccoli. Each ounce of chicken contains 110 calories (c), 5 grams (g) of fat, and 2 milligrams (mg) of iron; each ounce of rice contains 200 c, 6 g of fat, and 5 mg of iron; each ounce of broccoli contains 40 c, no fat, and 6 mg of iron. If the dietitian desires a meal containing 1200 c, 38 g of fat, and 47 mg of iron, how many ounces of each ingredient should he use?

52. Three pumps running together can fill a swimming pool in

4 hours. The first two pumps running together (with the third pump shut off) can fill the pool in 6 hours. If all three pumps run for one hour, and then the first pump is shut off, the other two pumps can finish filling the pool in 5 more hours. Find the time required for each pump, running alone, to fill the pool.

53. Recall that the equation of a parabola with axis parallel to the y-axis has the form $y = ax^2 + bx + c$. Given that $(-1, 8)$, $(0, 4)$, and $(1, 6)$ lie on such a parabola, find a, b, and c.

54. In electrical circuits, the formula

$$\frac{1}{R} = \frac{1}{R_1} + \frac{1}{R_2}$$

gives the total resistance R when two resistors R_1 and R_2 are connected in parallel. Given three resistors A, B, and C, suppose that the total resistance is 20 ohms if A and B are connected in parallel; 40 ohms if B and C are connected in parallel; and 30 ohms if A and C are connected in parallel. Find A, B, and C.

9.2 MATRICES AND SYSTEMS OF LINEAR EQUATIONS

The analysis of problems in a variety of disciplines often gives rise to systems containing many equations and many variables. For example, consider the following problem:

A company runs four oil refineries. Each refinery produces four petroleum-based products: kerosene, motor oil, diesel fuel, and gasoline. From one barrel of petroleum the first refinery produces 4 gallons of kerosene, 5 gallons of motor oil, 10 gallons of diesel fuel, and 15 gallons of gasoline. The other three refineries produce different amounts of these products as described in the table below.

Gallons of	Refinery 1 R_1	Refinery 2 R_2	Refinery 3 R_3	Refinery 4 R_4
Kerosene	4	5	4	3
Motor Oil	5	6	7	6
Diesel Fuel	10	8	10	9
Gasoline	15	12	9	15

If the company needs 298 gallons of kerosene, 450 gallons of motor oil, 682 gallons of diesel fuel, and 924 gallons of gasoline, how many barrels of petroleum are needed by each of these refineries to meet the demand?

If we let x_i be the number of barrels of petroleum used by the ith refinery R_i, then the total amount (in gallons) of each product produced is given by the following linear expressions:

$$\text{Kerosene: } 4x_1 + 5x_2 + 4x_3 + 3x_4$$
$$\text{Motor Oil: } 5x_1 + 6x_2 + 7x_3 + 6x_4$$
$$\text{Diesel Fuel: } 10x_1 + 8x_2 + 10x_3 + 9x_4$$
$$\text{Gasoline: } 15x_1 + 12x_2 + 9x_3 + 15x_4$$

To find the number of barrels of petroleum needed by each of the refineries to meet the demand for the four products, we equate each of the above linear expressions to the demand for the corresponding petroleum product. This gives us the system

$$\begin{cases} 4x_1 + 5x_2 + 4x_3 + 3x_4 = 298 \\ 5x_1 + 6x_2 + 7x_3 + 6x_4 = 450 \\ 10x_1 + 8x_2 + 10x_3 + 9x_4 = 682 \\ 15x_1 + 12x_2 + 9x_3 + 15x_4 = 924. \end{cases}$$

(1)

It is tedious to solve system (1) by the methods we have developed so far. However, the method of elimination, which was introduced in Section 9.1, leads to an efficient way to solve a system of linear equations such as system (1) using matrix methods.

Matrices

A **matrix** (plural: **matrices**) is a rectangular array of numbers written within brackets (or parentheses). Some examples of matrices are shown below.

$$\begin{bmatrix} 2 & -1 \\ 1 & 0 \end{bmatrix} \qquad \begin{bmatrix} 3 \\ 1 \\ 2 \\ 4 \end{bmatrix} \qquad [-1, 1, 2, 4, 5]$$

$$\begin{bmatrix} 1 & 0 & 0 \\ 0 & 1 & 0 \\ 0 & 0 & 1 \end{bmatrix} \qquad \begin{bmatrix} -2 & 1 & 3 & 0 \\ 0 & 1 & 2 & 1 \end{bmatrix}$$

Each number in a matrix is called an **element**, or **entry**, of the matrix. The elements that appear next to each other horizontally form a **row**, while those that appear next to each other vertically form a **column**. For example, the matrix

$$A = \begin{bmatrix} 1 & 2 & -3 & -7 \\ 0 & -1 & 2 & -4 \\ -6 & 5 & -2 & 6 \end{bmatrix}$$

consists of three rows and four columns. Each element of a matrix lies in a unique row and column. In matrix A above, for example, the element -3 lies in row 1 and column 3.

If the number of the row in which an element lies is the same as the number of the column in which it lies, the element is said to lie on the **main diagonal** of the matrix. In matrix A above, the main diagonal elements are 1, -1, and -2. Below, we identify the main diagonal elements of three matrices using broken diagonal lines.

$$\begin{bmatrix} 2 & 1 \\ 1 & 3 \end{bmatrix} \qquad \begin{bmatrix} 1 & 0 & 0 \\ 0 & 1 & 0 \\ 0 & 0 & 1 \end{bmatrix} \qquad \begin{bmatrix} -1 & 2 & -3 & -7 \\ 0 & -1 & 2 & -4 \\ -6 & 5 & -2 & 6 \end{bmatrix}$$

Main Diagonal Elements of a Matrix

Recall that the coefficients of the variables and the constant terms in a system of linear equations play a central role when the system is solved by the elimination method. An idea basic to the matrix methods in this section is that the operations performed in solving systems of linear equations do not affect the variables of the system, but affect only their coefficients and the constant terms. (The variables serve as placeholders only.) Hence, when we solve a system of linear equations, we can focus our attention on the coefficients of the variables and the constant terms in that system.

Consider the system

$$\begin{cases} x - 2y - z = 0 \\ 2x - 3y - 4z = 8 \\ 3x + y + 2z = -1. \end{cases} \qquad (2)$$

This system is said to be in **standard form** since the same variables appear in vertical columns and the constants appear on the right side of each equation.

The **coefficient matrix** of system (2) is the matrix

$$\begin{bmatrix} 1 & -2 & -1 \\ 2 & -3 & -4 \\ 3 & 1 & 2 \end{bmatrix},$$

which is the matrix whose elements are the coefficients of x, y, and z written in the same relative position as they appear in the system. The matrix

$$\begin{bmatrix} 1 & -2 & -1 & \bigg| & 0 \\ 2 & -3 & -4 & \bigg| & 8 \\ 3 & 1 & 2 & \bigg| & -1 \end{bmatrix},$$

formed from the coefficient matrix by listing the constant terms in an additional column on the right, is called the **augmented matrix** of the system.

Echelon Form of a Matrix

In Example 7 of Section 9.1, we used the elimination method to solve system (2). In this process we arrived at the equivalent system

$$\begin{cases} x - 2y - z = 0 \\ \quad\quad\ y - 2z = 8 \\ \quad\quad\quad\quad\ z = -3. \end{cases}$$

The latter system can be written as

$$\begin{cases} x - 2y - z = 0 \\ 0x + y - 2z = 8 \\ 0x + 0y + z = -3, \end{cases}$$

where zero is used as the coefficient of each variable that does not appear in an equation. The augmented matrix for the last system is

$$\begin{bmatrix} 1 & -2 & -1 & \bigg| & 0 \\ 0 & 1 & -2 & \bigg| & 8 \\ 0 & 0 & 1 & \bigg| & -3 \end{bmatrix}.$$

This matrix is said to be in *echelon form*.

In general, a matrix is said to be in echelon form provided it satisfies the following conditions:

Echelon Form of a Matrix

A matrix is in **echelon form** provided

(i)   the first nonzero entry (counting from left to right) in any nonzero row is 1;

(ii) the first nonzero entry in any nonzero row appears in a column to the right of the first nonzero entry of any previous row; and

(iii) any rows consisting entirely of zeros are found at the bottom of the matrix.

While each of the matrices

$$A = \begin{bmatrix} 1 & 0 \\ 0 & 1 \end{bmatrix}, \qquad B = \begin{bmatrix} 1 & 2 & 0 \\ 0 & 1 & 1 \\ 0 & 0 & 1 \end{bmatrix}, \qquad \text{and} \qquad C = \begin{bmatrix} 1 & 0 & 1 & -1 \\ 0 & 0 & 1 & 4 \\ 0 & 0 & 0 & 1 \end{bmatrix}$$

is in echelon form, none of the following matrices is in echelon form:

$$D = \begin{bmatrix} 0 & 1 \\ 1 & 0 \end{bmatrix}$$
The first nonzero entry in the second row appears in a column to the *left* of the first nonzero entry of the first row.

$$E = \begin{bmatrix} 1 & 2 & 0 \\ 0 & 0 & 0 \\ 0 & 1 & 1 \end{bmatrix}$$
Any row consisting entirely of zeros must be at the bottom of the matrix.

$$F = \begin{bmatrix} 1 & -1 & 1 & 4 \\ 0 & 1 & 1 & 0 \\ 0 & 0 & 2 & 5 \end{bmatrix}$$
The first nonzero entry of row 3 is not 1.

Matrices A, B, and C above illustrate an important property of matrices in echelon form: namely, any column containing the first (counting from left to right) nonzero entry of a row contains only zero entries below this element.

Starting with a system of linear equations, it is always possible to find an equivalent system whose augmented matrix is in echelon form. Recall from Theorem 9.1 that for a given system of equations, an equivalent system is obtained when any of the following operations is performed:

  (i)   two equations are interchanged;

 (ii)   an equation is multiplied by a nonzero constant; or

(iii)   a constant multiple of one equation is added to another.

Since the rows of the augmented matrix of a system of linear equations represent the equations in the system, the corresponding operations for matrices are called **elementary row operations**. Two augmented matrices A and B are said to be **row-equivalent**, denoted by $A \sim B$, if they are the augmented matrices of equivalent systems of equations. It follows immediately from Theorem 9.1 that a row-equivalent matrix is obtained when an elementary row operation is performed on an augmented matrix.

THEOREM 9.2

For a given augmented matrix, a row-equivalent matrix is obtained if:

 (i) two rows are interchanged;

 (ii) a row is multiplied by a nonzero constant; or

(iii) a constant multiple of one row is added to another row.

Using Theorem 9.2, it can be shown that every augmented matrix is row-equivalent to an augmented matrix in echelon form.

Gaussian Elimination

We now use Theorem 9.2 to find the solution set for system (2) on page 426. Compare the computations here with those given in the solution of Example 7 of Section 9.1.

We begin by writing the augmented matrix for the system:

$$\left[\begin{array}{ccc|c} 1 & -2 & -1 & 0 \\ 2 & -3 & -4 & 8 \\ 3 & 1 & 2 & -1 \end{array}\right]. \tag{3}$$

Our objective is to use Theorem 9.2 to transform matrix (3) into a row-equivalent matrix in echelon form. We will then write the system of equations corresponding to this matrix and use back substitution to find the solution set.

The first nonzero element of the first row of matrix (3) is 1, which occurs in the first column. We must now get a 0 as the first element in both the second and third rows. To get 0 as the first element in the second row, we multiply row 1 by -2 and add the result to row 2 (row 1 remains unchanged).

$$\left[\begin{array}{ccc|c} 1 & -2 & -1 & 0 \\ 0 & 1 & -2 & 8 \\ 3 & 1 & 2 & -1 \end{array}\right]. \tag{4}$$

To get a 0 as the first element of the third row, we multiply row 1 by -3 and add to row 3 (again, row 1 remains unchanged).

$$\left[\begin{array}{ccc|c} 1 & -2 & -1 & 0 \\ 0 & 1 & -2 & 8 \\ 0 & 7 & 5 & -1 \end{array}\right]. \tag{5}$$

In matrix (5), observe that the first nonzero element of the second row is 1, which occurs in the second column. Hence, the first nonzero element of the second row occurs in a column to the right of the first nonzero element of the first row. To get a 0 in the third row and second column, we multiply the second row by -7 and add the result to the third row (row 2 remains unchanged).

$$\left[\begin{array}{ccc|c} 1 & -2 & -1 & 0 \\ 0 & 1 & -2 & 8 \\ 0 & 0 & 19 & -57 \end{array}\right]. \tag{6}$$

To obtain 1 as the first nonzero element in row 3 of matrix (6), we multiply row 3 by $\frac{1}{19}$ obtaining the matrix

$$\left[\begin{array}{ccc|c} 1 & -2 & -1 & 0 \\ 0 & 1 & -2 & 8 \\ 0 & 0 & 1 & -3 \end{array}\right]. \tag{7}$$

Matrix (7) is in echelon form and, by Theorem 9.2, is row-equivalent to matrix (3). The system of equations corresponding to matrix (7) is

$$\begin{cases} x - 2y - z = 0 \\ y - 2z = 8 \\ z = -3. \end{cases}$$

Back substitution leads to the solution set $\{(1, 2, -3)\}$.

The process of using matrices to solve a system of linear equations, as outlined in this section, is called **Gaussian elimination** in honor of German mathematician Karl Friedrich Gauss.

We now summarize the guidelines for using Gaussian elimination with back substitution to solve a system of linear equations.

Gaussian Elimination with Back Substitution

(i)   Write the augmented matrix of the system of linear equations.

(ii) Perform elementary row operations on the augmented matrix to obtain a row-equivalent augmented matrix in echelon form.

(iii) Write the system of linear equations corresponding to the augmented matrix in echelon form.

(iv) Use back substitution to find the solution set.

In the examples that follow, we solve systems of linear equations using Gaussian elimination with back substitution. When using Gaussian elimination, it is important that the system of equations be in standard form. This allows us to transform the augmented matrix of the given system into an equivalent matrix in echelon form by working systematically in columns, from left to right, using elementary row operations.

For convenience we use the following symbols to represent the elementary row operations:

1. $R_i \leftrightarrow R_j$ Interchange row i and row j

2. $kR_i \rightarrow R_i$ Multiply row i by the constant k

3. $kR_j + R_i \rightarrow R_i$ Multiply row j by the constant k and add the result to row i

For example, the notation

$$
\begin{matrix} & (A) & \\ \begin{bmatrix} 2 & 1 & 3 \\ 4 & 1 & -1 \\ 0 & -5 & 6 \end{bmatrix} \end{matrix}
\quad -2R_1 + R_2 \rightarrow R_2 \quad
\begin{matrix} & (B) & \\ \begin{bmatrix} 2 & 1 & 3 \\ 0 & -1 & -7 \\ 0 & -5 & 6 \end{bmatrix} \end{matrix}
$$

means that we obtain the row-equivalent matrix B by multiplying row 1 of matrix A by -2 and adding the result to row 2 of matrix A.

E X A M P L E 10

Use Gaussian elimination to find the solution set of the system

$$\begin{cases} 2x - 5y + z = 0 \\ x - y + 2z = 3 \\ 4x - 6y + 3z = -1. \end{cases}$$

Solution   We write the augmented matrix of the system and then use elementary row operations to reduce it to a row-equivalent matrix in echelon form.

$$
\begin{bmatrix} 2 & -5 & 1 & | & 0 \\ 1 & -1 & 2 & | & 3 \\ 4 & -6 & 3 & | & -1 \end{bmatrix}
\quad R_1 \leftrightarrow R_2 \quad
\begin{bmatrix} 1 & -1 & 2 & | & 3 \\ 2 & -5 & 1 & | & 0 \\ 4 & -6 & 3 & | & -1 \end{bmatrix}
$$

$$-2R_1 + R_2 \to R_2 \quad \begin{bmatrix} 1 & -1 & 2 & | & 3 \\ 0 & -3 & -3 & | & -6 \\ 4 & -6 & 3 & | & -1 \end{bmatrix}$$

$$-4R_1 + R_3 \to R_3 \quad \begin{bmatrix} 1 & -1 & 2 & | & 3 \\ 0 & -3 & -3 & | & -6 \\ 0 & -2 & -5 & | & -13 \end{bmatrix}$$

$$-\tfrac{1}{3}R_2 \to R_2 \quad \begin{bmatrix} 1 & -1 & 2 & | & 3 \\ 0 & 1 & 1 & | & 2 \\ 0 & -2 & -5 & | & -13 \end{bmatrix}$$

$$2R_2 + R_3 \to R_3 \quad \begin{bmatrix} 1 & -1 & 2 & | & 3 \\ 0 & 1 & 1 & | & 2 \\ 0 & 0 & -3 & | & -9 \end{bmatrix}$$

$$-\tfrac{1}{3}R_3 \to R_3 \quad \begin{bmatrix} 1 & -1 & 2 & | & 3 \\ 0 & 1 & 1 & | & 2 \\ 0 & 0 & 1 & | & 3 \end{bmatrix}$$

The last matrix is in echelon form and corresponds to the system

$$\begin{cases} x - y + 2z = 3 \\ \quad\;\; y + z = 2 \\ \qquad\quad\;\; z = 3. \end{cases}$$

From the third equation we see that $z = 3$. Substituting 3 for z in the second equation yields

$$y + 3 = 2, \quad \text{or} \quad y = -1.$$

Finally, substituting -1 for y and 3 for z in the first equation gives

$$x - (-1) + 2(3) = 3, \quad \text{or} \quad x = -4.$$

Hence, the solution set is $\{(-4, -1, 3)\}$.

EXAMPLE 11

Find the solution set for the following system using Gaussian elimination.

$$\begin{cases} x + y - z = -1 \\ x + 2y + 3z = 3 \\ 3x + 2y - 7z = -7. \end{cases} \tag{8}$$

Solution We proceed as follows:

$$\begin{bmatrix} 1 & 1 & -1 & | & -1 \\ 1 & 2 & 3 & | & 3 \\ 3 & 2 & -7 & | & -7 \end{bmatrix} \quad -R_1 + R_2 \to R_2 \quad \begin{bmatrix} 1 & 1 & -1 & | & -1 \\ 0 & 1 & 4 & | & 4 \\ 3 & 2 & -7 & | & -7 \end{bmatrix}$$

$$-3R_1 + R_3 \to R_3 \quad \begin{bmatrix} 1 & 1 & -1 & | & -1 \\ 0 & 1 & 4 & | & 4 \\ 0 & -1 & -4 & | & -4 \end{bmatrix}$$

$$R_2 + R_3 \to R_3 \quad \begin{bmatrix} 1 & 1 & -1 & | & -1 \\ 0 & 1 & 4 & | & 4 \\ 0 & 0 & 0 & | & 0 \end{bmatrix}$$

The last matrix is in echelon form and the corresponding system of equations is

$$\begin{cases} x + y - z = -1 \\ y + 4z = 4 \\ 0 = 0. \end{cases} \qquad (9)$$

Note that the third equation in system (9) gives us no information concerning the variables. Hence, system (9) is equivalent to the system

$$\begin{cases} x + y - z = -1 \\ y + 4z = 4. \end{cases} \qquad (10)$$

Solving the second equation in system (10) for y in terms of z, we get

$$y = 4 - 4z.$$

Substituting $4 - 4z$ for y in the first equation yields

$$x + (4 - 4z) - z = -1, \qquad \text{or} \\ x = -5 + 5z.$$

Finally, letting $z = t$, we see that every ordered triple of the form $(-5 + 5t, 4 - 4t, t)$, where t is a real number, is a solution of the system. Hence, the solution set is $\{(-5 + 5t, 4 - 4t, t) \mid t \in \mathbb{R}\}$, and system (8) is consistent and dependent.

E X A M P L E 12

Determine the solution set of the following system using Gaussian elimination.

$$\begin{cases} x - 2y + 3z = 1 \\ -2x + 3y - 4z = 2 \\ x - 3y + 5z = 10. \end{cases}$$

Solution Using Gaussian elimination, we proceed as follows:

$$\begin{bmatrix} 1 & -2 & 3 & | & 1 \\ -2 & 3 & -4 & | & 2 \\ 1 & -3 & 5 & | & 10 \end{bmatrix} \quad \underset{\sim}{2R_1 + R_2 \to R_2} \quad \begin{bmatrix} 1 & -2 & 3 & | & 1 \\ 0 & -1 & 2 & | & 4 \\ 1 & -3 & 5 & | & 10 \end{bmatrix}$$

$$\underset{\sim}{-R_1 + R_3 \to R_3} \quad \begin{bmatrix} 1 & -2 & 3 & | & 1 \\ 0 & -1 & 2 & | & 4 \\ 0 & -1 & 2 & | & 9 \end{bmatrix}$$

$$\underset{\sim}{-R_2 \to R_2} \quad \begin{bmatrix} 1 & -2 & 3 & | & 1 \\ 0 & 1 & -2 & | & -4 \\ 0 & -1 & 2 & | & 9 \end{bmatrix}$$

$$\underset{\sim}{R_2 + R_3 \to R_3} \quad \begin{bmatrix} 1 & -2 & 3 & | & 1 \\ 0 & 1 & -2 & | & -4 \\ 0 & 0 & 0 & | & 5 \end{bmatrix}$$

Even though the last matrix is not in echelon form, we stop the procedure here since the equation corresponding to the third row of the last matrix is

$$0 = 5. \qquad \text{False}$$

This tells us that the given system is inconsistent, and its solution set is $\varnothing$.

In general, if any row of an augmented matrix has its only nonzero element in the last position, the system of linear equations corresponding to this matrix is inconsistent.

Up to this point, most of the systems of equations we have investigated contained the same number of variables as equations. It is not uncommon for a system of equations to contain more variables than equations or more equations than variables. In either case, Gaussian elimination can still be used as shown below.

EXAMPLE 13

Determine if the following system is consistent or inconsistent. If it is consistent, find the solution set.

$$\begin{cases} x - y = 4 \\ x + 3y = 8 \\ x - 5y = 12. \end{cases}$$

Solution Using Gaussian elimination, we have

$$\begin{bmatrix} 1 & -1 & | & 4 \\ 1 & 3 & | & 8 \\ 1 & -5 & | & 12 \end{bmatrix} \quad -R_1 + R_2 \to R_2 \quad \begin{bmatrix} 1 & -1 & | & 4 \\ 0 & 4 & | & 4 \\ 1 & -5 & | & 12 \end{bmatrix}$$

$$-R_1 + R_3 \to R_3 \quad \begin{bmatrix} 1 & -1 & | & 4 \\ 0 & 4 & | & 4 \\ 0 & -4 & | & 8 \end{bmatrix}$$

$$\tfrac{1}{4}R_2 \to R_2 \quad \begin{bmatrix} 1 & -1 & | & 4 \\ 0 & 1 & | & 1 \\ 0 & -4 & | & 8 \end{bmatrix}$$

$$4R_2 + R_3 \to R_3 \quad \begin{bmatrix} 1 & -1 & | & 4 \\ 0 & 1 & | & 1 \\ 0 & 0 & | & 12 \end{bmatrix}$$

Corresponding to the third row of the last matrix is the equation $0 = 12$, which is false. Therefore, this system is inconsistent. ———————

EXAMPLE 14

Determine by inspection whether the following system is consistent or inconsistent. If it is consistent, find the solution set.

$$\begin{cases} x - y - z = 0. \\ x + 2y + 2z = 0. \end{cases}$$

Solution Notice that the constant term in each of the equations in this system is 0. Hence, the ordered triple $(0, 0, 0)$ satisfies each equation and the system is consistent. If $(0, 0, 0)$ is the only solution, the system is independent. However, the system may have other solutions. We now use Gaussian elimination to determine the solution set.

$$\begin{bmatrix} 1 & -1 & -1 & | & 0 \\ 1 & 2 & 2 & | & 0 \end{bmatrix} \quad -R_1 + R_2 \to R_2 \quad \begin{bmatrix} 1 & -1 & -1 & | & 0 \\ 0 & 3 & 3 & | & 0 \end{bmatrix}$$

$$\tfrac{1}{3}R_2 \to R_2 \quad \begin{bmatrix} 1 & -1 & -1 & | & 0 \\ 0 & 1 & 1 & | & 0 \end{bmatrix}.$$

The system of equations corresponding to the last matrix is

$$\begin{cases} x - y - z = 0 \\ y + z = 0. \end{cases}$$

Solving the last equation for y in terms of z, we have

$$y = -z.$$

Substituting $-z$ for y in the first equation and solving for x yields

$$x - (-z) - z = 0, \quad \text{or} \quad x = 0.$$

Finally, if we let $z = t$ where t is any real number, the solution set for this system can be expressed as $S = \{(0, -t, t) \mid t \in \mathbb{R}\}$. Therefore the system is consistent and dependent. _____

Any system of linear equations in which the constant term in each equation is 0, such as the system in Example 14, is called **homogeneous**. Every homogeneous system is consistent, and the solution in which each variable is 0 is called the **trivial solution**. Any other solution is called a **nontrivial solution**.

Gauss-Jordan Elimination

A matrix in echelon form is said to be in **reduced echelon form** provided that any column containing the first nonzero entry of any row has zero for *all* other entries in that column. For example, the matrix

$$A = \begin{bmatrix} 1 & 0 & 2 & 0 & 0 \\ 0 & 1 & -1 & 0 & 3 \\ 0 & 0 & 0 & 1 & 1 \end{bmatrix}$$

is in reduced echelon form.

The method of solving a system of linear equation by transforming the augmented matrix into reduced echelon form is called the **Gauss-Jordan elimination method** in honor of Karl Friedrich Gauss and German geodesist Wilhelm Jordan (1842–1899). In Section 9.4, we will see that Gauss-Jordan elimination can be used to find the inverse of a matrix (if it exists).

EXAMPLE 15

Use Gauss-Jordan elimination to find the solution set of the following system of linear equations:

$$\begin{cases} x + 8z = 4 \\ x + 2y + 3z = -2 \\ 2x + 5y + 3z = 1. \end{cases}$$

Solution   We use elementary row operations to transform the augmented matrix of this system to reduced echelon form as follows:

$$\begin{bmatrix} 1 & 0 & 8 & | & 4 \\ 1 & 2 & 3 & | & -2 \\ 2 & 5 & 3 & | & 1 \end{bmatrix} \quad \begin{array}{c} -R_1 + R_2 \to R_2 \\ \curvearrowright \end{array} \quad \begin{bmatrix} 1 & 0 & 8 & | & 4 \\ 0 & 2 & -5 & | & -6 \\ 2 & 5 & 3 & | & 1 \end{bmatrix}$$

$$\begin{array}{c} -2R_1 + R_3 \to R_3 \\ \curvearrowright \end{array} \quad \begin{bmatrix} 1 & 0 & 8 & | & 4 \\ 0 & 2 & -5 & | & -6 \\ 0 & 5 & -13 & | & -7 \end{bmatrix}$$

$$\frac{1}{2}R_2 \to R_2 \quad \begin{bmatrix} 1 & 0 & 8 & | & 4 \\ 0 & 1 & -\frac{5}{2} & | & -3 \\ 0 & 5 & -13 & | & -7 \end{bmatrix}$$

$$-5R_2 + R_3 \to R_3 \quad \begin{bmatrix} 1 & 0 & 8 & | & 4 \\ 0 & 1 & -\frac{5}{2} & | & -3 \\ 0 & 0 & -\frac{1}{2} & | & 8 \end{bmatrix}$$

$$-2R_3 \to R_3 \quad \begin{bmatrix} 1 & 0 & 8 & | & 4 \\ 0 & 1 & -\frac{5}{2} & | & -3 \\ 0 & 0 & 1 & | & -16 \end{bmatrix}$$

$$\frac{5}{2}R_3 + R_2 \to R_2 \quad \begin{bmatrix} 1 & 0 & 8 & | & 4 \\ 0 & 1 & 0 & | & -43 \\ 0 & 0 & 1 & | & -16 \end{bmatrix}$$

$$-8R_3 + R_1 \to R_1 \quad \begin{bmatrix} 1 & 0 & 0 & | & 132 \\ 0 & 1 & 0 & | & -43 \\ 0 & 0 & 1 & | & -16 \end{bmatrix}$$

The last matrix is in reduced echelon form. By inspection, we see that $x = 132$, $y = -43$, and $z = -16$. Hence, the solution set is $\{(132, -43, -16)\}$.

EXAMPLE 16

To make 500 pounds of fertilizer that is 20% nitrogen, 10% phosphorus, and 10% potassium, a horticulturist has available three mixtures A, B, and C whose compositions are shown in the table below. How many pounds of each mixture should she use?

	Nitrogen	Phosphorus	Potassium
A	35%	4%	4%
B	10%	20%	8%
C	10%	8%	20%

Solution   We begin by letting

  x denote the number of pounds of mixture A needed,

  y denote the number of pounds of mixture B needed, and

  z denote the number of pounds of mixture C needed.

The total amount of nitrogen is 20% of 500 pounds, or 100 pounds. Since mixture A is 35% nitrogen, $0.35x$ pounds of nitrogen is provided by that mixture. Similarly $0.1y$ pounds of nitrogen and $0.1z$ pounds of nitrogen are provided by mixtures B and C, respectively. Hence

$$0.35x + 0.1y + 0.1z = 100. \qquad \text{Nitrogen equation}$$

Reasoning in a similar manner, we have

$$0.04x + 0.2y + 0.08z = 50, \qquad \text{Phosphorus equation}$$
$$0.04x + 0.08y + 0.2z = 50. \qquad \text{Potassium equation}$$

Thus, we see that a system of three linear equations in x, y, and z arises in the solution of this problem. Using Gaussian elimination, it can be shown that the solution of the above mentioned system is $\{(200, 150, 150)\}$. Hence, the horticulturist should use 200 pounds of mixture A and 150 pounds each of mixtures B and C.

EXERCISES 9.2

In Exercises 1–6, write the coefficient matrix and the augmented matrix for each system.

1. $\begin{cases} x - y + z = 4 \\ 2x - 3z = -8 \\ x + y - 5z = 0 \end{cases}$ **2.** $\begin{cases} -2u - 3w = 0 \\ 4u + v + w = 5 \\ -u - w = 6 \end{cases}$

3. $\begin{cases} 6s - 4u + 5v = 10 \\ 4t - v = 0 \\ -2s + 5u = 3 \\ s - t + u - 3v = -4 \end{cases}$

4. $\begin{cases} s - 2t + u - v = 0 \\ t + u + 4v = 0 \\ s - u = 0 \\ t - v = 0 \\ s - t + u + v = 0 \end{cases}$

5. $\begin{cases} 6u - 8v + 10w - 4x = 5 \\ 4u - w - 6x = 10 \end{cases}$

6. $\begin{cases} x - 2y = 5 \\ 2x + y = 6 \\ -x - y = 4 \\ 4x + y = 0 \end{cases}$

In Exercises 7–12, determine if the given matrix is in echelon form. If the matrix fails to be in echelon form, explain why.

7. $\begin{bmatrix} 1 & 2 & 3 & | & 0 \\ 0 & 0 & 2 & | & 1 \\ 0 & 0 & 0 & | & 4 \end{bmatrix}$ **8.** $\begin{bmatrix} 1 & 0 & 0 & 1 & 4 & | & 0 \\ 0 & 0 & 1 & 0 & 0 & | & 2 \\ 0 & 1 & 0 & 1 & 1 & | & 4 \end{bmatrix}$

9. $\begin{bmatrix} 0 & 0 & 1 & 0 & | & 2 \\ 0 & 0 & 0 & 1 & | & 0 \\ 0 & 0 & 0 & 0 & | & 0 \end{bmatrix}$ **10.** $\begin{bmatrix} 1 & 2 & 3 & 4 & | & 5 \\ 0 & 0 & 0 & 0 & | & -1 \\ 0 & 1 & 1 & 0 & | & 4 \end{bmatrix}$

11. $\begin{bmatrix} 1 & 0 & 0 & 0 & | & 2 \\ 0 & 1 & 1 & -1 & | & -5 \\ 0 & 0 & 0 & 0 & | & 0 \\ 0 & 0 & 0 & 0 & | & 1 \end{bmatrix}$

12. $\begin{bmatrix} 0 & 0 & 0 & 0 & 1 & | & 2 \\ 0 & 0 & 0 & 1 & 0 & | & -1 \\ 0 & 0 & 0 & 0 & 2 & | & 0 \end{bmatrix}$

In Exercises 13–18, use elementary row operations to write the matrix in echelon form.

13. $\begin{bmatrix} 2 & -1 & | & -4 \\ 1 & 2 & | & 3 \\ 3 & -1 & | & -1 \end{bmatrix}$ **14.** $\begin{bmatrix} 2 & -1 & 3 & | & 4 \\ 1 & 2 & 1 & | & 2 \\ 0 & -5 & 1 & | & 0 \end{bmatrix}$

15. $\begin{bmatrix} 2 & 1 & -3 & | & 0 \\ 3 & 2 & -4 & | & 2 \\ 1 & -1 & -3 & | & -6 \end{bmatrix}$

16. $\begin{bmatrix} 1 & 3 & -1 & 2 & | & 1 \\ -2 & -6 & 2 & 0 & | & 2 \\ 0 & 0 & 1 & -1 & | & 7 \\ 2 & 6 & -2 & 2 & | & 0 \end{bmatrix}$

17. $\begin{bmatrix} 1 & 1 & 3 & -4 & | & 1 \\ -5 & 2 & -1 & -3 & | & 0 \\ 7 & -2 & -6 & 1 & | & 5 \end{bmatrix}$

18. $\begin{bmatrix} 1 & 1 & 2 & 1 & 5 & | & 2 \\ 2 & 0 & 1 & 0 & 4 & | & 3 \\ 1 & 1 & 0 & 5 & 2 & | & -1 \\ 6 & -1 & -1 & 2 & 4 & | & -3 \\ -1 & 1 & -1 & 0 & 1 & | & 5 \end{bmatrix}$

In Exercises 19–24, find the solution set for the system of linear equations corresponding to each augmented matrix in echelon form.

19. $\begin{bmatrix} 1 & 0 & 0 & | & 2 \\ 0 & 1 & 0 & | & 0 \\ 0 & 0 & 1 & | & 1 \end{bmatrix}$ **20.** $\begin{bmatrix} 1 & 1 & 4 & | & 1 \\ 0 & 1 & 1 & | & 1 \\ 0 & 0 & 1 & | & 2 \end{bmatrix}$

21. $\begin{bmatrix} 1 & 0 & | & 1 \\ 0 & 1 & | & 1 \\ 0 & 0 & | & 2 \end{bmatrix}$ **22.** $\begin{bmatrix} 1 & 2 & | & 1 \\ 0 & 1 & | & -2 \\ 0 & 0 & | & 2 \end{bmatrix}$

23. $\begin{bmatrix} 1 & 1 & 1 & 0 & | & 1 \\ 0 & 1 & 1 & 1 & | & 0 \\ 0 & 0 & 1 & 2 & | & 1 \end{bmatrix}$ **24.** $\begin{bmatrix} 1 & 0 & 1 & 0 & | & 3 \\ 0 & 1 & 0 & 1 & | & 2 \\ 0 & 0 & 1 & 1 & | & -4 \end{bmatrix}$

In Exercises 25–50, use Gaussian elimination to find the solution set for each system. Indicate whether the system is consistent and independent, consistent and dependent, or inconsistent. If the system is consistent and dependent, express the solution set using the parameter t.

25. $\begin{cases} x - y - z = 0 \\ 2x + y - 3z = 1 \\ -x + 4y + 5z = 11 \end{cases}$ **26.** $\begin{cases} x + z = 3 \\ y - z = -2 \\ x - y = 1 \end{cases}$

27. $\begin{cases} 2x + 2y + z = 1 \\ 4x + 4y - 3z = 1 \\ 6x + 2y - 5z = 1 \end{cases}$

28. $\begin{cases} -3x - y + 2z = -1 \\ x - z = 0 \\ -4x + 3y + z = -3 \end{cases}$

29. $\begin{cases} -x + y - z = -6 \\ x + 3y + 4z = 21 \\ 2x - y - 2z = -8 \end{cases}$

30. $\begin{cases} -x - y + z = 6 \\ 3x - 2y + 4z = 4 \\ 5x + 3y - 2z = 21 \end{cases}$

31. $\begin{cases} x + y + z = 2 \\ x + 3y - 3z = 5 \\ x + 2y - z = 4 \end{cases}$

32. $\begin{cases} 2x - 5y + z = 1 \\ 4x - 23y - 3z = 5 \\ x + 4y + 3z = 4 \end{cases}$

33. $\begin{cases} 2x - y - z = 1 \\ -3x + 4y - 2z = 2 \\ 11y - z = 1 \end{cases}$

34. $\begin{cases} x + 2y + 3z = 5 \\ 3x - y - 2z = 0 \\ 2x + 2y - z = 3 \end{cases}$

35. $\begin{cases} 2x - y - 5z = -5 \\ 2x + 2y - z = -3 \\ -2x + 3y + 2z = 12 \end{cases}$

36. $\begin{cases} 3x + y + z = -7 \\ x - y + 4z = -19 \\ x + 3y - 2z = 21 \end{cases}$

37. $\begin{cases} \frac{1}{2}x + y + z = -\frac{1}{4} \\ x - 2y + 4z = -3 \\ -\frac{1}{10}x + y - \frac{4}{5}z = 1 \end{cases}$

38. $\begin{cases} \frac{3}{2}x - \frac{1}{2}y + 2z = 2 \\ \frac{1}{6}x + \frac{2}{3}y - z = \frac{1}{2} \\ \frac{3}{2}x - \frac{8}{3}y + \frac{17}{3}z = \frac{25}{6} \end{cases}$

39. $\begin{cases} x + y - 3z = 0 \\ x - y + 2z = 3 \\ x - y - z = 12 \end{cases}$

40. $\begin{cases} x - y - z = 0 \\ 2x + z = 0 \\ x + y + 2z = 0 \end{cases}$

41. $\begin{cases} x - 2y + 3z = -4 \\ 3x - 6y + 10z = 14 \\ 5x - 8y + 19z = -21 \\ 2x - 4y + 7z = -9 \end{cases}$

42. $\begin{cases} 2x - y + 3z = 2 \\ x - 2y - z = 3 \\ 3x + 5y - 2z = 2 \\ 4x + 3y + z = 6 \end{cases}$

43. $\begin{cases} s - 3u + 4v + 2w = 1 \\ 2s - 3u + 5v - 2w = -1 \\ -s + 2u - 3v + w = 4 \end{cases}$

44. $\begin{cases} s - u - v + w = 0 \\ 3s + 2u - v + 2w = -2 \\ -s - u + 4v + 3w = 1 \end{cases}$

45. $\begin{cases} s + v = 0 \\ -2s - u - v + w = 0 \\ s - 2u + v = -2 \\ u - v + w = 4 \end{cases}$

46. $\begin{cases} p - q - r + s = 0 \\ p + 2r - 3s = 6 \\ 2q + 4s = -6 \\ q - r - s = -1 \end{cases}$

47. $\begin{cases} x - y - 3z = 5 \\ -3x + 2y + z = -9 \end{cases}$

48. $\begin{cases} 6x + 5y - 2z = -2 \\ 5x - y - z = 5 \end{cases}$

49. $\begin{cases} u - v = 1 \\ 4u - 5v = 7 \\ 5u - 2v = 4 \end{cases}$

50. $\begin{cases} 3r - 2s = 2 \\ -r + 4s = -3 \\ 5r - 6s = -2 \end{cases}$

In Exercises 51–56, use Gauss-Jordan elimination to find the solution set for each system. Indicate whether the system is consistent and independent, consistent and dependent, or inconsistent. If the system is consistent and dependent, express the solution set using the parameter t.

51. $\begin{cases} x + 2y + z = 8 \\ -x + 3y - 2z = 1 \\ 3x + 4y - 7z = 10 \end{cases}$

52. $\begin{cases} 2x - y + z = 1 \\ x + y + z = 1 \\ -2x + 4y - z = 5 \end{cases}$

53. $\begin{cases} x + y = 2 \\ y + z = 3 \\ x - z = 1 \end{cases}$

54. $\begin{cases} p + q + r - s = 1 \\ 2p - 3q + r - 2s = -4 \\ 3p + q - 2r + s = 4 \end{cases}$

55. $\begin{cases} 3s - u - v + 4w = 2 \\ 6s + 3u - v + 4w = 3 \\ 9s + u - 8w = 6 \end{cases}$

56. $\begin{cases} 2r - 3s + u - v + w = 0 \\ 4r - 6s + 2u - 3v - w = -5 \\ -2r + 3s - 2u + 2v - w = 3 \end{cases}$

57. Recall that the defining equation for a quadratic function f has the form $f(x) = ax^2 + bx + c$ where $a \neq 0$. Determine the defining equation for the quadratic function f given that $f(-3) = -15$, $f(1) = -3$, and $f(2) = -5$.

58. Prove that there is no quadratic function f such that $f(-2) = -14$, $f(3) = 16$, and $f(5) = 28$.

59. In calculus, it is shown that an equation of the form
$$z = ax + by + c \qquad (1)$$
is the equation of a plane. Using equation (1) above, find an equation of the plane determined by the points $(1, 1, -1)$, $(4, 0, -2)$, and $(6, -2, 0)$.

60. Using equation (1) given in Exercise 59, find an equation of the plane determined by the points $(5, 1, -2)$, $(-2, 3, 0)$, and $(1, 0, 2)$.

61. Consider the system
$$\begin{cases} x - y - 3z = k \\ 2x - 3y + 4z = 0 \\ 3x - 4y + z = 1. \end{cases}$$

a. Find a value of k for which the system has infinitely many solutions. Give two of the solutions.

b. Is there a value of k for which the system has a unique solution? Justify your answer.

62. Find all the values of k for which the following system has a unique solution:

$$\begin{cases} x + y + z = k \\ kx + y + 2z = 2 \\ x - ky + z = 4. \end{cases}$$

63. A small company borrowed $500,000 for capital improvements. Some of the money was borrowed at an (annual) interest rate of 8%, some at a rate of 10%, and some at 12%. How much was borrowed at each rate, if the annual interest was $48,000 and the amount borrowed at 8% was the amount borrowed at 10%?

64. Find the solution of the problem concerning the oil company posed at the beginning of this section on page 425.

65. The perimeter of a certain triangle is 74 inches. Find the length of each side given that the longest side is 6 inches longer than the next longest and 10 inches longer than the shortest side.

66. The following table shows the protein, fat, and carbohydrate content of three food groups:

GRAMS PER 100 GRAMS

Food Groups	Protein	Fat	Carbohydrates
I	12	6	8
II	5	5	2
III	6	1	4

A nutritionist wishes to prepare two diets from these three food groups. The first diet must contain 47 grams of protein, 20 grams of fat, and 30 grams of carbohydrates. The second diet requires 53 grams of protein, 25 grams of fat, and 34 grams of carbohydrates. How many grams of each food group are required in the first diet, and how many grams of each food group are required in the second diet?

67. In the downtown section of a certain city, two sets of one-way streets intersect as shown in the accompanying figure.

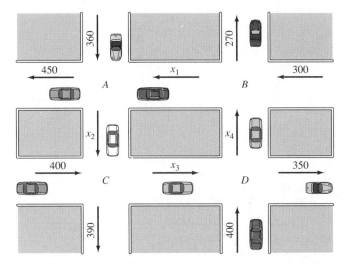

The numbers in the figure denote the average number of vehicles per hour that travel in the direction shown. At each intersection the number of automobiles entering must be the same as the number leaving. For example, at intersection A,

$$x_1 + 360 = x_2 + 450.$$

a. Find a system of linear equations in the variables x_1, x_2, x_3, and x_4 that describes the traffic flow at the four intersections in the figure.

b. Find the values of x_1, x_2, x_3, and x_4 in part (a).

68. A company has a taxable income of $5,000,000. The federal income tax is 30% of the portion of the income that remains after the state and local taxes have been deducted. The state income tax is 10% of the portion of income that remains after the federal and local taxes have been deducted, and the local tax is 5% of the portion of income that remains after the federal and state taxes have been deducted. Find the company's federal, state, and local income taxes.

9.3 MATRIX ALGEBRA

In Section 9.2, we found that matrices afford us an efficient way of solving a system of linear equations. The theory of matrices is powerful, and its applications are numerous in such fields as mechanics, aerodynamics, astronomy, nuclear physics, and economics. For this reason, we shall develop more matrix algebra in this section than is needed to solve systems of linear equations.

Let us begin by considering the matrix

$$A = \begin{bmatrix} -1 & 1 & 0 \\ 0 & 1 & 3 \end{bmatrix}.$$

Notice that matrix A has two rows and three columns. We say that A is a 2×3 (read "2 by 3") matrix. In general, a matrix having m rows and n columns is said to have **dimension $m \times n$**.

A matrix having only one row is called a **row matrix**, while a matrix having only one column is called a **column matrix**. For example, the 1×5 matrix

$$A = [2 \quad -1 \quad 4 \quad 6 \quad 9]$$

is a row matrix and the 3×1 matrix

$$B = \begin{bmatrix} 1 \\ 1 \\ 2 \end{bmatrix}$$

is a column matrix. A matrix that has the same number of rows as columns, such as the matrix

$$C = \begin{bmatrix} 1 & -1 \\ 4 & 2 \end{bmatrix}$$

is called a **square matrix**. Thus, a square matrix has dimension $n \times n$ for some positive integer n and is said to be of **order n**.

Each element of a matrix A lies in a particular row and a particular column of A. Consider, for example, the 2×4 matrix

$$A = \begin{bmatrix} 2 & 1 & 0 & -2 \\ -3 & 6 & 5 & -1 \end{bmatrix}.$$

Observe that 5 is the element of A that lies in the second row and third column of A. We denote the element 5 by the symbol a_{23}. Notice that we use a capital letter to denote a matrix and the corresponding lowercase letter to denote its elements. In the notation a_{ij}, the first subscript gives the row in which the element lies, and the second subscript gives the column in which it lies. It is often convenient to denote an $m \times n$ matrix A by the symbol

$$A = [a_{ij}].$$

The following definition tells us when two matrices are equal:

DEFINITION 9.1   Equality of Matrices

Two matrices are equal if they have the same dimension and their corresponding elements are equal. Thus, two $m \times n$ matrices $A = [a_{ij}]$ and $B = [b_{ij}]$ are equal provided $a_{ij} = b_{ij}$ for $1 \le i \le m$ and $1 \le j \le n$.

Note, for example, that

$$\begin{bmatrix} 1 & 3 \\ -2 & 4 \end{bmatrix} = \begin{bmatrix} 1 & \sqrt{9} \\ -2 & \sqrt{16} \end{bmatrix},$$

while

$$\begin{bmatrix} 6 & 5 \\ 4 & -3 \end{bmatrix} \ne \begin{bmatrix} 6 & -\sqrt{25} \\ 4 & -3 \end{bmatrix}$$

since $-\sqrt{25} \ne 5$.

Photo by Elliott and Fry.
THE LATE PROFESSOR ARTHUR CAYLEY.

North Wind Picture Archives

ARTHUR CAYLEY
(1821–1895)

English mathematician Arthur Cayley graduated at the top of his class at Trinity College, Cambridge in 1842. He spent the next three years on a fellowship that allowed him to travel, paint, read, and study architecture, as well as write and study mathematics. Unable to find a position in mathematics when his fellowship ended, Cayley studied law and through his practice met James Joseph Sylvester (1814–1897), also a lawyer and a mathematician. Together, they defined the operation of matrix multiplication and developed the theory of matrices. Their work proved invaluable in such areas as quantum mechanics and the theory of relativity. During the fourteen years he practiced law, Cayley published between two hundred and three hundred papers. In 1863 Cayley abandoned his lucrative practice to accept a professorship at Cambridge in order to devote all of his time to mathematics. A most prolific writer, Cayley ranks third behind Euler and Cauchy (1789–1857) in number of papers.

Matrix Addition and Scalar Multiplication

Two matrices can be added provided they have the same dimension.

> **DEFINITION 9.2**   Addition of Matrices
>
> Let $A = [a_{ij}]$ and $B = [b_{ij}]$ be $m \times n$ matrices. Then $A + B = C$ where $C = [c_{ij}]$ is the $m \times n$ matrix in which $c_{ij} = a_{ij} + b_{ij}$ for $1 \le i \le m$ and $1 \le j \le n$.

From Definition 9.2, we see that we add two matrices having the same dimension by adding their corresponding elements. For example,

$$\overset{3 \times 2}{\begin{bmatrix} -2 & -2 \\ -3 & 0 \\ 1 & 4 \end{bmatrix}} + \overset{3 \times 2}{\begin{bmatrix} 2 & -2 \\ 0 & 2 \\ 5 & 3 \end{bmatrix}} = \overset{3 \times 2}{\begin{bmatrix} 0 & -4 \\ -3 & 2 \\ 6 & 7 \end{bmatrix}},$$

while the sum of

$$A = \overset{2 \times 2}{\begin{bmatrix} 1 & 2 \\ -1 & 5 \end{bmatrix}} \text{ and } B = \overset{2 \times 3}{\begin{bmatrix} 6 & 6 & 5 \\ 1 & -3 & 2 \end{bmatrix}}$$

is not defined because these two matrices do not have the same dimension.

The operation of multiplying each element of a matrix by a real number, or a **scalar**, is called **scalar multiplication**.

> **DEFINITION 9.3**   Scalar Multiplication
>
> Let $A = [a_{ij}]$ be an $m \times n$ matrix and c be a scalar (real number). Then $cA = [ca_{ij}]$ is the $m \times n$ matrix formed by multiplying each element of A by c.

EXAMPLE 17

If $A = \begin{bmatrix} 3 & 1 & 1 \\ 5 & 2 & -1 \end{bmatrix}$, find each of the following:

a. $4A$ **b.** $(-1)A$ **c.** $(0)A$

Solution

a. $4A = \begin{bmatrix} 12 & 4 & 4 \\ 20 & 8 & -4 \end{bmatrix}$

b. $(-1)A = \begin{bmatrix} -3 & -1 & -1 \\ -5 & -2 & 1 \end{bmatrix}$

c. $(0)A = \begin{bmatrix} 0 & 0 & 0 \\ 0 & 0 & 0 \end{bmatrix}$

Below we list properties of matrix addition and scalar multiplication. You will, no doubt, observe the similarity of these properties to the corresponding properties of addition and multiplication of real numbers.

THEOREM 9.3   Properties of Matrix Addition and Scalar Multiplication

Let A, B, and C be $m \times n$ matrices and let c and d be scalars. Then

(i)	$A + B = B + A$	Commutative property of addition
(ii)	$A + (B + C) = (A + B) + C$	Associative property of addition
(iii)	$(cd)A = c(dA)$	Associative property of scalar multiplication
(iv)	$c(A + B) = cA + cB$	Distributive properties
(v)	$(c + d)A = cA + dA$	

It should be noted that, just as in the case of real numbers, the associative property of addition of matrices allows us to write expressions such as

$$A + B + C \qquad \text{and} \qquad A + B + C + D$$

without using parentheses. We illustrate properties of matrix addition and scalar multiplication in the following example:

EXAMPLE 18

Let $A = \begin{bmatrix} 2 & -1 \\ 3 & 4 \end{bmatrix}$, $B = \begin{bmatrix} 0 & 5 \\ 1 & 6 \end{bmatrix}$, $C = \begin{bmatrix} -4 & -5 \\ 2 & 3 \end{bmatrix}$, $c = -2$ and $d = 5$.

Verify that

a. $(A + B) + C = A + (B + C)$.   **b.** $c(dC) = (cd)C$.

Solution

a.
$$(A + B) + C = \left(\begin{bmatrix} 2 & -1 \\ 3 & 4 \end{bmatrix} + \begin{bmatrix} 0 & 5 \\ 1 & 6 \end{bmatrix} \right) + \begin{bmatrix} -4 & -5 \\ 2 & 3 \end{bmatrix}$$

$$= \begin{bmatrix} 2 & 4 \\ 4 & 10 \end{bmatrix} + \begin{bmatrix} -4 & -5 \\ 2 & 3 \end{bmatrix}$$

$$= \begin{bmatrix} -2 & -1 \\ 6 & 13 \end{bmatrix}$$

$$A + (B + C) = \begin{bmatrix} 2 & -1 \\ 3 & 4 \end{bmatrix} + \left(\begin{bmatrix} 0 & 5 \\ 1 & 6 \end{bmatrix} + \begin{bmatrix} -4 & -5 \\ 2 & 3 \end{bmatrix} \right)$$

$$= \begin{bmatrix} 2 & -1 \\ 3 & 4 \end{bmatrix} + \begin{bmatrix} -4 & 0 \\ 3 & 9 \end{bmatrix}$$

$$= \begin{bmatrix} -2 & -1 \\ 6 & 13 \end{bmatrix}$$

Thus, $(A + B) + C = A + (B + C)$.

b. $c(dC) = -2(5C)$

$$= -2\left(5\begin{bmatrix} -4 & -5 \\ 2 & 3 \end{bmatrix}\right) = -2\begin{bmatrix} -20 & -25 \\ 10 & 15 \end{bmatrix} = \begin{bmatrix} 40 & 50 \\ -20 & -30 \end{bmatrix}$$

$$(cd)C = (-2 \cdot 5)\begin{bmatrix} -4 & -5 \\ 2 & 3 \end{bmatrix} = -10\begin{bmatrix} -4 & -5 \\ 2 & 3 \end{bmatrix}$$

$$= \begin{bmatrix} 40 & 50 \\ -20 & -30 \end{bmatrix}$$

Therefore, $c(dC) = (cd)C$.

Zero Matrix

In Example 17, we found that $0A = \begin{bmatrix} 0 & 0 & 0 \\ 0 & 0 & 0 \end{bmatrix}$. The matrix $\begin{bmatrix} 0 & 0 & 0 \\ 0 & 0 & 0 \end{bmatrix}$ is called the 2×3 zero matrix and is denoted by $[0]_{2,3}$. In general, we have

---

DEFINITION 9.4   Zero Matrix

The $m \times n$ matrix in which every entry is 0 is called the $m \times n$ **zero matrix** and is denoted by $[0]_{m,n}$.

From Definition 9.4, we see that there is not a unique zero matrix, and in fact, each of the following is a zero matrix:

$$\begin{bmatrix} 0 & 0 \\ 0 & 0 \end{bmatrix}, \qquad \begin{bmatrix} 0 & 0 \\ 0 & 0 \\ 0 & 0 \end{bmatrix}, \qquad \text{and} \qquad \begin{bmatrix} 0 & 0 & 0 & 0 \\ 0 & 0 & 0 & 0 \\ 0 & 0 & 0 & 0 \end{bmatrix}.$$

However, for a fixed dimension, $m \times n$, the matrix $[0]_{m,n}$ is unique and plays the same role in matrix addition as the real number zero plays in the addition of real numbers. For example,

$$\begin{bmatrix} 0 & 0 \\ 0 & 0 \end{bmatrix} + \begin{bmatrix} -1 & 2 \\ -7 & 5 \end{bmatrix} = \begin{bmatrix} -1 & 2 \\ -7 & 5 \end{bmatrix},$$

and

$$\begin{bmatrix} 0 & 0 \\ 0 & 0 \\ 0 & 0 \end{bmatrix} + \begin{bmatrix} 2 & 3 \\ 4 & 5 \\ 6 & 7 \end{bmatrix} = \begin{bmatrix} 2 & 3 \\ 4 & 5 \\ 6 & 7 \end{bmatrix}.$$

Recall that for any real number a, $-a$ denotes the real number such that $a + (-a) = 0$. We call $-a$ the additive inverse of a. Similarly, if A is an $m \times n$ matrix, $-A$ is the $m \times n$ matrix such that $A + (-A) = [0]_{m,n}$. As in the case of real numbers, we call $-A$ the **additive inverse** of the matrix A. Each entry of the matrix $-A$ is the additive inverse of the corresponding entry of A, and thus, $-A = (-1)A$.

For real numbers a and b, $a - b$ is defined to be $a + (-b)$. Similarly, we define **subtraction** for two $m \times n$ matrices A and B as follows:

$$A - B = A + (-B)$$

For example, if $A = \begin{bmatrix} 2 & -1 \\ 3 & 4 \end{bmatrix}$ and $B = \begin{bmatrix} -1 & 5 \\ 2 & -3 \end{bmatrix}$,

$$A - B = A + (-B) = \begin{bmatrix} 2 & -1 \\ 3 & 4 \end{bmatrix} + \begin{bmatrix} 1 & -5 \\ -2 & 3 \end{bmatrix} = \begin{bmatrix} 3 & -6 \\ 1 & 7 \end{bmatrix}.$$

Matrix Multiplication

We now consider the operation of matrix multiplication. Based on the definition of the addition of two matrices, it is natural to guess that we multiply two matrices by multiplying their corresponding elements. However, such a definition does not yield a matrix that is useful to us. Before we give the definition of the product of two matrices, we define the dot product of a row matrix with a column matrix. (The dot product is also called the **scalar product** or the **inner product**.)

D E F I N I T I O N 9.5   Dot Product of a Row Matrix with a Column Matrix

If $A = [a_1\ a_2\ a_3 \ldots a_n]$ is a $1 \times n$ row matrix and

$$B = \begin{bmatrix} b_1 \\ b_2 \\ b_3 \\ \vdots \\ b_n \end{bmatrix} \text{ is an } n \times 1 \text{ column matrix,}$$

then the **dot product** $A \cdot B$ is the scalar defined by

$$A \cdot B = a_1 b_1 + a_2 b_2 + a_3 b_3 + \cdots + a_n b_n.$$

From Definition 9.5, we see that for $A \cdot B$ to be defined, the number of columns in the row matrix A must be equal to the number of rows in the column matrix B. Moreover, we see that $A \cdot B$ is a scalar, *not a matrix*. To illustrate,

$$[2\quad 1\quad 0] \cdot \begin{bmatrix} -1 \\ 2 \\ 4 \end{bmatrix} = 2(-1) + 1(2) + 0(4) = -2 + 2 + 0 = 0,$$

while the dot product

$$[-9\quad 4] \cdot \begin{bmatrix} -1 \\ 2 \\ 4 \end{bmatrix}$$

is not defined. (Why?)

The dot product plays a central role in the definition of the product of two matrices A and B. The element in the ith row and the jth column of the product

AB is the dot product of the ith row of A and the jth column of B. Hence, for the product AB to be defined, each row of A must have the same number of elements as each column of B. Also, the product AB has the same number of rows as A and the same number of columns as B. We summarize these results in the diagram below.

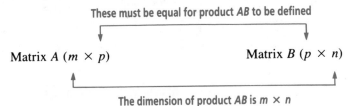

These must be equal for product AB to be defined

Matrix A $(m \times p)$ Matrix B $(p \times n)$

The dimension of product AB is $m \times n$

We now formally define matrix multiplication.

D E F I N I T I O N   9.6   Matrix Multiplication

If A is an $m \times p$ matrix and B is a $p \times n$ matrix, then the **product** of A and B is the $m \times n$ matrix $AB = C = [c_{ij}]$ where c_{ij} is the dot product of the ith row of A and the jth column of B.

E X A M P L E   19

If possible, compute AB and BA in each case.

a. $A = \begin{bmatrix} 2 & 1 \\ -1 & 3 \\ 3 & 4 \\ 2 & -5 \end{bmatrix}$, $B = \begin{bmatrix} -1 & 1 & 4 \\ 2 & 0 & 6 \end{bmatrix}$

b. $A = \begin{bmatrix} -1 & 0 & 1 \\ -5 & 1 & 3 \\ 7 & -1 & -4 \end{bmatrix}$, $B = \begin{bmatrix} 1 & -1 & 1 \\ -1 & 3 & -2 \\ 2 & -1 & 1 \end{bmatrix}$

c. $A = \begin{bmatrix} 1 & 0 & 8 \\ 1 & 2 & 3 \\ 2 & 5 & 3 \end{bmatrix}$, $B = \begin{bmatrix} 9 & -40 & 16 \\ -3 & 13 & -5 \\ -1 & 5 & -2 \end{bmatrix}$

Solution

must be equal

4×2 2×3

a. $AB = \begin{bmatrix} 2 & 1 \\ -1 & 3 \\ 3 & 4 \\ 2 & -5 \end{bmatrix} \begin{bmatrix} -1 & 1 & 4 \\ 2 & 0 & 6 \end{bmatrix}$

$$4 \times 3$$

$$= \begin{bmatrix} 2(-1) + 1(2) & 2(1) + 1(0) & 2(4) + 1(6) \\ -1(-1) + 3(2) & -1(1) + 3(0) & -1(4) + 3(6) \\ 3(-1) + 4(2) & 3(1) + 4(0) & 3(4) + 4(6) \\ 2(-1) + (-5)(2) & 2(1) + (-5)(0) & 2(4) + (-5)(6) \end{bmatrix}$$

$$= \begin{bmatrix} 0 & 2 & 14 \\ 7 & -1 & 14 \\ 5 & 3 & 36 \\ -12 & 2 & -22 \end{bmatrix}$$

not equal

$$4 \times 2$$

$$2 \times 3$$

$$BA = \begin{bmatrix} -1 & 1 & 4 \\ 2 & 0 & 6 \end{bmatrix} \begin{bmatrix} 2 & 1 \\ -1 & 3 \\ 3 & 4 \\ 2 & -5 \end{bmatrix}$$

is not defined since the number of columns in B is not equal to the number of rows in A.

must be equal

$$3 \times 3 \qquad\qquad 3 \times 3 \qquad\qquad 3 \times 3$$

b. $$AB = \begin{bmatrix} -1 & 0 & 1 \\ -5 & 1 & 3 \\ 7 & -1 & -4 \end{bmatrix} \begin{bmatrix} 1 & -1 & 1 \\ -1 & 3 & -2 \\ 2 & -1 & 1 \end{bmatrix} = \begin{bmatrix} 1 & 0 & 0 \\ 0 & 5 & -4 \\ 0 & -6 & 5 \end{bmatrix}$$

$$3 \times 3 \qquad\qquad 3 \times 3 \qquad\qquad 3 \times 3$$

$$BA = \begin{bmatrix} 1 & -1 & 1 \\ -1 & 3 & -2 \\ 2 & -1 & 1 \end{bmatrix} \begin{bmatrix} -1 & 0 & 1 \\ -5 & 1 & 3 \\ 7 & -1 & -4 \end{bmatrix} = \begin{bmatrix} 11 & -2 & -6 \\ -28 & 5 & 16 \\ 10 & -2 & -5 \end{bmatrix}$$

c. $$AB = \begin{bmatrix} 1 & 0 & 8 \\ 1 & 2 & 3 \\ 2 & 5 & 3 \end{bmatrix} \begin{bmatrix} 9 & -40 & 16 \\ -3 & 13 & -5 \\ -1 & 5 & -2 \end{bmatrix} = \begin{bmatrix} 1 & 0 & 0 \\ 0 & 1 & 0 \\ 0 & 0 & 1 \end{bmatrix}$$

$$BA = \begin{bmatrix} 9 & -40 & 16 \\ -3 & 13 & -5 \\ -1 & 5 & -2 \end{bmatrix} \begin{bmatrix} 1 & 0 & 8 \\ 1 & 2 & 3 \\ 2 & 5 & 3 \end{bmatrix} = \begin{bmatrix} 1 & 0 & 0 \\ 0 & 1 & 0 \\ 0 & 0 & 1 \end{bmatrix}$$

In part (a) of Example 19, we are given two matrices A and B for which the product AB is defined, while the product BA is undefined. Furthermore, notice that $AB \neq BA$ in part (b) of Example 19, even though both AB and BA are defined. This example illustrates the fact that, in general, matrix multiplication is not commutative.

Matrix multiplication, like matrix addition, has some general properties similar to the properties of real numbers. These properties will be useful when we solve matrix equations in Section 9.4. We note that there are two distributive laws and two special multiplication properties, since matrix multiplication is not commutative.

///

THEOREM 9.4   **Properties of Matrix Multiplication**

In the following statements, we are assuming all indicated products and sums are defined:

(i) $A(BC) = (AB)C$ \hspace{2cm} Associative property

(ii) $A(B + C) = AB + AC$ \hspace{1cm}⎫
(iii) $(B + C)A = BA + CA$ \hspace{1cm}⎬ Left & right distributive laws

(iv) If $A = B$, then $CA = CB$ \hspace{0.5cm}⎫ Left & right
(v) If $A = B$, then $AC = BC$ \hspace{0.5cm}⎬ multiplication properties

(vi) If c is a scalar, then $c(AB) = (cA)B = A(cB)$.

In Example 18, we illustrate the associative property of matrix multiplication and the right distributive law.

EXAMPLE 20

Let $A = \begin{bmatrix} 4 & 3 \\ -5 & 2 \end{bmatrix}$, $B = \begin{bmatrix} -5 & 6 \\ 1 & -1 \end{bmatrix}$, and $C = \begin{bmatrix} 7 & 1 & 3 \\ -1 & -2 & 4 \end{bmatrix}$.

Verify each of the following:

a. $(AB)C = A(BC)$ \hspace{1cm} **b.** $(A + B)C = AC + BC$

Solution

a. We begin by computing the product AB.

$$AB = \begin{bmatrix} 4 & 3 \\ -5 & 2 \end{bmatrix} \begin{bmatrix} -5 & 6 \\ 1 & -1 \end{bmatrix} = \begin{bmatrix} -17 & 21 \\ 27 & -32 \end{bmatrix}.$$

Thus,

$$(AB)C = \begin{bmatrix} -17 & 21 \\ 27 & -32 \end{bmatrix} \begin{bmatrix} 7 & 1 & 3 \\ -1 & -2 & 4 \end{bmatrix} = \begin{bmatrix} -140 & -59 & 33 \\ 221 & 91 & -47 \end{bmatrix}.$$

We now find the product BC.

$$BC = \begin{bmatrix} -5 & 6 \\ 1 & -1 \end{bmatrix} \begin{bmatrix} 7 & 1 & 3 \\ -1 & -2 & 4 \end{bmatrix} = \begin{bmatrix} -41 & -17 & 9 \\ 8 & 3 & -1 \end{bmatrix}.$$

Hence,

$$A(BC) = \begin{bmatrix} 4 & 3 \\ -5 & 2 \end{bmatrix} \begin{bmatrix} -41 & -17 & 9 \\ 8 & 3 & -1 \end{bmatrix} = \begin{bmatrix} -140 & -59 & 33 \\ 221 & 91 & -47 \end{bmatrix}.$$

Therefore, $(AB)C = A(BC)$.

b. We first compute $A + B$, getting

$$A + B = \begin{bmatrix} 4 & 3 \\ -5 & 2 \end{bmatrix} + \begin{bmatrix} -5 & 6 \\ 1 & -1 \end{bmatrix} = \begin{bmatrix} -1 & 9 \\ -4 & 1 \end{bmatrix}.$$

Hence,

$$(A + B)C = \begin{bmatrix} -1 & 9 \\ -4 & 1 \end{bmatrix} \begin{bmatrix} 7 & 1 & 3 \\ -1 & -2 & 4 \end{bmatrix} = \begin{bmatrix} -16 & -19 & 33 \\ -29 & -6 & -8 \end{bmatrix}.$$

Now

$$AC = \begin{bmatrix} 4 & 3 \\ -5 & 2 \end{bmatrix} \begin{bmatrix} 7 & 1 & 3 \\ -1 & -2 & 4 \end{bmatrix} = \begin{bmatrix} 25 & -2 & 24 \\ -37 & -9 & -7 \end{bmatrix},$$

and

$$BC = \begin{bmatrix} -5 & 6 \\ 1 & -1 \end{bmatrix} \begin{bmatrix} 7 & 1 & 3 \\ -1 & -2 & 4 \end{bmatrix} = \begin{bmatrix} -41 & -17 & 9 \\ 8 & 3 & -1 \end{bmatrix}.$$

Thus,

$$AC + BC = \begin{bmatrix} 25 & -2 & 24 \\ -37 & -9 & -7 \end{bmatrix} + \begin{bmatrix} -41 & -17 & 9 \\ 8 & 3 & -1 \end{bmatrix}$$

$$= \begin{bmatrix} -16 & -19 & 33 \\ -29 & -6 & -8 \end{bmatrix}.$$

Therefore $(A + B)C = AC + BC$.

Identity Matrix

In the solution of part (c) of Example 19 on page 445, we found that both matrix products AB and BA produced the matrix

$$C = \begin{bmatrix} 1 & 0 & 0 \\ 0 & 1 & 0 \\ 0 & 0 & 1 \end{bmatrix}.$$

Observe that each entry of C on the main diagonal is 1 while all other entries are 0. The matrix C is an example of an identity matrix.

In general, we have

//

DEFINITION 9.7   Identity Matrix

The $n \times n$ matrix in which each entry on the main diagonal is 1 and all other entries are 0 is called the $n \times n$ **identity matrix**. This square matrix is denoted by I_n.

To illustrate,

$$I_2 = \begin{bmatrix} 1 & 0 \\ 0 & 1 \end{bmatrix} \quad \text{and} \quad I_4 = \begin{bmatrix} 1 & 0 & 0 & 0 \\ 0 & 1 & 0 & 0 \\ 0 & 0 & 1 & 0 \\ 0 & 0 & 0 & 1 \end{bmatrix}.$$

are the identity matrices of order 2 and 4, respectively.

In the multiplication of *square* matrices of order n, the identity matrix I_n plays the same role as that played by 1 in multiplication of real numbers.

E X A M P L E   21

Verify that I_2 is the multiplicative identity for the set of 2×2 matrices. That is, verify that if

$$A = \begin{bmatrix} a_{11} & a_{12} \\ a_{21} & a_{22} \end{bmatrix}, \quad \text{then } I_2 A = A I_2 = A.$$

Solution

$$I_2 A = \begin{bmatrix} 1 & 0 \\ 0 & 1 \end{bmatrix} \begin{bmatrix} a_{11} & a_{12} \\ a_{21} & a_{22} \end{bmatrix} = \begin{bmatrix} (1)a_{11} + (0)a_{21} & (1)a_{12} + (0)a_{22} \\ (0)a_{11} + (1)a_{21} & (0)a_{12} + (1)a_{22} \end{bmatrix}$$

$$= \begin{bmatrix} a_{11} & a_{12} \\ a_{21} & a_{22} \end{bmatrix}.$$

Also,

$$A I_2 = \begin{bmatrix} a_{11} & a_{12} \\ a_{21} & a_{22} \end{bmatrix} \begin{bmatrix} 1 & 0 \\ 0 & 1 \end{bmatrix} = \begin{bmatrix} a_{11}(1) + a_{12}(0) & a_{11}(0) + a_{12}(1) \\ a_{21}(1) + a_{22}(0) & a_{21}(0) + a_{22}(1) \end{bmatrix}$$

$$= \begin{bmatrix} a_{11} & a_{12} \\ a_{21} & a_{22} \end{bmatrix}.$$

Therefore, $I_2 A = A I_2$, and hence, I_2 is the multiplicative identity for the set of 2×2 matrices.

In general, if A is an $m \times n$ matrix, then $I_m A = A$ and $A I_n = A$. (Note that if $m \neq n$, $I_m \neq I_n$.) For example, if $A = \begin{bmatrix} 2 & 1 & 3 \\ -1 & -1 & 2 \end{bmatrix}$, then

$$I_2 A = \begin{bmatrix} 1 & 0 \\ 0 & 1 \end{bmatrix} \begin{bmatrix} 2 & 1 & 3 \\ -1 & -1 & 2 \end{bmatrix} = \begin{bmatrix} 2 & 1 & 3 \\ -1 & -1 & 2 \end{bmatrix} = A, \quad \text{and}$$

$$A I_3 = \begin{bmatrix} 2 & 1 & 3 \\ -1 & -1 & 2 \end{bmatrix} \begin{bmatrix} 1 & 0 & 0 \\ 0 & 1 & 0 \\ 0 & 0 & 1 \end{bmatrix} = \begin{bmatrix} 2 & 1 & 3 \\ -1 & -1 & 2 \end{bmatrix} = A.$$

We have previously noted that, unlike the multiplication of real numbers, matrix multiplication is not commutative. There are other properties of multiplication of real numbers that do not hold for matrix multiplication. Recall that if a and b are real numbers for which ab is zero, then a is zero or b is zero. This is not the case for matrix multiplication as we now illustrate. If

$$A = \begin{bmatrix} 1 & 0 \\ 0 & 0 \end{bmatrix} \quad \text{and} \quad B = \begin{bmatrix} 0 & 0 \\ 1 & 0 \end{bmatrix},$$

then neither A nor B is the 2×2 zero matrix $[0]_{2,2}$, yet

$$AB = \begin{bmatrix} 1 & 0 \\ 0 & 0 \end{bmatrix} \begin{bmatrix} 0 & 0 \\ 1 & 0 \end{bmatrix} = \begin{bmatrix} 0 & 0 \\ 0 & 0 \end{bmatrix}.$$

E X A M P L E   22

A small bakery makes three main items: breads, cakes, and pies. The following matrix P, which shows the number of eggs and the number of cups of the other main ingredients needed for these items, is called the **production matrix**.

$$P = \begin{array}{c} \\ \\ \end{array} \begin{bmatrix} 1 & 4 & \frac{1}{2} & \frac{1}{2} & 1 \\ 3 & 4 & 2 & 1 & 1 \\ 1 & 1 & 1 & \frac{1}{2} & 1 \end{bmatrix} \begin{array}{c} \text{Breads} \\ \text{Cakes} \\ \text{Pies} \end{array}$$

Eggs Flour Sugar Shortening Milk

The costs, in cents, per egg or per cup for each ingredient when purchased either in small lots or in large lots is given in the **cost matrix** C.

Cost: Small Lot Large Lot

$$C = \begin{bmatrix} 7 & 5 \\ 10 & 8 \\ 14 & 10 \\ 10 & 6 \\ 13 & 10 \end{bmatrix} \begin{array}{c} \text{Eggs} \\ \text{Flour} \\ \text{Sugar} \\ \text{Shortening} \\ \text{Milk} \end{array}$$

Use matrix multiplication to find a matrix whose elements give the comparative costs per item for the two purchase options.

Solution Since the first row of the production matrix P gives the number of eggs and the number of cups of the other main ingredients needed for making bread, and the first column of the cost matrix C gives the cost, in cents, for buying these items in small lots, the dot product

$$[1 \quad 4 \quad \tfrac{1}{2} \quad \tfrac{1}{2} \quad 1] \cdot \begin{bmatrix} 7 \\ 10 \\ 14 \\ 10 \\ 13 \end{bmatrix} = 72$$

gives the cost, in cents, for making bread when the ingredients used are bought in small lots. Similarly, the dot product

$$[1 \quad 4 \quad \tfrac{1}{2} \quad \tfrac{1}{2} \quad 1] \cdot \begin{bmatrix} 5 \\ 8 \\ 10 \\ 6 \\ 10 \end{bmatrix} = 55$$

gives the cost for making bread when the ingredients are bought in large lots. Thus, we see that the matrix PC gives us the comparative costs per item for the two purchase options, and we have

$$PC = \begin{bmatrix} 1 & 4 & \frac{1}{2} & \frac{1}{2} & 1 \\ 3 & 4 & 2 & 1 & 1 \\ 1 & 1 & 1 & \frac{1}{2} & 1 \end{bmatrix} \begin{bmatrix} 7 & 5 \\ 10 & 8 \\ 14 & 10 \\ 10 & 6 \\ 13 & 10 \end{bmatrix}$$

Small Lot Large Lot

$$= \begin{bmatrix} 72 & 55 \\ 112 & 83 \\ 49 & 36 \end{bmatrix} \begin{array}{c} \text{Breads} \\ \text{Cakes} \\ \text{Pies} \end{array}$$

From matrix PC we find, for example, that it costs 13 cents more per pie when the ingredients are bought in small lots.

EXERCISES 9.3

In Exercises 1–4, find x and y.

1. $\begin{bmatrix} 3 & x \\ -y & 4 \end{bmatrix} = \begin{bmatrix} 3 & -5 \\ 2 & 4 \end{bmatrix}$

2. $\begin{bmatrix} -1 & 2 \\ x & -y \end{bmatrix} = \begin{bmatrix} -1 & 2 \\ -4 & y+2 \end{bmatrix}$

3. $\begin{bmatrix} 1 & -1 & 2 & 3 \\ 6 & -4 & -1 & 0 \\ -1 & 0 & 2 & -8 \end{bmatrix}$

$= \begin{bmatrix} 1 & -1 & 4x & 3 \\ y+2 & -4 & -1 & 0 \\ -1 & 0 & 2 & -16x \end{bmatrix}$

4. $\begin{bmatrix} 2x+3 & 1 & 2 & 0 \\ 1 & y-4 & -1 & 3 \\ -9 & 0 & -5 & -1 \\ -1 & -1 & -1 & -y+3 \end{bmatrix}$

$= \begin{bmatrix} x+1 & 1 & 2 & 0 \\ 1 & 2y+5 & -1 & 3 \\ -9 & 0 & 2x-1 & -1 \\ -1 & -1 & -1 & 12 \end{bmatrix}$

In Exercises 5–10, find, if possible, $A + B$, $A - B$, $3A$, and $3A - 4B$.

5. $A = \begin{bmatrix} 4 & 5 \\ -6 & 2 \end{bmatrix}$ $B = \begin{bmatrix} -1 & 0 \\ -7 & 10 \end{bmatrix}$

6. $A = \begin{bmatrix} -1 & 1 & 1 \\ 2 & 0 & 1 \\ -1 & 0 & 2 \end{bmatrix}$ $B = \begin{bmatrix} 1 & -1 \\ 6 & 2 \\ 0 & -4 \end{bmatrix}$

7. $A = \begin{bmatrix} 2 & 1 & 6 & 5 \\ 1 & 2 & -4 & 3 \end{bmatrix}$ $B = \begin{bmatrix} 4 & 6 \\ 0 & -1 \\ -2 & -2 \\ 0 & 1 \end{bmatrix}$

8. $A = \begin{bmatrix} 6 & 2 & 1 & 5 \\ 1 & 0 & 1 & -1 \\ 2 & 5 & 0 & 1 \end{bmatrix}$ $B = \begin{bmatrix} 2 & 0 & 0 & 0 \\ 0 & 1 & 0 & 0 \\ 0 & 0 & 0 & 4 \end{bmatrix}$

9. $A = \begin{bmatrix} 1 & 6 & -1 & -7 \\ 0 & 1 & 1 & -2 \\ 0 & 0 & 0 & 8 \\ 0 & 0 & 0 & -1 \end{bmatrix}$ $B = \begin{bmatrix} 2 & 0 & 0 & 0 \\ 0 & 1 & 0 & 0 \\ 0 & 0 & -3 & 0 \\ 0 & 0 & 0 & 4 \end{bmatrix}$

10. $A = \begin{bmatrix} 4 & 0 \\ 0 & 2 \\ 6 & 5 \\ -1 & -4 \\ 10 & -9 \end{bmatrix}$ $B = \begin{bmatrix} 2 & 4 \\ -5 & 0 \\ 1 & 1 \\ -3 & 3 \\ 6 & 5 \end{bmatrix}$

In Exercises 11–16, find the indicated dot product.

11. $[1 \quad 0 \quad 0] \cdot \begin{bmatrix} 0 \\ 1 \\ 0 \end{bmatrix}$

12. $[0 \quad 1 \quad 0] \cdot \begin{bmatrix} 0 \\ 0 \\ 1 \end{bmatrix}$

13. $[1 \quad -1 \quad 1 \quad 1] \cdot \begin{bmatrix} 1 \\ -1 \\ 1 \\ 1 \end{bmatrix}$

14. $[2 \quad -1 \quad 4 \quad 5] \cdot \begin{bmatrix} 3 \\ -1 \\ 0 \\ 6 \end{bmatrix}$

15. $[3 \quad 4 \quad 5 \quad -3] \cdot \begin{bmatrix} -2 \\ 3 \\ -3 \\ -3 \end{bmatrix}$

16. $[\frac{1}{2} \quad \frac{1}{2} \quad \frac{1}{2} \quad \frac{1}{2}] \cdot \begin{bmatrix} \frac{1}{2} \\ \frac{1}{2} \\ \frac{1}{2} \\ \frac{1}{2} \end{bmatrix}$

In Exercises 17–26, find AB and BA if possible.

17. $A = \begin{bmatrix} 1 & -1 \\ 2 & -2 \\ -3 & 3 \end{bmatrix}$ $B = \begin{bmatrix} 2 & 1 \\ 0 & 1 \end{bmatrix}$

18. $A = [1 \quad 2 \quad 3 \quad 4]$ $B = \begin{bmatrix} -3 & 1 & 2 \\ 0 & -1 & -3 \\ 1 & 1 & -1 \end{bmatrix}$

19. $A = \begin{bmatrix} 1 & 2 & 0 & 1 \\ 0 & -1 & 2 & 3 \\ 0 & 0 & -1 & 4 \end{bmatrix}$ $B = \begin{bmatrix} 1 & 0 & 0 & 0 \\ 0 & 1 & 0 & 0 \\ 0 & 0 & 1 & 0 \\ 0 & 0 & 0 & 1 \end{bmatrix}$

20. $A = \begin{bmatrix} 2 & -1 & 1 \\ 0 & 0 & -2 \\ 0 & 0 & 4 \end{bmatrix}$ $B = \begin{bmatrix} 1 & 0 & 4 \\ 0 & -1 & 3 \\ 0 & 0 & 1 \end{bmatrix}$

21. $A = \begin{bmatrix} 5 & 6 & 0 & 7 \\ -1 & 2 & -3 & 4 \end{bmatrix}$ $B = \begin{bmatrix} -1 & 0 \\ 1 & 0 \\ 0 & 1 \\ 1 & -1 \\ -1 & 1 \end{bmatrix}$

22. $A = \begin{bmatrix} 2 & 1 & 3 & 4 \\ 1 & -1 & 1 & 0 \end{bmatrix}$ $B = \begin{bmatrix} 1 & 1 & -1 \\ 0 & -1 & 0 \\ 1 & 0 & 1 \end{bmatrix}$

23. $A = \begin{bmatrix} 1 & 1 & 1 \\ 0 & 1 & 2 \\ 0 & 0 & 1 \end{bmatrix}$ $B = \begin{bmatrix} 1 & 0 & -1 \\ 1 & -1 & 2 \\ -1 & 1 & -1 \end{bmatrix}$

24. $A = \begin{bmatrix} 2 & 1 & 1 \\ 1 & 0 & -1 \\ 1 & 3 & 2 \end{bmatrix}$ $B = \begin{bmatrix} \frac{1}{2} & \frac{1}{6} & -\frac{1}{6} \\ \frac{1}{2} & \frac{1}{2} & \frac{1}{2} \\ \frac{1}{2} & \frac{5}{6} & -\frac{1}{6} \end{bmatrix}$

25. $A = \begin{bmatrix} 2 & 1 & 0 & 5 \\ -1 & 2 & 1 & 1 \\ 6 & -1 & 0 & -1 \end{bmatrix}$ $B = \begin{bmatrix} 1 & 0 & 1 \\ 1 & 1 & 0 \\ 1 & 1 & -1 \\ -1 & 0 & 1 \end{bmatrix}$

26. $A = \begin{bmatrix} 1 & -1 & 1 & -1 & -4 \\ -1 & 2 & 2 & -2 & 1 \\ 0 & 1 & 1 & 1 & 0 \end{bmatrix}$ $B = \begin{bmatrix} 1 & 0 & 1 \\ 0 & 1 & 0 \\ 0 & 0 & 1 \end{bmatrix}$

In Exercises 27–30, solve for the matrix $\bar{X}$ given that

$$A = \begin{bmatrix} 1 & -2 \\ 3 & 0 \\ -4 & 1 \end{bmatrix} \quad \text{and} \quad B = \begin{bmatrix} 3 & 0 \\ -2 & 1 \\ 0 & -5 \end{bmatrix}$$

27. $\bar{X} = -A - 3B$

28. $3\bar{X} = 2A - B$

29. $2\bar{X} + 4B = 5A$

30. $4\bar{X} - A + 2B = [0]_{3,2}$

In Exercises 31–34, verify each statement for the following matrices.

$$A = \begin{bmatrix} 1 & -2 \\ 0 & 0 \end{bmatrix} \quad B = \begin{bmatrix} 2 & 1 \\ -1 & 0 \end{bmatrix} \quad C = \begin{bmatrix} 2 & 2 \\ -1 & \frac{1}{2} \end{bmatrix}$$

31. The matrix $A \neq \begin{bmatrix} 0 & 0 \\ 0 & 0 \end{bmatrix}$; $AB = AC$; however $B \neq C$.

Compare your results with a similar property for real numbers.

32. $(AB)^2 \neq A^2B^2$ where $A^2 = AA$

33. $A^2 - B^2 \neq (A + B)(A - B)$

34. $A^2 + 2AB + B^2 \neq (A + B)(A + B)$

In Exercises 35–38, verify that $AB = BA = I_n$ where A and B are the given $n \times n$ matrices.

35. $A = \begin{bmatrix} 2 & -2 \\ 3 & -2 \end{bmatrix}$ $B = \begin{bmatrix} -1 & 1 \\ -\frac{3}{2} & 1 \end{bmatrix}$

36. $A = \begin{bmatrix} 1 & 3 & 4 \\ 2 & 8 & 6 \\ 5 & 1 & 35 \end{bmatrix}$ $B = \begin{bmatrix} 137 & -\frac{101}{2} & -7 \\ -20 & \frac{15}{2} & 1 \\ -19 & 7 & 1 \end{bmatrix}$

37. $A = \begin{bmatrix} 0 & \frac{1}{10} & \frac{1}{5} \\ \frac{1}{14} & \frac{4}{35} & -\frac{1}{5} \\ \frac{1}{14} & -\frac{13}{70} & \frac{1}{5} \end{bmatrix}$ $B = \begin{bmatrix} 2 & 8 & 6 \\ 4 & 2 & -2 \\ 3 & -1 & 1 \end{bmatrix}$

38. $A = \begin{bmatrix} 1 & 0 & 0 & -\frac{1}{2} \\ 0 & 0 & \frac{1}{2} & 0 \\ 0 & 1 & 0 & 0 \\ 0 & 0 & 0 & \frac{1}{2} \end{bmatrix}$ $B = \begin{bmatrix} 1 & 0 & 0 & 1 \\ 0 & 0 & 1 & 0 \\ 0 & 2 & 0 & 0 \\ 0 & 0 & 0 & 2 \end{bmatrix}$

In Exercises 39–44, find matrices A, $\bar{X}$, and B such that the given system of linear equations can be written in the form $A\bar{X} = B$.

39. $\begin{cases} x + y = 6 \\ x - 2y = 5 \end{cases}$

40. $\begin{cases} 2x - y = -7 \\ -3x + 2y = 4 \end{cases}$

41. $\begin{cases} t \quad - v = -! \\ u + 4v = 2 \\ 3t \quad - 5v = 6 \end{cases}$

42. $\begin{cases} t - u - v + w = 0 \\ 2u + v = 0 \end{cases}$

43. $\begin{cases} x + y = 4 \\ 2x + 4y = 2 \\ x - 5y = -8 \end{cases}$

44. $\begin{cases} t + v - w + x = 0 \\ v - x = 0 \\ t + v + 4w = 0 \\ v + w = 0 \\ t - w = 0 \end{cases}$

45. Let $A = \begin{bmatrix} a & b \\ c & d \end{bmatrix}$. The **transpose** of A, denoted by A^T, is the matrix obtained by interchanging the rows and columns of A. Using the given matrices

$$A = \begin{bmatrix} 2 & -9 \\ 10 & 4 \end{bmatrix} \quad \text{and} \quad B = \begin{bmatrix} -7 & -3 \\ 6 & 1 \end{bmatrix},$$

verify each of the following statements:

a. $(A + B)^T = A^T + B^T$

b. $(A^T)^T = A$

c. $(AB)^T = B^TA^T$

46. Let $A = \begin{bmatrix} a & b \\ c & d \end{bmatrix}$, $B = \begin{bmatrix} e & f \\ g & h \end{bmatrix}$, and $C = \begin{bmatrix} i & j \\ k & l \end{bmatrix}$.

Prove each of the following:

a. $A(BC) = A(BC)$ **b.** $(A + B)C = AC + BC$

47. In Section 8.5, we learned if the x- and y-axes are rotated through a positive angle θ, a new $x'y'$ coordinate system is formed. Using the equations of rotation given in Section 8.5, show that if a point P has coordinates (x, y) in the original system and coordinates (x', y') in the new system, then the coordinates of P are related as follows:

$$\begin{cases} x \cos \theta + y \sin \theta = x' \\ -x \sin \theta + y \cos \theta = y'. \end{cases}$$

The system above can be written in the following form:

$$\begin{bmatrix} \cos \theta & \sin \theta \\ -\sin \theta & \cos \theta \end{bmatrix} \begin{bmatrix} x \\ y \end{bmatrix} = \begin{bmatrix} x' \\ y' \end{bmatrix}.$$

If the coordinates of the point P in the original coordinate system are $(-2, 3)$, find the coordinates of P relative to the $x' - y'$ system given that

a. $\theta = 135°$ **b.** $\theta = \frac{7}{6}\pi$ **c.** $\theta = 50°$

48. In the matrix given below, hospitals in a certain country are classified by type and number of beds.

NUMBER OF BEDS

6–24	25–74	75 or more		
299	1894	3850	General	
22	130	350	Psychiatric	Type of
0	10	35	Chronic	Hospital
25	125	150	Others	

How many hospitals are

a. general hospitals?

b. psychiatric hospitals with 75 or more beds?

c. hospitals for chronically ill with at least 25 beds?

d. are not for the chronically ill?

49. A furniture manufacturer makes two types of dining tables from wood, plastic, and glass. The number of units of each raw material required for each type of table is given below.

UNITS REQUIRED

	Wood	Plastic	Glass
Table *A*	60	5	0
Table *B*	10	2	10

The manufacturer has three factories where the tables are built. The unit costs for each of the raw materials at the three plants are given below.

UNIT COSTS

	Factory 1	Factory 2	Factory 3
Wood	60	5	0
Plastic	10	2	10
Glass	12	10	13

a. Find the production matrix, P (see Example 22).
b. Find the cost matrix, C.
c. Using the production matrix and the cost matrix, find the total cost of producing each of the tables at each of the factories.

9.4 INVERSE OF A MATRIX

In Section 9.2, we solved systems of linear equations using either Gaussian elimination or Gauss-Jordan elimination. As we have seen, matrices play an integral part in both methods. In this section, we use matrices to develop yet another method for solving certain systems of linear equations in which the number of equations is equal to the number of variables.

Consider the system

$$\begin{cases} x + 2y - 4z = 3 \\ -x - y + 3z = 1 \\ -x - 2y + 5z = 4. \end{cases} \quad (1)$$

If we let

$$A = \begin{bmatrix} 1 & 2 & -4 \\ -1 & -1 & 3 \\ -1 & -2 & 5 \end{bmatrix}, \quad X = \begin{bmatrix} x \\ y \\ z \end{bmatrix}, \quad \text{and} \quad B = \begin{bmatrix} 3 \\ 1 \\ 4 \end{bmatrix},$$

then system (1) can be written as the **matrix equation**

$$AX = B. \quad (2)$$

Note that matrix equation (2) has the same form as the linear equation

$$ax = b, \quad (3)$$

where a and b are real numbers. Recall that if $a \neq 0$, we can solve equation (3) by multiplying both sides by a^{-1}, the multiplicative inverse of a to get

$$\begin{aligned} a^{-1}(ax) &= a^{-1}b \\ (a^{-1}a)x &= a^{-1}b \\ 1x &= a^{-1}b \\ x &= a^{-1}b. \end{aligned}$$

In solving equation (3), we used the fact that for $a \neq 0$, $a^{-1}(a) = 1$. We can solve matrix equation (2) in a similar fashion if we can find a matrix C such that $CA = I_3$. A matrix C of order 3 having the property that

$$AC = CA = I_3$$

is also called the *inverse* of matrix A. In general, we have

DEFINITION 9.8 Inverse of a Matrix

Let A be a square matrix of order n. If there exists an $n \times n$ matrix C such that

$$AC = CA = I_n,$$

then C is called the **inverse** of A and is denoted by A^{-1}.

If a matrix A has an inverse, then A is said to be **nonsingular** or **invertible**; otherwise, A is said to be **singular**. Notice that Definition 9.8 requires that A be a square matrix. To see why a nonsquare matrix A of dimension $m \times n$ $(m \neq n)$ cannot have an inverse, we simply note that if C is an $n \times m$ matrix, then the matrix AC has dimension $m \times m$ while the matrix CA has dimension $n \times n$. Thus, matrix AC cannot be equal to matrix CA. As we shall see, there are square matrices that do not have inverses. However, if a square matrix has an inverse, its inverse is unique (see Exercise 29.) Moreover, it can be shown that if C is a matrix such that $AC = I_n$, then $CA = I_n$ and hence, $C = A^{-1}$. Similarly, if $CA = I_n$, it follows that $AC = I_n$ and therefore, $C = A^{-1}$.

Finding the Inverse of a Matrix

Consider the matrix

$$A = \begin{bmatrix} 1 & 2 & -4 \\ -1 & -1 & 3 \\ -1 & -2 & 5 \end{bmatrix}$$

from matrix equation (2). Matrix A will have an inverse if there exists a matrix

$$C = \begin{bmatrix} c_{11} & c_{12} & c_{13} \\ c_{21} & c_{22} & c_{23} \\ c_{31} & c_{32} & c_{33} \end{bmatrix}$$

such that $AC = I_3$.

If $AC = I_3$, then

$$\begin{bmatrix} 1 & 2 & -4 \\ -1 & -1 & 3 \\ -1 & -2 & 5 \end{bmatrix} \begin{bmatrix} c_{11} & c_{12} & c_{13} \\ c_{21} & c_{22} & c_{23} \\ c_{31} & c_{32} & c_{33} \end{bmatrix} = \begin{bmatrix} 1 & 0 & 0 \\ 0 & 1 & 0 \\ 0 & 0 & 1 \end{bmatrix}.$$

Hence, from the definition of the product of matrices, we have

$$\begin{bmatrix} 1 & 2 & -4 \\ -1 & -1 & 3 \\ -1 & -2 & 5 \end{bmatrix} \begin{bmatrix} c_{11} \\ c_{21} \\ c_{31} \end{bmatrix} = \begin{bmatrix} 1 \\ 0 \\ 0 \end{bmatrix}, \tag{4}$$

$$\begin{bmatrix} 1 & 2 & -4 \\ -1 & -1 & 3 \\ -1 & -2 & 5 \end{bmatrix} \begin{bmatrix} c_{12} \\ c_{22} \\ c_{32} \end{bmatrix} = \begin{bmatrix} 0 \\ 1 \\ 0 \end{bmatrix}, \quad \text{and} \tag{5}$$

$$\begin{bmatrix} 1 & 2 & -4 \\ -1 & -1 & 3 \\ -1 & -2 & 5 \end{bmatrix} \begin{bmatrix} c_{13} \\ c_{23} \\ c_{33} \end{bmatrix} = \begin{bmatrix} 0 \\ 0 \\ 1 \end{bmatrix}. \tag{6}$$

To solve matrix equations (4), (5), and (6), we could use Gauss-Jordan elimination to transform the augmented matrices

$$
\left[\begin{array}{ccc|c}
1 & 2 & -4 & 1 \\
-1 & -1 & 3 & 0 \\
-1 & -2 & 5 & 0
\end{array}\right],
\left[\begin{array}{ccc|c}
1 & 2 & -4 & 0 \\
-1 & -1 & 3 & 1 \\
-1 & -2 & 5 & 0
\end{array}\right], \quad \text{and}
$$

$$
\left[\begin{array}{ccc|c}
1 & 2 & -4 & 0 \\
-1 & -1 & 3 & 0 \\
-1 & -2 & 5 & 1
\end{array}\right] \tag{7}
$$

into reduced echelon form. We note, however, that the matrix to the left of the vertical bar in all three of these augmented matrices is the matrix A. Therefore, the same row operations can be used to transform each of these augmented matrices into reduced echelon form. Thus, it is more efficient to combine these three augmented matrices into the single matrix

$$
\left[\begin{array}{ccc|ccc}
1 & 2 & -4 & 1 & 0 & 0 \\
-1 & -1 & 3 & 0 & 1 & 0 \\
-1 & -2 & 5 & 0 & 0 & 1
\end{array}\right]
$$

and transform it into reduced echelon form as follows:

$$
\left[\begin{array}{ccc|ccc}
1 & 2 & -4 & 1 & 0 & 0 \\
-1 & -1 & 3 & 0 & 1 & 0 \\
-1 & -2 & 5 & 0 & 0 & 1
\end{array}\right]
$$

$$
R_1 + R_2 \to R_2
\left[\begin{array}{ccc|ccc}
1 & 2 & -4 & 1 & 0 & 0 \\
0 & 1 & -1 & 1 & 1 & 0 \\
-1 & -2 & 5 & 0 & 0 & 1
\end{array}\right]
$$

$$
R_1 + R_3 \to R_3
\left[\begin{array}{ccc|ccc}
1 & 2 & -4 & 1 & 0 & 0 \\
0 & 1 & -1 & 1 & 1 & 0 \\
0 & 0 & 1 & 1 & 0 & 1
\end{array}\right]
$$

$$
R_3 + R_2 \to R_2
\left[\begin{array}{ccc|ccc}
1 & 2 & -4 & 1 & 0 & 0 \\
0 & 1 & 0 & 2 & 1 & 1 \\
0 & 0 & 1 & 1 & 0 & 1
\end{array}\right]
$$

$$
4R_3 + R_1 \to R_1
\left[\begin{array}{ccc|ccc}
1 & 2 & 0 & 5 & 0 & 4 \\
0 & 1 & 0 & 2 & 1 & 1 \\
0 & 0 & 1 & 1 & 0 & 1
\end{array}\right]
$$

$$
-2R_2 + R_1 \to R_1
\left[\begin{array}{ccc|ccc}
1 & 0 & 0 & 1 & -2 & 2 \\
0 & 1 & 0 & 2 & 1 & 1 \\
0 & 0 & 1 & 1 & 0 & 1
\end{array}\right] \tag{8}
$$

Matrix (8) is in reduced echelon form with the identity matrix I_3 appearing to the left of the vertical bar. We now reverse our earlier step of combining augmented matrices (7) and write matrix (8) as three augmented matrices:

$$
\left[\begin{array}{ccc|c}
1 & 0 & 0 & 1 \\
0 & 1 & 0 & 2 \\
0 & 0 & 1 & 1
\end{array}\right], \tag{9}
$$

$$
\left[\begin{array}{ccc|c}
1 & 0 & 0 & -2 \\
0 & 1 & 0 & 1 \\
0 & 0 & 1 & 0
\end{array}\right], \quad \text{and} \tag{10}
$$

$$\left[\begin{array}{ccc|c} 1 & 0 & 0 & 2 \\ 0 & 1 & 0 & 1 \\ 0 & 0 & 1 & 1 \end{array}\right]. \tag{11}$$

From matrix (9), we conclude that $c_{11} = 1$, $c_{21} = 2$, and $c_{31} = 1$. Similarly, from matrices (10) and (11), we find that $c_{12} = -2$, $c_{22} = 1$, and $c_{32} = 0$, while $c_{13} = 2$, $c_{23} = 1$, and $c_{33} = 1$. Thus, if

$$C = \begin{bmatrix} 1 & -2 & 2 \\ 2 & 1 & 1 \\ 1 & 0 & 1 \end{bmatrix},$$

$AC = I_3$. It follows that $CA = I_3$ and that the inverse of A is

$$C = A^{-1} = \begin{bmatrix} 1 & -2 & 2 \\ 2 & 1 & 1 \\ 1 & 0 & 1 \end{bmatrix}.$$

Notice that A^{-1} appears to the right of the vertical bar in matrix (8).

We can now solve matrix equation (2) using A^{-1} found above. We have

$$AX = B$$
$$A^{-1}(AX) = A^{-1}B$$
$$(A^{-1}A)X = A^{-1}B$$
$$I_3 X = A^{-1}B$$
$$X = A^{-1}B. \tag{12}$$

Since $A^{-1} = \begin{bmatrix} 1 & -2 & 2 \\ 2 & 1 & 1 \\ 1 & 0 & 1 \end{bmatrix}$ and $B = \begin{bmatrix} 3 \\ 1 \\ 4 \end{bmatrix}$, from equation (12),

$$X = \begin{bmatrix} 1 & -2 & 2 \\ 2 & 1 & 1 \\ 1 & 0 & 1 \end{bmatrix}\begin{bmatrix} 3 \\ 1 \\ 4 \end{bmatrix}$$

$$= \begin{bmatrix} 9 \\ 11 \\ 7 \end{bmatrix}.$$

The procedure we used to find A^{-1} can be used to find the inverse of any nonsingular matrix A. In general, if the $n \times n$ matrix A is nonsingular, we may find A^{-1} by adjoining the identity matrix I_n to A and transforming the resulting matrix

$$[A \mid I_n]$$

into reduced echelon form. The procedure is outlined below.

Finding the Inverse of a Nonsingular Matrix

Let A be an $n \times n$ matrix.

(i)   Form the matrix $[A \mid I_n]$.

(ii) Transform the matrix $[A \mid I_n]$ into reduced echelon form.

(iii) If the reduced echelon form of $[A \mid I_n]$ is $[I_n \mid C]$, then $A^{-1} = C$. Otherwise, A is singular.

EXAMPLE 23

Find the inverse (if it exists) of the matrix

$$A = \begin{bmatrix} 1 & 0 & -1 \\ 1 & -1 & 2 \\ -1 & 1 & -1 \end{bmatrix}.$$

Solution   We begin by forming the matrix

$$[A \mid I_3] = \begin{bmatrix} 1 & 0 & -1 & \vline & 1 & 0 & 0 \\ 1 & -1 & 2 & \vline & 0 & 1 & 0 \\ -1 & 1 & -1 & \vline & 0 & 0 & 1 \end{bmatrix}.$$

Transforming the augmented matrix above into reduced echelon form, we have

$$\begin{bmatrix} 1 & 0 & -1 & \vline & 1 & 0 & 0 \\ 1 & -1 & 2 & \vline & 0 & 1 & 0 \\ -1 & 1 & -1 & \vline & 0 & 0 & 1 \end{bmatrix}$$

$$-R_1 + R_2 \to R_2 \quad \begin{bmatrix} 1 & 0 & -1 & \vline & 1 & 0 & 0 \\ 0 & -1 & 3 & \vline & -1 & 1 & 0 \\ -1 & 1 & -1 & \vline & 0 & 0 & 1 \end{bmatrix}$$

$$R_1 + R_3 \to R_3 \quad \begin{bmatrix} 1 & 0 & -1 & \vline & 1 & 0 & 0 \\ 0 & -1 & 3 & \vline & -1 & 1 & 0 \\ 0 & 1 & -2 & \vline & 1 & 0 & 1 \end{bmatrix}$$

$$-R_2 \to R_2 \quad \begin{bmatrix} 1 & 0 & -1 & \vline & 1 & 0 & 0 \\ 0 & 1 & -3 & \vline & 1 & -1 & 0 \\ 0 & 1 & -2 & \vline & 1 & 0 & 1 \end{bmatrix}$$

$$-R_2 + R_3 \to R_3 \quad \begin{bmatrix} 1 & 0 & -1 & \vline & 1 & 0 & 0 \\ 0 & 1 & -3 & \vline & 1 & -1 & 0 \\ 0 & 0 & 1 & \vline & 0 & 1 & 1 \end{bmatrix}$$

$$3R_3 + R_2 \to R_2 \quad \begin{bmatrix} 1 & 0 & -1 & \vline & 1 & 0 & 0 \\ 0 & 1 & 0 & \vline & 1 & 2 & 3 \\ 0 & 0 & 1 & \vline & 0 & 1 & 1 \end{bmatrix}$$

$$R_3 + R_1 \to R_1 \quad \begin{bmatrix} 1 & 0 & 0 & \vline & 1 & 1 & 1 \\ 0 & 1 & 0 & \vline & 1 & 2 & 3 \\ 0 & 0 & 1 & \vline & 0 & 1 & 1 \end{bmatrix}$$

Since the last matrix is in reduced echelon form with I_3 appearing to the left of the vertical bar, the matrix to the right of the vertical bar is A^{-1}. Thus,

$$A^{-1} = \begin{bmatrix} 1 & 1 & 1 \\ 1 & 2 & 3 \\ 0 & 1 & 1 \end{bmatrix}.$$

You should verify that

$$\begin{bmatrix} 1 & 0 & -1 \\ 1 & -1 & 2 \\ -1 & 1 & -1 \end{bmatrix} \begin{bmatrix} 1 & 1 & 1 \\ 1 & 2 & 3 \\ 0 & 1 & 1 \end{bmatrix}$$

$$= \begin{bmatrix} 1 & 1 & 1 \\ 1 & 2 & 3 \\ 0 & 1 & 1 \end{bmatrix} \begin{bmatrix} 1 & 0 & -1 \\ 1 & -1 & 2 \\ -1 & 1 & -1 \end{bmatrix} = \begin{bmatrix} 1 & 0 & 0 \\ 0 & 1 & 0 \\ 0 & 0 & 1 \end{bmatrix}.$$

EXAMPLE 24

Find the inverse (if it exists) of the matrix

$$A = \begin{bmatrix} -2 & 3 & 3 \\ 3 & -4 & 1 \\ -5 & 7 & 2 \end{bmatrix}.$$

Solution   Proceeding as in Example 23, we form the matrix

$$[A \mid I_3] = \begin{bmatrix} -2 & 3 & 3 & \vline & 1 & 0 & 0 \\ 3 & -4 & 1 & \vline & 0 & 1 & 0 \\ -5 & 7 & 2 & \vline & 0 & 0 & 1 \end{bmatrix}.$$

Using elementary row operations, we transform $[A \mid I_3]$ into reduced echelon form as follows:

$$[A \mid I_3] = \begin{bmatrix} -2 & 3 & 3 & \vline & 1 & 0 & 0 \\ 3 & -4 & 1 & \vline & 0 & 1 & 0 \\ -5 & 7 & 2 & \vline & 0 & 0 & 1 \end{bmatrix}$$

$$-\tfrac{1}{2}R_1 \rightarrow R_1 \quad \begin{bmatrix} 1 & -\tfrac{3}{2} & -\tfrac{3}{2} & \vline & -\tfrac{1}{2} & 0 & 0 \\ 3 & -4 & 1 & \vline & 0 & 1 & 0 \\ -5 & 7 & 2 & \vline & 0 & 0 & 1 \end{bmatrix}$$

$$-3R_1 + R_2 \rightarrow R_2 \quad \begin{bmatrix} 1 & -\tfrac{3}{2} & -\tfrac{3}{2} & \vline & -\tfrac{1}{2} & 0 & 0 \\ 0 & \tfrac{1}{2} & \tfrac{11}{2} & \vline & \tfrac{3}{2} & 1 & 0 \\ -5 & 7 & 2 & \vline & 0 & 0 & 1 \end{bmatrix}$$

$$5R_1 + R_3 \rightarrow R_3 \quad \begin{bmatrix} 1 & -\tfrac{3}{2} & -\tfrac{3}{2} & \vline & -\tfrac{1}{2} & 0 & 0 \\ 0 & \tfrac{1}{2} & \tfrac{11}{2} & \vline & \tfrac{3}{2} & 1 & 0 \\ 0 & -\tfrac{1}{2} & -\tfrac{11}{2} & \vline & -\tfrac{5}{2} & 0 & 1 \end{bmatrix}$$

$$R_2 + R_3 \rightarrow R_3 \quad \begin{bmatrix} 1 & -\tfrac{3}{2} & -\tfrac{3}{2} & \vline & -\tfrac{1}{2} & 0 & 0 \\ 0 & \tfrac{1}{2} & \tfrac{11}{2} & \vline & \tfrac{3}{2} & 1 & 0 \\ 0 & 0 & 0 & \vline & -1 & 1 & 1 \end{bmatrix}$$

Notice that in the last matrix each entry in the third row to the left of the vertical bar is zero. Thus, it is impossible to obtain the identity matrix I_3 on the left side of the vertical bar. Hence, we conclude that A has no inverse and is, therefore, singular.

EXAMPLE 25

Find the solution set of each of the following systems:

a. $\begin{cases} x_1 & - & x_3 = 2 \\ x_1 - x_2 + 2x_3 = -1 \\ -x_1 + x_2 - & x_3 = 4 \end{cases}$ b. $\begin{cases} x_1 & - & x_3 = 0 \\ x_1 - x_2 + 2x_3 = 0 \\ -x_1 + x_2 - & x_3 = 0 \end{cases}$

Solution   Each of these systems can be expressed as a matrix equation in the form $AX = B$ where

$$A = \begin{bmatrix} 1 & 0 & -1 \\ 1 & -1 & 2 \\ -1 & 1 & -1 \end{bmatrix}.$$

From Example 23, we know that A is nonsingular and

$$A^{-1} = \begin{bmatrix} 1 & 1 & 1 \\ 1 & 2 & 3 \\ 0 & 1 & 1 \end{bmatrix}.$$

Hence, each system can be solved as follows:

a. $X = A^{-1}B = \begin{bmatrix} 1 & 1 & 1 \\ 1 & 2 & 3 \\ 0 & 1 & 1 \end{bmatrix} \begin{bmatrix} 2 \\ -1 \\ 4 \end{bmatrix} = \begin{bmatrix} 5 \\ 12 \\ 3 \end{bmatrix}.$

Therefore $x_1 = 5$, $x_2 = 12$, and $x_3 = 3$. It follows that the solution set is $\{(5, 12, 3)\}$.

b. $X = A^{-1}B = \begin{bmatrix} 1 & 1 & 1 \\ 1 & 2 & 3 \\ 0 & 1 & 1 \end{bmatrix} \begin{bmatrix} 0 \\ 0 \\ 0 \end{bmatrix} = \begin{bmatrix} 0 \\ 0 \\ 0 \end{bmatrix}.$

Thus, $\{(0, 0, 0)\}$ is the solution set.

An Application of Matrices: Coding Theory

Coding theory offers an interesting application of the inverse of a matrix. A **cryptogram** is a message written using a secret code. The intelligence agencies of world powers have long used sophisticated methods to encode and decode messages for security purposes.

We now outline a simple way to use matrix multiplication to encode and decode messages. We begin by assigning a positive integer to each letter in the alphabet, and for convenience, to a blank space.

A = 1	B = 2	C = 3
D = 4	E = 5	F = 6
G = 7	H = 8	I = 9
J = 10	K = 11	L = 12
M = 13	N = 14	O = 15
P = 16	Q = 17	R = 18
S = 19	T = 20	U = 21
V = 22	W = 23	X = 24
Y = 25	Z = 26	— = 27

Consider the message:

I LIKE MATHEMATICS.

We use blanks between adjacent words and break the message into groups of three letters.

I _ L I K E _ M A T H E M A T I C S

To each group of three letters there corresponds a 1×3 matrix, as indicated below.

I _ L   I K E   _ M A   T H E   M A T   I C S
[9 27 12] [9 11 5] [27 13 1] [20 8 5] [13 1 20] [9 3 19]

To encode this message, we select any 3×3 nonsingular matrix A and multiply each 1×3 matrix above on the right by A. If we choose the matrix

$$A = \begin{bmatrix} 1 & 0 & -1 \\ 1 & -1 & 2 \\ -1 & 1 & -1 \end{bmatrix}$$

whose inverse is

$$A^{-1} = \begin{bmatrix} 1 & 1 & 1 \\ 1 & 2 & 3 \\ 0 & 1 & 1 \end{bmatrix},$$

as determined in Example 23, we can encode the message I LIKE MATHE-MATICS as follows:

$$[9 \quad 27 \quad 12] \begin{bmatrix} 1 & 0 & -1 \\ 1 & -1 & 2 \\ -1 & 1 & -1 \end{bmatrix} = [24 \quad -15 \quad 33]$$

$$[9 \quad 11 \quad 5] \begin{bmatrix} 1 & 0 & -1 \\ 1 & -1 & 2 \\ -1 & 1 & -1 \end{bmatrix} = [15 \quad -6 \quad 8]$$

$$[27 \quad 13 \quad 1] \begin{bmatrix} 1 & 0 & -1 \\ 1 & -1 & 2 \\ -1 & 1 & -1 \end{bmatrix} = [39 \quad -12 \quad -2]$$

$$[20 \quad 8 \quad 5] \begin{bmatrix} 1 & 0 & -1 \\ 1 & -1 & 2 \\ -1 & 1 & -1 \end{bmatrix} = [23 \quad -3 \quad -9]$$

$$[13 \quad 1 \quad 20] \begin{bmatrix} 1 & 0 & -1 \\ 1 & -1 & 2 \\ -1 & 1 & -1 \end{bmatrix} = [-6 \quad 19 \quad -31]$$

$$[9 \quad 3 \quad 19] \begin{bmatrix} 1 & 0 & -1 \\ 1 & -1 & 2 \\ -1 & 1 & -1 \end{bmatrix} = [-7 \quad 16 \quad -22].$$

Hence, the sequence of encoded matrices that represents the message I LIKE MATHEMATICS is

$$[24 \quad -15 \quad 33][15 \quad -6 \quad 8][39 \quad -12 \quad -2]$$
$$[23 \quad -3 \quad -9][-6 \quad 19 \quad -31][-7 \quad 16 \quad -22].$$

Removing the matrix notation, we get

$$24 \quad -15 \quad 33 \quad 15 \quad -6 \quad 8 \quad 39 \quad -12 \quad -2$$
$$23 \quad -3 \quad -9 \quad -6 \quad 19 \quad -31 \quad -7 \quad 16 \quad -22.$$

You can, no doubt, appreciate the difficulty in attempting to decode the above message without knowing the matrix A. In Example 26, we illustrate a method for decoding a message.

EXAMPLE 26

Decode the message

$$26 \quad -7 \quad 4 \quad 12 \quad 2 \quad -15 \quad 15 \quad -7 \quad 26 \quad 0 \quad 1 \quad 17$$
$$-3 \quad 8 \quad 6 \quad 7 \quad 0 \quad 8 \quad 4 \quad 0 \quad 23,$$

given that it was encoded using the matrix

$$A = \begin{bmatrix} 1 & 0 & -1 \\ 1 & -1 & 2 \\ -1 & 1 & -1 \end{bmatrix}.$$

Solution   We begin by breaking the message into groups of three letters to form the matrices

[26   − 7   4][12   2   − 15][15   − 7   26][0   1   17]

[− 3   8   6][7   0   8][4   0   23].

From our earlier work, recall that

$$A^{-1} = \begin{bmatrix} 1 & 1 & 1 \\ 1 & 2 & 3 \\ 0 & 1 & 1 \end{bmatrix}.$$

Multiplying each encoded row matrix by A^{-1} on the right we obtain the decoded row matrices as follows:

$$[26 \quad -7 \quad 4] \begin{bmatrix} 1 & 1 & 1 \\ 1 & 2 & 3 \\ 0 & 1 & 1 \end{bmatrix} = [19 \quad 16 \quad 9]$$

$$[12 \quad 2 \quad -15] \begin{bmatrix} 1 & 1 & 1 \\ 1 & 2 & 3 \\ 0 & 1 & 1 \end{bmatrix} = [14 \quad 1 \quad 3]$$

$$[15 \quad -7 \quad 26] \begin{bmatrix} 1 & 1 & 1 \\ 1 & 2 & 3 \\ 0 & 1 & 1 \end{bmatrix} = [8 \quad 27 \quad 20]$$

$$[0 \quad 1 \quad 17] \begin{bmatrix} 1 & 1 & 1 \\ 1 & 2 & 3 \\ 0 & 1 & 1 \end{bmatrix} = [1 \quad 19 \quad 20]$$

$$[-3 \quad 8 \quad 6] \begin{bmatrix} 1 & 1 & 1 \\ 1 & 2 & 3 \\ 0 & 1 & 1 \end{bmatrix} = [5 \quad 19 \quad 27]$$

$$[7 \quad 0 \quad 8] \begin{bmatrix} 1 & 1 & 1 \\ 1 & 2 & 3 \\ 0 & 1 & 1 \end{bmatrix} = [7 \quad 15 \quad 15]$$

$$[4 \quad 0 \quad 23] \begin{bmatrix} 1 & 1 & 1 \\ 1 & 2 & 3 \\ 0 & 1 & 1 \end{bmatrix} = [4 \quad 27 \quad 27]$$

Replacing the numbers by their corresponding letter, we obtain the decoded message:

[19   16   9][14   1   3][8   27   20][1   19   20][5   19   27]
  S   P   I N   A   C H __   T A   S   T E   S   __

[7   15   15][4   27   27]
 G   O   O D   __   __

Notice that in order to encode the message SPINACH TASTES GOOD using 1×3 matrices, it was necessary to use two blanks to complete the last matrix [4   27   27].

EXERCISES 9.4

In Exercises 1–4, verify that $AB = BA = I_n$.

1. $A = \begin{bmatrix} 4 & -3 \\ 3 & -2 \end{bmatrix}$, $B = \begin{bmatrix} -2 & 3 \\ -3 & 4 \end{bmatrix}$

2. $A = \begin{bmatrix} 3 & 1 \\ -1 & 2 \end{bmatrix}$, $B = \begin{bmatrix} \frac{2}{7} & -\frac{1}{7} \\ \frac{1}{7} & \frac{3}{7} \end{bmatrix}$

3. $A = \begin{bmatrix} 2 & 1 & 1 \\ 1 & 3 & 1 \\ -1 & 4 & 0 \end{bmatrix}$, $B = \begin{bmatrix} 2 & -2 & 1 \\ \frac{1}{2} & -\frac{1}{2} & \frac{1}{2} \\ -\frac{7}{2} & \frac{9}{2} & -\frac{5}{2} \end{bmatrix}$

4. $A = \begin{bmatrix} 1 & 1 & 1 & 1 \\ 0 & 1 & 1 & 1 \\ 0 & 0 & 1 & 1 \\ 0 & 0 & 0 & 1 \end{bmatrix}$, $B = \begin{bmatrix} 1 & -1 & 0 & 0 \\ 0 & 1 & -1 & 0 \\ 0 & 0 & 1 & -1 \\ 0 & 0 & 0 & 1 \end{bmatrix}$

In Exercises 5–20, find the inverse (if it exists) of the matrix.

5. $\begin{bmatrix} 6 & 1 \\ -4 & 2 \end{bmatrix}$

6. $\begin{bmatrix} -1 & 3 \\ 2 & -6 \end{bmatrix}$

7. $\begin{bmatrix} 1 & 1 \\ -1 & 2 \end{bmatrix}$

8. $\begin{bmatrix} 5 & 5 \\ 2 & -3 \end{bmatrix}$

9. $\begin{bmatrix} 1 & 2 & 1 \\ -1 & 4 & 10 \\ 1 & 1 & 3 \end{bmatrix}$

10. $\begin{bmatrix} 1 & 0 & 1 \\ 0 & 1 & 2 \\ 3 & 5 & 4 \end{bmatrix}$

11. $\begin{bmatrix} 1 & -1 & 2 \\ 0 & 4 & 8 \\ 0 & 0 & 1 \end{bmatrix}$

12. $\begin{bmatrix} 1 & -5 & 6 \\ 0 & 0 & -1 \\ 0 & 0 & -7 \end{bmatrix}$

13. $\begin{bmatrix} 1 & -1 & 1 \\ 1 & 1 & -2 \\ -1 & 3 & 4 \end{bmatrix}$

14. $\begin{bmatrix} 1 & 0 & 0 \\ 2 & 1 & 0 \\ 1 & 5 & 2 \end{bmatrix}$

15. $\begin{bmatrix} 1 & 0 & 1 \\ 1 & 1 & 3 \\ 0 & 1 & 3 \end{bmatrix}$

16. $\begin{bmatrix} 1 & 2 & 1 \\ -1 & 0 & 1 \\ 1 & 4 & 3 \end{bmatrix}$

17. $\begin{bmatrix} 1 & 0 & 1 & 1 \\ -1 & 1 & 0 & 1 \\ 1 & 0 & 2 & -1 \\ 1 & 1 & -1 & 1 \end{bmatrix}$

18. $\begin{bmatrix} 1 & 0 & 0 & 0 \\ 1 & 3 & 0 & 0 \\ -1 & 1 & 2 & 0 \\ 1 & 2 & 1 & 1 \end{bmatrix}$

19. $\begin{bmatrix} 0.1 & 0.1 & 0.1 \\ -0.2 & 0.2 & -0.2 \\ 0.1 & 0.2 & 0.3 \end{bmatrix}$

20. $\begin{bmatrix} 0.1 & 0.2 & 0.3 \\ -0.1 & -0.3 & 0.1 \\ 0.2 & 0.1 & 0.4 \end{bmatrix}$

In Exercises 21 and 22, use the inverse matrix you computed in Exercise 5 to find the solution set for the given system of equations.

21. $\begin{cases} 6x + y = 4 \\ -4x + 2y = -3 \end{cases}$

22. $\begin{cases} 6x + y = 1 \\ -4x + 2y = 5 \end{cases}$

In Exercises 23 and 24, use the inverse matrix you computed in Exercise 13 to find the solution set for the given system of equations.

23. $\begin{cases} x - y + z = -1 \\ x + y - 2z = 0 \\ -x + 3y + 4z = 2 \end{cases}$

24. $\begin{cases} x - y + z = 2 \\ x + y - 2z = 1 \\ -x + 3y + 4z = 2 \end{cases}$

In Exercises 25 and 26, use the inverse matrix you computed in Exercise 17 to find the solution set for each system of equations.

25. $\begin{cases} x_1 \quad\quad + x_3 + x_4 = 1 \\ -x_1 + x_2 \quad\quad + x_4 = 2 \\ x_1 \quad\quad + 2x_3 - x_4 = 0 \\ x_1 + x_2 - x_3 + x_4 = 1 \end{cases}$

26. $\begin{cases} x_1 \quad\quad + x_3 + x_4 = 4 \\ -x_1 + x_2 \quad\quad + x_4 = 1 \\ x_1 \quad\quad + 2x_3 - x_4 = -1 \\ x_1 + x_2 - x_3 + x_4 = 0 \end{cases}$

In Exercises 27 and 28, use the inverse matrix you computed in Exercise 19 to find the solution set for each system of equations.

27. $\begin{cases} 0.1x + 0.1y + 0.1z = 0.7 \\ -0.2x + 0.2y - 0.2z = 0.3 \\ 0.1x + 0.2y + 0.3z = 0.1 \end{cases}$

28. $\begin{cases} 0.1x + 0.1y + 0.1z = 0.2 \\ -0.2x + 0.2y - 0.2z = 0.1 \\ 0.1x + 0.2y + 0.3z = 0.1 \end{cases}$

29. Suppose A is a nonsingular matrix of order n. Prove that the inverse of A is unique. [*Hint*: Assume that B and C are matrices such that $AB = BA = I_n$ and $AC = CA = I_n$. Prove that $B = C$.]

30. Give an example of two matrices A and B, each of order 2, such that $AB = [O]_{2,2}$, but $A \neq [O]_{2,2}$ and $B \neq [O]_{2,2}$.

31. Let A and B be square matrices such that A^{-1} exists. Prove that if $AB = [O]_{n,n}$, then $B = [O]_{n,n}$.

32. Give an example of two nonsingular matrices A and B, each of order 2, such that $(AB)^{-1} \neq A^{-1}B^{-1}$.

33. Let A and B be nonsingular matrices of order n. Prove that $(AB)^{-1} = B^{-1}A^{-1}$.

34. Let A be a nonsingular matrix. Prove that $(A^{-1})^{-1} = A$.

35. Give an example of two nonsingular matrices A and B, each of order n, such that $(A + B)^{-1} \neq A^{-1} + B^{-1}$.

36. Let A be a nonsingular matrix. Prove that
    a. If $AB = AC$, then $B = C$.
    b. If $BA = CA$, then $B = C$.

37. The domain of the function f defined by $f(x) = ax^2 + bx + c$ is $\mathbb{R}$. Hence, the graph of f contains points of the form $(-1, k_1)$, $(0, k_2)$ and $(1, k_3)$. Determine a, b, and c given that

a. $k_1 = 1$, $k_2 = -1$ and $k_3 = 2$.

b. $k_1 = -3$, $k_2 = 4$, and $k_3 = -1$.

38. Let $A = \begin{bmatrix} a & b \\ c & d \end{bmatrix}$, where $ad - bc \neq 0$. Verify that A is nonsingular by showing that the matrix

$$C = \frac{1}{ad - bc} \begin{bmatrix} d & -b \\ -c & a \end{bmatrix}$$

is the inverse of A.

39. Encode each of the following messages using the matrix

$$A = \begin{bmatrix} 3 & 3 & -1 \\ -2 & -2 & 1 \\ -4 & -5 & 2 \end{bmatrix}.$$

a. THE PRESIDENT IS HERE.

b. LONG LIVE FREEDOM.

40. a. Find the inverse of the matrix

$$A = \begin{bmatrix} 2 & 5 & 4 \\ 1 & 4 & 3 \\ 1 & -3 & -2 \end{bmatrix}.$$

b. Decode the following message given that it was encoded using matrix A in part (a):

47	9	15	34	21	21	31	58	46	52	10	15
64	40	39	53	117	94	92	140	117	29	3	7
75	177	141	82	191	152	56	76	63	20	36	29
36	67	55	92	122	103						

9.5 DETERMINANTS AND CRAMER'S RULE

Associated with each square matrix A is a real number called the **determinant** of A. To see how the idea of a determinant arises in connection with systems of linear equations, we consider the system

$$\begin{cases} a_{11}x_1 + a_{12}x_2 = b_1 \\ a_{21}x_1 + a_{22}x_2 = b_2. \end{cases} \tag{1}$$

Using the elimination method, let us find the values of x_1 and x_2 that satisfy system (1). We begin by multiplying the first equation by a_{22} and the second by $-a_{12}$ to obtain the equivalent system

$$\begin{cases} a_{11}a_{22}x_1 + a_{12}a_{22}x_2 = b_1a_{22} \\ -a_{12}a_{21}x_1 - a_{12}a_{22}x_2 = -b_2a_{12}. \end{cases} \tag{2}$$

Adding the first equation in system (2) to the second, we obtain the system

$$\begin{cases} a_{11}a_{22}x_1 + a_{12}a_{22}x_2 = b_1a_{22} \\ (a_{11}a_{22} - a_{12}a_{21})x_1 = b_1a_{22} - b_2a_{12}. \end{cases} \tag{3}$$

If $a_{11}a_{22} - a_{12}a_{21} \neq 0$, we see from the second equation that

$$x_1 = \frac{b_1a_{22} - b_2a_{12}}{a_{11}a_{22} - a_{12}a_{21}}. \tag{4}$$

Substituting this value of x_1 into the first equation in system (3) and solving for x_2, we find that

$$x_2 = \frac{b_2a_{11} - b_1a_{21}}{a_{11}a_{22} - a_{12}a_{21}}. \tag{5}$$

Equations (4) and (5) give us formulas for the solution of system (1) provided that $a_{11}a_{22} - a_{12}a_{21} \neq 0$. However, these formulas are difficult to remember as stated. Notice that the denominator in each of the formulas is $a_{11}a_{22} - a_{12}a_{21}$. Observing that the coefficient matrix for system (1) is

$$A = \begin{bmatrix} a_{11} & a_{12} \\ a_{21} & a_{22} \end{bmatrix},$$

we see that the number $a_{11}a_{22} - a_{12}a_{21}$ is the product of the elements on the main diagonal of A minus the product of the elements off the main diagonal. We call the real number $a_{11}a_{22} - a_{12}a_{21}$ the determinant of the coefficient matrix A.

In general, we have the following definition:

DEFINITION 9.9   **Determinant of a 2 × 2 matrix**

The **determinant** of the 2 × 2 matrix

$$A = \begin{bmatrix} a_{11} & a_{12} \\ a_{21} & a_{22} \end{bmatrix}$$

is the real number $a_{11}a_{22} - a_{12}a_{21}$. The determinant of A is denoted by det A or by $|A|$.

Cramer's Rule for a System of Two Linear Equations

Now that we have defined the determinant of a 2 × 2 matrix, we can express the solution of system (1) using determinants. First, we form the matrix

$$A_1 = \begin{bmatrix} b_1 & a_{12} \\ b_2 & a_{22} \end{bmatrix}$$

by replacing the elements a_{11} and a_{21} in the first column of the coefficient matrix A by the constants b_1 and b_2. Next we form the matrix

$$A_2 = \begin{bmatrix} a_{11} & b_1 \\ a_{21} & b_2 \end{bmatrix}$$

by replacing the elements a_{12} and a_{22} in the second column of A by the constants b_1 and b_2. Since det $A_1 = b_1a_{22} - a_{12}b_2$ and det $A_2 = a_{11}b_2 - b_1a_{21}$, we see from equations (4) and (5) that the solution of system (1) can be written in the form

$$x_1 = \frac{\det A_1}{\det A} \quad \text{and} \quad x_2 = \frac{\det A_2}{\det A},$$

provided det $A \neq 0$. The technique just described for finding the solution of a system of two linear equations in two variables is called **Cramer's Rule**, in honor of the Swiss mathematician Gabriel Cramer (1704–1752). It is important to note that we can use Cramer's Rule only when the determinant of the coefficient matrix is not zero.

E X A M P L E 27

Use Cramer's Rule, if possible, to find the solution set of the system

$$\begin{cases} 4x_1 - x_2 = 10 \\ 3x_1 + 7x_2 = 8. \end{cases}$$

Solution   The determinant of the coefficient matrix A is

$$\det A = \det \begin{bmatrix} 4 & -1 \\ 3 & 7 \end{bmatrix} = 4(7) - (-1)(3) = 31.$$

Since det $A \neq 0$, we can use Cramer's Rule. Forming the matrices A_1 and A_2 using the process outlined above, we have

$$A_1 = \begin{bmatrix} 10 & -1 \\ 8 & 7 \end{bmatrix} \quad \text{and} \quad A_2 = \begin{bmatrix} 4 & 10 \\ 3 & 8 \end{bmatrix}.$$

Since det $A_1 = \det \begin{bmatrix} 10 & -1 \\ 8 & 7 \end{bmatrix} = 10(7) - (-1)(8) = 78$, and det $A_2 = \det \begin{bmatrix} 4 & 10 \\ 3 & 8 \end{bmatrix} = 4(8) - 10(3) = 2$, it follows that, $x_1 = \frac{78}{31}$ and $x_2 = \frac{2}{31}$. Hence, the solution set is $\{(\frac{78}{31}, \frac{2}{31})\}$.

Minors and Cofactors

It is possible to define the determinant of any square matrix. The determinant of the 1×1 matrix $A = [a_{11}]$ is defined to be the real number a_{11}. In defining the determinant of a matrix A of order greater than 2, it is convenient to make the following definition:

DEFINITION 9.10   **Minors and Cofactors**

Let $A = [a_{ij}]$ be a square matrix.

  (i) The **minor** M_{ij} of the element a_{ij} is the determinant of the matrix that remains when the ith row and the jth column of matrix A are removed.
  (ii) The **cofactor** A_{ij} of the element a_{ij} is $A_{ij} = (-1)^{i+j} M_{ij}$.

From part (ii) of Definition 9.10 we see that the cofactor of a_{ij} is the same as its minor if $i + j$ is even, and is the negative of the minor if $i + j$ is odd.

EXAMPLE 28

Let $A = \begin{bmatrix} 2 & -1 & 3 \\ 0 & 1 & 4 \\ -1 & -2 & 5 \end{bmatrix}$. Find the minor and cofactor of each element in row one.

Solution   To find M_{11} and A_{11}, we first delete the first row and first column of the matrix

$$A = \begin{bmatrix} \cancel{2} & \cancel{-1} & \cancel{3} \\ \cancel{0} & 1 & 4 \\ \cancel{-1} & -2 & 5 \end{bmatrix},$$

to obtain the matrix

$$\begin{bmatrix} 1 & 4 \\ -2 & 5 \end{bmatrix}.$$

Hence, $M_{11} = \det\begin{bmatrix} 1 & 4 \\ -2 & 5 \end{bmatrix} = 5 - (-8) = 13$, and $A_{11} = (-1)^{1+1}M_{11}$

$= M_{11} = 13$. To obtain M_{12} and A_{12}, we delete the first row and second column of

$$A = \begin{bmatrix} 2 & -1 & 3 \\ 0 & 1 & 4 \\ -1 & -2 & 5 \end{bmatrix},$$

to get the matrix

$$\begin{bmatrix} 0 & 4 \\ -1 & 5 \end{bmatrix}.$$

Thus,

$$M_{12} = \det\begin{bmatrix} 0 & 4 \\ -1 & 5 \end{bmatrix} = 0 - (-4) = 4 \qquad \text{and}$$

$$A_{12} = (-1)^{1+2}M_{12} = -M_{12} = -4.$$

To compute M_{13} and A_{13}, we delete the first row and third column of

$$A = \begin{bmatrix} 2 & -1 & 3 \\ 0 & 1 & 4 \\ -1 & -2 & 5 \end{bmatrix}$$

to obtain the matrix

$$\begin{bmatrix} 0 & 1 \\ -1 & -2 \end{bmatrix}.$$

Therefore, $M_{13} = \det\begin{bmatrix} 0 & 1 \\ -1 & -2 \end{bmatrix} = 0 - (-1) = 1$, and $A_{13} = (-1)^{1+3}M_{13}$

$= 1$.

Determinant of a $n \times n$ Matrix

Now that we have defined the minor and cofactor of each entry in a square matrix A, we can define the determinant of that matrix.

DEFINITION 9.11   Determinant of an $n \times n$ Matrix

Let A be an $n \times n$ matrix where $n \geq 2$. The **determinant** of A is the sum of the n products obtained by multiplying each entry of any fixed row (or column) by its cofactor.

For example, consider the matrix

$$A = \begin{bmatrix} 2 & -1 & 3 \\ 0 & 1 & 4 \\ -1 & -2 & 5 \end{bmatrix}$$

given in Example 28. From Definition 9.11, if we select the first row of matrix

A, then det $A = 2A_{11} + (-1)A_{12} + 3A_{13}$. In Example 28, we found that $A_{11} = 13$, $A_{12} = -4$, and $A_{13} = 1$. Therefore, det $A = 2(13) + (-1)(-4) + 3(1) = 33$. We say that we have found the determinant of A by **expanding by the first row**. By Definition 9.11, we can find the value of det A by expanding by *any* row (or column). For example, expanding by the second row, we have

$$\text{det } A = 0 \cdot A_{21} + 1 \cdot A_{22} + 4 \cdot A_{23}.$$

Notice that $a_{21} = 0$ and hence, $a_{21}A_{21} = 0$. Thus, if we evaluate det A by expanding by the second row, we only need to determine A_{22} and A_{23}. Using Definition 9.10, it can be shown that $A_{22} = 13$ and $A_{23} = 5$. Thus

$$\begin{aligned} \text{det } A &= 0 \cdot A_{11} + 1 \cdot A_{22} + 4A_{23} \\ &= 0 + 1(13) + 4(5) \\ &= 33. \end{aligned}$$

In general, an efficient way of computing a determinant is to expand by the row (or column) containing the largest number of zero entries.

EXAMPLE 29

Compute the determinant of

$$A = \begin{bmatrix} 0 & -1 & 4 & 5 \\ 6 & 0 & 1 & 0 \\ -5 & 0 & 2 & -7 \\ -1 & 2 & 0 & -1 \end{bmatrix}.$$

Solution By inspection, we see that row 2 and column 2 have the largest number of zero entries. Expanding by column 2, we have

$$\text{det } A = -1 \cdot A_{12} + 0 \cdot A_{22} + 0 \cdot A_{32} + 2 \cdot A_{42}, \qquad \text{or}$$
$$\text{det } A = -1 \cdot A_{12} + 2 \cdot A_{42}. \tag{6}$$

Thus, we need only compute A_{12} and A_{42}. We have

$$\begin{aligned} A_{12} &= (-1)^{1+2} \det \begin{bmatrix} 6 & 1 & 0 \\ -5 & 2 & -7 \\ -1 & 0 & -1 \end{bmatrix} \\ &= -\det \begin{bmatrix} 6 & 1 & 0 \\ -5 & 2 & -7 \\ -1 & 0 & -1 \end{bmatrix} \\ &= -\left(6 \cdot (-1)^{1+1} \cdot \det \begin{bmatrix} 2 & -7 \\ 0 & -1 \end{bmatrix} \right. \\ &\qquad \left. + 1 \cdot (-1)^{1+2} \cdot \det \begin{bmatrix} -5 & -7 \\ -1 & -1 \end{bmatrix} \right) \quad \textbf{Expanding by the first row} \\ &= -[6(1)(-2) + 1(-1)(-2)] \\ &= 10. \end{aligned}$$

Computing A_{42}, we have

$$A_{42} = (-1)^{4+2} \cdot \det \begin{bmatrix} 0 & 4 & 5 \\ 6 & 1 & 0 \\ -5 & 2 & -7 \end{bmatrix}$$

$$= \det \begin{bmatrix} 0 & 4 & 5 \\ 6 & 1 & 0 \\ -5 & 2 & -7 \end{bmatrix}$$

$$= 4 \cdot (-1)^{1+2} \cdot \det \begin{bmatrix} 6 & 0 \\ -5 & -7 \end{bmatrix}$$

$$+ 5 \cdot (-1)^{1+3} \cdot \det \begin{bmatrix} 6 & 1 \\ -5 & 2 \end{bmatrix} \quad \text{Expanding by the first row}$$

$$= 4(-1)(-42) + 5(1)(17) = 253.$$

Substituting 10 for A_{12} and 253 for A_{42} in equation (6) yields

$$\det A = -1(10) + 2(253)$$
$$= 496.$$

In general, calculating the determinant of a matrix of fourth-order, or larger, can be arduous. However, in Example 29, the calculation of the determinant of the fourth-order matrix A is greatly simplified since A has a column (column 2) with two zero entries. Recall that in Section 9.2, we used elementary row operations to introduce zeros as entries of matrices. It is natural to ask how the value of the determinant of a matrix A is affected by performing an elementary row operation on A. This question is easily answered by the following theorem:

THEOREM 9.5

Let A and B be square matrices.

(i) If B is obtained from A by interchanging two rows (or columns) of A, then $\det B = -\det A$.

(ii) If B is obtained from A by adding a multiple of a row (or column) of A to another row (or column) of A, then $\det B = \det A$.

(iii) If B is obtained from A by multiplying a row (or column) of A by a nonzero constant c, then $\det B = c \det A$.

EXAMPLE 30

Evaluate the determinant of the matrix

$$A = \begin{bmatrix} 3 & -3 & 1 & 2 \\ -1 & 1 & 4 & 2 \\ 0 & 1 & -3 & 5 \\ -4 & 4 & 16 & 3 \end{bmatrix}.$$

Solution Using elementary row operations, we introduce zeros in the first column as follows:

$$\det A = \det \begin{bmatrix} 3 & -3 & 1 & 2 \\ -1 & 1 & 4 & 2 \\ 0 & 1 & -3 & 5 \\ -4 & 4 & 16 & 3 \end{bmatrix}$$

$$(-1)R_2 \rightarrow R_2 \\ = -\det \begin{bmatrix} 3 & -3 & 1 & 2 \\ 1 & -1 & -4 & -2 \\ 0 & 1 & -3 & 5 \\ -4 & 4 & 16 & 3 \end{bmatrix}$$ Theorem 9.5, part (iii)

$$R_1 \leftrightarrow R_2 \\ = \det \begin{bmatrix} 1 & -1 & -4 & -2 \\ 3 & -3 & 1 & 2 \\ 0 & 1 & -3 & 5 \\ -4 & 4 & 16 & 3 \end{bmatrix}$$ Theorem 9.5, part (i)

$$-3R_1 + R_2 \rightarrow R_2 \\ = \det \begin{bmatrix} 1 & -1 & -4 & -2 \\ 0 & 0 & 13 & 8 \\ 0 & 1 & -3 & 5 \\ -4 & 4 & 16 & 3 \end{bmatrix}$$ Theorem 9.5, part (ii)

$$4R_1 + R_4 \rightarrow R_4 \\ = \det \begin{bmatrix} 1 & -1 & -4 & -2 \\ 0 & 0 & 13 & 8 \\ 0 & 1 & -3 & 5 \\ 0 & 0 & 0 & -5 \end{bmatrix}$$ Theorem 9.5, part (ii)

$$= 1(-1)^{1+1} \det \begin{bmatrix} 0 & 13 & 8 \\ 1 & -3 & 5 \\ 0 & 0 & -5 \end{bmatrix}$$ Expanding by the first column

$$= \det \begin{bmatrix} 0 & 13 & 8 \\ 1 & -3 & 5 \\ 0 & 0 & -5 \end{bmatrix}$$

$$= 1 \cdot (-1)^{2+1} \det \begin{bmatrix} 13 & 8 \\ 0 & -5 \end{bmatrix}$$ Expanding by the first column

$$= -(-65) = 65.$$

Cramer's Rule

Now that we have defined the determinant of an $n \times n$ matrix, we can generalize Cramer's Rule to solve systems containing n linear equations in n variables, x_1, $x_2, \ldots, x_n$.

THEOREM 9.6   Cramer's Rule

If a system of n linear equations in the n variables, $x_1, x_2, \ldots, x_n$, has coefficient matrix A, with $\det A \neq 0$, then the solution of the system is given by

$$x_1 = \frac{\det A_1}{\det A}, \quad x_2 = \frac{\det A_2}{\det A}, \ldots, \quad x_n = \frac{\det A_n}{\det A},$$

where the matrix A_i is formed by replacing the ith column of A by the column of constants in the given system.

 CAUTION: Cramer's Rule does not apply if the number of equations in a system is not the same as the number of variables in that system.

From Cramer's Rule it follows that a system of n linear equations in n variables, having a coefficient matrix with a nonzero determinant, has a unique solution. However, if the determinant of the coefficient matrix is zero, Cramer's Rule does not apply. Such systems are either inconsistent or are consistent and dependent. It is worth noting that Cramer's Rule is not an efficient way to solve a system of linear equations that contains more than three equations since the determinants of so many matrices must be evaluated.

EXAMPLE 31

Use Cramer's Rule, if possible, to find the solution set of the system

$$\begin{cases} x_1 - x_2 + 2x_3 = 4 \\ x_1 + 3x_2 - x_3 = 1 \\ -x_1 + x_2 - x_3 = -5. \end{cases}$$

Solution The coefficient matrix of this system is

$$A = \begin{bmatrix} 1 & -1 & 2 \\ 1 & 3 & -1 \\ -1 & 1 & -1 \end{bmatrix}.$$

Using the definition of the determinant of a matrix, we find that det $A = 4$. Hence, Cramer's Rule is applicable. We now form the matrices A_1, A_2, and A_3 as defined in Cramer's Rule.

$$A_1 = \begin{bmatrix} 4 & -1 & 2 \\ 1 & 3 & -1 \\ -5 & 1 & -1 \end{bmatrix},$$

$$A_2 = \begin{bmatrix} 1 & 4 & 2 \\ 1 & 1 & -1 \\ -1 & -5 & -1 \end{bmatrix}, \quad \text{and} \quad A_3 = \begin{bmatrix} 1 & -1 & 4 \\ 1 & 3 & 1 \\ -1 & 1 & -5 \end{bmatrix}.$$

You should verify that det $A_1 = 18$, det $A_2 = -6$, and det $A_3 = -4$. From Cramer's Rule, it follows that

$$x_1 = \frac{\det A_1}{\det A} = \frac{18}{4} = \frac{9}{2}$$

$$x_2 = \frac{\det A_2}{\det A} = \frac{-6}{4} = -\frac{3}{2}$$

$$x_3 = \frac{\det A_3}{\det A} = \frac{-4}{4} = -1.$$

Hence, the solution set is $\{(\frac{9}{2}, -\frac{3}{2}, -1)\}$.

EXERCISES 9.5

In Exercises 1–4 , find the minor and cofactor for each entry in the given matrix.

1. $\begin{bmatrix} 1 & 1 & -1 \\ 0 & -1 & 1 \\ -1 & 2 & -1 \end{bmatrix}$

2. $\begin{bmatrix} 1 & -1 & 3 \\ 2 & 1 & 0 \\ 1 & -2 & -1 \end{bmatrix}$

3. $\begin{bmatrix} 2 & 0 & 0 & 0 \\ 0 & -3 & 0 & 0 \\ 0 & 0 & 4 & 0 \\ 0 & 0 & 0 & 5 \end{bmatrix}$

4. $\begin{bmatrix} -1 & 0 & 0 & 0 \\ 0 & -7 & 0 & 0 \\ 1 & 1 & -5 & 0 \\ 0 & 0 & 0 & 1 \end{bmatrix}$

In Exercises 5–20, find the determinant of each matrix.

5. $\begin{bmatrix} 6 & -1 \\ 4 & 5 \end{bmatrix}$

6. $\begin{bmatrix} 11 & -6 \\ 2 & 7 \end{bmatrix}$

7. $\begin{bmatrix} \sin x & \cos x \\ \cos x & -\sin x \end{bmatrix}$

8. $\begin{bmatrix} e^x & xe^x \\ e^x & xe^x + e^x \end{bmatrix}$

9. $\begin{bmatrix} 1 & 1 & -1 \\ 0 & -1 & 1 \\ -1 & 2 & -1 \end{bmatrix}$

10. $\begin{bmatrix} -6 & 1 & 4 \\ 2 & -1 & 5 \\ -1 & -1 & 1 \end{bmatrix}$

11. $\begin{bmatrix} 3 & 2 & 1 \\ 1 & 1 & -1 \\ 4 & 2 & 5 \end{bmatrix}$

12. $\begin{bmatrix} -6 & 1 & 4 \\ -2 & 1 & -5 \\ -1 & -1 & 1 \end{bmatrix}$

13. $\begin{bmatrix} 1 & 1 & 2 \\ 4 & 3 & 6 \\ 1 & -1 & -2 \end{bmatrix}$

14. $\begin{bmatrix} 2 & -1 & 2 \\ 1 & 3 & 2 \\ 5 & 1 & 6 \end{bmatrix}$

15. $\begin{bmatrix} 2 & 0 & 0 & 0 \\ 0 & -3 & 0 & 0 \\ 0 & 0 & 4 & 0 \\ 0 & 0 & 0 & 5 \end{bmatrix}$

16. $\begin{bmatrix} -1 & 0 & 0 & 0 \\ 0 & -7 & 0 & 0 \\ 1 & 1 & -5 & 0 \\ 2 & -2 & 0 & 1 \end{bmatrix}$

17. $\begin{bmatrix} i & j & k \\ 1 & 2 & -1 \\ 1 & 3 & 0 \end{bmatrix}$

18. $\begin{bmatrix} i & j & k \\ 2 & 1 & 0 \\ 1 & -1 & 1 \end{bmatrix}$

19. $\begin{bmatrix} 1 & 1 & -1 & 2 \\ 4 & 1 & 1 & 0 \\ 5 & -1 & 0 & 1 \\ 0 & 0 & 1 & -2 \end{bmatrix}$

20. $\begin{bmatrix} 2 & 4 & 0 & 1 \\ 0 & 1 & 2 & -1 \\ 1 & 1 & 0 & 2 \\ 1 & 0 & 1 & 0 \end{bmatrix}$

In Exercises 21–24, express the determinant as a polynomial in x.

21. $\begin{bmatrix} 2-x & 1 \\ -1 & -x \end{bmatrix}$

22. $\begin{bmatrix} 1-x & -3 \\ 2 & 3-x \end{bmatrix}$

23. $\begin{bmatrix} x-5 & -8 & -16 \\ -4 & x-1 & -8 \\ 4 & 4 & x+11 \end{bmatrix}$

24. $\begin{bmatrix} x-2 & 0 & -1 & -2 \\ 0 & x-2 & 1 & -3 \\ 0 & 0 & x+3 & -1 \\ 0 & 0 & 0 & x-4 \end{bmatrix}$

25. Determine the zeros of the polynomial found in
 a. Exercise 21. **b.** Exercise 22.
 c. Exercise 23. **d.** Exercise 24.

26. If $A = \begin{bmatrix} a_{11} & a_{12} & a_{13} & \cdots & a_{1n} \\ 0 & a_{22} & a_{23} & \cdots & a_{2n} \\ 0 & 0 & a_{33} & \cdots & a_{3n} \\ \cdot & \cdot & \cdot & \cdot & \cdot \\ \cdot & \cdot & \cdot & \cdot & \cdot \\ \cdot & \cdot & \cdot & \cdot & \cdot \\ 0 & 0 & 0 & \cdots & a_{nn} \end{bmatrix}$,

show that $\det A = a_{11} \cdot a_{22} \cdot \ldots \cdot a_{nn}$, which is the product of the elements on the main diagonal.

In Exercises 27–30, use the result stated in Exercise 26 to evaluate the determinant of each matrix.

27. $\begin{bmatrix} 4 & 1 & 1 & 2 & 3 \\ 0 & 5 & 1 & 0 & 2 \\ 0 & 0 & -1 & 1 & 0 \\ 0 & 0 & 0 & -3 & 1 \\ 0 & 0 & 0 & 0 & 7 \end{bmatrix}$

28. $\begin{bmatrix} 1 & 2 & 1 & 1 & 4 \\ 0 & -1 & 1 & 0 & 1 \\ 0 & 0 & -4 & 2 & 1 \\ 0 & 0 & 0 & 5 & 1 \\ 0 & 0 & 0 & 0 & -2 \end{bmatrix}$

29. $\begin{bmatrix} 1 & 0 & 0 & 0 \\ 0 & 1 & 0 & 0 \\ 0 & 0 & 1 & 0 \\ 0 & 0 & 0 & 1 \end{bmatrix}$

30. $\begin{bmatrix} 4 & 0 & 0 & 0 \\ 0 & 4 & 0 & 0 \\ 0 & 0 & 4 & 0 \\ 0 & 0 & 0 & 4 \end{bmatrix}$

31. Let $A = \begin{bmatrix} 3 & 0 & 0 & 0 \\ 1 & 2 & 0 & 0 \\ 4 & 0 & 2 & 0 \\ 3 & -1 & 1 & 3 \end{bmatrix}$.

 a. Examine Exercise 26 and use your intuition to make an educated guess for the value of $\det A$.
 b. Use Definition 9.11 on page 465 to find $\det A$.
 c. Find a formula for

$$\det \begin{bmatrix} a_{11} & 0 & 0 & \cdots & 0 \\ a_{21} & a_{22} & 0 & \cdots & 0 \\ a_{31} & a_{32} & a_{33} & \cdots & 0 \\ \cdot & \cdot & \cdot & \cdot & \cdot \\ \cdot & \cdot & \cdot & \cdot & \cdot \\ \cdot & \cdot & \cdot & \cdot & \cdot \\ a_{n1} & a_{n2} & a_{n3} & & a_{nn} \end{bmatrix}.$$

32. Let A be any square matrix. Show that $\det A = 0$ if A satisfies any one of the following conditions.
 a. A row (or column) of A has 0 as each of its entries.

b. Two rows (or columns) of A are identical.

c. One row (or column) of A is a multiple of another row (or column) of A.

In Exercises 33–38, state the property of determinants given in Theorem 9.5 or Exercise 32 that justifies the equation.

33. $\det \begin{bmatrix} 4 & -3 \\ 0 & 0 \end{bmatrix} = 0$

34. $\det \begin{bmatrix} 4 & -3 \\ -8 & 6 \end{bmatrix} = 0$

35. $\det \begin{bmatrix} 1 & 1 & -1 \\ 2 & 2 & -1 \\ 3 & -3 & 0 \end{bmatrix} = \det \begin{bmatrix} 1 & 1 & -1 \\ 2 & 2 & -1 \\ 0 & -6 & 3 \end{bmatrix}$

36. $\det \begin{bmatrix} 2 & -1 & -4 \\ -3 & 1 & 2 \\ 6 & -1 & -1 \end{bmatrix} = \det \begin{bmatrix} 2 & -1 & -4 \\ 9 & -1 & 0 \\ 6 & -1 & -1 \end{bmatrix}$

37. $\det \begin{bmatrix} 9 & 3 & 15 \\ 3 & -3 & 6 \\ 0 & 12 & 9 \end{bmatrix} = 3^3 \det \begin{bmatrix} 3 & 1 & 5 \\ 1 & -1 & 2 \\ 0 & 4 & 3 \end{bmatrix}$

38. $\det \begin{bmatrix} 2 & 1 & 3 \\ -2 & -3 & 1 \\ 1 & 0 & 4 \end{bmatrix} = \frac{1}{4^3} \det \begin{bmatrix} 8 & 4 & 12 \\ -8 & -12 & 4 \\ 4 & 0 & 16 \end{bmatrix}$

In Exercises 39–48, use Cramer's Rule (if applicable) to find the solution set of the given system of equations. If Cramer's Rule does not apply, use Gaussian elimination to solve the system.

39. $\begin{cases} 6x - y = 4 \\ 4x + 5y = -7 \end{cases}$

40. $\begin{cases} 11x - 6y = -3 \\ 2x + 7y = 5 \end{cases}$

41. $\begin{cases} 8x - 9y = 14 \\ 15x + 8y = 5 \end{cases}$

42. $\begin{cases} 19x - 42y = 10 \\ 52x + 14y = -7 \end{cases}$

43. $\begin{cases} x + y - z = 2 \\ -y + z = 5 \\ -x + 2y - z = 0 \end{cases}$

44. $\begin{cases} x - y + 3z = 1 \\ 2x + y = 1 \\ x - 2y - z = 1 \end{cases}$

45. $\begin{cases} 5x + 6y - z = 1 \\ 6x - 8y - 4z = -3 \\ -3x + 4y + 2z = 0 \end{cases}$

46. $\begin{cases} 10x + 10y - 10z = 3 \\ 15x - 13y + 10z = -1 \\ \frac{5}{2}x + \frac{5}{2}y - \frac{5}{2}z = 4 \end{cases}$

47. $\begin{cases} -x + 3y - 4z = 8 \\ x + y - 6z = -1 \\ 4x + 6y - z = 2 \end{cases}$

48. $\begin{cases} 2x - y - z = 1 \\ x + 2y + z = 0 \\ 3x + y - z = 5 \end{cases}$

49. If r_1, r_2, r_3, and r_4 are the four fourth roots of unity, show that

$$\det \begin{bmatrix} r_1 & r_2 & r_3 & r_4 \\ r_2 & r_3 & r_4 & r_1 \\ r_3 & r_4 & r_1 & r_2 \\ r_4 & r_1 & r_2 & r_3 \end{bmatrix} = 0.$$

50. It can be shown that the area A of a triangle determined by three noncollinear points $P_1 = (x_1, y_1)$, $P_2 = (x_2, y_2)$, and $P_3 = (x_3, y_3)$ is given by the formula $A = \frac{1}{2}|\det M|$, where

$$M = \begin{bmatrix} x_1 & y_1 & 1 \\ x_2 & y_2 & 1 \\ x_3 & y_3 & 1 \end{bmatrix}.$$

Verify this formula for the triangle in the accompanying figure.

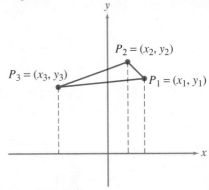

In Exercises 51–54, use the formula for the area of a triangle given in Exercise 50 to find the area of the triangle determined by the given points.

51. $P_1 = (4, 2)$, $P_2 = (-1, -10)$, $P_3 = (1, 6)$

52. $P_1 = (1, -1)$, $P_2 = (-6, 2)$, $P_3 = (4, 7)$

53. $P_1 = (\frac{1}{2}, -\frac{1}{3})$, $P_2 = (2, -\frac{1}{10})$, $P_3 = (4, \frac{2}{3})$

54. $P_1 = (0.4, -0.6)$, $P_2 = (1.6, 2.3)$, $P_3 = (4.1, -0.9)$

9.6 SYSTEMS OF NONLINEAR EQUATIONS

We know formulas for finding the area of certain regions in the plane such as circles, rectangles, triangles, and trapezoids. In calculus, techniques are developed that can be used to calculate the area of more general regions such as the region R_0 shown in Figure 9.6 on page 472.

From Figure 9.6, we see that region R_0 is bounded by the graphs of $x - y = 6$ and $x^2 - y = 8$. To calculate the area of R_0, it is necessary to determine the points of intersection of these graphs. While the techniques needed to calculate this area are beyond the scope of this course, we can, however, find the

FIGURE 9.6

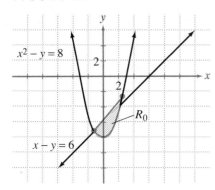

points of intersection of the two graphs. Since each of these points lies on both graphs, we can find each by solving the system of equations

$$\begin{cases} x - y = 6 \\ x^2 - y = 8. \end{cases} \qquad (1)$$

Solving Systems of Nonlinear Equations by Substitution

System (1) is called a **system of nonlinear equations** since it contains the equation $x^2 - y = 8$, which is not linear. Substitution is often an efficient method for finding the solution set of a system of nonlinear equations, especially if the system contains a linear equation. For example, we can find the solution set of system (1) using substitution as follows.

We begin by solving the first equation in system (1) for y to get $y = x - 6$. Substituting $x - 6$ for y in the second equation, we obtain

$$x^2 - (x - 6) = 8, \qquad \text{or}$$
$$x^2 - x - 2 = 0. \qquad (2)$$

Solving equation (2) by factoring, we get

$$(x - 2)(x + 1) = 0$$
$$x = 2 \qquad \text{or} \qquad x = -1.$$

Substituting these values of x into the equation $y = x - 6$, we find that

$$y = -4 \quad \text{when } x = 2 \qquad \text{and} \qquad y = -7 \quad \text{when } x = -1.$$

You should verify that the coordinates of the points $(2, -4)$ and $(-1, -7)$ satisfy system (1). It follows that the solution set for this system is $\{(2, -4), (-1, -7)\}$.

If a system of equations contains only two variables, and if the equations in the system can be readily graphed, we often graph each of them before we solve the system algebraically. By graphing the equations, we get an idea of how many solutions with real coordinates to expect, and we find the approximate location of each. While the coordinates of each solution of system (1) are real numbers, it is possible for the coordinates of a solution of a system of nonlinear equations to be nonreal complex numbers. Such a solution will not appear on the graphs of the equations that compose the system.

Before proceeding further, a word of caution is in order.

 CAUTION: Extraneous solutions sometimes arise in solving a system of nonlinear equations. It is, therefore, necessary to check each potential solution in the original system.

EXAMPLE 32

Find the solution set of the system

$$\begin{cases} -x + y = 1 \\ 4x^2 + y^2 = 4. \end{cases} \qquad (3)$$

Solution   The graphs of the two equations are the line and the ellipse shown in Figure 9.7. Since the graphs of these equations intersect at two points, we expect this system to have two solutions whose coordinates are real numbers.

Noting that the first equation in system (3) is linear, we use substitution to find the solution set. Solving the first equation in system (3) for x, we get

FIGURE 9.7

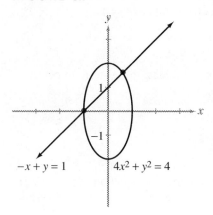

$-x + y = 1$ $4x^2 + y^2 = 4$

$$x = y - 1.$$

Substituting $y - 1$ for x in the second equation in system (3), we have

$$4(y - 1)^2 + y^2 = 4$$
$$4y^2 - 8y + 4 + y^2 = 4$$
$$5y^2 - 8y = 0$$
$$y(5y - 8) = 0$$
$$y = 0 \quad \text{or} \quad y = \frac{8}{5}. \tag{4}$$

Substituting these values of y into the equation $x = y - 1$, we get

$$x = -1 \quad \text{when } y = 0 \quad \text{and} \quad x = \frac{3}{5} \quad \text{when } y = \frac{8}{5}.$$

Since it can be verified that $(-1, 0)$ and $(\frac{3}{5}, \frac{8}{5})$ satisfy system (3), the solution set is $\{(-1, 0), (\frac{3}{5}, \frac{8}{5})\}$.

It is instructive for us to see what happens in the solution of Example 32 if we substitute the values of y given in equation (4) into the second equation of system (3). When $y = 0$, we get

$$4x^2 + 0^2 = 4$$
$$4x^2 = 4$$
$$x = \pm 1.$$

When $y = \frac{8}{5}$, we get

$$4x^2 + \left(\frac{8}{5}\right)^2 = 4$$
$$4x^2 = \frac{36}{25}$$
$$x = \pm \frac{3}{5}.$$

Thus, it appears that we have two additional solutions, namely $(1, 0)$ and $(-\frac{3}{5}, \frac{8}{5})$. However, you can show that neither of these points satisfies the equation $-x + y = 1$. Hence, $(1, 0)$ and $(-\frac{3}{5}, \frac{8}{5})$ are extraneous solutions of system (3).

Solving Systems of Nonlinear Equations by Elimination

Systems of nonlinear equations can often be solved using the elimination method as the following example illustrates.

EXAMPLE 33

Use the elimination method to find the solution set for the system

$$\begin{cases} x^2 + y^2 = 5 \\ 4x^2 + 9y^2 = 36. \end{cases} \tag{5}$$

Solution The graphs of the two equations are the circle and the ellipse shown in Figure 9.8. From Figure 9.8, we expect this system of equations to have four solutions with real numbers as their coordinates. Either variable can be elimi-

FIGURE 9.8

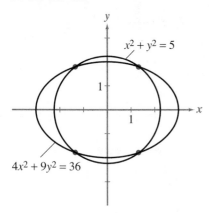

nated from system (5), and we arbitrarily choose to eliminate the variable x. If we multiply the first equation by -4, we get the equivalent system

$$\begin{cases} -4x^2 - 4y^2 = -20 \\ 4x^2 + 9y^2 = 36. \end{cases} \qquad (6)$$

By adding the two equations in system (6), we eliminate the x^2-term to get

$$5y^2 = 16.$$

Solving for y, we find that

$$y = \pm \frac{4\sqrt{5}}{5}.$$

We now substitute each of these values of y into the first equation in system (5). (We could also substitute these values of y into the second equation of system (5).) If $y = 4\sqrt{5}/5$, then

$$x^2 + \left(\frac{4\sqrt{5}}{5}\right)^2 = 5$$

$$x^2 + \frac{16}{5} = 5$$

$$x^2 = \frac{9}{5}$$

$$x = \pm \frac{3\sqrt{5}}{5}.$$

Similarly, if $y = -4\sqrt{5}/5$, $x = \pm 3\sqrt{5}/5$. You can verify that the solution set is

$$\left\{ \left(-\frac{3\sqrt{5}}{5}, -\frac{4\sqrt{5}}{5}\right), \left(-\frac{3\sqrt{5}}{5}, \frac{4\sqrt{5}}{5}\right), \right.$$
$$\left. \left(\frac{3\sqrt{5}}{5}, -\frac{4\sqrt{5}}{5}\right), \left(\frac{3\sqrt{5}}{5}, \frac{4\sqrt{5}}{5}\right) \right\}.$$

EXAMPLE 34

Find the solution set for the system

$$\begin{cases} 2x^2 - y^2 = 1 \\ xy = 1. \end{cases} \qquad (7)$$

Solution   We begin by graphing each of these hyperbolas in Figure 9.9. Figure 9.9 suggests that there are two solutions whose coordinates are real numbers. We will solve this system by substitution. Solving the equation $xy = 1$ for y, we get $y = 1/x$. Substituting $1/x$ for y in the first equation in the system yields

$$2x^2 - \left(\frac{1}{x}\right)^2 = 1$$

$$2x^2 - \frac{1}{x^2} = 1$$

$$2x^4 - x^2 - 1 = 0$$

$$(x^2 - 1)(2x^2 + 1) = 0.$$

It follows from the last equation that

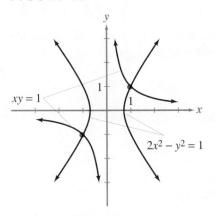

FIGURE 9.9

$$x = \pm 1 \quad \text{or} \quad x = \pm \frac{\sqrt{2}}{2}i.$$

Substituting these values of x into the equation $y = 1/x$ gives the solutions $(-1, -1)$, $(1, 1)$, $\left(-\frac{\sqrt{2}}{2}i, \sqrt{2}i\right)$, and $\left(\frac{\sqrt{2}}{2}i, -\sqrt{2}i\right)$. You can check to see that all four of these points satisfy system (7). Hence, all four points are solutions, although only the two points whose coordinates are real numbers appear as the points of intersection of the graphs shown in Figure 9.9. Therefore, the solution set of system (7) is

$$\left\{(-1, -1), (1, 1), \left(-\frac{\sqrt{2}}{2}i, \sqrt{2}i\right)\left(\frac{\sqrt{2}}{2}i, -\sqrt{2}i\right)\right\}.$$

EXAMPLE 35

Find the solution set of the system

$$\begin{cases} x^2 + y^2 = 25 \\ 9x^2 + 4y^2 = 40. \end{cases} \qquad (8)$$

FIGURE 9.10

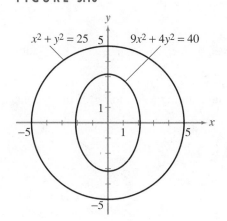

Solution   In Figure 9.10 we see that the graphs of the equations in system (8) do not intersect. Therefore, any solution of this system must have complex numbers as coordinates. We will solve system (8) using the elimination method. Multiplying the first equation by -4, we obtain the equivalent system

$$\begin{cases} -4x^2 - 4y^2 = -100 \\ 9x^2 + 4y^2 = 40. \end{cases} \qquad (9)$$

Adding the two equations in system (9) we get the equation

$$5x^2 = -60,$$

whose solutions are $x = -2\sqrt{3}i$ and $x = 2\sqrt{3}i$. Substituting $-2\sqrt{3}i$ for x in the equation $x^2 + y^2 = 25$, we find that $y = \pm\sqrt{37}$. Similarly, if $x = 2\sqrt{3}i$, $y = \pm\sqrt{37}$. You can verify that the solution set for system (8) is

$$\{(-2\sqrt{3}i, -\sqrt{37}), (-2\sqrt{3}i, \sqrt{37}), (2\sqrt{3}i, -\sqrt{37}), (2\sqrt{3}i, \sqrt{37})\}.$$

Additional Techniques for Solving Systems of Nonlinear Equations

We now illustrate other techniques for solving systems of nonlinear equations.

EXAMPLE 36

Find the solution set of the system

$$\begin{cases} \dfrac{1}{x^2} + \dfrac{2}{y^2} = 7 \\ \dfrac{3}{x^2} - \dfrac{1}{y^2} = 0. \end{cases} \qquad (10)$$

Solution   If we let $u = 1/x^2$ and $v = 1/y^2$, then system (10) becomes the system of linear equations

$$\begin{cases} u + 2v = 7 \\ 3u - v = 0. \end{cases} \qquad (11)$$

You should verify that the values of u and v that satisfy system (11) are $u = 1$ and $v = 3$. Substituting 1 for u in the equation $u = 1/x^2$ yields

$$\frac{1}{x^2} = 1, \qquad \text{or} \qquad x = \pm 1.$$

Similarly, substituting 3 for v in the equation $v = 1/y^2$ yields

$$\frac{1}{y^2} = 3, \qquad \text{or} \qquad y = \frac{\pm\sqrt{3}}{3}.$$

You can verify that the solution set for system (10) is

$$\left\{ \left(-1, -\frac{\sqrt{3}}{3} \right), \left(-1, \frac{\sqrt{3}}{3} \right), \left(1, -\frac{\sqrt{3}}{3} \right), \left(1, \frac{\sqrt{3}}{3} \right) \right\}.$$

EXAMPLE 37

Find the solution set of the system

$$\begin{cases} 6x^2 + xy - 12y^2 = 0 \\ 3x^2 + 2xy = 1. \end{cases} \qquad (12)$$

Solution   The first equation can be solved by factoring. Since

$$(2x + 3y)(3x - 4y) = 0,$$

it follows that $y = -\frac{2}{3}x$ or $y = \frac{3}{4}x$. Substituting $-\frac{2}{3}x$ for y in the second equation of system (12), we have

$$3x^2 + 2x\left(-\frac{2}{3}x \right) = 1$$

$$3x^2 - \frac{4}{3}x^2 = 1$$

$$\frac{5}{3}x^2 = 1$$

$$x = \pm\frac{\sqrt{15}}{5}.$$

Since $y = -\frac{2}{3}x$, it follows that $y = \mp 2\sqrt{15}/15$. Substituting $\frac{3}{4}x$ for y in the equation $3x^2 + 2xy = 1$ yields

$$3x^2 + 2x\left(\frac{3}{4}x \right) = 1$$

$$\frac{9}{2}x^2 = 1$$

$$x = \pm\frac{\sqrt{2}}{3}.$$

Since $y = \frac{3}{4}x$, we have $y = \pm(\sqrt{2}/4)$. You should verify that the solution set for system (12) is

$$\left\{\left(-\frac{\sqrt{15}}{5}, \frac{2\sqrt{15}}{15}\right)\left(\frac{\sqrt{15}}{5}, -\frac{2\sqrt{15}}{15}\right), \left(-\frac{\sqrt{2}}{3}, -\frac{\sqrt{2}}{4}\right)\left(\frac{\sqrt{2}}{3}, \frac{\sqrt{2}}{4}\right)\right\}.$$

EXAMPLE 38

Find the points of intersection of the graphs of the equations $y = 2^x$ and $y = 2^{2x} - 6$.

FIGURE 9.11

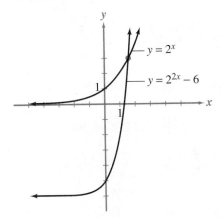

Solution   From the graphs of the two equations sketched in Figure 9.11, we expect to find only one point of intersection. This point of intersection (x, y) must satisfy the system

$$\begin{cases} y = 2^x \\ y = 2^{2x} - 6. \end{cases} \qquad (13)$$

In particular, the point of intersection will be the solution of system (13), which has real numbers as its coordinates. Using the laws of exponents we can express the second equation in system (13) in the form

$$y = (2^x)^2 - 6. \qquad (14)$$

From the first equation in system (13), we know that $y = 2^x$. Substituting y for 2^x in equation (14) yields

$$y = y^2 - 6$$
$$y^2 - y - 6 = 0$$
$$(y + 2)(y - 3) = 0.$$

From the last equation, we see that $y = -2$ or $y = 3$. If $y = -2$, the first equation in system (13) becomes $-2 = 2^x$. Since 2^x must be positive, we discard $y = -2$. On the other hand, if $y = 3$, we get $3 = 2^x$. Solving this last equation for x, we find that

$$\ln 3 = \ln 2^x = x \ln 2, \qquad \text{or}$$
$$x = \frac{\ln 3}{\ln 2}.$$

Therefore, the point of intersection of the two graphs is

$$\left(\frac{\ln 3}{\ln 2}, 3\right) \approx (1.58, 3).$$

EXAMPLE 39

A civic club adopted a project that would cost $1020. Before the project was completed, eight new members joined the club and agreed to pay their share of the cost of the project. The cost per member was thereby reduced by $2. Find the original number of members and the original cost per member for this project.

Solution   We begin by letting x denote the original number of members, and y denote the original cost, in dollars, per member. Then $x + 8$ denotes the total number of members, while $y - 2$ denotes the (new) cost per member. Since the total cost of the project was $1020, we know that

$$xy = \$1020$$

and we also know that

$$(x + 8)(y - 2) = \$1020.$$

Hence, to solve the problem, we must solve the system

$$\begin{cases} xy = 1020 \\ (x + 8)(y - 2) = 1020. \end{cases} \tag{15}$$

To solve system (15), we begin by writing the second equation as

$$xy - 2x + 8y - 16 = 1020.$$

Since $xy = 1020$, we have

$$1020 - 2x + 8y - 16 = 1020.$$

Solving the last equation for x yields

$$2x = 8y - 16, \quad \text{or}$$
$$x = 4y - 8. \tag{16}$$

Substituting $4y - 8$ for x in the first equation in system (15), we get

$$(4y - 8)y = 1020, \quad \text{or}$$
$$4y^2 - 8y = 1020.$$

Simplifying the last equation and then solving for y, we have

$$4y^2 - 8y - 1020 = 0$$
$$y^2 - 2y - 255 = 0$$
$$(y - 17)(y + 15) = 0$$
$$y = 17 \quad \text{or} \quad y = -15.$$

We discard $y = -15$ since the cost per member must be positive. If $y = 17$, by equation (16), we know that $x = 60$. You can verify that $(60, 17)$ is a solution of system (15). Thus, there were originally 60 members in the club, and the original cost per member was \$17.

EXERCISES 9.6

In Exercises 1–12, graph each equation in the given system and then find the solution set of the system.

1. $\begin{cases} x + y - 4 = 0 \\ x - y^2 + 1 = 0 \end{cases}$

2. $\begin{cases} x^2 + y - 3 = 0 \\ 2x - y - 1 = 0 \end{cases}$

3. $\begin{cases} y = \sqrt{x - 1} \\ y = x - 3 \end{cases}$

4. $\begin{cases} x + (y - 2)^2 = 0 \\ -x + y = -1 \end{cases}$

5. $\begin{cases} x^2 + (y - 1)^2 = 1 \\ (x + 1)^2 + y^2 = 1 \end{cases}$

6. $\begin{cases} (x - 2)^2 + (y + 1)^2 = 16 \\ (x - 2)^2 + y = 0 \end{cases}$

7. $\begin{cases} x - y^3 = 0 \\ x - y^2 = 0 \end{cases}$

8. $\begin{cases} 4x^2 + y = 4 \\ x^4 - y = 1 \end{cases}$

9. $\begin{cases} xy = 4 \\ 4x + y = -8 \end{cases}$

10. $\begin{cases} x - y = 0 \\ \dfrac{(x - 1)^2}{16} - \dfrac{(y - 6)^2}{9} = -1 \end{cases}$

11. $\begin{cases} x - y^4 = 0 \\ x + y^4 = 2 \end{cases}$

12. $\begin{cases} x + y = 0 \\ x^2 - 2xy + y^2 = 10 \end{cases}$

In Exercises 13–20, graph each pair of equations and find the point(s) of intersection, if any.

13. $\begin{cases} y = x^3 \\ y = 32\sqrt{x} \end{cases}$

14. $\begin{cases} x - y^2 = 0 \\ 2x - y^2 = 6 \end{cases}$

15. $\begin{cases} x^2 - y = 0 \\ x^{2/3} - y = 0 \end{cases}$

16. $\begin{cases} y = x^3 \\ y = 2x^3 + x^2 - 2x \end{cases}$

17. $\begin{cases} y = e^x \\ y = e^{2x} - 1 \end{cases}$

18. $\begin{cases} y = 2^x \\ y = 4^x - 6 \end{cases}$

19. $\begin{cases} y = \ln x \\ y = \ln(10 - x) \end{cases}$ **20.** $\begin{cases} y = \ln|2x| \\ y = \ln|x + 1| \end{cases}$

In Exercises 21–40, find the solution set of the given system of equations.

21. $\begin{cases} x + 3y^2 = 0 \\ x + y^2 - 12y = 5 \end{cases}$ **22.** $\begin{cases} x - y^2 = 0 \\ x + y^2 = 32 \end{cases}$

23. $\begin{cases} xy - y^2 = 0 \\ x^2 + 2xy + y^2 = 10 \end{cases}$ **24.** $\begin{cases} x^2 - xy = 0 \\ x^2 - xy + 2y^2 = 4 \end{cases}$

25. $\begin{cases} x^2 + y^2 = 10 \\ 16x^2 + 9y^2 = 144 \end{cases}$ **26.** $\begin{cases} x^2 + y^2 = 4 \\ xy = 1 \end{cases}$

27. $\begin{cases} \dfrac{3}{x} - \dfrac{4}{y} = 2 \\ \dfrac{5}{x} + \dfrac{7}{y} = 1 \end{cases}$ **28.** $\begin{cases} \dfrac{2}{x} - \dfrac{6}{y} = 3 \\ -\dfrac{3}{x} + \dfrac{5}{y} = 2 \end{cases}$

29. $\begin{cases} \dfrac{4}{x^2} + \dfrac{3}{y^2} = 26 \\ \dfrac{3}{x^2} - \dfrac{11}{y^2} = -7 \end{cases}$ **30.** $\begin{cases} \dfrac{1}{x^2} + \dfrac{1}{y^2} = 2 \\ \dfrac{2}{x^2} + \dfrac{3}{y^2} = 9 \end{cases}$

31. $\begin{cases} 2x^2 + y^2 = 4 \\ x^2 - y^2 = 16 \end{cases}$ **32.** $\begin{cases} x^2 - y = 6 \\ x^2 + y^2 = 1 \end{cases}$

33. $\begin{cases} x^2 - y^2 = 1 \\ \dfrac{x^2}{4} + y^2 = 1 \end{cases}$ **34.** $\begin{cases} \dfrac{x^2}{9} + y^2 = 1 \\ x^2 + \dfrac{y^2}{9} = 1 \end{cases}$

35. $\begin{cases} x^2 - xy + y^2 = 10 \\ xy - y^2 = 6 \end{cases}$ **36.** $\begin{cases} x^2 - xy + y^2 = 2 \\ x^2 + y^2 = 3 \end{cases}$

37. $\begin{cases} x + \log_3(y + 1) = 3 \\ x - \log_3(y + 1) = 1 \end{cases}$ **38.** $\begin{cases} x + \log_4(y + 6) = 0 \\ x + 5\log_4(y + 6) = 4 \end{cases}$

39. $\begin{cases} e^{2x} - ye^x = 9 \\ 2e^x + y = 0 \end{cases}$ **40.** $\begin{cases} 3^x + y = 6 \\ -3^x + 3y = 2 \end{cases}$

41. The difference in the area of two circles is 40π square inches. The difference of their radii is 4 inches. What is the radius of the smaller circle?

42. The product of two numbers is 28. Twice the square of the first number is four more than four times the second number. Find the two numbers.

43. The area of a right triangle is 12 square meters. If the length of the hypotenuse of the triangle is $2\sqrt{13}$ meters, find the length of each leg.

44. Find the radius of each circle shown in the accompanying figure given that the total area enclosed by the two circles is $\frac{29}{2}\pi$ square meters.

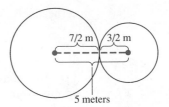

7/2 m   3/2 m

5 meters

45. A travel club in Georgia chartered a plane for a trip to Cancun, Mexico. The total cost for the flight, \$31,500, was to be equally shared by all people on the flight. After all the arrangements were made, five additional people chose to go along and the cost per person was reduced by \$15. How many people were originally scheduled for the flight and what was the original cost per person?

46. A right circular cylinder has a volume of 224 cubic inches. If the height of the cylinder is 2 inches less than four times the radius of its base, find the height of the cylinder and the radius of its base.

47. A rectangle is inscribed in a circle of radius $\sqrt{10}$ centimeters. If the area of the rectangle is 12 square centimeters, find the dimensions of the rectangle.

48. A right circular cylinder is inscribed in a sphere of radius 5 centimeters. If the volume of the cylinder is 288π cubic centimeters, find the height of the cylinder and radius of the base.

49. The point $(2, 5)$ lies on the parabola $y = x^2 + 1$. Find the value of m for which the line $y - 5 = m(x - 2)$ and the parabola intersect only at the point $(2, 5)$.

50. The point $(-1, 3)$ lies on the parabola $y = x^2 - 2x$. Find the value of m for which the line $y - 3 = m(x + 1)$ and the parabola intersect only at the point $(-1, 3)$.

9.7 SYSTEMS OF LINEAR INEQUALITIES— LINEAR PROGRAMMING

In Section 1.6, we discussed linear inequalities in one variable. In this section, we extend our discussion of linear inequalities to those containing two variables. We note that the techniques we developed for solving inequalities in one variable also apply to inequalities in two variables. Later in this section we discuss systems of linear inequalities and show how they can be used in an area of mathematics called **linear programming**.

A **linear inequality in the variables x and y** is an inequality that can be written in the form

$$ax + by + c < 0 \quad \text{or} \quad ax + by + c \leq 0,$$

where a, b, and c are real numbers and at least one of a and b is not zero. It follows that inequalities of the form

$$ax + by + c > 0 \qquad \text{or} \qquad ax + by + c \geq 0$$

are linear inequalities since they are equivalent to the inequalities

$$-ax - by - c < 0 \qquad \text{or} \qquad -ax - by - c \leq 0,$$

respectively. The following are examples of linear inequalities:

$$x - y - 1 < 0; \qquad x - y - 1 \leq 0;$$
$$2x - y \geq 3; \qquad x + y + 2 > 0.$$

An ordered pair of real numbers (x_0, y_0) is a **solution** of a given linear inequality in x and y provided that a true statement is obtained when x and y are replaced by x_0 and y_0, respectively. For example, the ordered pair $(-2, 3)$ is a solution of the inequality $x - y - 1 < 0$ since

$$-2 - 3 - 1 = -6 < 0.$$

Graphing Linear Inequalities

The **graph of a linear inequality in x and y** is the collection of all the points in the Cartesian plane whose coordinates satisfy the inequality. We now illustrate a procedure for graphing a linear inequality in two variables.

Consider the inequality

$$x - y - 1 < 0. \qquad (1)$$

If we replace the inequality symbol in (1) by an equality sign, we obtain

$$x - y - 1 = 0,$$

whose graph is the line l shown in Figure 9.12(a). In Figure 9.12(a), we see that the line l separates the plane into three disjoint sets: region R_1, region R_2, and the set of points on the line l. Region R_1 and region R_2 are called **open half-planes**, and the line l is called the **boundary** of each of these half-planes.

Since the coordinates of the points on line l do not satisfy inequality (1), we need only consider the points in region R_1 and region R_2. Let $P = (x_0, y_0)$ be any point not on the line l. We call the point P a **test point**. It follows from the definition of a solution of a linear inequality that the point $P = (x_0, y_0)$ is a solution of inequality (1) if and only if

$$x_0 - y_0 - 1 < 0,$$

or equivalently, $\qquad\qquad x_0 - 1 < y_0.$

Next, we choose $Q = (x_0, x_0 - 1)$ to be the point on the line l having the same x-coordinate as our test point P. From Figure 9.12(b), we see that the point P lies in the open half-plane R_1 if and only if the y-coordinate of Q is less than the y-coordinate of P. That is, the test point P lies in region R_1 if and only if $x_0 - 1 < y_0$ (or $x_0 - y_0 - 1 < 0$). It follows that the test point P satisfies the inequality $x - y - 1 < 0$ if and only if it lies in the open half-plane R_1. Therefore, the graph of the inequality $x - y - 1 < 0$ is the open half-plane R_1 shown in Figure 9.13. The line l is represented as a broken line in Figure 9.13 to indicate that the points on l are not included in the solution.

If the original inequality had been $x - y - 1 \leq 0$, the solution would also have included the points on line l. In this case, we represent line l with a solid line to indicate that the points on l *are* included in the solution (see Figure 9.14).

FIGURE 9.12
a.

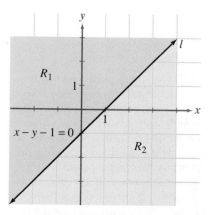

b.

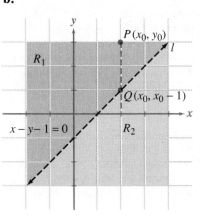

FIGURE 9.13
The graph of $x - y - 1 < 0$

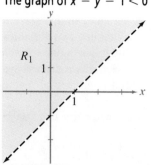

FIGURE 9.14

The graph of $x - y - 1 \le 0$

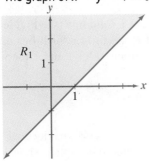

We call the solution set for the inequality $x - y - 1 \le 0$ a **closed half-plane** since it contains the boundary line l.

In general, the graph of the linear inequality $ax + by + c < 0$ is an open half-plane, while the graph of the linear inequality $ax + by + c \le 0$ is the corresponding closed half-plane. Below, we outline the procedure used for determining the graph of the linear inequality $x - y - 1 < 0$.

Procedure for Graphing Linear Inequalities

The graph of the linear inequality $ax + by + c < 0$ can be drawn using the following steps.

(i) Graph the associated equation $ax + by + c = 0$ using a broken line to indicate that the points on this line are not part of the solution.

(ii) Choose a test point not on the graph of the associated line. (The origin $(0, 0)$ is a convenient test point, provided it does not lie on the associated line $ax + by + c = 0$.)

(iii) The graph of the inequality is the open half-plane containing the test point if the inequality is satified by that point. Otherwise, the graph is the other open half-plane.

The graph of the linear inequality $ax + by + c \le 0$ is the closed half-plane formed by the union of the graph of $ax + by + c < 0$ and the graph of the associated equation $ax + by + c = 0$. In this case, the boundary line is shown on the graph as a solid line.

EXAMPLE 40

Sketch the graph of the linear inequality

$$2x - 3y > 4. \qquad (2)$$

Solution   We begin by sketching the graph of the associated equation $2x - 3y = 4$ using a broken line as shown in Figure 9.15(a). As our test point, we choose the origin $(0, 0)$, which lies in the open half-plane above the line $2x - 3y = 4$. Substituting 0 for x and y in inequality (2), we get $2(0) - 3(0) > 4$ or $0 > 4$. Since the last inequality is false, the point $(0, 0)$ does not satisfy inequality (2), and thus, the graph of inequality (2) is the open half-plane lying below the line as shown in Figure 9.15(b).

FIGURE 9.15

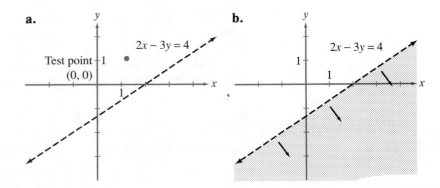

a.   Test point $(0, 0)$ $2x - 3y = 4$

b. $2x - 3y = 4$

EXAMPLE 41

Graph the inequality $3x + y \geq 0$.

Solution   First, we graph the associated equation using a solid line (see Figure 9.16(a)). Next, we choose as our test point, the point $P = (0, 1)$, which lies in the open half plane above the line $3x + y = 0$. Substituting 0 for x and 1 for y in the inequality $3x + y \geq 0$ yields

$$3(0) + 1 \geq 0 \qquad \text{or}$$
$$1 \geq 0.$$

Since the last inequality is true, the test point $(0, 1)$ satisfies the original inequality. Hence, the graph of the inequality $3x + y \geq 0$ is the closed half-plane shown in Figure 9.16(b).

FIGURE 9.16

a.

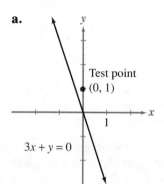

b.

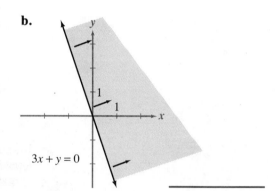

Graphing a System of Linear Inequalities

A **system of linear inequalities** is a system such as

$$\begin{cases} 2x - 3y > 4 \\ 3x + y \geq 0 \end{cases} \tag{3}$$

or

$$\begin{cases} x - y \geq -5 \\ x - y \leq 1 \\ 2x + y \geq 8 \\ x \leq 5. \end{cases} \tag{4}$$

Such systems often arise in the analysis of problems in business, economics, engineering, and other fields.

A **solution** of a system of linear inequalities in the variables x and y is an ordered pair of real numbers (x_0, y_0) that satisfies each inequality in the system. For example, the ordered pair $(2, 5)$ is a solution of system (4). To graph a system of inequalities in two variables, we sketch the graph of each inequality in the system in the same coordinate system and take the intersection of all the graphs.

EXAMPLE 42

Sketch the graph of the system of inequalities

$$\begin{cases} 2x - 3y > 4 \\ 3x + y \geq 0. \end{cases} \tag{5}$$

Solution We graphed the first inequality of system (5) in Example 40 and the second in Example 41. These two graphs are reproduced in parts (a) and (b) of Figure 9.17. By definition, the graph of system (5) is the intersection of these two graphs as shown in part (c) of Figure 9.17. (Note that the point $(\frac{4}{11}, -\frac{12}{11})$ is not part of the solution.)

FIGURE 9.17 **a.** The graph of $2x - 3y > 4$ **b.** The graph of $3x + y \geq 0$ **c.** The graph of the system

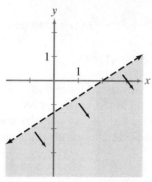
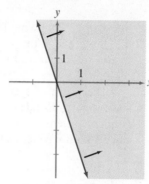
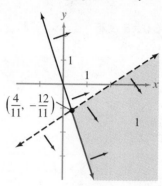

EXAMPLE 43

FIGURE 9.18

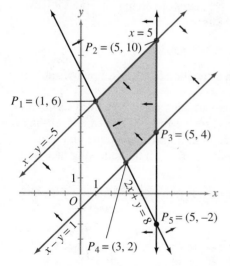

Sketch the graph of the system of linear inequalities

$$\begin{cases} x - y \geq -5 \\ x - y \leq 1 \\ 2x + y \geq 8 \\ x \leq 5. \end{cases} \quad (6)$$

Solution We begin by graphing the associated equation for each inequality in the given system. Using the origin as our test point in each inequality, we find that the graph of this system is the trapezoidal region shown in Figure 9.18.

We emphasize that the solution set for the system of linear inequalities in Example 43 consists of the points in the shaded region in Figure 9.18 including line segments $\overline{P_1P_2}$, $\overline{P_2P_3}$, $\overline{P_3P_4}$, and $\overline{P_1P_4}$. Moreover, we call the points P_1, P_2, P_3, and P_4 **vertices** or **corner points** of the solution set. In general, a **vertex** of a solution set is a point in the solution set that is the intersection of two boundary lines. (Notice that the point $P_5 = (5, -2)$ is not a vertex of the solution set since it is not in the solution set.)

The solution set of system (6) is an example of a convex set. In general, a set S in the plane is called **convex** if for any points P and Q in S, the line segment $\overline{PQ}$ lies in S. See Figure 9.19.

FIGURE 9.19 **a.** Set is convex **b.** Set is not convex

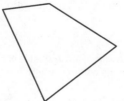
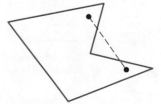

EXAMPLE 44

Sketch the graph of the following system of linear inequalities and find the vertices of the solution set.

$$\begin{cases} x - y \geq -5 \\ x + y \geq -1 \\ x - y \leq 0. \end{cases} \tag{7}$$

FIGURE 9.20

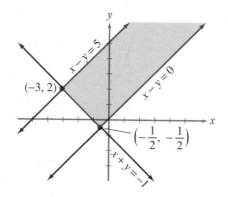

Solution   Following the procedure we have developed in this section, we find that the graph is the shaded region shown in Figure 9.20. The vertices for this region are the points $(-3, 2)$ and $(-\frac{1}{2}, -\frac{1}{2})$. _____

The graph of system (7), shown in Figure 9.20, is said to be **unbounded** since it extends indefinitely far in a particular direction. On the other hand, the graph of system (6) shown in Figure 9.18, is said to be **bounded** since the region can be wholly contained within some circle with sufficiently large radius.

Linear Programming

Business professionals are often faced with the problem of maximizing profits while minimizing both the cost and use of resources. In practice, such problems involve maximizing or minimizing a function of several variables defined on a set determined by a system of linear inequalities. Solutions to such problems can often be found using an area of mathematics known as **linear programming**. In this section, we restrict our attention to problems involving functions of two variables whose defining equations have the form

$$f(x, y) = ax + by + c,$$

where a, b, and c are real numbers. Such a function is called a **linear function in the variables x and y.**

The domain of a linear function in the variables x and y is always a subset of the Cartesian plane. It can be shown that such a function, when defined on a closed, convex, and bounded set S, attains both a maximum value and a minimum value on S.* This result is stated, without proof, in the following theorem:

THEOREM 9.7

Let f be a linear function defined by $f(x, y) = ax + by + c$. If f is defined on a closed, convex, and bounded set S, then f attains both a maximum value and a minimum value on S. Moreover, each of these values of f is attained at a vertex of S.

We apply Theorem 9.7 in the following example:

_____

*Although the concept of a closed set is beyond the scope of this text, it can be shown that the solution set of each system of linear inequalities in the remainder of this section is a closed set.

EXAMPLE 45

Let f be the function defined by

$$f(x, y) = 3x - 2y.$$

Show that f has a maximum value and a minimum value on the solution set S of Example 43 and tell where these values occur.

Solution   We have previously noted that the solution set S in Example 43 is both convex and bounded. By Theorem 9.7, f attains both a maximum and a minimum value on S, and each of these values occurs at a vertex of S. From Figure 9.18 on page 483, we see that the vertices of S are the points $(1, 6)$, $(3, 2)$, $(5, 4)$, and $(5, 10)$. The values of f at these vertices are given in Table 9.1.

TABLE 9.1

Vertex	Value of $f(x, y) = 3x - 2y$
$(1, 6)$	-9
$(3, 2)$	5
$(5, 4)$	7
$(5, 10)$	-5

From Table 9.1, we see that f attains its maximum value 7 at the vertex $(5, 4)$ and its minimum value -9 at the vertex $(1, 6)$.

The following example will serve as a vehicle for introducing some new terminology. It will also lead to a general procedure for solving linear programming problems that involve a linear function in two variables.

EXAMPLE 46

A candy producer sells two types of boxes of chocolate candies. Each box of type A contains 20 pieces with creamy centers and 10 pieces with nutty centers. On the other hand, each box of type B contains 10 pieces with creamy centers and 20 pieces with nutty centers. The profit on each box of type A is $2.50, while the profit on each box of type B is $1.50. If the candy maker has 15,000 pieces of candy with creamy centers and 9000 pieces with nutty centers available to be used in the two types of boxes, how many boxes of each type should be made to maximize his profit? What is his maximum profit?

Solution   First, we let x denote the number of type A boxes to be produced and y the number of type B boxes to be produced. Then the profit function P is given by

$$P(x) = 2.50x + 1.50y.$$

Our objective is to maximize the function P, and for this reason, we call P the **objective function**.

In maximizing P, we are bound by certain restrictions, or **constraints**, which are placed on the variables x and y. Since x and y denote numbers of boxes of candy to be produced, we know that

$$x \geq 0 \quad \text{and} \quad y \geq 0.$$

In addition, since each type A box contains 20 pieces with creamy centers and each type B box contains 10 pieces with creamy centers and since there are only 15,000 pieces with creamy centers, we know that

$$20x + 10y \leq 15,000.$$

Similarly, since there are only 9000 pieces with nutty centers, and since each type A box contains 10 such pieces while each type B box contains 20 pieces, we have

$$10x + 20y \leq 9,000.$$

We now have a mathematical model for this problem:

$$\text{Maximize:} \quad P(x, y) = 2.50x + 1.50y$$

$$\text{Subject to the constraints:} \begin{cases} x \geq 0 \\ y \geq 0 \\ 20x + 10y \leq 15,000 \\ 10x + 20y \leq 9,000 \end{cases}$$

An ordered pair (x, y) of real numbers that satisfies all the constraints given in this problem is called a **feasible solution** of the problem. The set of all feasible solutions S is graphed in Figure 9.21.

From Figure 9.21, we see that the set S is bounded and convex. Since the set S is also closed, it follows from Theorem 9.7 that the profit function P has a maximum value on S and that the maximum value occurs at a vertex of S. In Table 9.2, we list the values of the function P at each vertex of S. From Table 9.2, we see that the candy producer should make 700 boxes of type A and 100 boxes of type B. His profit will be $1900.00.

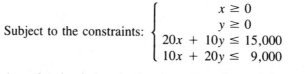

Vertex	Value of $P(x, y) = 2.50x + 1.50y$
(0, 0)	$ 0.00
(0, 450)	$ 675.00
(700, 100)	$1,900.00
(750, 0)	$1,875.00

EXERCISES 9.7

In Exercises 1–8, sketch the graph of the given linear inequality.

1. $x \leq 4$

2. $y \geq -1$

3. $x - 5y > 0$

4. $x - y \leq 0$

5. $4x + 3y \leq 7$

6. $5x - 2y - 6 \geq 0$

7. $10x + y > 15$

8. $15x + 14y \leq 10$

In Exercises 9–22, graph each system of linear inequalities and give all vertices. Is the region bounded? Is the region convex?

9. $\begin{cases} x \geq 0 \\ y \geq 1 \\ 2x + y \geq 4 \\ 3x + 2y \leq 12 \end{cases}$

10. $\begin{cases} x \geq 50 \\ x \leq 100 \\ y \geq 100 \\ x + y - 230 \geq 0 \end{cases}$

11. $\begin{cases} x \geq 0 \\ x \leq 5 \\ y \leq 6 \\ 3x + 4y \geq 12 \\ 2x - y \leq 8 \end{cases}$

12. $\begin{cases} x \geq 0 \\ x \leq 10 \\ y \geq 10 \\ x - y \leq 0 \\ x + y - 50 \leq 0 \end{cases}$

13. $\begin{cases} x \geq 0 \\ y \geq 0 \\ 4x + 7y \leq 500 \\ 11x + 3y \leq 400 \end{cases}$

14. $\begin{cases} x \geq 0 \\ y \geq 0 \\ 2x + 10y \leq 16 \\ 2x + 3y \leq 9 \end{cases}$

15. $\begin{cases} x \geq 0 \\ y \geq 0 \\ x + 2y \geq 3 \\ 2x + y \geq 4 \end{cases}$ **16.** $\begin{cases} x \leq 7 \\ y \leq 4 \\ 2x + 3y \leq 12 \\ 3x + y \leq 11 \end{cases}$

17. $\begin{cases} x \geq 0 \\ x + 3y \leq 60 \\ x + y \geq 10 \\ x - y \leq 0 \end{cases}$ **18.** $\begin{cases} x \geq 0 \\ 5x + y \geq 70 \\ x + 3y \geq 105 \\ 3x + 2y \geq 96 \end{cases}$

19. $\begin{cases} x \geq 0 \\ y \leq 15 \\ 5x + y \geq 10 \\ 5x + y \leq 35 \end{cases}$ **20.** $\begin{cases} x \geq 0 \\ y \geq 0 \\ x + y \geq 2 \\ x - 3y \geq 0 \end{cases}$

In Exercises 21–24, find the maximum value and minimum value of each function on the region determined in Exercise 9.

21. $f(x, y) = 2x + y$ **22.** $f(x, y) = x + 4y$

23. $f(x, y) = 2x + 4y$ **24.** $f(x, y) = 4x + 5y$

In Exercises 25–28, find the maximum value and minimum value of each function on the region determined in Exercise 11.

25. $f(x, y) = x - y$ **26.** $f(x, y) = 6x - y$

27. $f(x, y) = -4x + y$ **28.** $f(x, y) = -x - 2y$

29. Explain why the function defined by $f(x, y) = 6x + 4y$ has no maximum value on the solution set in Exercise 15. Does this contradict Theorem 9.7? Why?

30. Explain why the function defined by $f(x, y) = 15y - 10x$ has no minimum value on the solution set in Exercise 18. Does this contradict Theorem 9.7? Why?

In Exercises 31–34, give a system of linear inequalities that determines the region within the given boundary.

31. The rectangle with vertices $(1, 6)$, $(4, -6)$, $(5, 7)$, and $(8, -5)$.

32. The parallelogram with vertices $(1, 6)$, $(3, 8)$, $(4, -6)$, and $(6, -4)$.

33. The trapezoid with vertices $(-1, -3)$, $(-1, 4)$, $(3, -7)$, and $(3, 10)$.

34. The pentagon with vertices $(-3, 4)$, $(-1, -2)$, $(1, 10)$, $(4, -6)$, and $(7, 3)$.

35. A candidate for governor in California wishes to use a combination of radio and television advertisements in her campaign. Research has shown that each 1-minute spot on television reaches 0.08 million people and each 1-minute spot on radio reaches 0.01 million. The candidate feels that she must reach at least 3.10 million people, and that she should buy a total of no more than 100 minutes of advertisement. How many minutes of each medium should be used to minimize costs if television costs $3000 per minute and radio costs $400 per minute?

36. An investor has a maximum of $10,000 to invest in two types of bonds. The return on bond A is 9.6%, while the return on bond B is 12%. Since bond B is not as safe as bond A, the investor decides that her investment in bond B will not exceed 40% of her investment in bond A by more

than $200. How much should she invest at each rate to obtain the maximum interest per year?

37. The Green Forest Lumber Company owns timberland on which it produces pine and fir trees. From these trees, the company produces pulpwood, resin, and lumber. An acre of pine produces 6 units of pulpwood, 4 units of resin, and 2 units of lumber. An acre of fir will produce 2 units of pulpwood, 2 units of resin, and 4 units of lumber. Green Forest estimates that it costs $800 to harvest an acre of pine and $1500 to harvest an acre of fir. How many acres of each type of tree should the lumber company harvest to minimize the cost and to fill contracts that require the production of at least 12 units of pulpwood, 10 units of resin, and 4 units of lumber?

38. Bill has two types of food he is taking on a hiking trip. Each ounce of type A food costs $0.40, contains 150 calories, and supplies 21 units of vitamins. Each ounce of type B food costs $0.35, contains 60 calories, and supplies 42 units of vitamins. Bill has determined that he will need at least 3000 calories and 1260 units of vitamins per day. However, he wishes for each day's food supply to be no greater than 60 ounces. How many ounces of each type of food should be allocated per day to minimize his costs?

39. A manufacturer makes two types of birdhouses, A-frames and bungalows, using cedar and plywood. Each A-frame requires 2 square feet of cedar and three square feet of plywood, while each bungalow requires 3 square feet of cedar and 4 square feet of plywood. The manufacturer has on hand 180 square feet of cedar and 250 square feet of plywood, and wishes to make no more than 75 birdhouses. If

Photos © Fredrik D. Bodin

his profit on an A-frame is $4.00 and his profit on a bungalow is $5.00, how many birdhouses of each type should be manufactured to maximize his profit? What is his maximum profit?

40. Repco Electronics has a contract to supply a retail outlet with up to 60 video recorders and up to 40 television sets. Four worker-hours are required to produce a video recorder and 6 worker-hours are required to produce a television. The electronics company wishes to use no more than 420 worker-hours to fulfill this contract. If the company's profit is $35 on each video recorder and $50 on each television, determine the number of each that should be produced to maximize the company's profit. What is the maximum profit?

41. A U.S. government agency desires to study an endangered species of turtles. The government contracted with a large university for its marine biology department to furnish professors and research assistants to conduct the study. Each professor will spend three hours per week collecting data in the field and five hours per week analyzing data in a laboratory center. Each research assistant will spend four hours in the field and three hours per week in the laboratory. The minimum weekly labor-hour requirements are 25 hours in the field and 27 hours in the laboratory. The contract specifies that no more than 8 professors and research assistants are to be employed in this project. If the cost per week is $500 for each professor and $400 for each re-

search assistant, how many of each should be hired to minimize the costs? What is this minimum cost?

42. Star Oil Company produces gasoline and diesel fuel. While the company has access to all the crude oil it can use, the company's refinery can produce no more than 2 million barrels of gasoline a day and no more than 2.5 million barrels of diesel fuel per day. In addition, government regulations prevent Star from producing more than 3 million barrels per day of gasoline and diesel fuel together. The company's profit per barrel of gasoline is $2.20 and $2.00 per barrel of diesel fuel. If Star Oil can sell all the gasoline and diesel fuel it produces per day, how many barrels of gasoline and diesel fuel should it produce each day to maximize its profit? What is its maximum profit?

43. The director of libraries at a large university wants to extend the library hours during the week of final examinations. She wants the staff to work the additional hours in two types of shifts. A type-1 shift will require that two senior librarians and one student assistant work for three hours. A type-2 shift will require that one senior librarian and five student assistants work for two hours. The director wants to use no more than 36 employees. There will be at least 24 additional hours in the expanded schedule. If the cost of a type-1 shift is $30 per hour, and the cost of a type-2 shift is $25 per hour, determine how many shifts of each type the director should schedule to minimize the labor costs. Find the minimum labor costs.

CHAPTER 9 REVIEW EXERCISES

In Exercises 1–10, use the substitution method to find the solution set of each system.

1. $\begin{cases} x - y = 4 \\ 4x + y = -6 \end{cases}$

2. $\begin{cases} 4x - y = 4 \\ 2x + 3y = 5 \end{cases}$

3. $\begin{cases} 3x + 8y = 0 \\ -5x + y - 6 = 0 \end{cases}$

4. $\begin{cases} a - b + 6 = 0 \\ 3a + 3b - 5 = 0 \end{cases}$

5. $\begin{cases} x - y = 1 \\ 3x^2 + y = 13 \end{cases}$

6. $\begin{cases} x^2 - y^2 = 2 \\ x - y^2 = 0 \end{cases}$

7. $\begin{cases} x - 2y = 0 \\ x^2 - 3y^2 = 9 \end{cases}$

8. $\begin{cases} 4c - 3d = 0 \\ 3cd = 1 \end{cases}$

9. $\begin{cases} x^2 + y^2 - 25 = 0 \\ x^2 - 6y + 2 = 0 \end{cases}$

10. $\begin{cases} x^2 + y^2 - 25 = 0 \\ x^2 - y^2 + 7 = 0 \end{cases}$

In Exercises 11–20, use the elimination method to find the solution set of each system.

11. $\begin{cases} 3x + y = 2 \\ 2x + y = 3 \end{cases}$

12. $\begin{cases} 2x - 3y = -1 \\ 3x - 7y = 1 \end{cases}$

13. $\begin{cases} \frac{2}{3}x + \frac{3}{4}y = \frac{7}{2} \\ \frac{1}{4}x - \frac{2}{5}y = \frac{11}{2} \end{cases}$

14. $\begin{cases} 0.5a + 0.2b = 1.3 \\ 0.3a + 0.2b = 1.1 \end{cases}$

15. $\begin{cases} x - 2y^2 = 0 \\ x^2 - 3y^2 = 9 \end{cases}$

16. $\begin{cases} x^2 + y^2 = 34 \\ x^2 + y = 8 \end{cases}$

17. $\begin{cases} \dfrac{3}{x} + \dfrac{3}{y} = \dfrac{4}{5} \\ \dfrac{3}{x} - \dfrac{1}{y} = \dfrac{2}{5} \end{cases}$

18. $\begin{cases} \dfrac{6}{x} + \dfrac{4}{y} = 1 \\ \dfrac{1}{x} - \dfrac{1}{y} = \dfrac{1}{16} \end{cases}$

19. $\begin{cases} x + y + 3z = 2 \\ x + z = 3 \\ 3x + y = -2 \end{cases}$

20. $\begin{cases} r - s + 3t = 2 \\ 4r + 7s - t = 7 \\ r + 4s - 2t = 3 \end{cases}$

In Exercises 21–26, use Gaussian elimination to find the solution set of each system.

21. $\begin{cases} x + 2y - z = 6 \\ -2x + y - 3z = 13 \\ -3x + 2y - 3z = 16 \end{cases}$

22. $\begin{cases} 3x - 2y + 4z = 6 \\ 2x + 3y - 5z = -8 \\ -5x + 4y - 3z = -7 \end{cases}$

23. $\begin{cases} x + y - z = 100 \\ 10x + 4y + 5z = 680 \\ x - 2y = 0 \end{cases}$

24. $\begin{cases} a + b + c = 140 \\ 3a + 4b + 8c = 840 \\ a - 2b = 0 \end{cases}$

25. $\begin{cases} r + 2s - 3t + 2u = 0 \\ 2r + 5s - 8t + 6u = 5 \\ 3r + 4s - 5t + 2u = 4 \end{cases}$

26. $\begin{cases} x + 2y + 2z = 2 \\ 3x - y - z = 5 \\ 2x - 5y + 3z = -4 \\ x + 4y + 6z = 0 \end{cases}$

In Exercises 27–30, find AB and BA, if possible.

27. $A = \begin{bmatrix} 2 & -1 & 0 & 1 \\ 1 & 1 & -1 & 0 \end{bmatrix}$ $B = \begin{bmatrix} 5 & -4 \\ 2 & 1 \\ 3 & -6 \end{bmatrix}$

28. $A = \begin{bmatrix} 1 & 2 & 1 \\ -1 & -1 & 4 \\ 4 & 5 & 5 \end{bmatrix}$ $B = \begin{bmatrix} 2 & 1 & 0 & 4 \\ 1 & -1 & 1 & -3 \\ 0 & 1 & 0 & 1 \end{bmatrix}$

29. $A = \begin{bmatrix} 1 & 2 & 1 & 0 \\ 0 & 1 & 1 & -1 \\ 2 & -1 & 0 & 7 \\ 1 & 1 & -1 & 4 \end{bmatrix}$ $B = \begin{bmatrix} 1 & 2 \\ 3 & 4 \\ -1 & -5 \\ 2 & 3 \end{bmatrix}$

30. $A = \begin{bmatrix} 1 & 2 & 1 & 0 \\ 0 & 1 & 1 & 0 \\ 0 & 0 & 1 & 2 \\ 0 & 0 & 0 & 1 \end{bmatrix}$

$B = \begin{bmatrix} 2 & 0 & 0 & 0 \\ -1 & 2 & 0 & 0 \\ 1 & 0 & 2 & 0 \\ 0 & -1 & 0 & 2 \end{bmatrix}$

In Exercises 31–38, find the inverse (if it exists) of the matrix.

31. $\begin{bmatrix} 1 & 2 \\ -1 & 10 \end{bmatrix}$

32. $\begin{bmatrix} 4 & 6 \\ -3 & 8 \end{bmatrix}$

33. $\begin{bmatrix} 1 & -1 & 3 \\ -1 & 0 & -1 \\ 2 & -1 & 4 \end{bmatrix}$

34. $\begin{bmatrix} 3 & 2 & 5 \\ 1 & -3 & 2 \\ 2 & 2 & -1 \end{bmatrix}$

35. $\begin{bmatrix} 1 & 2 & 0 \\ 0 & 1 & 1 \\ 0 & 0 & 4 \end{bmatrix}$

36. $\begin{bmatrix} -3 & 2 & 1 \\ 0 & 1 & 3 \\ 4 & -6 & 0 \end{bmatrix}$

37. $\begin{bmatrix} 1 & 1 & -1 & 4 \\ 1 & 0 & 4 & -1 \\ 6 & 0 & 0 & 5 \\ 2 & 1 & -1 & 3 \end{bmatrix}$

38. $\begin{bmatrix} 5 & 4 & 2 & 1 \\ 2 & 3 & 1 & -2 \\ -5 & -7 & -3 & 9 \\ 1 & -2 & -1 & 4 \end{bmatrix}$

In Exercises 39–44, find the determinant of each matrix.

39. $\begin{bmatrix} 10 & 15 \\ 14 & -13 \end{bmatrix}$

40. $\begin{bmatrix} -4 & 1 \\ 0 & 3 \end{bmatrix}$

41. $\begin{bmatrix} 1 & -1 & 3 \\ 4 & 6 & 9 \\ 8 & 7 & 0 \end{bmatrix}$

42. $\begin{bmatrix} 1 & -1 & 4 \\ 10 & -8 & 2 \\ 0 & 1 & -9 \end{bmatrix}$

43. $\begin{bmatrix} 1 & 2 & -1 \\ 0 & 4 & -5 \\ 0 & 0 & -3 \end{bmatrix}$

44. $\begin{bmatrix} 9 & -2 & 6 \\ 3 & 4 & -1 \\ 1 & 2 & -3 \end{bmatrix}$

In Exercises 45–50, use Cramer's Rule, if applicable, to find the solution set of each system. If Cramer's Rule does not apply, use Gaussian elimination to solve the system.

45. $\begin{cases} 3x - 4y = 8 \\ 5x + 7y = 15 \end{cases}$

46. $\begin{cases} 7x - 9y = 14 \\ -3x + 10y = 5 \end{cases}$

47. $\begin{cases} x + y + z = 14 \\ 2x - z = 5 \\ 3x + 2y = -6 \end{cases}$

48. $\begin{cases} 6x + 2y - z = 7 \\ 5x - 3y + 2z = -4 \\ 4x - 8y + 5z = 3 \end{cases}$

49. $\begin{cases} x - z = 2 \\ y + 2z = 4 \\ 3x + 5z = -1 \end{cases}$

50. $\begin{cases} u - 3v + w = 4 \\ -2u - v + 3w = -5 \\ 6u + 3v - w = 2 \end{cases}$

In Exercises 51–56, graph the solution of each system of inequalities. Determine if the set is bounded and find the vertices.

51. $\begin{cases} 2x - y \geq 6 \\ 3x + 2y \leq 12 \end{cases}$

52. $\begin{cases} 7x + 2y - 14 < 0 \\ 5x + y \geq 4 \end{cases}$

53. $\begin{cases} x \geq 0 \\ y \geq 0 \\ x + y \geq 1 \\ x - y \leq 4 \end{cases}$

54. $\begin{cases} x \geq 0 \\ y \geq 0 \\ x + y \geq 4 \\ x - y \leq -6 \end{cases}$

55. $\begin{cases} x \geq 0 \\ y \geq 0 \\ 2x - y \geq -2 \\ 2x + 5y \leq 22 \\ x - y \leq 4 \end{cases}$

56. $\begin{cases} x \geq 0 \\ y \geq 0 \\ 2x - y \geq -2 \\ 3x + y \leq 9 \\ 6x + y \leq 15 \end{cases}$

57. Find the maximum value of the function defined by $f(x, y) = 5x + 2y$ subject to the constraints given in Exercise 55.

58. Find the minimum value of the function defined by $f(x, y) = 2x - y$ subject to the constraints given in Exercise 56.

59. The sum of the numerator and denominator of a fraction is 12. If the numerator is decreased by 2 and the denominator is increased by 1, then the product of the resulting fraction and the given fraction is $\frac{7}{6}$. Find the numerator and denominator of the given fraction assuming both are positive.

60. The area of the ground covered by a rectangular garden is 450 square feet. It is divided into four equal parts by two paths, which are 3 feet wide and are parallel to the sides of the garden (see figure). If the area of the paths that pass through the center of the garden is 126 square feet, what are the length and width of the garden?

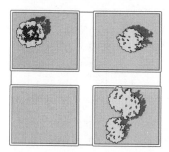

61. The area of a lot at the corner where two streets meet is 6000 square feet. A sidewalk 5 feet wide runs along the front of the lot and down one side. If the area of the sidewalk is 975 square feet, how long and how wide is the lot?

62. Suppose the height of an object is a function of time t and has defining equation $f(t) = at^2 + bt + c$ where t is measured in seconds and the height $f(t)$ is measured in feet. Assume a, b, and c are constants. If the heights of the object after 1, 2, and 3 seconds are 544 feet, 544 feet, and 222 feet, respectively, find the time at which the object is at ground level. That is, find the value of $t(t > 0)$ at which $f(t) = 0$.

63. Find the equation of the circle passing through the points $(-1, 1)$, $(0, 2)$, and $(3, 3)$. Express your answer in the form $Ax^2 + Ay^2 + Bx + Cy + D = 0$.

64. A chemist needs 20 liters of a 25% acid solution. A 20% solution, a 15% solution, and a 40% solution are available in the laboratory. How many liters of each solution should be combined to obtain the desired result if the chemist uses half as much 20% solution as 15% solution?

65. A wholesale distributor of art manufactures two types of wooden frames. To cut, assemble, and finish the first type of frame, it takes 20, 10, and 10 minutes, respectively. To cut, assemble, and finish the second type of frame it takes 10, 20, and 60 minutes, respectively. The first type of frame costs the distributor $3.60 each while the second type costs her $10.80 each. If each production run requires at least 8 hours of cutting, at least 10 hours of assembling, and at least 18 hours of finishing, how many frames of each type should she manufacture to minimize her cost? What is the minimum cost?

CHAPTER 10

Additional Topics

A sequence is a function whose domain is the set of positive integers. In the first three sections of this chapter, we study sequences, with an emphasis on arithmetic and geometric sequences. Our study of sequences concludes with a brief introduction to geometric series—a topic studied in depth in calculus. This chapter also includes a discussion of the binomial theorem and an important method of proof known as the Principle of Mathematical Induction. In the concluding section, we describe and study a procedure known as partial fraction decomposition.

North Wind Picture Archives

SIR ISAAC NEWTON (1642–1727)

Born on Christmas Day to a pros-
perous English farm family, Isaac
Newton showed little academic
promise as a young boy. He did,
however, exhibit a remarkable tal-
ent for working with mechanical
devices such as clocks, sundials,
and water wheels. Persuaded by
an uncle who recognized his un-
usual abilities, Newton entered
Trinity College in Cambridge University in 1661. Newton, guided
by the gifted mathematics teacher
Isaac Barrow, immersed himself in
the study of mathematics and science but he graduated without
distinction. Following his return
home to Woolstroke, Newton, in
an eighteen-month period, in-
vented differential and integral
calculus, proved the binomial
theorem, and determined that
white light is composed of all col-
ors from red to violet. Newton returned to Trinity for his master's
degree, and in 1669 he succeeded
his mentor, Isaac Barrow, to the
prestigious Lucasian Chair of
Mathematics. While at Trinity, he
formulated the law of gravitation
and the basic theories of light,
thermodynamics, and hydro-
dynamics. He later became master
of the London mint and was
knighted in 1705.

10.1 SEQUENCES AND SUMMATION NOTATION

Most of tne functions we have studied thus far have intervals of real numbers as
their domains. In this section we will discuss functions that have the set of positive integers as their domain.

Definition of Sequence

A **sequence** is a function whose domain is the set of positive integers.

The nth Term of a Sequence

Since every sequence has the set of positive integers as its domain, it follows
that a given sequence is completely determined by its range. For convenience we
will represent the range elements

$$f(1), f(2), f(3), \ldots, f(n), \ldots$$

of a sequence using the subscript notation

$$a_1, a_2, a_3, \ldots, a_n, \ldots.$$

We then say that $a_1 = f(1)$ is the **first term** of the sequence, $a_2 = f(2)$ is the
second term, and so on. The sequence is then completely determined by its
terms $a_1, a_2, a_3, \ldots, a_n, \ldots$, and can be represented by its **nth term**, or
general term, a_n, using the notation $\{a_n\}$. We use braces to distinguish the sequence
$\{a_n\}$ from its nth term a_n. Consider, for example, the sequence f defined by
$f(n) = \dfrac{1}{n}$, for $n = 1, 2, 3, \ldots$. This sequence, which is denoted by $\left\{\dfrac{1}{n}\right\}$, has

the following terms:

$$1, \frac{1}{2}, \frac{1}{3}, \ldots, \frac{1}{n}, \ldots$$

Since a sequence is a function, it has a graph. It is interesting to compare the
graph of the sequence $\left\{\dfrac{1}{n}\right\}$ with the graph of $f(x) = \dfrac{1}{x}$, where $x > 0$. See
Figure 10.1.

FIGURE 10.1

a. $f(x) = \dfrac{1}{x}, x > 0$

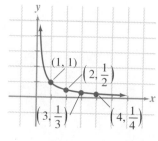

b. $f(n) = \dfrac{1}{n}, n = 1, 2, 3, \ldots$

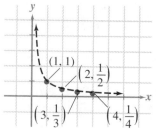

EXAMPLE 1

Find the first three terms and the twentieth term of each sequence.

a. $\{3n - 2\}$ **b.** $\left\{\dfrac{n + 1}{2n - 5}\right\}$ **c.** $\{\cos n\pi\}$

Solution

a. Since the nth term of this sequence is $a_n = 3n - 2$, we can find the first three terms by substituting 1, 2, and 3 for n to get

$$a_1 = 3(1) - 2 = 1,$$
$$a_2 = 3(2) - 2 = 4, \quad \text{and}$$
$$a_3 = 3(3) - 2 = 7.$$

Similarly, substituting 20 for n gives us the twentieth term

$$a_{20} = 3(20) - 2 = 58.$$

b. Substituting 1, 2, 3, and 20 for n in the equation $a_n = \dfrac{n + 1}{2n - 5}$, we find that

$$a_1 = \frac{1 + 1}{2(1) - 5} = -\frac{2}{3},$$

$$a_2 = \frac{2 + 1}{2(2) - 5} = -3,$$

$$a_3 = \frac{3 + 1}{2(3) - 5} = 4, \quad \text{and}$$

$$a_{20} = \frac{20 + 1}{2(20) - 5} = \frac{21}{35}.$$

c. Noting that $a_n = \cos n\pi$, we proceed as in parts (a) and (b).

$$a_1 = \cos \pi = -1,$$
$$a_2 = \cos 2\pi = 1,$$
$$a_3 = \cos 3\pi = -1, \quad \text{and}$$
$$a_{20} = \cos 20\pi = 1.$$

Defining a Sequence Recursively

A sequence can also be defined using a formula that expresses each term in terms of one or more of the previous terms. A sequence defined in this manner is said to be defined **recursively** and a formula that determines the terms of the sequence is called a **recursion formula**. We illustrate this method of defining a sequence in our next example.

EXAMPLE 2

The following two sequences are defined recursively. List the first four terms of each sequence.

a. $a_1 = 4; a_n = \dfrac{2a_{n-1} + 1}{12}, \quad$ for $n \geq 2$

b. $a_1 = \dfrac{1}{2}, a_n = (-1)^{n-1}a_{n-1} + 4, \quad$ for $n \geq 2$

Solution

a. We are given the first term $a_1 = 4$. Substituting 4 for a_1 in the recursion formula, we find the second term

$$a_2 = \frac{2a_1 + 1}{12} = \frac{2(4) + 1}{12} = \frac{9}{12} = \frac{3}{4}.$$

In like manner, we use the second term to find the third and the third to find the fourth.

$$a_3 = \frac{2a_2 + 1}{12} = \frac{2(\frac{3}{4}) + 1}{12} = \frac{\frac{5}{2}}{12} = \frac{5}{24}$$

$$a_4 = \frac{2a_3 + 1}{12} = \frac{2(\frac{5}{24}) + 1}{12} = \frac{\frac{17}{12}}{12} = \frac{17}{144}.$$

Hence, the first four terms of this sequence are 4, $\frac{3}{4}$, $\frac{5}{24}$, and $\frac{17}{144}$.

b. Again, we are given the first term $a_1 = \frac{1}{2}$. Substituting $\frac{1}{2}$ for a_1 in the recursion formula, we get the second term

$$a_2 = (-1)^1 a_1 + 4 = -\frac{1}{2} + 4 = \frac{7}{2}.$$

We can use a_2 to find a_3, and then use a_3 to find a_4 as follows:

$$a_3 = (-1)^2 a_2 + 4 = \frac{7}{2} + 4 = \frac{15}{2}$$

$$a_4 = (-1)^3 a_3 + 4 = -\frac{15}{2} + 4 = -\frac{7}{2}.$$

Thus, $\frac{1}{2}$, $\frac{7}{2}$, $\frac{15}{2}$, and $-\frac{7}{2}$ are the first four terms of this sequence.

We see from the solution of Example 2 that the recursive method of defining a sequence has a distinct disadvantage. We cannot find the *n*th term of a sequence that is defined recursively without first finding the $n - 1$ terms that precede it. This disadvantage, however, is far outweighed by the ease with which the recursive method of defining sequences can be adapted to computer methods.

Factorials

Some important sequences in mathematics are defined using special types of products known as factorials.

DEFINITION 10.1

If n is a positive integer, *n* **factorial** is defined by

$$n! = n(n - 1)(n - 2) \ldots 3 \cdot 2 \cdot 1.$$

For convenience, we also define **zero factorial** to be 1 and write

$$0! = 1.$$

From Definition 10.1, it follows that

$$1! = 1$$
$$2! = 2 \cdot 1 = 2$$
$$3! = 3 \cdot 2 \cdot 1 = 6$$
$$4! = 4 \cdot 3 \cdot 2 \cdot 1 = 24$$
$$5! = 5 \cdot 4 \cdot 3 \cdot 2 \cdot 1 = 120.$$
$$\vdots$$

As the following example illustrates, we often make use of the fact that

$$n! = n(n - 1)!, \qquad \text{for } n \geq 1,$$
$$n! = n(n - 1)(n - 2)!, \qquad \text{for } n \geq 2, \text{ and so on,}$$

when we simplify expressions involving factorials.

EXAMPLE 3

Evaluate each of the following expressions:

a. $\dfrac{(9 - 2)!}{6!}$ **b.** $\dfrac{12!}{2! \cdot 10!}$

Solution

a. Since $(9 - 2)! = 7!$ and since $7! = 7 \cdot 6!$, it follows that

$$\frac{(9 - 2)!}{6!} = \frac{7 \cdot 6!}{6!} = 7.$$

b. Writing $12!$ as $12 \cdot 11 \cdot 10!$, we have

$$\frac{12!}{2! \cdot 10!} = \frac{12 \cdot 11 \cdot 10!}{(2 \cdot 1) \cdot 10!} = 66.$$

EXAMPLE 4

Consider the sequence $\{a_n\}$ where $a_n = (2n)!$.

a. Find a_1, a_2, a_3, and a_k.   **b.** Simplify the expression $\dfrac{a_{k+1}}{a_k}$.

Solution

a. Substituting $1, 2, 3$, and k, in succession, for n in the equation $a_n = (2n)!$, we get

$$a_1 = (2 \cdot 1)! = 2! = 2$$
$$a_2 = (2 \cdot 2)! = 4! = 24$$
$$a_3 = (2 \cdot 3)! = 6! = 720$$
$$a_k = (2k)!$$

b. Since $a_{k+1} = [2 \cdot (k + 1)]! = (2k + 2)!$, and since $(2k + 2)! = (2k + 2)(2k + 1)(2k)!$, it follows that

$$\frac{a_{k+1}}{a_k} = \frac{(2k + 2)(2k + 1)(2k)!}{(2k)!} = (2k + 2)(2k + 1).$$

Summation Notation

We sometimes need to find the sum of the first n terms of a sequence $\{a_n\}$. Using the Greek letter Σ (capital sigma) as a summation symbol, we can represent this sum by writing

$$a_1 + a_2 + a_3 + \cdots + a_n = \sum_{i=1}^{n} a_i.$$

The notation $\sum_{i=1}^{n} a_i$ is read "the sum of (the terms) a_i, as i goes from 1 to n." In this **summation notation**, the letter i is called the **index of summation** and the numbers 1 and n indicate the first and last values of i in the sum.

EXAMPLE 5

Express each of the following sums using summation notation:

a. $2 + 4 + 8 + 16 + 32$

b. $2 + 2 + \frac{4}{3} + \cdots + \dfrac{2^n}{n!}$

c. $x - x^3 + x^5 - x^7 + \cdots + x^{25}$

Solution

a. We observe that the terms of the sum $2 + 4 + 8 + 16 + 32$ have the form $a_i = 2^i$, where i ranges from 1 to 5. We therefore write

$$2 + 4 + 8 + 16 + 32 = \sum_{i=1}^{5} 2^i.$$

b. Since the last term is $a_n = \dfrac{2^n}{n!}$, we know that the terms of this sum have the form $a_i = \dfrac{2^i}{i!}$. Thus

$$2 + 2 + \frac{4}{3} + \cdots + \frac{2^n}{n!} = \sum_{i=1}^{n} \frac{2^i}{i!}.$$

c. We note that the sum $x - x^3 + x^5 - x^7 + \cdots + x^{25}$ involves powers of x that are odd positive integers and that the terms of the sum alternate in sign. Since odd positive integers have the form $2i - 1$, where i is a positive integer, we can represent these powers of x by x^{2i-1}. We can obtain the alternating signs by placing a factor of the form $(-1)^i$ or $(-1)^{i+1}$ in the ith term. Since the first term has a positive coefficient, we use the factor $(-1)^{i+1}$. (When the first term has a negative coefficient, we use $(-1)^i$.) Thus

$$x - x^3 + x^5 - x^7 + \cdots + x^{25} = \sum_{i=1}^{13} (-1)^{i+1} x^{2i-1}.$$

As the following example illustrates, we can use summation notation with letters other than i as the index of summation, and we can begin the sum with numbers other than 1.

EXAMPLE 6

Rewrite each of the following sums without using summation notation and find the value of each sum:

a. $\displaystyle\sum_{j=3}^{7} (3j^2 + 1)$ **b.** $\displaystyle\sum_{k=0}^{3} (-1)^k \left(4 - \frac{1}{2^k}\right)$

Solution

a. Since the index of summation j starts with 3 and ends with 7, we replace j in the general term $3j^2 + 1$ successively by $3, 4, 5, 6,$ and 7, and add the results. Thus

$$\sum_{j=3}^{7} (3j^2 + 1) = (3 \cdot 3^2 + 1) + (3 \cdot 4^2 + 1)$$
$$+ (3 \cdot 5^2 + 1) + (3 \cdot 6^2 + 1) + (3 \cdot 7^2 + 1)$$
$$= 28 + 49 + 76 + 109 + 148$$
$$= 410.$$

b. In this sum the index of summation k starts with 0 and ends with 3. Thus we replace k successively by $0, 1, 2,$ and 3 in the general term

$(-1)^k \left(4 - \dfrac{1}{2^k}\right)$ to obtain

$$\sum_{k=0}^{3} (-1)^k \left(4 - \frac{1}{2^k}\right) = (-1)^0 \left(4 - \frac{1}{1}\right) + (-1)^1 \cdot \left(4 - \frac{1}{2}\right)$$
$$+ (-1)^2 \left(4 - \frac{1}{4}\right) + (-1)^3 \cdot \left(4 - \frac{1}{8}\right)$$
$$= 3 - \frac{7}{2} + \frac{15}{4} - \frac{31}{8}$$
$$= -\frac{5}{8}.$$

The following theorem gives some useful properties of sums:

THEOREM 10.1

Let $\{a_n\}$ and $\{b_n\}$ be sequences and let c be any real number. Then

(i) $\displaystyle\sum_{i=1}^{n} (a_i + b_i) = \sum_{i=1}^{n} a_i + \sum_{i=1}^{n} b_i$

(ii) $\displaystyle\sum_{i=1}^{n} (a_i - b_i) = \sum_{i=1}^{n} a_i - \sum_{i=1}^{n} b_i$

(iii) $\displaystyle\sum_{i=1}^{n} ca_i = c\sum_{i=1}^{n} a_i$

(iv) $\displaystyle\sum_{i=1}^{n} c = n \cdot c$

Proof   We will prove parts (i) and (iv) and leave the proofs of parts (ii) and (iii) as an exercise (see Exercise 70). To prove part (i) we write the sum $\sum_{i=1}^{n} (a_i + b_i)$ without using summation notation and use the associative and commutative properties of addition.

$$
\begin{aligned}
\sum_{i=1}^{n} (a_i + b_i) &= (a_1 + b_1) + (a_2 + b_2) + \cdots + (a_n + b_n) \\
&= a_1 + b_1 + a_2 + b_2 + \cdots + a_n + b_n \\
&= (a_1 + a_2 + \cdots + a_n) + (b_1 + b_2 + \cdots + b_n) \\
&= \sum_{i=1}^{n} a_i + \sum_{i=1}^{n} b_i.
\end{aligned}
$$

To prove part (iv), we first write the sum $\sum_{i=1}^{n} c$ without using summation notation.

$$
\sum_{i=1}^{n} c = c + c + \cdots + c. \qquad (n \text{ terms})
$$

Since each of the n terms in the sum is the real number c, it follows that

$$
\sum_{i=1}^{n} c = n \cdot c.
$$

EXAMPLE 7

Use Theorem 10.1 to find the value of the following sum:

$$
\sum_{k=1}^{5} (13k - 10).
$$

Solution   We first rewrite the sum using parts (ii) and (iii) of Theorem 10.1.

$$
\begin{aligned}
\sum_{k=1}^{5} (13k - 10) &= \sum_{k=1}^{5} 13k - \sum_{k=1}^{5} 10 \\
&= 13 \sum_{k=1}^{5} k - \sum_{k=1}^{5} 10.
\end{aligned}
$$

Since $\sum_{k=1}^{5} k = 1 + 2 + 3 + 4 + 5 = 15$, and since $\sum_{k=1}^{5} 10 = 5 \cdot 10 = 50$, by part (iv) of Theorem 10.1, it follows that

$$
\sum_{k=1}^{5} (13k - 10) = 13(15) - 50 = 145.
$$

EXERCISES 10.1

In Exercises 1–20, find the first three terms and the tenth term of each sequence.

1. $\{3n - 5\}$

2. $\{2n^2 + 1\}$

3. $\{\pi\}$

4. $\{\frac{5}{4}\}$

5. $\left\{1 - \dfrac{1}{2^n}\right\}$

6. $\left\{\dfrac{1}{3^n} - 2\right\}$

7. $\left\{\dfrac{n^2 - 2}{n + 1}\right\}$

8. $\left\{\dfrac{3n^2}{2n - 1}\right\}$

9. $\{(n - 1)(n + 1)\}$

10. $\{n(2n + 1)\}$

11. $\left\{\left(1 + \dfrac{1}{n}\right)^n\right\}$

12. $\left\{\left(2 - \dfrac{2}{n}\right)^n\right\}$

13. $\left\{\dfrac{(-1)^n}{2n - 3}\right\}$

14. $\left\{\dfrac{(-1)^{n-1}}{\sqrt{n}}\right\}$

15. $\left\{\dfrac{\sin\left(\dfrac{n\pi}{2}\right)}{n^2}\right\}$

16. $\left\{\dfrac{\cos(n\pi)}{2^n}\right\}$

17. $\left\{\dfrac{(n + 2)!}{(n - 1)!}\right\}$

18. $\left\{\dfrac{(2n + 2)!}{(2n)!}\right\}$

19. $\left\{\dfrac{\log n}{\sqrt{n}}\right\}$

20. $\left\{\dfrac{\ln n}{e^n}\right\}$

In Exercises 21–22, evaluate each expression.

21. $\dfrac{20!}{(20 - 3)! \cdot 3!}$

22. $\dfrac{15!}{5! \cdot 10!}$

In Exercises 23–24, simplify the expression $\dfrac{a_{k+1}}{a_k}$ for each sequence $\{a_n\}$.

23. $\{(n + 1)!\}$

24. $\{(2n - 1)!\}$

In each of Exercises 25–34, a sequence $\{a_n\}$ is defined recursively. List the first five terms of each sequence.

25. $a_1 = 2; a_n = 2a_{n-1} - 1, \quad$ for $n \geq 2$

26. $a_1 = 5; a_n = 3a_{n-1} - 1, \quad$ for $n \geq 2$

27. $a_1 = \frac{1}{2}; a_n = \dfrac{a_{n-1}}{2}, \quad$ for $n \geq 2$

28. $a_1 = \frac{2}{3}; a_n = \dfrac{a_{n-1}}{3}, \quad$ for $n \geq 2$

29. $a_1 = 4; a_n = (a_{n-1})^{1/2}, \quad$ for $n \geq 2$

30. $a_1 = -1; a_n = (a_{n-1})^n, \quad$ for $n \geq 2$

31. $a_1 = -1; a_n = (-1)^{n-1}a_{n-1} - 5, \quad$ for $n \geq 2$

32. $a_1 = 3; a_n = (-1)^n a_{n-1} - 5, \quad$ for $n \geq 2$

33. $a_1 = 1, a_2 = 3; a_n = \dfrac{a_{n-1} + a_{n-2}}{2}, \quad$ for $n \geq 3$

34. $a_1 = 1, a_2 = 2; a_n = (-1)^{n+1}\dfrac{a_{n-2}}{a_{n-1}}, \quad$ for $n \geq 3$

In Exercises 35–44, express each sum using summation notation.

35. $2 + 4 + 6 + 8 + \cdots + 100$

36. $1 + \frac{1}{4} + \frac{1}{9} + \frac{1}{16} + \cdots + \frac{1}{100}$

37. $1 + 2x + 3x^2 + 4x^3 + 5x^4$

38. $1 + \dfrac{1}{x} + \dfrac{1}{x^2} + \dfrac{1}{x^3} + \cdots + \dfrac{1}{x^n}$

39. $1 + 4 + 27 + 256 + \cdots + n^n$

40. $1 - \frac{2}{3} + \frac{4}{9} - \frac{8}{27} + \frac{16}{81}$

41. $\dfrac{e}{2} - \dfrac{e^2}{3} + \dfrac{e^3}{4} - \dfrac{e^4}{5} + \cdots + (-1)^{n+1}\dfrac{e^n}{n + 1}$

42. $\log 1 + \dfrac{\log 2}{2} + \dfrac{\log 3}{4} + \dfrac{\log 4}{8} + \cdots + \dfrac{\log 10}{512}$

43. $\dfrac{1}{1 - \frac{1}{2}} + \dfrac{2}{2 - \frac{1}{3}} + \dfrac{3}{3 - \frac{1}{4}} + \cdots + \dfrac{10}{10 - \frac{1}{11}}$

44. $\dfrac{1}{2 \cdot 3} + \dfrac{1}{4 \cdot 5} + \dfrac{1}{6 \cdot 7} + \cdots + \dfrac{1}{20 \cdot 21}$

In Exercises 45–54, write each sum without using summation notation and find the value of each sum.

45. $\displaystyle\sum_{i=1}^{6} (i + 4)$

46. $\displaystyle\sum_{i=1}^{5} (i - 3)$

47. $\displaystyle\sum_{i=1}^{4} (2i - 1)^2$

48. $\displaystyle\sum_{i=1}^{3} (3i^2 - 7)$

49. $\displaystyle\sum_{k=1}^{5} \dfrac{1}{2^{k-1}}$

50. $\displaystyle\sum_{k=0}^{4} \left(\dfrac{2}{5}\right)^k$

51. $\displaystyle\sum_{k=0}^{6} \sin\left(\dfrac{k\pi}{3}\right)$

52. $\displaystyle\sum_{k=0}^{8} \cos\left(\dfrac{k\pi}{2}\right)$

53. $\displaystyle\sum_{k=1}^{5} (-1)^{k-1}\ln e^{k-1}$

54. $\displaystyle\sum_{k=1}^{4} (-1)^k e^{\ln k}$

In Exercises 55–60, evaluate each sum using Theorem 10.1 whenever possible.

55. $\displaystyle\sum_{k=1}^{300} \pi$

56. $\displaystyle\sum_{k=1}^{100} e^2$

57. $\displaystyle\sum_{i=1}^{5} 50i^2$

58. $\displaystyle\sum_{i=1}^{6} \dfrac{i}{100}$

59. $\displaystyle\sum_{j=1}^{8} (5j + 32{,}000)$

60. $\displaystyle\sum_{j=1}^{7} (3j - 1700)$

61. Find the first ten terms of the sequence $\{a_n\}$ defined recursively as follows:

$$a_1 = 1, a_2 = 1; a_n = a_{n-1} + a_{n-2}, \quad \text{for } n \geq 3.$$

This sequence, which has numerous applications in the natural and physical sciences, is called the **Fibonacci sequence** in honor of its discoverer, the Italian mathematician Leonardo Fibonacci (1170–1250).

62. Sir Isaac Newton devised a method for approximating the square root of a positive real number N. The method involves a sequence defined recursively as follows:

$$a_1 = \dfrac{N}{2}; a_n = \dfrac{1}{2}\left(a_{n-1} + \dfrac{N}{a_{n-1}}\right), \quad \text{for } n \geq 2.$$

Approximate $\sqrt{5}$ by finding a_5, the fifth term of this sequence, and compare the result to the value of $\sqrt{5}$ given by your calculator. Round each result to the nearest thousandth.

63. Find the first six terms of the sequence defined recursively by

$$a_1 = 1; a_n = na_{n-1}, \quad \text{for } n \geq 2.$$

Then evaluate $1 + \displaystyle\sum_{i=1}^{6} \dfrac{1}{a^i}$ and compare the result to the value of the irrational number e given by your calculator. Round each result to the nearest thousandth.

64. Consider the sum $\sum_{k=1}^{n} \left(\frac{1}{k} - \frac{1}{k-1} \right)$. Writing the sum without using summation notation and simplifying, we find that

$$\sum_{k=1}^{n} \left(\frac{1}{k} - \frac{1}{k-1} \right) = \left(1 - \frac{1}{2} \right) + \left(\frac{1}{2} - \frac{1}{3} \right)$$

$$+ \left(\frac{1}{3} - \frac{1}{4} \right) + \left(\frac{1}{4} - \frac{1}{5} \right) + \cdots + \left(\frac{1}{n-1} - \frac{1}{n} \right)$$

$$+ \left(\frac{1}{n} - \frac{1}{n+1} \right) = 1 - \frac{1}{n+1} = \frac{n}{n+1}.$$

A sum such as this one is called a **telescoping sum**. Find the value of the telescoping sum

$$\sum_{k=2}^{20} \left[\frac{1}{(k-1)^2} - \frac{1}{k^2} \right].$$

65. A $10,000 deposit is made in an account that earns 7.5% interest compounded quarterly. The amount (in dollars) in the account after n years is given by the formula

$$a_n = 10,000 \left(1 + \frac{0.075}{4} \right)^{4n}.$$

   a. Find the first four terms of the sequence $\{a_n\}$.
   b. Find the amount in the account at the end of the tenth year.

66. Assume that each year an automobile depreciates by 25% of its value at the beginning of the year. Find the value of a new $20,000 automobile after it has depreciated for four years.

67. A teacher obtains a job with a starting salary of $25,000. If he receives a 6% raise each year for five consecutive years, determine his salary after the raise at the end of the fifth year.

68. An executive is offered a starting salary of $70,000 with the promise of a 10% raise at the end of each year she remains with the company. Determine the executive's salary after the raise has been added at the end of the fourth year.

69. Find specific sums $\sum_{i=1}^{n} a_i$ and $\sum_{i=1}^{n} b_i$ which show that, in general, $\sum_{i=1}^{n} a_i b_i \neq \sum_{i=1}^{n} a_i \cdot \sum_{i=1}^{n} b_i$.

70. Prove parts (ii) and (iii) of Theorem 10.1.

10.2 ARITHMETIC SEQUENCES

A grocery clerk has 300 cans of peaches that are to be stacked into a triangular display. If each row after the bottom row has one less can than the previous row, and the top row is to contain a single can, how many cans should be placed in the bottom row? We can solve this problem using a special type of sequence known as an arithmetic sequence.

Arithmetic Sequence

A sequence $\{a_n\}$ is called an **arithmetic sequence** (or **arithmetic progression**) if there is a real number d such that

$$a_n = a_{n-1} + d, \quad \text{for all } n > 1.$$

The number d is called the **common difference**.

We see from this definition that each term of an arithmetic sequence, after the first, can be obtained by adding the common difference d to the previous term. In other words, an arithmetic sequence $\{a_n\}$ having common difference d can be defined recursively by the formula $a_n = a_{n-1} + d$, for $n > 1$.

EXAMPLE 8

Write the first four terms of an arithmetic sequence whose first term is 3 and whose common difference is 7.

Solution We are given the first term $a_1 = 3$. Using the recursion formula $a_n = a_{n-1} + d$, where $d = 7$, we can obtain the second, third, and fourth terms as follows:

$$a_2 = 3 + 7 = 10$$
$$a_3 = 10 + 7 = 17$$
$$a_4 = 17 + 7 = 24.$$

EXAMPLE 9

Show that the sequence $\{5n - 2\}$ is an arithmetic sequence, and find the common difference d.

Solution We must show that, for some real number d, the terms of this sequence satisfy the equation

$$a_n = a_{n-1} + d \qquad \text{or}$$
$$d = a_n - a_{n-1}.$$

Since $a_n = 5n - 2$ and $a_{n-1} = 5(n - 1) - 2 = 5n - 7$, we have

$$d = (5n - 2) - (5n - 7)$$
$$= 5n - 2 - 5n + 7$$
$$= 5.$$

Hence, the sequence $\{5n - 2\}$ is an arithmetic sequence with common difference 5.

The nth Term of an Arithmetic Sequence

Using the recursion formula that defines an arithmetic sequence, we can easily determine a formula for the nth term of the sequence. To see this, we write the terms of an arithmetic sequence $\{a_n\}$ whose common difference is d as follows:

$$a_1 = a_1 + 0d$$
$$a_2 = a_1 + d = a_1 + 1d$$
$$a_3 = a_2 + d = (a_1 + d) + d = a_1 + 2d$$
$$a_4 = a_3 + d = (a_1 + 2d) + d = a_1 + 3d$$
$$\vdots$$

We note that, for each term a_n, the coefficient of d is equal to $n - 1$. Hence, we have the following result:

nth Term of an Arithmetic Sequence

The nth term of an arithmetic sequence $\{a_n\}$, whose common difference is d, is given by

$$a_n = a_1 + (n - 1)d, \quad \text{for } n \geq 1.$$

EXAMPLE 10

Find the fortieth term of the arithmetic sequence 8, 11, 14, ...

Solution   We see from the first three terms that the common difference is 3. To find a_{40}, we substitute 40 for n, 8 for a_1, and 3 for d in the equation

$$a_n = a_1 + (n - 1)d,$$

to obtain
$$a_{40} = 8 + 39 \cdot 3 = 125.$$

EXAMPLE 11

Find the general term of an arithmetic sequence $\{a_n\}$ whose first term is 1.8 and whose common difference is -3.7.

Solution   Substituting 1.8 for a_1 and -3.7 for d in the equation

$$a_n = a_1 + (n - 1)d$$

we find that

$$
\begin{aligned}
a_n &= 1.8 + (n - 1)(-3.7) \\
&= 1.8 - 3.7n + 3.7 \\
&= -3.7n + 5.5.
\end{aligned}
$$

Thus the general term of this sequence is $a_n = -3.7n + 5.5$.

EXAMPLE 12

Find the first term and the general term of an arithmetic sequence whose sixth term is 20 and whose eleventh term is 36.

Solution   When $n = 6$, the equation

$$a_n = a_1 + (n - 1)d$$

becomes
$$a_6 = a_1 + 5d,$$

and, since $a_6 = 20$,

$$20 = a_1 + 5d \qquad\qquad (1)$$

Similarly, when $n = 11$, we have

$$a_{11} = a_1 + 10d,$$

and since $a_{11} = 36$,

$$36 = a_1 + 10d. \qquad\qquad (2)$$

Equations (1) and (2) above comprise a system of two equations in the two unknowns a_1 and d:

$$
\begin{cases}
a_1 + 5d = 20 \\
a_1 + 10d = 36.
\end{cases}
$$

Subtracting the first equation from the second, we have

$$5d = 16, \qquad \text{or} \qquad d = \frac{16}{5}.$$

Substituting $\frac{16}{5}$ for d in the first equation in the system, we find the first term of the sequence.

$$a_1 + 5\left(\frac{16}{5}\right) = 20, \quad \text{or}$$

$$a_1 = 4.$$

Substituting 4 for a_1 and $\frac{16}{5}$ for d in the equation

$$a_n = a_1 + (n - 1)d$$

gives us

$$a_n = 4 + (n - 1)\left(\frac{16}{5}\right)$$

$$= 4 + \left(\frac{16}{5}\right)n - \frac{16}{5}$$

$$= \frac{16}{5}n + \frac{4}{5}.$$

Thus, the general term of this arithmetic sequence is

$$a_n = \frac{16}{5}n + \frac{4}{5}, \quad \text{or}$$

$$a_n = \frac{16n + 4}{5}.$$

EXAMPLE 13

An exercise program calls for walking 15 minutes each day for the first week. Each week thereafter, the time spent walking each day increases by 5 minutes. In how many weeks will participants in this program be walking for one hour each day?

Solution   Let $\{a_n\}$ be an arithmetic sequence whose first term a_1 and common difference d are defined as follows:

$a_1 = 15$, the time (in minutes) spent walking each day of the first week,

and

$d = 5$, the weekly increase in walking time.

To find the number of weeks required for the daily walking time to increase to 60 minutes, we must find the value of n for which a_n is equal to 60. Using the equation

$$a_n = a_1 + (n - 1)d,$$

we find that

$$60 = 15 + (n - 1)5$$
$$60 = 15 + 5n - 5$$
$$5n = 50$$
$$n = 10.$$

Thus, the time spent walking each week will increase to one hour in 10 weeks.

The Sum of the First n Terms of an Arithmetic Sequence

The method used by Karl Gauss (see box at the bottom of this page) to find the sum of the first 100 positive integers can be generalized to obtain a formula for the sum of the first n terms of an arithmetic sequence $\{a_n\}$ whose common difference is d. By expressing the nth term a_n in the form $a_1 + (n - 1)d$, for $n \geq 1$, this sum, which we denote by S_n, becomes

$$S_n = a_1 + (a_1 + d) + \cdots + [a_1 + (n - 2)d] + [a_1 + (n - 1)d].$$

We now write the terms in the sum S_n in reverse order and add the two equations as shown below.

$$
\begin{array}{rcccccc}
S_n = & a_1 & + & (a_1 + d) & + \cdots + & [a_1 + (n - 2)d] & + & [a_1 + (n - 1)d] \\
S_n = & [a_1 + (n - 1)d] & + & [a_1 + (n - 2)d] & + \cdots + & (a_1 + d) & + & a_1 \\
\hline
2S_n = & 2a_1 + (n - 1)d & + & 2a_1 + (n - 1)d & + \cdots + & 2a_1 + (n - 1)d & + & 2a_1 + (n - 1)d
\end{array}
$$

Since the right side of the equation

$$2S_n = [2a_1 + (n - 1)d] + [2a_1 + (n - 1)d] + \cdots$$
$$+ [2a_1 + (n - 1)d] + [2a_1 + (n - 1)d]$$

contains n summands of the form $2a_1 + (n - 1)d$, we see that

$$2S_n = n[2a_1 + (n - 1)d], \qquad \text{or} \tag{3}$$
$$2S_n = 2na_1 + n(n - 1)d.$$

Dividing both sides of the latter equation by 2, we obtain the following convenient formula for S_n:

$$S_n = na_1 + \frac{n(n - 1)}{2}\, d.$$

From equation (3) above, we can obtain an alternate formula for the sum of the first n terms of an arithmetic sequence. We begin by rearranging terms on the right side of equation (3).

As noted in the previous section, we are often interested in finding the sum of the first n terms of a given sequence. We thus turn our attention to the problem of finding the sum of the first n terms of an arithmetic sequence. As a step in this direction, we recall a story that is told about the legendary German mathematician Karl Gauss as a young boy. According to the story, ten-year-old Gauss and his classmates were instructed by their teacher to add the integers from 1 to 100. Believing that this problem would occupy the class for quite a while, the teacher planned to use the time to catch up on some overdue work of his own.

Before he had time to collect his thoughts, however, the young Gauss appeared at his desk with the correct solution scrawled on his slate. As the story goes, Gauss obtained the sum of the first 100 positive integers as shown below.

$$
\begin{array}{rcccccccc}
1 & + & 2 & + & 3 & + \cdots + & 98 & + & 99 & + & 100 \\
100 & + & 99 & + & 98 & + \cdots + & 3 & + & 2 & + & 1 \\
\hline
101 & + & 101 & + & 101 & + \cdots + & 101 & + & 101 & + & 101
\end{array}
$$

This sum, which consists of 100 summands—each equal to 101, is twice the desired sum. Thus, the sum of the first 100 positive integers is $\frac{1}{2}(100)(101) = 5050$.

$$2S_n = n[2a_1 + (n - 1)d]$$
$$= n\{a_1 + [a_1 + (n - 1)d]\}.$$

Since $a_1 + (n - 1)d = a_n$, we have

$$2S_n = n(a_1 + a_n), \qquad \text{or}$$

$$S_n = \frac{n}{2}(a_1 + a_n).$$

We summarize our results as follows:

Sum of the First n Terms of an Arithmetic Sequence

Let $\{a_n\}$ be an arithmetic sequence with common difference d. The sum S_n of the first n terms of this sequence is given by

(i) $$S_n = na_1 + \frac{n(n - 1)}{2}d,$$

or equivalently by

(ii) $$S_n = \frac{n}{2}(a_1 + a_n).$$

Using the formula

$$S_n = na_1 + \frac{n(n - 1)}{2}d, \tag{4}$$

we can solve the grocery clerk's problem that was posed at the beginning of this section. Recall that the clerk wishes to stack 300 cans of peaches in a triangular display. He wants each row, after the bottom row, to contain one less can than the previous row and the top row to contain a single can.

To find the solution, we first number the rows of the display starting at the top, with row 1 containing 1 can, row 2 containing 2 cans, etc. We then define a sequence $\{a_n\}$, in which, for $i \geq 1$, the ith term a_i is equal to the number of cans in row i of the display. The sequence $\{a_n\}$ is thus an arithmetic sequence whose first term is 1 and whose common difference is 1. Noting that the total number of cans in the display is 300, we substitute 300 for the sum S_n, and 1 for both a_1 and d in equation (4) above.

$$S_n = na_1 + \frac{n(n - 1)}{2}d$$

$$300 = n(1) + \frac{n(n - 1)}{2} \tag{1}$$

$$300 = n + \frac{n^2 - n}{2}$$

$$600 = 2n + n^2 - n$$

$$n^2 + n - 600 = 0.$$

The latter quadratic equation can be solved by factoring.

$$n^2 + n - 600 = 0$$
$$(n + 25)(n - 24) = 0$$
$$n = -25 \quad \text{or} \quad n = 24.$$

Since the number of cans cannot be negative, it follows that the clerk should place 24 cans in the bottom row of the display. We can verify our result by substituting 1 for a_1 and 24 for both n and a_n in the formula

$$S_n = \frac{n}{2}(a_1 + a_n),$$

to obtain

$$S_{24} = \frac{24}{2}(1 + 24) = 12 \cdot 25 = 300.$$

E X A M P L E 14

Find the sum of the first 20 terms of an arithmetic sequence whose first term is 14 and whose common difference is $-\frac{7}{3}$.

Solution Since we are given a_1, n, and d, we use the equation

$$S_n = na_1 + \frac{n(n - 1)}{2}d$$

to find the sum of the first 20 terms of this arithmetic sequence. Substituting 20 for n, 14 for a_1, and $-\frac{7}{3}$ for d, we find that

$$S_{20} = (20)(14) + \frac{(20)(19)}{2}\left(-\frac{7}{3}\right)$$

$$= 280 + \left(-\frac{1330}{3}\right)$$

$$= -\frac{490}{3}.$$

E X A M P L E 15

If the first and eighteenth terms of an arithmetic sequence are 3 and 122, respectively, find the common difference d and the sum of the first 30 terms.

Solution To find d, we substitute 3 for a_1 and 122 for $a_n = a_{18}$ in the equation $a_n = a_1 + (n - 1)d$. This gives us

$$122 = 3 + (18 - 1)d$$
$$119 = 17d$$
$$d = 7.$$

Since we know the values of a_1, n, and d, we can find the sum of the first 30 terms of this sequence using the formula

$$S_n = na_1 + \frac{n(n - 1)}{2}d.$$

Substituting 30 for n, 3 for a_1, and 7 for d, we get

$$S_{30} = (30)(3) + \frac{(30)(29)}{2}(7) = 3135.$$

E X A M P L E 16

Verify the following formula for the sum of the first n positive integers:

$$\sum_{i=1}^{n} i = \frac{n(n+1)}{2}.$$

Solution The positive integers 1, 2, 3, . . . form an arithmetic sequence whose first term is 1 and whose common difference is 1. To find the sum of the first n terms of this sequence, we can use the formula

$$S_n = na_1 + \frac{n(n-1)}{2}d,$$

where $a_1 = 1$ and $d = 1$. Hence,

$$S_n = n(1) + \frac{n^2 - n}{2}(1)$$

$$= \frac{2n + n^2 - n}{2}$$

$$= \frac{n^2 + n}{2}$$

$$= \frac{n(n+1)}{2}.$$

Thus, we have

$$\sum_{i=1}^{n} i = \frac{n(n+1)}{2}.$$

Arithmetic Means

If a and b are real numbers, their average, $\dfrac{a+b}{2}$, is also called the **arithmetic mean** of a and b. Since the numbers a, $\dfrac{a+b}{2}$, and b have a common difference, they can be thought of as the first three terms of an arithmetic sequence. In general, if m_1, m_2, m_3, . . . , m_k are numbers such that

$$a, m_1, m_2, \ldots, m_k, b$$

are the first $k + 2$ terms of an arithmetic sequence, then the numbers m_1, m_2, m_3, . . . , m_k are called the **k arithmetic means of a and b**. The process of determining the numbers m_1, m_2, m_3, . . . , m_k is called **inserting k arithmetic means between a and b**. We illustrate this process in the following example.

E X A M P L E 17

Insert 4 arithmetic means between 3 and 12.

Solution We must find numbers m_1, m_2, m_3, and m_4 such that 3, m_1, m_2,

m_3, m_4, 12 are the first six terms of an arithmetic sequence. We can find the common difference d by substituting 3 for a_1 and 12 for $a_n = a_6$ in the equation

$$a_n = a_1 + (n - 1)d.$$

This gives us

$$12 = 3 + (6 - 1)d$$
$$9 = 5d$$
$$d = \frac{9}{5}.$$

Using this common difference, we find that the 4 arithmetic means between 3 and 12 are

$$m_1 = a_1 + \frac{9}{5} = 3 + \frac{9}{5} = \frac{24}{5},$$

$$m_2 = m_1 + \frac{9}{5} = \frac{24}{5} + \frac{9}{5} = \frac{33}{5},$$

$$m_3 = m_2 + \frac{9}{5} = \frac{33}{5} + \frac{9}{5} = \frac{42}{5}, \quad \text{and}$$

$$m_4 = m_3 + \frac{9}{5} = \frac{42}{5} + \frac{9}{5} = \frac{51}{5}.$$

EXERCISES 10.2

In Exercises 1–10, determine if the given sequence is an arithmetic sequence. For each sequence that is an arithmetic sequence, find the common difference d.

1. $\{4n - 1\}$
2. $\{2n + 1\}$
3. $\{n(n + 1)\}$
4. $\{3n^2 - 2\}$
5. $\{\frac{3}{2}n + 7\}$
6. $\{7 - \frac{2}{5}n\}$
7. $\{e^{n-1}\}$
8. $\{1 + \ln n\}$
9. $\{\log 10^n\}$
10. $\{\ln 2^n\}$

In Exercises 11–20, find the general term a_n of each arithmetic sequence.

11. $4, 7, 10, \ldots$
12. $2, 7, 12, \ldots$
13. $5, 1, -3, \ldots$
14. $1, -4, -9, \ldots$
15. $2, \frac{5}{2}, 3, \ldots$
16. $0, \frac{1}{2}, 1, \ldots$
17. $-\frac{1}{3}, \frac{1}{3}, 1, \ldots$
18. $-\frac{7}{4}, \frac{1}{2}, \frac{11}{4}, \ldots$
19. $\ln 2, \ln 4, \ln 8, \ldots$
20. $\log 5, \log 25, \log 125, \ldots$

In Exercises 21–30, find the fortieth term of each arithmetic sequence.

21. $\{7n - 1\}$
22. $\{\frac{2}{3}n + 4\}$
23. $\{5 + \frac{1}{2}n\}$
24. $\{12 - 1.5n\}$
25. $11, 16, 21, \ldots$
26. $-4, 3, 10, \ldots$
27. $-2, \frac{3}{2}, 5, \ldots$
28. $-\frac{5}{3}, 2, \frac{17}{3}, \ldots$
29. $a_1 = -\frac{12}{5}; d = \frac{3}{5}$
30. $a_1 = \frac{3}{4}; d = -\frac{9}{4}$
31. $a_3 = -1; a_5 = -9$
32. $a_2 = 2; a_6 = -10$

In Exercises 33–44, find the sum of the first ten terms of each arithmetic sequence.

33. $2, 8, 14, \ldots$
34. $-12, -8, -4, \ldots$
35. $1.3, 0.6, -0.1, \ldots$
36. $4, 5.2, 6.4, \ldots$
37. $\{5n - 2\}$
38. $\{\frac{3}{2}n - 4\}$
39. $a_1 = -4; d = \frac{3}{2}$
40. $a_1 = 7; d = -\frac{8}{5}$
41. $a_1 = \log 2; d = \log 2$
42. $a_1 = \ln 100; d = \ln 10$
43. $a_3 = 5\pi/2, a_{10} = 13\pi$
44. $a_2 = 2\sqrt{3}, a_7 = \frac{19\sqrt{3}}{2}$

In Exercises 45–52, insert the indicated number of arithmetic means between the given pair of numbers.

45. 3 arithmetic means between 4 and 28
46. 4 arithmetic means between 3 and 16
47. 8 arithmetic means between 2 and 7
48. 6 arithmetic means between 1 and 15
49. 5 arithmetic means between $\frac{1}{3}$ and 4
50. 7 arithmetic means between $\frac{5}{2}$ and 12
51. 4 arithmetic means between -11 and 2
52. 5 arithmetic means between -8 and 3
53. Determine whether or not 1000 is a term of the arithmetic sequence $2, 5, 8, \ldots$.
54. Determine whether or not 270 is a term of the arithmetic sequence $-3, 4, 11, \ldots$.

55. The sum of the first six terms of an arithmetic sequence is 27. If the first term is -8, find the general term a_n.

56. The sum of the first ten terms of an arithmetic sequence is -30. If the first term is 6, find the general term a_n.

57. The sum of three consecutive terms of an arithmetic sequence is 30 and their product is 510. Find the three terms. [*Hint*: Let $x - d$, x, and $x + d$ represent the three terms.]

58. The sum of three consecutive terms of an arithmetic sequence is 6 and the sum of their squares is 44. Find the three terms. (See Exercise 57 for a hint.)

59. An exercise program calls for jogging $\frac{1}{4}$ mile the first day and increasing the distance by $\frac{1}{8}$ mile each day thereafter. In how many days will the jogging distance be equal to 2 miles?

60. A fitness program calls for cycling on a stationary bicycle for 10 minutes each day of the first week. Each week thereafter the daily cycling regimen is increased by 5 minutes. In how many weeks will participants be cycling for 90 minutes each day?

61. The first row of seats in a theater has 18 seats and each row after the first has 3 more seats than the previous row. If there are 24 rows altogether, determine the total number of seats in the theater.

62. Logs are placed in a pile so that each row, starting at the bottom, has one less log than the previous row. If the bottom row has 48 logs and the top row has 16 logs, how many logs are in the pile?

63. Let n be a positive integer. Find a formula for the sum of the first n odd positive integers. [*Hint*: Odd positive integers have the form $2n - 1$, where $n = 1, 2, 3, \ldots$.]

Photo Francene Keery © Stock, Boston, Inc. 1991.
All Rights Reserved.

64. Let a_k and a_m, where $k < m$, be terms of an arithmetic sequence $\{a_n\}$. Show that the common difference d of the sequence is given by

$$d = \frac{a_m - a_k}{m - k}.$$

10.3 GEOMETRIC SEQUENCES AND SERIES

The multiplicative analogs of arithmetic sequences are called *geometric sequences*.

Geometric Sequence

A sequence $\{a_n\}$ is called a **geometric sequence** (or **geometric progression**) if there is a nonzero real number r such that

$$a_n = ra_{n-1}, \quad \text{for all } n > 1.$$

The number r is called the **common ratio**.

We see from this definition that each term of a geometric sequence, after the first, is obtained by multiplying the previous term by the common ratio r. The number r is called the common ratio since the ratio of any term to the previous term is always r—that is, $\dfrac{a_n}{a_{n-1}} = r$, for $n > 1$.

EXAMPLE 18

If the first term of a geometric sequence $\{a_n\}$ is $\frac{1}{3}$ and the common ratio is $-\frac{2}{5}$, find the second, third, and fourth terms of the sequence.

Solution   From the equation $a_n = ra_{n-1}$, where $r = -\frac{2}{5}$ and $n = 2, 3,$ and 4 in succession, we find that

$$a_2 = ra_1 = \left(-\frac{2}{5}\right)\left(\frac{1}{3}\right) = -\frac{2}{15},$$

$$a_3 = ra_2 = \left(-\frac{2}{5}\right)\left(-\frac{2}{15}\right) = \frac{4}{75}, \quad \text{and}$$

$$a_4 = ra_3 = \left(-\frac{2}{5}\right)\left(\frac{4}{75}\right) = -\frac{8}{375}.$$

EXAMPLE 19

Determine which of the following sequences are geometric sequences. If a given sequence is a geometric sequence, find its common ratio r.

a. $\left\{\dfrac{1}{2^{n-1}}\right\}$ **b.** $\left\{\dfrac{1}{n+1}\right\}$ **c.** $\left\{\left(-\dfrac{\pi}{3}\right)^n\right\}$

Solution

a. Let a_k and a_{k+1} be consecutive terms of the sequence $\{1/2^{n-1}\}$. Then $a_k = 1/2^{k-1}$ and $a_{k+1} = 1/2^{(k+1)-1} = 1/2^k$. Calculating the ratio a_{k+1}/a_k, we find that

$$\frac{a_{k+1}}{a_k} = \frac{1/2^k}{1/2^{k-1}} = \frac{1}{2^k} \cdot 2^{k-1} = 2^{-1} = \frac{1}{2}.$$

Since the ratio a_{k+1}/a_k is a constant, the sequence $\{1/2^{n-1}\}$ is a geometric sequence. Its common ratio is $\frac{1}{2}$.

b. Let $a_k = \dfrac{1}{k+1}$ and $a_{k+1} = \left[\dfrac{1}{(k+1)+1}\right] = \dfrac{1}{k+2}$ be consecutive terms of the sequence $\left\{\dfrac{1}{n+1}\right\}$. Then

$$\frac{a_{k+1}}{a_k} = \frac{\dfrac{1}{k+2}}{\dfrac{1}{k+1}} = \frac{k+1}{k+2}.$$

Since the ratio a_{k+1}/a_k is not a constant, this sequence is not a geometric sequence.

c. As in parts (a) and (b), let $a_k = (-\pi/3)^k$ and $a_{k+1} = (-\pi/3)^{k+1}$ be consecutive terms of the sequence $\{(-\pi/3)^n\}$. Then $a_{k+1}/a_k = \dfrac{(-\pi/3)^{k+1}}{(-\pi/3)^k} = -\pi/3$. Since the ratio a_{k+1}/a_k is a constant, this sequence is a geometric sequence. Its common ratio is $r = -\pi/3$.

The nth Term of a Geometric Sequence

To determine a formula for the nth term of a geometric sequence $\{a_n\}$ with common ratio r, we note that

$$a_1 = a_1 r^0$$
$$a_2 = a_1 r^1$$
$$a_3 = a_2 r = (a_1 r)r = a_1 r^2$$
$$a_4 = a_3 r = (a_1 r^2)r = a_1 r^3$$
$$\cdot$$
$$\cdot$$
$$\cdot$$

Since, for $n \geq 1$, the nth term a_n is the product of a_1 and r^{n-1}, we have the following result:

nth Term of a Geometric Sequence

If $\{a_n\}$ is a geometric sequence with common ratio r, its nth term a_n is given by the formula

$$a_n = a_1 r^{n-1}, \quad \text{for } n \geq 1.$$

EXAMPLE 20

Find the tenth term of the geometric sequence $\frac{4}{3}, -\frac{8}{9}, \frac{16}{27}, \ldots$.

Solution We can find the common ratio by dividing the second term by the first. Thus

$$r = \frac{-\dfrac{8}{9}}{\dfrac{4}{3}} = -\frac{8}{9} \cdot \frac{3}{4} = -\frac{2}{3}.$$

Using the equation $a_n = a_1 r^{n-1}$, where $n = 10$, we find that

$$a_{10} = \frac{4}{3} \cdot \left(-\frac{2}{3}\right)^9 = -\frac{2048}{59049}.$$

EXAMPLE 21

Find the first term of a geometric sequence whose third term is 8 and whose sixth term is 27.

Solution Using the formula $a_n = a_1 r^{n-1}$, with $n = 3$ and $n = 6$, respectively, we obtain

$$a_3 = a_1 r^2 \quad \text{and} \quad a_6 = a_1 r^5.$$

Since $a_3 = 8$ and $a_6 = 27$, we have the following system of equations:

$$\begin{cases} 8 = a_1 r^2 \\ 27 = a_1 r^5. \end{cases} \tag{1}$$

Solving the first equation for a_1, we find that $a_1 = \dfrac{8}{r^2}$. Substituting $\dfrac{8}{r^2}$ for a_1 in the second equation gives us

$$27 = \frac{8}{r^2} \cdot r^5, \qquad \text{or} \qquad 8r^3 = 27.$$

Hence,
$$r = \sqrt[3]{\frac{27}{8}} = \frac{3}{2}.$$

To find a_1, we substitute $\frac{3}{2}$ for r in the first equation of system (1) above:

$$8 = a_1 \left(\frac{3}{2}\right)^2$$
$$8 = \frac{9}{4} a_1$$
$$a_1 = \frac{32}{9}.$$

Thus, the first term of this geometric sequence is $\frac{32}{9}$.

Geometric Means

As with arithmetic sequences, we can find numbers $m_1, m_2, \ldots m_k$ between two real numbers a and b so that $a, m_1, m_2, \ldots, m_k, b$ are the first $k + 2$ terms of a geometric sequence. The process of finding the numbers $m_1, m_2, \ldots m_k$ is called **inserting k geometric means between a and b**.

EXAMPLE 22

Insert 3 geometric means between 2 and 162.

Solution Let m_1, m_2, and m_3 be three numbers such that

$$2, m_1, m_2, m_3, 162$$

are the first five terms of a geometric sequence. Since we know the first and fifth terms of this geometric sequence, we can find the common ratio r using the formula $a_n = a_1 r^{n-1}$, with $n = 5$. Thus

$$a_5 = a_1 r^4$$
$$162 = 2r^4$$
$$r^4 = 81$$
$$r = \pm 3.$$

Hence, there are two geometric sequences that satisfy the stated conditions. If $r = 3$, we have the sequence

$$2, 6, 18, 54, 162, \ldots.$$

On the other hand, if $r = -3$, we have the sequence

$$2, -6, 18, -54, 162, \ldots.$$

Thus, when $r = 3$, the three geometric means between 2 and 162 are 6, 18, and 54, and when $r = -3$, the three geometric means are -6, 18, and -54.

The Sum of the First *n* Terms of a Geometric Sequence

We will now determine a formula for the sum of the first n terms of a geometric sequence $\{a_n\}$ whose common ratio is r. If S_n denotes this sum, then

$$S_n = a_1 + a_2 + a_3 + \cdots + a_n,$$

or, since $a_n = a_1 r^{n-1}$, for $n \geq 1$,

$$S_n = a_1 + a_1 r + a_1 r^2 + \cdots + a_1 r^{n-1}. \tag{2}$$

Multiplying both sides of equation (2) by r, we obtain the equation

$$rS_n = a_1 r + a_1 r^2 + a_1 r^3 + \cdots + a_1 r^n. \tag{3}$$

If we subtract equation (3) from equation (2), we get

$$
\begin{array}{rcllll}
S_n = & a_1 + a_1 r + a_1 r^2 + \cdots + a_1 r^{n-1} \\
-rS_n = & -a_1 r - a_1 r^2 - \cdots - a_1 r^{n-1} - a_1 r^n \\
\hline
S_n - rS_n = & a_1 + 0 \quad + 0 \quad + \cdots + 0 \quad - a_1 r^n
\end{array}
$$

Thus, $\quad S_n - rS_n = a_1 - a_1 r^n, \qquad$ or

$$(1 - r)S_n = a_1 (1 - r^n).$$

If $r \neq 1$, we can divide both sides of the latter equation by $1 - r$ to obtain

$$S_n = \frac{a_1(1 - r^n)}{1 - r}.$$

In summary, we have

Sum of the First *n* Terms of a Geometric Sequence

Let $\{a_n\}$ be a geometric sequence with common ratio r. The sum S_n of the first n terms of this sequence is given by the formula

$$S_n = \frac{a_1(1 - r^n)}{1 - r}, \quad \text{if } r \neq 1.$$

EXAMPLE 23

Find the value of each of the following sums:

a. $\displaystyle\sum_{i=1}^{8} 5 \left(\frac{1}{2}\right)^{i-1}$ **b.** $\displaystyle\sum_{i=0}^{10} (-3)^i$

Solution

a. Expanding the sum $\displaystyle\sum_{i=1}^{8} 5 \left(\frac{1}{2}\right)^{i-1}$, we obtain

$$\sum_{i=1}^{8} 5 \left(\frac{1}{2}\right)^{i-1} = 5 + 5\left(\frac{1}{2}\right) + 5\left(\frac{1}{2}\right)^2 + \cdots + 5\left(\frac{1}{2}\right)^7.$$

We thus have the sum of the first 8 terms of a geometric sequence whose first term is $a_1 = 5$ and whose common ratio is $r = \frac{1}{2}$. Using the formula

$$S_n = \frac{a_1(1 - r^n)}{1 - r},$$

where $n = 8$, we get

$$S_8 = \frac{5[1 - (\frac{1}{2})^8]}{1 - \frac{1}{2}} = \frac{5(1 - \frac{1}{256})}{\frac{1}{2}} = \frac{5(\frac{255}{256})}{\frac{1}{2}} = \frac{1275}{128}.$$

b. Since $\sum_{i=0}^{10}(-3)^i = 1 + (-3) + (-3)^2 + \cdots + (-3)^{10}$, we have the sum of the first 11 terms of a geometric sequence with first term $a_1 = 1$ and common ratio $r = -3$. Using the formula

$$S_n = \frac{a_1(1 - r^n)}{1 - r},$$

where $n = 11$, the sum of these terms is

$$S_{11} = \frac{1[1 - (-3)^{11}]}{1 - (-3)} = \frac{1 - (-177,147)}{4} = 44,287.$$

E X A M P L E   24

A deposit of $100 is made on the first day of each month for five years into an account that pays 6% interest compounded monthly. How much money is in the account at the end of five years?

Solution   In Section 4.1 we found that a principal of P dollars invested at interest rate r, compounded n times per year for t years, produces an amount A, which is given by the formula

$$A = P\left(1 + \frac{r}{n}\right)^{nt}$$

To solve this problem, we will consider the 60 monthly deposits made during the 5 years as 60 individual investments. Since the first deposit of $100 earns interest for all 60 months, the amount it produces (in dollars) is given by

$$A = 100\left(1 + \frac{0.06}{12}\right)^{12\cdot5} \qquad \text{or}$$

$$\doteq 100(1.005)^{60}.$$

Similarly, the second $100 deposit earns interest for 59 months, and produces the amount

$$A = 100(1.005)^{59}.$$

We continue in this manner until we reach the final monthly deposit, which produces the amount

$$A = 100(1.005)^1.$$

Since the total amount A in the account after 5 years is the sum of these 60 amounts, we have

$$A = 100(1.005)^1 + 100(1.005)^2 + 100(1.005)^3 + \cdots + 100(1.005)^{60}.$$

Thus, A is the sum of the first 60 terms of a geometric sequence with first term $a_1 = 100(1.005)^1 = 100.5$, and common ratio $r = 1.005$. Using the formula

$$S_n = \frac{a_1(1 - r^n)}{1 - r},$$

where $n = 60$, we find that

$$S_{60} = \frac{100.5[1 - (1.005)^{60}]}{1 - 1.005} \approx 7011.89.$$

Hence the amount in the account after 5 years (rounded to the nearest cent) is $7011.89.

Geometric Series

As we have seen, the formula $S_n = \dfrac{a_1(1 - r^n)}{1 - r}$ enables us to determine the sum of the first n terms of any geometric sequence. As we shall now see, it is sometimes possible to determine the sum of *all* the terms of a geometric sequence. Consider, for example, a geometric sequence $\{a_n\}$ whose common ratio is r. To indicate the sum of all the terms of this sequence we write

$$a_1 + a_2 + a_3 + \cdots,$$

or, using summation notation,

$$\sum_{i=1}^{\infty} a_i.$$

The indicated sum of all the terms of a geometric sequence is called a **geometric series**. In general, the indicated sum of all the terms of a sequence is called an **infinite series** or simply a **series**. While a thorough study of series requires concepts developed in calculus, we can give an intuitive argument to describe the behavior of geometric series.

Consider, for example, the following two geometric sequences:

(1) $\{2^n\}$ **(2)** $\{(\frac{1}{3})^n\}$

The indicated sum of the terms of the first sequence is the geometric series $\sum_{i=1}^{\infty} 2^i$, while that of the second sequence is the series $\sum_{i=1}^{\infty} (\frac{1}{3})^i$. To determine (if possible) the values of these sums, we will consider the values of S_n, the sum of the first n terms of each sequence, for increasing values of n. For each sequence, the values of S_n for $n = 1, 2, \ldots, 5$ and 10 are displayed in Table 10.1 on page 516.

We see from Table 10.1 that the values of S_n for the series $\sum_{i=1}^{\infty} 2^i$ are getting larger and larger as the value of n increases. Using terminology developed in Section 3.6, we can write

$$S_n \to \infty \quad \text{as} \quad n \to \infty.$$

We conclude, therefore, that the sum $\sum_{i=1}^{\infty} 2^i$ is infinite, and say that the series $\sum_{i=1}^{\infty} 2^i$ has no sum. On the other hand, we see from the table that the values of S_n for the series $\sum_{i=1}^{\infty} (\frac{1}{3})^i$ appear to be getting closer and closer to $0.5 = \frac{1}{2}$ as n gets larger and larger. In this case, since

$$S_n \to \frac{1}{2} \quad \text{as} \quad n \to \infty,$$

TABLE 10.1

n	$S_n = \sum_{i=1}^{n} 2i$	$S_n = \sum_{i=1}^{n} (\frac{1}{3})^i$
1	2	$\frac{1}{3} = 0.333\ldots$
2	$2 + 4 = 6$	$\frac{1}{3} + \frac{1}{9} = \frac{4}{9} = 0.444\ldots$
3	$2 + 4 + 8 = 14$	$\frac{1}{3} + \frac{1}{9} + \frac{1}{27} = \frac{13}{27} = 0.481481\ldots$
4	$2 + 4 + 8 + 16 = 30$	$\frac{1}{3} + \frac{1}{9} + \frac{1}{27} + \frac{1}{81} = \frac{40}{81} \approx 0.49382716$
5	$2 + 4 + 8 + 16 + 32 = 62$	$\frac{1}{3} + \frac{1}{9} + \frac{1}{27} + \frac{1}{81} + \frac{1}{243} = \frac{121}{243} \approx 0.49794239$
$\vdots$	$\vdots$	$\vdots$
10	$2 + 4 + 8 + \cdots + 1024 = 2046$	$\frac{1}{3} + \frac{1}{9} + \frac{1}{27} + \cdots + \frac{1}{59049} = \frac{29524}{59049} \approx 0.49999153$

we conclude that the value of the sum $\sum_{i=1}^{\infty} (\frac{1}{3})^i$ is $\frac{1}{2}$. Theorem 10.2, which can be verified for *any* sequence $\{a_n\}$, generalizes our findings to this point:

THEOREM 10.2

Let $\{a_n\}$ be a geometric sequence with common ratio r, and let S_n denote the sum of the first n terms of the sequence.

(i) If $S_n \rightarrow S(S \in \mathbb{R})$, as $n \rightarrow \infty$, then $\sum_{i=1}^{\infty} a_i = S$, and we say that the series $\sum_{i=1}^{\infty} a_i$ has sum S.

(ii) If $S_n \rightarrow \infty$, or $S_n \rightarrow -\infty$, as $n \rightarrow \infty$, we say that the series $\sum_{i=1}^{\infty} a_i$ has no sum.

If a geometric series has a sum S, we can find it by investigating the behavior of $S_n = \dfrac{a_1(1 - r^n)}{1 - r}$, as $n \rightarrow \infty$. Since a_1 and r are constants, we see that the value of S_n, as $n \rightarrow \infty$, depends on the value of r^n, as $n \rightarrow \infty$. The following result, which can be verified for any real number $r \neq 1$, describes the behavior of r^n, as $n \rightarrow \infty$.

Value of r^n, as $n \rightarrow \infty$

Let r be any real number other than 1.

(i) If $|r| < 1$, $r^n \rightarrow 0$, as $n \rightarrow \infty$.
(ii) If $|r| > 1$, $|r^n| \rightarrow \infty$, as $n \rightarrow \infty$.

In Table 10.2, we illustrate this result using $r = \frac{1}{3}$ and $r = 2$, where $n = 1, 2, \ldots, 5$, and 10. Thus, when $|r| < 1$, the fact that $r^n \rightarrow 0$ as $n \rightarrow \infty$, tells us that

$$S_n = \frac{a_1(1 - r^n)}{1 - r} \rightarrow \frac{a_1(1 - 0)}{1 - r} \rightarrow \frac{a_1}{1 - r}, \quad \text{as } n \rightarrow \infty.$$

TABLE 10.2

n	1	2	3	4	5	...	10
$(\frac{1}{3})^n$	0.333 $\cdots$	0.111 $\cdots$	0.037037 $\cdots$	0.012346	0.004115	$\cdots$	0.000017
2^n	2	4	8	16	32	$\cdots$	1024

This means that the sum S of the series $\sum_{i=1}^{\infty} a_i$ is $\dfrac{a_1}{1-r}$, whenever $|r| < 1$. On the other hand, when $|r| > 1$, it can be shown that $|S_n| \to \infty$, as $n \to \infty$. It follows that the series $\sum_{i=1}^{\infty} a_i$ has no sum when $|r| > 1$. In summary we have:

Sum of a Geometric Series

Let $\{a_n\}$ be a geometric sequence with common ratio $r \neq 1$.

(i) If $|r| < 1$, the geometric series $\sum_{i=1}^{\infty} a_i$ has sum $S = \dfrac{a_1}{1-r}$.

(ii) If $|r| > 1$, the geometric series $\sum_{i=1}^{\infty} a_i$ has no sum.

If we apply this result to the series $\sum_{i=1}^{\infty} 2^i$ and $\sum_{i=1}^{\infty} (\frac{1}{3})^i$, which were discussed earlier, we see that $\sum_{i=1}^{\infty} 2^i$ has no sum since $|r| = 2 > 1$, while $\sum_{i=1}^{\infty} (\frac{1}{3})^i$ has sum $S = \dfrac{a_1}{1-r} = \dfrac{\frac{1}{3}}{1-\frac{1}{3}} = \frac{1}{2}$, since $|r| = \frac{1}{3} < 1$. We note that these results agree with our earlier findings.

EXAMPLE 25

Find the sum of each of the following geometric series or tell why the series has no sum.

a. $16 - 2 + \frac{1}{4} - \frac{1}{32} + \cdots$ **b.** $1 + \frac{5}{3} + \frac{25}{9} + \cdots$

Solution

a. To find the common ratio r we divide the second term by the first. Thus $r = -\frac{2}{16} = -\frac{1}{8}$. Since $|r| = \frac{1}{8} < 1$, the sum of the series $16 - 2 + \frac{1}{4} - \frac{1}{32} + \cdots$ is

$$\frac{a_1}{1-r} = \frac{16}{1-(-\frac{1}{8})} = 16 \cdot \frac{8}{9} = \frac{128}{9}.$$

b. Since $|r| = \frac{5}{3} > 1$, the series $1 + \frac{5}{3} + \frac{25}{9} + \cdots$ has no sum.

In Section 1.2 we demonstrated a method for expressing a repeating decimal number as a quotient of two integers. In the following example we show how this can be done using a geometric series.

EXAMPLE 26

Express the rational number $0.53161616\ldots$ as a quotient of two integers.

Solution   Since the first two digits are not part of the repeating cycle of digits, we write

$$0.53161616\ldots = 0.53 + 0.00161616\ldots.$$

The number $0.00161616\ldots$ can be written as

$$0.00161616\ldots = 0.0016 + 0.000016 + 0.00000016 + \cdots,$$

which is the sum of the terms of the geometric sequence whose first term is $a_1 = 0.0016$ and whose common ratio is $r = 0.01$. Since $|r| = 0.01 < 1$, the sum of this series is

$$\frac{a_1}{1 - r} = \frac{0.0016}{1 - 0.01} = \frac{0.0016}{0.99} = \frac{16}{9900}.$$

Thus $0.53161616\ldots = 0.53 + \frac{16}{9900} = \frac{53}{100} + \frac{16}{9900} = \frac{5263}{9900}.$

We note that this is the same result we obtained in Example 4 of Section 1.2.

EXAMPLE 27

Consider the geometric sequence $1, \dfrac{2x}{3}, \dfrac{4x^2}{9}, \ldots$, where x is a real number. Determine all values of x for which the geometric series $\displaystyle\sum_{i=1}^{\infty} \left(\frac{2x}{3}\right)^{i-1}$ has a sum S, where $S \in \mathbb{R}$.

Solution   Since $r = \dfrac{2x}{3}$ is the common ratio of the geometric sequence $1, \dfrac{2x}{3},$ $\dfrac{4x^2}{9}, \ldots$, the series $\displaystyle\sum_{i=1}^{\infty} \left(\frac{2x}{3}\right)^{i-1}$ has sum S provided $|r| = \left|\dfrac{2x}{3}\right| < 1$. Solving this inequality for x, we find that

$$\left|\frac{2x}{3}\right| < 1$$
$$|2x| < 3$$
$$-3 < 2x < 3$$
$$-\frac{3}{2} < x < \frac{3}{2}.$$

Thus the series $\displaystyle\sum_{i=1}^{\infty} \left(\frac{2x}{3}\right)^{i-1}$ has a sum S for all values of x in the interval $\left(-\frac{3}{2}, \frac{3}{2}\right)$.

EXERCISES 10.3

In Exercises 1–10, determine whether or not the given sequence is a geometric sequence. If it is, find its common ratio r.

1. $\{(\frac{1}{4})^n\}$
2. $\{(-\frac{1}{3})^n\}$
3. $\{1/2n\}$
4. $\{n^2\}$
5. $\{(-\frac{2}{5})^{n-1}\}$
6. $\{(\frac{3}{4})^n\}$

7. $\{e^{nx}\}$
8. $\{\pi^{(n-1)x}\}$
9. $\{\log 2^{2^n}\}$
10. $\{\ln 10^{n^2}\}$

In Exercises 11–20, find the general term of each geometric sequence.

11. $3, 6, 12, \ldots$
12. $2, -1, \frac{1}{2}, \ldots$
13. $-4, 3, -\frac{9}{4}, \ldots$
14. $5, 3, \frac{9}{5}, \ldots$

15. 1.2, 0.24, 0.048, . . . **16.** $-5, 1, -0.2, \ldots$

17. $1, -x^2, x^4, \ldots$ **18.** $1, x/3, x^2/9, \ldots$

19. ln 10, ln 100, ln 10,000, . . .

20. log 5, log 25, log 625, . . .

21. Find the third term of a geometric sequence whose first term is 3 and whose fifth term is 48.

22. Find the fourth term of a geometric sequence whose first term is 7 and whose fifth term is 112.

23. Find the fourth term of a geometric sequence whose second term is -6 and whose fifth term is $\frac{81}{4}$.

24. Find the third term of a geometric sequence whose second term is 4 and whose sixth term is $\frac{4}{81}$.

25. Find the second term of a geometric sequence whose third term is $\frac{4}{3}$ and whose sixth term is $-\frac{32}{81}$.

26. Find the fourth term of a geometric sequence whose second term is -4 and whose sixth term is $-\frac{64}{81}$.

In Exercises 27–36, find the indicated sum of terms for each geometric sequence.

27. $8, 4, 2, \ldots; S_{10}$ **28.** $7, 14, 28, \ldots; S_8$

29. $300, -30, 3, \ldots; S_{12}$ **30.** $162, -54, 18, \ldots; S_{10}$

31. $\frac{1}{2}, \frac{3}{8}, \frac{9}{32}, \ldots; S_6$ **32.** $4, 3, \frac{9}{4}, \ldots; S_{12}$

33. ln 4, ln 2, ln $\sqrt{2}$, . . .; S_{10}

34. log 27, log 3, log $\sqrt[3]{3}$, . . .; S_5

35. $10, 10^{x+1}, 10^{2x+1}, \ldots; S_8$ **36.** $1, -x/3, x^2/9, \ldots; S_7$

In Exercises 37–42, insert the indicated number of geometric means between the given pair of numbers.

37. 3 geometric means between 2 and 162

38. 3 geometric means between 10 and 20

39. 4 geometric means between -27 and $\frac{32}{9}$

40. 4 geometric means between 2 and 0.00064

41. 3 geometric means between 4 and 0.0324

42. 3 geometric means between -24 and $-\frac{243}{2}$

In Exercises 43–52, find the sum of each geometric series or tell why the series has no sum.

43. $6 + 4 + \frac{8}{3} + \cdots$ **44.** $12 + 9 + \frac{27}{4} + \cdots$

45. $1 - \frac{2}{3} + \frac{4}{9} - \frac{8}{27} + \cdots$

46. $1 - \frac{5}{6} + \frac{25}{36} - \frac{125}{216} + \cdots$

47. $4 + 1.2 + 0.36 + \cdots$

48. $0.3 + 0.06 + 0.012 + \cdots$

49. $-\sqrt{3} + 3 - 3\sqrt{3} + \cdots$

50. $-\sqrt{2} + 2 - 2\sqrt{2} + \cdots$

51. $1/e + 1/e^2 + 1/e^3 + \cdots$

52. $2 + \pi/2 + \pi^2/8 + \cdots$

In Exercises 53–60, express each rational number as a quotient of two integers.

53. $0.141414 \ldots$ **54.** $0.272727 \ldots$

55. $2.632632 \ldots$ **56.** $3.158158 \ldots$

57. $10.50219219 \ldots$ **58.** $1.36445445 \ldots$

59. $3.20482048 \ldots$ **60.** $12.06930693 \ldots$

61. The product of three consecutive terms of a geometric sequence is -216 and their sum is 7. Find the common ratio r. [Hint: Let $\frac{a}{r}$, a, and ar denote the three terms.]

62. The product of three consecutive terms of a geometric sequence is 8 and their sum is $\frac{26}{3}$. Find the common ratio r.

63. For 12 consecutive years a $500 deposit is made on January 1 into an account that pays 8% interest compounded annually. Find the value of the account at the end of the twelfth year.

64. On the first day of each month for six years, a deposit of $50 is made into an account that pays 9% interest compounded monthly. How much money is in the account at the end of six years?

65. A company pays $24,000 for a copy machine that depreciates at the rate of 25% per year. Find the depreciated value of the copy machine 8 years later. [Hint: The depreciated value of the machine at the end of each year is 75% of its value at the beginning of that year.]

66. If a new automobile costs $36,000 and depreciates in value by 20% each year, find its depreciated value at the end of 10 years.

Courtesy of Mazda Corporation of America

67. A ball is dropped from a height of 20 feet. On each bounce, the ball rises to a height that is one-half its previous height. Find the total distance the ball travels. [Hint: Find the sum of the "downward" distances $(20 + 10 + 5 + \cdots)$ and the upward distances $(10 + 5 + \frac{5}{2} + \cdots)$.]

68. A ball dropped from a height of 60 feet bounces 12 feet high on its first bounce. On each successive bounce, the ball bounces to a height that is $\frac{2}{3}$ the height of the previous bounce. Find the total distance the ball travels.

69. Let $\{a_n\}$ be a geometric sequence of positive terms whose common ratio is r. Show that the sequence $\{\ln a_n\}$ is an arithmetic sequence and find its common difference d.

70. The product $P_n = a_1 \cdot a_2 \cdot a_3 \cdots a_n$ is called the **nth partial product** of the sequence $\{a_n\}$. Find the nth partial product of a geometric sequence $\{a_n\}$ whose common ratio is r.

71. Consider the geometric sequence $1, 2x, 4x^2, \ldots$, where x is a real number. Determine all values of x for which the series $\sum_{i=1}^{\infty} (2x)^{i-1}$ has a sum S, where $S \in \mathbb{R}$.

72. Consider the geometric sequence $1, \dfrac{3x}{5}, \dfrac{9x^2}{25}, \ldots$, where x is a real number. Determine all values of x for which the series $\sum_{i=1}^{\infty} \left(\dfrac{3x}{5}\right)^{i-1}$ has a sum S, where $S \in \mathbb{R}$.

73. If square $ABCD$ has sides of length 1 (see figure), and the

pattern of shaded regions continues indefinitely, find the total area of the shaded regions.

74. Assume that square $ABCD$ has sides of length 1 and that square $EFGH$ is formed by connecting the midpoints of the sides of square $ABCD$ (see figure). If the pattern of shaded regions continues indefinitely, find the total area of the shaded regions.

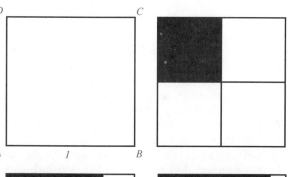

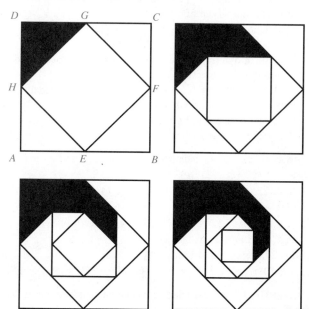

10.4 MATHEMATICAL INDUCTION

There are many mathematical statements that can be shown to be true for every positive integer n. For example, in Section 10.2, we found the following formula for the sum of the first n positive integers:

$$1 + 2 + 3 + \cdots + n = \frac{n(n+1)}{2}. \qquad (1)$$

As the following table shows, statement (1) is true for $n = 1, 2, 3,$ and 4:

n	$1 + 2 + 3 + \cdots + n$	$\dfrac{n(n+1)}{2}$
1	1	$\dfrac{1(1+1)}{2} = 1$
2	$1 + 2 = 3$	$\dfrac{2(2+1)}{2} = 3$
3	$1 + 2 + 3 = 6$	$\dfrac{3(3+1)}{2} = 6$
4	$1 + 2 + 3 + 4 = 10$	$\dfrac{4(4+1)}{2} = 10$

As verified in the table, statement (1) is true when $n = 1, 2, 3,$ and 4. However, since the set of positive integers is infinite, it is impossible when using this approach to prove that statement (1) is true for *every* positive integer n. We can, however, prove that this statement is true for every positive integer n by using a type of reasoning known as the principle of mathematical induction.

Principle of Mathematical Induction

Let S be a set of positive integers that satisfies the following two conditions:

(i)   1 is an element of S;

(ii)   If the positive integer k is in S, then $k + 1$ is in S.
        Then S is the set of positive integers; that is, $S = \mathbb{N}$.

To see how the principle of mathematical induction works, observe that 1 is in S by condition (i). Since 1 is in S, $1 + 1 = 2$ is in S by condition (ii). Since 2 is in S, $2 + 1 = 3$ is in S by condition (ii). From repeated applications of condition (ii), it follows that $3 + 1 = 4$ is in S, $4 + 1 = 5$ is in S, $5 + 1 = 6$ is in S, and so on.

Before using the principle of mathematical induction to prove that statement (1) is true for every positive integer n, we should point out that in a typical proof that is based on the principle of mathematical induction, we begin by defining the set S. The proof then has two steps. In the first step, we verify that condition (i) holds by showing that 1 is in S. In the second step, we verify that condition (ii) holds, by assuming that k is in S and then showing that $k + 1$ is in S.

EXAMPLE 28

Use the principle of mathematical induction to prove that, for every positive integer n,

$$1 + 2 + 3 + \cdots + n = \frac{n(n + 1)}{2}.$$

Solution   We begin by defining the set S as follows:

$$S = \left\{ n \text{ in } \mathbb{N} \mid 1 + 2 + 3 + \cdots + n = \frac{n(n + 1)}{2} \right\}.$$

Using the principle of mathematical induction, we proceed as follows. First, we must show that 1 is in S. Since $1 = \dfrac{1(1 + 1)}{2}$, 1 is in S.

Next, we must show that if k is in S, then $k + 1$ is in S. If we assume that k is in S, then we know that

$$1 + 2 + 3 + \cdots + k = \frac{k(k + 1)}{2}. \tag{2}$$

To show that $k + 1$ is in S, we must show that

$$1 + 2 + 3 + \cdots + k + (k + 1) = \frac{(k + 1)(k + 2)}{2}. \tag{3}$$

Starting with the left side of equation (3), we obtain the right side as follows:

$$1 + 2 + 3 + \cdots + k + (k + 1)$$
$$= (1 + 2 + 3 + \cdots + k) + (k + 1)$$
$$= \left[\frac{k(k + 1)}{2} \right] + (k + 1) \qquad \text{By equation (2)}$$
$$= \frac{k(k + 1) + 2(k + 1)}{2} \qquad \text{Adding fractions}$$
$$= \frac{(k + 1)(k + 2)}{2}.$$

Therefore, $k + 1$ is in S.

Since both conditions of the principle of mathematical induction have been verified, it follows that $S = \mathbb{N}$; that is, equation (1) is true for every positive integer n.

In Section 10.1, we defined n factorial where n is a nonnegative integer as follows:

If $n > 0$, then $n! = n(n - 1)(n - 2) \cdots 3 \cdot 2 \cdot 1$;

If $n = 0$, then $0! = 1$.

We also noted that if $n > 1$, $n! = n(n - 1)!$ In our next example, we use the principle of mathematical induction to verify a statement that involves n factorial.

EXAMPLE 29

Use the principle of mathematical induction to prove that, for every positive integer n,

$$\frac{1}{2!} + \frac{2}{3!} + \frac{3}{4!} + \cdots + \frac{n}{(n + 1)!} = 1 - \frac{1}{(n + 1)!}.$$

Solution   We begin by defining the set S as follows:

$$S = \left\{ n \text{ in } \mathbb{N} \mid \frac{1}{2!} + \frac{2}{3!} + \frac{3}{4!} + \cdots + \frac{n}{(n + 1)!} = 1 - \frac{1}{(n + 1)!} \right\}.$$

Using the principle of mathematical induction, we will show that $S = \mathbb{N}$.

First, we show that 1 is in S. To do this, we must show that

$$\frac{1}{2!} = 1 - \frac{1}{(1 + 1)!}.$$

Since $(1 + 1)! = 2! = 2 \cdot 1 = 2$, it follows that

$$1 - \frac{1}{(1 + 1)!} = 1 - \frac{1}{2} = \frac{1}{2} = \frac{1}{2!}.$$

Hence, 1 is in S.

Next, we show that if k is in S, then $k + 1$ is in S. If we assume that k is in S, then we know that

$$\frac{1}{2!} + \frac{2}{3!} + \frac{3}{4!} + \cdots + \frac{k}{(k + 1)!} = 1 - \frac{1}{(k + 1)!}. \qquad (4)$$

To show that $k + 1$ is in S, we must show that

$$\frac{1}{2!} + \frac{2}{3!} + \frac{3}{4!} + \cdots + \frac{k}{(k + 1)!} + \frac{k + 1}{(k + 2)!} = 1 - \frac{1}{(k + 2)!}. \quad (5)$$

We start with the left side of equation (5) and obtain the right side as follows:

$$\frac{1}{2!} + \frac{2}{3!} + \frac{3}{4!} + \cdots + \frac{k}{(k + 1)!} + \frac{k + 1}{(k + 2)!}$$

$$= \left(\frac{1}{2!} + \frac{2}{3!} + \frac{3}{4!} + \cdots + \frac{k}{(k + 1)!} \right) + \frac{k + 1}{(k + 2)!}$$

$$= \left[1 - \frac{1}{(k + 1)!} \right] + \frac{k + 1}{(k + 2)!} \qquad \textbf{By equation (4)}$$

$$= 1 - \frac{1}{(k + 1)!} + \frac{k + 1}{(k + 2)(k + 1)!} \qquad (k + 2)! = (k + 2)(k + 1)!$$

$$= 1 + \frac{-(k + 2) + (k + 1)}{(k + 2)(k + 1)!} \qquad \textbf{Adding fractions}$$

$$= 1 + \frac{-k - 2 + k + 1}{(k + 2)!} \qquad (k + 2)(k + 1)! = (k + 2)!$$

$$= 1 - \frac{1}{(k + 2)!}.$$

Therefore, $k + 1$ is in S. Based on the principle of mathematical induction, we conclude that $S = \mathbb{N}$. Hence, for every possible integer n,

$$\frac{1}{2!} + \frac{2}{3!} + \frac{3}{4!} + \cdots + \frac{n}{(n + 1)!} = 1 - \frac{1}{(n + 1)!}.$$

Our next example involves a recursively defined sequence of positive integers.

EXAMPLE 30

For every positive integer n, the integer x_n is defined recursively as follows:

(1) $x_1 = 3$;
(2) If $n \geq 2$, $x_n = x_{n-1} + (2n + 1)$.

Use the principle of mathematical induction to prove that for every positive integer n, $x_n = n(n + 2)$.

Solution First, we let

$$S = \{n \text{ in } \mathbb{N} \mid x_n = n(n + 2)\}.$$

Since $x_1 = 3 = 1(1 + 2)$, we see that 1 is in S.

If we assume that the positive integer k is in S, then we know that

$$x_k = k(k + 2). \quad (6)$$

In order to show that $k + 1$ is in S, we must verify that

$$x_{k+1} = (k + 1)[(k + 1) + 2],$$

or that
$$x_{k+1} = (k + 1)(k + 3).$$

Since $k + 1 \geq 2$, it follows from the recursive definition of x_n that

$$
\begin{aligned}
x_{k+1} &= x_{(k+1)-1} + [2(k + 1) + 1] \\
&= x_k + (2k + 3) \\
&= k(k + 2) + (2k + 3) \qquad \text{By equation (6)} \\
&= k^2 + 4k + 3 \\
&= (k + 1)(k + 3).
\end{aligned}
$$

Since $k + 1$ is in S, it follows that $S = \mathbb{N}$. Thus, $x_n = n(n + 2)$, for every positive integer n.

E X A M P L E   31

Use the principle of mathematical induction to prove that, for every positive integer n, $n < 2^n$.

Solution   We begin by letting $S = \{n \text{ in } \mathbb{N} \mid n < 2^n\}$. Since $1 < 2 = 2^1$, it follows that 1 is in S. If we assume that k is in S, then we know that

$$k < 2^k. \tag{7}$$

In order to show that $k + 1$ is in S, we must show that

$$k + 1 < 2^{k+1}.$$

Multiplying both sides of inequality (7) by 2, we obtain

$$2k < 2 \cdot 2^k, \qquad \text{or} \qquad 2k < 2^{k+1}.$$

Since $1 \leq k$, it follows that

$$k + 1 \leq k + k = 2k.$$

Thus, we have

$$k + 1 \leq 2k \qquad \text{and} \qquad 2k < 2^{k+1}.$$

By the transitive law of order, it follows that $k + 1 < 2^{k+1}$. Thus, $k + 1$ is in S, and $S = \mathbb{N}$. Hence, by the principle of mathematical induction, $n < 2^n$, for every positive integer n.

Notice that in each of the previous examples in this section we have considered statements that are true for every positive integer n. In the next example we see how the principle of mathematical induction can be modified when we are considering a statement that is false for some values of n, but true for all values of n greater than or equal to some fixed positive integer.

E X A M P L E   32

Use the principle of mathematical induction to prove that

$$3^n > (n + 1)^2,$$

for every positive integer n such that $n \geq 3$.

Solution   First, we let $S = \{n \text{ in } \mathbb{N} \mid 3^n > (n + 1)^2\}$. To show that 3 is in S, we must show that

$$3^3 > (3 + 1)^2.$$

Since $27 > 16$, it follows that 3 is in S.

If we assume that k is in S, then for some positive integer $k \geq 3$,

$$3^k > (k + 1)^2. \tag{8}$$

To show that $k + 1$ is in S, we must show that

$$3^{k+1} > [(k + 1) + 1]^2,$$

or that

$$3^{k+1} > (k + 2)^2.$$

Multiplying both sides of inequality (8) by 3, we obtain

$$3(3^k) > 3(k + 1)^2 = 3k^2 + 6k + 3 = k^2 + 4k + (2k^2 + 2k + 3),$$

or

$$3^{k+1} > k^2 + 4k + (2k^2 + 2k + 3). \tag{9}$$

Since we know that $k \geq 3$, it follows that

$$2k^2 + 2k + 3 > 4. \tag{10}$$

Adding $k^2 + 4k$ to both sides of inequality (10) yields

$$k^2 + 4k + (2k^2 + 2k + 3) > k^2 + 4k + 4. \tag{11}$$

Using inequalities (9) and (11) and the transitive law of order, we can write

$$3^{k+1} > k^2 + 4k + 4, \quad \text{or} \quad 3^{k+1} > (k + 2)^2.$$

This shows that $k + 1$ is in S. By the principle of mathematical induction, it follows that S contains every positive integer $n \geq 3$.

EXERCISES 10.4

In Exercises 1–32, use the principle of mathematical induction to prove that each statement is true for every positive integer n.

1. $\dfrac{1}{1 \cdot 2} + \dfrac{1}{2 \cdot 3} + \dfrac{1}{3 \cdot 4} + \cdots + \dfrac{1}{n(n + 1)} = \dfrac{n}{n + 1}$

2. $1 \cdot 2 + 2 \cdot 3 + 3 \cdot 4 + \cdots + n(n + 1)$
   $= \dfrac{n(n + 1)(n + 2)}{3}$

3. $1 + 4 + 7 + \cdots + (3n - 2) = \dfrac{n(3n - 1)}{2}$

4. $2^1 + 2^2 + 2^3 + \cdots + 2^n = 2^{n+1} - 2$

5. $1^2 + 3^2 + 5^2 + \cdots + (2n - 1)^2$
   $= \dfrac{n(2n + 1)(2n - 1)}{3}$

6. $1^3 + 2^3 + 3^3 + \cdots + n^3 = \dfrac{n^2(n + 1)^2}{4}$

7. $1 \cdot 1! + 2 \cdot 2! + 3 \cdot 3! + \cdots + n \cdot n! = (n + 1)! - 1$

8. $5 + 10 + 15 + \cdots + 5n = \dfrac{5n(n + 1)}{2}$

9. $3^n > 2^n$ 10. If $b \neq 0$, $\left(\dfrac{a}{b}\right)^n = \dfrac{a^n}{b^n}$.

11. $n + 4 < 6n^2$

12. $1 + 27 + 125 + \cdots + (2n - 1)^3 = n^2(2n^2 - 1)$

13. $\dfrac{1}{1 \cdot 3} + \dfrac{1}{3 \cdot 5} + \dfrac{1}{5 \cdot 7} + \cdots + \dfrac{1}{(2n - 1)(2n + 1)}$
    $= \dfrac{n}{2n + 1}$

14. $2^3 + 4^3 + 6^3 + \cdots + (2n)^3 = 2n^2(n + 1)^2$

15. $n^2 + 1 > n$

16. $1 \cdot 2^2 + 2 \cdot 3^2 + 3 \cdot 4^2 + \cdots + n(n + 1)^2$
    $= \dfrac{n(n + 1)(n + 2)(3n + 5)}{12}$

17. $1 + 5 + 9 + \cdots + (4n - 3) = n(2n - 1)$

18. $1^2 + 4^2 + 7^2 + \cdots + (3n - 2)^2 = \dfrac{n(6n^2 - 3n - 1)}{2}$

19. $1 \cdot 2 \cdot 3 + 2 \cdot 3 \cdot 4 + 3 \cdot 4 \cdot 5 + \cdots + n(n + 1)(n + 2)$
    $= \dfrac{n(n + 1)(n + 2)(n + 3)}{4}$

20. $3 + 3^2 + 3^3 + \cdots + 3^n = \dfrac{3^{n+1} - 3}{2}$

21. $1 \cdot 2 + 3 \cdot 4 + 5 \cdot 6 + \cdots + (2n - 1)(2n)$
    $= \dfrac{n(n + 1)(4n - 1)}{3}$

22. $(a \cdot b)^n = a^n \cdot b^n$

23. $8 + 16 + 24 + \cdots + 8n < (2n + 1)^2$

24. $[r \operatorname{cis} \theta]^n = r^n \operatorname{cis} (n\theta)$ (DeMoivre's theorem, Section 7.3.

25. $\sin(x + n\pi) = (-1)^n \sin x$

26. $\cos(x + n\pi) = (-1)^n \cos x$

27. 2 is a factor of $n(n + 1)$.

28. 2 is a factor of $n^2 - n + 2$.

29. 3 is a factor of $n(n + 1)(n + 2)$.

30. 2 is a factor of $5^n - 3^n$.   **31.**   3 is a factor of $4^n - 1$.

32. 8 is a factor of $3^{2n} - 1$.

33. For every positive integer n, the integer x_n is defined recursively as follows:

(1) $x_1 = 2$;   (2)   If $n \geq 2$, $x_n = x_{n-1} + 2n$.

Use the principle of mathematical induction to prove that, for every positive integer n, $x_n = n(n + 1)$.

34. Let $S = \left\{ n \in \mathbb{N} \mid 2 + 4 + 6 + \cdots + 2n = \dfrac{(2n + 1)^2}{4} \right\}$.

a. Show that if $k \in S$, for some positive integer k, then $k + 1 \in S$.

b. Does it follow that $S = \mathbb{N}$? Explain.

In Exercises 35–40, prove that each statement is true for the indicated integer values of n. (See Example 32.)

35. $4^n > n^4$,   for $n \geq 5$ **36.** $n! > 2^n$,   for $n \geq 4$

37. $3^n > 2^n + 1$,   for $n \geq 2$ **38.** $2^n > n^2$,   for $n \geq 5$

39. $2^{n+3} < (n + 1)!$,   for $n \geq 5$

40. $4^n < n!$,   for $n \geq 9$

41. The **Fibonacci numbers** f_n are defined recursively as follows:

$$f_1 = 1;$$
$$f_2 = 1;$$
$$\text{If } n \geq 1, f_{n+2} = f_{n+1} + f_n.$$

Use the principle of mathematical induction to prove that each of the following statements about the Fibonacci numbers is true for every positive integer integer n.

a. $f_1 + f_3 + f_5 + \cdots + f_{2n-1} = f_{2n}$

b. $f_2 + f_4 + f_6 + \cdots + f_{2n} = f_{2n+1} - 1$

c. $f_1 + f_2 + f_3 + \cdots + f_n = f_{n+2} - 1$

d. $f_n \cdot f_{n+3} - f_{n+1} \cdot f_{n+2} = (-1)^{n+1}$

42. For every positive integer n, the integer x_n is defined recursively as follows:

(1) $x_1 = 1$;   (2)   If $n \geq 2$, $x_n = x_{n-1} + 2$.

Use the principle of mathematical induction to prove that, for every positive integer n, $x_n = 2n - 1$.

43. For every positive integer n, the integer x_n is defined recursively as follows:

(1) $x_1 = 2$;   (2)   If $n \geq 2$, $x_n = 4x_{n-1}$.

Use the principle of mathematical induction to prove that, for every positive integer n, $x_n = 2^{2n-1}$.

44. Use the principle of mathematical induction to prove that if $x \neq 1$, then $x - 1$ is a factor of $x^n - 1$, for every positive integer n.

45. Use the principle of mathematical induction to prove that, for every positive integer n, $1^3 + 2^3 + 3^3 + \cdots + n^3 = (1 + 2 + 3 + \cdots + n)^2$.

10.5   BINOMIAL THEOREM

If a and b are real numbers, we know that

$$(a + b)^1 = a + b$$

and that

$$(a + b)^2 = (a + b)(a + b) = a^2 + 2ab + b^2.$$

In this section we discuss a theorem that will enable us to find the expansion of $(a + b)^n$, for any positive integer n. To this end, consider the following expansions of $(a + b)^n$ where $n = 3, 4,$ and 5:

$$(a + b)^3 = (a + b)(a + b)^2 = a^3 + 3a^2b + 3ab^2 + b^3,$$
$$(a + b)^4 = (a + b)(a + b)^3$$
$$= a^4 + 4a^3b + 6a^2b^2 + 4ab^3 + b^4, \text{ and}$$
$$(a + b)^5 = (a + b)(a + b)^4 = a^5 + 5a^4b + 10a^3b^2$$
$$+ 10a^2b^3 + 5ab^4 + b^5.$$

Based on the expansions of $(a + b)^n$ shown above, we observe the patterns described in Table 10.3. As illustrated in Table 10.3, patterns (1)–(6) hold for the expansion of $(a + b)^4$. You should check that these patterns also hold for the other expansions shown.

Binomial Coefficients

Next, we will consider the coefficients of the terms in the expansion of $(a + b)^n$.

TABLE 10.3

General Pattern for $(a + b)^n$	Illustration Using $(a + b)^4$
	$(a + b)^4 = a^4 + 4a^3b + 6a^2b^2 + 4ab^3 + b^4$
(1) There are $n + 1$ terms.	There are $4 + 1 = 5$ terms.
(2) The first term is a^n, and the last term is b^n.	The first term is a^4, and the last term is b^4.
(3) For any term, the sum of the exponent of a and the exponent of b is n.	For any term, the sum of the exponent of a and the exponent of b is 4.
(4) The exponents of a begin with n in the first term, decrease by 1 in each subsequent term, and end with 0 in the last term.	The exponents of a begin with 4 in the first term, decrease by 1 in each subsequent term, and end with 0 in the last term.
(5) The exponents of b begin with 0 in the first term, increase by 1 in each subsequent term, and end with n in the last term.	The exponents of b begin with 0 in the first term, increase by 1 in each subsequent term, and end with 4 in the last term.
(6) The variable parts of the terms have the pattern $a^n, a^{n-1}b, a^{n-2}b^2, \cdots, a^2b^{n-2}, ab^{n-1}, b^n$	The variable parts of the terms have the pattern $a^4, a^3b, a^2b^2, ab^3, b^4$

As we learned in Section 10.1, n factorial, where n is a nonnegative integer, is defined as follows:

$$\text{If } n > 0, \ n! = (n)(n - 1)(n - 2) \ \cdots \cdots (3)(2)(1); \quad \text{and}$$
$$\text{If } n = 0, \ 0! = 1.$$

It can be shown that the coefficient of the term in the expansion of $(a + b)^n$ which involves $a^{n-r}b^r$ is

$$\frac{n!}{r! \, (n - r)!}.$$

We call the number $\dfrac{n!}{r! \, (n - r)!}$ a binomial coefficient and denote it by $\dbinom{n}{r}$.

Binomial Coefficient

If n and r are integers such that $0 \le r \le n$, then the **binomial coefficient** $\dbinom{n}{r}$ is defined as follows:

$$\binom{n}{r} = \frac{n!}{r! \, (n - r)!}.$$

It is important to note that, although the binomial coefficient $\dbinom{n}{r} = \dfrac{n!}{r! \, (n - r)!}$ is written as a fraction, it can be shown that it is always a positive integer. Also, notice that $\dbinom{n}{0} = 1$ and that $\dbinom{n}{n} = 1$ (see Exercise 37).

EXAMPLE 33

Verify that the coefficients in the expansion of $(a + b)^4$ are given by $\binom{4}{0}$, $\binom{4}{1}$, $\binom{4}{2}$, $\binom{4}{3}$, and $\binom{4}{4}$.

Solution   First, we find the binomial coefficients:

$$\binom{4}{0} = \frac{4!}{0!\,(4 - 0)!} = \frac{4!}{4!} = 1,$$

$$\binom{4}{1} = \frac{4!}{1!\,(4 - 1)!} = \frac{4!}{3!} = \frac{4 \cdot 3!}{3!} = 4,$$

$$\binom{4}{2} = \frac{4!}{2!\,(4 - 2)!} = \frac{4 \cdot 3 \cdot 2 \cdot 1}{(2 \cdot 1)(2 \cdot 1)} = 6,$$

$$\binom{4}{3} = \frac{4!}{3!\,(4 - 3)!} = \frac{4 \cdot 3!}{3!\,1!} = 4,$$

$$\binom{4}{4} = \frac{4!}{4!\,(4 - 4)!} = \frac{4!}{4!\,0!} = 1.$$

Since, as noted earlier in this section,

$$(a + b)^4 = a^4 + 4a^3b + 6a^2b^2 + 4ab^3 + b^4,$$

we see that the coefficients in the expansion of $(a + b)^4$ are, indeed, $\binom{4}{0}$, $\binom{4}{1}$, $\binom{4}{2}$, $\binom{4}{3}$, and $\binom{4}{4}$.

The Binomial Theorem

As the following theorem states, the patterns we have discussed for the expansion of $(a + b)^n$ hold for every positive integer n:

Binomial Theorem

If a and b are real (or complex) numbers and n is a positive integer, then

$$(a + b)^n = \binom{n}{0} a^n + \binom{n}{1} a^{n-1}b + \binom{n}{2} a^{n-2}b^2 + \cdots$$

$$+ \binom{n}{r} a^{n-r}b^r + \cdots + \binom{n}{n-1} ab^{n-1} + \binom{n}{n} b^n.$$

(For an outline of the proof of the binomial theorem using mathematical induction, see Exercise 38.)

We conclude this section with several examples that involve applications of the binomial theorem.

EXAMPLE 34

Use the binomial theorem to expand $(a + b)^7$.

Solution   By the binomial theorem, we have

$$(a + b)^7 = \binom{7}{0} a^7 + \binom{7}{1} a^6b + \binom{7}{2} a^5b^2 + \binom{7}{3} a^4b^3$$

$$+ \binom{7}{4} a^3b^4 + \binom{7}{5} a^2b^5 + \binom{7}{6} ab^6 + \binom{7}{7} b^7.$$

Calculating the binomial coefficients, we find that

$$(a + b)^7 = a^7 + 7a^6b + 21a^5b^2 + 35a^4b^3$$
$$+ 35a^3b^4 + 21a^2b^5 + 7ab^6 + b^7.$$

EXAMPLE 35

Expand $(3x + 4y)^5$ using the binomial theorem.

Solution   Using the binomial theorem, with $a = 3x$, $b = 4y$, and $n = 5$, we have

$$(3x + 4y)^5 = (3x)^5 + \binom{5}{1} (3x)^4(4y)$$

$$+ \binom{5}{2} (3x)^3(4y)^2 + \binom{5}{3} (3x)^2(4y)^3$$

$$+ \binom{5}{4}(3x)(4y)^4 + (4y)^5$$

$$= 243x^5 + (5)(81x^4)(4y) + (10)(27x^3)(16y^2)$$
$$+ (10)(9x^2)(64y^3) + (5)(3x)(256y^4) + 1024y^5$$

$$= 243x^5 + 1620x^4y + 4320x^3y^2$$
$$+ 5760x^2y^3 + 3840xy^4 + 1024y^5.$$

EXAMPLE 36

Use the binomial theorem to expand $(2x - 3y)^4$.

Solution   By the binomial theorem, we have

$$(2x - 3y)^4 = [2x + (-3y)]^4 = (2x)^4$$

$$+ \binom{4}{1} (2x)^3(-3y) + \binom{4}{2} (2x)^2(-3y)^2$$

$$+ \binom{4}{3} (2x)(-3y)^3 + (-3y)^4$$

$$= 16x^4 + (4)(8x^3)(-3y) + (6)(4x^2)(9y^2)$$
$$+ (4)(2x)(-27y^3) + 81y^4$$

$$= 16x^4 - 96x^3y + 216x^2y^2$$
$$- 216xy^3 + 81y^4.$$

Finding a Particular Term in a Binomial Expansion

Sometimes we are interested in finding a particular term in a binomial expansion rather than the complete expansion. From the binomial theorem we see that in the expansion of $(a + b)^n$:

the first term is $\binom{n}{0} a^n,$

the second term is $\binom{n}{1} a^{n-1}b,$

the third term is $\binom{n}{2} a^{n-2}b^2,$

and in general, for $1 \le k \le n + 1$, it can be shown that the kth term is

$$\binom{n}{k-1} a^{n-(k-1)}b^{k-1}.$$

In summary, we have

Formula for the kth Term of a Binomial Expression

The kth term in the expansion of $(a + b)^n$ is given by the formula

$$\binom{n}{k-1} a^{n-(k-1)}b^{k-1}, \quad \text{where } 1 \le k \le n + 1.$$

EXAMPLE 37

Find the sixth term in the expansion of $\left(\dfrac{2}{\sqrt{x}} - 3\sqrt{x} \right)^{12}.$

Solution   Using the formula $\binom{n}{k-1} a^{n-(k-1)}b^{k-1}$ with $n = 12$ and $k = 6$, the

sixth term in the expansion of $(a + b)^{12}$ is given by

$$\binom{12}{5} a^{12-5}b^5, \quad \text{or} \quad 792a^7b^5.$$

Since $a = 2/\sqrt{x}$ and $b = -3\sqrt{x}$, it follows that the sixth term in the expansion

of $\left(\dfrac{2}{\sqrt{x}} - 3\sqrt{x} \right)^{12}$ is

$$792 \left(\frac{2}{\sqrt{x}} \right)^7 (-3\sqrt{x})^5 = (792)(128x^{-7/2})(-243x^{5/2})$$

$$= -24{,}634{,}368x^{-1}$$

$$= -\frac{24{,}634{,}368}{x}.$$

EXAMPLE 38

Find the term in the expansion of $(x^3 + 2y^2)^{10}$ that involves x^{12}.

Solution   We begin by letting $a = x^3$, $b = 2y^2$, and $n = 10$. Since $x^{12} = (x^3)^4 = a^4$, we must find the term in the expansion of $(a + b)^{10}$ that involves a^4. Since the sum of the exponents of a and b must be 10, it follows that the term we seek involves b^6. From the formula $\binom{n}{k-1} a^{n-(k-1)}b^{k-1}$ for the kth term in the expansion of $(a + b)^n$, we see that the exponent of b is $k - 1$. Hence,

$$k - 1 = 6, \quad \text{or} \quad k = 7.$$

This means that we must find the seventh term in the expansion of $(a + b)^{10}$, which is given by

$$\binom{10}{6} a^{10-6} b^6, \quad \text{or} \quad 210 a^4 b^6.$$

Since $a = x^3$ and $b = 2y^2$, the desired term is

$$210(x^3)^4 (2y^2)^6 = 13{,}440 x^{12} y^{12}.$$

EXERCISES 10.5

In Exercises 1–6, find each binomial coefficient.

1. $\binom{8}{3}$ **2.** $\binom{7}{4}$

3. $\binom{12}{8}$ **4.** $\binom{13}{10}$

5. $\binom{n}{n-1}$ **6.** $\binom{10}{2}$

In Exercises 7–18, use the binomial theorem to expand each expression.

7. $(a + b)^8$ **8.** $(a - b)^8$
9. $(x - 2y)^5$ **10.** $(2x - y)^6$

11. $\left(\sqrt{x} - \dfrac{2}{\sqrt{x}} \right)^4$ **12.** $\left(3\sqrt{x} - \dfrac{4}{\sqrt{x}} \right)^5$

13. $\left(\dfrac{x}{2} - 4y \right)^3$ **14.** $(5x + 4y)^4$

15. $\left(6x + \dfrac{y}{3} \right)^6$ **16.** $\left(3y - \dfrac{x}{6} \right)^3$

17. $(x^{-1} - 2x)^5$ **18.** $(4x - 2x^{-2})^5$

In Exercises 19–22, use the binomial theorem to find the first three terms (only) of each expansion.

19. $(x + 2y)^{10}$ **20.** $(x - 4y)^{12}$
21. $(2x - 3y)^{15}$ **22.** $(2x + 3y)^{13}$

In Exercises 23–36, use the binomial theorem to find the indicated term (only) of each expansion.

23. $(2x - y)^{12}$;   third term
24. $(3x - 2y)^{10}$;   eighth term

25. $\left(\dfrac{x}{3} - 6y \right)^{11}$;   fourth term

26. $\left(2x + \dfrac{y}{4} \right)^9$;   fourth term

27. $\left(\sqrt{x} + \dfrac{10}{\sqrt{x}} \right)^9$;   sixth term

28. $\left(\sqrt{x} - \dfrac{3}{\sqrt{x}} \right)^7$;   third term

29. $(5x + y)^{13}$;   fifth term   **30.** $(6x + 2y)^8$;   fifth term
31. $(x^3 - 3y^2)^{11}$;   the term involving y^6
32. $(x^2 - 3y^3)^9$;   the term involving x^8
33. $(2x^2 + 4y^3)^8$;   the term involving x^8
34. $(2x^3 - 3y^2)^{15}$;   the term involving y^6
35. $(\sqrt{y} - \sqrt{x})^{10}$;   the term involving y^3
36. $(\sqrt{x} + \sqrt{y})^8$;   the term involving y^2
37. Let n and r be integers where $0 \le r \le n$.

   a.   Show that $\dbinom{n}{r} = \dbinom{n}{n-r}$.

   b.   Show that $\dbinom{n}{0} = 1$.

   c.   Show that $\dbinom{n}{n} = 1$.

   d.   Show that $\dbinom{n}{r} + \dbinom{n}{r-1} = \dbinom{n+1}{r}$.

38. For each positive integer n, consider the following equation from the binomial theorem:

$$(a + b)^n = \binom{n}{0} a^n + \binom{n}{1} a^{n-1}b + \binom{n}{2} a^{n-2}b^2 + \cdots$$
$$+ \binom{n}{r} a^{n-r}b^r + \cdots + \binom{n}{n} b^n. \quad (1)$$

Let S be the set of all positive integers for which equation (1) is true.

   a.   Show that 1 is in S; that is, show that the following statement is true:

$$(a + b)^1 = \binom{1}{0} a^1 + \binom{1}{1} b^1.$$

   b.   Let k be a positive integer, and assume that k is in S; that is, assume that the following statement is true:

$$(a + b)^k = \binom{k}{0} a^k + \binom{k}{1} a^{k-1}b + \cdots$$
$$+ \binom{k}{k-1} ab^{k-1} + \binom{k}{k} b^k. \quad (2)$$

Now, prove that $k + 1$ is in S; that is, show that the following statement is true:

$$(a + b)^{k+1} = \binom{k+1}{0} a^{k+1} + \binom{k+1}{1} a^k b + \cdots$$
$$+ \binom{k+1}{k} ab^k + \binom{k+1}{k+1} b^{k+1}.$$

[*Hint*: Multiply both sides of equation (2) by $a + b$ and simplify the right side of the resulting equation. By parts (b) and (c) of Exercise 37, we can replace

$$\binom{k}{0} \text{ by } \binom{k+1}{0} \text{ (since both are equal to 1), and}$$

$$\binom{k}{k} \text{ by } \binom{k+1}{k+1} \text{ (again, since both are equal to 1).}$$

You will also need to use part (d) of Exercise 37.] Using the principle of mathematical induction we conclude that $S = \mathbb{N}$, so that equation (1) is true for every positive integer n.

39. Let n be a positive integer. Show that

$$\binom{n}{0} + \binom{n}{1} + \binom{n}{2} + \cdots + \binom{n}{n} = 2^n.$$

[*Hint*: Use the binomial theorem with $a = b = 1$.]

40. Show that if n is an integer such that $n \geq 2$ and if $0 < r < n$, then

$$\binom{n}{r} = \binom{n-1}{r-1} + \binom{n-1}{r}.$$

In Exercises 41–46, use the binomial theorem to find the indicated power of each complex number.

41. $(2 + 3i)^5$

42. $(1 - 4i)^6$

43. $(2 - 3i)^4$

44. $(5 + 2i)^5$

45. $(\sqrt{2} + \sqrt{2}i)^8$

46. $(\sqrt{2} - \sqrt{2}i)^8$

The following triangular array of numbers, which is called **Pascal's triangle**, can be used to obtain binomial coefficients.

$$1$$
$$1 \quad 1$$
$$1 \quad 2 \quad 1$$
$$1 \quad 3 \quad 3 \quad 1$$
$$1 \quad 4 \quad 6 \quad 4 \quad 1$$
$$1 \quad 5 \quad 10 \quad 10 \quad 5 \quad 1$$
$$\cdots\cdots\cdots\cdots\cdots\cdots$$

The number in the first row of the array is the coefficient in the expansion of $(a + b)^0 = 1$. The numbers in the second row are the coefficients in the expansion of $(a + b)^1$. The numbers in the third row are the coefficients in the expansion of $(a + b)^2$, and so on. In any row after the first, the following pattern holds: Each number in the row that is different from 1 is the sum of the two numbers in the previous row, which are above and immediately to the left and right of that number. For example, consider the fourth and fifth rows of the array:

$$1 \quad 3 \quad 3 \quad 1$$
$$1 \quad 4 \quad 6 \quad 4 \quad 1.$$

The second entry in the fifth row, 4, is the sum of the entries 1 and 3 from the fourth row, which appear above and immediately to the left and right of 4. Similarly, $6 = 3 + 3$, and $4 = 3 + 1$. Exercises 47–48 involve Pascal's triangle.

47. Find the eighth row of Pascal's triangle, and then use the result in the expansion of $(2x - y)^7$.

48. Find the ninth row of Pascal's triangle, and then use the result in the expansion of $(x - 2y)^8$.

10.6 PARTIAL FRACTIONS

Recall that a rational expression is an expression of the form $\dfrac{P(x)}{Q(x)}$, where $P(x)$ and $Q(x)$ are polynomials and $Q(x) \neq 0$. If the degree of $P(x)$ is less than the degree of $Q(x)$, we say that the rational expression is **proper**. For example, each of the following is a proper rational expression:

$$\frac{2x + 1}{x^2 - 3}, \quad \frac{4x^5 - 2x^3 + 6}{5x^7 + 3x^3 - 7}, \quad \frac{3x^4 + 2x^3 + 5x^2 - 6x}{6x^5 + 8x^4 - x^2 + 9}.$$

In calculus, it is sometimes convenient to express a rational expression as the sum of two or more simpler rational expressions. For example, it is easy to verify that

$$\frac{2x}{x^2 - 4} = \frac{1}{x - 2} + \frac{1}{x + 2}. \quad (1)$$

The rational expressions

$$\frac{1}{x - 2} \quad \text{and} \quad \frac{1}{x + 2}$$

that appear on the right side of equation (1) are called partial fractions, and their sum

$$\frac{1}{x - 2} + \frac{1}{x + 2}$$

is called the partial fraction decomposition of $\dfrac{2x}{x^2 - 4}$.

In general, it can be shown that if $\dfrac{P(x)}{Q(x)}$ is a proper rational expression, then

$$\frac{P(x)}{Q(x)} = F_1(x) + F_2(x) + \cdots + F_n(x),$$

where each function $F_i(x)$ has one of the forms:

$$\frac{A}{(ax + b)^m} \qquad\qquad (2)$$

or $$\frac{Bx + C}{(ax^2 + bx + c)^n}, \qquad\qquad (3)$$

for some real numbers A, B, and C and positive integers m and n. The sum $F_1(x) + F_2(x) + \cdots + F_n(x)$ is called the **partial fraction decomposition** of $\dfrac{P(x)}{Q(x)}$, and each function $F_i(x)$ is called a **partial fraction**.

It is important to note that for partial fractions of the form (3), the quadratic polynomial $ax^2 + bx + c$ must be **irreducible** over $\mathbb{R}$. A quadratic polynomial is irreducible over $\mathbb{R}$ when the associated quadratic equation $ax^2 + bx + c = 0$ has no real roots; that is, when the discriminant $b^2 - 4ac$ is negative.

The factorization of the polynomial $Q(x)$ determines the exact number of terms of forms (2) and (3) which appear in the partial fraction decomposition of $\dfrac{P(x)}{Q(x)}$. It can be shown that every polynomial of positive degree that has real coefficients can be factored into linear or irreducible (over $\mathbb{R}$) quadratic factors. The following result is fundamental in our work:

Partial Fraction Decomposition

Consider the proper rational expression $\dfrac{P(x)}{Q(x)}$.

Case 1 (*Linear factors*) For each distinct factor of $Q(x)$ having the form $(ax + b)^m$, the partial fraction decomposition contains a sum of the form

$$\frac{A_1}{ax + b} + \frac{A_2}{(ax + b)^2} + \cdots + \frac{A_m}{(ax + b)^m},$$

where $A_1, A_2, \ldots, A_m$ are constants.

Case 2 (*Irreducible quadratic factors*) For each distinct factor of $Q(x)$ having the form $(ax^2 + bx + c)^n$, where $b^2 - 4ac < 0$, the partial fraction decomposition contains a sum of the form

$$\frac{B_1 x + C_1}{ax^2 + bx + c} + \frac{B_2 x + C_2}{(ax^2 + bx + c)^2} + \cdots + \frac{B_n x + C_n}{(ax^2 + bx + c)^n}$$

where $B_1, B_2, \ldots, B_n$ and $C_1, C_2, \ldots, C_n$ are constants.

EXAMPLE 39

Find the partial fraction decomposition of

$$\frac{4x - 1}{2x^2 - x - 1}.$$

Solution   First, we note that

$$\frac{4x - 1}{2x^2 - x - 1}$$

is a proper rational expression. Next, we factor the denominator $2x^2 - x - 1$ as follows:

$$2x^2 - x - 1 = (x - 1)(2x + 1). \tag{4}$$

Each factor shown on the right side of equation (4) has the form $(ax + b)^m$, as described in case 1, with $m = 1$. Thus, for the factor $x - 1$, we introduce the term

$$\frac{A}{x - 1},$$

and, for the factor $2x + 1$, we introduce the term

$$\frac{B}{2x + 1}.$$

The partial fraction decomposition thus has the form

$$\frac{4x - 1}{(x - 1)(2x + 1)} = \frac{A}{x - 1} + \frac{B}{2x + 1}, \tag{5}$$

where A and B are constants. To find A and B, we multiply both sides of equation (5) by $(x - 1)(2x + 1)$ to obtain

$$4x - 1 = A(2x + 1) + B(x - 1). \tag{6}$$

Because equation (6) is an identity, we can substitute *any* convenient value of x to help us determine the values of A and B. In particular, we can choose the values of x that cause the factors $x - 1$ and $2x + 1$ to be zero, namely 1 and $-\frac{1}{2}$.

If $x = 1$, then

$$4(1) - 1 = A(3) + B(0)$$
$$3 = 3A, \quad \text{or}$$
$$A = 1.$$

If $x = -\frac{1}{2}$, then

$$4\left(-\frac{1}{2}\right) - 1 = A(0) + B\left(-\frac{3}{2}\right)$$

$$-3 = -\frac{3}{2}B$$

$$B = 2.$$

Therefore, $$\frac{4x - 1}{(x - 1)(2x + 1)} = \frac{1}{x - 1} + \frac{2}{2x + 1}.$$

EXAMPLE 40

Find the partial fraction decomposition of

$$\frac{6x^2 + 19x + 17}{(x^2 + 3x + 2)(x + 2)}.$$

Solution Factoring the quadratic polynomial that appears in the denominator of

$$\frac{6x^2 + 19x + 17}{(x^2 + 3x + 2)(x + 2)},$$

we find that

$$x^2 + 3x + 2 = (x + 1)(x + 2).$$

Thus, we must find the partial fraction decomposition of

$$\frac{6x^2 + 19x + 17}{(x + 1)(x + 2)^2}.$$

For the factor $x + 1$, we introduce the term

$$\frac{A}{x + 1},$$

and, for the factor $(x + 2)^2$, we introduce the sum

$$\frac{B}{x + 2} + \frac{C}{(x + 2)^2}.$$

Hence, $$\frac{6x^2 + 19x + 17}{(x + 1)(x + 2)^2} = \frac{A}{x + 1} + \frac{B}{x + 2} + \frac{C}{(x + 2)^2}. \qquad (7)$$

Multiplying both sides of equation (7) by $(x + 1)(x + 2)^2$, we have

$$6x^2 + 19x + 17 = A(x + 2)^2 + B(x + 1)(x + 2) + C(x + 1). \qquad (8)$$

Substituting -1 for x in equation (8) yields

$$6(1) + 19(-1) + 17 = A(1) + B(0) + C(0)$$
$$4 = A, \qquad \text{or}$$
$$A = 4.$$

Substituting -2 for x in equation (8), we have

$$6(4) + 19(-2) + 17 = A(0) + B(0) + C(-1)$$
$$3 = -C, \qquad \text{or}$$
$$C = -3.$$

There are no other values of x that cause a factor of $(x + 1)(x + 2)^2$ to be zero. However, we can find B by substituting 4 for A, -3 for C, and any convenient value for x, say 1, in equation (8). Thus, we have

$$6(1) + 19(1) + 17 = 4(3)^2 + B(2)(3) + (-3)(2)$$
$$42 = 36 + 6B - 6$$
$$12 = 6B$$
$$B = 2.$$

It follows that the decomposition is

$$\frac{6x^2 + 19x + 17}{(x + 1)(x + 2)^2} = \frac{4}{x + 1} + \frac{2}{x + 2} - \frac{3}{(x + 2)^2}.$$

The procedure we used to find the constants A, B, and C in the last two examples works well when the factors of the denominator of $\dfrac{P(x)}{Q(x)}$ are all linear. However, when $Q(x)$ has an irreducible quadratic factor, we use an alternate approach, which is illustrated in the next two examples.

EXAMPLE 41

Find the partial fraction decomposition of

$$\frac{5x^3 - 6x^2 + 15x - 25}{x^4 + 5x^2}.$$

Solution   We begin by factoring the denominator of

$$\frac{5x^3 - 6x^2 + 15x - 25}{x^4 + 5x^2}.$$

as follows:

$$x^4 + 5x^2 = x^2(x^2 + 5).$$

For the factor x^2, we introduce the sum

$$\frac{A}{x} + \frac{B}{x^2}.$$

Since the factor $x^2 + 5$ has the form $(ax^2 + bx + c)^n$ with $n = 1$, we introduce the term

$$\frac{Cx + D}{x^2 + 5}.$$

Hence, $\quad \dfrac{5x^3 - 6x^2 + 15x - 25}{x^2(x^2 + 5)} = \dfrac{A}{x} + \dfrac{B}{x^2} + \dfrac{Cx + D}{x^2 + 5}.$ $\qquad$ *(9)*

Multiplying both sides of equation (9) by $x^2(x^2 + 5)$ yields

$5x^3 - 6x^2 + 15x - 25$
$$= A(x)(x^2 + 5) + B(x^2 + 5) + (Cx + D)(x^2). \quad \textbf{(10)}$$

Substituting 0 for x in equation (10), we have

$$-25 = A(0) + B(5) + D(0)$$
$$-25 = 5B$$
$$B = -5.$$

Since there are no other real values of x that cause a factor of $x^2(x^2 + 5)$ to be zero, we multiply and rearrange terms on the right side of equation (10) to obtain

$$5x^3 - 6x^2 + 15x - 25 = (A + C)x^3 + (B + D)x^2 + 5Ax + 5B.$$

Substituting -5 for B in the latter equation yields

$$5x^3 - 6x^2 + 15x - 25 = (A + C)x^3 + (D - 5)x^2 + 5Ax - 25. \quad \textbf{(11)}$$

Since two polynomials are equal if and only if the coefficients of like terms are equal, we obtain the system of linear equations

$$\begin{cases} A + C = 5 \\ D - 5 = -6 \\ 5A = 15. \end{cases} \qquad (12)$$

Solving system (12) we find that $A = 3$, $C = 2$, and $D = -1$. Hence,

$$\frac{5x^3 - 6x^2 + 15x - 25}{x^4 + 5x^2} = \frac{3}{x} - \frac{5}{x^2} + \frac{2x - 1}{x^2 + 5}.$$

E X A M P L E 42

Find the partial fraction decomposition of

$$\frac{10x^2 + 8x + 11}{(2x^2 + x + 3)^2}.$$

Solution   Since $2x^2 + x + 3$ is irreducible over $\mathbb{R}$, we write

$$\frac{10x^2 + 8x + 11}{(2x^2 + x + 3)^2} = \frac{Ax + B}{2x^2 + x + 3} + \frac{Cx + D}{(2x^2 + x + 3)^2}. \qquad (13)$$

Multiplying both sides of equation (13) by $(2x^2 + x + 3)^2$, we have

$$10x^2 + 8x + 11 = (Ax + B)(2x^2 + x + 3) + Cx + D. \qquad (14)$$

Since there are no real values of x that make a factor of $(2x^2 + x + 3)^2$ zero, we multiply and rearrange terms on the right side of equation (14) to obtain

$$10x^2 + 8x + 11 = 2Ax^3 + (A + 2B)x^2 \\ + (3A + B + C)x + (3B + D). \qquad (15)$$

Equating coefficients of the like powers of x on each side of equation (15), we obtain the system of linear equations

$$\begin{cases} 2A = 0 \\ A + 2B = 10 \\ 3A + B + C = 8 \\ 3B + D = 11. \end{cases} \qquad (16)$$

Solving system (16), we find that $A = 0$, $B = 5$, $C = 3$, and $D = -4$. Therefore,

$$\frac{10x^2 + 8x + 11}{(2x^2 + x + 3)^2} = \frac{5}{2x^2 + x + 3} + \frac{3x - 4}{(2x^2 + x + 3)^2}.$$

To find the partial fraction decomposition of an improper rational expression, we first divide the numerator by the denominator and then work with the remainder term. This process is illustrated in our next example.

E X A M P L E 43

Find the partial fraction decomposition of

$$\frac{3x^5 + x^4 + 20x^3 - x^2 + 15x - 25}{x^4 + 5x^2}.$$

Solution   Since the rational expression

$$\frac{3x^5 + x^4 + 20x^3 - x^2 + 15x - 25}{x^4 + 5x^2}$$

is improper, we first perform the long division

$$
\begin{array}{r}
3x + 1 \\
x^4 + 5x^2 \overline{\smash{\big)}\ 3x^5 + x^4 + 20x^3 - x^2 + 15x - 25} \\
\underline{3x^5 + 15x^3} \\
x^4 + 5x^3 - x^2 + 15x - 25 \\
\underline{x^4 + 5x^2} \\
5x^3 - 6x^2 + 15x - 25
\end{array}
$$

Hence, we can write

$$\frac{3x^5 + x^4 + 20x^3 - x^2 + 15x - 25}{x^4 + 5x^2}$$

$$= (3x + 1) + \frac{5x^3 - 6x^2 + 15x - 25}{x^4 + 5x^2}. \quad (17)$$

Since the remainder term

$$\frac{5x^3 - 6x^2 + 15x - 25}{x^4 + 5x^2}$$

from equation (17) is a proper rational expression, we can find its partial fraction decomposition using the methods illustrated earlier in this section. Using the result of Example 41, we can write

$$\frac{3x^5 + x^4 + 20x^3 - x^2 + 15x - 25}{x^4 + 5x^2} = (3x + 1) + \frac{3}{x} - \frac{5}{x^2} + \frac{2x - 1}{x^2 + 5}.$$

EXERCISES 10.6

In Exercises 1–34, find the partial fraction decomposition of each rational expression.

1. $\dfrac{4}{x^2 - 4}$

2. $\dfrac{3x - 3}{x^2 - 9}$

3. $\dfrac{-x - 7}{x^2 - x - 6}$

4. $\dfrac{5x}{x^2 + x - 6}$

5. $\dfrac{x + 9}{x^2 - 3x - 10}$

6. $\dfrac{3x - 13}{x^2 - 2x - 3}$

7. $\dfrac{x^2 + 7x + 1}{(x - 1)(x^2 + x - 2)}$

8. $\dfrac{2x^2 + 14x + 15}{(x + 2)(x^2 - x - 6)}$

9. $\dfrac{7x^2 - 17x + 3}{(x - 2)(x^2 - x - 2)}$

10. $\dfrac{4x^2 + 42x + 98}{(x + 5)(x^2 + 4x - 5)}$

11. $\dfrac{9x^2 - 43x + 30}{x^3 - 10x^2}$

12. $\dfrac{x^2 + x - 3}{x^4 - x^3}$

13. $\dfrac{5x^2 - 6x + 15}{x^3 + 3x}$

14. $\dfrac{-2x^2 + 5x - 2}{x^3 + x}$

15. $\dfrac{-2x^2 - 5}{x^4 + x^2}$

16. $\dfrac{-x^2 + 2x + 6}{x^4 + 2x^2}$

17. $\dfrac{-5x^2 + 4}{x^4 + 2x^2}$

18. $\dfrac{3x^3 - 8x^2 + 5x - 6}{(x - 1)^2(x^2 + 1)}$

19. $\dfrac{x^2 + 5x - 5}{(x^2 + 6)^2}$

20. $\dfrac{2x^3 + 3x^2 + 9x + 11}{(x^2 + 4)^2}$

21. $\dfrac{2x^3 - x^2 + 2x - 1}{(2x^2 + x + 1)^2}$

22. $\dfrac{x^3 + 3x^2 + 10x + 1}{(x^2 + 3x + 10)^2}$

23. $\dfrac{3x^3 - x^2 + 13x - 2}{x^4 + 9x^2 + 20}$

24. $\dfrac{2x^3 + 6x + 4}{x^4 + 6x^2 + 8}$

25. $\dfrac{3x^4 + 4x^3 + 21x^2 + 38x + 16}{x^3 + 8x}$

26. $\dfrac{2x^4 - x^3 + 17x^2 - 10x + 18}{x^3 + 9x}$

27. $\dfrac{2x^3 - x^2 - 14x - 13}{x^2 - x - 6}$

28. $\dfrac{3x^3 + 2x^2 - 14x + 6}{x^2 + x - 6}$

29. $\dfrac{x^2 + 2x + 3}{(x + 2)^3}$ **30.** $\dfrac{4x + 11}{(x + 3)^3}$ **33.** $\dfrac{5x^2 + 3x + 7}{x^3 + 3x^2 + 6x + 4}$ **34.** $\dfrac{9x^2 - 7x - 32}{2x^3 - 9x^2 + 7x + 6}$

31. $\dfrac{x^2 - 2x + 7}{x^3 - x^2 + 5x - 5}$ **32.** $\dfrac{2x^2 - 5x + 24}{x^3 - 2x^2 + 7x - 14}$

CHAPTER 10 REVIEW EXERCISES

In Exercises 1–6, find the first four terms and the eighth term of each sequence $\{a_n\}$.

1. $\left\{\dfrac{n + 1}{2n - 1}\right\}$ **2.** $\left\{\dfrac{(-1)^{n-1}}{2^n}\right\}$

3. $\left\{\dfrac{\cos \dfrac{n\pi}{2}}{n^2}\right\}$ **4.** $\left\{\left(-\dfrac{3}{2}\right)^n \left(1 - \dfrac{1}{n}\right)\right\}$

5. $\left\{\dfrac{(-1)^n x^n}{n!}\right\}$ **6.** $\left\{\dfrac{2n! \ln n}{(n - 1)!}\right\}$

In each of Exercises 7–10, a sequence $\{a_n\}$ is defined recursively. Find the second, third, and fourth terms of each sequence.

7. $a_1 = 3;\ a_n = \dfrac{a_{n-1}}{2} + 1,\quad$ for $n \geq 2$

8. $a_1 = \dfrac{5}{2};\ a_n = \dfrac{(-1)^n a_{n-1}}{n + 1},\quad$ for $n \geq 2$

9. $a_1 = x;\ a_n = \dfrac{(-1)^n (a_{n-1})^2}{n^2},$ for $n \geq 2$

10. $a_1 = 1,\ a_2 = 3;\ a_n = 5a_{n-2} - \dfrac{a_{n-1}}{2},\quad$ for $n \geq 3$

In Exercises 11–14, express each sum using summation notation.

11. $1 - \frac{1}{2} + \frac{1}{4} - \frac{1}{8} + \cdots + \frac{1}{256}$

12. $(1 - \frac{1}{3})^2 + (1 - \frac{1}{5})^3 + (1 - \frac{1}{7})^4 + \cdots + (1 - \frac{1}{21})^{11}$

13. $\dfrac{\ln 2}{1 \cdot 2} + \dfrac{\ln 3}{2 \cdot 3} + \dfrac{\ln 4}{3 \cdot 4} + \dfrac{\ln 5}{4 \cdot 5} + \cdots + \dfrac{\ln 100}{99 \cdot 100}$

14. $(x - 2) - \dfrac{(x - 2)^2}{2} + \dfrac{(x - 2)^3}{6} - \dfrac{(x - 2)^4}{24} + \cdots$
$+ \dfrac{(-1)^{n-1}(x - 2)^n}{n!}$

In Exercises 15–18, find the value of each sum.

15. $\displaystyle\sum_{i=1}^{5} \left(\dfrac{2}{3}\right)^{i-1}$ **16.** $\displaystyle\sum_{k=1}^{10} \dfrac{(2k - 1)^2}{k}$

17. $\displaystyle\sum_{k=0}^{8} \dfrac{\sin k\pi}{k!}$ **18.** $\displaystyle\sum_{i=1}^{8} (3i - 4)$

19. Find the value of the telescoping sum

$$\sum_{k=1}^{10} \left[\dfrac{1}{(k + 1)^2} - \dfrac{1}{k^2}\right].$$

20. An accountant takes a job with a starting salary of $30,000. If she receives an 8% raise at the end of each year for six consecutive years, find her salary after the raise at the end of the sixth year.

In Exercises 21–26, find the general term a_n of each arithmetic sequence.

21. $5, 11, 17, \ldots$ **22.** $-\frac{4}{5}, -\frac{7}{5}, -2, \ldots$

23. $8, 5.3, 2.6, \ldots$ **24.** $\frac{7}{3}, \frac{11}{4}, \frac{19}{6}, \ldots$

25. $\dfrac{3\pi}{2}, \dfrac{5\pi}{3}, \dfrac{11\pi}{6}, \ldots$ **26.** $\ln 4, \ln 16, \ln 64, \ldots$

In Exercises 27–30, find the indicated term of each arithmetic sequence.

27. $-\frac{3}{2}, 2, \frac{11}{2}, \ldots$; twentieth term

28. $3.2, 0.8, -1.6, \ldots$; tenth term

29. $\log 2, \log 4, \log 8, \ldots$; twelfth term

30. $x + 1, x + 3, x + 5, \ldots$; fiftieth term

In Exercises 31–34, find the sum of the first ten terms of each arithmetic sequence.

31. $30, 24, 18, \ldots$ **32.** $-8.1, -6.3, -4.5, \ldots$

33. $\frac{5}{2}, \frac{13}{4}, 4, \ldots$ **34.** $\ln 2, \ln 4, \ln 8, \ldots$

In Exercises 35–36, insert the indicated number of arithmetic means between each given pair of numbers.

35. 3 arithmetic means between -5 and 13

36. 4 arithmetic means between $\frac{3}{4}$ and $\frac{17}{2}$

37. The sum of the first five terms of an arithmetic sequence is 55. If the fifth term is 15, find the general term a_n.

38. A college student plans to get lots of sleep during his three-month summer vacation. Each night he plans to sleep 15 minutes more than he slept the night before. In how many days will the student be sleeping 24 hours a day if he starts with 8 hours of sleep the first night?

In Exercises 39–44, find the general term a_n of each geometric sequence.

39. $\frac{3}{5}, \frac{9}{25}, \frac{27}{125}, \ldots$ **40.** $-8, 14, -\frac{49}{2}, \ldots$

41. $3.2, 0.8, 0.2, \ldots$ **42.** $\dfrac{\sqrt{3}}{2}, -\dfrac{3}{4}, \dfrac{3\sqrt{3}}{8}, \ldots$

43. $\ln 3, \ln 9, \ln 81, \ldots$ **44.** $\dfrac{1}{x^2}, \dfrac{1}{x^5}, \dfrac{1}{x^8}, \ldots$

In Exercises 45–48, find the indicated sum of terms for each geometric sequence.

45. $1, \frac{1}{10}, \frac{1}{100}, \ldots$; S_6 **46.** $1.2, 1.8, 2.7, \ldots$; S_8

47. $15, -20, \frac{80}{3}, \ldots$; S_8

48. $\ln 4, \ln 2, \ln \sqrt{2}, \ldots$; S_6

In Exercises 49–50, insert the indicated number of geometric means between each given pair of numbers.

49. 3 geometric means between 4 and 12

50. 4 geometric means between -8 and $\frac{243}{128}$

In Exercises 51–54, find the sum of each geometric series or tell why the series has no sum.

51. $-1 + \frac{1}{3} - \frac{1}{9} + \cdots$

52. $2 + 2\sqrt{2} + 4 + \cdots$

53. $\frac{5}{6} - \frac{5}{4} + \frac{15}{8} - \frac{45}{16} + \cdots$

54. $e/2 + e/4 + e/8 + \cdots$

55. Express the rational number $2.3161616\ldots$ as a quotient of two integers.

56. Find the first term of a geometric sequence whose second term is $\frac{3}{2}$ and whose fifth term is $\frac{32}{243}$.

57. On the first day of each month for 5 years, a deposit of $100 is made into an account that pays 9% interest compounded monthly. How much money is in this account at the end of 5 years?

58. A ball dropped from a height of 50 feet bounces 10 feet high on its first bounce. On each successive bounce, the ball bounces to a height that is $\frac{3}{5}$ the height of its previous bounce. Find the total distance the ball travels.

In Exercises 59–66, use the principle of mathematical induction to prove that each statement is true for the indicated values of the positive integer n.

59. $3 \cdot 1^2 + 3 \cdot 2^2 + 3 \cdot 3^2$
$$+ \cdots + 3 \cdot n^2 = \frac{n(n+1)(2n+1)}{2}$$

60. $2(-1) + 2(-1)^2 + 2(-1)^3$
$$+ \cdots + 2(-1)^n = (-1)^n - 1$$

61. $2^n + 10n < 3^n$, for all $n \geq 4$

62. 2 is a factor of $5^n - 1$.

63. 2 is a factor of $7^n - 3^n$, for all n

64. $\dfrac{1}{(2)(5)} + \dfrac{1}{(5)(8)} + \dfrac{1}{(8)(11)} + \cdots +$
$$\dfrac{1}{(3n-1)(3n+2)} = \dfrac{n}{2(3n+2)}, \quad \text{for all } n$$

65. $\ln 1 + \ln 2 + \ln 3 + \cdots + \ln n = \ln(n!)$, for all n

66. $2^n > 3n$, for all $n \geq 4$

In Exercises 67–70, use the binomial theorem to expand each expression.

67. $(2x - y)^5$

68. $(3x + 1)^6$

69. $\left(\sqrt{x} + \dfrac{2}{\sqrt{x}}\right)^4$

70. $(x^{-1} + 3y)^4$

In Exercises 71–74, use the binomial theorem to find the indicated term (only) of each expansion.

71. $(3x + y)^{10}$; third term

72. $(2x - 3y)^{11}$; fifth term

73. $(x^2 - 2y^3)^9$; term involving y^{12}

74. $(x^3 + 4y^2)^6$; term involving x^9

In Exercises 75–84, find the partial fraction decomposition of each rational expression.

75. $\dfrac{11x - 8}{x^2 - x - 6}$

76. $\dfrac{-x - 37}{x^2 + 2x - 15}$

77. $\dfrac{5x^2 - 41x + 98}{(x - 5)(x^2 - 4x - 5)}$

78. $\dfrac{5x^2 + 25x + 31}{(x + 3)(x^2 + 5x + 6)}$

79. $\dfrac{2x^3 + x^2 + 33x + 15}{(x^2 + 16)^2}$

80. $\dfrac{x^3 + 2x^2 + 6x + 1}{(x^2 + 2x + 7)^2}$

81. $\dfrac{2x^2 + 5x + 7}{(x + 2)^3}$

82. $\dfrac{5x + 1}{(x - 1)^3}$

83. $\dfrac{2x^5 + 3x^3 + 4x^2 + 6x + 4}{x^3 + x}$

84. $\dfrac{2x^2 + 3x + 1}{x^3 + 2x - x^2 - 2}$

APPENDIX A

Algebraic Review

In this section, we review some basic algebraic concepts including exponents, radicals, polynomials, factoring, and rational expressions. We begin with a discussion of exponentiation.

EXPONENTS AND RADICALS

If x is a real number and n is a positive integer, the **nth power of x**, denoted by x^n, is defined by

$$x^n = \underbrace{x \cdot x \cdot x \ldots x.}_{n \text{ factors}}$$

The number x is called the **base** and the integer n is called the **exponent** or **power**. As we shall now see, this concept can readily be extended to include zero and negative integers as exponents.

If x is a nonzero real number, and n is a positive integer, then

$$x^0 = 1 \qquad \text{and}$$

$$x^{-n} = \frac{1}{x^n}.$$

For example, $\pi^0 = 1$ and $(-2)^{-4} = 1/(-2)^4 = \frac{1}{16}$, while expressions such as 0^0 and 0^{-2} are undefined.

Arithmetic operations with integral exponents are summarized in the following laws:

Laws of Exponents

Let x and y be nonzero real numbers and let m and n be integers. Then

(i) $x^m \cdot x^n = x^{m+n}$;

(ii) $\dfrac{x^m}{x^n} = x^{m-n}$;

(iii) $(x^m)^n = x^{m \cdot n}$;

(iv) $(xy)^n = x^n \cdot y^n$;

(v) $\left(\dfrac{x}{y}\right)^n = \dfrac{x^n}{y^n}$.

EXAMPLE 1

Use the laws of exponents to write each of the following expressions in simplest form. Leave no zero exponents and no negative exponents.

a. $\left(\dfrac{5ab^7}{2a^2}\right)^3$ b. $(-\tfrac{3}{4}m^2n^{-3})^{-2}$ c. $(-8p^{-3}q^3)^2(13p^4q^{-6})$

Solution

a. $\left(\dfrac{5ab^7}{2a^2}\right)^3 = \dfrac{(5ab^7)^3}{(2a^2)^3} = \dfrac{5^3a^3(b^7)^3}{2^3(a^2)^3} = \dfrac{125a^3b^{21}}{8a^6} = \dfrac{125}{8}a^{-3}b^{21}$

$= \dfrac{125}{8}\left(\dfrac{1}{a^3}\right)b^{21} = \dfrac{125b^{21}}{8a^3}.$

b. $(-\tfrac{3}{4}m^2n^{-3})^{-2} = (-\tfrac{3}{4})^{-2}(m^2)^{-2}(n^{-3})^{-2} = \dfrac{1}{(-\tfrac{3}{4})^2}\dfrac{1}{(m^2)^2}n^6$

$= \dfrac{1}{(\tfrac{9}{16})}\dfrac{n^6}{m^4} = \dfrac{16n^6}{9m^4}.$

c. $(-8p^{-3}q^3)^2(13p^4q^{-6}) = [(-8)^2(p^{-3})^2(q^3)^2](13p^4q^{-6})$

$= (64p^{-6}q^6)(13p^4q^{-6}) = 832p^{-2}q^0 = \dfrac{832}{p^2}.$

In order to extend the concept of exponents to include rational numbers, we must first define the principal nth root of a real number.

Principal nth Root of a Real Number

(i) If n is an even positive integer, and if a is a nonnegative real number, the **principal nth root of a** is the nonnegative real number b such that $b^n = a$.

(ii) If n is an odd positive integer, and if a is a real number, the **principal nth root of a** is the real number b such that $b^n = a$.

In either case, the principal nth root of a is denoted by $\sqrt[n]{a}$.

In the expression $\sqrt[n]{a}$, which is called a **radical**, a is called the **radicand**, n is called the **index**, and the symbol $\sqrt{}$ is called the **radical sign**. When the index is 2, it is customary to write $\sqrt{a}$ rather than $\sqrt[2]{a}$. To illustrate the definition of principal nth root, we see that

$\sqrt{49} = 7$, since $7^2 = 49$; $\sqrt[3]{-125} = -5$, since $(-5)^3 = -125$;
and $\sqrt[4]{81} = 3$, since $3^4 = 81$.

It is important to note that, when the index n is even, $\sqrt[n]{a}$ is not defined for $a < 0$. Thus, expressions such as $\sqrt{-1}$, $\sqrt{-9}$, and $\sqrt[4]{-16}$ do not exist in the set $\mathbb{R}$. Operations with radicals are governed by the following laws:

Laws of Radicals

Let x and y be real numbers and let m and n be positive integers. Then

(i) $\sqrt[n]{xy} = \sqrt[n]{x}\,\sqrt[n]{y}$;

(ii) $\sqrt[n]{\dfrac{x}{y}} = \dfrac{\sqrt[n]{x}}{\sqrt[n]{y}}$ $\quad y \neq 0$;

(iii) $\sqrt[m]{\sqrt[n]{x}} = \sqrt[mn]{x}$;

(iv) $\sqrt[n]{x^n} = \begin{cases} |x|, & \text{if } n \text{ is even} \\ x, & \text{if } n \text{ is odd} \end{cases}$;

provided each of the indicated roots exists.

CAUTION: It is incorrect to write $\sqrt{x^2} = x$, when $x < 0$. For example, if $x = -2$, then $\sqrt{x^2} = \sqrt{(-2)^2} = \sqrt{4} = 2 \neq x$. The correct result, according to the laws of radicals, is $\sqrt{x^2} = |x|$. Thus, if $x = -2$, $\sqrt{x^2} = |x| = |-2| = 2$.

Using the laws of radicals, we can simplify radicals by removing as many factors as possible from the radicand. As a rule, no exponents larger than or equal to the index should remain under the radical sign.

EXAMPLE 2

Simplify each of the following expressions. Assume all variables represent positive real numbers.

a. $\dfrac{\sqrt[3]{81y^7}}{\sqrt[3]{-3y}}$ **b.** $\sqrt[4]{8m^3n^4}\,\sqrt[4]{2mn^{13}}$ **c.** $\sqrt{\sqrt[3]{64t^{17}}}$ **d.** $\sqrt{\dfrac{1}{4p}}$

Solution

a. $\dfrac{\sqrt[3]{81y^7}}{\sqrt[3]{-3y}} = \sqrt[3]{\dfrac{81y^7}{-3y}} = \sqrt[3]{-27y^6} = -3y^2$

b. $\sqrt[4]{8m^3n^4}\,\sqrt[4]{2mn^{13}} = \sqrt[4]{16m^4n^{17}} = \sqrt[4]{(16m^4n^{16})n} = 2\,mn^4\,\sqrt[4]{n}$

c. $\sqrt{\sqrt[3]{64t^{17}}} = \sqrt[6]{64t^{17}} = \sqrt[6]{(64t^{12})t^5} = 2t^2\,\sqrt[6]{t^5}$

d. $\sqrt{\dfrac{1}{4p}} = \dfrac{\sqrt{1}}{\sqrt{4p}} = \dfrac{1}{2\sqrt{p}}$

Notice that the solution of part (d) of the preceding example is a fraction in which a radical appears in the denominator. It is possible to express this fraction as an equivalent fraction with no radical in the denominator by using a process known as **rationalizing the denominator**. Using the fact that $(\sqrt{p})^2 = p$, we can rationalize the denominator of the fraction $\dfrac{1}{2\sqrt{p}}$ by multiplying by 1 written in the form $\dfrac{\sqrt{p}}{\sqrt{p}}$. Thus

$$\frac{1}{2\sqrt{p}} = \frac{1}{2\sqrt{p}} \cdot \frac{\sqrt{p}}{\sqrt{p}} = \frac{\sqrt{p}}{2p}.$$

Variations of this procedure are illustrated in the following example.

EXAMPLE 3

Rationalize the denominator of each expression. Assume all variables represent positive real numbers.

a. $\dfrac{5a}{\sqrt{b} + \sqrt{c}}$ **b.** $\dfrac{2x}{\sqrt[3]{3y^2}}$

Solution

a. To rationalize the denominator, we multiply numerator and denominator by $\sqrt{b} - \sqrt{c}$, which is called the **conjugate** of the expression $\sqrt{b} + \sqrt{c}$.

$$\frac{5a}{\sqrt{b} + \sqrt{c}} = \frac{5a}{\sqrt{b} + \sqrt{c}} \cdot \frac{\sqrt{b} - \sqrt{c}}{\sqrt{b} - \sqrt{c}} = \frac{5a(\sqrt{b} - \sqrt{c})}{b - c}.$$

b. To eliminate the cube root in the denominator, we must obtain a perfect cube in the radicand. Since $(3y^2)(9y) = 27y^3$ is a perfect cube, we multiply numerator and denominator by $\sqrt[3]{9y}$. Thus,

$$\frac{2x}{\sqrt[3]{3y^2}} = \frac{2x}{\sqrt[3]{3y^2}} \cdot \frac{\sqrt[3]{9y}}{\sqrt[3]{9y}} = \frac{2x\sqrt[3]{9y}}{\sqrt[3]{27y^3}} = \frac{2x\sqrt[3]{9y}}{3y}.$$

Sometimes in calculus we must rationalize the numerator of an expression. The process is analogous to rationalizing the denominator, as the next example illustrates.

EXAMPIE 4

Rationalize the numerator of each expression. Assume all variables represent positive real numbers.

a. $\dfrac{2\sqrt{t} - 5}{3t}$ **b.** $\dfrac{\sqrt{x + h} - \sqrt{x}}{h}$

Solution

a. Multiplying numerator and denominator by $2\sqrt{t} + 5$, the conjugate of the numerator, we get

$$\frac{2\sqrt{t} - 5}{3t} = \frac{2\sqrt{t} - 5}{3t} \cdot \frac{2\sqrt{t} + 5}{2\sqrt{t} + 5} = \frac{4t - 25}{3t(2\sqrt{t} + 5)}.$$

b. Multiplying numerator and denominator by the conjugate of the numerator, we have

$$\frac{\sqrt{x+h} - \sqrt{x}}{h} = \frac{\sqrt{x+h} - \sqrt{x}}{h} \cdot \frac{\sqrt{x+h} + \sqrt{x}}{\sqrt{x+h} + \sqrt{x}}$$

$$= \frac{(x+h) - x}{h(\sqrt{x+h} + \sqrt{x})} = \frac{h}{h(\sqrt{x+h} + \sqrt{x})} = \frac{1}{\sqrt{x+h} + \sqrt{x}}.$$

As noted earlier, we can use the concept of principal nth root to help us define x^r, where r is a rational number.

Definition of Rational Exponent

Let m and n be integers such that $n > 0$. Assume also that m and n have no common factors other than 1. If x is a real number such that $\sqrt[n]{x}$ exists, then

$$x^{m/n} = (\sqrt[n]{x})^m = \sqrt[n]{x^m}.$$

Before we illustrate this definition with an example, we note that the concept of negative exponent extends naturally from integer exponents to rational exponents. Thus

$$x^{-r} = \frac{1}{x^r}, \quad \text{where } r \text{ is any rational number and } x \text{ is any nonzero real number.}$$

EXAMPLE 5

Evaluate each of the following expressions:

a. $(-27)^{4/3}$ **b.** $(\frac{4}{25})^{-3/2}$ **c.** $(-2)^{2/3}$

Solution

a. $(-27)^{4/3} = (\sqrt[3]{-27})^4 = (-3)^4 = 81.$

b. $(\frac{4}{25})^{-3/2} = \frac{1}{(\frac{4}{25})^{3/2}} = \frac{1}{(\sqrt{\frac{4}{25}})^3} = \frac{1}{(\frac{2}{5})^3} = \frac{125}{8}$

c. $(-2)^{2/3} = \sqrt[3]{(-2)^2} = \sqrt[3]{4}.$

It can be shown that the laws of exponents, which were given for integral exponents, are also valid for rational exponents. When $m = 1$, we see from the definition of $x^{m/n}$, that $x^{1/n} = \sqrt[n]{x}$. Thus, any radical having index n can be written in terms of rational exponents. For example, $\sqrt{x} = x^{1/2}$, $\sqrt[3]{x} = x^{1/3}$, $\sqrt[4]{x} = x^{1/4}$, etc. It follows therefore, that the laws of radicals can be expressed in terms of rational exponents. As an illustration, the law $\sqrt[n]{ab} = \sqrt[n]{a} \cdot \sqrt[n]{b}$, can be written as $(ab)^{1/n} = a^{1/n} \cdot b^{1/n}$.

E X A M P L E   6

Simplify each of the following expressions. Assume all variables represent positive real numbers.

a. $\dfrac{x^{5/6}}{x^{1/2}}$ **b.** $(4k^{4/3}m^{-8/3})^{3/4}$ **c.** $\left(\dfrac{8m^8n^2}{m^5n^{11}}\right)^{2/3}$

Solution

a. $\dfrac{x^{5/6}}{x^{1/2}} = x^{5/6-1/2} = x^{1/3}$.

b. $(4k^{4/3}m^{-8/3})^{3/4} = 4^{3/4}(k^{4/3})^{3/4}(m^{-8/3})^{3/4} = \sqrt[4]{4^3}\, km^{-2} = \dfrac{\sqrt[4]{64}\, k}{m^2}$

$$= \dfrac{\sqrt[4]{16\cdot 4}\, k}{m^2} = \dfrac{2\sqrt[4]{4}\, k}{m^2}.$$

c. $\left(\dfrac{8m^8n^2}{m^5n^{11}}\right)^{2/3} = (8m^3n^{-9})^{2/3} = 8^{2/3}(m^3)^{2/3}\,(n^{-9})^{2/3}$

$$= (\sqrt[3]{8})^2 m^2 n^{-6} = \dfrac{4m^2}{n^6}$$

POLYNOMIALS

Expressions such as $3x^2 - 5x^7 + 1$, $\sqrt{13}abc^2$, $\sqrt[5]{8xy} - 2\sqrt[3]{x^2y}$, $\dfrac{\frac{7}{3}z - z^{2/5}}{1 - z^2}$, and

$-\frac{4}{9}t^{-3} + \dfrac{1}{\pi}$ are called *algebraic expressions*. An **algebraic expression** is any combination of constants and variables resulting from the arithmetic operations of addition, subtraction, multiplication, division, and exponentiation (including both powers and roots). Whenever we write an algebraic expression, we assume that the variables involved only assume values for which the expression has meaning. We thus assume that no denominators are equal to zero and that all implied roots exist. One relatively simple type of algebraic expression is called a *polynomial*.

Polynomial in *x* of Degree *n*

Let *x* be a real number and let *n* be a nonnegative integer. An expression of the form

$$a_nx^n + a_{n-1}x^{n-1} + \cdots + a_1x + a_0$$

where $a_0, a_1, \ldots, a_n$ are constants and $a_n \neq 0$, is called a **polynomial in *x* of degree *n*.**

For example, $2x^3 - 3x + 5$, $1 - \sqrt{2}x^{12}$, $\frac{3}{4}x + 5$, -8, and πx^4 are polynomials in *x* of degrees 3, 12, 1, 0, and 4, respectively, while expressions such as $3\sqrt{x} - 1$, $\dfrac{1}{x^2} - 5$, and $\sqrt[3]{2x + 11}$ are not polynomials. A polynomial in x is a sum of expressions of the form $a_k x^k$, which are called **terms**. In the term $a_k x^k$, the constant a_k is called the **coefficient**, and if $a_k \neq 0$, we say that the term has **degree k**. If the coefficients $a_1, a_2, \ldots, a_n$ are all zero, we have a polynomial consisting of only the constant term a_0. Such a polynomial is called a **constant polynomial** and its degree is 0 if $a_0 \neq 0$. If all the coefficients of a polynomial are equal to zero, it is called the **zero polynomial**. It is customary not to assign a degree to the zero polynomial. Consider, for example, the polynomial $5x^2 - \sqrt{2}x^7 + \frac{3}{4}x - 1$. This polynomial has degree 7, and consists of the four terms described in Table A.1.

TABLE A.1

Term	Coefficient	Degree
$5x^2$	5	2
$-\sqrt{2}x^7$	$-\sqrt{2}$	7
$\frac{3}{4}x$	$\frac{3}{4}$	1
-1	-1	0

Before discussing the arithmetic of polynomials, we should point out that some polynomials are given special names. A **monomial** is a polynomial having exactly one term, a **binomial** has exactly two terms, and a **trinomial** has exactly three. In addition, polynomials of degree one are called **linear polynomials,** while polynomials of degree two are called **quadratic polynomials**. Terms having the same degree, such as $\sqrt{3}x^4$ and $-5x^4$, are called **like terms**. Polynomials are added or subtracted by adding or subtracting their like terms. To illustrate,

$$\left(13x^7 - 3\sqrt{2}x^5 + \frac{2}{3}x + 5\right) - (x^7 - 2x^5 + x^2 + 4x - 2)$$

$$= 13x^7 - 3\sqrt{2}x^5 + \frac{2}{3}x + 5 - x^7 + 2x^5 - x^2 - 4x + 2$$

$$= 12x^7 + (2 - 3\sqrt{2})x^5 - x^2 - \frac{10}{3}x + 7.$$

To multiply two polynomials, we multiply every term of one of the polynomials by every term of the other, and combine like terms. This process requires repeated usage of the exponent law $x^m \cdot x^n = x^{m+n}$, as we see in the next example.

EXAMPLE 7

Find each of the following products:

a. $(y - 7)(y^2 - 2y + 1)$ **b.** $(5x^2 - 2)^3$

Solution

a. $(y - 7)(y^2 - 2y + 1) = y(y^2 - 2y + 1) - 7(y^2 - 2y + 1)$
$$= y^3 - 2y^2 + y - 7y^2 + 14y - 7$$
$$= y^3 - 9y^2 + 15y - 7$$

b. $(5x^2 - 2)^3 = (5x^2 - 2)(5x^2 - 2)(5x^2 - 2)$
$$= (5x^2 - 2)(25x^4 - 10x^2 - 10x^2 + 4)$$
$$= (5x^2 - 2)(25x^4 - 20x^2 + 4)$$
$$= 125x^6 - 100x^4 + 20x^2 - 50x^4 + 40x^2 - 8$$
$$= 125x^6 - 150x^4 + 60x^2 - 8$$

Often we must deal with polynomials that involve more than one variable. A **polynomial in two variables** is a sum of terms of the form $kx^m y^n$, where k is a constant, x and y are real numbers, and m and n are nonnegative integers. We define polynomials of three or more variables in a similar manner. A polynomial in more than one variable is usually referred to as a **polynomial in several variables**.

To find the **degree** of a monomial in several variables, we add the exponents of all the variables. For example, the monomial $-3p^5 qr^6$ has degree 12. The **degree** of a polynomial in several variables is the degree of its term with highest degree. For example, the degree of the polynomial $12x^2 y^4 - 3xy^6 + 13x - 8y^5$ is 7.

As we see in the following example, arithmetic operations on polynomials in several variables are analogous to those on polynomials in one variable. **Like terms** are terms such as $3a^3 bc^2$ and $-\frac{5}{2}c^2 a^3 b$ that contain the same powers of the same variables.

EXAMPLE 8

Perform the indicated operations.

a. $(x^4 + 4x^3 y + 6x^2 y^2 - y^4) - (2x^3 y - 7x^2 y^2 - x^3 y^2 + 3y^4)$

b. $(3m + 2n)^2 + (3m - 2n)^2$

c. $(2a^2 - 3a + 1)(a^3 + 3a^2 - 5)$

Solution

a. $(x^4 + 4x^3 y + 6x^2 y^2 - y^4) - (2x^3 y - 7x^2 y^2 - x^3 y^2 + 3y^4)$
$$= x^4 + 4x^3 y + 6x^2 y^2 - y^4 - 2x^3 y + 7x^2 y^2 + x^3 y^2 - 3y^4$$
$$= x^4 + 2x^3 y + x^3 y^2 + 13x^2 y^2 - 4y^4$$

b. $(3m + 2n)^2 + (3m - 2n)^2 = (3m + 2n)(3m + 2n)$
$$+ (3m - 2n)(3m - 2n)$$
$$= 9m^2 + 6mn + 6mn + 4n^2 + 9m^2$$
$$- 6mn - 6mn + 4n^2$$
$$= 18m^2 + 8n^2$$

c. $(2a^2 - 3a + 1)(a^3 + 3a^2 - 5) = 2a^5 + 6a^4 - 10a^2 - 3a^4 - 9a^3$
$$+ 15a + a^3 + 3a^2 - 5$$
$$= 2a^5 + 3a^4 - 8a^3 - 7a^2 + 15a - 5$$

Sometimes a polynomial of a certain degree can be factored as a product of two or more polynomials of lower degree. The procedure for factoring the polynomial essentially reverses the process of multiplication and is often difficult to accomplish. To simplify matters, we will limit our discussion to polynomials with integer coefficients whose factors also have integer coefficients.

Suppose we wish to factor the trinomial $6x^3 + 21x^2 - 45x$. We begin by looking for factors that are common to all three terms. These **common factors**, when present, should be factored out before any additional factoring is attempted. Since $3x$ is a common factor here, we write

$$6x^3 + 21x^2 - 45x = 3x(2x^2 + 7x - 15).$$

We now attempt to factor the trinomial $2x^2 + 7x - 15$. Trinomials of the form $ax^2 + bx + c$ can sometimes be factored into two binomial factors. In this case, by trial and error, we find that

$$2x^2 + 7x - 15 = (2x - 3)(x + 5).$$

Hence, in factored form,

$$6x^3 + 21x^2 - 45x = 3x(2x - 3)(x + 5).$$

Some binomials can be factored by using one of the following formulas:

Special Factoring Formulas

$x^2 - y^2 = (x + y)(x - y)$	Difference of two squares
$x^3 - y^3 = (x - y)(x^2 + xy + y^2)$	Difference of two cubes
$x^3 + y^3 = (x + y)(x^2 - xy + y^2)$	Sum of two cubes

EXAMPLE 9

Factor each of the following polynomials completely:

a. $18a^3b - 50ab^3$ **b.** $27m^3 + 125$ **c.** $8t^3 - 1$

Solution

a. $18a^3b - 50ab^3 = 2ab(9a^2 - 25b^2) = 2ab(3a + 5b)(3a - 5b)$

b. $27m^3 + 125 = (3m)^3 + 5^3 = (3m + 5)(9m^2 - 15m + 25)$

c. $8t^3 - 1 = (2t)^3 - 1^3 = (2t - 1)(4t^2 + 2t + 1)$

Polynomials consisting of four or more terms can sometimes be factored by grouping pairs (or triples) of terms and then factoring each grouping. This method of factoring is appropriately called **factoring by grouping**.

EXAMPLE 10

Factor each of the following polynomials completely:

a. $x^3 - 4x + 3x^2 - 12$

b. $a^2x - b^2y + b^2x - a^2y$

c. $m^2 - 10m + 25 - 4n^2$

Solution

a. Grouping pairs of terms and factoring common factors from each pair, gives us

$$x^3 - 4x + 3x^2 - 12 = (x^3 - 4x) + (3x^2 - 12)$$
$$= x(x^2 - 4) + 3(x^2 - 4).$$

Since $x^2 - 4$ is a common factor, we can write

$$x(x^2 - 4) + 3(x^2 - 4) = (x^2 - 4)(x + 3).$$

Noting that $x^2 - 4$ is a difference of two squares, we have the complete factorization

$$x^3 - 4x + 3x^2 - 12 = (x + 2)(x - 2)(x + 3).$$

b. Rearranging the terms and grouping, we have

$$a^2x - b^2y + b^2x - a^2y = a^2x + b^2x - a^2y - b^2y$$
$$= (a^2x + b^2x) - (a^2y + b^2y).$$

Note the sign change in the second parentheses.

Factoring the common factors from each pair of terms yields

$$(a^2x + b^2x) - (a^2y + b^2y) = x(a^2 + b^2) - y(a^2 + b^2).$$

Since the common factor $a^2 + b^2$ is a sum of two squares, it cannot be factored. Thus, the complete factorization is

$$a^2x - b^2y + b^2x - a^2y = (a^2 + b^2)(x - y).$$

c. Grouping the first three terms, we have

$$m^2 - 10m + 25 - 4n^2 = (m^2 - 10m + 25) - 4n^2.$$

Since $m^2 - 10m + 25 = (m - 5)(m - 5) = (m - 5)^2$, the latter equation becomes

$$m^2 - 10m + 25 - 4n^2 = (m - 5)^2 - 4n^2.$$

Noting that $(m - 5)^2 - 4n^2 = (m - 5)^2 - (2n)^2$ is a difference of two squares, we have

$$m^2 - 10m + 25 - 4n^2 = (m - 5 + 2n)(m - 5 - 2n).$$

RATIONAL EXPRESSIONS

Quotients of algebraic expressions such as

$$\frac{x^2 - 1}{x^2 + 3x + 2}, \quad \frac{a^{-1} - b^{-1}}{\dfrac{1}{ab}}, \quad \frac{\dfrac{1}{3 + h} - \dfrac{1}{3}}{h}, \quad \text{and} \quad \frac{\dfrac{1}{x} - \dfrac{1}{y}}{\sqrt{x} + \sqrt{y}}$$

are called **algebraic fractions**. In particular, a quotient of two polynomials such as $\dfrac{y^3 - 16y^2 + 28}{1 - 5y^4}$ is called a **rational expression**. Whenever we write an algebraic fraction, we assume that the variables involved only assume values for which the expression has meaning. Thus, we assume $h \neq 0$ and $h \neq -3$ in the expression $\dfrac{\dfrac{1}{3 + h} - \dfrac{1}{3}}{h}$ and that x and y are both positive in the expression $\dfrac{\dfrac{1}{x} - \dfrac{1}{y}}{\sqrt{x} + \sqrt{y}}$.

Sometimes an algebraic fraction can be simplified by dividing both numerator and denominator by a common factor. When the numerator and denominator have no common factor other than 1, we say that the expression has been **reduced to lowest terms** or **simplified**. For example,

$$\frac{x^2 - 1}{x^2 + 3x + 2} = \frac{(x + 1)(x - 1)}{(x + 2)(x + 1)} = \frac{x - 1}{x + 2}, \quad \text{and}$$

$$\frac{m^4 - 1}{m^3 + m^2 + m + 1} = \frac{(m^2 + 1)(m^2 - 1)}{(m^3 + m^2) + (m + 1)}$$

$$= \frac{(m^2 + 1)(m + 1)(m - 1)}{m^2(m + 1) + (m + 1)}$$

$$= \frac{(m^2 + 1)(m + 1)(m - 1)}{(m^2 + 1)(m + 1)} = m - 1.$$

The following example illustrates that algebraic fractions can be multiplied or divided as if they were numerical fractions:

EXAMPLE 11

Perform the indicated operations and simplify.

a. $\dfrac{a + 3}{2a - 3b} \cdot \dfrac{4a^2 - 9b^2}{ab + 3b}$ **b.** $\dfrac{x^3 - y^3}{x + y} \div \dfrac{2x^2 - xy - y^2}{x^2 - y^2}$

Solution

a. Factoring (where possible) and multiplying, we have

$$\frac{a + 3}{2a - 3b} \cdot \frac{4a^2 - 9b^2}{ab + 3b} = \frac{(a + 3)(2a + 3b)(2a - 3b)}{(2a - 3b) \cdot b(a + 3)} = \frac{2a + 3b}{b}.$$

b. Inverting the divisor and multiplying, we obtain

$$\frac{x^3 - y^3}{x + y} \div \frac{2x^2 - xy - y^2}{x^2 - y^2} = \frac{x^3 - y^3}{x + y} \cdot \frac{x^2 - y^2}{2x^2 - xy - y^2}$$

$$= \frac{(x - y)(x^2 + xy + y^2)(x + y)(x - y)}{(x + y)(2x + y)(x - y)}$$

$$= \frac{(x - y)(x^2 + xy + y^2)}{2x + y}.$$

As in numerical fractions, we must have a common denominator in order to add algebraic fractions. The procedure for finding the **least common denominator** (LCD) and adding algebraic fractions is outlined below.

1. Factor each denominator completely. The LCD is the product of all the different factors appearing in any of the denominators, with each factor raised to the highest power to which it appears in any of the denominators.

2. Express each fraction as an equivalent fraction that has the LCD as its denominator. We do this by multiplying numerator and denominator by any factors that are present in the LCD, but missing from the denominator of that fraction.

3. Add the numerators and place the sum over the LCD. Make sure the final answer is in lowest terms.

E X A M P L E 12

Perform the indicated operations and simplify.

a. $\dfrac{3t}{t^2 - 2t + 1} + \dfrac{t + 1}{t^2 + t - 2}$ **b.** $\dfrac{1}{x^2 - 4} + \dfrac{3}{x - 2} - \dfrac{2}{x^2 + 2x}$

Solution

a. Factoring each denominator, we have

$$\frac{3t}{t^2 - 2t + 1} + \frac{t + 1}{t^2 + t - 2} = \frac{3t}{(t - 1)^2} + \frac{t + 1}{(t + 2)(t - 1)}.$$

Thus the LCD is $(t - 1)^2(t + 2)$. Expressing both fractions as equivalent fractions that have the LCD as their denominators, we obtain

$$\frac{3t}{(t - 1)^2} + \frac{t + 1}{(t + 2)(t - 1)} = \frac{3t}{(t - 1)^2} \cdot \frac{t + 2}{t + 2}$$

$$+ \frac{t + 1}{(t + 2)(t - 1)} \cdot \frac{t - 1}{t - 1}$$

$$= \frac{3t^2 + 6t}{(t - 1)^2(t + 2)} + \frac{t^2 - 1}{(t - 1)^2(t + 2)}.$$

Adding the numerators and placing the sum over the LCD, we find that

$$\frac{3t}{t^2 - 2t + 1} + \frac{t + 1}{t^2 + t - 2} = \frac{4t^2 + 6t - 1}{(t - 1)^2(t + 2)}.$$

We note that our answer is in lowest terms since the numerator $4t^2 + 6t - 1$ cannot be factored.

b. We proceed as in part (a).

$$\frac{1}{x^2 - 4} + \frac{3}{x - 2} - \frac{2}{x^2 + 2x} = \frac{1}{(x + 2)(x - 2)} + \frac{3}{x - 2} - \frac{2}{x(x + 2)}$$

$$\text{LCD} = x(x + 2)(x - 2)$$

$$= \frac{1}{(x + 2)(x - 2)} \cdot \frac{x}{x} + \frac{3}{x - 2} \cdot \frac{x(x + 2)}{x(x + 2)} - \frac{2}{x(x + 2)} \cdot \frac{x - 2}{x - 2}$$

$$= \frac{x}{x(x + 2)(x - 2)} + \frac{3x^2 + 6x}{x(x + 2)(x - 2)} - \frac{2x - 4}{x(x + 2)(x - 2)}$$

$$= \frac{3x^2 + 5x + 4}{x(x + 2)(x - 2)}.$$

E X A M P L E 13

Perform the indicated operations and simplify.

a. $\dfrac{y - \dfrac{4}{y}}{1 - \dfrac{2}{y}}$ **b.** $\dfrac{\dfrac{1}{\sqrt{x + h}} - \dfrac{1}{\sqrt{x}}}{h}$ **c.** $\dfrac{(x^2 - 1)^4(2x) + x^2(4)(x^2 - 1)^3(2x)}{(x^2 - 1)^8}$

Solution

a. Multiplying numerator and denominator by y (the LCD) and simplifying, we have

$$\frac{y - \dfrac{4}{y}}{1 - \dfrac{2}{y}} \cdot \frac{y}{y} = \frac{y^2 - 4}{y - 2} = \frac{(y + 2)(y - 2)}{y - 2} = y + 2.$$

b. Multiplying numerator and denominator by $\sqrt{x}(\sqrt{x + h})$, which is the LCD, we obtain

$$\frac{\dfrac{1}{\sqrt{x + h}} - \dfrac{1}{\sqrt{x}}}{h} = \frac{\dfrac{1}{\sqrt{x + h}} - \dfrac{1}{\sqrt{x}}}{h} \cdot \frac{\sqrt{x}(\sqrt{x + h})}{\sqrt{x}(\sqrt{x + h})}$$

$$= \frac{\dfrac{1}{\sqrt{x + h}} \cdot \sqrt{x}(\sqrt{x + h}) - \dfrac{1}{\sqrt{x}} \cdot \sqrt{x}(\sqrt{x + h})}{h \cdot \sqrt{x}(\sqrt{x + h})}$$

$$= \frac{\sqrt{x} - \sqrt{x + h}}{h\sqrt{x^2 + hx}}.$$

c. Factoring $2x(x^2 - 1)^3$ out of each term in the numerator and simplifying, we obtain

$$\frac{(x^2 - 1)^4(2x) + x^2(4)(x^2 - 1)^3(2x)}{(x^2 - 1)^8} = \frac{2x(x^2 - 1)^3[(x^2 - 1) + 4x^2]}{(x^2 - 1)^8}$$

$$= \frac{2x(5x^2 - 1)}{(x^2 - 1)^5}$$

$$= \frac{10x^3 - 2x}{(x^2 - 1)^5}.$$

EXERCISES A.1

In Exercises 1–12, simplify each expression, leaving no zero exponents and no negative exponents.

1. $(xy^2)(-2x^3y)^{-2}$

2. $\dfrac{8x^{-8}y^{-12}}{2x^{-2}y^{-6}}$

3. $\dfrac{3a^{-3}b^0c^2}{-6a^3bc^{-2}}$

4. $\left(\dfrac{2x^{-2}}{y^3}\right)^{-1}$

5. $(4a^2b)^4\left(\dfrac{-a^3}{2b}\right)^3$

6. $\dfrac{-12x^{-9}y^{10}}{4x^{-12}y^5}$

7. $\left(\dfrac{-5x^{-2}y}{2y^3}\right)^{-2}$

8. $\dfrac{(rs^5)(r^2s^{-2})^3}{(r^2s^{-4})^3}$

9. $(\frac{1}{2}x^{-4}y^3)^{-3}(-3x^{-5}y^0)^2$

10. $\left(\dfrac{2m^{-4}n^2p^0}{18m^3n^{-5}}\right)^{-2}$

11. $\dfrac{(x + y)^{-3}(x + y)^5}{(x + y)^{-1}}$

12. $\dfrac{b^{-1} + a^{-1}}{(ab)^{-1}}$

In Exercises 13–24, simplify each expression. Assume all variables represent positive real numbers.

13. $\sqrt[3]{16t^4u^3v^9}$

14. $\sqrt{\dfrac{8x^2y}{2x^4y^{-1}}}$

15. $\sqrt{\left(\dfrac{2a^{-2}}{b^{-3}}\right)^{-2}}$

16. $\sqrt[5]{-2m^4n^{-1}}\ \sqrt[5]{16m^{-9}n^4}$

17. $\sqrt[3]{\dfrac{54p^2qr^{-1}}{p^{-1}r^5}}$

18. $\sqrt[4]{(2a - 3b)^4}$

19. $\sqrt{\dfrac{t^3}{2tv^4}}\ \sqrt{\dfrac{t^{-2}v^3}{(3tv)^2}}$

20. $\sqrt[3]{(5 - 2y)^2}\ \sqrt[3]{(5 - 2y)^4}$

21. $\dfrac{\sqrt[6]{2p^{10}q^{12}r^{-5}}}{\sqrt[6]{128p^{-2}q^7r}}$

22. $\sqrt{a^3b}\left(\dfrac{1}{\sqrt{ab}} - \sqrt{ab}\right)$

23. $\sqrt{\sqrt[3]{x^7y^6z^{-12}}}$

24. $\sqrt{\sqrt{\dfrac{32ab^{-1}}{2a^5b^{-8}}}}$

In Exercises 25–30, rationalize the denominator of each expression. Assume all variables represent positive real numbers.

25. $\dfrac{5}{\sqrt{3}}$

26. $\dfrac{\sqrt[3]{3x}}{\sqrt[3]{4x^2}}$

27. $\dfrac{2}{3-\sqrt{2}}$

28. $\dfrac{4-\sqrt{3}}{4+\sqrt{3}}$

29. $\dfrac{2x-1}{\sqrt{x}-2}$

30. $\dfrac{\sqrt{a}+\sqrt{b}}{\sqrt{a}-\sqrt{b}}$

In Exercises 31–34, rationalize the numerator of each expression. Assume all variables represent positive real numbers.

31. $\dfrac{\sqrt{a}-2}{3\sqrt{a}}$

32. $\dfrac{1-\sqrt{2x}}{\sqrt{5x}}$

33. $\dfrac{\sqrt{m}+\sqrt{n}}{2mn}$

34. $\dfrac{\sqrt{2x+2h+1}-\sqrt{2x+1}}{h}$

In Exercises 35–46, simplify each expression. Assume all variables represent positive real numbers.

35. $\left(\dfrac{9}{16}\right)^{-3/2}$

36. $a^{4/3}a^{-3/2}a^{1/6}$

37. $(-27x^{-3}y^{12})^{2/3}$

38. $\dfrac{r^2s^{-1/2}t^{1/3}}{r^{1/3}s^{-3/4}t^3}$

39. $(p^2q^{-3})^{1/3}(p^{-2}q)^{-1/2}$

40. $(-8m^{5/2}n^{-6}p^0)^{4/3}$

41. $\left(\dfrac{x^3y^7z^{-3}}{x^2y^{-5}z^{-10}}\right)^{2/5}$

42. $(\tfrac{1}{4})^{3/2}(-\tfrac{1}{8})^{2/3}$

43. $\left(\dfrac{a^{-2/3}b^{1/6}}{a^{1/2}b^{-5/6}}\right)^6$

44. $\dfrac{(-27u^3v^0w^6)^{4/3}}{(125u^{3/2}v^6w^{-3})^{2/3}}$

45. $\dfrac{(2x-1)^{2/5}(2x-1)^{-1/3}}{(2x-1)^{-2/3}}$

46. $\left(\dfrac{x^{-2}y^3}{x^4y^{-3}}\right)^{-1/2}\left(\dfrac{x^4y^{-5}}{xy^4}\right)^{-1/3}$

In Exercises 47–60, perform the indicated operations and simplify.

47. $(2x^3+\tfrac{1}{2}x^2+4x)-(3x^3-\tfrac{1}{2}x^2+2x-4)$

48. $(2x^3y^2-5xy+x^2y^3)+(3xy-x^2y^3)-(x^3y^2+2xy)$

49. $3x^2(2x-1+15x^3)$

50. $(4m-3)(3m+7)$

51. $(7y+1)(-2y+9)$

52. $(ax-y)(ax+2y)$

53. $(r+2)(3r^2-12r+4)$

54. $(3u-2v)^2$

55. $(2x-1)(4x^2+1)$

56. $(x-y)(x^2+xy+y^2)$

57. $(2x^2-3)(4x^2-6x+1)$

58. $(2m-3)^3$

59. $(x-3y+1)^2$

60. $(2x+1)(3x-1)(4x+3)$

In Exercises 61–76, factor each expression completely.

61. a^2-2a-3

62. $6x^2-4x-16$

63. $16-9z^2$

64. $a^2-3ab-4b^2$

65. $6t^2+10t+4$

66. $3x^5-12x^3$

67. y^3-125

68. $2m^3+16$

69. $x^3-2x^2y-x+2y$

70. $5ru+10rv+2ut+4vt$

71. $4u^2+12uv+9v^2-9$

72. $x^2+2xy+y^2-16$

73. y^4-16

74. $4x^{10}-5x^5-6$

75. $36a^2+60ab^2+25b^4$

76. $6r^5s-3r^3s^2-30rs^3$

In Exercises 77–92, simplify each expression.

77. $\dfrac{4a^2+12a+9}{2a^2+3a}$

78. $\dfrac{5v^2+10v+5}{v^3+v^2-v-1}$

79. $\dfrac{x^2-1}{3x^3}\cdot\dfrac{x^2-5x}{x^3-1}$

80. $\dfrac{3a+12}{a^2+8a+16}\cdot\dfrac{a^2-16}{3a^2-11a-4}$

81. $\dfrac{9m^2-18m-16}{6m+4}\div\dfrac{9m^2-64}{6m+16}$

82. $\dfrac{x^2+x-6}{5x^2-3x-2}\div\dfrac{x^2-9}{5x^2+2x}$

83. $\dfrac{1}{a^2-5a+6}-\dfrac{1}{a^2-4}$

84. $\dfrac{x+1}{x^2+x}+\dfrac{3x}{x^2-1}$

85. $\dfrac{\dfrac{1}{a^2}-\dfrac{1}{b^2}}{\dfrac{a}{b}-\dfrac{b}{a}}$

86. $\dfrac{\dfrac{1}{x}-\dfrac{2}{x^2}+1}{\dfrac{1}{x}-\dfrac{3}{x^2}+2}$

87. $\dfrac{\dfrac{1}{(x+h)^2}-\dfrac{1}{x^2}}{h}$

88. $\dfrac{(x+h)^{-3}-x^{-3}}{h}$

89. $(3x^2+1)^3(7)(5x^2+2)^6(10x)$ $+(5x^2+2)^7(3)\cdot(3x^2+1)^2(6x)$

90. $(6x+2)^{1/2}(4)(3x^2-2x+1)^3(6x-2)$ $+(3x^2-2x+1)^4(\tfrac{1}{2})(6x+2)^{-1/2}(6)$

91. $\dfrac{(3x-1)^{1/3}(5)(3x^2-1)^4(6x)-(3x^2-1)^5(\tfrac{1}{3})(3x-1)^{-2/3}(3)}{[(3x-1)^{1/3}]^2}$

92. $(3)\left(\dfrac{5-2x^2}{4x+1}\right)^2\cdot\dfrac{(4x+1)(-4x)-(5-2x^2)(4)}{(4x+1)^2}$

APPENDIX B

Tables

TABLE 1 Common Logarithms

N	0	1	2	3	4	5	6	7	8	9
1.0	.0000	.0043	.0086	.0128	.0170	.0212	.0253	.0294	.0334	.0374
1.1	.0414	.0453	.0492	.0531	.0569	.0607	.0645	.0682	.0719	.0755
1.2	.0792	.0828	.0864	.0899	.0934	.0969	.1004	.1038	.1072	.1106
1.3	.1139	.1173	.1206	.1239	.1271	.1303	.1335	.1367	.1399	.1430
1.4	.1461	.1492	.1523	.1553	.1584	.1614	.1644	.1673	.1703	.1732
1.5	.1761	.1790	.1818	.1847	.1875	.1903	.1931	.1959	.1987	.2014
1.6	.2041	.2068	.2095	.2122	.2148	.2175	.2201	.2227	.2253	.2279
1.7	.2304	.2330	.2355	.2380	.2405	.2430	.2455	.2480	.2504	.2529
1.8	.2553	.2577	.2601	.2625	.2648	.2672	.2695	.2718	.2742	.2765
1.9	.2788	.2810	.2833	.2856	.2878	.2900	.2923	.2945	.2967	.2989
2.0	.3010	.3032	.3054	.3075	.3096	.3118	.3139	.3160	.3181	.3201
2.1	.3222	.3243	.3263	.3284	.3304	.3324	.3345	.3365	.3385	.3404
2.2	.3424	.3444	.3464	.3483	.3502	.3522	.3541	.3560	.3579	.3598
2.3	.3617	.3636	.3655	.3674	.3692	.3711	.3729	.3747	.3766	.3784
2.4	.3802	.3820	.3838	.3856	.3874	.3892	.3909	.3927	.3945	.3962
2.5	.3979	.3997	.4014	.4031	.4048	.4065	.4082	.4099	.4116	.4133
2.6	.4150	.4166	.4183	.4200	.4216	.4232	.4249	.4265	.4281	.4298
2.7	.4314	.4330	.4346	.4362	.4378	.4393	.4409	.4425	.4440	.4456
2.8	.4472	.4487	.4502	.4518	.4533	.4548	.4564	.4579	.4594	.4609
2.9	.4624	.4639	.4654	.4669	.4683	.4698	.4713	.4728	.4742	.4757
3.0	.4771	.4786	.4800	.4814	.4829	.4843	.4857	.4871	.4886	.4900
3.1	.4914	.4928	.4942	.4955	.4969	.4983	.4997	.5011	.5024	.5038
3.2	.5051	.5065	.5079	.5092	.5105	.5119	.5132	.5145	.5159	.5172
3.3	.5185	.5198	.5211	.5224	.5237	.5250	.5263	.5276	.5289	.5302
3.4	.5315	.5328	.5340	.5353	.5366	.5378	.5391	.5403	.5416	.5428
3.5	.5441	.5453	.5465	.5478	.5490	.5502	.5514	.5527	.5539	.5551
3.6	.5563	.5575	.5587	.5599	.5611	.5623	.5635	.5647	.5658	.5670
3.7	.5682	.5694	.5705	.5717	.5729	.5740	.5752	.5763	.5775	.5786
3.8	.5798	.5809	.5821	.5832	.5843	.5855	.5866	.5877	.5888	.5899
3.9	.5911	.5922	.5933	.5944	.5955	.5966	.5977	.5988	.5999	.6010
4.0	.6021	.6031	.6042	.6053	.6064	.6075	.6085	.6096	.6107	.6117
4.1	.6128	.6138	.6149	.6160	.6170	.6180	.6191	.6201	.6212	.6222
4.2	.6232	.6243	.6253	.6263	.6274	.6284	.6294	.6304	.6314	.6325
4.3	.6335	.6345	.6355	.6365	.6375	.6385	.6395	.6405	.6415	.6425
4.4	.6435	.6444	.6454	.6464	.6474	.6484	.6493	.6503	.6513	.6522
4.5	.6532	.6542	.6551	.6561	.6571	.6580	.6590	.6599	.6609	.6618
4.6	.6628	.6637	.6646	.6656	.6665	.6675	.6684	.6693	.6702	.6712
4.7	.6721	.6730	.6739	.6749	.6758	.6767	.6776	.6785	.6794	.6803
4.8	.6812	.6821	.6830	.6839	.6848	.6857	.6866	.6875	.6884	.6893
4.9	.6902	.6911	.6920	.6928	.6937	.6946	.6955	.6964	.6972	.6981
5.0	.6990	.6998	.7007	.7016	.7024	.7033	.7042	.7050	.7059	.7067
5.1	.7076	.7084	.7093	.7101	.7110	.7118	.7126	.7135	.7143	.7152
5.2	.7160	.7168	.7177	.7185	.7193	.7202	.7210	.7218	.7226	.7235
5.3	.7243	.7251	.7259	.7267	.7275	.7284	.7292	.7300	.7308	.7316
5.4	.7324	.7332	.7340	.7348	.7356	.7364	.7372	.7380	.7388	.7396
5.5	.7404	.7412	.7419	.7427	.7435	.7443	.7451	.7459	.7466	.7474
5.6	.7482	.7490	.7497	.7505	.7513	.7520	.7528	.7536	.7543	.7551
5.7	.7559	.7566	.7574	.7582	.7589	.7597	.7604	.7612	.7619	.7627
5.8	.7634	.7642	.7649	.7657	.7664	.7672	.7679	.7686	.7694	.7701
5.9	.7709	.7716	.7723	.7731	.7738	.7745	.7752	.7760	.7767	.7774
6.0	.7782	.7789	.7796	.7803	.7810	.7818	.7825	.7832	.7839	.7846
6.1	.7853	.7860	.7868	.7875	.7882	.7889	.7896	.7903	.7910	.7917
6.2	.7924	.7931	.7938	.7945	.7952	.7959	.7966	.7973	.7980	.7987
6.3	.7993	.8000	.8007	.8014	.8021	.8028	.8035	.8041	.8048	.8055
6.4	.8062	.8069	.8075	.8082	.8089	.8096	.8102	.8109	.8116	.8122
6.5	.8129	.8136	.8142	.8149	.8156	.8162	.8169	.8176	.8182	.8189
6.6	.8195	.8202	.8209	.8215	.8222	.8228	.8235	.8241	.8248	.8254
6.7	.8261	.8267	.8274	.8280	.8287	.8293	.8299	.8306	.8312	.8319
6.8	.8325	.8331	.8338	.8344	.8351	.8357	.8363	.8370	.8376	.8382
6.9	.8388	.8395	.8401	.8407	.8414	.8420	.8426	.8432	.8439	.8445
7.0	.8451	.8457	.8463	.8470	.8476	.8482	.8488	.8494	.8500	.8506
7.1	.8513	.8519	.8525	.8531	.8537	.8543	.8549	.8555	.8561	.8567
7.2	.8573	.8579	.8585	.8591	.8597	.8603	.8609	.8615	.8621	.8627
7.3	.8633	.8639	.8645	.8651	.8657	.8663	.8669	.8675	.8681	.8686
7.4	.8692	.8698	.8704	.8710	.8716	.8722	.8727	.8733	.8739	.8745
7.5	.8751	.8756	.8762	.8768	.8774	.8779	.8785	.8791	.8797	.8802
7.6	.8808	.8814	.8820	.8825	.8831	.8837	.8842	.8848	.8854	.8859
7.7	.8865	.8871	.8876	.8882	.8887	.8893	.8899	.8904	.8910	.8915
7.8	.8921	.8927	.8932	.8938	.8943	.8949	.8954	.8960	.8965	.8971
7.9	.8976	.8982	.8987	.8993	.8998	.9004	.9009	.9015	.9020	.9025
8.0	.9031	.9036	.9042	.9047	.9053	.9058	.9063	.9069	.9074	.9079
8.1	.9085	.9090	.9096	.9101	.9106	.9112	.9117	.9122	.9128	.9133
8.2	.9138	.9143	.9149	.9154	.9159	.9165	.9170	.9175	.9180	.9186
8.3	.9191	.9196	.9201	.9206	.9212	.9217	.9222	.9227	.9232	.9238
8.4	.9243	.9248	.9253	.9258	.9263	.9269	.9274	.9279	.9284	.9289
8.5	.9294	.9299	.9304	.9309	.9315	.9320	.9325	.9330	.9335	.9340
8.6	.9345	.9350	.9355	.9360	.9365	.9370	.9375	.9380	.9385	.9390
8.7	.9395	.9400	.9405	.9410	.9415	.9420	.9425	.9430	.9435	.9440
8.8	.9445	.9450	.9455	.9460	.9465	.9469	.9474	.9479	.9484	.9489
8.9	.9494	.9499	.9504	.9509	.9513	.9518	.9523	.9528	.9533	.9538
9.0	.9542	.9547	.9552	.9557	.9562	.9566	.9571	.9576	.9581	.9586
9.1	.9590	.9595	.9600	.9605	.9609	.9614	.9619	.9624	.9628	.9633
9.2	.9638	.9643	.9647	.9652	.9657	.9661	.9666	.9671	.9675	.9680
9.3	.9685	.9689	.9694	.9699	.9703	.9708	.9713	.9717	.9722	.9727
9.4	.9731	.9736	.9741	.9745	.9750	.9754	.9759	.9763	.9768	.9773
9.5	.9777	.9782	.9786	.9791	.9795	.9800	.9805	.9809	.9814	.9818
9.6	.9823	.9827	.9832	.9836	.9841	.9845	.9850	.9854	.9859	.9863
9.7	.9868	.9872	.9877	.9881	.9886	.9890	.9894	.9899	.9903	.9908
9.8	.9912	.9917	.9921	.9926	.9930	.9934	.9939	.9943	.9948	.9952
9.9	.9956	.9961	.9965	.9969	.9974	.9978	.9983	.9987	.9991	.9996

TABLE 3   Natural Logarithms

n	0.0	0.1	0.2	0.3	0.4	0.5	0.6	0.7	0.8	0.9
0*		7.697	8.391	8.796	9.084	9.307	9.489	9.643	9.777	9.895
1	0.000	0.095	0.182	0.262	0.336	0.405	0.470	0.531	0.588	0.642
2	0.693	0.742	0.788	0.833	0.875	0.916	0.956	0.993	1.030	1.065
3	1.099	1.131	1.163	1.194	1.224	1.253	1.281	1.308	1.335	1.361
4	1.386	1.411	1.435	1.459	1.482	1.504	1.526	1.548	1.569	1.589
5	1.609	1.629	1.649	1.668	1.686	1.705	1.723	1.740	1.758	1.775
6	1.792	1.808	1.825	1.841	1.856	1.872	1.887	1.902	1.917	1.932
7	1.946	1.960	1.974	1.988	2.001	2.015	2.028	2.041	2.054	2.067
8	2.079	2.092	2.104	2.116	2.128	2.140	2.152	2.163	2.175	2.186
9	2.197	2.208	2.219	2.230	2.241	2.251	2.262	2.272	2.282	2.293
10	2.303	2.313	2.322	2.332	2.342	2.351	2.361	2.370	2.380	2.389

* Subtract 10 if $n < 1$; for example, $\ln 0.3 \approx 8.796 - 10 = -1.204$.

TABLE 2   Natural Exponential Function Values

x	e^x	e^{-x}	x	e^x	e^{-x}
0.00	1.0000	1.0000	2.50	12.182	0.0821
0.05	1.0513	0.9512	2.60	13.464	0.0743
0.10	1.1052	0.9048	2.70	14.880	0.0672
0.15	1.1618	0.8607	2.80	16.445	0.0608
0.20	1.2214	0.8187	2.90	18.174	0.0550
0.25	1.2840	0.7788	3.00	20.086	0.0498
0.30	1.3499	0.7408	3.10	22.198	0.0450
0.35	1.4191	0.7047	3.20	24.533	0.0408
0.40	1.4918	0.6703	3.30	27.113	0.0369
0.45	1.5683	0.6376	3.40	29.964	0.0334
0.50	1.6487	0.6065	3.50	33.115	0.0302
0.55	1.7333	0.5769	3.60	36.598	0.0273
0.60	1.8221	0.5488	3.70	40.447	0.0247
0.65	1.9155	0.5220	3.80	44.701	0.0224
0.70	2.0138	0.4966	3.90	49.402	0.0202
0.75	2.1170	0.4724	4.00	54.598	0.0183
0.80	2.2255	0.4493	4.10	60.340	0.0166
0.85	2.3396	0.4274	4.20	66.686	0.0150
0.90	2.4596	0.4066	4.30	73.700	0.0136
0.95	2.5857	0.3867	4.40	81.451	0.0123
1.00	2.7183	0.3679	4.50	90.017	0.0111
1.10	3.0042	0.3329	4.60	99.484	0.0101
1.20	3.3201	0.3012	4.70	109.95	0.0091
1.30	3.6693	0.2725	4.80	121.51	0.0082
1.40	4.0552	0.2466	4.90	134.29	0.0074
1.50	4.4817	0.2231	5.00	148.41	0.0067
1.60	4.9530	0.2019	6.00	403.43	0.0025
1.70	5.4739	0.1827	7.00	1096.6	0.0009
1.80	6.0496	0.1653	8.00	2981.0	0.0003
1.90	6.6859	0.1496	9.00	8103.1	0.0001
2.00	7.3891	0.1353	10.00	22026.0	0.00005
2.10	8.1662	0.1225			
2.20	9.0250	0.1108			
2.30	9.9742	0.1003			
2.40	11.0232	0.0907			

TABLE 4 Values of the Trigonometric Functions

t	degrees	sin t	cos t	tan t	cot t	sec t	csc t	degrees	t
.0000	0°00'	.0000	1.0000	.0000	—	1.000	—	90°00'	1.5708
.0029	10	.0029	1.0000	.0029	343.8	1.000	343.8	50	1.5679
.0058	20	.0058	1.0000	.0058	171.9	1.000	171.9	40	1.5650
.0087	30	.0087	1.0000	.0087	114.6	1.000	114.6	30	1.5621
.0116	40	.0116	.9999	.0116	85.94	1.000	85.95	20	1.5592
.0145	50	.0145	.9999	.0145	68.75	1.000	68.76	10	1.5563
.0175	1°00'	.0175	.9998	.0175	57.29	1.000	57.30	89°00'	1.5533
.0204	10	.0204	.9998	.0204	49.10	1.000	49.11	50	1.5504
.0233	20	.0233	.9997	.0233	42.96	1.000	42.98	40	1.5475
.0262	30	.0262	.9997	.0262	38.19	1.000	38.20	30	1.5446
.0291	40	.0291	.9996	.0291	34.37	1.000	34.38	20	1.5417
.0320	50	.0320	.9995	.0320	31.24	1.001	31.26	10	1.5388
.0349	2°00'	.0349	.9994	.0349	28.64	1.001	28.65	88°00'	1.5359
.0378	10	.0378	.9993	.0378	26.43	1.001	26.45	50	1.5330
.0407	20	.0407	.9992	.0407	24.54	1.001	24.56	40	1.5301
.0436	30	.0436	.9990	.0437	22.90	1.001	22.93	30	1.5272
.0465	40	.0465	.9989	.0466	21.47	1.001	21.49	20	1.5243
.0495	50	.0494	.9988	.0495	20.21	1.001	20.23	10	1.5213
.0524	3°00'	.0523	.9986	.0524	19.08	1.001	19.11	87°00'	1.5184
.0553	10	.0552	.9985	.0553	18.07	1.002	18.10	50	1.5155
.0582	20	.0581	.9983	.0582	17.17	1.002	17.20	40	1.5126
.0611	30	.0610	.9981	.0612	16.35	1.002	16.38	30	1.5097
.0640	40	.0640	.9980	.0641	15.60	1.002	15.64	20	1.5068
.0669	50	.0669	.9978	.0670	14.92	1.002	14.96	10	1.5039
.0698	4°00'	.0698	.9976	.0699	14.30	1.002	14.34	86°00'	1.5010
.0727	10	.0727	.9974	.0729	13.73	1.003	13.76	50	1.4981
.0756	20	.0756	.9971	.0758	13.20	1.003	13.23	40	1.4952
.0785	30	.0785	.9969	.0787	12.71	1.003	12.75	30	1.4923
.0814	40	.0814	.9967	.0816	12.25	1.003	12.29	20	1.4893
.0844	50	.0843	.9964	.0846	11.83	1.004	11.87	10	1.4864
.0873	5°00'	.0872	.9962	.0875	11.43	1.004	11.47	85°00'	1.4835
.0902	10	.0901	.9959	.0904	11.06	1.004	11.10	50	1.4806
.0931	20	.0929	.9957	.0934	10.71	1.004	10.76	40	1.4777
.0960	30	.0958	.9954	.0963	10.39	1.005	10.43	30	1.4748
.0989	40	.0987	.9951	.0992	10.08	1.005	10.13	20	1.4719
.1018	50	.1016	.9948	.1022	9.788	1.005	9.839	10	1.4690
.1047	6°00'	.1045	.9945	.1051	9.514	1.006	9.567	84°00'	1.4661
.1076	10	.1074	.9942	.1080	9.255	1.006	9.309	50	1.4632
.1105	20	.1103	.9939	.1110	9.010	1.006	9.065	40	1.4603
.1134	30	.1132	.9936	.1139	8.777	1.006	8.834	30	1.4573
.1164	40	.1161	.9932	.1169	8.556	1.007	8.614	20	1.4544
.1193	50	.1190	.9929	.1198	8.345	1.007	8.405	10	1.4515
.1222	7°00'	.1219	.9925	.1228	8.144	1.008	8.206	83°00'	1.4486
		cos t	sin t	cot t	tan t	csc t	sec t	degrees	t

t	degrees	sin t	cos t	tan t	cot t	sec t	csc t	degrees	t
.1222	7°00'	.1219	.9925	.1228	8.144	1.008	8.206	83°00'	1.4486
.1251	10	.1248	.9922	.1257	7.953	1.008	8.016	50	1.4457
.1280	20	.1276	.9918	.1287	7.770	1.008	7.834	40	1.4428
.1309	30	.1305	.9914	.1317	7.596	1.009	7.661	30	1.4399
.1338	40	.1334	.9911	.1346	7.429	1.009	7.496	20	1.4370
.1367	50	.1363	.9907	.1376	7.269	1.009	7.337	10	1.4341
.1396	8°00'	.1392	.9903	.1405	7.115	1.010	7.185	82°00'	1.4312
.1425	10	.1421	.9899	.1435	6.968	1.010	7.040	50	1.4283
.1454	20	.1449	.9894	.1465	6.827	1.011	6.900	40	1.4254
.1484	30	.1478	.9890	.1495	6.691	1.011	6.765	30	1.4224
.1513	40	.1507	.9886	.1524	6.561	1.012	6.636	20	1.4195
.1542	50	.1536	.9881	.1554	6.435	1.012	6.512	10	1.4166
.1571	9°00'	.1564	.9877	.1584	6.314	1.012	6.392	81°00'	1.4137
.1600	10	.1593	.9872	.1614	6.197	1.013	6.277	50	1.4108
.1629	20	.1622	.9868	.1644	6.084	1.013	6.166	40	1.4079
.1658	30	.1650	.9863	.1673	5.976	1.014	6.059	30	1.4050
.1687	40	.1679	.9858	.1703	5.871	1.014	5.955	20	1.4021
.1716	50	.1708	.9853	.1733	5.769	1.015	5.855	10	1.3992
.1745	10°00'	.1736	.9848	.1763	5.671	1.015	5.759	80°00'	1.3963
.1774	10	.1765	.9843	.1793	5.576	1.016	5.665	50	1.3934
.1804	20	.1794	.9838	.1823	5.485	1.016	5.575	40	1.3904
.1833	30	.1822	.9833	.1853	5.396	1.017	5.487	30	1.3875
.1862	40	.1851	.9827	.1883	5.309	1.018	5.403	20	1.3846
.1891	50	.1880	.9822	.1914	5.226	1.018	5.320	10	1.3817
.1920	11°00'	.1908	.9816	.1944	5.145	1.019	5.241	79°00'	1.3788
.1949	10	.1937	.9811	.1974	5.066	1.019	5.164	50	1.3759
.1978	20	.1965	.9805	.2004	4.989	1.020	5.089	40	1.3730
.2007	30	.1994	.9799	.2035	4.915	1.020	5.016	30	1.3701
.2036	40	.2022	.9793	.2065	4.843	1.021	4.945	20	1.3672
.2065	50	.2051	.9787	.2095	4.773	1.022	4.876	10	1.3643
.2094	12°00'	.2079	.9781	.2126	4.705	1.022	4.810	78°00'	1.3614
.2123	10	.2108	.9775	.2156	4.638	1.023	4.745	50	1.3584
.2153	20	.2136	.9769	.2186	4.574	1.024	4.682	40	1.3555
.2182	30	.2164	.9763	.2217	4.511	1.024	4.620	30	1.3526
.2211	40	.2193	.9757	.2247	4.449	1.025	4.560	20	1.3497
.2240	50	.2221	.9750	.2278	4.390	1.026	4.502	10	1.3468
.2269	13°00'	.2250	.9744	.2309	4.331	1.026	4.445	77°00'	1.3439
.2298	10	.2278	.9737	.2339	4.275	1.027	4.390	50	1.3410
.2327	20	.2306	.9730	.2370	4.219	1.028	4.336	40	1.3381
.2356	30	.2334	.9724	.2401	4.165	1.028	4.284	30	1.3352
.2385	40	.2363	.9717	.2432	4.113	1.029	4.232	20	1.3323
.2414	50	.2391	.9710	.2462	4.061	1.030	4.182	10	1.3294
.2443	14°00'	.2419	.9703	.2493	4.011	1.031	4.134	76°00'	1.3265
		cos t	sin t	cot t	tan t	csc t	sec t	degrees	t

(Continued)

TABLE 4 (Continued)

t	degrees	sin t	cos t	tan t	cot t	sec t	csc t	degrees	t
.3665	21°00′	.3584	.9336	.3839	2.605	1.071	2.790	69°00′	1.2043
.3694	10	.3611	.9325	.3872	2.583	1.072	2.769	50	1.2014
.3723	20	.3638	.9315	.3906	2.560	1.074	2.749	40	1.1985
.3752	30	.3665	.9304	.3939	2.539	1.075	2.729	30	1.1956
.3782	40	.3692	.9293	.3973	2.517	1.076	2.709	20	1.1926
.3811	50	.3719	.9283	.4006	2.496	1.077	2.689	10	1.1897
.3840	22°00′	.3746	.9272	.4040	2.475	1.079	2.669	68°00′	1.1868
.3869	10	.3773	.9261	.4074	2.455	1.080	2.650	50	1.1839
.3898	20	.3800	.9250	.4108	2.434	1.081	2.632	40	1.1810
.3927	30	.3827	.9239	.4142	2.414	1.082	2.613	30	1.1781
.3956	40	.3854	.9228	.4176	2.394	1.084	2.595	20	1.1752
.3985	50	.3881	.9216	.4210	2.375	1.085	2.577	10	1.1723
.4014	23°00′	.3907	.9205	.4245	2.356	1.086	2.559	67°00′	1.1694
.4043	10	.3934	.9194	.4279	2.337	1.088	2.542	50	1.1665
.4072	20	.3961	.9182	.4314	2.318	1.089	2.525	40	1.1636
.4102	30	.3987	.9171	.4348	2.300	1.090	2.508	30	1.1606
.4131	40	.4014	.9159	.4383	2.282	1.092	2.491	20	1.1577
.4160	50	.4041	.9147	.4417	2.264	1.093	2.475	10	1.1548
.4189	24°00′	.4067	.9135	.4452	2.246	1.095	2.459	66°00′	1.1519
.4218	10	.4094	.9124	.4487	2.229	1.096	2.443	50	1.1490
.4247	20	.4120	.9112	.4522	2.211	1.097	2.427	40	1.1461
.4276	30	.4147	.9100	.4557	2.194	1.099	2.411	30	1.1432
.4305	40	.4173	.9088	.4592	2.177	1.100	2.396	20	1.1403
.4334	50	.4200	.9075	.4628	2.161	1.102	2.381	10	1.1374
.4363	25°00′	.4226	.9063	.4663	2.145	1.103	2.366	65°00′	1.1345
.4392	10	.4253	.9051	.4699	2.128	1.105	2.352	50	1.1316
.4422	20	.4279	.9038	.4734	2.112	1.106	2.337	40	1.1286
.4451	30	.4305	.9026	.4770	2.097	1.108	2.323	30	1.1257
.4480	40	.4331	.9013	.4806	2.081	1.109	2.309	20	1.1228
.4509	50	.4358	.9001	.4841	2.066	1.111	2.295	10	1.1199
.4538	26°00′	.4384	.8988	.4877	2.050	1.113	2.281	64°00′	1.1170
.4567	10	.4410	.8975	.4913	2.035	1.114	2.268	50	1.1141
.4596	20	.4436	.8962	.4950	2.020	1.116	2.254	40	1.1112
.4625	30	.4462	.8949	.4986	2.006	1.117	2.241	30	1.1083
.4654	40	.4488	.8936	.5022	1.991	1.119	2.228	20	1.1054
.4683	50	.4514	.8923	.5059	1.977	1.121	2.215	10	1.1025
.4712	27°00′	.4540	.8910	.5095	1.963	1.122	2.203	63°00′	1.0996
.4741	10	.4566	.8897	.5132	1.949	1.124	2.190	50	1.0966
.4771	20	.4592	.8884	.5169	1.935	1.126	2.178	40	1.0937
.4800	30	.4617	.8870	.5206	1.921	1.127	2.166	30	1.0908
.4829	40	.4643	.8857	.5243	1.907	1.129	2.154	20	1.0879
.4858	50	.4669	.8843	.5280	1.894	1.131	2.142	10	1.0850
.4887	28°00′	.4695	.8829	.5317	1.881	1.133	2.130	62°00′	1.0821
		cos t	sin t	cot t	tan t	csc t	sec t	degrees	t

t	degrees	sin t	cos t	tan t	cot t	sec t	csc t	degrees	t
.2443	14°00′	.2419	.9703	.2493	4.011	1.031	4.134	76°00′	1.3265
.2473	10	.2447	.9696	.2524	3.962	1.031	4.086	50	1.3235
.2502	20	.2476	.9689	.2555	3.914	1.032	4.039	40	1.3206
.2531	30	.2504	.9681	.2586	3.867	1.033	3.994	30	1.3177
.2560	40	.2532	.9674	.2617	3.821	1.034	3.950	20	1.3148
.2589	50	.2560	.9667	.2648	3.776	1.034	3.906	10	1.3119
.2618	15°00′	.2588	.9659	.2679	3.732	1.035	3.864	75°00′	1.3090
.2647	10	.2616	.9652	.2711	3.689	1.036	3.822	50	1.3061
.2676	20	.2644	.9644	.2742	3.647	1.037	3.782	40	1.3032
.2705	30	.2672	.9636	.2773	3.606	1.038	3.742	30	1.3003
.2734	40	.2700	.9628	.2805	3.566	1.039	3.703	20	1.2974
.2763	50	.2728	.9621	.2836	3.526	1.039	3.665	10	1.2945
.2793	16°00′	.2756	.9613	.2867	3.487	1.040	3.628	74°00′	1.2915
.2822	10	.2784	.9605	.2899	3.450	1.041	3.592	50	1.2886
.2851	20	.2812	.9596	.2931	3.412	1.042	3.556	40	1.2857
.2880	30	.2840	.9588	.2962	3.376	1.043	3.521	30	1.2828
.2909	40	.2868	.9580	.2994	3.340	1.044	3.487	20	1.2799
.2938	50	.2896	.9572	.3026	3.305	1.045	3.453	10	1.2770
.2967	17°00′	.2924	.9563	.3057	3.271	1.046	3.420	73°00′	1.2741
.2996	10	.2952	.9555	.3089	3.237	1.047	3.388	50	1.2712
.3025	20	.2979	.9546	.3121	3.204	1.048	3.356	40	1.2683
.3054	30	.3007	.9537	.3153	3.172	1.049	3.326	30	1.2654
.3083	40	.3035	.9528	.3185	3.140	1.049	3.295	20	1.2625
.3113	50	.3062	.9520	.3217	3.108	1.050	3.265	10	1.2595
.3142	18°00′	.3090	.9511	.3249	3.078	1.051	3.236	72°00′	1.2566
.3171	10	.3118	.9502	.3281	3.047	1.052	3.207	50	1.2537
.3200	20	.3145	.9492	.3314	3.018	1.053	3.179	40	1.2508
.3229	30	.3173	.9483	.3346	2.989	1.054	3.152	30	1.2479
.3258	40	.3201	.9474	.3378	2.960	1.056	3.124	20	1.2450
.3287	50	.3228	.9465	.3411	2.932	1.057	3.098	10	1.2421
.3316	19°00′	.3256	.9455	.3443	2.904	1.058	3.072	71°00′	1.2392
.3345	10	.3283	.9446	.3476	2.877	1.059	3.046	50	1.2363
.3374	20	.3311	.9436	.3508	2.850	1.060	3.021	40	1.2334
.3403	30	.3338	.9426	.3541	2.824	1.061	2.996	30	1.2305
.3432	40	.3365	.9417	.3574	2.798	1.062	2.971	20	1.2275
.3462	50	.3393	.9407	.3607	2.773	1.063	2.947	10	1.2246
.3491	20°00′	.3420	.9397	.3640	2.747	1.064	2.924	70°00′	1.2217
.3520	10	.3448	.9387	.3673	2.723	1.065	2.901	50	1.2188
.3549	20	.3475	.9377	.3706	2.699	1.066	2.878	40	1.2159
.3578	30	.3502	.9367	.3739	2.675	1.068	2.855	30	1.2130
.3607	40	.3529	.9356	.3772	2.651	1.069	2.833	20	1.2101
.3636	50	.3557	.9346	.3805	2.628	1.070	2.812	10	1.2072
.3665	21°00′	.3584	.9336	.3839	2.605	1.071	2.790	69°00′	1.2043
		cos t	sin t	cot t	tan t	csc t	sec t	degrees	t

Table (28°–35° / 55°–62°)

t	degrees	sin t	cos t	tan t	cot t	sec t	csc t	degrees	t
.4887	28°00'	.4695	.8829	.5317	1.881	1.133	2.130	62°00'	1.0821
.4916	10	.4720	.8816	.5354	1.868	1.134	2.118	50	1.0792
.4945	20	.4746	.8802	.5392	1.855	1.136	2.107	40	1.0763
.4974	30	.4772	.8788	.5430	1.842	1.138	2.096	30	1.0734
.5003	40	.4797	.8774	.5467	1.829	1.140	2.085	20	1.0705
.5032	50	.4823	.8760	.5505	1.816	1.142	2.074	10	1.0676
.5061	29°00'	.4848	.8746	.5543	1.804	1.143	2.063	61°00'	1.0647
.5091	10	.4874	.8732	.5581	1.792	1.145	2.052	50	1.0617
.5120	20	.4899	.8718	.5619	1.780	1.147	2.041	40	1.0588
.5149	30	.4924	.8704	.5658	1.767	1.149	2.031	30	1.0559
.5178	40	.4950	.8689	.5696	1.756	1.151	2.020	20	1.0530
.5207	50	.4975	.8675	.5735	1.744	1.153	2.010	10	1.0501
.5236	30°00'	.5000	.8660	.5774	1.732	1.155	2.000	60°00'	1.0472
.5265	10	.5025	.8646	.5812	1.720	1.157	1.990	50	1.0443
.5294	20	.5050	.8631	.5851	1.709	1.159	1.980	40	1.0414
.5323	30	.5075	.8616	.5890	1.698	1.161	1.970	30	1.0385
.5352	40	.5100	.8601	.5930	1.686	1.163	1.961	20	1.0356
.5381	50	.5125	.8587	.5969	1.675	1.165	1.951	10	1.0327
.5411	31°00'	.5150	.8572	.6009	1.664	1.167	1.942	59°00'	1.0297
.5440	10	.5175	.8557	.6048	1.653	1.169	1.932	50	1.0268
.5469	20	.5200	.8542	.6088	1.643	1.171	1.923	40	1.0239
.5498	30	.5225	.8526	.6128	1.632	1.173	1.914	30	1.0210
.5527	40	.5250	.8511	.6168	1.621	1.175	1.905	20	1.0181
.5556	50	.5275	.8496	.6208	1.611	1.177	1.896	10	1.0152
.5585	32°00'	.5299	.8480	.6249	1.600	1.179	1.887	58°00'	1.0123
.5614	10	.5324	.8465	.6289	1.590	1.181	1.878	50	1.0094
.5643	20	.5348	.8450	.6330	1.580	1.184	1.870	40	1.0065
.5672	30	.5373	.8434	.6371	1.570	1.186	1.861	30	1.0036
.5701	40	.5398	.8418	.6412	1.560	1.188	1.853	20	1.0007
.5730	50	.5422	.8403	.6453	1.550	1.190	1.844	10	.9977
.5760	33°00'	.5446	.8387	.6494	1.540	1.192	1.836	57°00'	.9948
.5789	10	.5471	.8371	.6536	1.530	1.195	1.828	50	.9919
.5818	20	.5495	.8355	.6577	1.520	1.197	1.820	40	.9890
.5847	30	.5519	.8339	.6619	1.511	1.199	1.812	30	.9861
.5876	40	.5544	.8323	.6661	1.501	1.202	1.804	20	.9832
.5905	50	.5568	.8307	.6703	1.492	1.204	1.796	10	.9803
.5934	34°00'	.5592	.8290	.6745	1.483	1.206	1.788	56°00'	.9774
.5963	10	.5616	.8274	.6787	1.473	1.209	1.781	50	.9745
.5992	20	.5640	.8258	.6830	1.464	1.211	1.773	40	.9716
.6021	30	.5664	.8241	.6873	1.455	1.213	1.766	30	.9687
.6050	40	.5688	.8225	.6916	1.446	1.216	1.758	20	.9657
.6080	50	.5712	.8208	.6959	1.437	1.218	1.751	10	.9628
.6109	35°00'	.5736	.8192	.7002	1.428	1.221	1.743	55°00'	.9599
		cos t	sin t	cot t	tan t	csc t	sec t	degrees	t

Table (35°–42° / 48°–55°)

t	degrees	sin t	cos t	tan t	cot t	sec t	csc t	degrees	t
.6109	35°00'	.5736	.8192	.7002	1.428	1.221	1.743	55°00'	.9599
.6138	10	.5760	.8175	.7046	1.419	1.223	1.736	50	.9570
.6167	20	.5783	.8158	.7089	1.411	1.226	1.729	40	.9541
.6196	30	.5807	.8141	.7133	1.402	1.228	1.722	30	.9512
.6225	40	.5831	.8124	.7177	1.393	1.231	1.715	20	.9483
.6254	50	.5854	.8107	.7221	1.385	1.233	1.708	10	.9454
.6283	36°00'	.5878	.8090	.7265	1.376	1.236	1.701	54°00'	.9425
.6312	10	.5901	.8073	.7310	1.368	1.239	1.695	50	.9396
.6341	20	.5925	.8056	.7355	1.360	1.241	1.688	40	.9367
.6370	30	.5948	.8039	.7400	1.351	1.244	1.681	30	.9338
.6400	40	.5972	.8021	.7445	1.343	1.247	1.675	20	.9308
.6429	50	.5995	.8004	.7490	1.335	1.249	1.668	10	.9279
.6458	37°00'	.6018	.7986	.7536	1.327	1.252	1.662	53°00'	.9250
.6487	10	.6041	.7969	.7581	1.319	1.255	1.655	50	.9221
.6516	20	.6065	.7951	.7627	1.311	1.258	1.649	40	.9192
.6545	30	.6088	.7934	.7673	1.303	1.260	1.643	30	.9163
.6574	40	.6111	.7916	.7720	1.295	1.263	1.636	20	.9134
.6603	50	.6134	.7898	.7766	1.288	1.266	1.630	10	.9105
.6632	38°00'	.6157	.7880	.7813	1.280	1.269	1.624	52°00'	.9076
.6661	10	.6180	.7862	.7860	1.272	1.272	1.618	50	.9047
.6690	20	.6202	.7844	.7907	1.265	1.275	1.612	40	.9018
.6720	30	.6225	.7826	.7954	1.257	1.278	1.606	30	.8988
.6749	40	.6248	.7808	.8002	1.250	1.281	1.601	20	.8959
.6778	50	.6271	.7790	.8050	1.242	1.284	1.595	10	.8930
.6807	39°00'	.6293	.7771	.8098	1.235	1.287	1.589	51°00'	.8901
.6836	10	.6316	.7753	.8146	1.228	1.290	1.583	50	.8872
.6865	20	.6338	.7735	.8195	1.220	1.293	1.578	40	.8843
.6894	30	.6361	.7716	.8243	1.213	1.296	1.572	30	.8814
.6923	40	.6383	.7698	.8292	1.206	1.299	1.567	20	.8785
.6952	50	.6406	.7679	.8342	1.199	1.302	1.561	10	.8756
.6981	40°00'	.6428	.7660	.8391	1.192	1.305	1.556	50°00'	.8727
.7010	10	.6450	.7642	.8441	1.185	1.309	1.550	50	.8698
.7039	20	.6472	.7623	.8491	1.178	1.312	1.545	40	.8668
.7069	30	.6494	.7604	.8541	1.171	1.315	1.540	30	.8639
.7098	40	.6517	.7585	.8591	1.164	1.318	1.535	20	.8610
.7127	50	.6539	.7566	.8642	1.157	1.322	1.529	10	.8581
.7156	41°00'	.6561	.7547	.8693	1.150	1.325	1.524	49°00'	.8552
.7185	10	.6583	.7528	.8744	1.144	1.328	1.519	50	.8523
.7214	20	.6604	.7509	.8796	1.137	1.332	1.514	40	.8494
.7243	30	.6626	.7490	.8847	1.130	1.335	1.509	30	.8465
.7272	40	.6648	.7470	.8899	1.124	1.339	1.504	20	.8436
.7301	50	.6670	.7451	.8952	1.117	1.342	1.499	10	.8407
.7330	42°00'	.6691	.7431	.9004	1.111	1.346	1.494	48°00'	.8378
		cos t	sin t	cot t	tan t	csc t	sec t	degrees	t

(Continued)

TABLE 4 (Continued)

t	degrees	sin t	cos t	tan t	cot t	sec t	csc t	degrees	t
.7330	42°00'	.6691	.7431	.9004	1.111	1.346	1.494	48°00'	.8378
.7359	10	.6713	.7412	.9057	1.104	1.349	1.490	50	.8348
.7389	20	.6734	.7392	.9110	1.098	1.353	1.485	40	.8319
.7418	30	.6756	.7373	.9163	1.091	1.356	1.480	30	.8290
.7447	40	.6777	.7353	.9217	1.085	1.360	1.476	20	.8261
.7476	50	.6799	.7333	.9271	1.079	1.364	1.471	10	.8232
.7505	43°00'	.6820	.7314	.9325	1.072	1.367	1.466	47°00'	.8203
.7534	10	.6841	.7294	.9380	1.066	1.371	1.462	50	.8174
.7563	20	.6862	.7274	.9435	1.060	1.375	1.457	40	.8145
.7592	30	.6884	.7254	.9490	1.054	1.379	1.453	30	.8116
.7621	40	.6905	.7234	.9545	1.048	1.382	1.448	20	.8087
.7650	50	.6926	.7214	.9601	1.042	1.386	1.444	10	.8058
.7679	44°00'	.6947	.7193	.9657	1.036	1.390	1.440	46°00'	.8029
.7709	10	.6967	.7173	.9713	1.030	1.394	1.435	50	.7999
.7738	20	.6988	.7153	.9770	1.024	1.398	1.431	40	.7970
.7767	30	.7009	.7133	.9827	1.018	1.402	1.427	30	.7941
.7796	40	.7030	.7112	.9884	1.012	1.406	1.423	20	.7912
.7825	50	.7050	.7092	.9942	1.006	1.410	1.418	10	.7883
.7854	45°00'	.7071	.7071	1.0000	1.0000	1.414	1.414	45°00'	.7854
		cos t	sin t	cot t	tan t	csc t	sec t	degrees	t

TABLE 5   Trigonometric Functions of Real Numbers

t	sin t	cos t	tan t	cot t	sec t	csc t
.00	.0000	1.0000	.0000	—	1.000	—
.01	.0100	1.0000	.0100	99.997	1.000	100.00
.02	.0200	.9998	.0200	49.993	1.000	50.00
.03	.0300	.9996	.0300	33.323	1.000	33.34
.04	.0400	.9992	.0400	24.987	1.001	25.01
.05	.0500	.9988	.0500	19.983	1.001	20.01
.06	.0600	.9982	.0601	16.647	1.002	16.68
.07	.0699	.9976	.0701	14.262	1.002	14.30
.08	.0799	.9968	.0802	12.473	1.003	12.51
.09	.0899	.9960	.0902	11.081	1.004	11.13
.10	.0998	.9950	.1003	9.967	1.005	10.02
.11	.1098	.9940	.1104	9.054	1.006	9.109
.12	.1197	.9928	.1206	8.293	1.007	8.353
.13	.1296	.9916	.1307	7.649	1.009	7.714
.14	.1395	.9902	.1409	7.096	1.010	7.166
.15	.1494	.9888	.1511	6.617	1.011	6.692
.16	.1593	.9872	.1614	6.197	1.013	6.277
.17	.1692	.9856	.1717	5.826	1.015	5.911
.18	.1790	.9838	.1820	5.495	1.016	5.586
.19	.1889	.9820	.1923	5.200	1.018	5.295
.20	.1987	.9801	.2027	4.933	1.020	5.033
.21	.2085	.9780	.2131	4.692	1.022	4.797
.22	.2182	.9759	.2236	4.472	1.025	4.582
.23	.2280	.9737	.2341	4.271	1.027	4.386
.24	.2377	.9713	.2447	4.086	1.030	4.207
.25	.2474	.9689	.2553	3.916	1.032	4.042
.26	.2571	.9664	.2660	3.759	1.035	3.890
.27	.2667	.9638	.2768	3.613	1.038	3.749
.28	.2764	.9611	.2876	3.478	1.041	3.619
.29	.2860	.9582	.2984	3.351	1.044	3.497
.30	.2955	.9553	.3093	3.233	1.047	3.384
.31	.3051	.9523	.3203	3.122	1.050	3.278
.32	.3146	.9492	.3314	3.018	1.053	3.179
.33	.3240	.9460	.3425	2.920	1.057	3.086
.34	.3335	.9428	.3537	2.827	1.061	2.999
.35	.3429	.9394	.3650	2.740	1.065	2.916
.36	.3523	.9359	.3764	2.657	1.068	2.839
.37	.3616	.9323	.3879	2.578	4.073	2.765
.38	.3709	.9287	.3994	2.504	1.077	2.696
.39	.3802	.9249	.4111	2.433	1.081	2.630
.40	.3894	.9211	.4228	2.365	1.086	2.568
.41	.3986	.9171	.4346	2.301	1.090	2.509
.42	.4078	.9131	.4466	2.239	1.095	2.452
.43	.4169	.9090	.4586	2.180	1.100	2.399
.44	.4259	.9048	.4708	2.124	1.105	2.348
.45	.4350	.9004	.4831	2.070	1.111	2.299
.46	.4439	.8961	.4954	2.018	1.116	2.253
.47	.4529	.8916	.5080	1.969	1.122	2.208
.48	.4618	.8870	.5206	1.921	1.127	2.166
.49	.4706	.8823	.5334	1.875	1.133	2.125
.50	.4794	.8776	.5463	1.830	1.139	2.086
.51	.4882	.8727	.5594	1.788	1.146	2.048
.52	.4969	.8678	.5726	1.747	1.152	2.013
.53	.5055	.8628	.5859	1.707	1.159	1.978
.54	.5141	.8577	.5994	1.668	1.166	1.945
.55	.5227	.8525	.6131	1.631	1.173	1.913
.56	.5312	.8473	.6269	1.595	1.180	1.883
.57	.5396	.8419	.6410	1.560	1.188	1.853
.58	.5480	.8365	.6552	1.526	1.196	1.825
.59	.5564	.8309	.6696	1.494	1.203	1.797
.60	.5646	.8253	.6841	1.462	1.212	1.771
.61	.5729	.8196	.6989	1.431	1.220	1.746
.62	.5810	.8139	.7139	1.401	1.229	1.721
.63	.5891	.8080	.7291	1.372	1.238	1.697
.64	.5972	.8021	.7445	1.343	1.247	1.674
.65	.6052	.7961	.7602	1.315	1.256	1.652
.66	.6131	.7900	.7761	1.288	1.266	1.631
.67	.6210	.7838	.7923	1.262	1.276	1.610
.68	.6288	.7776	.8087	1.237	1.286	1.590
.69	.6365	.7712	.8253	1.212	1.297	1.571
.70	.6442	.7648	.8423	1.187	1.307	1.552
.71	.6518	.7584	.8595	1.163	1.319	1.534
.72	.6594	.7518	.8771	1.140	1.330	1.517
.73	.6669	.7452	.8949	1.117	1.342	1.500
.74	.6743	.7385	.9131	1.095	1.354	1.483
.75	.6816	.7317	.9316	1.073	1.367	1.467
.76	.6889	.7248	.9505	1.052	1.380	1.452
.77	.6961	.7179	.9697	1.031	1.393	1.437
.78	.7033	.7109	.9893	1.011	1.407	1.422
.79	.7104	.7038	1.009	.9908	1.421	1.408

(Continued)

TABLE 5 (Continued)

t	sin t	cos t	tan t	cot t	sec t	csc t
1.20	.9320	.3624	2.572	.3888	2.760	1.073
1.21	.9356	.3530	2.650	.3773	2.833	1.069
1.22	.9391	.3436	2.733	.3659	2.910	1.065
1.23	.9425	.3342	2.820	.3546	2.992	1.061
1.24	.9458	.3248	2.912	.3434	3.079	1.057
1.25	.9490	.3153	3.010	.3323	3.171	1.054
1.26	.9521	.3058	3.113	.3212	3.270	1.050
1.27	.9551	.2963	3.224	.3102	3.375	1.047
1.28	.9580	.2867	3.341	.2993	3.488	1.044
1.29	.9608	.2771	3.467	.2884	3.609	1.041
1.30	.9636	.2675	3.602	.2776	3.738	1.038
1.31	.9662	.2579	3.747	.2669	3.878	1.035
1.32	.9687	.2482	3.903	.2562	4.029	1.032
1.33	.9711	.2385	4.072	.2456	4.193	1.030
1.34	.9735	.2288	4.256	.2350	4.372	1.027
1.35	.9757	.2190	4.455	.2245	4.566	1.025
1.36	.9779	.2092	4.673	.2140	4.779	1.023
1.37	.9799	.1994	4.913	.2035	5.014	1.021
1.38	.9819	.1896	5.177	.1931	5.273	1.018
1.39	.9837	.1798	5.471	.1828	5.561	1.017
1.40	.9854	.1700	5.798	.1725	5.883	1.015
1.41	.9871	.1601	6.165	.1622	6.246	1.013
1.42	.9887	.1502	6.581	.1519	6.657	1.011
1.43	.9901	.1403	7.055	.1417	7.126	1.010
1.44	.9915	.1304	7.602	.1315	7.667	1.009
1.45	.9927	.1205	8.238	.1214	8.299	1.007
1.46	.9939	.1106	8.989	.1113	9.044	1.006
1.47	.9949	.1006	9.887	.1011	9.938	1.005
1.48	.9959	.0907	10.983	.0910	11.029	1.004
1.49	.9967	.0807	12.350	.0810	12.390	1.003
1.50	.9975	.0707	14.101	.0709	14.137	1.003
1.51	.9982	.0608	16.428	.0609	16.458	1.002
1.52	.9987	.0508	19.670	.0508	19.695	1.001
1.53	.9992	.0408	24.498	.0408	24.519	1.001
1.54	.9995	.0308	32.461	.0308	32.476	1.000
1.55	.9998	.0208	48.078	.0208	48.089	1.000
1.56	.9999	.0108	92.620	.0108	92.626	1.000
1.57	1.0000	.0008	1255.8	.0008	1255.8	1.000

t	sin t	cos t	tan t	cot t	sec t	csc t
.80	.7174	.6967	1.030	.9712	1.435	1.394
.81	.7243	.6895	1.050	.9520	1.450	1.381
.82	.7311	.6822	1.072	.9331	1.466	1.368
.83	.7379	.6749	1.093	.9146	1.482	1.355
.84	.7446	.6675	1.116	.8964	1.498	1.343
.85	.7513	.6600	1.138	.8785	1.515	1.331
.86	.7578	.6524	1.162	.8609	1.533	1.320
.87	.7643	.6448	1.185	.8437	1.551	1.308
.88	.7707	.6372	1.210	.8267	1.569	1.297
.89	.7771	.6294	1.235	.8100	1.589	1.287
.90	.7833	.6216	1.260	.7936	1.609	1.277
.91	.7895	.6137	1.286	.7774	1.629	1.267
.92	.7956	.6058	1.313	.7615	1.651	1.257
.93	.8016	.5978	1.341	.7458	1.673	1.247
.94	.8076	.5898	1.369	.7303	1.696	1.238
.95	.8134	.5817	1.398	.7151	1.719	1.229
.96	.8192	.5735	1.428	.7001	1.744	1.221
.97	.8249	.5653	1.459	.6853	1.769	1.212
.98	.8305	.5570	1.491	.6707	1.795	1.204
.99	.8360	.5487	1.524	.6563	1.823	1.196
1.00	.8415	.5403	1.557	.6421	1.851	1.188
1.01	.8468	.5319	1.592	.6281	1.880	1.181
1.02	.8521	.5234	1.628	.6142	1.911	1.174
1.03	.8573	.5148	1.665	.6005	1.942	1.166
1.04	.8624	.5062	1.704	.5870	1.975	1.160
1.05	.8674	.4976	1.743	.5736	2.010	1.153
1.06	.8724	.4889	1.784	.5604	2.046	1.146
1.07	.8772	.4801	1.827	.5473	2.083	1.140
1.08	.8820	.4713	1.871	.5344	2.122	1.134
1.09	.8866	.4625	1.917	.5216	2.162	1.128
1.10	.8912	.4536	1.965	.5090	2.205	1.122
1.11	.8957	.4447	2.014	.4964	2.249	1.116
1.12	.9001	.4357	2.066	.4840	2.295	1.111
1.13	.9044	.4267	2.120	.4718	2.344	1.106
1.14	.9086	.4176	2.176	.4596	2.395	1.101
1.15	.9128	.4085	2.234	.4475	2.448	1.096
1.16	.9168	.3993	2.296	.4356	2.504	1.091
1.17	.9208	.3902	2.360	.4237	2.563	1.086
1.18	.9246	.3809	2.427	.4120	2.625	1.082
1.19	.9284	.3717	2.498	.4003	2.691	1.077

Answers to Selected Exercises

CHAPTER 1

Exercises 1.1

1. $A \cup B = \{-3, -1, 0, \frac{1}{2}, 2, 7\}$; $A \cap B = \{-3, 7\}$

3. $\varnothing, \{-1\}, \{0\}, \{1\}, \{-1, 0\}, \{-1, 1\}, \{0, 1\}, \{-1, 0, 1\}$

5. $A' = \{1, 2, 3, \dots, 49\}$

7. T **9.** F **11.** F **13.** F **15.** T

17. (a) 4 (b) 2 (c) 1 (d) 2
  (e) 1024 (f) 2^{41} (g) 16 (h) 2^{26}

19.

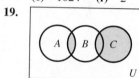

$(A \cap B) \cup C$

21.

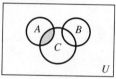

$A \cap (B \cup C)$

23.

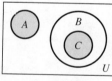

$A \cup (B \cap C)$

25.

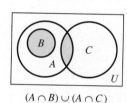

$(A \cap B) \cup (A \cap C)$

Exercises 1.2

1. T **3.** F **5.** T **7.** T **9.** F

11. $\frac{209}{100}$ **13.** $\frac{9}{11}$ **15.** $\frac{67}{9}$ **17.** $\frac{8506}{9990}$

19. Distributive property

21. Identity property of addition

23. Inverse property of multiplication

25. Associative property of addition

27. Inverse property of addition

29. Distributive property

31. Associative property of addition

33. Identity property of addition

35. Inverse property of multiplication

37. 2 **39.** $r - 1$ **41.** $-y$

43. $\{x \mid -2 \le x < 1\}$

45. $\{x \mid x \ge -4\}$

47. $\{x \mid x \le -\sqrt{2}\}$

49. $\{x \mid x < 0\} \cup \{x \mid 1 < x < \frac{13}{8}\}$

51. 7 **53.** $\frac{17}{9}$ **55.** $5\pi/4$

Exercises 1.3

1.

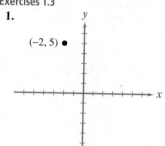

$(-2, 5)$

3.

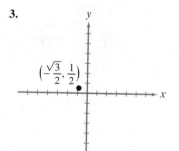

$\left(-\frac{\sqrt{3}}{2}, \frac{1}{2}\right)$

5.

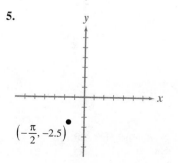

$\left(-\frac{\pi}{2}, -2.5\right)$

7. $2\sqrt{5}$ 9. $\sqrt{5}$ 11. $\sqrt{21}/4$ 13. $2\sqrt{2}a$
15. $(0, -1)$ 17. $(2.6, -1)$ 19. $(\sqrt{13}/2, 0)$
21. $(a, 0)$ 23. $(\frac{7}{2}, \frac{27}{2})$ 25. $5\sqrt{2} + 2\sqrt{17}$
27. $(-3, 0)$ and $(3, 0)$ 29. $\pm\sqrt{3}$ 31. Inside the circle
33. On the circle 35. $A = 16\sqrt{3}$ 37. Equilateral
39. Neither 41. Isosceles 45. Not collinear
47. Collinear 51. $A = 30$

Exercises 1.4

1.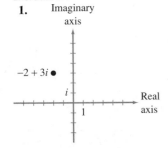
3.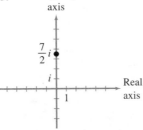
5.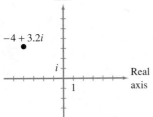
7.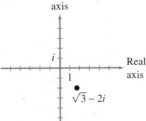

9. $9i$ 11. $-4\sqrt{2}i$ 13. -12 15. 6 17. 3
19. $7 - 4i$ 21. $\frac{33}{4} - \frac{7}{2}i$ 23. $-16 - 38i$
25. $-3 - 4i$ 27. $-110 + 74i$ 29. $-10i$
31. $\frac{1}{17} + \frac{4}{17}i$ 33. $-6 + 7i$
35. $\dfrac{1 - \pi^2}{1 + \pi^2} - \dfrac{2\pi}{1 + \pi^2}\,i$
37. $2 - 4i$ 39. $\frac{1}{2}i$ 41. $\frac{1}{10} + \frac{3}{10}i$ 43. $\sqrt{2}/9 - \frac{4}{9}i$
45. 1 47. $-i$ 49. -1 51. i 53. 1
55. 1 57. $\sqrt{58}$ 59. $\sqrt{13}/4$ 61. 16 63. 25
65. 1 67. $\frac{1}{3}$ 69. $6\sqrt{5}$
71. $x = -2, y = -5$

Exercises 1.5
1. $\{-5\}$ 3. $\{-4\}$ 5. $\{\frac{96}{5}\}$ 7. $\{0\}$
9. $L = \dfrac{P - 2W}{2}$ 11. $W = \dfrac{A}{L}$ 13. $r = \dfrac{C}{2\pi}$
15. $F = \frac{9}{5}C + 32$ 17. $h = \dfrac{V}{\pi r^2}$
19. $y = -\dfrac{A}{B}x - \dfrac{C}{B}$
21. 45 student tickets and 135 adult tickets 23. $9
25. $520 27. 14 days 29. $\{-\frac{3}{2}, \frac{1}{3}\}$ 31. $\{-21, 2\}$
33. $\{-\frac{1}{2}, \frac{4}{3}\}$ 35. $\left\{\dfrac{2 \pm \sqrt{10}}{2}\right\}$ 37. $\left\{2 \pm \dfrac{\sqrt{3}i}{3}\right\}$

39. $\{\frac{3}{2}\}$ 41. $\left\{\dfrac{2 \pm \sqrt{2}}{2}\right\}$ 43. $\{1 \pm i\}$ 45. $\{-\frac{4}{3}\}$
47. 20; 2 real roots 49. 0; 1 real (double) root
51. -16; 2 complex conjugate roots
53. $\{-\frac{1}{3}, 4\}$ 55. $\left\{\dfrac{7 \pm \sqrt{57}}{4}\right\}$ 57. $\{0, \frac{4}{5}\}$
59. $\{\frac{5}{3}\}$
61. Since $r \geq 0$, $r = \dfrac{\sqrt{A\pi}}{\pi}$. 63. $b = \pm\sqrt{D + 4ac}$
65. $h = \dfrac{3 \pm \sqrt{16A + 89}}{8}$ 69. 15 m, 36 m
71. 12 ft wide by 30 ft long; Perimeter = 84 ft 73. 5 and 7
75. 5 seconds after it is thrown 77. 7 m by 13 m 79. 5

Exercises 1.6
1. $(-\infty, -4]$ 3. $(\frac{16}{5}, \infty)$
5. $(-4, 3]$ 7. $(-\infty, -35)$
9. $(3, \infty)$ 11. $[-\frac{5}{3}, 2)$
13. $[-\frac{21}{4}, \infty)$ 15. $(-\infty, 4]$
17. $[-2, -\frac{1}{2}]$ 19. $[\frac{19}{2}, \infty)$
21. $[-4, 3]$ 23. $(-\infty, -\frac{1}{2}] \cup [\frac{3}{2}, \infty)$
25. $[2 - \sqrt{6}, 2 + \sqrt{6}]$ $2 - \sqrt{6} \approx -0.4$ $2 + \sqrt{6} \approx 4.4$
27. $\left(\dfrac{-1 - \sqrt{2}}{3}, \dfrac{-1 + \sqrt{2}}{3}\right)$ $\dfrac{-1 - \sqrt{2}}{3} \approx -0.8$ $\dfrac{-1 + \sqrt{2}}{3} \approx 0.1$
29. $\varnothing$ 31. $(-\infty, -4) \cup (3, \infty)$
33. $\varnothing$ 35. $(-\infty, -2) \cup (8, \infty)$
37. $(-\infty, -\frac{5}{3}] \cup [4, \infty)$ 39. $\mathbb{R}$

41. The temperature F, in degrees Fahrenheit, has the average annual range $-4 \le F \le 86$.

43. The mental age M satisfies. $9 \le M \le 15$.

45. At least 51 pots.   **47.** (a) $(0, 5.5)$ (b) $[2.5, 3]$

Exercises 1.7

1. $\{28\}$ **3.** $\{-\frac{6}{5}\}$ **5.** $\varnothing$ **7.** $\{4\}$ **9.** $\varnothing$

11. $\{2, 3\}$ **13.** $\{0\}$ **15.** $\{1\}$ **17.** $\{4\}$

19. $\{-\frac{27}{8}, 1\}$ **21.** $\{\pm 2, \pm 3i\}$

23. $\left\{ -2, -1, 1 \pm \sqrt{3}i, \frac{1}{2} \pm \frac{\sqrt{3}}{2}i \right\}$ **25.** $\{-\frac{4}{5}\}$

27. $\{\pm\sqrt{3}, \pm\sqrt{3}i\}$ **29.** $\{-27, 64\}$ **31.** $\varnothing$

33. $\{-\frac{5}{2}, 4\}$ **35.** $\{\frac{1}{29}\}$ **37.** $\{-5, \frac{3}{4}\}$ **39.** $\{\frac{1}{2}\}$

41. $\{-2, 10\}$

43. $(-\infty, -\frac{1}{5}] \cup [1, \infty)$ **45.** $(\frac{5}{7}, 1)$

47. $\{-3, \frac{1}{4}\}$ **49.** $(-\infty, \frac{1}{3}] \cup [2, \infty)$

51. $(-2, \frac{10}{3})$ **53.** $(-\infty, -1] \cup [1, \infty)$

55. $(-4, 16)$ **57.** $(-\infty, -28] \cup [52, \infty)$

59. $[1, 6)$ **61.** $(-\infty, -2] \cup [1, 2)$

63. $(-4, -1] \cup (2, 3]$

65. $(-\infty, -2) \cup (-1, 1) \cup (2, \infty)$

67. $[-2, 1) \cup [2, \infty)$

Chapter 1 Review Exercises

1. (a) K (b) $\varnothing$ (c) H (d) H (e) H (f) H

3. $\varnothing, \{-2\}, \{0\}, \{2\}, \{-2, 0\}, \{-2, 2\}, \{0, 2\}, \{-2, 0, 2\}$

5.
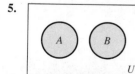
$A \cup B$

7.
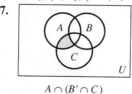
$A \cap (B' \cap C)$

9.

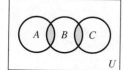
$(A \cap B) \cup (B \cap C)$

11. $\frac{2135}{999}$ **13.** Associative property of addition

15. Identity property of addition

17. Identity property of multiplication   **19.** $\pi - 3.14$

21. (a) $\{x \mid -2 < x \le 4\}$

(b) $\{x \mid x \ge -1\}$

(c) $\{x \mid -3 < x < -1 \text{ or } x \ge 1\}$

(d) $\{x \mid x \le -\pi \text{ or } 0 < x \le \frac{5}{4}\}$

23.

25. $d(A, B) = \frac{\sqrt{265}}{4}, M = (\frac{7}{8}, 1)$

27. $(-1, -2)$

29.
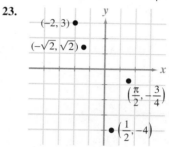

31. $-3 + 4i$ **33.** $7 + i$ **35.** $-\frac{1}{2} + \frac{5}{2}i$

37. $\sqrt{13}$ **39.** $-i$ **41.** 1 **43.** $2 - 6i$

45. $-\frac{2}{13} - \frac{3}{13}i$ **47.** $\{-\frac{1}{6}\}$ **49.** $\{-\frac{11}{4}\}$ **51.** $\varnothing$

53. $B = -\dfrac{Ax + C}{y} \ (y \ne 0)$ **55.** $b_1 = \dfrac{2A - b_2 h}{h} \ (h \ne 0)$

57. $x = \dfrac{2 \pm \sqrt{2 + 2y}}{2}$

59. $[\frac{16}{9}, \infty)$

61. $(1, 3]$

63. -11; two complex conjugate roots

65. 25; two distinct real roots   **67.** $\{-\frac{4}{3}, \frac{3}{2}\}$ **69.** $\{1\}$

71. $\left\{ \frac{1}{3} \pm \frac{\sqrt{17}}{3}i \right\}$ **73.** $\left\{ \dfrac{3 \pm \sqrt{33}}{6} \right\}$

75. $\{\frac{1}{2} \pm \frac{1}{2}i\}$ **77.** $\left\{ \dfrac{-1 \pm \sqrt{21}}{4} \right\}$ **79.** $\{2 \pm \sqrt{5}\}$

81. $(-2, 3)$

83. $(-\infty, -\frac{1}{3}) \cup (2, \infty)$

85. $(-2 - \sqrt{5}, -2 + \sqrt{5})$ $-2 - \sqrt{5} \approx -4.2$ $-2 + \sqrt{5} \approx 0.2$

87. $\mathbb{R}$

89. $\{-1, 2\}$

91. $(-\infty, 1] \cup [2, \infty)$

93. $[0, \frac{6}{5}]$

95. $(-\infty, -9) \cup (11, \infty)$

97. $\varnothing$ **99.** $\{9\}$ **101.** $\{0\}$ **103.** $\{\pm i, \pm 2i\}$

105. $\{4, 16\}$ **107.** $\{2, 4\}$

109. $(-\infty, 1] \cup (2, \infty)$

111. $(-\infty, -3) \cup (-2, 2) \cup (3, \infty)$

113. $\varnothing$ **115.** 600 adult tickets, 800 student tickets

117. \$9.50   **119.** 25 ft $\times$ 60 ft

CHAPTER 2

Exercises 2.1

1. dom $(A) = \{0, 1, 2, 3\}$; ran $(A) = \{0, 1, 2\}$

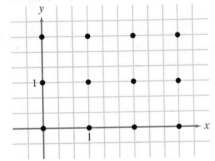

3. dom $(C) = \{-3, -2, 0, 1, 4\}$;
ran $(C) = \{-\frac{3}{2}, -1, 0, \frac{1}{2}, 2\}$

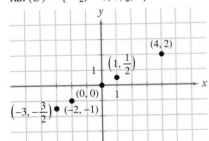

5. Only C **7.** $(x - 3)^2 + (y + 4)^2 = 100$

9. $(x - 2)^2 + (y + 1)^2 = 45$

11. $(x - 1)^2 + (y + 2)^2 = 1$

13. $(x - 2)^2 + (y - 2)^2 = 10$

15. Center: $(0, 0)$; Radius: 6

17. Center: $(1, 1)$; Radius: 1

19. Center: $(\frac{1}{3}, 0)$; Radius: $\sqrt{226}/3$

21. Center: $(\frac{1}{2}, 1)$; Radius: 2

23. Domain: $(-\infty, 0]$
Range: $\mathbb{R}$
Not a function

25. Domain: $[-3, 3]$
Range: $[-3, 3]$
Not a function

27. Domain: $(-\infty, 0) \cup (0, \infty)$
Range: $\{-1, 1\}$
A function

29. $f(0) = -1; f(-1) = 3; f(2) = 9$

31. $f(0) = 0; f(2) = -\frac{1}{2}; f(3)$ is undefined

33. (a) $\sqrt{6a - 5}$ (b) $\sqrt{6a + 6h - 5}$

(c) $\dfrac{\sqrt{6a + 6h - 5} - \sqrt{6a - 5}}{h}$ (d) $\sqrt{\dfrac{6 + 5a}{a}}$

(e) $\dfrac{1}{\sqrt{6a - 5}}$ (f) $\sqrt{6a^2 - 5}$

35. (a) $\dfrac{a + 1}{a^3 + 4a}$ (b) $\dfrac{a + h + 1}{(a + h)^3 + 4(a + h)}$

(c) $\dfrac{-(2a^3 + 3a^2 + 4 + ah^2 + 3a^2h + 3ah + h^2)}{[(a + h)^3 + 4(a + h)](a^3 + 4a)}$

(d) $\dfrac{a^2 + a^3}{1 + 4a^2}$ (e) $\dfrac{a^3 + 4a}{a + 1}$ (f) $\dfrac{a^2 + 1}{a^6 + 4a^2}$

37. $4x + 2h - 1$ **39.** $\dfrac{1}{\sqrt{x + h} + \sqrt{x}}$ **41.** $\dfrac{-10}{x^2 + hx}$

43. $\mathbb{R}$ **45.** $[-\frac{3}{2}, \infty)$ **47.** $(-\infty, -\frac{1}{2}) \cup (-\frac{1}{2}, \infty)$

49. $\mathbb{R}$ **51.** $A(x) = \dfrac{100x - 2x^2}{2}$

53. $A(x) = 2x\sqrt{a^2 - x^2}$ **55.** A, E **57.** A, B, E

63.

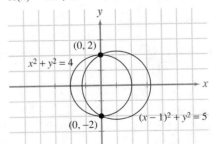

Exercises 2.2

1. Slope: 3
x-intercept: $(-2, 0)$
y-intercept: $(0, 6)$

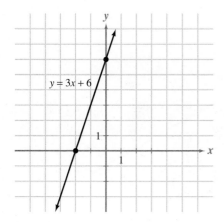

3. Slope: -6
x-intercept: $(\frac{5}{6}, 0)$
y-intercept: $(0, 5)$

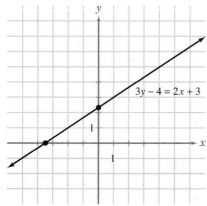

5. Slope: $\frac{2}{3}$
x-intercept: $(-\frac{7}{2}, 0)$
y-intercept: $(0, \frac{7}{3})$

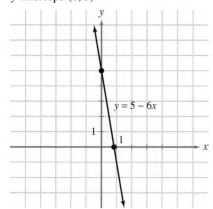

7. Slope: undefined
x-intercept: $(8, 0)$
y-intercept: none

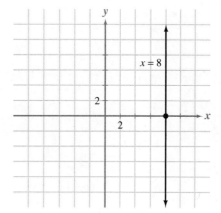

9. $f(x) = -x + 1$

11. $f(x) = -\frac{2}{3}x - 3$ **13.** $f(x) = -\frac{4}{7}x + \frac{41}{7}$

15. $f(x) = x + \sqrt{2} + 1$

17. $y = -2x + 3$; x-intercept: $(\frac{3}{2}, 0)$

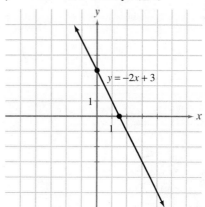

19. $y = \dfrac{\sqrt{3}}{3}x$; x-intercept: $(0, 0)$

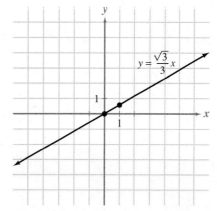

21. $y = \frac{3}{2}x + \frac{5}{2}$
Slope: $\frac{3}{2}$
y-intercept: $(0, \frac{5}{2})$

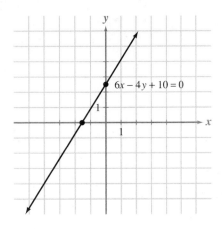

23. $y = -2$
Slope: 0
y-intercept: $(0, -2)$

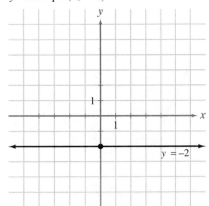

25. $y = -\frac{1}{3}x + \frac{5}{3}$
Slope: $-\frac{1}{3}$
y-intercept: $(0, \frac{5}{3})$

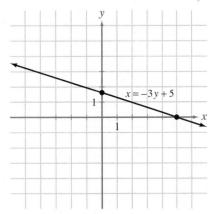

27. 3
31. (a) $R(C) = \frac{1}{3}(4C + 200)$ (b) $466.67 (c) $700
33. (a) $V(t) = -5000t + 100,000$ (b) 11.5 years
35. **6 pounds**

37.

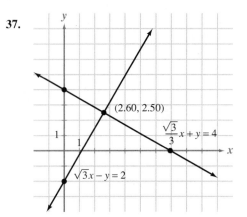

Exercises 2.3
1. (a) Minimum value of f is -3
(b) Vertex: $(-1, -3)$
Axis of symmetry: $x = -1$
x-intercepts: $\left(\dfrac{-2 - \sqrt{6}}{2}, 0\right)$, $\left(\dfrac{-2 + \sqrt{6}}{2}, 0\right)$
y-intercept: $(0, -1)$
(c) Range: $[-3, \infty)$

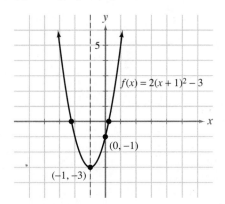

3. (a) Minimum value of f is $\frac{159}{16}$

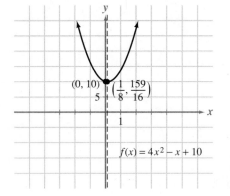

(b) Vertex: $(\frac{1}{8}, \frac{159}{16})$
Axis of symmetry: $x = \frac{1}{8}$
x-intercepts: none
y-intercepts: $(0, 10)$
(c) Range: $[\frac{159}{16}, \infty)$

5. (a) Maximum value of f is 11
(b) Vertex: $(-4, 11)$
Axis of symmetry: $x = -4$
x-intercepts: $(-4 - \sqrt{11}, 0), (-4 + \sqrt{11}, 0)$
y-intercept: $(0, -5)$
(c) Range: $(-\infty, 11]$

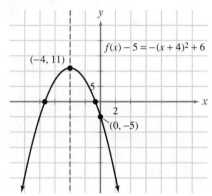

7. (a) Minimum value of f is $-\frac{5}{12}$
(b) Vertex: $(-\frac{1}{2}, -\frac{5}{12})$
Axis of symmetry: $x = -\frac{1}{2}$
x-intercepts: $\left(\dfrac{-1 - \sqrt{5}}{2}, 0\right), \left(\dfrac{-1 + \sqrt{5}}{2}, 0\right)$
y-intercept: $(0, -\frac{1}{3})$
(c) Range: $[-\frac{5}{12}, \infty)$

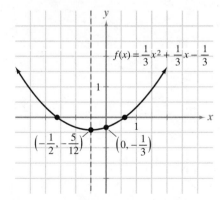

9. (a) Maximum value of f is 0
(b) Vertex: $(\frac{2}{3}, 0)$
Axis of symmetry: $x = \frac{2}{3}$
x-intercept: $(\frac{2}{3}, 0)$
y-intercept: $(0, -4)$
(c) Range: $(-\infty, 0]$

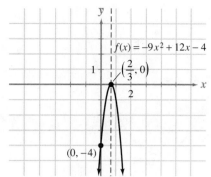

11. (a) Minimum value of f is -1
(b) Vertex: $(8, -1)$
Axis of symmetry: $x = 8$
x-intercepts: $(8 - \sqrt{3}, 0), (8 + \sqrt{3}, 0)$
y-intercept: $(0, \frac{61}{3})$
(c) Range: $[-1, \infty)$

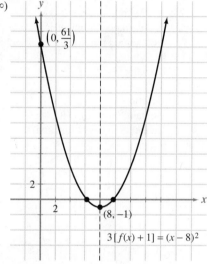

13. $(0, 4)$ 15. $(-\infty, -3] \cup [1, \infty)$
17. $(-\infty, -5) \cup (-5, \infty)$ 19. $(-\infty, -2] \cup [5, \infty]$
21. $(-1, 8), (3, 8)$ 23. $f(x) = -\frac{1}{4}x^2 + 2x$
25. $f(x) = -\frac{4}{9}x^2 - \frac{8}{9}x + \frac{32}{9}$ 26. 4
27. \$1450   29. (a) 22.06 ft   (b) 2.11 sec
31. (a) $C(p) = 800,000 - 2000p$
(b) $R(p) = 12,000p - 40p^2$
(d) \$72 and \$278.00

33.

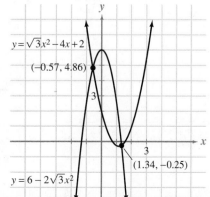

35. (a)

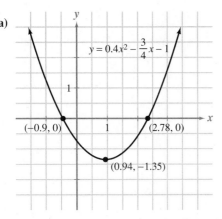

(b) Vertex: $(0.94, -1.35)$ **(c)** $[-0.90, 2.78]$

Exercises 2.4

1. Odd; symmetric with respect to the origin
3. Neither odd nor even
5. Even; graph is symmetric with respect to the y-axis
7. Neither odd nor even   9. Neither odd nor even
11. Function   13. Not a function
    Domain: $\mathbb{R}$ Domain: $\mathbb{R}$
    Range: $(-\infty, -1]$ Range: $(-2, \infty)$
15. Function
    Domain: $(-\infty, 0) \cup (0, \infty)$
    Range: $\mathbb{R}$; an even function

17.

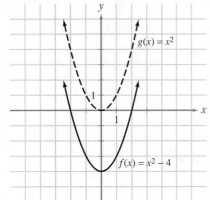

19.

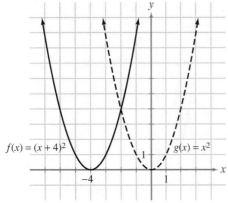

21.

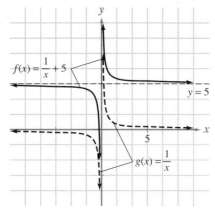

23.

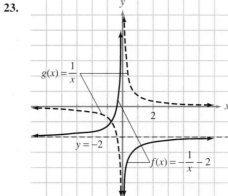

25. Domain: $[-10, \infty)$
    Range: $[0, \infty)$

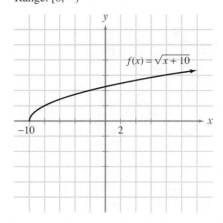

27. Domain: $\mathbb{R}$
Range: $\mathbb{R}$

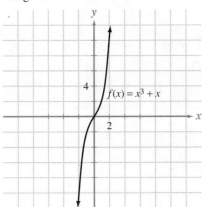

$f(x) = x^3 + x$

29. Domain: $\mathbb{R}$
Range: $x^2 \mid x$ is an integer

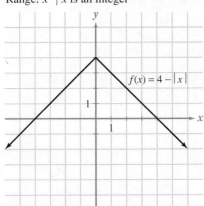

$f(x) = 4 - |x|$

31. Domain: $\mathbb{R}$
Range: $[-3, \infty)$

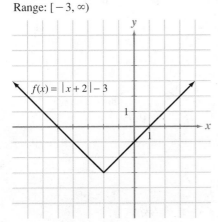

$f(x) = |x + 2| - 3$

33. Domain: $\mathbb{R}$
Range: $\mathbb{R}$

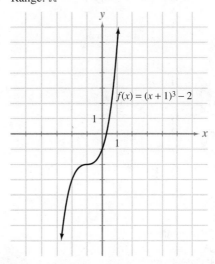

$f(x) = (x + 1)^3 - 2$

35. Domain: $(-\infty, 0) \cup (0, \infty)$
Range: $(-2, \infty)$

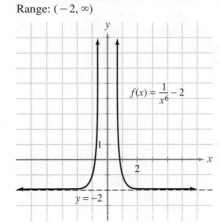

$f(x) = \dfrac{1}{x^6} - 2$

$y = -2$

37. Domain: $\mathbb{R}$
Range: $\{-2, 1\}$

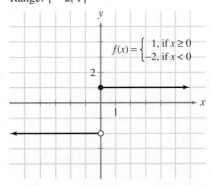

$f(x) = \begin{cases} 1, & \text{if } x \geq 0 \\ -2, & \text{if } x < 0 \end{cases}$

39. Domain: $\mathbb{R}$
Range: $\{x \mid x$ is a nonnegative integer$)$

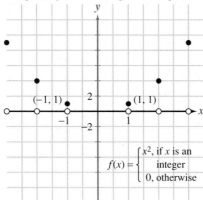

$$f(x) = \begin{cases} x^2, \text{ if } x \text{ is an integer} \\ 0, \text{ otherwise} \end{cases}$$

41. Domain: $\mathbb{R}$
Range: $\{x \mid x$ is an integer$\}$

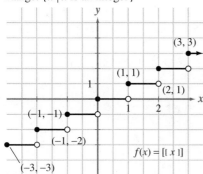

$f(x) = [\![x]\!]$

43.

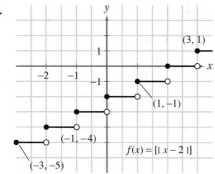

$f(x) = [\![x - 2]\!]$

45.

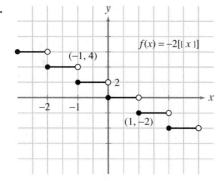

$f(x) = -2[\![x]\!]$

51.

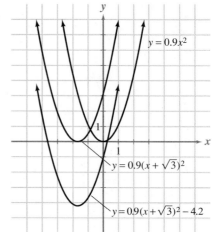

$y = 0.9x^2$

$y = 0.9(x + \sqrt{3})^2$

$y = 0.9(x + \sqrt{3})^2 - 4.2$

Exercises 2.5

1. (a) $(f + g)(x) = 4$; dom $(f + g) = \mathbb{R}$
(b) $(f - g)(x) = 2x$; dom $(f - g) = \mathbb{R}$
(c) $(fg)(x) = 4 - x^2$; dom $(fg) = \mathbb{R}$
(d) $\left(\dfrac{f}{g}\right)(x) = \dfrac{2 + x}{2 - x}$; dom $\left(\dfrac{f}{g}\right) = (-\infty, 2) \cup (2, \infty)$

3. (a) $(f + g)(x) = 21 - x$; dom $(f + g) = \mathbb{R}$
(b) $(f - g)(x) = 2x^2 - x + 9$; dom $(f - g) = \mathbb{R}$
(c) $(fg)(x) = -x^4 + x^3 - 9x^2 - 6x + 90$; dom $(fg) = \mathbb{R}$
(d) $\left(\dfrac{f}{g}\right)(x) = \dfrac{x^2 - x + 15}{6 - x^2}$;
dom $\left(\dfrac{f}{g}\right) = (-\infty, -\sqrt{6}) \cup (-\sqrt{6}, \sqrt{6}) \cup (\sqrt{6}, \infty)$

5. (a) $(f + g)(x) = \dfrac{x^2 + 4x + 10}{5x}$;
dom $(f + g) = (-\infty, 0) \cup (0, \infty)$
(b) $(f - g)(x) = \dfrac{x^2 + 4x - 10}{5x}$;
dom $(f - g) = (-\infty, 0) \cup (0, \infty)$
(c) $(fg)(x) = \dfrac{2x + 8}{5x}$; dom $(fg) = (-\infty, 0) \cup (0, \infty)$
(d) $\left(\dfrac{f}{g}\right)(x) = \dfrac{x^2 + 4x}{10}$; dom $\left(\dfrac{f}{g}\right) = (-\infty, 0) \cup (0, \infty)$

7. (a) $(f + g)(x) = x^5 + x^{1/5}$; dom $(f + g) = \mathbb{R}$
(b) $(f - g)(x) = x^5 - x^{1/5}$; dom $(f - g) = \mathbb{R}$
(c) $(fg)(x) = x^{26/5}$; dom $(fg) = \mathbb{R}$
(d) $\left(\dfrac{f}{g}\right)(x) = x^{24/5}$; dom $\left(\dfrac{f}{g}\right) = (-\infty, 0) \cup (0, \infty)$

9. (a) $(f + g)(x) = \sqrt{x + 3} + |x - 2|$;
dom $(f + g) = [-3, \infty)$
(b) $(f - g)(x) = \sqrt{x + 3} - |x - 2|$;
dom $(f - g) = [-3, \infty)$
(c) $(fg)(x) = \sqrt{x + 3} \, (|x - 2|)$; dom $(fg) = [-3, \infty)$

(d) $\left(\dfrac{f}{g}\right)(x) = \dfrac{\sqrt{x+3}}{|x-2|}$; dom$\left(\dfrac{f}{g}\right) = [-3, 2) \cup (2, \infty)$

11. **(a)** $(f+g)(x) = \dfrac{x^2 + 6x + 4}{x^2 - 4}$;

dom$(f+g) = (-\infty, -2) \cup (-2, 2) \cup (2, \infty)$

(b) $(f-g)(x) = \dfrac{x^2 + 4x + 8}{x^2 - 4}$;

dom$(f-g) = (-\infty, -2) \cup (-2, 2) \cup (2, \infty)$

(c) $(fg)(x) = \dfrac{x+3}{x^2 - 4}$;

dom$(fg) = (-\infty, -2) \cup (-2, 2) \cup (2, \infty)$

(d) $\left(\dfrac{f}{g}\right)(x) = \dfrac{x^2 + 5x + 6}{x - 2}$;

dom$\left(\dfrac{f}{g}\right) = (-\infty, -2) \cup (-2, 2) \cup (2, \infty)$

13. **(a)** 1 **(b)** 7 **(c)** -12 **(d)** $-\frac{4}{3}$

15. **(a)** **(i)** $-1 - 2x$ **(ii)** $13 - 2x$

(b) dom$(f \circ g) = $ dom$(g \circ f) = \mathbb{R}$

17. **(a)** **(i)** $1 - 4x^2$ **(ii)** $-8 + 24x - 16x^2$

(b) dom$(f \circ g) = $ dom$(g \circ f) = \mathbb{R}$

19. **(a)** **(i)** $\dfrac{1}{1+x}$ **(ii)** $\dfrac{x+1}{x}$

(b) dom$(f \circ g) = $ dom$(g \circ f) = (-\infty, -1) \cup (-1, 0)$
$\cup (0, \infty)$

21. **(a)** **(i)** $6 - \left|\dfrac{x^2 - 1}{x^2}\right|$ **(ii)** $\dfrac{35 - 12|x| + x^2}{36 - 12|x| + x^2}$

(b) dom$(f \circ g) = (-\infty, 0) \cup (0, \infty)$
dom$(g \circ f) = (-\infty, -6) \cup (-6, 6) \cup (6, \infty)$

23. **(a)** **(i)** $\sqrt{x^4 + 3x^2}$ **(ii)** $x^2 + 3x$

(b) dom$(f \circ g) = \mathbb{R}$; dom$(g \circ f) = [0, \infty)$

25. **(a)** $3 - \frac{4}{3}x + \dfrac{x^2}{9}$ **(b)** $3 - \dfrac{x^2}{9}$

(c) $\dfrac{7 - x^2}{3}$ **(d)** $\dfrac{3 - 4x + x^2}{3}$

(e) $\dfrac{x^2}{81} - 1$ **(f)** $x^2 - 1$

27. **(a)** -4 **(b)** π

29. **(a)** 4 **(b)** 4 **(c)** 3 **(d)** 3 **(e)** 16 **(f)** -6

In Exercises 30–37, the functions *f* and *g* are not unique. We give one possible pair for each exercise.

31. $f(x) = \sqrt{x}$; $g(x) = x + 3$

33. $f(x) = 5/x^3$; $g(x) = x - 3$

35. $f(x) = \sqrt[5]{x}$; $g(x) = (x-2)^4$

37. $f(x) = x^{-5}$; $g(x) = \dfrac{10x}{8 - x}$

39. $f(x) = \begin{cases} -x^2 + 2x - 1, & \text{if } x < 1 \\ x^2 - 2x + 1, & \text{if } x \geq 1 \end{cases}$

41. If $r(t)$ is the radius of the sphere at time t, then the volume is $V = \frac{4}{3}\pi r^3(t)$ and the surface area is $SA = 4\pi r^2(t)$.

Exercises 2.6

1. **(a)**

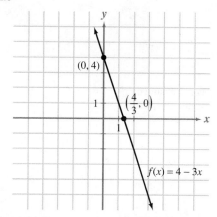

$(0, 4)$
$\left(\frac{4}{3}, 0\right)$
$f(x) = 4 - 3x$

(b) *f* is decreasing on $(-\infty, \infty)$ **(c)** *f* is one-to-one

3. **(a)**

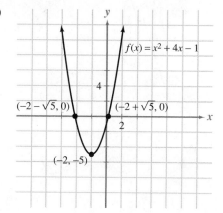

$f(x) = x^2 + 4x - 1$
$(-2 - \sqrt{5}, 0)$
$(-2 + \sqrt{5}, 0)$
$(-2, -5)$

(b) *f* is increasing on $[-2, \infty)$ and decreasing on $(-\infty, -2]$

(c) *f* is not one-to-one

5. **(a)**

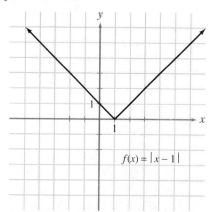

$f(x) = |x - 1|$

(b) *f* is increasing on $[1, \infty)$ and decreasing on $(-\infty, 1]$
(c) *f* is not one-to-one

7. (a)

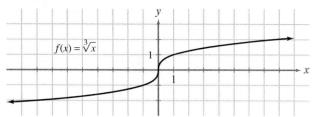

(b) f is increasing on $(-\infty, \infty)$
(c) f is one-to-one

9. (a)

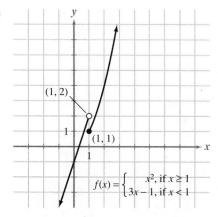

(b) f is increasing on $(-\infty, 1)$ and $[1, \infty)$
(c) f is not one-to-one

11.

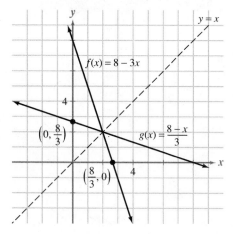

13.

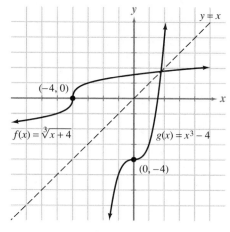

15.

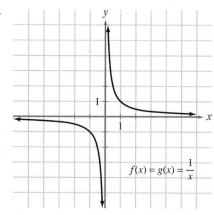

17. $f^{-1}(x) = \dfrac{x+10}{9}$

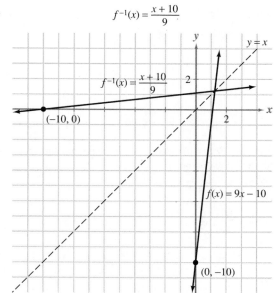

19.

$$f^{-1}(x) = x^{5/3}$$

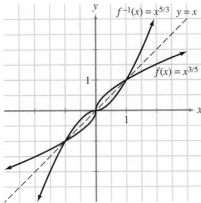

21. $f^{-1}(x) = (x-2)^2$ for $x \ge 2$

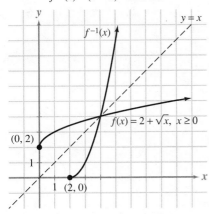

23. $f^{-1}(x) = \sqrt{3+x} - 2,\ x \ge -3$

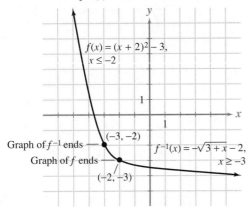

25. (a) $f^{-1}(x) = 2 - 5x$

(b) $f^{-1}(1) = -3,\ 1/f(1) = 5$

27. $f^{-1}(x) = \dfrac{x-b}{a}$

29. A one-to-one function that contains at least two ordered pairs cannot be even.

33. (a)

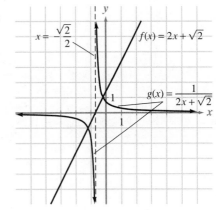

(b) No

Chapter 2 Review Exercises

1. Not a function
Domain: $[-4, 4]$
Range: $[-2, 2]$

3. Function
Domain: $\mathbb{R}$
Range: $(0, \infty)$

5. Not a function
Domain: $[-1, 1]$
Range: $[-1, 1]$

7. Function
Domain: $\mathbb{R}$
Range: $[-2, \infty)$

9. Center: $(0, 0)$
Radius: 7

11. Center: $(5, -2)$
Radius: $\sqrt{12}$

13. Center: $(5, 5)$
Radius: $\sqrt{15}$

15. Center: $(0.7, 0.1)$
Radius: 2

17. Domain: $\mathbb{R}$

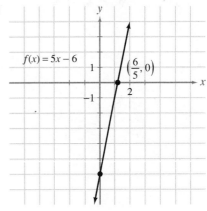

19. Domain: $\mathbb{R}$
Symmetric with respect to y-axis

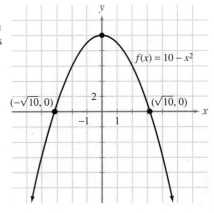

21. Domain: $(-\infty, -3) \cup (-3, \infty)$

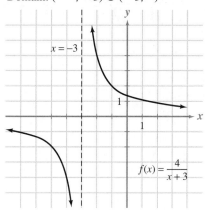

23. Domain: $\mathbb{R}$
Symmetric with respect to y-axis

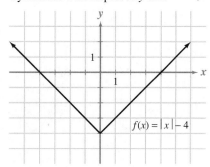

25. Domain: $[-5, \infty)$

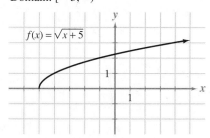

27. Domain: $\mathbb{R}$

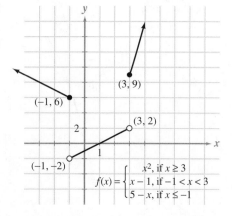

29. Domain: $\mathbb{R}$

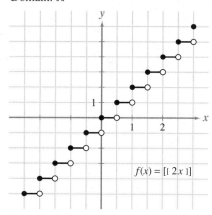

31. **(a)** 1   **(b)** 13   **(c)** $1 - 7a - a^2$
(d) $-3 - 2a - h$
(e) $\dfrac{11a^2 - 3a - 1}{a^2}$

33. **(a)** $\frac{1}{4}$ **(b)** $-\frac{1}{2}$ **(c)** $\dfrac{1}{2a + 4}$
(d) $-\dfrac{1}{2a^2 + 2ah}$ **(e)** $\dfrac{a}{2}$

35. Slope: 3

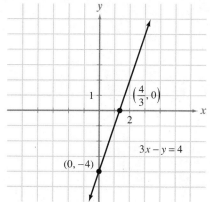

37. Slope: undefined

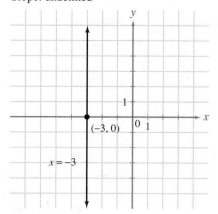

39. Slope: $-\frac{5}{2}$

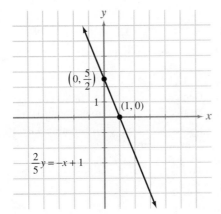

41. $3x + 4y - 17 = 0$

43. $2x + 3y - 15 = 0$

45. Minimum value: -9

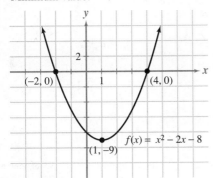

47. Maximum value: 4

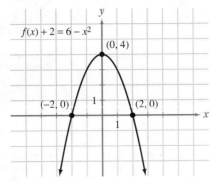

49. Minimum value: -1

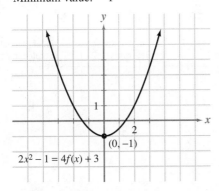

51. **(a)** Even function

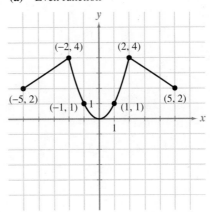

(b) Odd function

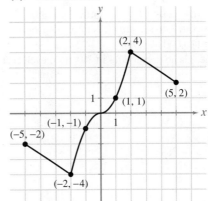

53. Increasing on $(-\infty, -1]$; decreasing on $[1, \infty)$

55. Increasing on $(-\infty, -1)$ and on $(-1, 0]$; decreasing on $[0, 1)$ and on $(1, \infty)$

57. $\frac{1}{2}$ **59.** $\frac{7}{2}$

61. $(f \circ g)(x) = \dfrac{x^2 - x - 4}{3}$; dom $(f \circ g) = \mathbb{R}$

63. $(f \circ g)(x) = \dfrac{1}{\sqrt{x^2 - 6x - 4}}$;

dom $(f \circ g) = (-\infty, -2) \cup (-2, 0] \cup [6, 8) \cup (8, \infty)$

65. One-to-one   **67.** One-to-one

69. $f^{-1}(x) = x^5$

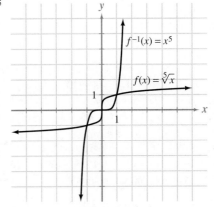

71. $f^{-1}(x) = -\sqrt{9 - x^2}, \quad 0 \le x \le 3$

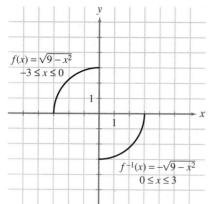

$f(x) = \sqrt{9-x^2}$
$-3 \le x \le 0$

$f^{-1}(x) = -\sqrt{9-x^2}$
$0 \le x \le 3$

73. $A(x) = 10x - x^2; 25$
75. (a) 64 ft   (b) 2 sec after it is thrown upward
(c) 0.54 sec and 3.46 sec after it is thrown upward
79. (a) $f(p) = 110{,}000 - 200p$ (b) \$450
(c) No. There is no demand for the VCRs if each cost
\$600.

CHAPTER 3
Exercises 3.2
1. $3m - 2; 5$ **3.** $y^3 + 2y^2 + 4y + 8; 17$
5. $a^3 - 3a^2 + a - 3; 4a - 4$
7. $-\frac{1}{2}x^2 - \frac{11}{4}x + \frac{15}{8}; -\frac{51}{8}x + \frac{11}{4}$
9. $x^4 + ix^3; 0$ **11.** $3x^2 - 6x + 4; -7$
13. $y^3 - 5y - 13; -55$ **15.** $x^2 - 3x + 2; 0$
17. $x^6 + x^5 + x^4 + x^3 + 1; 1$
19. $x^5 + x^4 + x^3 + x^2 + x + 1; 0$
21. $-x^4 - x^3 - x^2 - x - 1; 0$
23. $x^3 + (a^3 + a)x^2 + (a^4 + 2a^2)x + a^5 + 3a^3;$
$a^6 + 4a^4$
25. $8x^2 - 2; 0$ **27.** $x^2 + \frac{1}{2}x + \frac{1}{4}; 0$
29. $x^2 + (-1 - 2i)x + 2i; 0$
31. $x^3 + (1 - i)x^2 - 3ix - 3 - 3i; -5$
33. $x^4 + ix^3 - 2x^2 - 2ix + 2; 2i$
35. $2y^3 - 6iy^2 - 4y; 0$
37. $m^4 - 2im^3 + (-3 - i)m^2 + (-2 + 6i)m + 12 + 4i;$
$6 - 23i$
49. $-\frac{3}{2}, 2$ **51.** $-2, -1$ **53.** $\pm\sqrt{3}$ **57.** -8

Exercises 3.3
1. 0   **3.** 6   **5.** $2 + i$ **7.** $-1 + 2\sqrt{2}$
9. $-3 + 2i$ **11.** Yes   **13.** Yes
15. No   **17.** Yes   **19.** No   **21.** No
23. $x^3 - 4x^2 + x + 6$ **25.** $x^3 + 3x^2 - 10x$
27. $x^4 + x^3 - x^2 + x - 2$
29. $8x^4 - 18x^3 + x^2 + 6x$
31. $x^5 - 2x^4 - 3x^3 - 12x^2 - 28x - 16$
33. 2 is a root of multiplicity 3

35. $-2, 1$ **37.** $1, -\frac{1}{2} \pm \dfrac{\sqrt{3}}{2}i$ **39.** $\pm\frac{1}{2}$
41. $\pm i$ **45.** $-1, 3$ **49.** $x^3 - 5x^2 + 2x + 8$

Exercises 3.4
1. Degree 4; 1, 5, $-\frac{3}{4}$ (multiplicity 2)
3. Degree 5; $\pm 2i, \sqrt{3}$ (multiplicity 3)
5. Degree 8; $-1 \pm \sqrt{3}i$, 1 (multiplicity 2),
2 (multiplicity 4)
13. $x^4 - 6x^3 + 22x + 15$
15. $x^5 + 5x^4 + \frac{13}{4}x^3 - \frac{15}{2}x^2 + \frac{9}{4}x$
17. $\pm 1, \pm 5, \pm\frac{1}{2}, \pm\frac{5}{2}$
19. $\pm 1, \pm 2, \pm 4, \pm 8, \pm\frac{1}{2}, \pm\frac{1}{4}, \pm\frac{1}{8}, \pm\frac{1}{16}$
21. $\pm 1, \pm 2, \pm 3, \pm 4, \pm 6, \pm 9, \pm 18, \pm 36, \pm\frac{1}{2}, \pm\frac{3}{2}, \pm\frac{9}{2}$
23. $0, \pm 1, \pm\frac{1}{2}$ **31.** 2, 3   **33.** $\frac{1}{2}, 2$ (multiplicity 3)
35. -2 (multiplicity 2), 1 (multiplicity 3)
37. $-\frac{1}{3}, \frac{1}{2}, \frac{5}{2}$ **39.** $1, 2 \pm \sqrt{2}$
41. $2, \frac{1}{3}, \dfrac{-1 \pm \sqrt{5}}{2}$ **43.** $-\frac{2}{3}, -1 \pm 2i$
45. $-\frac{2}{3}, \frac{1}{2}, 2$ **47.** $9(x + 1)(x - \frac{1}{3})(x - \frac{2}{3})$
49. $3(x + 2)(x - \frac{2}{3})(x + \sqrt{2}i)(x - \sqrt{2}i)$
53. $-2, -1, 0, 3$ **55.** $-2, -\frac{1}{2},$ and $\frac{1}{3}$
57. $(\frac{1}{2}, -\frac{3}{8}), (2, 15)$ **59.** $(-1, -7), (2, 53)$

Exercises 3.5

1.

No. of positive real roots	No. of negative real roots	No. of nonreal complex roots
1	2	0
1	0	2

3.

No. of positive real roots	No. of negative real roots	No. of nonreal complex roots
5	0	0
3	0	2
1	0	4

5.

No. of positive real roots	No. of negative real roots	No. of nonreal complex roots
2	1	2
0	1	4

7.

No. of positive real roots	No. of negative real roots	No. of nonreal complex roots
2	1	0
0	1	2

9. $-1, 9$ **11.** $-2, 2$ **13.** $-1, 3$ **15.** $-3, 2$

19. $x^4 - 3x^2 - 4$ **21.** $x^4 - \frac{1}{2}x^3 - 9x^2 - 17x - 12$

23. $x^5 - 2x^4 + \frac{69}{4}x^3 - \frac{1}{2}x^2 + \frac{17}{4}x$

25. $x^3 - \left(1 + \frac{\sqrt{3}}{2}\right)x^2 + \left(1 + \frac{\sqrt{3}}{2}\right)x - \frac{\sqrt{3}}{2}$

27. $\frac{1}{2}, 1 - 2\sqrt{2} \approx -1.8, 1 + 2\sqrt{2} \approx 3.8$

29. $-1, 1.5$ **31.** -1.2 **33.** 1.63 **39.** 220

41. 1.2 sec.   **45.** 1.1 **47.** 1.227 **49.** 1.836

Exercises 3.6
1. Yes   **3.** No   **5.** Yes   **7.** No

9.

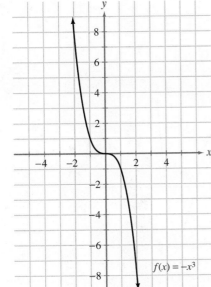

11.

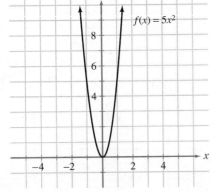

13.

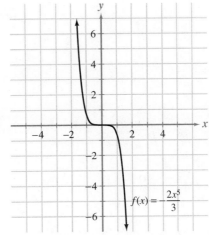

15.

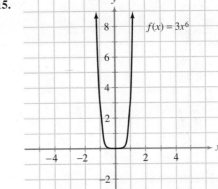

17.

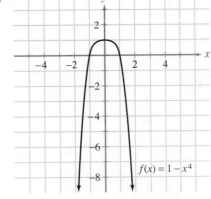

19.

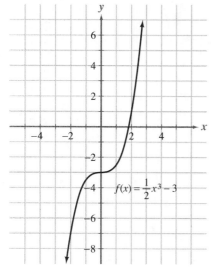

$$f(x) = \frac{1}{2}x^3 - 3$$

25.

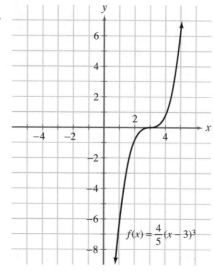

$$f(x) = \frac{4}{5}(x-3)^3$$

27. (e)   **29.** (c)   **31.** (f)

21.

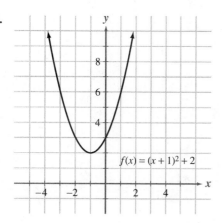

$$f(x) = (x+1)^2 + 2$$

33.

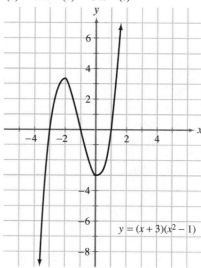

$$y = (x+3)(x^2 - 1)$$

23.

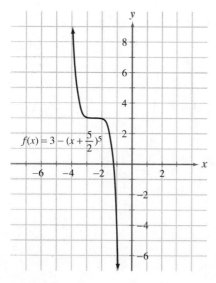

$$f(x) = 3 - (x + \frac{5}{2})^5$$

35.

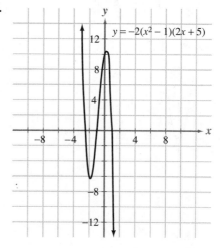

$$y = -2(x^2 - 1)(2x + 5)$$

37.

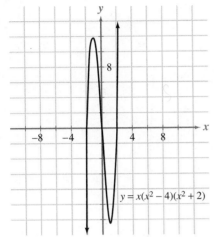

$y = x(x^2 - 4)(x^2 + 2)$

39. $y = \dfrac{1}{4}(x^2 - 9)(x + 5)(x - 1)$

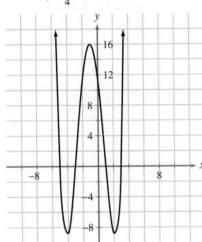

41.

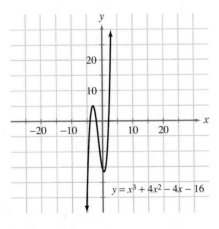

$y = x^3 + 4x^2 - 4x - 16$

43.

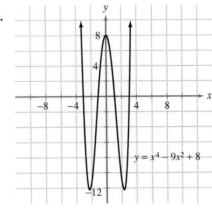

$y = x^4 - 9x^2 + 8$

45. $y = -2x^3 - 8x^2 + 3x + 12$

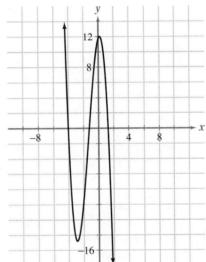

47. $y = x^4 - 2x^3 - x^2 + 6x - 8$

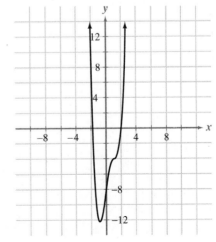

49. $y = -x^4 + 4x^3 - 3x^2 - x + 3$

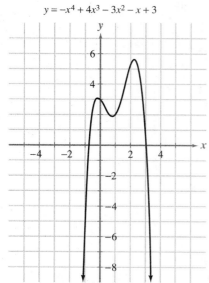

51. $y = \frac{1}{2}x^5 - \frac{3}{2}x^3 - 2x$

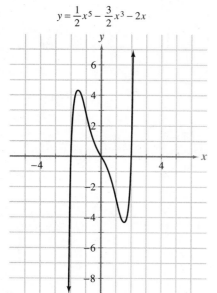

53.   2 real zeros

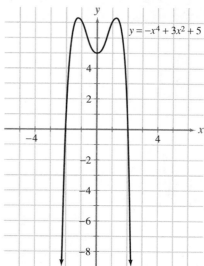

$y = -x^4 + 3x^2 + 5$

55.   3 real zeros

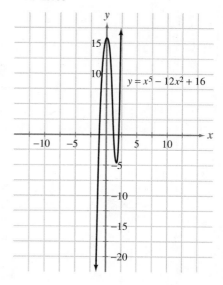

$y = x^5 - 12x^2 + 16$

57. $(-1.4, -0.3)$

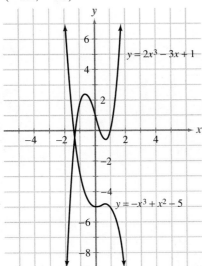

$y = 2x^3 - 3x + 1$

$y = -x^3 + x^2 - 5$

1.

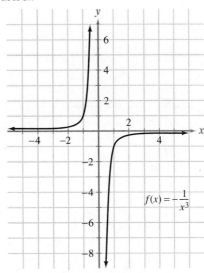

$f(x) = -\dfrac{1}{x^3}$

59. Increases on $(-\infty, -1) \cup (3, \infty)$;
decreases on $(-1, 3)$

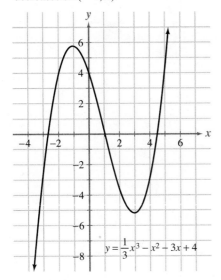

$y = \dfrac{1}{3}x^3 - x^2 - 3x + 4$

3.

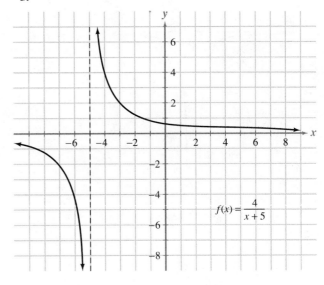

$f(x) = \dfrac{4}{x + 5}$

5.

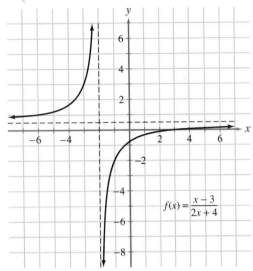

$$f(x) = \frac{x - 3}{2x + 4}$$

7.

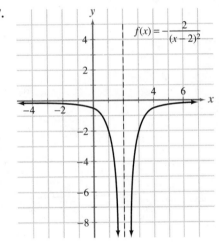

$$f(x) = -\frac{2}{(x - 2)^2}$$

9.

$$f(x) = \frac{1}{(x + 3)(x - 2)}$$

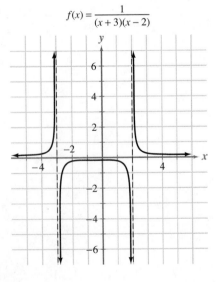

11.

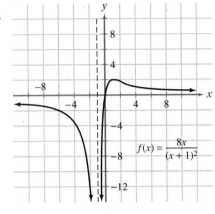

$$f(x) = \frac{8x}{(x + 1)^2}$$

13.

$$f(x) = \frac{x^2 - 1}{x(x + 4)}$$

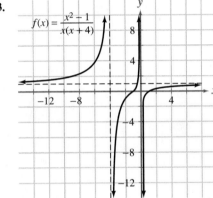

15.

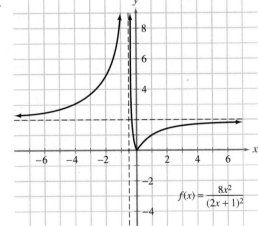

$$f(x) = \frac{8x^2}{(2x + 1)^2}$$

17.

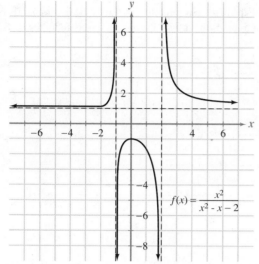

$$f(x) = \frac{x^2}{x^2 - x - 2}$$

19.

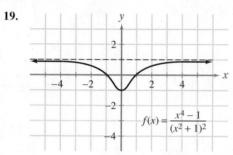

$$f(x) = \frac{x^4 - 1}{(x^2 + 1)^2}$$

21. $f(x) = x - 1, \quad x \neq -1$

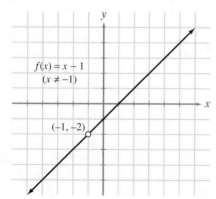

$f(x) = x - 1$
$(x \neq -1)$

$(-1, -2)$

23. $f(x) = x + 2, \quad x \neq 3$

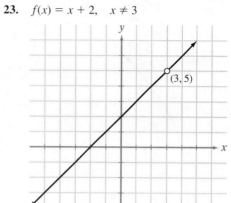

$(3, 5)$

25. $f(x) = x^2 + 2x + 4, \quad x \neq 2$

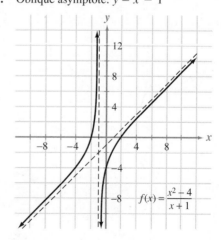

$(2, 12)$

27. Oblique asymptote: $y = x - 1$

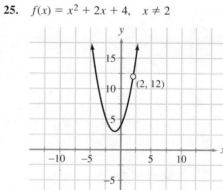

$$f(x) = \frac{x^2 - 4}{x + 1}$$

29.   Oblique asymptote: $y = x + 3$

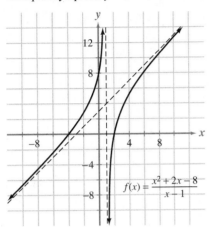

$$f(x) = \frac{x^2 + 2x - 8}{x - 1}$$

31.   Oblique asymptote: $y = x - 3$

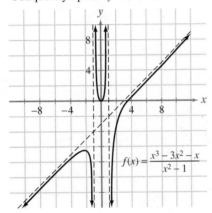

$$f(x) = \frac{x^3 - 3x^2 - x}{x^2 - 1}$$

33. $\overline{C}(x) = x + 2 + \dfrac{15}{x}; x > 0$

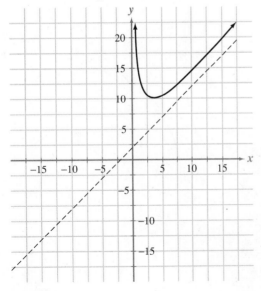

35.   The horizontal asymptote $y = 5$ tells the researcher that all but 5 mg of the drug is eventually absorbed by the body.

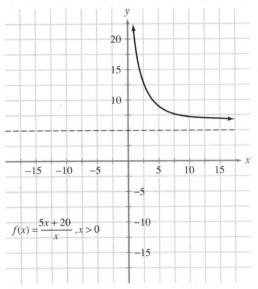

$$f(x) = \frac{5x + 20}{x}, x > 0$$

37.   No.

39.

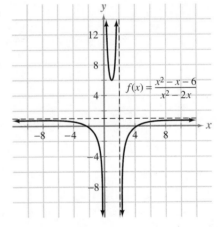

$$f(x) = \frac{x^2 - x - 6}{x^2 - 2x}$$

Chapter 3 Review Exercises

1. $x - 2; 11$ **3.** $x + 2; 2x - 1$

5. $3x^3 - 7x^2 + 21x - 67; 200$ **7.** $6x^2 + x + \frac{7}{3}; \frac{25}{9}$

9. $x^2 + ix + 1; -4 + i$

11. $x^3 + (1 - i)x^2 - ix - 1 - i; -1$

17. -6 **19.** 14 **21.** 57 **23.**   Yes   **25.**   No

27. $x^3 - 3x^2 - x + 3$ **29.** $x^4 - 2x^3 - 5x^2 + 8x + 4$

31. $-\frac{1}{2}, \frac{1}{3}$ **33.** $3i, 2$

37. 2 **39.**   Degree 4; $-\frac{4}{3}, -1, \frac{1}{2}$ (multiplicity 2)

41.   Degree 3; $0, \pm\sqrt{3}\,i$

43. $x^4 - 4x^3 - 3x^2 + 14x - 8$

45. $x^3 - \frac{2}{3}x^2 - \frac{7}{12}x - \frac{1}{12}$ **47.** $\pm 1, \pm 3, \pm\frac{1}{3}$

49. $\pm 1, \pm 2, \pm 3, \pm 4, \pm 6, \pm 12, \pm\frac{1}{2}, \pm\frac{1}{3}, \pm\frac{1}{6},$
$\pm\frac{1}{9}, \pm\frac{1}{18}, \pm\frac{2}{3}, \pm\frac{2}{9}, \pm\frac{3}{2}, \pm\frac{4}{3}, \pm\frac{4}{9}$

51. $-\frac{4}{3}, -1, \frac{1}{2}, 1$ **53.** None   **55.** None
59. -1 and 1

61.

No. of positive real roots	No. of negative real roots	No. of nonreal complex roots
3	1	0
1	1	2

63.

No. of positive real roots	No. of negative real roots	No. of nonreal complex roots
1	1	4

65. $-2, 5$ **67.** $-1, 2$ **69.** 0.6 **71.** $2, -1.2$
73. 1.71 **75.** 3 sec
77.

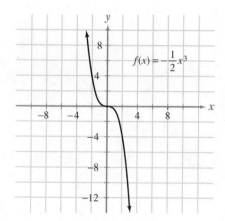

79.

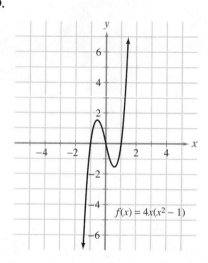

81. $f(x) = -\frac{5}{2}(x+1)^2(x-1)$

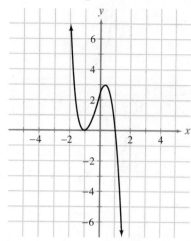

83. $f(x) = 2x^3 - 5x^2 - 4x + 3$

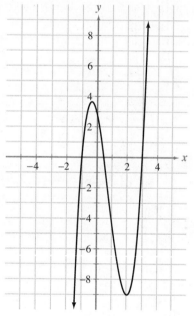

85.

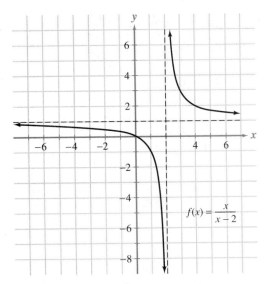

$$f(x) = \frac{x}{x-2}$$

87.

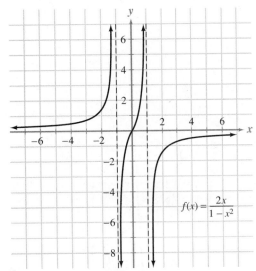

$$f(x) = \frac{2x}{1-x^2}$$

89.

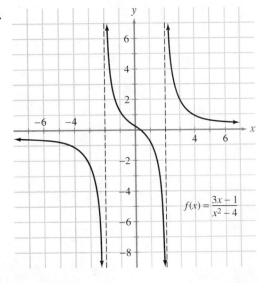

$$f(x) = \frac{3x-1}{x^2-4}$$

91.

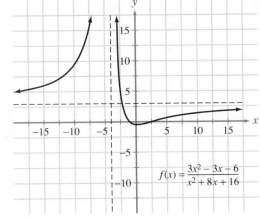

$$f(x) = \frac{3x^2 - 3x - 6}{x^2 + 8x + 16}$$

93. Oblique asymptote: $y = x$

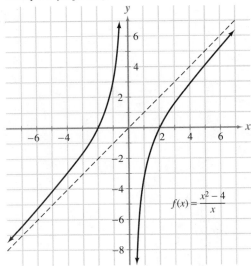

$$f(x) = \frac{x^2 - 4}{x}$$

95. Oblique asymptote: $y = x - 1$

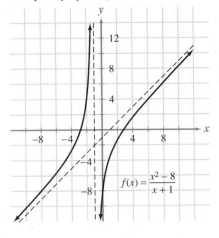

$$f(x) = \frac{x^2 - 8}{x + 1}$$

97. $\overline{C}(x) = \frac{1}{2}x + 2 + \dfrac{5}{x}; \; x > 0$

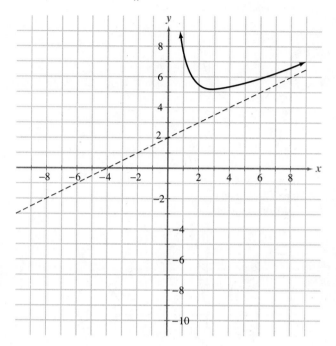

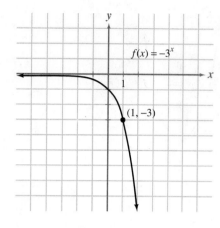

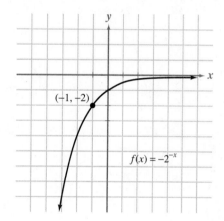

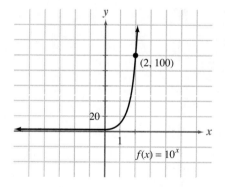

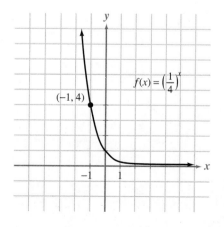

9. dom $(f) = \mathbb{R}$
ran $(f) = [1, \infty)$

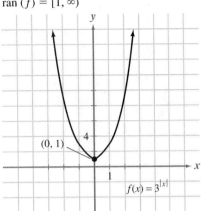

$f(x) = 3^{|x|}$
$(0, 1)$

11. dom $(f) = \mathbb{R}$
ran $(f) = (0, \infty)$
Horizontal asymptote: $y = 0$

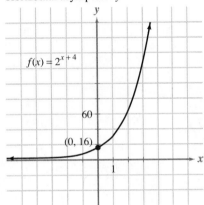

$f(x) = 2^{x+4}$
$(0, 16)$

13. dom $(f) = \mathbb{R}$
ran $(f) = (2, \infty)$
Horizontal asymptote: $y = 2$

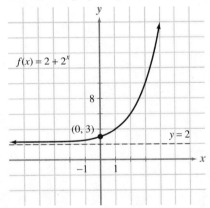

$f(x) = 2 + 2^x$
$(0, 3)$
$y = 2$

15. dom $(f) = \mathbb{R}$
ran $(f) = (0, \infty)$
Horizontal asymptote: $y = 0$

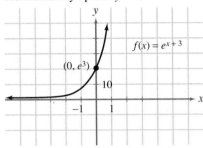

$f(x) = e^{x+3}$
$(0, e^3)$

17. dom $(f) = \mathbb{R}$
ran $(f) = (3, \infty)$
Horizontal asymptote: $y = 3$

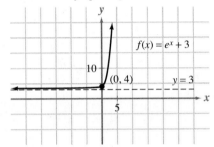

$f(x) = e^x + 3$
$(0, 4)$
$y = 3$

19. dom $(f) = \mathbb{R}$
ran $(f) = \mathbb{R}$

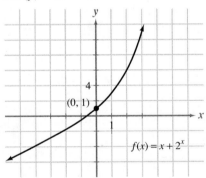

$(0, 1)$
$f(x) = x + 2^x$

21. dom $(f) = (-\infty, 0) \cup (0, \infty)$
ran $(f) = (0, 1) \cup (1, \infty)$
Horizontal asymptote: $y = 1$

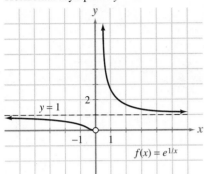

23. dom $(f) = \mathbb{R}$
ran $(f) = (0, \infty)$
Horizontal asymptote: $y = 0$

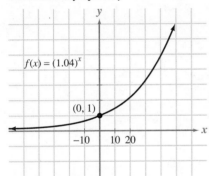

25. dom $(f) = \mathbb{R}$
ran $(f) = (0, 16]$
Horizontal asymptote: $y = 0$

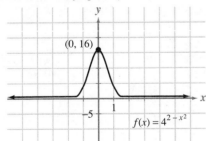

27.

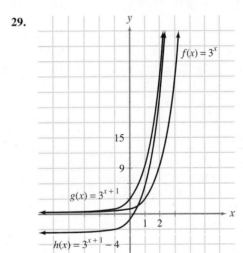

29.

31.

x	3	3.1	3.14	3.141
3^x	27	30.13533	31.48914	31.52375

3.1415	3.14159	π
31.54107	31.54419	31.54428

As x approaches π, 3^x approaches 3^π.

33. dom (f) = ran (f^{-1}) = $\mathbb{R}$
ran (f) = dom (f^{-1}) = $(0, \infty)$

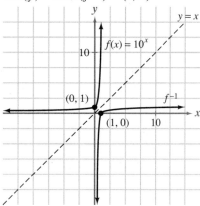

35. dom (f) = ran (f^{-1}) = $\mathbb{R}$
ran (f) = dom (f^{-1}) = $(0, \infty)$

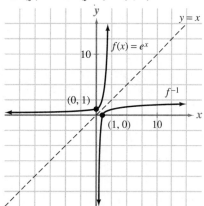

39. \$2821.20   **41.**   \$6867.22
43. Mary earns approximately 7 cents more than Janice.
45. First year: \$101.52
Second year: \$336.17
Third year: \$592.85
47. (a) $-\frac{1}{5}, 0$ (b) $-3, 1$ (c)   3
49.

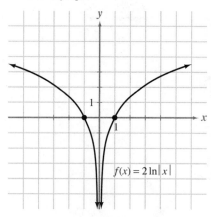

51.

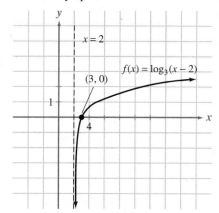

Exercises 4.2
1. $\log_6 216 = 3$ **3.** $\log_{1/3} 9 = -2$ **5.** $\log_5 \sqrt{5} = \frac{1}{2}$
7. $\ln 10 = 4t$ **9.** $4^{-1} = \frac{1}{4}$ **11.** $e^{1/4} = \sqrt[4]{e}$
13. $x = 6^a$ **15.** 7   **17.** 4   **19.** $-\frac{4}{3}$ **21.** 25
23. dom $(f) = (2, \infty)$
Vertical asymptote: $x = 2$

25. dom $(f) = (-\infty, 0) \cup (0, \infty)$
Vertical asymptote: $x = 0$

27. dom $(f) = (2, \infty)$
Vertical asymptote: $x = 2$

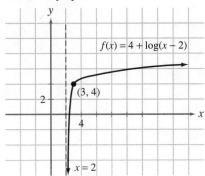

29. dom $(f) = (-\infty, 0)$
Vertical asymptote: $x = 0$

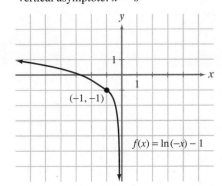

31. dom $(f) = [1, \infty)$

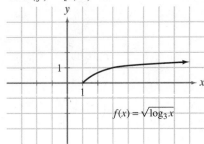

33. dom $(f) = [1, \infty)$

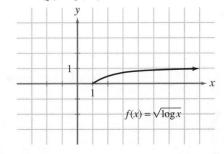

35. 10   **37.** $\frac{1}{9}$ **39.** ± 5 **41.** 1, 4   **43.** $\pm \sqrt{6}$

45. **a.** $(-\frac{1}{3}, \infty)$ **b.** $(-\infty, \frac{8}{3})$ **c.** $(0, \infty)$
     d. $(-\infty, 0) \cup (0, \infty)$

In Exercises 47–54, the solutions are not unique. We give one possible pair of functions for each exercise.

47. $g(x) = \log_6 x$; $h(x) = 3x + 1$
49. $g(x) = \ln |x|$; $h(x) = x + 2$
51. $g(x) = \ln x$; $h(x) = x^2$
53. $g(x) = e^x$; $h(x) = x^2 + x - 1$
55. **(a)** pH $= 2.92$ **(b)** 5.01×10^{-8} **(c)** 8.3 (a base)
57.

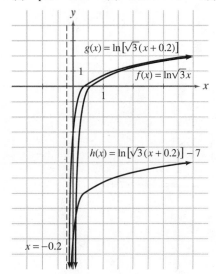

59. **(a)**

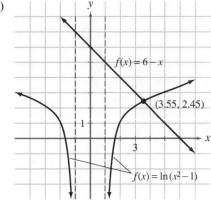

     (b) (3.55, 2.45)

Exercises 4.3
1. 1   **3.** 1   **5.** 0   **7.** 4   **9.** b
11. **a.** 2.29   **b.** -0.49 **c.** 0.278   **d.** 0.647
13. $\ln(81x^3)$ **15.** $\log_6 \frac{16}{3}$ **17.** $\log_b \frac{(x + y)^2}{8z}$
19. $\log_5 \dfrac{x}{\sqrt{2x^2 + x - 3}}$
21. $2\log_b x + 2\log_b y - 3\log_b z$

23. $\frac{1}{5}(2\log_2 x - \log_2 y - \log_2 z)$

25. $\ln(2 - x) - \ln(x + 1) - 2\ln(2x - 1)$

27. $\log_b x + \log_b y + \frac{1}{2}\log_b z$

29. $\dfrac{\ln 10}{\ln 2} \approx 3.32$ **31.** $\dfrac{2\ln 4 + 3\ln 5}{\ln 4 + \ln 5} \approx 2.54$

33. $\frac{20001}{9960} \approx 2.01$ **35.** $\pm\sqrt{23} \approx \pm 4.80$

37. $-5 + 2\sqrt{10} \approx 1.32$ **39.** $\pm\sqrt{13} \approx 3.61, \pm 5\sqrt{i}$

41. $\dfrac{\log 6}{\log 5}$ **43.** $\dfrac{\log x}{\log 3}$ **45.** $\dfrac{\log(x - 1)}{\log e}$

47. $\dfrac{\ln 15}{\ln 10}$ **49.** $\dfrac{\ln(x^2 - 1)}{\ln b}$

51. $\ln(y \pm \sqrt{y^2 - 1})$ **53.** $\frac{1}{2}\ln\left(\dfrac{y + 1}{y - 1}\right)$

63.

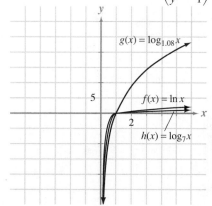

Exercises 4.4

1. $1419.07   **3.** Approximately $50

5. $k \approx -4.33 \times 10^{-4}$

7. Approximately 13,000 years old

9. **(a)** $11,250   **(b)** $7750   **(c)** $5339

11. Approximately every 11 hours

13. **(a)** Approximately 28 days
     (b) Approximately 46 days

15. 4,449,372   **17.** 14.69

19. **(a)** 1000
     (b) Approximately 4526

     (c)

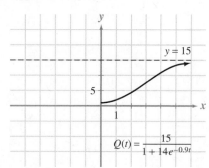

     (d) 15,000

21. **(a)** 21.05   **(b)** 28.99

Chapter 4 Review Exercises

1. dom $(f) = \mathbb{R}$
ran $(f) = (0, \infty)$
Horizontal asymptote: $y = 0$

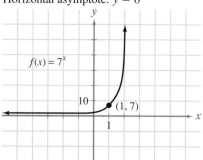

3. dom $(f) = (0, \infty)$
ran $(f) = \mathbb{R}$
Vertical asymptote: $x = 0$

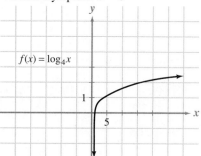

5. dom $(f) = \mathbb{R}$
ran $(f) = [1, \infty)$

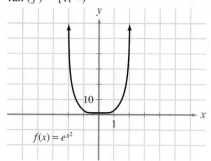

7. dom $(f) = (-4, \infty)$
ran $(f) = \mathbb{R}$
Vertical asymptote: $x = -4$

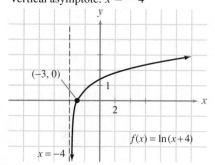

9. dom $(f) = \mathbb{R}$
ran $(f) = (0, \infty)$
Horizontal asymptote: $y = 0$

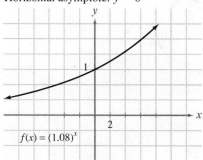

$f(x) = (1.08)^x$

11. $\{(\ln \frac{49}{64})(\ln \frac{16}{7})^{-1}\}$ **13.** $\{\frac{299}{95}\}$ **15.** $\{-6, 2\}$

17. $\{\frac{3}{2}\}$ **19.** $\frac{1}{3}$ **21.** 2 **23.** 2

25. $3 \log x + \frac{1}{4}(\log z - 3 \log y)$

27. **(a)** $\dfrac{\log I - \log I_0}{\log b}$ **(b)** 4.19

29. 114.77 **31.** **(a)** $\dfrac{2}{3 \ln 2}$ **(b)** 0

33. Chose second plan since its yield is approximately $9000 greater than that of the first plan.

35. **(a)** $11,234.21 **(b)** $11,233.22 **37.** 43.34 billion

39. **(a)** 20 **(b)** 11.5 days

CHAPTER 5

Exercises 5.1

1. $P(5\pi) = (-1, 0)$

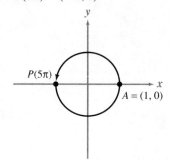

3. $P(-2\pi/3) = (-\frac{1}{2}, -\sqrt{3}/2)$

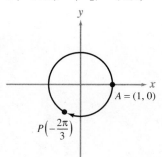

5. $P(-7\pi/6) = (-\sqrt{3}/2, \frac{1}{2})$

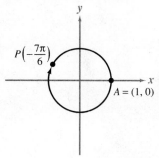

7. $P(-7\pi/4) = (\sqrt{2}/2, \sqrt{2}/2)$

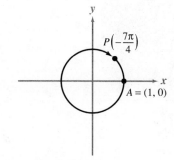

9. $P(23\pi/3) = (\frac{1}{2}, -\sqrt{3}/2)$

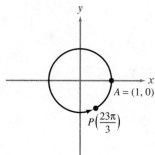

11. $P(50\pi/6) = P(25\pi/3) = (\frac{1}{2}, \sqrt{3}/2)$

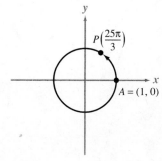

13. $P(-4\pi/3) = (-\frac{1}{2}, \sqrt{3}/2)$

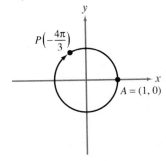

15. $P(22\pi/8) = P(11\pi/4) = (-\sqrt{2}/2, \sqrt{2}/2)$

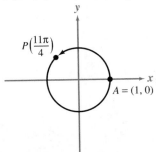

17. $P(-25\pi/6) = (\sqrt{3}/2, -\frac{1}{2})$

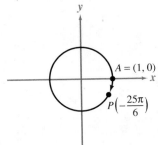

19.

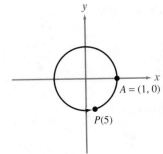

21.

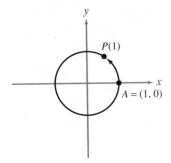

23.

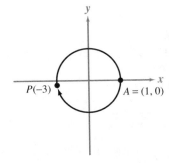

25.

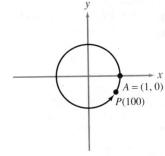

27.

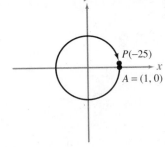

29.

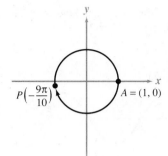

$P\left(-\dfrac{9\pi}{10}\right)$ $A = (1, 0)$

31.

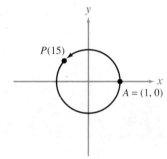

$P(15)$ $A = (1, 0)$

33.

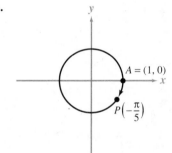

$A = (1, 0)$ $P\left(-\dfrac{\pi}{5}\right)$

35. $4\pi/3$ and $5\pi/3$ **37.** 0 and 2π **39.** $\pi/2$ and $3\pi/2$
41. $-2\sqrt{5}/5$ **43.** $2\sqrt{2}/3$ **45.** $\sqrt{41}$
47. $-\sqrt{10}/10$

Exercises 5.2

1. Undefined   **3.** $-\sqrt{2}$ **5.** $-\sqrt{3}/2$ **7.** $-\frac{1}{2}$
9. Undefined   **11.** 1   **13.** Undefined
15. $\sqrt{2}/2$ **17.** 2   **19.** Undefined   **21.** 0

	sin t	cos t	tan t	cot t	sec t	csc t
23.	$-\frac{4}{5}$	$-\frac{3}{5}$	$\frac{4}{3}$	$\frac{3}{4}$	$-\frac{5}{3}$	$-\frac{5}{4}$
25.	$5/\sqrt{41}$	$-4/\sqrt{41}$	$-\frac{5}{4}$	$-\frac{4}{5}$	$-\sqrt{41}/4$	$\sqrt{41}/5$
27.	$\sqrt{7}/4$	$\frac{3}{4}$	$\sqrt{7}/3$	$3/\sqrt{7}$	$\frac{4}{3}$	$4/\sqrt{7}$
29.	$-\sqrt{22}/5$	$\sqrt{3}/5$	$-\sqrt{22}/\sqrt{3}$	$-\sqrt{3}/\sqrt{22}$	$5/\sqrt{3}$	$-5/\sqrt{22}$

31. 0.6374   **33.** -1.1918 **35.** 0.9650   **37.** -1.2711
39. -0.7713 **41.** 0.5038

51. (a) $-3\sqrt{11}/10$ (b) $3\sqrt{11}$
53. (a) $5/\sqrt{26}$ (b) $-\sqrt{26}$
55. (a) $10/\sqrt{19}$ (b) $\sqrt{19}/9$
57. (a) $-3\sqrt{7}/8$ (b) $3\sqrt{7}$
59. (a) $5/\sqrt{61}$ (b) $-6/\sqrt{61}$
61. (a) $2\sqrt{6}/5$ (b) $\frac{1}{5}$
63.

c	sin t	cos t	tan t	cot t	sec t	csc t
$\sqrt{3}$	$\sqrt{6}/3$	$-\sqrt{3}/3$	$-\sqrt{2}$	$-1/\sqrt{2}$	$-\sqrt{3}$	$\sqrt{3}/\sqrt{2}$

73.

t	0.1	0.01	0.001	0.0001
cos t	0.9950	0.99995	0.9999995	0.999999995
$\dfrac{1-\cos t}{t}$	0.0500	0.00500	0.0005	0.00005

t	0.00001	0.000001
cos t	0.999999999	1
$\dfrac{1-\cos t}{2}$	0.000005	0

As t gets closer and closer to 0, $\dfrac{1-\cos t}{t}$ is approaching 0.

Exercises 5.3

1. $f(x) = \cot x$

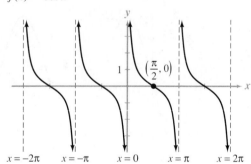

$\left(\dfrac{\pi}{2}, 0\right)$

$x = -2\pi$ $x = -\pi$ $x = 0$ $x = \pi$ $x = 2\pi$

3. $f(x) = \csc x$

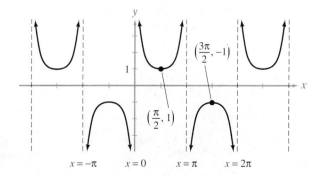

$\left(\dfrac{3\pi}{2}, -1\right)$ $\left(\dfrac{\pi}{2}, 1\right)$

$x = -\pi$ $x = 0$ $x = \pi$ $x = 2\pi$

5. $y = 2 \sin x$; period = 2π, amplitude = 2, phase shift = 0, range = $[-2, 2]$

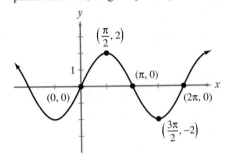

7. $y = \cos 3x$; period = $2\pi/3$, amplitude = 1, phase shift = 0, range = $[-1, 1]$

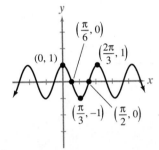

9. $y = -4 \sin(-x)$; period = 2π, amplitude = 4, phase shift = 0, range = $[-4, 4]$

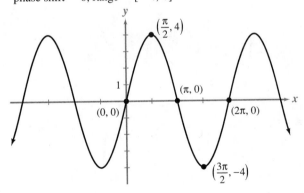

11. $y = 2 \sin 3x$; period = $2\pi/3$, amplitude = 2, phase shift = 0, range = $[-2, 2]$

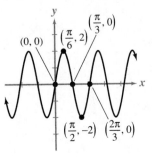

13. $y = -\pi/2 \cos(-x)$; period = 2π, amplitude = $\pi/2$, phase shift = 0, range = $[-\pi/2, \pi/2]$

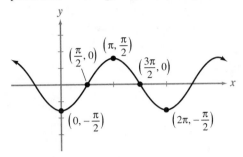

15. $y = 2 \cos(\pi - x)$; period = 2π, amplitude = 2, phase shift = π, range = $[-2, 2]$

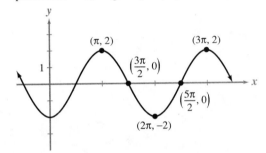

17. $y = \frac{1}{3} \sin\left(\dfrac{x}{3} + 1\right)$; period = 6π, amplitude = $\frac{1}{3}$, phase shift = -3, range = $[-\frac{1}{3}, \frac{1}{3}]$

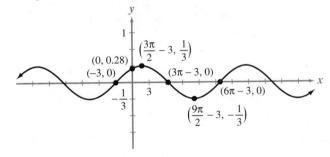

19. $y = -\frac{4}{3}\cos(4 - x)$; period = 2π, amplitude = $\frac{4}{3}$, phase shift = 4, range = $[-\frac{4}{3}, \frac{4}{3}]$

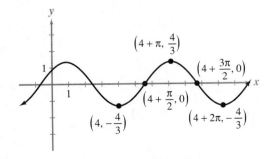

21. $y = -5\sin(2x - \pi)$; period $= \pi$, amplitude $= 5$,
phase shift $= \pi/2$, range $= [-5, 5]$

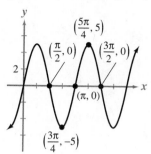

23. $y = \frac{1}{2}\cos\left(\frac{x}{2} + 3\right)$; period $= 4\pi$, amplitude $= \frac{1}{2}$,
phase shift $= -6$, range $= [-\frac{1}{2}, \frac{1}{2}]$

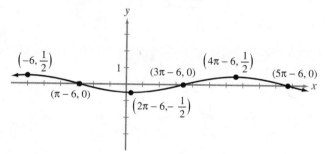

25. $y = -\sin(\pi - 3x)$; period $= 2\pi/3$, amplitude $= 1$,
phase shift $= \pi/3$, range $= [-1, 1]$

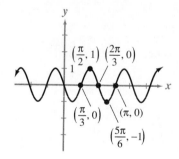

27. $y = 2\sin(2x + 2\pi)$; period $= \pi$, amplitude $= 2$,
phase shift $= -\pi$, range $= [-2, 2]$

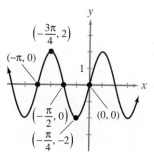

29. $y = -3\cos(\pi x + 2\pi)$; period $= 2$, amplitude $= 3$,
phase shift $= -2$, range $= [-3, 3]$

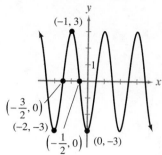

31. $y = \pi\sin\left(1 - \frac{x}{2}\right)$; period $= 4\pi$, amplitude $= \pi$,
phase shift $= 2$, range $= [-\pi, \pi]$

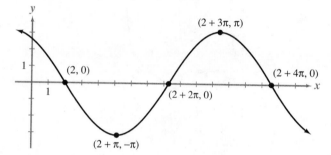

33. $y = -\cot x$; period $= \pi$, phase shift $= 0$,
range $= \mathbb{R}$

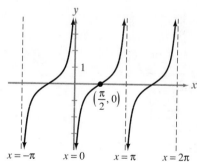

35. $y = \tan 2x$; period $= \pi/2$, phase shift $= 0$,
range $= \mathbb{R}$

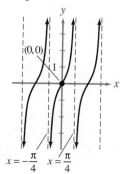

37. $y = -\tan(2x - 1)$; period $= \pi/2$, phase shift $= \frac{1}{2}$, range $= \mathbb{R}$

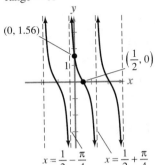

39. $y = \cot(1 - x)$; period $= \pi$, phase shift $= 1$, range $= \mathbb{R}$

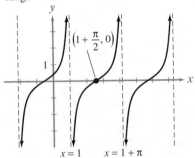

41. $y = -\tan(1 - 2x)$; period $= \pi/2$, phase shift $= \frac{1}{2}$, range $= \mathbb{R}$

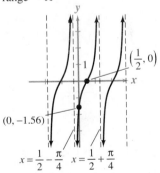

43. $y = \sec 2x$; period $= \pi$, phase shift $= 0$, range $= (-\infty, -1] \cup [1, \infty)$

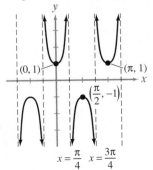

45. $y = \sec(x + \pi)$; period $= 2\pi$, phase shift $= -\pi$, range $= (-\infty, -1] \cup [1, \infty)$

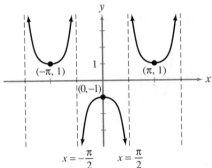

47. $y = -2 \csc\left(x - \dfrac{\pi}{2}\right)$; period $= 2\pi$, phase shift $= \pi/2$, range $= (-\infty, -2] \cup [2, \infty)$

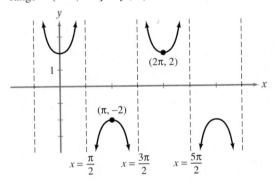

49. $y = \frac{1}{2} \tan\left(\dfrac{x}{2} + 1\right)$; period $= 2\pi$, phase shift $= -2$, range $= \mathbb{R}$

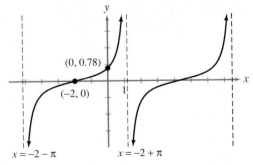

51. (a) $y = \cos\left(x + \dfrac{\pi}{2}\right)$

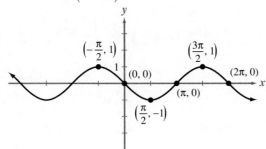

(b) $y = -\sin x$

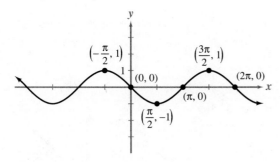

The graphs of $y = \cos\left(x + \dfrac{\pi}{2}\right)$ and $y = -\sin x$ are identical.

53. (a) $y = \tan\left(x + \dfrac{\pi}{2}\right)$

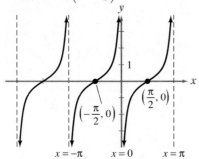

(b) $y = -\cot x$

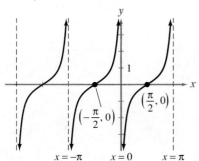

The graphs of $y = \tan\left(x + \dfrac{\pi}{2}\right)$ and $y = -\cot x$ are identical.

55. (a) $\frac{1}{125}$ **(b)** 3 **(c)** $t_0/250$
    (d) 125 cycles per sec **(e)** $t_0/250$ sec

57. (a) $\frac{1}{5}$ **(b)** 100 **(c)** $\dfrac{1}{5\pi}$
    (d) 5 cycles per sec **(e)** $(3\pi + 4)/20\pi$ sec

59. (a) $y = \sqrt{2}\,\sin(\sqrt{3}x - 1)$

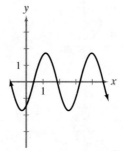

(b) $y = -\sqrt{3}\,\cos(\sqrt{5}x + 1)$

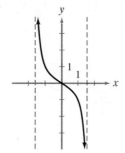

61. (a) $y = \tan(-x)$

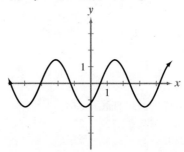

(b) $y = -\tan x$

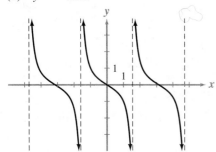

The graphs of $y = \tan(-x)$ and $y = -\tan x$ are identical.

Exercises 5.4

1. $-\pi/4$ **3.** $\pi/6$ **5.** $-\pi/6$ **7.** $-1/\sqrt{3}$
9. $-\pi/6$ **11.** $-\pi/6$ **13.** -25 **15.** $\frac{4}{5}$
17. $1/\sqrt{5}$ **19.** $\sqrt{6}/3$ **21.** Undefined   **23.** 1.5
25. 100 **27.** $\sqrt{2501}$ **29.** 1.737 **31.** -0.8481
33. 1.057 **35.** 0.3802
37. $f(x) = -\sin^{-1} x$

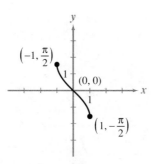

39. $f(x) = 2 + \cos^{-1} x$

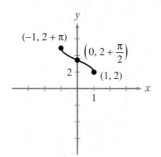

41. $f(x) = 1 - \tan^{-1} x$

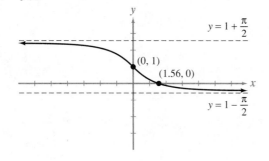

43. $f(x) = -\cos^{-1} x$

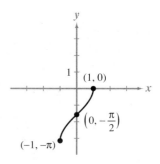

45. $f(x) = 1 - \sin^{-1} x$

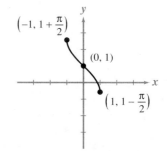

47. $f(x) = \cos^{-1}(\cos x)$

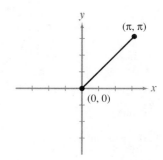

49. $f(x) = \tan(\arctan x)$

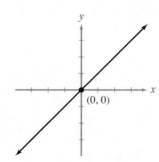

55. $y = \sec^{-1} x$, where y is in $[0, \pi/2) \cup [\pi, 3\pi/2)$

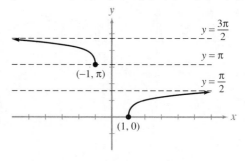

57. $y = \cot^{-1} x$, where y is in $(0, \pi)$

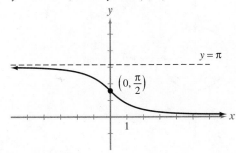

59. No, since the cosine function is not one-to-one on this interval.

61. Since the inverse tangent function is an odd function (see Exercise 51), it follows that its graph is symmetric with respect to the origin.

63. $x = \frac{1}{3} \cos\left(\dfrac{y - 4}{5}\right) + 3$ **65.** $x = -\sin\left(\dfrac{y + 3}{4}\right) + 9$

67. $x = \dfrac{2}{\pi} - \dfrac{1}{\pi} \tan(y - 3)$

69. $f(x) = \sin x$ and $g(x) = \ln x$ (other solutions are possible)

71. $f(x) = \cos x$ and $g(x) = \tan^{-1} x$ (other solutions are possible)

73. $f(x) = e^x$ and $g(x) = \sec x$ (other solutions are possible)

75. **(a)** $y = -2 \sin^{-1}(x - 1) + 1$

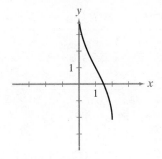

(b) $y = 2 \cos^{-1}(2x + 1) - 1$

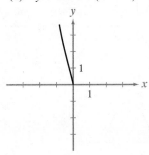

Chapter 5 Review Exercises

1. $P(-9\pi) = (-1, 0)$

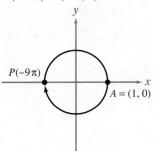

3. $P(17\pi/3) = (\frac{1}{2}, -\sqrt{3}/2)$

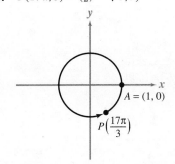

5. $P(-25\pi/4) = (\sqrt{2}/2, -\sqrt{2}/2)$

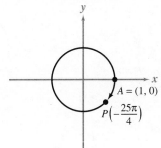

7. $P(-17\pi/6) = (-\sqrt{3}/2, -\frac{1}{2})$

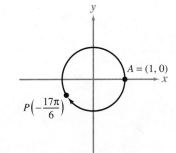

9. $P(25\pi/2) = (0, 1)$

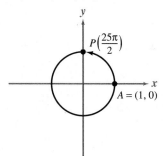

11.

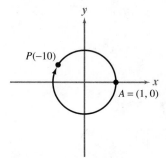

13.

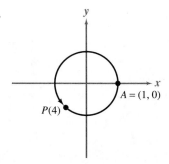

15.

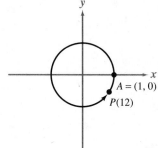

17.

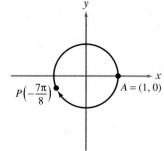

19.

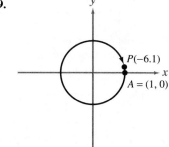

21. 0 and π **23.** No solutions **25.** $\pi/6$ and $11\pi/6$
27. $-\sqrt{22}/5$ **29.** $-3\sqrt{26}$
31. Undefined **33.** $2/\sqrt{3}$ **35.** $-\sqrt{2}/2$
37. $\sqrt{3}/2$ **39.** -1 **41.** $-1/\sqrt{3}$

	sin t	cos t	tan t	cot t	sec t	csc t
43.	$1/\sqrt{3}$	$-\sqrt{6}/3$	$-1/\sqrt{2}$	$-\sqrt{2}$	$-3/\sqrt{6}$	$\sqrt{3}$
45.	$-\sqrt{3}/5$	$\sqrt{22}/5$	$-\sqrt{3}/\sqrt{22}$	$-\sqrt{22}/\sqrt{3}$	$5/\sqrt{22}$	$-5/\sqrt{3}$
47.	$-\frac{2}{3}$	$-\sqrt{5}/3$	$2/\sqrt{5}$	$\sqrt{5}/2$	$-3/\sqrt{5}$	$-\frac{3}{2}$

49. 1.538 **51.** -0.8796 **53.** -0.6359
55. -1.826 **57.** -0.2915 **59.** $\sqrt{5}/2$
61. $-7/4\sqrt{3}$ **63.** $1/\sqrt{15}$
65.

c	sin t	cos t	tan t	cot t	sec t	csc t
-4	$-\sqrt{11}/4$	$-\sqrt{5}/4$	$\sqrt{11}/\sqrt{5}$	$\sqrt{5}/\sqrt{11}$	$-4/\sqrt{5}$	$-4/\sqrt{11}$

69. $y = 5\cos x$; period $= 2\pi$, amplitude $= 5$,
phase shift $= 0$, range $= [-5, 5]$

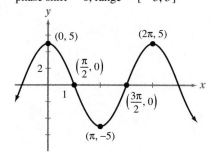

71. $y = \sin 4x$; period $= \pi/2$, amplitude $= 1$,
phase shift $= 0$, range $= [-1, 1]$

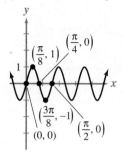

73. $y = -3 \cos(x/3)$; period $= 6\pi$, amplitude $= 3$,
phase shift $= 0$, range $= [-3, 3]$

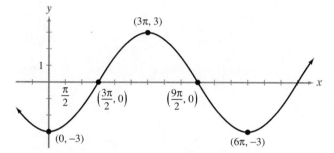

75. $y = 2 \sin(3x/2)$; period $= 4\pi/3$, amplitude $= 2$,
phase shift $= 0$, range $= [-2, 2]$

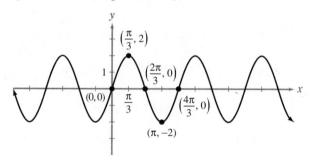

77. $y = \frac{1}{2} \cos(x/2)$; period $= 4\pi$, amplitude $= \frac{1}{2}$,
phase shift $= 0$, range $= [-\frac{1}{2}, \frac{1}{2}]$

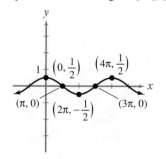

79. $y = -2 \cos(3x + 1)$; period $= 2\pi/3$, amplitude $= 2$, phase
shift $= -\frac{1}{3}$, range $= [-2, 2]$

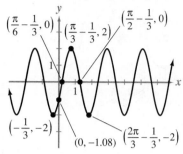

81. $y = 4 \sin(2 - 3x)$; period $= 2\pi/3$, amplitude $= 4$,
phase shift $= \frac{2}{3}$, range $= [-4\ 4]$

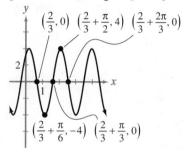

83. $y = \cot(-x)$; period $= \pi$, phase shift $= 0$,
range $= \mathbb{R}$

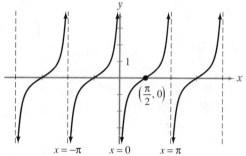

85. $y = 2 \tan(x/2)$; period $= 2\pi$, phase shift $= 0$,
range $= \mathbb{R}$

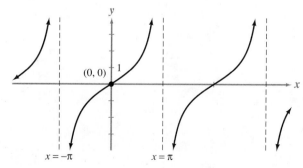

87. $y = \sec(2 - x)$; period $= 2\pi$, phase shift $= 2$,
range $= (-\infty, -1] \cup [1, \infty)$

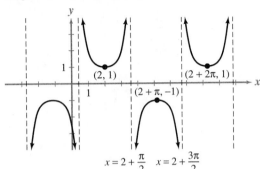

89. $y = -2\csc(2 - 4x)$; period $= \pi/2$, phase shift $= \frac{1}{2}$,
range $= (-\infty, -2] \cup [2, \infty)$

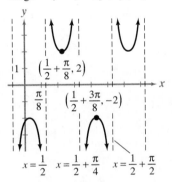

91. (a) $\frac{1}{60}$ (b) 50   (c) $\frac{1}{120}$
   (d) 60 cycles per sec   (e) $\frac{1}{80}$ sec
93. $-\pi/3$ **95.** $\pi/4$ **97.** $\pi/2$ **99.** $-\pi/4$
101. $-\pi/3$ **103.** $-\pi$ **105.** $\pi/3$
107. $\sqrt{7}/4$ **109.** $\sqrt{26}$ **111.** -0.5944
113. 1.531   **115.** 2.786
117. $f(x) = -\cos^{-1} x$

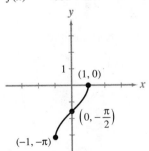

119. $f(x) = 1 + \tan^{-1} x$

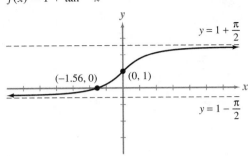

121. $f(x) = 2 + \sin^{-1}(-x)$

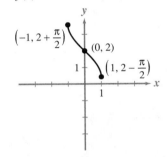

123. $x = -\frac{12}{11} - \frac{1}{11}\sin\left(\dfrac{y - 9}{10}\right)$

125. $x = \dfrac{1}{\pi}\tan\left(\dfrac{y + 5}{6}\right) - \dfrac{2}{\pi}$

127. $x = \frac{1}{4} + \frac{1}{8}\cos\left(\dfrac{y - 4}{7}\right)$

CHAPTER 6

Exercises 6.1
   1. 75.537°   **3.** 187.253°   **5.** 88°7′23″
   7. 286°53′49″
   9. 1.66 radians

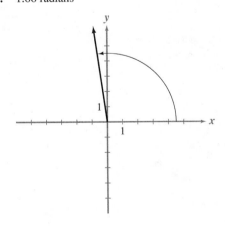

11. $\pi^2/90$ radians

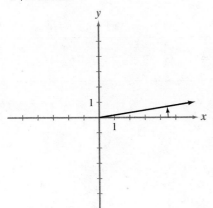

17. $-7\pi/4$ radians

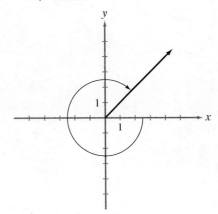

13. $-4\pi/3$ radians

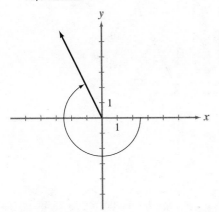

19. $-162°$

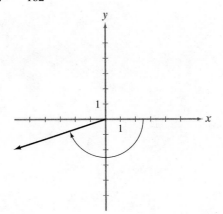

15. $19\pi/6$ radians

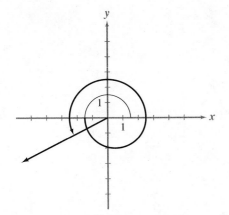

21. $286.479°$

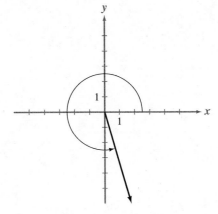

23. $-330°$

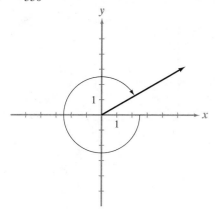

25. $-405°$

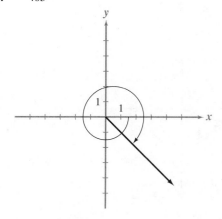

27. $-82.5°$

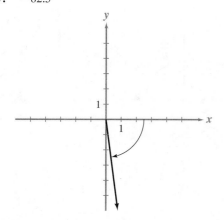

29. $s = 40\pi/9$ cm, $A = 200\pi/9$ cm^2
31. $s = 2\pi^2/45$ in., $A = 8\pi^2/45$ in.^{2}
33. $s = 60$ ft, $A = 90$ ft^2 **35.** $-\sqrt{3}/2$ **37.** $\sqrt{2}/2$

39. $-\sqrt{3}$ **41.** $\sqrt{3}$

43.

sin θ	cos θ	tan θ	cot θ	sec θ	csc θ
$-\dfrac{4}{\sqrt{17}}$	$-\dfrac{1}{\sqrt{17}}$	4	$\dfrac{1}{4}$	$-\sqrt{17}$	$-\dfrac{\sqrt{17}}{4}$

45.

sin θ	cos θ	tan θ	cot θ	sec θ	csc θ
$\dfrac{5}{\sqrt{26}}$	$-\dfrac{1}{\sqrt{26}}$	-5	$-\dfrac{1}{5}$	$-\sqrt{26}$	$\dfrac{\sqrt{26}}{5}$

47. $\sin\theta = \sqrt{119}/12$ $\tan\theta = \sqrt{119}/5$
$\cot\theta = 5/\sqrt{119}$ $\sec\theta = \frac{12}{5}$
$\csc\theta = 12/\sqrt{119}$

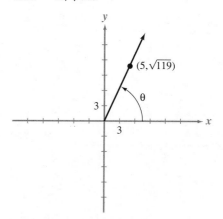

49. $\sin\theta = -2\sqrt{2}/3$ $\cos\theta = \frac{1}{3}$ $\tan\theta = -2\sqrt{2}$
$\cot\theta = -1/(2\sqrt{2})$ $\csc\theta = -3/(2\sqrt{2})$

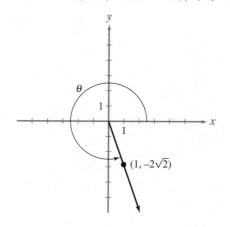

51. $\cos \theta = -4\sqrt{2}/9$ $\tan \theta = 7/4\sqrt{2}$ $\cot \theta = 4\sqrt{2}/7$
$\sec \theta = -9/4\sqrt{2}$ $\csc \theta = -\frac{9}{7}$

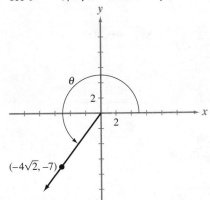

53. $2640/\pi$ **55.** $25/198$ **57.** 15 ft
59. $35\pi/18$ in.   **61.** π radians   **63.** $315/\pi$ cm
65. Angular velocity: 2800π radians per min;
linear velocity: 8400π in. per min
67. Angular velocity: 200π radians per min;
linear velocity: 1000π ft per min
69. $176/(13\pi)$ **71.** $\frac{24}{5}$ **73.** 3.2

Exercises 6.2
1. -1 **3.** 2   **5.** $\csc \theta$ **7.** $\csc x$
9. $\tan^4 v$ **11.** $-\csc^2 \theta$
79. If $x = 7\pi/6$, then $-\frac{1}{2} \neq \frac{1}{2}$.
81. If $x = \pi/6$, then $\sqrt{3} \neq 1/\sqrt{3}$.
83. If $x = \pi/4$, then $\tan(\cot x) = \tan 1 \approx 1.56 \neq 1$.
85. If $x = \pi/4$, then $2 \neq 1$.
87. If $x = \pi/3$, then $0.69 \neq -1.44$ **89.** $a \cos \theta$
91. $(1/a) \cos \theta$ **93.** $\frac{2}{3} \cot \theta$ **95.** $\frac{1}{3} \cos \theta$

97. **(a)** $y = \dfrac{\sin^2 2x}{\cos^2 2x}$

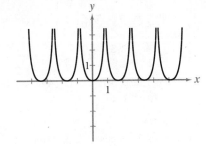

(b) $y = \tan^2 2x$

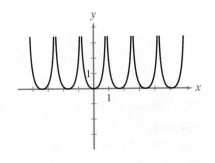

Over the interval $(-\pi/2, \pi/2)$, the graphs of
$y = \sin^2 2x/\cos^2 2x$ and $y = \tan^2 2x$ are identical.

Exercises 6.3
1. $\dfrac{\sqrt{6} + \sqrt{2}}{4}$ **3.** $\dfrac{\sqrt{3} - 1}{\sqrt{3} + 1}$ **5.** $\sqrt{6} + \sqrt{2}$
7. $\dfrac{-\sqrt{2} - \sqrt{6}}{4}$ **9.** $\dfrac{\sqrt{2} - \sqrt{6}}{4}$
11. $\dfrac{-\sqrt{6} - \sqrt{2}}{4}$ **13.** 0   **15.** $\dfrac{\sqrt{2}}{2}$ **17.** -1
19. $\dfrac{\sqrt{3}}{2}$ **21.** 0   **23.** $-\sqrt{3}$ **25.** 0
27. **(a)** $\dfrac{-\sqrt{2} + \sqrt{105}}{12}$ **(b)** $\dfrac{\sqrt{30} + \sqrt{7}}{12}$
(c) $\dfrac{-\sqrt{2} - \sqrt{105}}{12}$ **(d)** $\dfrac{\sqrt{30} - \sqrt{7}}{12}$
(e) $\dfrac{\sqrt{105} - \sqrt{2}}{\sqrt{30} + \sqrt{7}}$ **(f)** $\dfrac{\sqrt{2} + \sqrt{105}}{\sqrt{7} - \sqrt{30}}$
(g) Q I   **(h)** Q IV
29. **(a)** $\dfrac{1 + 2\sqrt{35}}{6\sqrt{5}}$ **(b)** $\dfrac{-2 + \sqrt{35}}{6\sqrt{5}}$
(c) $\dfrac{1 - 2\sqrt{35}}{6\sqrt{5}}$ **(d)** $\dfrac{-2 - \sqrt{35}}{6\sqrt{5}}$
(e) $\dfrac{2\sqrt{35} + 1}{\sqrt{35} - 2}$ **(f)** $\dfrac{2\sqrt{35} - 1}{\sqrt{35} + 2}$
(g) Q I   **(h)** Q III
31. **(a)** $\dfrac{2 + 6\sqrt{2}}{3\sqrt{13}}$ **(b)** $\dfrac{-4\sqrt{2} + 3}{3\sqrt{13}}$ **(c)** $\dfrac{2 - 6\sqrt{2}}{3\sqrt{13}}$
(d) $\dfrac{-4\sqrt{2} - 3}{3\sqrt{13}}$ **(e)** $\dfrac{2 + 6\sqrt{2}}{3 - 4\sqrt{2}}$
(f) $\dfrac{6\sqrt{2} - 2}{4\sqrt{2} + 3}$ **(g)** Q II   **(h)** Q III
33. $-\sin \theta$ **35.** $-\sin t$
37. $-\sin u$ **39.** $-\tan v$ **41.** $-\cos w$
43. $\dfrac{\sqrt{3} \tan u + 1}{\sqrt{3} - \tan u}$ **45.** $\dfrac{-4 - 2\sqrt{2}}{3\sqrt{17}} \approx -0.55$
47. $\dfrac{3\sqrt{5} + 2\sqrt{7}}{12} \approx 1.00$ **49.** $\dfrac{4 + 6\sqrt{6}}{8\sqrt{6} - 3} \approx 1.13$
69. **(a)** $y = \sin(\pi - x)$

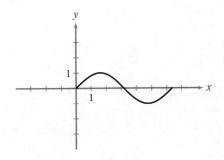

(b) $y = \sin x$

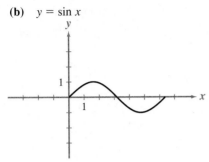

Over the interval $[0, 2\pi)$, the graphs of the functions $y = \sin(\pi - x)$ and $y = \sin x$ are identical.

71. **(a)** $y = \cos\left(\dfrac{\pi}{6} - x\right)$

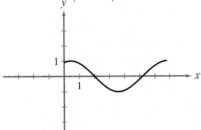

(b) $y = \sin\left(\dfrac{\pi}{3} + x\right)$

Over the interval $[0, 2\pi)$, the graphs of the functions $y = \cos\left(\dfrac{\pi}{6} - x\right)$ and $y = \sin\left(\dfrac{\pi}{3} + x\right)$ are identical.

Exercises 6.4

1. $\sin 24$ **3.** $\cos 100°$ **5.** $\cos 5t$ **7.** $\tan 25x$

9. $\tan 50°$

11. $\sin 2v = -2\sqrt{66}/25;\ \cos 2v = \frac{19}{25};$
$\tan 2v = -2\sqrt{66}/19;$ Q IV

13. $\sin 2v = -\frac{4}{5};\ \cos 2v = \frac{3}{5};\ \tan 2v = -\frac{4}{3};$ Q IV

15. $\sin 2v = 4\sqrt{6}/25;\ \cos 2v = -\frac{23}{25};$
$\tan 2v = -4\sqrt{6}/23;$ Q II

17. $\sin 2v = \frac{24}{25};\ \cos 2v = -\frac{7}{25};$
$\tan 2v = -\frac{24}{7};$ Q II

19. $\sin 2v = \sqrt{35}/18;\ \cos 2v = \frac{17}{18};$
$\tan 2v = \sqrt{35}/17;$ Q I

21. $-\dfrac{\sqrt{2 - \sqrt{2}}}{2}$ **23.** $\dfrac{\sqrt{2 - \sqrt{2 - \sqrt{2}}}}{2}$

25. $-\sqrt{\dfrac{2 + \sqrt{3}}{2 - \sqrt{3}}}$ **27.** $\dfrac{\sqrt{2 - \sqrt{3}}}{2}$

29. $-\sqrt{\dfrac{2 + \sqrt{2}}{2 - \sqrt{2}}}$ **31.** $-\dfrac{\sqrt{2 + \sqrt{2}}}{2}$

33. $-\dfrac{\sqrt{2 - \sqrt{2}}}{2}$ **35.** $\sqrt{\dfrac{2 - \sqrt{3}}{2 + \sqrt{3}}}$

37. $\sin\dfrac{v}{2} = \sqrt{\dfrac{6 - \sqrt{35}}{12}};$

$\cos\dfrac{v}{2} = -\sqrt{\dfrac{6 + \sqrt{35}}{12}};$

$\tan\dfrac{v}{2} = -\sqrt{\dfrac{6 - \sqrt{35}}{6 + \sqrt{35}}}$

39. $\sin\dfrac{v}{2} = \dfrac{\sqrt{11}}{4};\ \cos\dfrac{v}{2} = \dfrac{\sqrt{5}}{4};\ \tan\dfrac{v}{2} = \dfrac{\sqrt{55}}{5}$

41. $\sin\dfrac{v}{2} = \sqrt{\dfrac{\sqrt{10} + 1}{2\sqrt{10}}};$

$\cos\dfrac{v}{2} = -\sqrt{\dfrac{\sqrt{10} - 1}{2\sqrt{10}}};$

$\tan\dfrac{v}{2} = -\sqrt{\dfrac{\sqrt{10} + 1}{\sqrt{10} - 1}}$

43. $\sin\dfrac{v}{2} = \sqrt{\dfrac{3}{7}};\ \cos\dfrac{v}{2} = \dfrac{2}{\sqrt{7}};\ \tan\dfrac{v}{2} = \dfrac{\sqrt{3}}{2}$

45. $4\sqrt{6}/25$ **47.** -1 **49.** $-\frac{12}{5}$

51. $-12\sqrt{13}/49$

Exercises 6.5

1. $\frac{1}{2}(\cos 5x + \cos x)$ **3.** $\frac{1}{2}(\cos 30° - \cos 70°)$

5. $\frac{1}{2}(\sin 8\theta - \sin 2\theta)$ **7.** $\frac{1}{2}(\sin 18t - \sin 2t)$

9. $5(\cos 50t + \cos 10t)$ **11.** $2\sin(9\theta/2)\cos(3\theta/2)$

13. $2\sin 37° \sin 25°$ **15.** $2\cos(15x/2)\cos(x/2)$

17. $-2\cos 16v \sin 4v$ **19.** $-2\sin(y/2)\cos(5y/2)$

21. $-\sqrt{2}/2$ **23.** $-\sqrt{2}/2$ **25.** $(\sqrt{3} - 2)/4$

27. $(\sqrt{3} + 2)/4$ **29.** $-\sqrt{2}/2$ **31.** $\frac{1}{4}$ **33.** $\sqrt{2}/4$

53. $\frac{1}{2}\sin(2ax + 2b)$

Exercises 6.6

1. $-\sin 20°$ **3.** $-\cos 76°$ **5.** $-\tan 1.36$

7. $\{x \mid x = \pi/6 + 2k\pi$ or $x = 5\pi/6 + 2k\pi$, where k is any integer$\}$

9. $\{x \mid x = \pi/4 + k\pi$, where k is any integer$\}$

11. $\{x \mid x = 5\pi/4 + 2k\pi$ or $x = 7\pi/4 + 2k\pi$, where k is any integer$\}$

13. $\{x \mid x = 5\pi/6 + 2k\pi$ or $x = 7\pi/6 + 2k\pi$, where k is any integer$\}$

15. $\{x \mid x = 5\pi/6 + k\pi$, where k is any integer$\}$

17. $\{x \mid x = \pi/2 + 2k\pi$, where k is any integer$\}$

19. $\{x \mid x = 3\pi/4 + 2k\pi$ or $x = 5\pi/4 + 2k\pi$, where k is any integer$\}$

21. $\{2\pi/3, 4\pi/3\}$ 23. $\{3\pi/8, 7\pi/8, 11\pi/8, 15\pi/8\}$

25. $\{\pi/3, 2\pi/3\}$ 27. $\{\pi/4, 3\pi/4, 5\pi/4, 7\pi/4\}$

29. $\{5\pi/6, 7\pi/6\}$ 31. $\{240°, 300°\}$ 33. $\{45°, 315°\}$

35. $\{30°, 120°, 210°, 300°\}$ 37. $\{105°, 165°, 285°, 345°\}$

39. $\{45°, 90°, 135°, 270°\}$ 41. $\pi/2, 7\pi/6, 11\pi/6$

43. $\pi/2, 3\pi/2$ 45. $0, \pi, \pi/4, 7\pi/4$

47. $\pi/6, \pi/2, 5\pi/6, 3\pi/2$ 49. $0, \pi, 7\pi/6, 11\pi/6$

51. $2\pi/3, 4\pi/3$ 53. $7\pi/6, 11\pi/6$

55. $\pi/6, 5\pi/6, 3.9, 5.5$ 57. $0, \pi/4, 3\pi/4, \pi, 5\pi/4, 7\pi/4$

59. $0.2, 2.9$ 61. $0.2, 1.0, 3.4, 4.1$ 63. $11\pi/6$

65. $\pi/2, 11\pi/6$ 67. $\pi/2, \pi$ 71. 0.01 sec

73. (a) 2.06 sec   (b) 2.19 sec

75. (a) 1.55 sec   (b) 5.04 sec

77. 24.3°   79. 76.6°

81. $y = 3 \cos x$ and $y = 2$

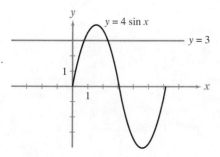

From the graphs, we estimate the x-coordinates of the points of intersection (to the nearest tenth) to be 0.9 and 5.6. Since the second linear equation solved in Example 41 is equivalent to $3 \cos x = 2$, these solutions should be identical to those we found there.

83. $y = 4 \sin x$ and $y = 3$

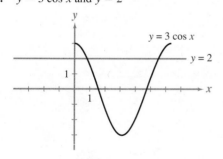

From the graphs, we estimate the x-coordinates of the points of intersection (to the nearest tenth) to be 0.8 and 2.3. These values of x approximate the solutions in $[0, 2\pi)$ to the equation $\sin x = \frac{3}{4}$.

85. $y = \cos x$ and $y = e^{-x/5}$

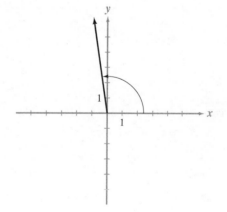

From the graphs, we estimate the x-coordinates of the points of intersection (to the nearest tenth) to be 0.4 and 5.1. These values of x approximate the solutions in $[0, 2\pi)$ to the equation $\cos x = e^{-x/5}$.

Chapter 6 Review Exercises

1. 202.725°

3. $5\pi/9$ radians

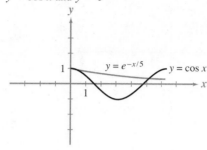

5. $-5\pi/4$ radians

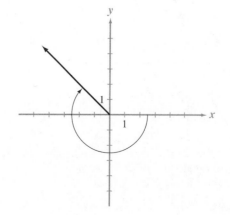

7. $-126°$

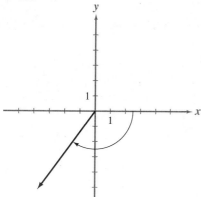

9. $97.5°$

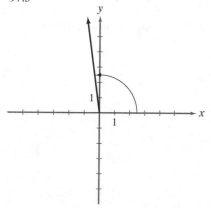

11. $s = 7\pi/3$ cm, $A = 14\pi$ cm^2 **13.** $\frac{1}{2}$ **15.** $\sqrt{2}/2$

17. $\sin\theta = -3/\sqrt{13}$ $\cos\theta = -2/\sqrt{13}$
$\sec\theta = -\sqrt{13}/2$
$\csc\theta = -\sqrt{13}/3$
$\cot\theta = \frac{2}{3}$

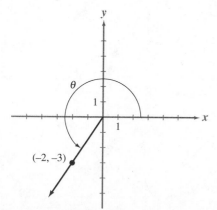

19. $\cos\theta = \sqrt{65}/9$ $\sec\theta = 9/\sqrt{65}$ $\csc\theta = -\frac{9}{4}$
$\tan\theta = -4/\sqrt{65}$ $\cot\theta = -\sqrt{65}/4$

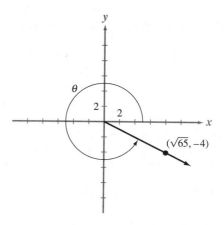

21. $1760/\pi$ **23.** $3\pi/2$ ft   **25.** $300/\pi$ cm

27. Angular velocity: 3000π radians per min, linear velocity: $18{,}000\pi$ in. per min

29. $44/13\pi$ **31.** -1 **33.** $\cos^4 t$

35. If $x = 2\pi/3$, then $-\frac{1}{2} \neq \frac{1}{2}$.   **37.** $3\sec\theta$

39. $(\sqrt{2} - \sqrt{6})/4$ **41.** $(\sqrt{2} - \sqrt{6})/4$

43. 1 **45.** $-\sqrt{2}/2$ **47.** 1

49. (a) $\dfrac{\sqrt{105} - \sqrt{2}}{12}$ (b) $\dfrac{\sqrt{7} + \sqrt{30}}{12}$

(c) $\dfrac{\sqrt{105} + \sqrt{2}}{12}$ (d) $\dfrac{\sqrt{7} - \sqrt{30}}{12}$

(e) $\dfrac{\sqrt{105} - \sqrt{2}}{\sqrt{7} + \sqrt{30}}$ (f) $\dfrac{\sqrt{105} + \sqrt{2}}{\sqrt{7} - \sqrt{30}}$

(g) Q I   (h) Q II

51. $-\cos\theta$ **53.** $\dfrac{\sqrt{3}\cos u + \sin u}{2}$

55. $\dfrac{4\sqrt{3} - 1}{2\sqrt{17}} \approx 0.72$

57. $\sin 50$ **59.** $\cos 46°$

61. $\sin 2v = -2\sqrt{46}/25$; $\cos 2v = \frac{21}{25}$;
$\tan 2v = -2\sqrt{46}/21$; Q IV

63. $\dfrac{\sqrt{2 + \sqrt{2}}}{2}$ **65.** $\sqrt{\dfrac{2 + \sqrt{2}}{2 - \sqrt{2}}}$

67. $\sin\dfrac{v}{2} = \dfrac{\sqrt{5}}{2\sqrt{3}}$; $\cos\dfrac{v}{2} = -\dfrac{\sqrt{7}}{2\sqrt{3}}$; $\tan\dfrac{v}{2} = -\dfrac{\sqrt{5}}{\sqrt{7}}$

69. $-\frac{1}{2}$ **71.** $\frac{1}{2}(\cos x - \cos 5x)$

73. $\frac{1}{2}(\sin 70° - \frac{1}{2})$ **75.** $2\cos(9\theta/2)\cos(3\theta/2)$

77. $2\cos 37° \sin 25°$ **79.** $\sqrt{6}/2$ **81.** $-\frac{1}{4}(2 + \sqrt{2})$

83. $\{x \mid x = 7\pi/6 + 2k\pi$ or $x = 11\pi/6 + 2k\pi$, where k is any integer$\}$

85. $\{x \mid x = 3\pi/4 + k\pi$, where k is any integer$\}$

87. $\{\pi/2, 3\pi/2\}$ **89.** $\{5\pi/4, 7\pi/4\}$

91. $\{75°, 105°, 195°, 225°, 315°, 345°\}$

93. $\{0°, 135°, 180°, 225°\}$ **95.** $\pi/2, 7\pi/6,$ and $11\pi/6$

97. $0, \pi, 7\pi/6,$ and $11\pi/6$ **99.** $\pi/3, 2\pi/3, 4\pi/3,$ and $5\pi/3$

101. 1.3, 2.0, 4.4, and 5.2 **103.** $7\pi/6$ and $11\pi/6$

105. 0.03 sec **107.** 1.16 sec **109.** 34.0°

CHAPTER 7

Exercises 7.1

1. $\beta = 18.6°, a \approx 30, b \approx 10$

3. $\beta \approx 8.0°, \alpha \approx 82.0°, a \approx 816$

5. $\alpha = 29.3°, \beta \approx 60.7°, c \approx 4.80$

7. $\alpha = 63°10', b \approx 1644, c \approx 3642$

9. $\alpha \approx 39°40', b \approx 3.79, c \approx 4.92$

11. $\alpha = 44.8°, a \approx 13{,}700, b \approx 13{,}800$

13. $\beta = 62°, \alpha \approx 1.4, c \approx 2.9$

15. $\alpha = 79.6°, a \approx 396, c \approx 402$

17. $\beta = 30°, a = 12\sqrt{3}, c = 24$

19. $\alpha = 60°, a = 144\sqrt{3}, c = 288$

21. $\alpha = 60°, \beta = 30°, b = 5$

23. $\beta = 45°, b = 17, c = 17\sqrt{2}$

25. 1110 ft **27.** 17°

29. Equal sides ≈ 58.6 cm, third side ≈ 70.8 cm

31. 29.6 cm **33.** 14.2° **37.** 8.7 cm

39. 58° **41.** 13,760 m **43.** 3990 mi

45. 960 ft **47.** 11.9° **49.** 5990 ft

51. 15.8 mi

53. Approximately 254 paces in the direction S28.2°E

55. 353 m **57.** 490 mi; 161°20′

Exercises 7.2

1. $\alpha = 57.5°, a \approx 32.4, b \approx 13.0$

3. $\beta_1 \approx 59°40', \alpha_1 \approx 83°10', a_1 \approx 1440; \beta_2 \approx 120°20',$
$\alpha_2 \approx 22°30', a_2 \approx 554$

5. $\beta_1 \approx 73°, \gamma_1 \approx 88°20', c_1 \approx 734; \beta_2 \approx 107°,$
$\gamma_2 \approx 54°20', c_2 \approx 597$

7. $\alpha \approx 16.7°, \beta \approx 111.5°, \gamma \approx 51.8°$

9. $\gamma = 95°55', a \approx 0.94, c \approx 9.89$

11. $\alpha_1 \approx 57°30', \gamma_1 \approx 90°40', c_1 \approx 14.2; \alpha_2 \approx 122°30',$
$\gamma_2 \approx 25°40', c_2 \approx 6.2$

13. $\alpha \approx 24°20', \gamma \approx 121°58', c \approx 25.00$

15. $c \approx 140.2, \alpha \approx 20°20', \beta \approx 39°10'$

17. $b \approx 3.56, \alpha \approx 36.8°, \gamma \approx 28.1°$ **19.** No triangle

21. $\alpha \approx 52°, \gamma \approx 70°10', c \approx 5.59$

23. $b \approx 67.0, \alpha \approx 19.4°, \gamma \approx 28.3°$ **25.** 1890

27. 715 **29.** 3,700,000 **31.** 0.48

33. 7360 **35.** 697,000 **37.** 53°10′

39. 42.3 ft **41.** 120 km; N88°10′E

43. 37.7° **45.** 13.4° **47.** 217 ft

49. 7250 m^2 **51.** 8750 in.^{2}

53. 6.9 mi; N3.3°W **55.** 1080 m **57.** 8.40 mi

59. 31.9 cm; 441 cm^2

61. 65°; 150 in.^{2} **63.** $\dfrac{12}{\sqrt[4]{3}}$ in.

Exercises 7.3

1. $2 \operatorname{cis} \dfrac{\pi}{3}$ **3.** $4\sqrt{2} \operatorname{cis} \dfrac{3\pi}{4}$ **5.** $5 \operatorname{cis} \dfrac{3\pi}{2}$ **7.** $\operatorname{cis} \dfrac{11\pi}{6}$

9. 0 **11.** $2\sqrt{3} \operatorname{cis} \dfrac{5\pi}{6}$

13. $\sqrt{13} \operatorname{cis} (360° - \tan^{-1} \tfrac{2}{3}) \approx 3.6 \operatorname{cis} 326°$

15. $2 \operatorname{cis} \dfrac{5\pi}{6}$

17. $2\sqrt{3} \operatorname{cis} (180° + \tan^{-1} \sqrt{2}) \approx 3.5 \operatorname{cis} 235°$

19. $13 \operatorname{cis} (180° - \tan^{-1} \tfrac{12}{5}) \approx 13 \operatorname{cis} 113°$

21. $\dfrac{\sqrt{3}}{2} \operatorname{cis} \pi$ **23.** $4 \operatorname{cis} \dfrac{\pi}{3}$ **25.** $-\sqrt{2} + \sqrt{2}i$

27. $\sqrt{5}i$ **29.** $-\dfrac{3\sqrt{2}}{2} + \dfrac{3\sqrt{6}}{2}i$

31. $3 - 3\sqrt{3}i$ **33.** $-6\sqrt{3} - 6i$

35. $-\tfrac{5}{2}$ **37.** $-\dfrac{15}{2} - \dfrac{15\sqrt{3}}{2}i$

39. $\dfrac{3}{4} - \dfrac{\sqrt{3}}{4}i$ **41.** $(\sqrt{6} - \sqrt{2}) + (\sqrt{6} + \sqrt{2})i$

43. $4\sqrt{2} - 4\sqrt{2}i$ **45.** $8 \operatorname{cis} \pi$ **47.** $\tfrac{5}{2} \operatorname{cis} \dfrac{7\pi}{6}$

49. $6 \operatorname{cis} \dfrac{\pi}{4}$ **51.** $\dfrac{\sqrt{3}}{2} \operatorname{cis} 210°$ **53.** $-\dfrac{15\sqrt{6}}{2} + \dfrac{15\sqrt{2}}{2}i$

55. $3\sqrt{2} - 3\sqrt{2}i$ **57.** $\dfrac{3\sqrt{3}}{4}i$

59. $-\dfrac{25}{4} - \dfrac{25\sqrt{3}}{4}i$

65.

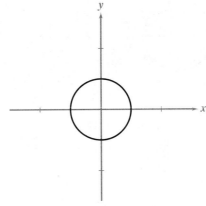

67. **(a)** $1, i, -1$

69. $z_1 \cdot z_2 \cdot z_3 = r_1 \cdot r_2 \cdot r_3 \operatorname{cis} (\theta_1 + \theta_2 + \theta_3)$

Exercises 7.4

1. $8 + 8\sqrt{3}i$ **3.** 27 **5.** $-\dfrac{\sqrt{2}}{2} - \dfrac{\sqrt{2}}{2}i$

7. $-16\sqrt{2}i$ **9.** $-\dfrac{81}{2} + \dfrac{81\sqrt{3}}{2}i$ **11.** $-\dfrac{27}{64}i$

13. $-8 - 8\sqrt{3}i$ **15.** $\dfrac{1}{2} - \dfrac{\sqrt{3}}{2}i$ **17.** $4 - 4i$

19. $-i$ **21.** -324 **23.** $524{,}288 + 524{,}288\sqrt{3}i$

25. 280 cis 140°   **27.** 825 cis 103°

29. 1728 cis 212°   **31.** 0.13 cis 225°

33. 169 cis 225°   **35.** 2197 cis 338°

37. $2^{1/5}$ cis 66°, $2^{1/5}$ cis 138°, $2^{1/5}$ cis 210°,
 $2^{1/5}$ cis 282°, $2^{1/5}$ cis 354°

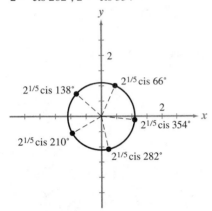

39. $2^{1/3}$ cis 75°, $2^{1/3}$ cis 195°, $2^{1/3}$ cis 315°

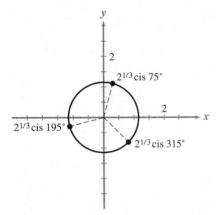

41. $4^{1/4}$ cis 30°, $4^{1/4}$ cis 120°, $4^{1/4}$ cis 210°, $4^{1/4}$ cis 300°

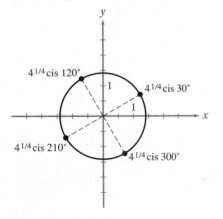

43. 4 cis 45°, 4 cis 225°

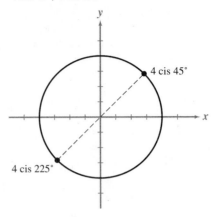

45. cis 0°, cis 60°, cis 120°, cis 180°, cis 240°, cis 300°

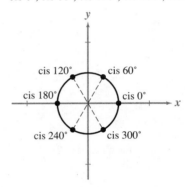

47. 2 cis 45°, 2 cis 135°, 2 cis 225°, 2 cis 315°

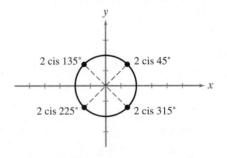

49. 2 cis 40°, 2 cis 160°, 2 cis 280°

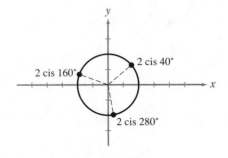

51. 3 cis 56.25°, 3 cis 146.25°, 3 cis 236.25°, 3 cis 326.25°

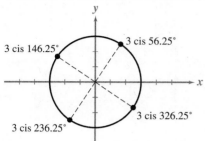

53. 2 cis 46°, 2 cis 118°, 2 cis 190°, 2 cis 262°, 2 cis 334°

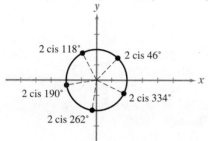

55. 3 cis 60°, 3 cis 180°, 3 cis 300°
57. cis 18°, cis 90°, cis 162°, cis 234°, cis 306°
59. $2^{1/8}$ cis 78.75°, $2^{1/8}$ cis 168.75°,
    $2^{1/8}$ cis 258.75°, $2^{1/8}$ cis 348.75°

61. $-\dfrac{1}{32} - \dfrac{\sqrt{3}}{32}\,i$ **63.** $-1 - i, 1 + i, 1 - i$

Exercises 7.5
 1. $\|v\| = \sqrt{13}$

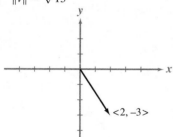

 3. $\|v\| = \sqrt{85}/4$

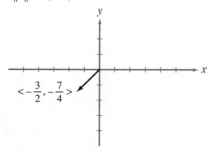

 5. $\|v\| = 6$

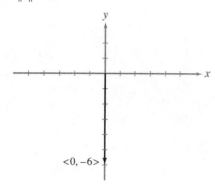

 7. $\|v\| = 2\sqrt{7}$

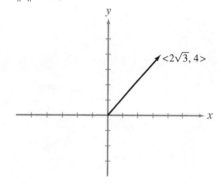

 9. $-2v = \langle 6, -12\rangle; \frac{1}{3}v = \langle -1, 2\rangle$

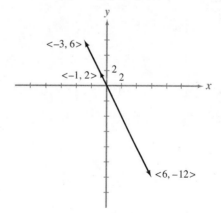

11. $-2v = \langle 4, \frac{15}{2} \rangle;\ \frac{1}{3}v = \langle -\frac{2}{3}, -\frac{5}{4} \rangle$

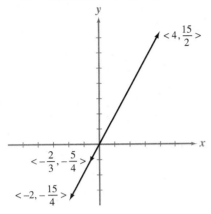

13. $\langle -1, -7 \rangle$

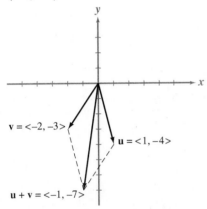

15. $\langle 1, -\frac{9}{2} \rangle$

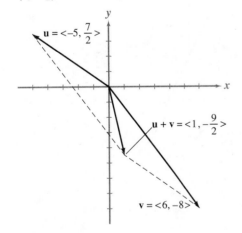

17. $\langle -\frac{33}{5}, 3 \rangle$

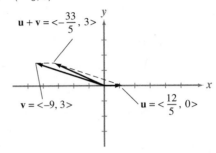

19. $\langle 7, -14 \rangle$ **21.** $\langle -\frac{7}{4}, 7 \rangle$ **23.** $\langle 2\sqrt{2}, -5\sqrt{2} \rangle$

25. $\langle -230, 680 \rangle$ **27.** $\langle \frac{3}{2}, -16 \rangle$

29. $\langle 20, -50 \rangle$ **31.** $315°$ **33.** $210°$

35. $180° - \tan^{-1}\frac{4}{3} \approx 127°$

37. $360° - \tan^{-1}\frac{5}{2} \approx 292°$

39. $\langle \frac{3}{2}, 3\sqrt{3}/2 \rangle$ **41.** $\langle -12\sqrt{2}, -12\sqrt{2} \rangle$

43. $\langle 9.52 \cos 137°, 9.52 \sin 137° \rangle \approx \langle -6.96, 6.49 \rangle$

45. 287 lb; N40.9°W **47.** 810 km/hr; N78°40'W

49. $\frac{37}{2}i + 6j$ **51.** $5i - \frac{123}{8}j$

53. $\left(\dfrac{4\sqrt{3} - 25}{8} \right)i + \left(\dfrac{15 - 96\sqrt{3}}{32} \right)j$

55. 544 mph; S53°E **57.** 239 mph; S58°W

59. $21°$ **61.** 1050 lb **63.** 330 lb

65. 315 lb; N0.2°W

67. (a) -11 (b) -4 (c) -15

69. (a) $138°$ (b) $97.1°$ (c) $152°$

71. (a) $\langle 8, -12 \rangle$ (b) $\langle \frac{79}{12}, \frac{5}{4} \rangle$

Exercises 7.6

1.

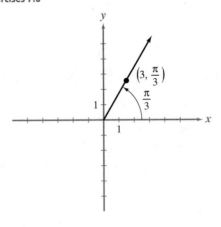

3.

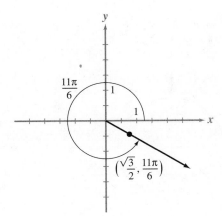

5.

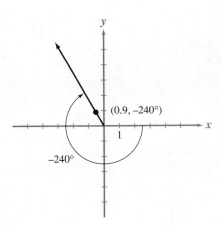

7.

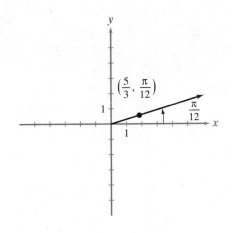

9.

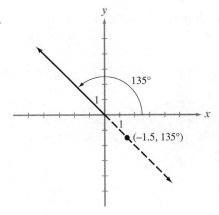

11.

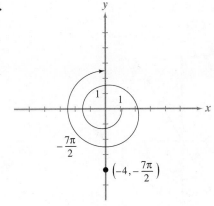

13. $(\sqrt{2}, \sqrt{2})$ **15.** $(\sqrt{5}/2, -\sqrt{15}/2)$

17. $(\sqrt{6} - \sqrt{2}, \sqrt{6} + \sqrt{2})$ **19.** $(0, 3)$

21. $(4\sqrt{2}/5, -4\sqrt{2}/5)$ **23.** $(4, \pi/6)$ **25.** $(4, \pi)$

27. $(5, 360° - \tan^{-1}\frac{4}{3}) \approx (5, 307°)$

29. $(\sqrt{6}/2, 180° + \tan^{-1}(\sqrt{2}/2)) \approx (1.2, 215°)$

31. $(4.5, 360° - \tan^{-1}\frac{4}{3}) \approx (4.5, 307°)$

33. $(5, 9\pi/4), (5, 17\pi/4), (-5, 5\pi/4), (-5, 13\pi/4)$

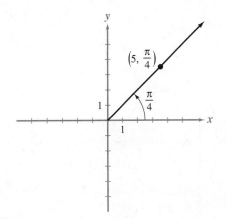

35. $(3, \pi/6), (3, 13\pi/6), (-3, 19\pi/6), (-3, 31\pi/6)$

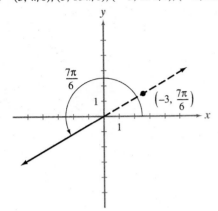

37. $(2.5, \pi/3), (2.5, 7\pi/3), (-2.5, -8\pi/3),$
$(-2.5, -14\pi/3)$

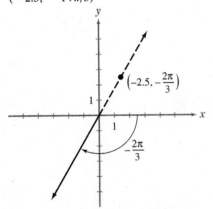

39. **(a)** yes   **(b)** no   **(c)** yes   **41.** $x^2 + y^2 = 9$
43. $y = -\sqrt{3}x$ **45.** $2x + y = 5$ **47.** $y = 4x^2 - 1$

49. $x^2 + y^2 - 8x - 9 = 0$ **51.** $r^2 = \dfrac{1}{\sin 2\theta}$

53. $r^2 = \dfrac{4}{1 + 3\sin^2 \theta}$ **55.** $r = 4\sin 2\theta$

57. $r^2 = \sin 2\theta$

59. $r = 3 + 2\cos\theta$
(limacon)

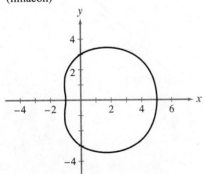

61. $r = 2\sin\theta - 3$
(limacon)

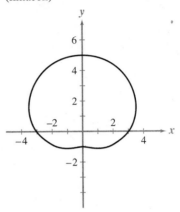

63. $r^2 = \cos 2\theta$
(lemniscate)

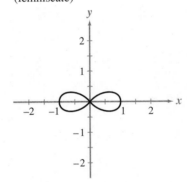

65. $r^2 = 9\sin 2\theta$
(lemniscate)

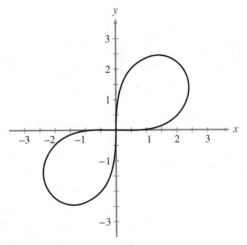

67. $r = 4 + 4 \sin \theta$
(cardioid)

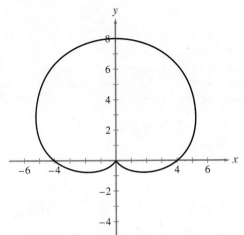

69. $r = 2 - 3 \cos \theta$
(limacon with loop)

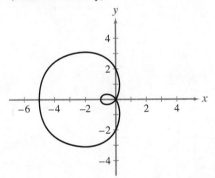

71. $r = 4 \sin 2\theta$
(4-leaf rose)

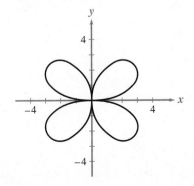

73. $r = 2 - 2 \cos \theta$
(cardioid)

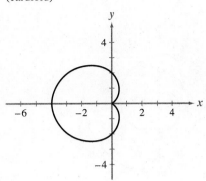

75. $r^2 = 4 \sin 4\theta$
(4-leaf rose)

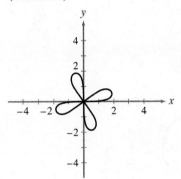

77. $r^2 = 8 \cos 2\theta$
(lemniscate)

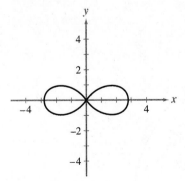

79. $r = 2\theta$
     (Spiral of Archimedes)

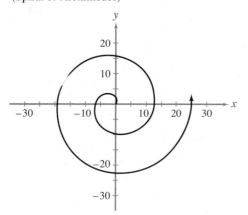

81. $r = \dfrac{2\theta}{\pi}$
     (Spiral of Archimedes)

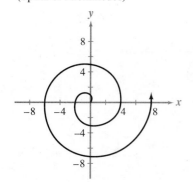

83. $r = 3 + \sec\theta$
     (conchoid)

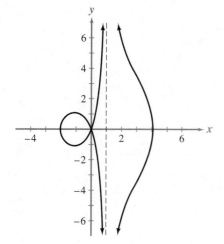

85. $r = 1/\theta$
     (reciprocal spiral)

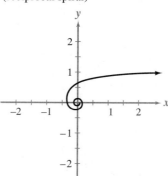

87. $r = \sin\theta\,\tan\theta$
     (cissoid)

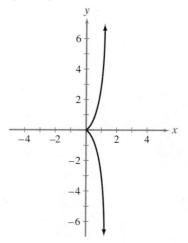

89. $r = e\,\theta$
     (logarithmic spiral)

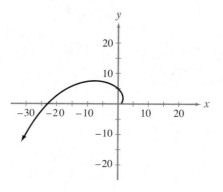

93. $\sqrt{29}$

95. $r = 1 - \sqrt{2} \cos \theta$
(limacon with loop)

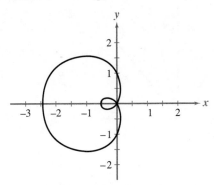

97. $r^2 = 3 \cos 2\theta$
(lemniscate)

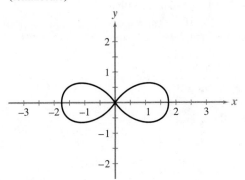

99.

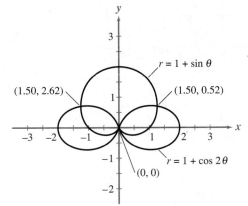

Chapter 7 Review Exercises

1. $b = 8, \alpha \approx 62°, \beta \approx 28°$

3. $\beta = 36°10', a \approx 107, c \approx 133$

5. $c \approx 1830, \alpha \approx 37.7°, \beta \approx 52.3°$

7. $\alpha = 72.9°, a \approx 894, c \approx 935$

9. $\beta = 142°10', a \approx 67.8, c \approx 42.9$

11. $\gamma \approx 58.1°, \alpha \approx 36.8°, a \approx 7.3$

13. $\alpha \approx 86.4°, \beta \approx 39.5°, \gamma \approx 54.1°$

15. $a \approx 13.8, \beta \approx 74°40', \gamma \approx 55°10'$ **17.**   No triangle

19. $c \approx 12.1, \beta \approx 84°20', \alpha \approx 53°$

21. $\gamma_1 \approx 24.9°, \beta_1 \approx 139.5°, b_1 \approx 2450; \gamma_2 \approx 155.1°,$
$\beta_2 \approx 9.3°, b_2 \approx 611$

23. 38,600 ft   **25.**   437 cm^2 **27.**   17 ft

29. 13.6 mi   **31.**   17 ft

33. 1,270,000 m^2 **35.** $2\sqrt{2} \operatorname{cis} \dfrac{7\pi}{4}$

37. $2 \operatorname{cis} \dfrac{7\pi}{6}$ **39.** $\dfrac{5}{2} \operatorname{cis} \dfrac{3\pi}{2}$

41. $\sqrt{5} \operatorname{cis} (360° - \tan^{-1} \tfrac{1}{2}) \approx 2.24 \operatorname{cis} 333°$

43. $-2 + 2\sqrt{3}i$ **45.** $-8\sqrt{2} - 8\sqrt{2}i$

47. $\dfrac{3}{4} - \dfrac{\sqrt{3}}{4}i$ **49.** $\dfrac{5(\sqrt{6} + \sqrt{2})}{4} + \dfrac{5(\sqrt{6} - \sqrt{2})}{4}i$

51. $45 \operatorname{cis} \dfrac{8\pi}{15}$ **53.** $\dfrac{\sqrt{2}}{2} \operatorname{cis} \dfrac{4\pi}{3}$

55. $9 - 3\sqrt{3}i$ **57.** $-\dfrac{7\sqrt{3}}{2} + \dfrac{7}{2}i$

59. $-125i$ **61.** $-\dfrac{1}{128} - \dfrac{\sqrt{3}}{128}i$ **63.**   1

65. $-\dfrac{\sqrt{3}}{2} - \dfrac{1}{2}i$ **67.** $-2,097,152i$

69. $\dfrac{256}{6561} \operatorname{cis} 256°$ **71.** $0.536 \operatorname{cis} 22.2°$

73. $28,561 \operatorname{cis} 90.4°$

75.

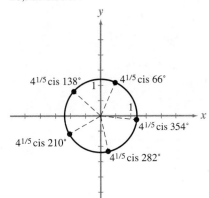

77.

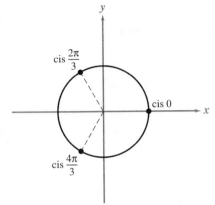

79.

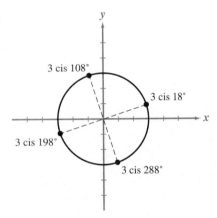

81. $\|\mathbf{v}\| = \sqrt{17}$

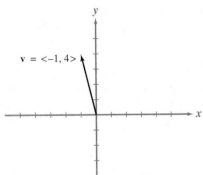

83. $\|\mathbf{v}\| = \sqrt{365}/4$

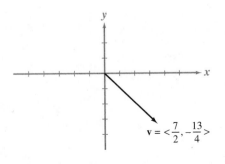

85. $\mathbf{u} + \mathbf{v} = \langle 2, 8 \rangle$

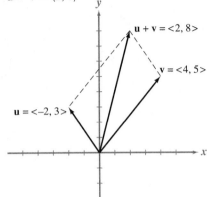

87. $\langle 10, -17 \rangle$ **89.** $\langle 28, -44 \rangle$ **91.** $5\pi/4$
93. $180° - \tan^{-1} \frac{3}{4} \approx 143°$ **95.** $\langle -5\sqrt{2}/2, 5\sqrt{2}/2 \rangle$
97. $\frac{19}{2}i - 12j$ **99.** 424 lb; S12.9°W
101. 603 lb   **103.** 546 mph, S19°W
105. $(\frac{3}{2}, 3\sqrt{3}/2)$ **107.** $(\frac{6}{5}, -6\sqrt{3}/5)$
109. $(4, 4\pi/3)$ **111.** $(5, 180° - \tan^{-1} \frac{4}{3}) \approx (5, 127°)$
113. $x^2 + y^2 = 25$ **115.** $y = 1/x$
117. $r^2 = \dfrac{1}{3 \sin \theta \cos \theta}$ **119.** $r^2 = \dfrac{9}{1 + 8 \sin^2 \theta}$
121. $r = 3 + 3 \sin \theta$
(cardioid)

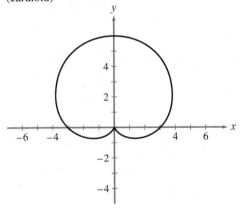

123. $r = 16 \cos 2\theta$
(4-leaf rose)

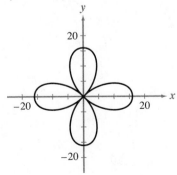

125. $r = 5 - 2 \cos \theta$
(limacon)

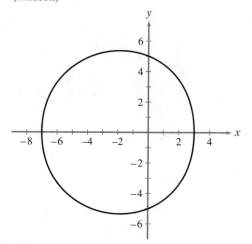

127. $r = 1 - 3 \sin \theta$
(limacon with loop)

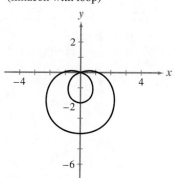

129. $r^2 = 16 \cos 2\theta$
(lemniscate)

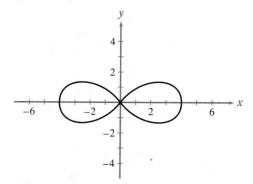

131. $r = \theta/2$
(Spiral of Archimedes)

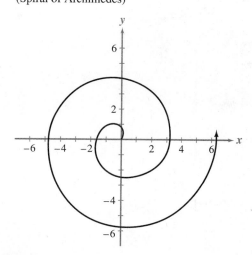

CHAPTER 8

Exercises 8.2

1. Vertex: $(0, 0)$
Focus: $(0, 4)$
Axis: y-axis
Directrix: $y = -4$

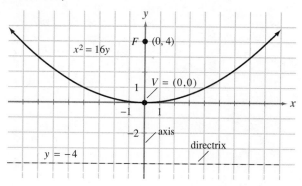

3. Vertex: $(0, 0)$
Focus: $(\frac{3}{2}, 0)$
Axis: x-axis
Directrix: $x = -\frac{3}{2}$

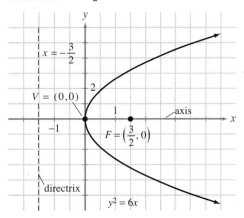

5. Vertex: $(0, 0)$
Focus: $(0, -\frac{5}{2})$
Axis: y-axis
Directrix: $y = \frac{5}{2}$

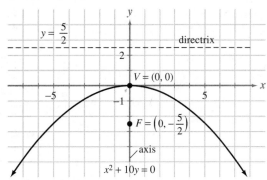

7. Vertex: $(0, 0)$
Focus: $(\frac{5}{8}, 0)$
Axis: x-axis
Directrix: $x = -\frac{5}{8}$

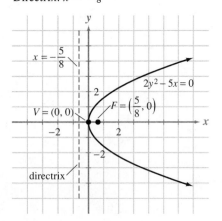

9. Vertex: $(1, 2)$
Focus: $(1, 0)$
Axis: $x = 1$
Directrix: $y = 4$

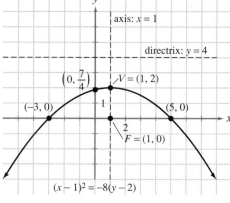

11. Vertex: $(3, 1)$
Focus: $(5, 1)$
Axis: $y = 1$
Directrix: $x = 1$

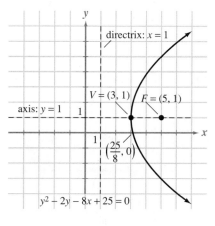

13. Vertex: $(\frac{5}{4}, \frac{29}{4})$
Focus: $(\frac{5}{4}, \frac{115}{16})$
Axis: $x = \frac{5}{4}$
Directrix: $y = \frac{117}{16}$

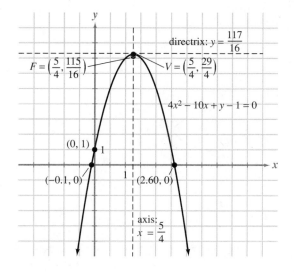

15. Vertex: $(0, 13)$
Focus: $\left(0, \frac{53}{4}\right)$
Axis: y-axis
Directrix: $y = \frac{51}{4}$

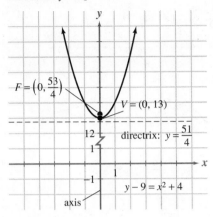

17. Vertex: $(-12, 3)$
Focus: $\left(-\frac{47}{4}, 3\right)$
Axis: $y = 3$
Directrix: $x = -\frac{49}{4}$

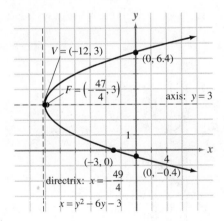

19. Vertex: $\left(-\frac{1}{2}, 1\right)$
Focus: $\left(-\frac{1}{2}, \frac{11}{8}\right)$
Axis: $x = -\frac{1}{2}$
Directrix: $y = \frac{5}{8}$

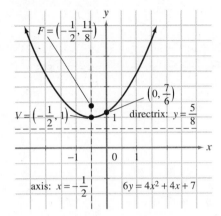

21. $x^2 = -y$ **23.** $(y - 3)^2 = -4(x + 2)$

25. $y^2 = 4(x + 1)$

27. $(x + 4)^2 = 12(y + 2)$ **29.** $y^2 = -8x$

33. On the axis $\frac{9}{4}$ ft from the vertex.

35. $2\sqrt{10} \approx 6.32$ ft from the vertical line passing through the end of the water pipe.

37. **(a)** $y = 2x - 2$

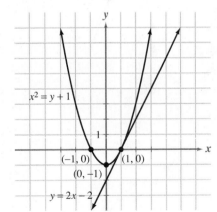

(b) $y = \frac{1}{2}x$

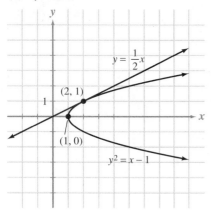

(c) $y = -2x + 7$

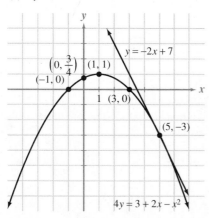

(d) $y = -x$

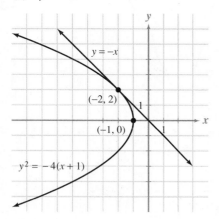

39. 55.6 ft

41. **(a)**

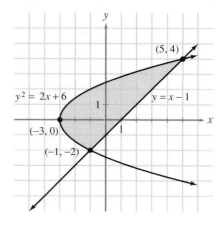

(b)

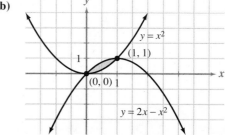

(c)

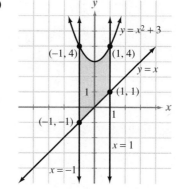

(d)

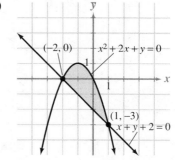

47. (a)

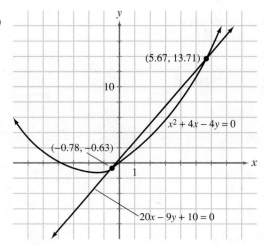

(b) $(-0.78, -0.63), (5.67, 13.71)$

Exercises 8.3

1. Center: $(0, 0)$
Foci: $(\pm\sqrt{11}, 0)$
Vertices: $(\pm 6, 0)$
Endpoints of minor axis: $(0, \pm 5)$

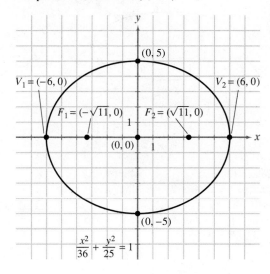

3. Center: $(0, 0)$
Foci: $(0, \pm\sqrt{11})$
Vertices: $(0, \pm 6)$
Endpoints of minor axis: $(\pm 5, 0)$

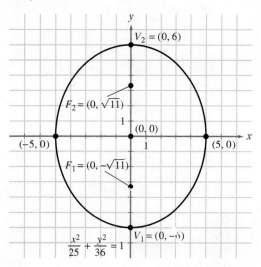

5. Center: $(0, 0)$
Foci: $(\pm 3, 0)$
Vertices: $(\pm 4, 0)$
Endpoints of
minor axis: $(0, \pm\sqrt{7})$

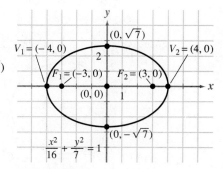

7. Center: $(-2, 3)$
Foci: $(-2 \pm \sqrt{5}, 3)$
Vertices: $(-5, 3), (1, 3)$
Endpoints of minor axis: $(-2, 1), (-2, 5)$

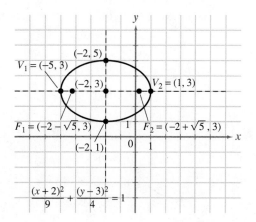

9. Center: $(0, -5)$
Foci: $(0, -6), (0, -4)$
Vertices: $(0, -5 - \sqrt{2}), (0, -5 + \sqrt{2})$
Endpoints of minor axis: $(\pm 1, -5)$

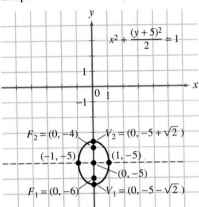

11. Center: $(0, 0)$
Foci: $(\pm \sqrt{7}/6, 0)$
Vertices: $(\pm \sqrt{3}/2, 0)$
Endpoints of minor axis: $(0, \pm \sqrt{5}/3)$

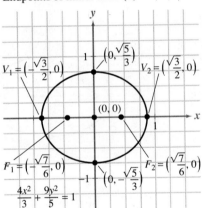

13. Center: $(0, 0)$
Foci: $(0, \pm 1)$
Vertices: $(0, \pm 2)$
Endpoints of minor axis: $(\pm \sqrt{3}, 0)$

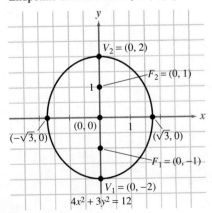

15. Center: $(0, 0)$
Foci: $(\pm \sqrt{55}/2, 0)$
Vertices: $(\pm 4, 0)$
Endpoints of minor axis: $(0, \pm \frac{3}{2})$

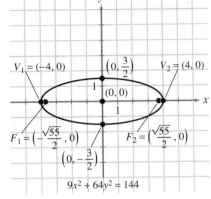

17. Center: $(-2, 1)$
Foci: $(-2 - 2\sqrt{5}, 1), (-2 + 2\sqrt{5}, 1)$
Vertices: $(-7, 1), (3, 1)$
Endpoints of minor axis: $(-2, 1 - \sqrt{5})$,
$(-2, 1 + \sqrt{5})$

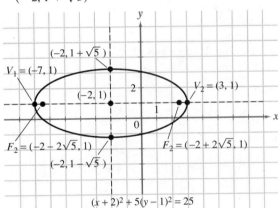

19. Center: $(2, -2)$
Foci: $(0, -2), (4, -2)$
Vertices: $(-1, -2), (5, -2)$
Endpoints of minor axis: $(2, -2 - \sqrt{5})$,
$(2, -2 + \sqrt{5})$

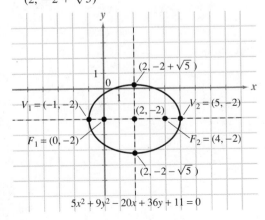

21. Center: $(2, 2)$
Foci: $(-1, 2), (5, 2)$
Vertices: $(2 - 3\sqrt{2}, 2), (2 + 3\sqrt{2}, 2)$
Endpoints of minor axis: $(2, -1), (2, 5)$

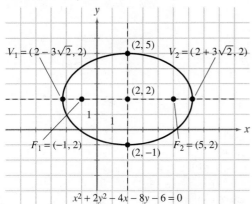

23. $\dfrac{x^2}{16} + \dfrac{y^2}{25} = 1$ **25.** $\dfrac{x^2}{9} + \dfrac{(y - 2)^2}{8} = 1$

27. $\dfrac{(x - 3)^2}{64} + \dfrac{(y - 2)^2}{9} = 1$

29. $\dfrac{(x - 4)^2}{16} + \dfrac{y^2}{25} = 1$ **31.** $\dfrac{(x + 1)^2}{4} + \dfrac{(y - 1)^2}{8} = 1$

35. $\dfrac{x^2}{900} + \dfrac{y^2}{400} = 1$, 17.32 ft **37.** Approximately 359 ft

43. (a)

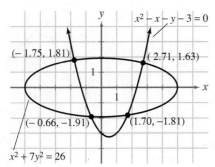

(b) $(-0.66, -1.91), (-1.75, 1.81), (1.70, -1.81),$
$(2.71, 1.63)$

Exercises 8.4

1. Center: $(0, 0)$
Vertices: $(\pm 4, 0)$
Foci: $(\pm 5, 0)$
Length of transverse axis: 8
Length of conjugate axis: 6

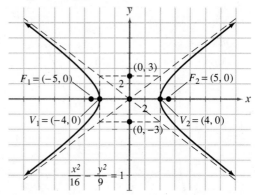

3. Center: $(0, 0)$
Vertices: $(0, \pm 5)$
Foci: $(0, \pm 13)$
Length of transverse axis: 10
Length of conjugate axis: 24

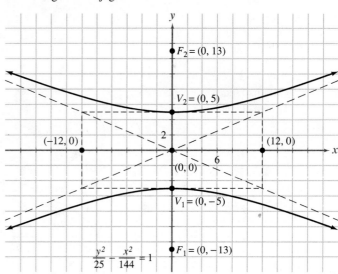

5.   Center: $(0, 0)$
Vertices: $(\pm\frac{5}{3}, 0)$
Foci: $(\pm 5\sqrt{13}/6, 0)$
Length of transverse axis: $\frac{10}{3}$
Length of conjugate axis: 5

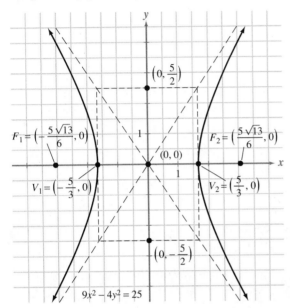

7.   Center: $(-3, 0)$
Vertices: $(-3 - \sqrt{6}, 0), (-3 + \sqrt{6}, 0)$
Foci: $(-3 - \sqrt{7}, 0), (-3 + \sqrt{7}, 0)$
Length of transverse axis: $2\sqrt{6}$
Length of conjugate axis: 2

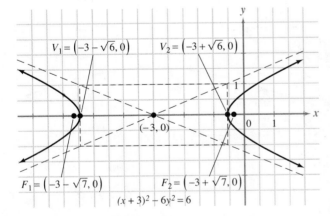

9.   Center: $(-2, 3)$
Vertices: $(-2, 4), (-2, 2)$
Foci: $(-2, 3 - \sqrt{5}), (-2, 3 + \sqrt{5})$
Length of transverse axis: 2
Length of conjugate axis: 4

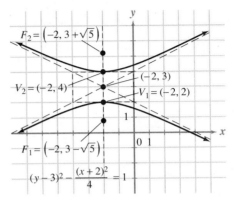

11.   Center: $(4, 2)$
Vertices: $(3, 2), (5, 2)$
Foci: $(4 - \sqrt{2}, 2), (4 + \sqrt{2}, 2)$
Length of transverse axis: 2
Length of conjugate axis: 2

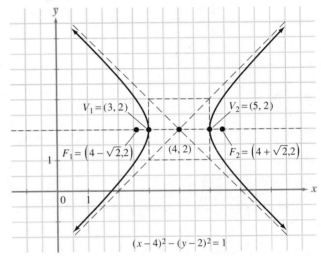

13.   Center: $(-2, 1)$
Vertices: $(-7, 1), (3, 1)$
Foci: $(-2 - \sqrt{29}, 1), (-2 + \sqrt{29}, 1)$
Length of transverse axis: 10
Length of conjugate axis: 4

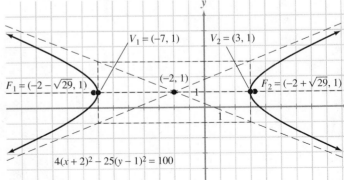

15. Center: $(-2, 1)$
Vertices: $(-2, -2), (-2, 4)$
Foci: $(-2, -4), (-2, 6)$
Length of transverse axis: 6
Length of conjugate axis: 8

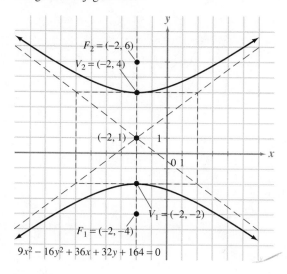

$9x^2 - 16y^2 + 36x + 32y + 164 = 0$

17. Center: $(2, 1)$
Vertices: $(2 - \sqrt{6}, 1), (2 + \sqrt{6}, 1)$
Foci: $(2 - \sqrt{15}, 1), (2 + \sqrt{15}, 1)$
Length of transverse axis: $2\sqrt{6}$
Length of conjugate axis: 6

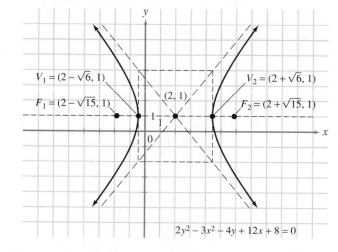

$2y^2 - 3x^2 - 4y + 12x + 8 = 0$

19. Center: $(0, 4)$
Vertices: $(\pm\sqrt{3}, 4)$
Foci: $(\pm\sqrt{15}, 4)$
Length of transverse axis: $2\sqrt{3}$
Length of conjugate axis: $4\sqrt{3}$

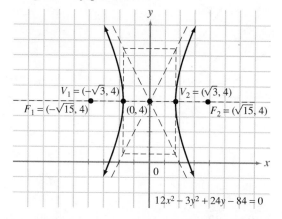

$12x^2 - 3y^2 + 24y - 84 = 0$

21. $\dfrac{y^2}{16} - \dfrac{x^2}{20} = 1$ **23.** $\dfrac{(y-3)^2}{4} - \dfrac{(x-2)^2}{12} = 1$

25. $\dfrac{(y-5)^2}{5} - \dfrac{(x-1)^2}{4} = 1$

27. $\dfrac{y^2}{5} - \dfrac{x^2}{20} = 1$

29. $(x+2)^2 - \dfrac{(y-4)^2}{9} = 1$ **31.** $\frac{7}{36}$ **33.** Parabola

35. Ellipse   **37.** Hyperbola   **39.** Circle

43. On one branch of the hyperbola with foci located at the radio towers R_1 and R_2 and $a = 25$. This hyperbola is defined by $\dfrac{y^2}{625} - \dfrac{x^2}{21,875} = 1$ by introducing a rectangular coordinate system with R_1 and R_2 appropriately placed on the y-axis.

45. $x^2 - 3y^2 - 40y - 48 = 0$

Exercise 8.5

1. Ellipse

$$\dfrac{(x')^2}{4} + \dfrac{(y')^2}{\frac{4}{3}} = 1$$

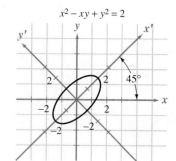

$x^2 - xy + y^2 = 2$

3. Parabola
$$(x')^2 = \frac{\sqrt{2}}{2}y'$$

$$x^2 + 2xy + y^2 + x - y = 0$$

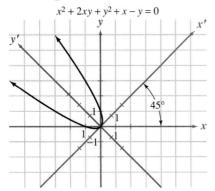

5. Hyperbola
$$\frac{(x')^2}{10} - \frac{(y')^2}{10} = 1$$

$$2x^2 + 3xy - 2y^2 = 25$$

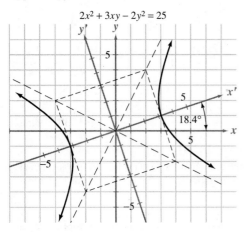

7. Ellipse
$$\frac{(x')^2}{12} + \frac{(y')^2}{4} = 1$$

$$x^2 - xy + y^2 = 6$$

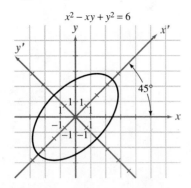

9. Ellipse
$$\frac{(x')^2}{\frac{1}{13}} + \frac{(y')^2}{\frac{1}{5}} = 1$$

$$11x^2 + 4\sqrt{3}xy + 7y^2 - 1 = 0$$

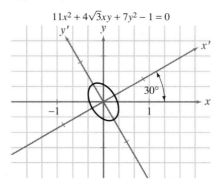

11. Parabola
$$(x')^2 = y'$$

$$x^2 + 2\sqrt{3}xy + 3y^2 + 2\sqrt{3}x - 2y = 0$$

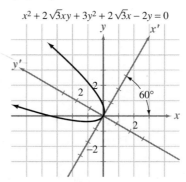

13. Hyperbola
$$(x' - 1)^2 - \frac{(y')^2}{\frac{1}{3}} = 1$$

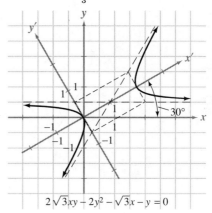

$$2\sqrt{3}xy - 2y^2 - \sqrt{3}x - y = 0$$

15. Parabola

$$\left(y' - \frac{\sqrt{2}}{12}\right)^2 = -\frac{\sqrt{2}}{6}\left(x' - \frac{85\sqrt{2}}{24}\right)$$

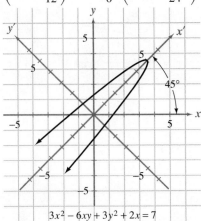

$$3x^2 - 6xy + 3y^2 + 2x = 7$$

17. Parabola
$$(y')^2 = 8x'$$

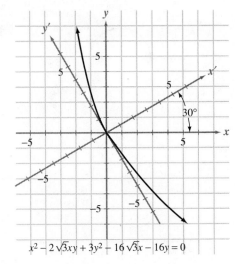

$$x^2 - 2\sqrt{3}xy + 3y^2 - 16\sqrt{3}x - 16y = 0$$

19. Hyperbola

$$\frac{\left(y' + \frac{3\sqrt{10}}{5}\right)^2}{\frac{91}{12}} - \frac{\left(x' + \frac{\sqrt{10}}{20}\right)^2}{\frac{91}{8}} = 1$$

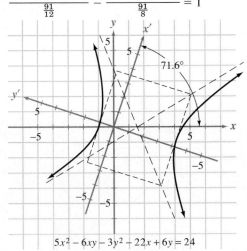

$$5x^2 - 6xy - 3y^2 - 22x + 6y = 24$$

23. Ellipse   **25.** Hyperbola   **27.** Parabola
29. Parabola   **31.** Hyperbola
33.

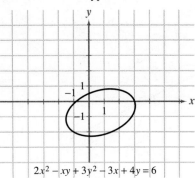

$$2x^2 - xy + 3y^2 - 3x + 4y = 6$$

35.

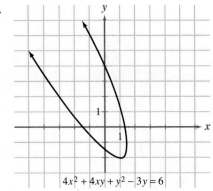

$$4x^2 + 4xy + y^2 - 3y = 6$$

37.

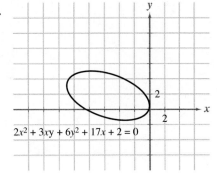

$$2x^2 + 3xy + 6y^2 + 17x + 2 = 0$$

39.

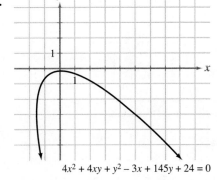

$$4x^2 + 4xy + y^2 - 3x + 145y + 24 = 0$$

Exercises 8.6

1.

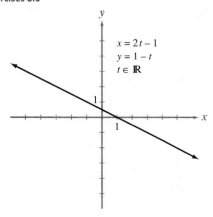

$$x = 2t - 1$$
$$y = 1 - t$$
$$t \in \mathbb{R}$$

3.

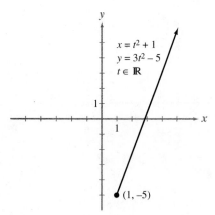

$$x = t^2 + 1$$
$$y = 3t^2 - 5$$
$$t \in \mathbb{R}$$

$(1, -5)$

5.

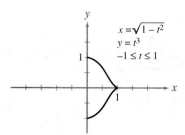

$$x = \sqrt{1 - t^2}$$
$$y = t^3$$
$$-1 \le t \le 1$$

7.

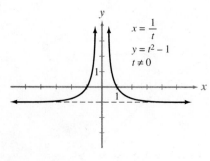

$$x = \frac{1}{t}$$
$$y = t^2 - 1$$
$$t \ne 0$$

9.

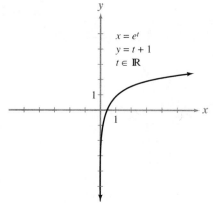

$$x = e^t$$
$$y = t + 1$$
$$t \in \mathbb{R}$$

11.

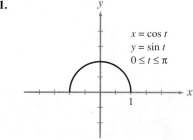

$$x = \cos t$$
$$y = \sin t$$
$$0 \le t \le \pi$$

13.

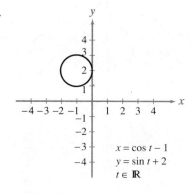

$$x = \cos t - 1$$
$$y = \sin t + 2$$
$$t \in \mathbb{R}$$

15.

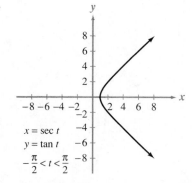

$$x = \sec t$$
$$y = \tan t$$
$$-\frac{\pi}{2} < t < \frac{\pi}{2}$$

17.

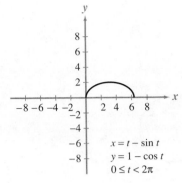

$x = t - \sin t$
$y = 1 - \cos t$
$0 \le t < 2\pi$

19.

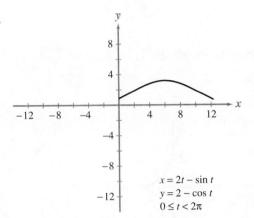

$x = 2t - \sin t$
$y = 2 - \cos t$
$0 \le t < 2\pi$

21. $y = -\frac{3}{2}x + \frac{11}{2}$

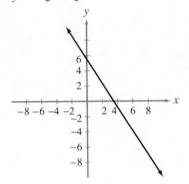

23. $y = 9x^2 + 9x - 5$

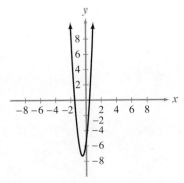

25. $y = x^{2/3} - 1$

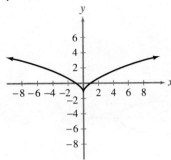

27. $x^2 + y^2 = 4$

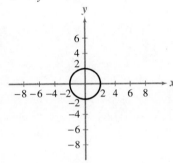

29. $y = 3x; \ x > 0$

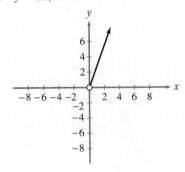

31. $y = \dfrac{1}{x} + 1; \ x > 0$

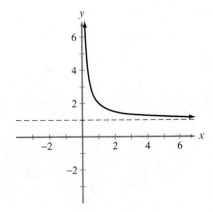

33. $\dfrac{x^2}{9} + \dfrac{y^2}{1} = 1$

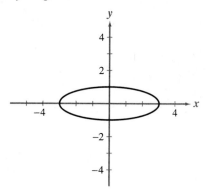

35. $\dfrac{x^2}{9} + \dfrac{(y+1)^2}{4} = 1$

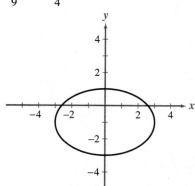

37. $y = \sqrt{x^2 - 1}; \ |x| \geq 1$

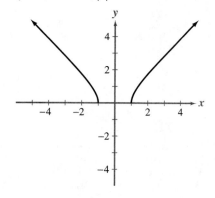

39. $(x - 1)^2 + y^2 = 1$

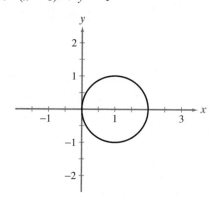

41. 37.4 sec; 41,040 ft **43.** $x = \dfrac{4p}{t^2}$ and $y = \dfrac{4p}{t}$

47. $x^{2/3} + y^{2/3} = 1$

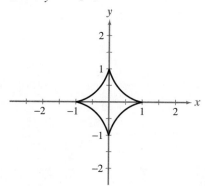

Chapter 8 Review Exercises

1. Parabola
Vertex: $(0, 0)$
Focus: $(0, \frac{1}{2})$
Directrix: $y = -\frac{1}{2}$

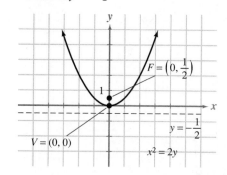

3. Hyperbola
Center: $(0, 0)$
Vertices: $(\pm 1, 0)$
Foci: $(\pm \sqrt{3}, 0)$
Endpoints of conjugate axis: $(0, \pm \sqrt{2})$

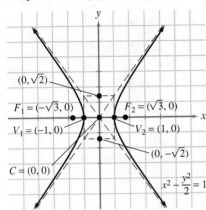

5. Ellipse
Center: $(0, 0)$
Vertex: $(\pm \sqrt{10}, 0)$
Foci: $(\pm 2, 0)$
Endpoints of minor axis: $(0, \pm \sqrt{6})$

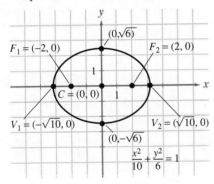

7. Circle
Center: $(0, 0)$
Radius: $\frac{3}{2}$

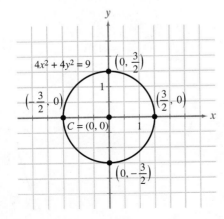

9. Hyperbola
Center: $(0, 0)$
Vertices: $(\pm \frac{1}{3}, 0)$
Foci: $(\pm \sqrt{10}/3, 0)$
Endpoints of the conjugate axis: $(0, \pm 1)$

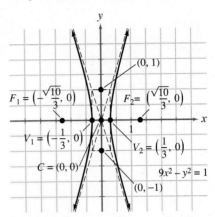

11. Hyperbola
Center: $(-2, 5)$
Vertices: $(-5, 5), (1, 5)$
Foci: $(-7, 5), (3, 5)$
Endpoints of conjugate axis: $(-2, 1)$ and $(-2, 9)$

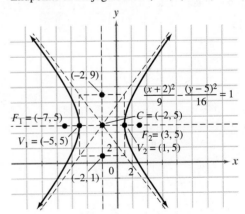

13. Parabola
Vertex: $(-2, 2)$
Focus: $(-2, 3)$
Directrix: $y = 1$

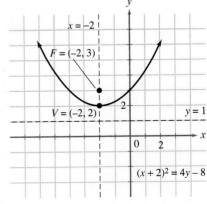

15. Ellipse
Center: $(3, 0)$
Vertices: $(1, 0)$, $(5, 0)$
Foci: $(3 - \sqrt{3}, 0)$, $(3 + \sqrt{3}, 0)$
Endpoints of minor axis: $(3, -1)$, $(3, 1)$

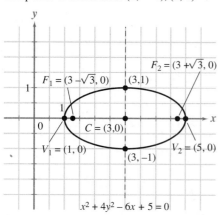

17. Hyperbola
Center: $(-1, -2)$
Vertices: $(-4, -2)$, $(2, -2)$
Foci: $(-1 - \sqrt{58}, -2)$, $(-1 + \sqrt{58}, -2)$
Endpoints of conjugate axis: $(-1, -9)$ and $(-1, 5)$

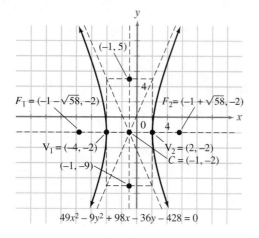

19. Hyperbola
Center: $(2, 2)$
Vertices: $(2, -1)$, $(2, 5)$
Foci: $(2, -4)$, $(2, 8)$
Endpoints of conjugate axis: $(2 - 3\sqrt{3}, 2)$,
$(2 + 3\sqrt{3}, 2)$

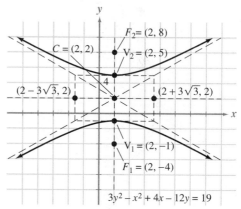

21. $(y + 1)^2 = 4(x + 4)$ **23.** $\dfrac{y^2}{4} - \dfrac{x^2}{12} = 1$

25. $\dfrac{(x + 1)^2}{25} + \dfrac{(y + 4)^2}{\frac{25}{4}} = 1$ **27.** $(x + 1)^2 = \frac{4}{7}(y + 1)$

29. Ellipse
$\dfrac{(x')^2}{\frac{3}{2}} + (y')^2 = 1$

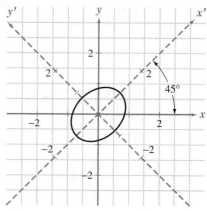

31. Hyperbola
$\dfrac{(x')^2}{\frac{182}{169}} - \dfrac{(y')^2}{\frac{182}{169}} = 1$

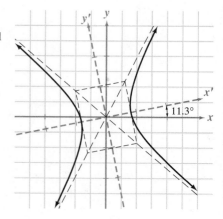

33. Two parallel lines
$y' = \pm 2/\sqrt{5}$

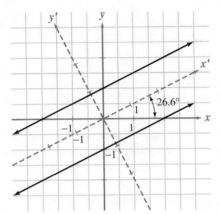

35. Hyperbola
$$\frac{\left(x' - \dfrac{\sqrt{2}}{2}\right)^2}{2} - \frac{\left(y' + \dfrac{3\sqrt{2}}{2}\right)^2}{2} = 1$$

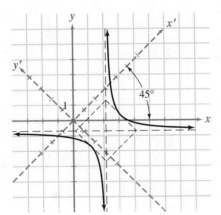

37. Parabola
$(x')^2 = -2y'$

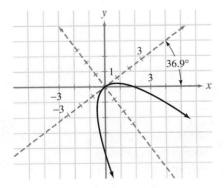

39. $x = t^2 - 1$
$y = \sqrt{t - 1}$
$t \geq 1$

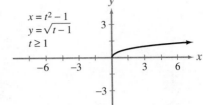

41.

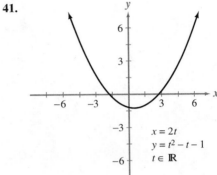
$x = 2t$
$y = t^2 - t - 1$
$t \in \mathbb{R}$

43. $x = \dfrac{1}{2} \sec t$
$y = 2 \tan t$
$-\dfrac{\pi}{2} < t < \dfrac{\pi}{2}$

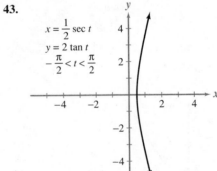

45. $y = 3x^2 - 2;\ x \geq 0$

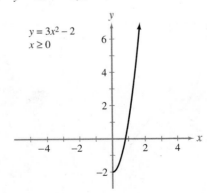

47. $y = 2/(x + 1)$

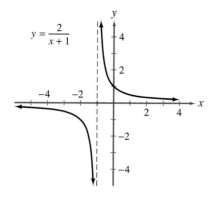

$$y = \frac{2}{x + 1}$$

49. $(x - 4)^2 + (y - 4)^2 = 16$

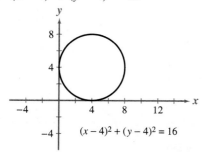

$(x - 4)^2 + (y - 4)^2 = 16$

51. 24.8 sec; 7550 ft   **53.** $(x - 2)^2 = -8(y + 3)$

55. $\dfrac{x^2}{5} + \dfrac{y^2}{30} = 1$

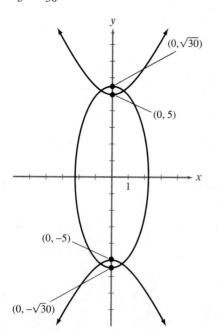

$(0, \sqrt{30})$

$(0, 5)$

$(0, -5)$

$(0, -\sqrt{30})$

CHAPTER 9

Exercises 9.1

1. $\{(3, -2)\}$ **3.** $\{(\frac{1}{4}, -\frac{7}{12})\}$ **5.** $\{(\frac{3}{7}, \frac{3}{7})\}$

7. $\{(\frac{1}{3}(-4t + 7), t) \mid t \in \mathbb{R}\}$

9. $\{(\frac{3}{8}, -\frac{3}{8})\}$ **11.** $\{(\frac{7}{9}, -\frac{1}{3})\}$

13. $\{(-1, \frac{3}{2})\}$ **15.** $\{(t, \frac{4}{3}t - 6) \mid t \in \mathbb{R}\}$

17. $\varnothing$ **19.** $\{(\frac{335}{91}, \frac{470}{91})\}$ **21.** $\varnothing$

23. $\left\{ \left(\dfrac{7k - 1}{8}, \dfrac{3k + 3}{16} \right) \;\middle|\; k \in \mathbb{R} \right\}$

25. $\{(-1, 2, 1)\}$ **27.** $\{(\frac{2}{5}, \frac{3}{5}, \frac{12}{5})\}$ **29.** $\varnothing$ **31.** $\varnothing$

33. $\{(\frac{34}{3}, -\frac{37}{6}, \frac{11}{3})\}$ **35.** $\{(2, 1, -4)\}$ **37.** $\{(-1, 2, 0)\}$

39. $\{(0, 0, 0)\}$ **41.** $\{(\frac{1}{2}, -1)\}$ **43.** $\{(\frac{1}{2}, -\frac{1}{2}, \frac{1}{3})\}$

45. 6.5 m, 12.5 m   **47.** 1, 1   **49.** 1 standard, 9 deluxe

51. 4 oz. chicken, 3 oz. rice, 4 oz. broccoli

53. $a = 3, b = -1, c = 4$

Exercises 9.2

1. $\begin{bmatrix} 1 & -1 & 1 \\ 2 & 0 & -3 \\ 1 & 1 & -5 \end{bmatrix}$; $\begin{bmatrix} 1 & -1 & 1 & 4 \\ 2 & 0 & -3 & -8 \\ 1 & 1 & -5 & 0 \end{bmatrix}$

3. $\begin{bmatrix} 6 & 0 & -4 & 5 \\ 0 & 4 & 0 & -1 \\ -2 & 0 & 5 & 0 \\ 1 & -1 & 1 & -3 \end{bmatrix}$;

$\begin{bmatrix} 6 & 0 & -4 & 5 & 10 \\ 0 & 4 & 0 & -1 & 0 \\ -2 & 0 & 5 & 0 & 3 \\ 1 & -1 & 1 & -3 & -4 \end{bmatrix}$

5. $\begin{bmatrix} 6 & -8 & 10 & -4 \\ 4 & 0 & -1 & -6 \end{bmatrix}$; $\begin{bmatrix} 6 & -8 & 10 & -4 & 5 \\ 4 & 0 & -1 & -6 & 10 \end{bmatrix}$

7. No. The first nonzero element in the second row is not one.

9. Yes.

11. No. The third row contains all zeros, while the fourth row contains a nonzero element.

13. $\begin{bmatrix} 1 & 2 & 3 \\ 0 & 1 & 2 \\ 0 & 0 & 1 \end{bmatrix}$

15. $\begin{bmatrix} 1 & -1 & -3 & -6 \\ 0 & 1 & 1 & 4 \\ 0 & 0 & 0 & 0 \end{bmatrix}$

17. $\begin{bmatrix} 1 & 1 & 3 & -4 & 1 \\ 0 & 1 & 2 & -\frac{23}{7} & \frac{5}{7} \\ 0 & 0 & 1 & \frac{4}{63} & -\frac{31}{63} \end{bmatrix}$

19. $\{(2, 0, 1)\}$ **21.** $\varnothing$

23. $\{(1 + t, -1 + t, 1 - 2t, t) \mid t \in \mathbb{R}\}$

25. $\{(3, 1, 2)\}$; consistent; independent

27. $\{(\frac{3}{10}, \frac{1}{10}, \frac{1}{5})\}$; consistent; independent

29. $\{(1, 0, 5)\}$; consistent; independent

31. $\emptyset$; inconsistent
33. $\{(0, 0, -1)\}$; consistent; independent
35. $\{(-4, 2, -1)\}$; consistent; independent
37. $\emptyset$; inconsistent
39. $\{(0, -9, -3)\}$; consistent; independent
41. $\emptyset$; inconsistent
43. $\{(14 - t, t + 7, t, 4) \mid t \in \mathbb{R}\}$; consistent; dependent
45. $\{(1, 1, -1, 2)\}$; consistent; independent
47. $\{(-5t - 1, -8t - 6, t) \mid t \in \mathbb{R}\}$; consistent; dependent
49. $\emptyset$; inconsistent
51. $\{(3, 2, 1)\}$; consistent; independent
53. $\emptyset$; inconsistent
55. $\{(\frac{32}{33}t + \frac{23}{33}, -\frac{8}{11}t - \frac{3}{11}, \frac{84}{11}t + \frac{4}{11}, t) \mid t \in \mathbb{R}\}$; consistent;
    dependent
57. $f(x) = -x^2 + x - 3$　　**59.** $z = -x - 2y + 2$
61. **(a)** $k = 1$　　**(b)** No
63. \$200,000 at 8%, \$200,000 at 10%, and \$100,000 at 12%
65. 20, 24, 30
67. **(a)** $\begin{cases} x_1 - x_2 = 90 \\ x_1 - x_4 = 30 \\ x_2 - x_3 = -10 \\ x_3 - x_4 = -50 \end{cases}$

    (b)
    $\{(t + 30, t - 60, t - 50, t) \mid t \text{ is a positive integer} \geq 60\}$

Exercises 9.3

1. $x = -5, y = -2$　　**3.** $x = \frac{1}{2}, y = 4$

5. $A + B = \begin{bmatrix} 3 & 5 \\ -13 & 12 \end{bmatrix}$, $A - B = \begin{bmatrix} 5 & 5 \\ 1 & -8 \end{bmatrix}$,

   $3A = \begin{bmatrix} 12 & 15 \\ -18 & 6 \end{bmatrix}$, $3A - 4B = \begin{bmatrix} 16 & 15 \\ 10 & -34 \end{bmatrix}$

7. $3A = \begin{bmatrix} 6 & 3 & 18 & 15 \\ 3 & 6 & -12 & 9 \end{bmatrix}$

9. $A + B = \begin{bmatrix} 3 & 6 & -1 & -7 \\ 0 & 2 & 1 & -2 \\ 0 & 0 & -3 & 8 \\ 0 & 0 & 0 & 3 \end{bmatrix}$,

   $A - B = \begin{bmatrix} -1 & 6 & -1 & -7 \\ 0 & 0 & 1 & -2 \\ 0 & 0 & 3 & 8 \\ 0 & 0 & 0 & -5 \end{bmatrix}$,

   $3A = \begin{bmatrix} 3 & 18 & -3 & -21 \\ 0 & 3 & 3 & -6 \\ 0 & 0 & 0 & 24 \\ 0 & 0 & 0 & -3 \end{bmatrix}$,

   $3A - 4B = \begin{bmatrix} -5 & 18 & -3 & -21 \\ 0 & -1 & 3 & -6 \\ 0 & 0 & 12 & 24 \\ 0 & 0 & 0 & -19 \end{bmatrix}$

11. 0　　**13.** 4　　**15.** 0

17. $AB = \begin{bmatrix} 2 & 0 \\ 4 & 0 \\ -6 & 0 \end{bmatrix}$

19. $AB = A = \begin{bmatrix} 1 & 2 & 0 & 1 \\ 0 & -1 & 2 & 3 \\ 0 & 0 & -1 & 4 \end{bmatrix}$

21. $BA = \begin{bmatrix} -5 & -6 & 0 & -7 \\ 5 & 6 & 0 & 7 \\ -1 & 2 & -3 & 4 \\ 6 & 4 & 3 & 3 \\ -6 & -4 & -3 & -3 \end{bmatrix}$

23. $AB = \begin{bmatrix} 1 & 0 & 0 \\ -1 & 1 & 0 \\ -1 & 1 & -1 \end{bmatrix}$, $BA = \begin{bmatrix} 1 & 1 & 0 \\ 1 & 0 & 1 \\ -1 & 0 & 0 \end{bmatrix}$

25. $AB = \begin{bmatrix} -2 & 1 & 7 \\ 1 & 3 & -1 \\ 6 & -1 & 5 \end{bmatrix}$, $BA = \begin{bmatrix} 8 & 0 & 0 & 4 \\ 1 & 3 & 1 & 6 \\ -5 & 4 & 1 & 7 \\ 4 & -2 & 0 & -6 \end{bmatrix}$

27. $\begin{bmatrix} -10 & 2 \\ 3 & -3 \\ 4 & 14 \end{bmatrix}$　　**29.** $\begin{bmatrix} -\frac{7}{2} & -5 \\ \frac{23}{2} & -2 \\ -10 & \frac{25}{2} \end{bmatrix}$

39. $A = \begin{bmatrix} 1 & 1 \\ 1 & -2 \end{bmatrix}$, $\overline{X} = \begin{bmatrix} x \\ y \end{bmatrix}$, $B = \begin{bmatrix} 6 \\ 5 \end{bmatrix}$

41. $A = \begin{bmatrix} 1 & 0 & -1 \\ 0 & 1 & 4 \\ 3 & 0 & -5 \end{bmatrix}$, $\overline{X} = \begin{bmatrix} t \\ u \\ v \end{bmatrix}$, $B = \begin{bmatrix} -1 \\ 2 \\ 6 \end{bmatrix}$

43. $A = \begin{bmatrix} 1 & 1 \\ 2 & 4 \\ 1 & -5 \end{bmatrix}$, $\overline{X} = \begin{bmatrix} x \\ y \end{bmatrix}$, $B = \begin{bmatrix} 4 \\ 2 \\ -8 \end{bmatrix}$

47. **(a)** $\left(\dfrac{5\sqrt{2}}{2}, -\dfrac{\sqrt{2}}{2} \right)$

    (b) $\left(\sqrt{3} - \dfrac{3}{2}, -1 - \dfrac{3\sqrt{3}}{2} \right)$

    (c) $(-2\cos 50° + 3\sin 50°, 2\sin 50° + 3\cos 50°)$
        $\approx (1.01, 3.46)$

49. **(a)** $P = \begin{bmatrix} 60 & 5 & 0 \\ 10 & 2 & 10 \end{bmatrix}$　　**(b)** $C = \begin{bmatrix} 60 & 5 & 0 \\ 10 & 2 & 10 \\ 12 & 10 & 13 \end{bmatrix}$

    (c) Table A: Factory 1; \$3650　　Table B: Factory 1; \$740
    　　　　　　Factory 2; \$310　　　　　　　　　Factory 2; \$154
    　　　　　　Factory 3; \$50　　　　　　　　　　Factory 3; \$150

Exercises 9.4

5. $\begin{bmatrix} \frac{1}{8} & -\frac{1}{16} \\ \frac{1}{4} & \frac{3}{8} \end{bmatrix}$　　**7.** $\begin{bmatrix} \frac{2}{3} & -\frac{1}{3} \\ \frac{1}{3} & \frac{1}{3} \end{bmatrix}$

9. $\begin{bmatrix} \frac{2}{23} & -\frac{5}{23} & \frac{16}{23} \\ \frac{13}{23} & \frac{2}{23} & -\frac{11}{23} \\ -\frac{5}{23} & \frac{1}{23} & \frac{6}{23} \end{bmatrix}$

11. $\begin{bmatrix} 1 & \frac{1}{4} & -4 \\ 0 & \frac{1}{4} & -2 \\ 0 & 0 & 1 \end{bmatrix}$

13. $\begin{bmatrix} \frac{5}{8} & \frac{7}{16} & \frac{1}{16} \\ -\frac{1}{8} & \frac{5}{16} & \frac{3}{16} \\ \frac{1}{4} & -\frac{1}{8} & \frac{1}{8} \end{bmatrix}$

15. $\begin{bmatrix} 0 & 1 & -1 \\ -3 & 3 & -2 \\ 1 & -1 & 1 \end{bmatrix}$

17. $\begin{bmatrix} \frac{1}{8} & -\frac{3}{8} & \frac{1}{8} & \frac{3}{8} \\ -\frac{1}{2} & \frac{1}{2} & \frac{1}{2} & \frac{1}{2} \\ \frac{1}{4} & \frac{1}{4} & \frac{1}{4} & -\frac{1}{4} \\ \frac{5}{8} & \frac{1}{8} & -\frac{3}{8} & -\frac{1}{8} \end{bmatrix}$

19. $\begin{bmatrix} 12.5 & -1.25 & -5.0 \\ 5.0 & 2.5 & 0 \\ -7.5 & -1.25 & 5.0 \end{bmatrix}$

21. $\{(\frac{11}{16}, -\frac{1}{8})\}$ **23.** $\{(-\frac{1}{2}, \frac{1}{2}, 0)\}$ **25.** $\{(-\frac{1}{4}, 1, \frac{1}{2}, \frac{3}{4})\}$

27. $\{(7.875, 4.25, -5.125)\}$

37. **(a)** $a = \frac{5}{2}, b = \frac{1}{2}, c = -1$
(b) $a = -6, b = 1, c = 4$

39. **(a)** $\begin{array}{cccccccc} 79 & -51 & -110 & 111 & -68 & -152 & 63 & -39 \\ -97 & 13 & -4 & -13 & 132 & -85 & -197 & 130 \\ -84 & -195 & 64 & -41 & -100 \end{array}$

(b) $\begin{array}{cccccccc} 67 & -40 & -95 & 90 & -56 & -139 & 88 & -57 \\ -136 & 81 & -48 & -102 & 26 & -16 & -37 & 57 \\ -29 & -71 \end{array}$

Exercises 9.5

1. $M_{11} = -1; M_{12} = 1; M_{13} = -1; M_{21} = 1; M_{22} = -2;$
$M_{23} = 3; M_{31} = 0; M_{32} = 1; M_{33} = -1;$
$A_{11} = -1; A_{12} = -1; A_{13} = -1; A_{21} = -1; A_{22} = -2;$
$A_{23} = -3; A_{31} = 0; A_{32} = -1; A_{33} = -1$

3. $M_{11} = -60; M_{12} = 0; M_{13} = 0; M_{14} = 0; M_{21} = 0;$
$M_{22} = 40; M_{23} = 0; M_{24} = 0;$
$A_{11} = -60; A_{12} = 0; A_{13} = 0; A_{14} = 0; A_{21} = 0;$
$A_{22} = 40; A_{23} = 0; A_{24} = 0;$
$M_{31} = 0; M_{32} = 0; M_{33} = -30; M_{34} = 0; M_{41} = 0;$
$M_{42} = 0; M_{43} = 0; M_{44} = -24;$
$A_{31} = 0; A_{32} = 0; A_{33} = -30; A_{34} = 0; A_{41} = 0;$
$A_{42} = 0; A_{43} = 0; A_{44} = -24$

5. 34   **7.** -1 **9.** -1 **11.** 1   **13.** 0
15. -120 **17.** $3i - j + k$ **19.** -9
21. $x^2 - 2x + 1$ **23.** $x^3 + 5x^2 + 3x - 9$
25. **(a)** 1   **(b)** $2 \pm \sqrt{5}i$ **(c)** $-3, 1$ **(d)** $-3, 2, 4$
27. 420   **29.** 1   **31.** **(b)** 36   **(c)** $a_{11} \cdot a_{22} \cdots a_{nn}$
33. If each element of a row of a square matrix A is zero, then $\det(A) = 0$.
35. Property II (Theorem 9.5)

37. Property III (Theorem 9.5)
39. $\{(\frac{13}{34}, -\frac{29}{17})\}$ **41.** $\{(\frac{157}{199}, -\frac{170}{199})\}$ **43.** $\{(7, 12, 17)\}$
45. $\varnothing$ **47.** $\{(-\frac{39}{16}, \frac{221}{112}, \frac{5}{56})\}$ **51.** 28   **53.** $\frac{41}{120}$

Exercises 9.6

1. $\left\{ \left(\dfrac{9 - \sqrt{21}}{2}, \dfrac{-1 + \sqrt{21}}{2} \right), \left(\dfrac{9 + \sqrt{21}}{2}, \dfrac{-1 - \sqrt{21}}{2} \right) \right\}$

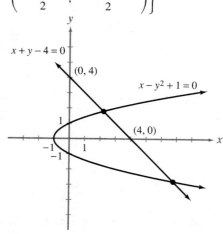

3. $\{(5, 2)\}$

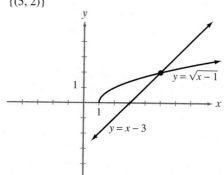

5. $\{(-1, 1), (0, 0)\}$

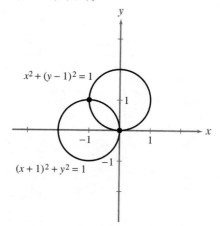

7. $\{(0, 0), (1, 1)\}$

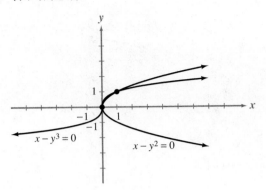

$x - y^3 = 0$

$x - y^2 = 0$

9. $\{(-1, -4)\}$

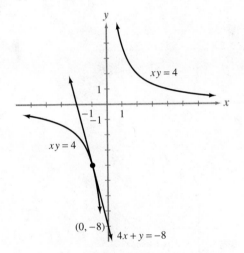

$xy = 4$

$xy = 4$

$(0, -8)$ $4x + y = -8$

11. $\{(1, -1), (1, 1), (1, -i), (1, i)\}$

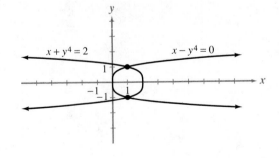

$x + y^4 = 2$ $x - y^4 = 0$

13. Points of intersection: $(0, 0)$, $(4, 64)$

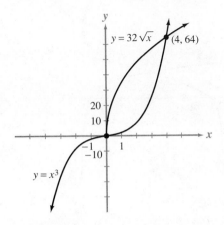

$y = 32\sqrt{x}$ $(4, 64)$

$y = x^3$

15. Points of intersection: $(-1, 1)$, $(0, 0)$, $(1, 1)$

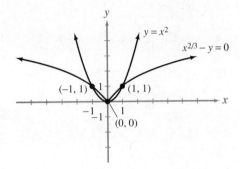

$y = x^2$

$x^{2/3} - y = 0$

$(-1, 1)$ $(1, 1)$

$(0, 0)$

17. Point of intersection: $\left(\ln\left(\dfrac{1 + \sqrt{5}}{2}\right), \dfrac{1 + \sqrt{5}}{2}\right)$

$\approx (.48, 1.62)$

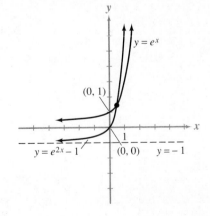

$y = e^x$

$(0, 1)$

$y = e^{2x} - 1$ $(0, 0)$ $y = -1$

19. Point of intersection: $(5, \ln 5) \approx (5, 1.61)$

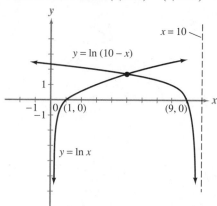

21. $\left\{ \left(\dfrac{-93 - 18\sqrt{26}}{2}, \dfrac{-6 - \sqrt{26}}{2} \right), \right.$
$\left. \left(\dfrac{-93 + 18\sqrt{26}}{2}, \dfrac{-6 + \sqrt{26}}{2} \right) \right\}$

23. $\{(-\sqrt{10}/2, -\sqrt{10}/2), (\sqrt{10}/2, \sqrt{10}/2),$
$(-\sqrt{10}, 0), (\sqrt{10}, 0)\}$

25. $\{(-\sqrt{(54/7)}, -4\sqrt{7}/7), (-\sqrt{(54/7)}, 4\sqrt{7}/7),$
$(\sqrt{(54/7)}, -4\sqrt{7}/7), (\sqrt{(54/7)}, 4\sqrt{7}/7)\}$

27. $\{(\frac{41}{18}, -\frac{41}{7})\}$

29. $\{(-\sqrt{\frac{1}{5}}, -\sqrt{\frac{1}{2}}), (-\sqrt{\frac{1}{5}}, \sqrt{\frac{1}{2}}), (\sqrt{\frac{1}{5}}, -\sqrt{\frac{1}{2}}),$
$(\sqrt{\frac{1}{5}}, \sqrt{\frac{1}{2}})\}$

31. $\{(-\sqrt{\frac{20}{3}}, -\sqrt{\frac{28}{3}}i), (-\sqrt{\frac{20}{3}}, \sqrt{\frac{28}{3}}i),$
$(\sqrt{\frac{20}{3}}, -\sqrt{\frac{28}{3}}i), (\sqrt{\frac{20}{3}}, \sqrt{\frac{28}{3}}i)\}$

33. $\{(-2\sqrt{\frac{2}{5}}, -\sqrt{\frac{3}{5}}), (-2\sqrt{\frac{2}{5}}, \sqrt{\frac{3}{5}}),$
$(2\sqrt{\frac{2}{5}}, -\sqrt{\frac{3}{5}}), (2\sqrt{\frac{2}{5}}, \sqrt{\frac{3}{5}})\}$

35. $\{(-4, -2 - \sqrt{2}i), (-4, -2 + \sqrt{2}i),$
$(4, 2 - \sqrt{2}i), (4, 2 + \sqrt{2}i)\}$

37. $\{(2, 2)\}$ **39.** $\{(\frac{1}{2}\ln 3, -2\sqrt{3})\}$ **41.** 3

43. 4 m, 6 cm   **45.** 100, \$315   **47.** 2 m, 6 cm   **49.** 4

Exercises 9.7

1.

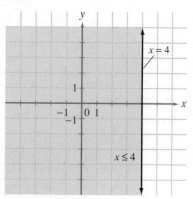

3.

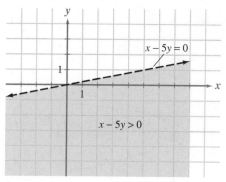

5.

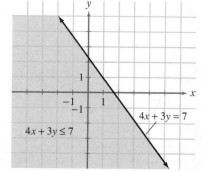

7.

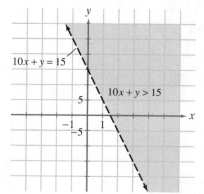

9. This region is bounded and convex.

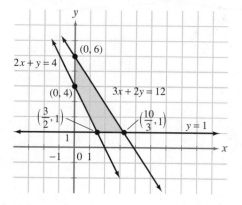

11. This region is bounded and convex.

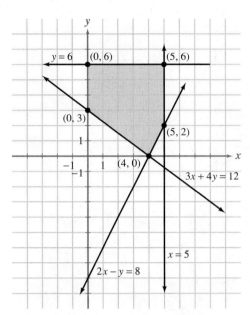

13. This region is bounded and convex.

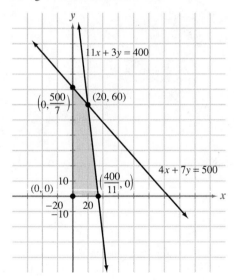

15. This region is unbounded and convex.

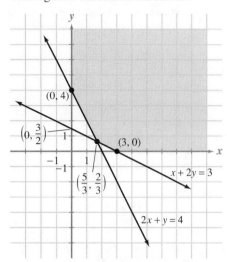

17. This region is bounded and convex.

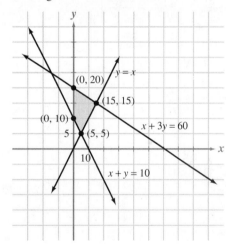

19. This region is unbounded and convex.

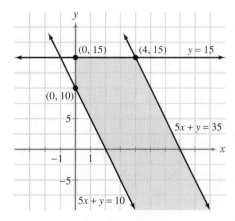

21. $\frac{23}{3}, 4$ **23.** $24, 7$ **25.** $4, -6$ **27.** $6, -18$

29. There is no upper bound for the values of x or the values of y. No. The solution set in Exercise 15 is not bounded.

31. $\begin{cases} 4x + y \geq 10 \\ 4x + y \leq 27 \\ x - 4y \geq -23 \\ x - 4y \leq 28 \end{cases}$ **33.** $\begin{cases} x \geq -1 \\ x + y \geq -4 \\ 3x - 2y \geq -11 \\ x \leq 3 \end{cases}$

35. 30 min of television; 70 min of radio

37. 2.5 acres of pine; 0 acres of fir

39. 50 A-frames and 25 bungalows; \$325

41. 3 professors and 4 assistants; \$3100

43. 6 type-1 shifts and 3 type-2 shifts; \$690

Chapter 9 Review Exercises

1. $\{(-\frac{2}{5}, -\frac{22}{5})\}$ **3.** $\{(-\frac{48}{43}, \frac{18}{43})\}$

5. $\{(-\frac{7}{3}, -\frac{10}{3}), (2, 1)\}$ **7.** $\{(-6, -3), (6, 3)\}$

9. $\{(-2\sqrt{14}i, -9), (2\sqrt{14}i, -9), (-4, 3), (4, 3)\}$

11. $\{(-1, 5)\}$ **13.** $\{(\frac{1326}{109}, -\frac{670}{109})\}$

15. $\left\{ \left(\dfrac{3 + \sqrt{153}}{4}, -\sqrt{\dfrac{3 + \sqrt{153}}{8}} \right), \right.$

$\left(\dfrac{3 - \sqrt{153}}{4}, \sqrt{\dfrac{3 + \sqrt{153}}{8}} \right),$

$\left(\dfrac{3 - \sqrt{153}}{4}, -\sqrt{\dfrac{\sqrt{153} - 3}{8}}\, i \right),$

$\left. \left(\dfrac{3 - \sqrt{153}}{4}, \sqrt{\dfrac{\sqrt{153} - 3}{8}}\, i \right) \right\}$

17. $\{(6, 10)\}$ **19.** $\{(1, -5, 2)\}$

21. $\{(-1, 2, -3)\}$ **23.** $\{(\frac{2360}{39}, \frac{1180}{39}, -\frac{120}{13})\}$ **25.** $\varnothing$

27. AB is undefined. $\; BA = \begin{bmatrix} 6 & -9 & 4 & 5 \\ 5 & -1 & -1 & 2 \\ 0 & -9 & 6 & 3 \end{bmatrix}$.

29. $AB = \begin{bmatrix} 6 & 5 \\ 0 & -4 \\ 13 & 21 \\ 13 & 23 \end{bmatrix}$. BA is undefined.

31. $\begin{bmatrix} \frac{5}{6} & -\frac{1}{6} \\ \frac{1}{12} & \frac{1}{12} \end{bmatrix}$ **33.** Singular

35. $\begin{bmatrix} 1 & -2 & \frac{1}{2} \\ 0 & 1 & -\frac{1}{4} \\ 0 & 0 & \frac{1}{4} \end{bmatrix}$ **37.** $\begin{bmatrix} -\frac{5}{11} & 0 & \frac{1}{11} & \frac{5}{11} \\ -\frac{21}{44} & \frac{1}{4} & -\frac{5}{11} & \frac{65}{44} \\ \frac{1}{4} & \frac{1}{4} & 0 & -\frac{1}{4} \\ \frac{6}{11} & 0 & \frac{1}{11} & -\frac{6}{11} \end{bmatrix}$

39. -340 **41.** -195 **43.** -12 **45.** $\{(\frac{116}{41}, \frac{5}{41})\}$

47. $\{(\frac{44}{3}, -25, \frac{73}{3})\}$ **49.** $\{(\frac{9}{8}, \frac{46}{8}, -\frac{7}{8})\}$

51. Unbounded; vertex: $(\frac{24}{7}, \frac{6}{7})$

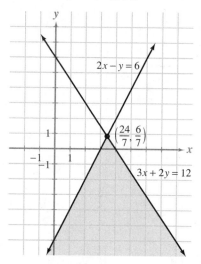

53. Unbounded; vertices: $(0, 1), (1, 0), (4, 0)$

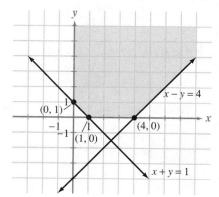

55. Bounded; vertices: $(0, 0), (0, 2), (1, 4), (4, 0), (6, 2)$

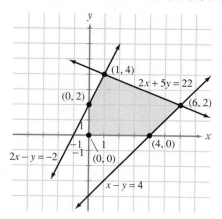

57. 34   **59.** 7, 5   **61.** 150 ft, 40 ft

63. $x^2 + y^2 - 6x + 4y - 12 = 0$

65. 36 of first type and 12 of the second type; \$259.20

CHAPTER 10

Exercises 10.1

1. $-2, 1, 4; 25$ **3.** $\pi, \pi, \pi, ; \pi$ **5.** $\frac{1}{2}, \frac{3}{4}, \frac{7}{8}, \frac{1023}{1024}$

7. $-\frac{1}{2}, \frac{2}{3}, \frac{7}{4}; \frac{98}{11}$ **9.** $0, 3, 8; 99$ **11.** $2, \frac{9}{4}, \frac{64}{27}; (1.1)^{10}$

13. $1, 1, -\frac{1}{3}; \frac{1}{17}$ **15.** $1, 0, -\frac{1}{9}; 0$ **17.** $6, 24, 60; 1320$

19. $0, \dfrac{\log 2}{\sqrt{2}}, \dfrac{\log 3}{\sqrt{3}}; \dfrac{\log 10}{\sqrt{10}}$ **21.** 1140 **23.** $k + 2$

25. $2, 3, 5, 9, 17$ **27.** $\frac{1}{2}, \frac{1}{4}, \frac{1}{8}, \frac{1}{16}, \frac{1}{32}$

29. $4, 2, 2^{1/2}, 2^{1/4}, 2^{1/8}$ **31.** $-1, -4, -9, 4, -1$

33. $1, 3, 2, 2.5, 2.25$ **35.** $\displaystyle\sum_{k=1}^{50} 2k$ **37.** $\displaystyle\sum_{k=1}^{5} kx^{k-1}$

39. $\displaystyle\sum_{k=1}^{n} k^k$ **41.** $\displaystyle\sum_{k=1}^{n} (-1)^{k+1} \dfrac{e^k}{k+1}$ **43.** $\displaystyle\sum_{k=1}^{10} \dfrac{k}{k - \dfrac{1}{k+1}}$

45. $(1 + 4) + (2 + 4) + (3 + 4) + (4 + 4) + (5 + 4)$
$+ (6 + 4) = 45$

47. $1^2 + 3^2 + 5^2 + 7^2 = 84$

49. $\dfrac{1}{2^0} + \dfrac{1}{2^1} + \dfrac{1}{2^2} + \dfrac{1}{2^3} + \dfrac{1}{2^4} = \dfrac{31}{16}$

51. $\sin 0 + \sin \dfrac{\pi}{3} + \sin \dfrac{2\pi}{3} + \sin \pi + \sin \dfrac{4\pi}{3}$
$+ \sin \dfrac{5\pi}{3} + \sin 2\pi = 0$

53. $(-1)^0 \ln e^0 + (-1)^1 \ln e^1 + (-1)^2 \ln e^2$
$+ (-1)^3 \ln e^3 + (-1)^4 \ln e^4 = 2$

55. 300π **57.** 2750 **59.** $256, 180$

61. $1, 1, 2, 3, 5, 8, 13, 21, 34, 55$

63. $1, 2, 6, 24, 120, 720; 2.718$

65. $10,000\left(1 + \dfrac{0.075}{4}\right)^4, 10,000\left(1 + \dfrac{0.075}{4}\right)^8,$
$10,000\left(1 + \dfrac{0.075}{4}\right)^{12}, 10,000\left(1 + \dfrac{0.075}{4}\right)^{16};$
$\$21,023.49$

67. $\$33,455.64$

Exercises 10.2

1. Yes; 4 **3.** No   **5.** Yes; $\frac{3}{2}$ **7.** No

9. Yes; 1 **11.** $3n + 1$ **13.** $9 - 4n$ **15.** $\dfrac{n+3}{2}$

17. $\frac{2}{3}n - 1$ **19.** $n \ln 2$ **21.** 279 **23.** 25

25. 206 **27.** $\frac{269}{2}$ **29.** 21 **31.** -149 **33.** 290

35. -18.5 **37.** 255 **39.** 27.5 **41.** $55 \log 2$

43. $\dfrac{125\pi}{2}$ **45.** $10, 16, 22$

47. $\frac{23}{9}, \frac{28}{9}, \frac{11}{3}, \frac{38}{9}, \frac{43}{9}, \frac{16}{3}, \frac{53}{9}, \frac{58}{9}$ **49.** $\frac{17}{18}, \frac{14}{9}, \frac{13}{6}, \frac{25}{9}, \frac{61}{18}$

51. $-\frac{42}{5}, -\frac{29}{5}, -\frac{16}{5}, -\frac{3}{5}$ **53.** No   **55.** $5n - 13$

57. $3, 10, 17$ **59.** 15 **61.** 1260 **63.** n^2

Exercises 10.3

1. Yes; $\frac{1}{4}$ **3.** No   **5.** Yes; $-\frac{2}{5}$ **7.** Yes; e^x

9. Yes; 2 **11.** $3(2^{n-1})$ **13.** $-4(-\frac{3}{4})^{n-1}$

15. $1.2(0.2)^{n-1}$ **17.** $(-x^2)^{n-1}$ **19.** $(\ln 10)\, 2^{n-1}$

21. 12 **23.** $-\frac{27}{2}$ **25.** -2

27. $\dfrac{1023}{64}$ **29.** $S_{12} \approx \dfrac{3000}{11}$ **31.** $\dfrac{3367}{2048}$

33. $\dfrac{1023 \ln 2}{256}$ **35.** $\dfrac{10(1 - 10^{8x})}{1 - 10^x}$

37. $6, 18, 54$ or $-6, 18, -54$ **39.** $18, -12, 8, -\frac{16}{3}$

41. $1.2, 0.36, 0.108$ or $-1.2, 0.36, -0.108$ **43.** 18

45. $\frac{3}{5}$ **47.** $\frac{40}{7}$

49. no sum; $|r| = \sqrt{3} > 1$ **51.** $\dfrac{1}{e-1}$ **53.** $\frac{14}{99}$

55. $\dfrac{2630}{999}$ **57.** $\dfrac{349,723}{33,300}$ **59.** $\dfrac{32,045}{9999}$

61. $-\frac{2}{3}$ or $-\frac{3}{2}$ **63.** $\$10,247.65$

65. $\$2,402.71$ **67.** 60 ft

69. $d = \ln r$ **71.** $x \in (-\frac{1}{2}, \frac{1}{2})$ **73.** 1 (total area)

Exercises 10.5

1. 56 **3.** 495 **5.** n

7. $a^8 + 8a^7b + 28a^6b^2 + 56a^5b^3 + 70a^4b^4 + 56a^3b^5$
$+ 28a^2b^6 + 8ab^7 + b^8$

9. $x^5 - 10x^4y + 40x^3y^2 - 80x^2y^3 + 80xy^4 - 32y^5$

11. $x^2 - 8x + 24 - \dfrac{32}{x} + \dfrac{16}{x^2}$

13. $\dfrac{x^3}{8} - 3x^2y + 24xy^2 - 64y^3$

15. $46,656x^6 + 15,552x^5y + 2,160x^4y^2 + 160x^3y^3 + \dfrac{20}{3}x^2y^4$
$+ \dfrac{4}{27}xy^5 + \dfrac{y^6}{729}$

17. $\dfrac{1}{x^5} - \dfrac{10}{x^3} + \dfrac{40}{x} - 80x + 80x^3 - 32x^5$

19. $x^{10} + 20x^9y + 180x^8y^2$

21. $32,768x^{15} - 737,280x^{14}y + 7,741,440x^{13}y^2$

23. $67,584x^{10}y^2$ **25.** $-\dfrac{440}{81}x^8y^3$ **27.** $\dfrac{12,600,000}{\sqrt{x}}$

29. $1,396,484,375x^9y^4$ **31.** $-4,455x^{24}y^6$

33. $286,720x^8y^{12}$ **35.** $210x^2y^3$

41. $122 - 597i$ **43.** $-119 + 120i$ **45.** 256

47. $128x^7 - 448x^6y + 672x^5y^2 - 560x^4y^3 + 280x^3y^4$
$- 84x^2y^5 + 14xy^6 - y^7$

Exercises 10.6

1. $\dfrac{1}{x - 2} - \dfrac{1}{x + 2}$ **3.** $\dfrac{1}{x + 2} - \dfrac{2}{x - 3}$

5. $\dfrac{-1}{x+2} + \dfrac{2}{x-5}$ **7.** $\dfrac{2}{x-1} + \dfrac{3}{(x-1)^2} - \dfrac{1}{x+2}$

9. $\dfrac{4}{x-2} - \dfrac{1}{(x-2)^2} + \dfrac{3}{x+1}$

11. $\dfrac{4}{x} - \dfrac{3}{x^2} + \dfrac{5}{x-10}$

13. $\dfrac{5}{x} - \dfrac{6}{x^2+3}$ **15.** $\dfrac{-5}{x^2} + \dfrac{3}{x^2+1}$

17. $\dfrac{2}{x^2} - \dfrac{7}{x^2+2}$ **19.** $\dfrac{1}{x^2+6} + \dfrac{5x-11}{(x^2+6)^2}$

21. $\dfrac{x-1}{2x^2+x+1} + \dfrac{2x}{(2x^2+x+1)^2}$

23. $\dfrac{x+2}{x^2+4} + \dfrac{2x-3}{x^2+5}$

25. $(3x+4) + \dfrac{2}{x} + \dfrac{-5x+6}{x^2+8}$

27. $(2x+1) - \dfrac{2}{x-3} + \dfrac{1}{x+2}$

29. $\dfrac{1}{x+2} - \dfrac{2}{(x+2)^2} + \dfrac{3}{(x+2)^3}$

31. $\dfrac{1}{x-1} - \dfrac{2}{x^2+5}$ **33.** $\dfrac{3}{x+1} + \dfrac{2x-5}{x^2+2x+4}$

Chapter 10 Review Exercises

1. $2, 1, \frac{4}{5}, \frac{5}{7}; \frac{3}{5}$ **3.** $0, -\frac{1}{4}, 0, \frac{1}{16}; \frac{1}{64}$

5. $-x, \dfrac{x^2}{2}, -\dfrac{x^3}{6}, \dfrac{x^4}{24}; \dfrac{x^8}{40,320}$ **7.** $\frac{5}{2}, \frac{9}{4}, \frac{17}{8}$

9. $\dfrac{x^2}{4}, -\dfrac{x^4}{144}, \dfrac{x^8}{331,776}$ **11.** $\displaystyle\sum_{k=1}^{9} (-\tfrac{1}{2})^{k-1}$

13. $\displaystyle\sum_{k=1}^{99} \dfrac{\ln(k+1)}{k(k+1)}$ **15.** $\frac{211}{81}$ **17.** 0 **19.** $-\frac{120}{121}$

21. $6n-1$ **23.** $-2.7n+10.7$ **25.** $\dfrac{\pi}{6}n + \dfrac{4\pi}{3}$

27. 65 **29.** $12 \log 2$ **31.** 30 **33.** $\frac{235}{4}$

35. $-\frac{1}{2}, 4, \frac{17}{2}$ **37.** $2n+5$ **39.** $(\frac{3}{5})^n$

41. $3.2(\frac{1}{4})^{n-1}$ **43.** $2^{n-1} \ln 3$ **45.** 1.11111

47. $-\dfrac{42,125}{729}$

49. $4(3)^{1/4}, 4(9)^{1/4}, 4(27)^{1/4}$ or $-4(3)^{1/4}, 4(9)^{1/4}, -4(27)^{1/4}$

51. $-\frac{3}{4}$ **53.** No sum since $|r| = \frac{3}{2} > 1$ **55.** $\frac{2293}{990}$

57. $\$7,598.98$

67. $32x^5 - 80x^4y + 80x^3y^2 - 40x^2y^3 + 10xy^4 - y^5$

69. $x^2 + 8x + 24 + \dfrac{32}{x} + \dfrac{16}{x^2}$

71. $295,245x^8y^2$ **73.** $2,016x^{10}y^{12}$

75. $\dfrac{5}{x-3} + \dfrac{6}{x+2}$

77. $\dfrac{1}{x-5} + \dfrac{3}{(x-5)^2} + \dfrac{4}{x+1}$

79. $\dfrac{2x+1}{x^2+16} + \dfrac{x-1}{(x^2+16)^2}$

81. $\dfrac{2}{x+2} - \dfrac{3}{(x+2)^2} + \dfrac{5}{(x+2)^3}$

83. $(2x^2+1) + \left(\dfrac{4}{x} + \dfrac{5}{x^2+1}\right)$

APPENDIX

Exercises A.1

1. $\dfrac{1}{4x^5}$ **3.** $-\dfrac{c^4}{2a^6b}$ **5.** $-32a^{17}b$

7. $\dfrac{4x^4y^4}{25}$ **9.** $\dfrac{72x^2}{y^9}$ **11.** $(x+y)^3$

13. $2tuv^3 \sqrt[3]{2t}$ **15.** $\dfrac{a^2}{2b^3}$ **17.** $\dfrac{3p\sqrt[3]{2q}}{r^2}$

19. $\dfrac{1}{3tv\sqrt{2v}}$ **21.** $\dfrac{p^2\sqrt[6]{q^5}}{2r}$ **23.** $\dfrac{xy\sqrt[6]{x}}{z^2}$

25. $\dfrac{5\sqrt{3}}{3}$ **27.** $\dfrac{6+2\sqrt{2}}{7}$ **29.** $\dfrac{(2x-1)\sqrt{x-2}}{x-2}$

31. $\dfrac{a-4}{3a+6\sqrt{a}}$ **33.** $\dfrac{m-n}{2mn(\sqrt{m}-\sqrt{n})}$

35. $\dfrac{64}{27}$ **37.** $\dfrac{9y^8}{x^2}$ **39.** $\dfrac{p^{5/3}}{q^{3/2}}$

41. $x^{2/5}y^{24/5}z^{14/5}$ **43.** $\dfrac{b^6}{a^7}$

45. $(2x-1)^{11/15}$ **47.** $-x^3 + x^2 + 2x + 4$

49. $6x^3 - 3x^2 + 45x^5$ **51.** $-14y^2 + 61y + 9$

53. $3r^3 - 6r^2 - 20r + 8$ **55.** $8x^3 - 4x^2 + 2x - 1$

57. $8x^4 - 12x^3 - 10x^2 + 18x - 3$

59. $x^2 + 9y^2 + 2x - 6y - 6xy + 1$

61. $(a+1)(a-3)$ **63.** $(4+3z)(4-3z)$

65. $2(3t+2)(t+1)$ **67.** $(y-5)(y^2+5y+25)$

69. $(x-2y)(x+1)(x-1)$

71. $(2u+3v+3)(2u+3v-3)$

73. $(y^2+4)(y+2)(y-2)$ **75.** $(6a+5b^2)^2$

77. $\dfrac{2a+3}{a}$ **79.** $\dfrac{(x+1)(x-5)}{3x^2(x^2+x+1)}$ **81.** 1

83. $\dfrac{5}{(a-3)(a^2-4)}$ **85.** $-\dfrac{1}{ab}$ **87.** $-\dfrac{2x+h}{x^2(x+h)^2}$

89. $2x(3x^2+1)^2(5x^2+2)^6(150x^2+53)$

91. $\dfrac{(3x^2-1)^4(87x^2-30x+1)}{(3x-1)^{4/3}}$

Index

GRAPHS OF THE TRIGONOMETRIC FUNCTIONS

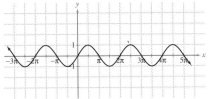

$y = \sin x$

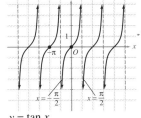

$y = \tan x$

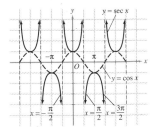

$y = \sec x$

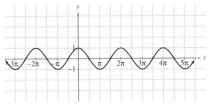

$y = \cos x$

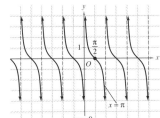

$y = \cot x$

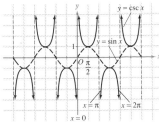

$y = \csc x$

GRAPHS OF THE INVERSE TRIGONOMETRIC FUNCTIONS

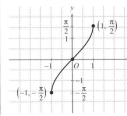

$y = \sin^{-1} x$

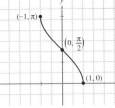

$y = \cos^{-1} x$

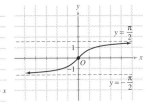

$y = \tan^{-1} x$

THE DOUBLE-ANGLE FORMULAS

$\sin 2v = 2 \sin v \cos v$

$\cos 2v = \cos^2 v - \sin^2 v = 1 - 2 \sin^2 v$
$\qquad = 2 \cos^2 v - 1$

$\tan 2v = \dfrac{2 \tan v}{1 - \tan^2 v}$

THE HALF-ANGLE FORMULAS

$\sin \dfrac{v}{2} = \pm \sqrt{\dfrac{1 - \cos v}{2}}$

$\cos \dfrac{v}{2} = \pm \sqrt{\dfrac{1 + \cos v}{2}}$

$\tan \dfrac{v}{2} = \pm \sqrt{\dfrac{1 - \cos v}{1 + \cos v}}$

The sign on the right side of each equation is determined by the quadrant in which $\dfrac{v}{2}$ lies.

SPECIAL VALUES OF THE TRIGONOMETRIC FUNCTIONS

θ (radians)	θ (degrees)	$\sin\theta$	$\cos\theta$	$\tan\theta$	$\cot\theta$	$\sec\theta$	$\csc\theta$
0	$0°$	0	1	0	undefined	1	undefined
$\dfrac{\pi}{6}$	$30°$	$\dfrac{1}{2}$	$\dfrac{\sqrt{3}}{2}$	$\dfrac{\sqrt{3}}{3}$	$\sqrt{3}$	$\dfrac{2\sqrt{3}}{3}$	2
$\dfrac{\pi}{4}$	$45°$	$\dfrac{\sqrt{2}}{2}$	$\dfrac{\sqrt{2}}{2}$	1	1	$\sqrt{2}$	$\sqrt{2}$
$\dfrac{\pi}{3}$	$60°$	$\dfrac{\sqrt{3}}{2}$	$\dfrac{1}{2}$	$\sqrt{3}$	$\dfrac{\sqrt{3}}{3}$	2	$\dfrac{2\sqrt{3}}{3}$
$\dfrac{\pi}{2}$	$90°$	1	0	undefined	0	undefined	1